Reaction Mechanisms in Organic Chemistry

Reaction Mechanisms in Organic Chemistry

Metin Balcı

Author

Prof. Dr. Metin Balcı
Middle East Technical University
Department of Chemistry
06800 Ankara
Turkey

Cover Image: © Gul Menges &
Metin Balcı

■ All books published by **WILEY-VCH** are carefully produced. Nevertheless, authors, editors, and publisher do not warrant the information contained in these books, including this book, to be free of errors. Readers are advised to keep in mind that statements, data, illustrations, procedural details or other items may inadvertently be inaccurate.

Library of Congress Card No.: applied for

British Library Cataloguing-in-Publication Data
A catalogue record for this book is available from the British Library.

Bibliographic information published by the Deutsche Nationalbibliothek
The Deutsche Nationalbibliothek lists this publication in the Deutsche Nationalbibliografie; detailed bibliographic data are available on the Internet at <http://dnb.d-nb.de>.

© 2022 WILEY-VCH GmbH, Boschstr. 12, 69469 Weinheim, Germany

All rights reserved (including those of translation into other languages). No part of this book may be reproduced in any form – by photoprinting, microfilm, or any other means – nor transmitted or translated into a machine language without written permission from the publishers. Registered names, trademarks, etc. used in this book, even when not specifically marked as such, are not to be considered unprotected by law.

Print ISBN: 978-3-527-34964-7
ePDF ISBN: 978-3-527-83460-0
ePub ISBN: 978-3-527-83459-4

Typesetting   Straive, Chennai, India
Printing and Binding   CPI Group (UK) Ltd, Croydon, CR0 4YY

Printed on acid-free paper

C9783527349647_120224

For my Wife Jale.

Contents

Preface *xv*
About the Author *xvii*
Abbreviations *xix*

1 Basic Concepts *1*
1.1 Introduction to Reaction Mechanisms *1*
1.2 Covalent Bonding and Hybridization *2*
1.2.1 sp^3-Hybridization of Carbon *4*
1.2.2 sp^2-Hybridization of Carbon *7*
1.2.3 sp-Hybridization of Carbon *10*
1.2.4 Bond Lengths *12*
1.3 Electrophiles and Nucleophiles *13*
1.3.1 Electrophiles (Electrophilic Compounds) *14*
1.3.2 Nucleophiles (Nucleophilic Compounds) *15*
1.4 Inductive and Mesomeric Effects *15*
1.4.1 Inductive Effect *15*
1.4.2 Mesomeric Effect (Resonance Structures) *18*
1.5 Formal Charge and Oxidation Number *24*
1.5.1 Formal Charge *24*
1.5.2 Oxidation Number *25*
1.6 Acids and Bases *28*
1.6.1 Arrhenius Acid–Base Theory *29*
1.6.2 Brønsted–Lowry Acid–Base Theory *29*
1.6.3 Lewis Acid–Base Theory *30*
1.6.4 Pearson Hard and Soft Acid–Base Theory *32*
1.6.4.1 Hard Acids and Bases *32*
1.6.4.2 Soft Acids and Bases *33*
1.6.5 pKa Values of Acids *34*
1.6.5.1 Factors Affecting the Acidity Strength of Organic Compounds *35*
1.6.6 pK_b Values of Bases *38*
1.6.7 Factors Affecting the Strengths of Bases in Nitrogen-Containing Compounds *38*
1.6.8 Heterocyclic Bases *40*
1.7 Reaction Kinetics and Energy Diagrams *41*
1.7.1 Thermodynamic vs. Kinetic Control of Reactions *43*
1.7.2 Reaction rate *44*
Problems *46*
References *48*

2 Nucleophilic Substitution Reaction *50*
2.1 Types of Chemical Reactions *50*

2.1.1	Polar Reactions 50	
2.1.2	Radical Reactions 50	
2.1.3	Pericyclic Reactions 50	
2.1.3.1	Relationship Between Nucleophilicity and Basicity 51	
2.1.3.2	Leaving Group 54	
2.2	Unimolecular Nucleophilic Substitution Reactions, S_N1 55	
2.2.1	Stereochemistry in S_N1 Reactions 58	
2.2.2	Optical Activity 59	
2.2.3	Other Factors Affecting S_N1 Reactions: Steric Factors 61	
2.3	Bimolecular Substitution Reactions, S_N2 62	
2.3.1	Stereochemistry of S_N2 Reactions 64	
2.3.2	Factors Affecting the S_N2 Reaction Mechanism 66	
2.3.2.1	The Structure of the Substrate 66	
2.3.2.2	Solvent Effect 68	
2.3.2.3	Leaving Group Effect 69	
2.3.2.4	Structure of the Nucleophile 70	
2.3.3	Nucleophilic Substitution in Allylic Systems: Allylic Rearrangement 71	
2.3.3.1	Stereochemistry in Allylic Substitution Reactions 73	
2.3.4	Internal Nucleophilic Substitution Reaction, S_Ni 74	
2.3.5	Neighboring Group Participation in Nucleophilic Substitution Reactions 75	
2.3.5.1	Reaction Rate 76	
2.3.5.2	Configuration Retention 77	
2.3.5.3	Molecular Rearrangement 78	
2.3.6	Ambident Nucleophiles 78	
2.3.7	Various Nucleophilic Substitution Reactions 80	
2.3.7.1	Williamson Ether Synthesis 80	
2.3.7.2	Ether Cleavage 81	
2.3.7.3	Reactions of Epoxides 82	
2.3.7.4	Substitution in Unsaturated Systems 83	
	Problems 83	
	References 85	
3	**Elimination Reactions** 87	
3.1	Unimolecular Elimination Reactions, E1 88	
3.1.1	E1 Reaction Mechanism 88	
3.1.1.1	Dehydration of Alcohols 91	
3.1.2	Factors Affecting the Ratio of E1 and S_N1 92	
3.2	Bimolecular Elimination Reactions, E2 94	
3.2.1	Kinetic Isotope Effect 98	
3.2.2	Stereochemistry in E2 Elimination Reactions 98	
3.2.2.1	Erythro- and Threo-Configurations 100	
3.2.3	E2 Elimination in the Cyclohexane System 105	
3.2.3.1	Conformation and Configuration in Cyclohexane 105	
3.2.3.2	syn-Elimination (cis-Elimination) 112	
3.3	Unimolecular Conjugate Base Elimination, E1cb 112	
3.4	Elimination Reaction in Synthesis 114	
3.4.1	Halogen Elimination 114	
3.4.2	Hofmann Elimination: Quaternary Ammonium Salts 116	
3.4.3	Pyrolytic Elimination: Intramolecular cis-Elimination Reactions 119	
3.4.4	Elimination at the Bridgehead: Bredt's Rule 123	
3.4.5	Grob Fragmentation 126	
	Problems 129	
	References 131	

4 Addition Reactions to Alkenes *133*
4.1 Halogen Addition to Alkenes: Halogenation *133*
4.1.1 Stereochemistry of Halogen Addition *136*
4.2 Addition of Hydrogen Halides to Alkenes: Markovnikov's Rule *138*
4.2.1 *Anti*-Markovnikov Addition of Hydrogen Halides to Alkenes *141*
4.3 Addition of Water and Alcohols to Alkenes *143*
4.3.1 Hydration *143*
4.3.2 Alkoxylation *144*
4.3.3 Formation of Halohydrins *145*
4.4 Hydration Alkenes: Oxymercuration and Demercuration *146*
4.5 Hydroboration of Alkenes: *anti*-Markovnikov Hydration *148*
4.6 Oxidation of Alkenes *152*
4.6.1 Epoxidation *152*
4.6.2 Dioxirane *156*
4.6.3 Epoxide Ring-Opening Reactions *157*
4.6.4 Vicinal *cis*-Dihydroxylation *159*
4.6.5 Dihydroxylation via PIFA *161*
4.6.6 Enzymatic Dihydroxylation *162*
4.6.7 Ozonolysis: Oxidative Cleavage of Alkenes *162*
4.7 Reduction of Alkenes *165*
4.7.1 Heterogeneous Catalytic Reduction *166*
4.7.2 Homogeneous Catalytic Reduction *168*
4.8 Addition to Conjugated Dienes *171*
Problems *176*
References *177*

5 Carbonyl Compounds *181*
5.1 Reactivity of the Carbonyl Group *181*
5.1.1 Structure–Reactivity Relationships *183*
5.2 Reactions of Carbonyl Compounds: Addition Reactions *186*
5.2.1 Hydration: Addition of Water to Carbonyl Groups *187*
5.2.2 Hemiacetal Formation: Addition of Alcohols to Carbonyl Groups *190*
5.2.2.1 Cyclic Acetals and Their Synthetic Application *191*
5.2.2.2 What is the Protecting Group? *192*
5.2.2.3 Protection of Diols *193*
5.2.2.4 Formation of Thioacetals and Their Synthetic Application *193*
5.2.2.5 Umpolung: Polarity Inversion of the Aldehyde Carbonyl Group *195*
5.2.3 Reactions of Aldehydes and Ketones with Amines *196*
5.2.3.1 Oximes *199*
5.2.3.2 Hydrazones *199*
5.3 Reduction of Carbonyl Groups *200*
5.3.1 Wolff–Kishner Reduction (Under Basic Conditions) *200*
5.3.2 Clemmensen Reduction (Under Acidic Conditions) *201*
5.3.3 Metal Hydride Reduction of the Carbonyl Groups *201*
5.3.3.1 Diisobutyl Aluminum Hydride (DIBAL) Reduction *203*
5.3.3.2 Sodium Borohydride ($NaBH_4$) Reduction *203*
5.3.3.3 Reduction of Carboxylic Acids *205*
5.3.3.4 Meerwein–Ponndorf–Verley Reduction and Oppenauer Oxidation *209*
5.4 Reaction of Carbonyl Groups with Organometallic Compounds *210*
5.4.1 Grignard Reagents *210*
5.4.2 Reactions with Active Hydrogen-Containing Compounds *213*
5.4.3 Reactions with Carbonyl Compounds *213*
5.4.4 Stereochemistry *214*

5.4.5	Reaction with Esters	*215*
5.4.6	Reactions with Different Functional Groups	*215*
5.5	Reaction of Carbonyl Groups with Ylides	*218*
5.5.1	Phosphonium Ylides and Wittig Reactions	*219*
5.5.2	Reaction Mechanism	*219*
5.5.3	Stable Ylides	*220*
5.5.4	Unstable Ylides	*220*
5.5.5	Wittig–Schlosser Reaction	*221*
5.5.6	Wittig–Horner Reaction	*222*
5.5.7	Horner–Wadsworth–Emmons Reaction (HWE Reaction)	*223*
5.5.8	Sulfur Ylides	*224*
5.5.9	Julia Olefination	*226*
5.5.10	Peterson Olefination	*227*
5.6	Reactivity of α-Carbon Atoms	*228*
5.6.1	Acidity of α-Hydrogens	*228*
5.6.2	Keto–Enol Tautomerism	*229*
5.6.3	Acid-Catalyzed Enolization	*232*
5.6.4	Base-Catalyzed Enolization	*233*
5.6.5	Kinetic and Thermodynamic Enolates	*234*
5.6.6	Enol and Enolate Reactions	*235*
5.6.6.1	Racemization of Chiral Ketones, α-Epimerization	*235*
5.6.6.2	α-Halogenation of Aldehydes and Ketones	*237*
5.6.7	α-Alkylation of Carbonyl Compounds	*241*
5.6.7.1	Stork Enamine Reaction	*243*
5.6.7.2	Enolates Derived from 1,3-Dicarbonyl Compounds	*245*
5.6.7.3	Enolates, Ambident Nucleophiles: C vs. O Alkylation	*249*
5.7	Condensation Reactions of Carbonyl Compounds	*251*
5.7.1	Aldol Condensation	*252*
5.7.2	Crossed Aldol Condensation	*254*
5.7.3	Robinson Annulation	*256*
5.7.4	Claisen Ester Condensation	*257*
5.7.5	Crossed Claisen Ester Condensation	*259*
5.7.6	Dieckmann Condensation	*260*
5.7.7	Knoevenagel Condensation	*261*
5.7.8	Perkin Condensation	*264*
5.7.9	Stobbe Condensation	*265*
5.7.10	Role of Condensation Reactions in Synthetic Chemistry	*266*
5.8	Ester Hydrolysis Reactions	*268*
5.8.1	Ester Hydrolysis	*269*
5.8.2	Ester Hydrolysis Under Basic Conditions	*270*
5.8.3	$B_{AL}2$ Mechanism	*271*
5.8.4	Ester Hydrolysis Under Acidic Conditions	*272*
5.8.5	Asymmetric Ester Hydrolysis	*273*
5.8.6	Transesterification	*273*
	Problems	*275*
	References	*278*
6	**Aromaticity** *281*	
6.1	Aromatic Compounds	*281*
6.1.1	Discovery and Structure of Benzene	*281*
6.1.2	Aromatic, Antiaromatic, and Nonaromatic Compounds	*284*
6.1.2.1	Aromatic Compounds	*284*

6.1.2.2	Antiaromatic Compounds	*285*
6.1.2.3	Nonaromatic Compounds	*286*
6.1.3	Determination of the Molecular Orbitals of Aromatic Compounds	*287*
6.1.4	What Are the Criteria for Aromaticity? How Does One Quantify Aromaticity?	*288*
6.1.4.1	Thermodynamic and Aromatic Resonance Stabilization Energy	*288*
6.1.4.2	Structural Evidence for Aromaticity	*289*
6.1.4.3	Magnetic Evidence for Aromaticity	*289*
6.1.5	Homoaromaticity	*291*
6.1.6	Möbius Aromaticity	*293*
6.2	Aromatic Ions	*294*
6.3	Annulenes	*303*
6.3.1	Cyclobutadiene	*303*
6.3.2	[10]Annulene	*305*
6.3.3	[12]Annulenes	*307*
6.3.4	[14] and Higher Annulenes	*308*
6.4	Aromaticity in Fused Systems	*311*
6.5	Aromaticity in Heterocyclic Compounds	*314*
6.5.1	Heteroaromatic Compounds with Three-Membered Ring	*314*
6.5.2	Heteroaromatic Compounds with a Five-Membered Ring	*315*
6.5.3	Heteroaromatic Compounds with Six-Membered Ring	*323*
6.5.3.1	Electrophilic Aromatic Substitution	*324*
6.5.3.2	Nucleophilic Aromatic Substitution	*326*
6.5.3.3	Pyridine *N*-Oxides	*327*
6.5.3.4	Electrophilic Substitution	*328*
6.5.3.5	Nucleophilic Substitution	*328*
6.5.3.6	Six-Membered Ring Heteroaromatic Compounds with Two Nitrogen Atoms	*329*
6.5.4	Heteroaromatic Compounds with a Seven-Membered Ring	*330*
6.5.4.1	Oxepine	*330*
6.5.4.2	1*H*-Azepine	*331*
6.5.4.3	Thiepine	*332*
6.6	Electrophilic Aromatic Substitution: Chemistry of Benzene	*333*
6.6.1	Halogenation of Benzene	*335*
6.6.2	Nitration	*336*
6.6.3	Sulfonation	*337*
6.6.4	Friedel–Crafts Acylation	*338*
6.6.5	Friedel–Crafts Alkylation	*339*
6.6.6	Clemmensen Reduction	*341*
6.6.7	Reactivity of Monosubstituted Benzene Derivatives	*341*
6.6.8	Directing Effects of Substituents: Activating Groups	*344*
6.6.9	Directing Effects of Substituents: Deactivating Groups	*349*
6.6.10	Electrophilic Aromatic Substitution on Disubstituted Benzenes	*352*
6.7	Functionalization of the Side-Chain Substituents of Benzene	*354*
6.7.1	Oxidation of the Side Chain of Alkylbenzenes	*354*
6.7.2	Halogenation of Side Chains of Alkylbenzenes	*355*
6.7.3	Arenediazonium Ion as an Electrophile	*356*
6.8	Nucleophilic Aromatic Substitution Reactions	*357*
6.8.1	Addition–Elimination Mechanism ($S_N Ar$ Mechanism)	*358*
6.8.2	Reaction of Arenediazonium Salts with Nucleophiles	*360*
6.8.2.1	Reductive Dediazonization	*360*
6.8.3	Elimination–Addition Mechanism	*363*
6.9	Polycyclic Aromatic Compounds	*365*
6.9.1	Naphthalene	*365*

6.9.2	Benzenoid Aromatic Compounds *368*	
	Problems *373*	
	References *375*	

7	**Reactive Intermediates: Carbocations** *381*
7.1	Structure and Stability of Carbocations *384*
7.2	Generation of Carbocations *387*
7.2.1	Ionization Mechanism *387*
7.2.2	Electrophilic Addition to π Bonds *388*
7.3	Detection of Carbocations *388*
7.4	Reactions of Carbocations: Rearrangements *389*
7.4.1	Reactions with Nucleophiles *389*
7.4.2	Double-bond Formation by Proton Elimination *389*
7.4.3	Rearrangement *389*
7.4.4	Carbocation Rearrangement *390*
7.4.5	Ethyl Carbocation *390*
7.4.6	Isopropyl Carbocation *390*
7.4.7	Cyclopentyl Carbocation *391*
7.4.8	The Wagner–Meerwein Rearrangement *391*
7.4.9	Pinacol Rearrangement *393*
7.4.10	Tiffeneau–Demjanov Rearrangement *395*
7.4.11	Dienone–Phenol Rearrangement *398*
7.4.12	Neighboring Group Participation in Molecular Rearrangement (Anchimeric Assistance) *398*
7.4.13	Nonclassical Carbocations *401*
7.4.14	Nametkin Rearrangement *406*
7.4.15	Hydride Shift *406*
7.4.16	Base-induced Nucleophilic Rearrangements *408*
7.4.17	Favorskii Rearrangement *408*
7.4.18	Ramberg–Bäcklund Rearrangement *411*
7.4.19	Benzil–Benzilic Acid Rearrangement *412*
7.5	Rearrangement to Electron-deficient Nitrogen *412*
7.5.1	Beckmann Rearrangement *413*
7.5.2	Neber Rearrangement *415*
7.5.3	Stieglitz Rearrangement *416*
7.6	Rearrangement to Electron-deficient Oxygen *416*
7.6.1	Baeyer–Villiger Rearrangement (or Oxidation) *417*
	Problems *419*
	References *421*

8	**Reactive Intermediates Carbanions, Carbenes, and Nitrenes** *424*
8.1	Carbanions: Electrophilic Rearrangements *424*
8.1.1	Stevens Rearrangement *425*
8.1.2	Sommelet–Hauser Rearrangement *427*
8.1.3	Wittig Rearrangement *429*
8.2	Carbenes *429*
8.2.1	Naming Carbenes *430*
8.2.2	Structure and Reactivity of Carbenes *430*
8.2.3	Inductive Effect *430*
8.2.4	Mesomeric Effect *430*
8.2.5	Carbene Generation *431*
8.2.5.1	Carbene Precursors: Synthesis of Diazo Compounds *432*
8.2.5.2	Bamford–Stevens Reaction *432*
8.2.5.3	N-Nitrosoalkyl Urea Compounds *433*

8.2.5.4	Diazirines	*434*
8.2.5.5	α-Elimination Method	*435*
8.2.5.6	Simmons–Smith Reaction	*437*
8.2.6	Carbene Reactions	*438*
8.2.6.1	Carbene Cycloaddition Reactions	*439*
8.2.6.2	Carbene Insertion Reactions	*445*
8.2.6.3	Carbene Rearrangements	*446*
8.3	Azides and Nitrenes	*453*
8.3.1	Nitrene Synthesis	*454*
8.3.1.1	Synthesis of Acyl Azides	*455*
8.3.2	Nitrene Rearrangements	*455*
8.3.2.1	Curtius Rearrangement	*455*
8.3.2.2	Schmidt Rearrangement	*457*
8.3.2.3	Lossen Rearrangement	*457*
8.3.2.4	Hofmann Rearrangement	*458*
8.3.3	Nitrene Cycloaddition Reactions	*458*
8.3.4	Nitrene Insertion Reactions	*460*
	Problems	*461*
	References	*465*
9	**Reactive Intermediates: Radicals and Singlet Oxygen**	*468*
9.1	Structure of Radicals and Their Stability	*468*
9.1.1	Generation of Radicals	*471*
9.1.2	Radical Reactions	*472*
9.1.2.1	Atom-Abstraction Reaction	*472*
9.1.2.2	Radical Combination and Disproportionation	*475*
9.1.2.3	Kolbe Electrolysis (Kolbe Reaction)	*476*
9.1.2.4	Hunsdiecker Reaction	*477*
9.1.2.5	Radical Addition to Alkenes	*478*
9.1.2.6	Manganese(III)-Mediated Oxidative Radical Additions to Alkenes	*480*
9.1.2.7	Birch Reduction	*481*
9.1.2.8	Di-π-methane Rearrangement	*483*
9.1.2.9	Diradicals Derived from Diazo Compounds	*486*
9.2	Singlet Oxygen	*487*
9.2.1	The Electronic Configuration of the Oxygen Molecule	*487*
9.2.2	Singlet Oxygen Generation	*489*
9.2.2.1	Generation of Photosensitized Singlet Oxygen	*489*
9.2.2.2	Chemical Sources for Singlet Oxygen	*490*
9.2.3	Reactions of Singlet Oxygen	*492*
9.2.3.1	Ene Reactions	*492*
9.2.3.2	[2 + 2] Cycloaddition Reactions	*493*
9.2.3.3	[4 + 2] Cycloaddition Reactions	*495*
9.2.3.4	Bicyclic Endoperoxides in Synthesis	*496*
	References	*497*
10	**Pericyclic Reactions**	*500*
10.1	Frontier Molecular Orbitals	*502*
10.2	Electrocyclic Reactions	*505*
10.2.1	Thermal Electrocyclic Reactions	*505*
10.2.2	Photochemical Electrocyclic Reactions	*506*
10.2.3	Application of the Woodward–Hoffman Rules to Electrocyclic Reactions	*508*
10.2.3.1	Neutral Compounds	*508*
10.2.3.2	Ionic Compounds	*510*

10.3	Correlation Diagrams	*512*
10.3.1	Symmetry Elements	*512*
10.4	Cycloaddition Reactions	*515*
10.4.1	Stereoselectivity in Cycloaddition Reactions: Secondary Orbital Interactions	*520*
10.4.2	Factors Affecting the Rates of Cycloadditions	*521*
10.4.3	Coefficients of the Frontier Orbitals: Application to Regioselectivity in Diels–Alder Reactions	*522*
10.5	Sigmatropic Reactions	*526*
10.6	Cope and Claisen Rearrangements	*530*
10.6.1	Oxy-Cope Rearrangement	*531*
10.6.2	Claisen Rearrangement	*532*
	Problems	*533*
	References	*535*
11	**Carbon–Carbon Coupling Reactions**	*537*
11.1	History	*537*
11.2	Mizoroki–Heck Coupling Reaction	*539*
11.2.1	Regioselectivity	*542*
11.3	Stille Coupling Reaction	*547*
11.3.1	Synthesis of Organotin Compounds	*550*
11.4	Suzuki–Miyaura Coupling Reaction	*552*
11.5	Negishi Coupling Reaction	*558*
11.6	Sonogashira Coupling Reaction	*562*
11.7	Kumada Coupling Reaction	*566*
11.8	Hiyama Coupling Reaction	*567*
11.8.1	Hiyama–Denmark Coupling	*569*
11.9	Buchwald–Hartwig Coupling Reaction	*570*
11.10	Tsuji–Trost Coupling Reaction	*571*
11.11	Palladium-Catalyzed Carbonylation Reactions	*573*
11.11.1	Carbonylative Coupling Reactions with Organometallic Reagents	*576*
11.11.2	$Mo(CO)_6$-Mediated Carbonylation	*577*
	References	*578*

Solutions *583*

Index *605*

Preface

For more than 40 years, I have taught courses on reaction mechanisms in organic chemistry at undergraduate and graduate level throughout my academic career. During my classes, I realized that many students do not like organic chemistry and most believe that it is a lesson to be memorized. I have searched for a way to change this idea. Organic chemistry is a systemic field like mathematics and hence memorizing is not necessary. Students need to know the reaction mechanisms in organic chemistry and understand their many common points. For example, there are many condensation reactions and students usually try to memorize them. However, they all have the same mechanisms and students must learn this. Starting from that point, I decided to write a book about these reaction mechanisms and emphasize the points common to them.

In my lectures, I always tell the students that chemical reactions occur because of a reaction between an acid and a base (a Lewis acid and a Lewis base). Therefore, they should learn the acid/base concepts very well in order to understand chemistry. For this, they need to determine in which part of a molecule the electron density is increased or decreased. They can achieve this by fully understanding the factors that influence electron density. In other words, they should have a firm grasp of the mesomeric effect and the inductive effect. A student who comprehends these concepts can predict what product will be formed as a result of a chemical reaction and will not rely on memorization. Therefore, I included these concepts in the first part of this book. I explained the mesomeric effect and inductive effect through detailed examples.

Another disadvantage faced by students is that most of them cannot perceive organic compounds in a three-dimensional structure. For this, the concept of hybridization needs to be explained clearly and understood fully. Then, the students can imagine the three-dimensional structure and understand how isomers are formed in some reactions. Therefore, I explained hybridization in detail in the introduction part.

A further aspect of organic chemistry that confuses students is oxidation/reduction. For this, they should know oxidation numbers well and not mix them up with formal charges. This subject is examined in detail in the book, along with examples.

After this introduction, the students begin to grasp the reactions quickly. I included substitution, elimination, and addition reactions as the essential ones. A student who has taken a one-semester organic chemistry course can easily comprehend these issues. Then, I examined the reactions derived from the carbonyl group. I emphasized that essential reactions are significant in this subject, which is very broad. With a good grasp of the carbonyl group's polarization and the acidity of the alpha proton, students can understand the carbonyl group and realize how simple condensation reactions are.

In the part on aromaticity, I first explained classical reactions after the concepts of aromaticity. Here too, I tried to find common points mechanistically. After these chapters, I moved on to more advanced topics that both undergraduate and graduate students can follow.

The chemistry of intermediates is extremely important mechanistically. A student who has learned the chemistry of the intermediate well can fully understand organic chemistry. For this reason, I focused extensively on carbocations, carbanions, radicals, and carbenes, which are involved as intermediates during the reactions. A student who can deal with these issues will have a full understanding of all the rearrangements.

The book's final chapters are devoted to the Woodward–Hoffmann rules and modern C–C couplings. These are subjects written for both undergraduate and graduate students.

In summary, in this book, I tried to discourage students from relying on memorization by drawing attention to reaction mechanisms' common points. I hope that this book will be useful to both undergraduate and graduate students at any time.

There may be one name on the front cover, but every book is the result of the hard work of many people. I gratefully acknowledge the following professors for their helpful critiques of this book at many stages during its development. They devoted a tremendous amount of time to review this book and made valuable suggestions.

Çağatay Dengiz	Middle East Technical University
Hamdullah Kılıç	Atatürk University
Nurullah Saraçoğlu	Atatürk University

I would also like to thank the following professors for reviewing the chapters and providing feedback on developing new features. I sincerely appreciate all of their suggestions.

Aliye Altundaş	Gazi University
İlker Avan	Anadolu University
Raşit Çalışkan	Süleymen Demirel University
Murat Çelik	Atatürk University
Yasin Çetinkaya	Atatürk University
Arif Daştan	Atatürk University
Dilem Doğan	Erciyes University
Serdal Kaya	Necmettin Erbakan University
Nurettin Menteş	Van Yüzüncü Yıl University
Hasan Seçen	Atatürk University
Nermin Şimşek Kuş	Mersin University
Meltem Tan	Van Yüzüncü Yıl University
Yavuz Taşkesenligil	Atatürk University

The author gratefully acknowledges the critical readings by the students Aslıcan Özdemir, Başak Karagöllü, Fevzi Can İnyurt, Flora Mammadova, Furkan Melih Günay, İpek Öktem, İpek Savaş, Kübra Erden, and Semin Özsinan, members of Dr. Çağatay Dengiz's group at the Middle East Technical University. Finally, I thank all my students, at both Middle East Technical University and Atatürk University, for their positive interactions over the years, which have guided me in creating this book. I also thank Mr. Russell Fraser for proofreading.

Although every effort has been made to eliminate all possible errors, it is likely that some typographical mistakes remain in the text. I would welcome comments from the readers, particularly those that point out mistakes so that they can be corrected, as well as suggestions for additions or other changes that will make the book more valuable to the reader.

This book of extensive scope could not have been produced without the excellent support I have had from many people at John Wiley-VCH. I would particularly like to thank the Editorial Director, Dr. Gudrun Walter, and the Executive Commissioning Editor, Dr. Elke Maase. They have been of great assistance at every stage of production. I am deeply grateful to my managing editor, Katherine Wong, who was always ready to do whatever was needed to make this book the best that it could be.

I would also thank Content Refinement Specialist Abisheka Santhoshini, and others involved in the typesetting.

I would also like to thank my daughter Gülşah, my son-in-law Onur, and son Berkay for their continued support and encouragement during the writing process.

Finally, my wife Jale offered unwavering patience and support and understood why I was writing. There is no adequate way to express my appreciation. Jale provided a comfortable environment that enabled me to write this book.

January 2021

Metin Balcı
Ankara, Turkey

About the Author

With his wife Jale

Metin Balcı was born in Erzurum, Turkey, in 1948. He received his "Diplom Chemiker" degree in 1972, followed by a PhD degree in 1976 from the University of Cologne, where he worked with Professor Emanuel Vogel. He did postdoctoral work with Professors Harald Günther (Siegen, Germany), Waldemar Adam (Puerto Rico), and W. M. Jones (Florida).

In 1980, he joined the Department of Chemistry at Atatürk University and he became a full professor there in 1987. He spent one year (1986) at the University of Cologne and one year (1996–1997) at Auburn University (USA) as a guest professor. In 1997, he moved to Middle East Technical University in Ankara because of its reputation.

Metin Balcı has received several prizes, including the 1983 "Junior Research Prize" and the "Scientific Award" in 1989 from the Scientific and Technical Research Council of Turkey. Furthermore, he was awarded the "Science Prize" by the Science and Technology Foundation in 1990 and the "Chemistry Prize" by the Chemistry Foundation and the "Science Prize" by the Ministry of Culture in 1991. He has received the Best Teacher of the Year Award (2000, 2003, and 2004) and the Distinguished Teaching Award at Middle East Technical University. His name was given to the NMR labs in the Department of Chemistry, Atatürk University, in 2010. He is an elected member of the Turkish Academy of Sciences.

His main research interest involves the synthesis of cyclitols, endoperoxides, cyclic strained compounds, bromine chemistry, and heterocyclic compounds. He has published 280 scientific papers and he retired in 2015. He has also published books entitled *Basic ^1H and ^{13}C NMR Spectroscopy* (426 pages, Elsevier, January 2005 and Turkish version in 2000 by METU Press). *The Reaction Mechanism in Organic Chemistry* (in Turkish, 2008) and an autobiography, *Science Rising from the East*, in 2019 (in Turkish) were published by the Turkish Academy of Sciences.

Metin Balcı has always maintained a strong interest in the great outdoors. He enjoys observing nature, hiking, skiing, and reading, as well as world travel.

Abbreviations

9-BBN	9-borabicyclo[3.3.1]nonane
ABCN	1,1-azobis(cyclohexanecarbonitril)
acac	acetylacetonate
AIBN	azobisisobutyronitrile
BH(Sia)$_2$	disiamylborane
BINAP	2,2'-bis(diphenylphosphino)-1,1'-binaphthyl
BINOL	binaphthol
Boc	*t-b*utyloxycarbonyl
COD	cyclooctadiene
COT	cyclooctatetraene
CSA	camphorsulfonic acid
DBA	bis(dibenzylideneacetone)
DBU	1,8-diazabicyclo-[5.4.0]undec-7-ene
DDQ	2,3-dichloro-5,6-dicyano-1,4-benzoquinone
DIBAL	diisobutyl aluminum hydride
DIOP	*O*-isopropylidene-2,3-dihydroxy-1,4-bis(diphenylphosphino)butane
DIPT	diisopropyl tartrate
DMDO	dimethyldioxirane
DMF	dimethylformamide
DMSO	dimethylsulfoxide
DPPF	[1,1'-bis(diphenylphosphino)ferrocene]dichloropalladium(II)
DPPP	bis(diphenylphosphino)propane
EPR	electron spin resonance spectroscopy
ESCA	electron spectroscopy for chemical analysis
HBpin	pinacolborane
HIA	hydride ion affinity
HMPA	hexamethylphosphoramide
HOMO	highest occupied molecular orbital
HSAB	hard and soft acids and bases
ISC	intersystem crossing
KHMDS	potassium bistrimethylsilylamide
LDA	lithium diisopropylamide
LHMDS	lithium bis(trimethylsilyl)amide
LTMP	lithium tetramethylpiperidine
LUMO	lowest unoccupied molecular orbital
m-CPBA	*meta*-chloroperbenzoic acid
MMPP	magnesium monoperoxyphthalate
MNDO	modified neglect of differential overlap
NBS	*N*-bromosuccunimide
n-BuLi	*n*-butyllithium
NICS	nucleus-independent chemical shift

NMO	*N*-methylmorpholine-*N*-oxide
NMP	*N*-methylpyrrolidone
PEPSI	pyridine enhanced precatalyst, preparation, stabilization, and inhibition
PIFA	phenyl-iodine(III) bis(trifluoroacetate)
PLE	pig liver esterase
PTAD	4-phenyl-1,2,4-triazoline-3,5-dione
RAMP	(*R*)-1-amino-2-methoxymethylpyrrolidine
SAMP	(*S*)-1-amino-2-methoxymethylpyrrolidine
SOMO	singly occupied molecular orbital
TASF	tris(diethylamino)-sulfonium difluorotrimethylsilicate
TBAF	tetrabutylammonium fluoride
TBDMS	*t*-butyldimethylsilyl
TEBA	triethylbenzylammonium salt
TFA	trifluoroaceticacid
TFPAA	trifluoroperaceticacid
THF	tetrahydrofuran
THP	tetrahydropyran
TIBSA	2,4,6-triisopropylbenzene sulfonyl azide
TMEDA	*N*,*N*,*N*′,*N*′-tetramethylenediamine
TMSOK	potassium trimethylsilanolate
PBDPSCl	*t*-butyl(chloro)diphenylsilane
TPP	tetraphenylporphine
XPS	X-ray photoelectron spectroscopy

1
Basic Concepts

1.1 Introduction to Reaction Mechanisms

Chemical reactions are the processes by which chemicals interact to form new chemicals with different compositions. A new compound formed as a result of a chemical reaction does not bear the properties of the starting compounds; it has its own unique properties. In order for a chemical reaction to begin, some conditions (temperature, pressure, catalyst, etc.) must exist. For some compounds, it is sufficient to bring them together to start a chemical reaction. For example, water and sodium (provided they do not come in contact with air) are normally stable. However, a very violent reaction happens when these two come together. Sodium metal reacts rapidly with water to form a colorless solution of sodium hydroxide (NaOH) and releases hydrogen gas.

$$2Na + 2H_2O \longrightarrow 2NaOH + H_2 \uparrow$$
$$\text{Reactants} \qquad\qquad \text{Products}$$

There is a common point regarding all chemical reactions, namely, transfer of electrons from one reactant to another (electron exchange). Therefore, one of the starting compounds should be capable of denoting electrons, while the other should accept electrons. In order for electron exchange to take place, some reactions require catalysts. A catalyst is a substance that can be added to a reaction to increase the reaction rate without being consumed in the process. The function of catalysts is to facilitate electron transfer by activating bond electrons and to lower activation energies (E_a). As a result of electron transfer, some bonds are broken, some bonds rearrange, and new bonds are formed.

Benzene is an unreactive compound in the presence of halogens (Cl_2, Br_2, or I_2) because they are not electrophilic enough to attack the benzene ring and disrupt its aromaticity. However, the halogens must be activated by Lewis acid catalysts such as $FeBr_3$ or $AlBr_3$. $FeBr_3$ accepts electron pairs from bromine and makes it much more electrophilic. For a detailed mechanism, see Section 6.6.

No reaction ⇍ Br_2 ⟵ ⬡ ⟶ ⬡–Br (Br_2, $FeBr_3$)

Let us roughly analyze the processes that occur during the reaction of sodium with water. We see that the hydrogen atom, which is bonded to oxygen, breaks off from oxygen and forms hydrogen gas by binding to another hydrogen atom; on the other hand, the neutral sodium atom is oxidized by donating an electron and forming an ionic bond with the hydroxyl anion. This reaction appears to be very simple. However, complex processes are involved. We need to understand how products are formed as a result of a reaction. When a reactant turns into a product, we should look at which intermediates are involved during this reaction.

If we have a good understanding of what is going on at the intermediate stages, then we can guess what kind of products are formed as a result of the reaction. The step-by-step sequence of the intermediate stages of a reaction until the product is formed is called the *reaction mechanism*.

We have to understand what is happening at the intermediate levels. Otherwise, if we try to learn about a chemical reaction by writing reactants on one side of the equation and products on the other side, we will not be able to master organic chemistry. If the reaction mechanism is learned well, then one can see that organic chemistry is as enjoyable and systematic

Reaction Mechanisms in Organic Chemistry, First Edition. Metin Balcı.
© 2022 WILEY-VCH GmbH. Published 2022 by WILEY-VCH GmbH.

as mathematics. This is because, once the reaction mechanism is known, it is possible to predict what reaction will occur between two reactants and what products will be formed. Otherwise, if the reactions are learned without examining the mechanism (a type of learning based on memorizing), it becomes clear that organic chemistry cannot be understood. Then, organic chemistry becomes extremely boring.

Let us go back to the beginning. We emphasized that all chemical reactions occur as a result of electron exchange. To decide whether a compound is an electron donor or not, it is necessary to examine its electronic structure. In other words, it is necessary to examine the bonds between atoms and how the bonds are polarized.

The elements share electrons so that each atom attains a noble gas configuration. For example, two chlorine atoms can each attain a filled second shell by sharing their unpaired valence electrons. A bond formed by sharing electrons is called a *covalent bond*. Similarly, the hydrogen molecule, H_2, can also form a covalent bond by sharing electrons. The atoms that share the bonding electrons in the H—H and Cl—Cl covalent bonds are identical. Such bonds are called nonpolar covalent bonds. In some compounds, the bonding electrons are shared equally between the atoms.

$$\ddot{H}\cdot + \cdot\ddot{H}\colon \longrightarrow \ddot{H}\colon\ddot{H}$$
$$\colon = \text{Nonpolar bonds}$$
$$\colon\ddot{Cl}\cdot + \cdot\ddot{Cl}\colon \longrightarrow \colon\ddot{Cl}\colon\ddot{Cl}\colon$$

In contrast, the bonding electrons between two different atoms are more attracted to one atom than to another because of different electronegativities (the ability of an atom in a molecule to attract electrons toward itself). The symmetrical distribution of the electrons between the two atoms is disrupted. This condition is called *bond polarization*.

Equal distribution of the bond electrons between A and B

A is more electronegative than B. Electrons are attracted toward, A

B is more electronegative than A. Electrons are attracted toward, B

If electrons are attracted more strongly by A, the electron density increases around atom A and decreases around B. Therefore, this polarization makes A and B atoms more reactive. In such a case, groups with high electron density prefer to bind to atom B, while those with low electron density prefer to bind to atom A. In order to understand the reaction mechanism, electron polarization between bonds must be known very well. In this chapter, we first discuss about bonds, and in the next section, we focus on bond polarization. Here, I would like to draw the attention of the reader to two concepts. The first is the *inductive effect* and the second is the *mesomeric effect*. For a student who knows and understands these two concepts well, it will be easier and more enjoyable to travel along the paths of organic chemistry that seem to be winding.

One of the other important points in organic chemistry, after learning bonding theory, is to think of molecules in a three-dimensional environment and estimate their true structures. This is an extremely simple thing. However, in order to understand it, it is necessary to work with simple organic models.

1.2 Covalent Bonding and Hybridization

In organic chemistry, unlike in inorganic chemistry, we deal with covalent bonds. Bonds are formed by overlapping of orbitals and the placement of electrons in these orbitals. Let us first look at how a hydrogen molecule is formed. The covalent bond between two hydrogen atoms is formed when the 1s orbital of one hydrogen atom overlaps with the 1s orbital of a second hydrogen atom as shown below. Overlapping orbitals can be pure orbitals as well as hybrid orbitals.

1.2 Covalent Bonding and Hybridization

We will begin the discussion with the simplest molecule in organic chemistry, methane (CH_4), with only one carbon atom. It is well known that the carbon atom is located in the center of the methane molecule, and it has four covalent C—H bonds. All the four bonds have the same length, and all bond angles are also the same (109.5°). This structure, which forms a smooth tetrahedron, is called a *tetrahedral* structure.

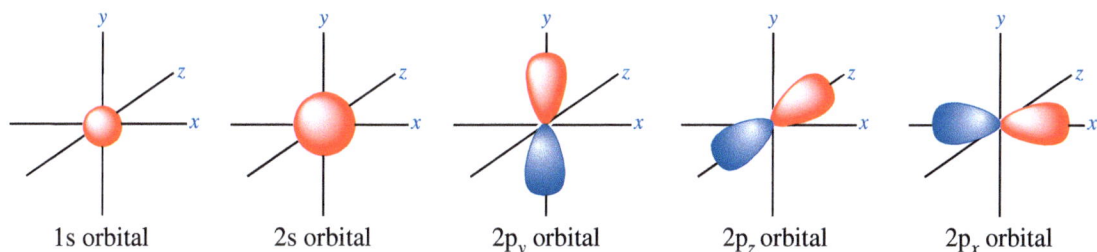

In the perspective formula, the hydrogen atoms shown with solid lines and the carbon atom are located in the paper plane. The bond represented as a dashed wedge projects behind the plane of the paper. The bond represented as solid wedge projects out of the paper (toward the viewer). To understand this structure of the methane molecule, it is necessary to first examine the electronic configuration of the carbon atom. The electronic configuration of the carbon atom is as follows:

$$1s^2 2s^2 2p_x^1 p_y^1 p_z^0$$

The atomic orbitals of the carbon atom are shown in Figure 1.1. The 2s orbital is drawn larger than the 1s orbital. Because the 2s orbital is in a location more remote from the core, it covers a larger area than the 1s orbital. The energy levels of these orbitals are different. The energy level of the 2s orbital is lower than that of the p orbitals. The energy levels of the 2p orbitals are equal to each other; in other words, they are degenerated.

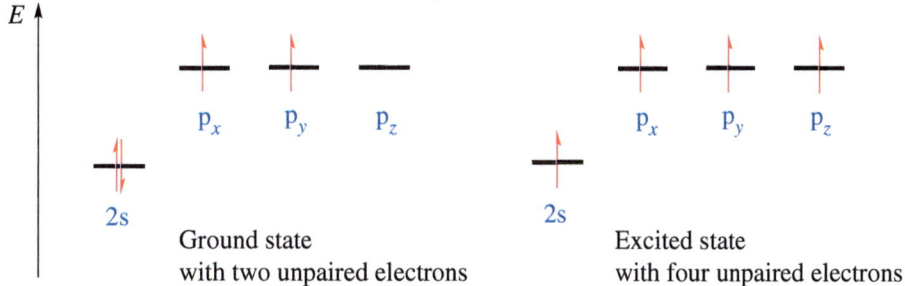

Figure 1.1 Atomic orbitals of the carbon atom.

The carbon atom has only two unpaired electrons in its electronic configuration. This indicates that this atom can form only two covalent bonds, but with two covalent bonds, it would not complete its octet configuration. However, we know that the carbon atom forms four covalent bonds. Now, we need an explanation why carbon forms four covalent bonds. Without unpaired electrons, this configuration does not allow four bonds to be made. With a small amount of energy, one electron from the 2s orbital can be promoted into the empty $2p_z$ orbital (Figure 1.2). This energy is readily compensated for by bond formation.

Figure 1.2 Electronic configuration of the carbon atom in the ground state and the excited state.

Now, this configuration of carbon can form four covalent bonds. If carbon uses an s and three p orbitals to form these four bonds, the bond formed with p orbitals will be different from the bond formed with an s orbital. On the other hand, we know from spectroscopic studies that methane has a tetrahedral structure and the four C—H bonds in methane are identical. We have to answer the question *How can the carbon atom form four identical bonds by using three p orbitals and one s orbital?* Carbon uses the hybrid orbitals. To be able to generate this geometry, the 2s orbital on carbon is mixed with all the three 2p orbitals to make four equivalent sp^3 orbitals with tetrahedral symmetry (Figure 1.3).

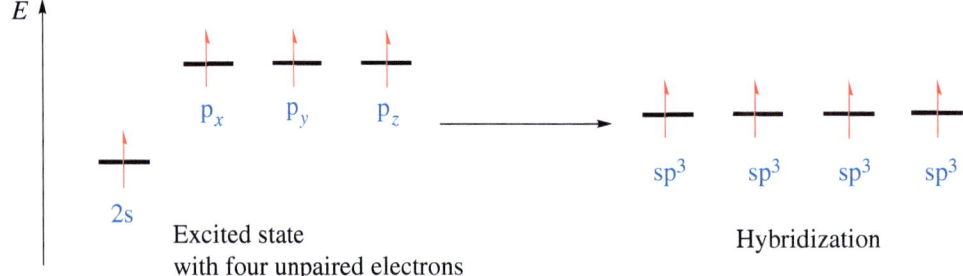

Figure 1.3 Electronic configuration of the carbon atom in the ground state and the hybridized state.

1.2.1 sp^3-Hybridization of Carbon

Now, we know why the bonds in methane are equal. One s orbital and three p orbitals of the carbon atom combine to form four new orbitals. The concept of combining was proposed by Linus Pauling in 1931. Because the number of combining orbitals is 4, the number of newly formed orbitals is also 4. The phenomenon of orbitals creating new orbitals by combining with each other is called *hybridization*. These new orbitals (hybrid orbitals) are formed by the combination of one s and three p orbitals, and so they are called sp^3 hybrid orbitals. The superscript 3 above the letter "p" indicates that three p orbitals were mixed with one s orbital to form hybrid orbitals. These orbitals are expressed as "s-p-three"; "s-p-cubed" is wrong. The hybridization involved is called sp^3 hybridization. The energies of these hybrid orbitals are equal. The hybrid orbitals formed create a pyramidal structure (tetrahedral) and the angles between the hybrid orbitals are 109.5° (Figure 1.4). The new sp^3 hybrid orbitals consist of two lobes such as the p orbitals. However, the lobes of an sp^3 orbital are not the same size. The larger lobe is used in covalent bond formation. Each sp^3 orbital has a 75% p character and 25% s character. This ratio is very important. As we explain some reactions later in the book, we will return to the s and p ratios of the hybrid orbitals.

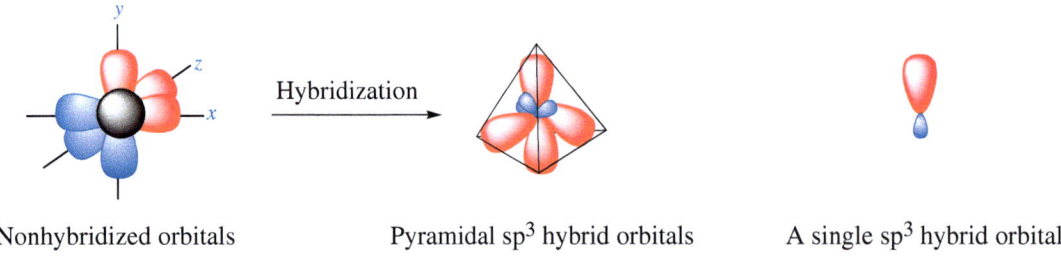

Figure 1.4 Shape of the sp^3 hybrid orbitals.

After explaining hybridization, we can go back to the methane molecule. The bonds between the carbon and hydrogen atoms in the methane molecule are not formed by the combination of pure s or p orbitals. The four sp^3 hybrid orbitals of carbon can combine with four hydrogen s orbitals forming in methane (CH_4). Because the sp^3 hybrid orbitals are equal, the bonds formed by the carbon atom are also equal, and the methane molecule has a pyramidal structure.

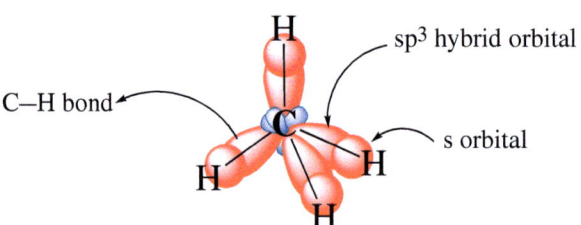

If two orbitals overlap along the bond axis (i.e. end-to-end overlap of atomic or molecular orbitals), the bond formed is called a σ bond. In saturated carbon compounds, the bonds that connect atoms are σ bonds, and the hybridization of the carbon atom is always sp^3. For example, the hybridization of the carbon atom in carbon tetrachloride (CCl$_4$) is sp^3. The structure is a pyramidal structure just like in methane, and all carbon–chlorine bonds are equal. In all compounds given below, the angle between the bonds is 109.5° as the substituents are equivalent.

$$CH_4, \ CCl_4, \ CBr_4, \ C(CH_3)_4, \ C(CH_2CH_3)_4, \ C(OR)_4$$

However, if different substituents are attached to the carbon atom, the hybridization of the carbon atom in saturated systems (tetravalent carbon) does not change, but there are some deviations from the ideal tetrahedral structure. The bond angles of some atoms and bond lengths vary according to the connected groups.

After examining the methane molecule, let us apply hybridization to the ethane molecule. In this case, we need to examine each carbon atom separately. Both carbon atoms in ethane are sp^3 hybridized and they are tetrahedral. These hybrid orbitals then combine and form σ-bonds. The ethane molecule has seven σ-bonds. Six of them are σ bonds that bind hydrogen atoms, and they are formed by overlapping of sp^3 hybrid orbitals of carbon atoms with s orbitals of hydrogen atoms. This reveals that σ bonds can also be formed by the combination of different orbitals. The C—C bond is formed by overlapping of one sp^3 orbital of one carbon with an sp^3 orbital of the other carbon (Figure 1.5).

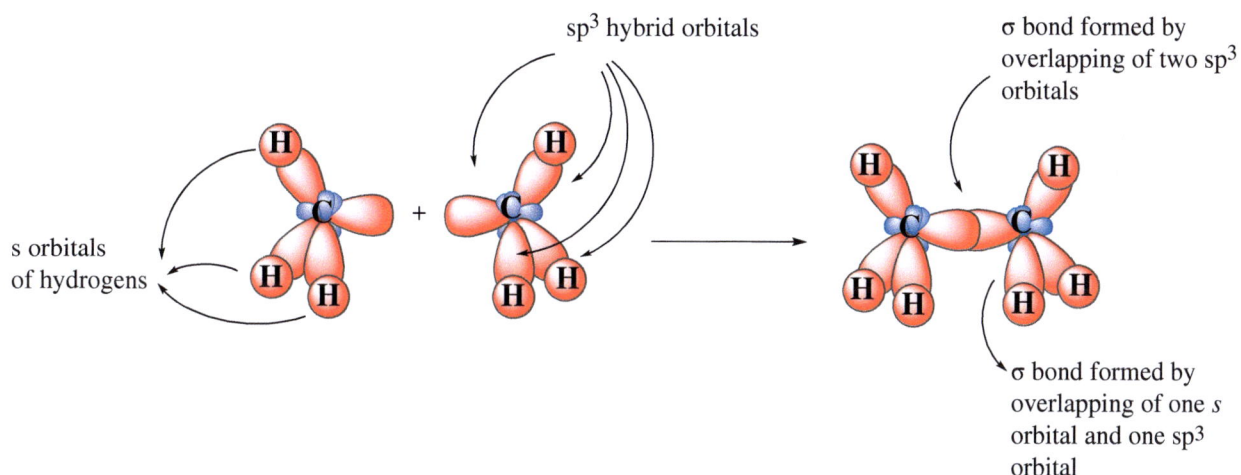

Figure 1.5 Overlap of two sp^3 orbitals to form the carbon–carbon bond in ethane.

What sort of orbitals describe the bonding in compounds such as ammonia and water having bonding and nonbonding pairs of electrons? Let us examine hybridization in ammonia (NH$_3$). The outer shell of the nitrogen atom consists of one s orbital and three p orbitals, as in the carbon atom. Nitrogen has five valence electrons. According to the basic electronic configuration of the nitrogen atom, the nitrogen has three unpaired electrons, which explains why nitrogen is trivalent, as three covalent bonds are needed for octet formation. Three p orbitals can be used for overlapping with the s orbitals of hydrogen atoms, leaving the nonbonding electrons in the 2s orbital. However, this arrangement does not minimize the repulsion between the electrons. The best solution is again sp^3-hybridization. During the formation of ammonia, one 2s orbital and three 2p orbitals of nitrogen combine to form four hybrid orbitals having equivalent energy, which is then considered an sp^3 type of hybridization just like the carbon atom (Figure 1.6).

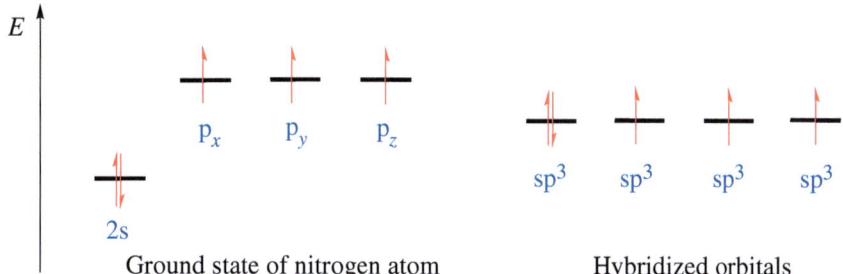

Figure 1.6 The electronic configuration of nitrogen atom in the ground state and the hybridized state.

Three of the sp³ orbitals are used to bond the hydrogen atoms. The lone electron pair settles in the remaining fourth sp³ hybrid orbital. This orbital is called a *nonbonding orbital* and the electrons in this orbital are called *nonbonding electrons*. Because the sp³ hybrid orbitals form a pyramidal structure, the hydrogen atoms attached to the nitrogen atom also form a pyramidal structure. The H—N—H bond angles are 107.3° in ammonia and it has an almost tetrahedral structure. Now, we can understand why ammonia does not have a planar structure (Figure 1.7).

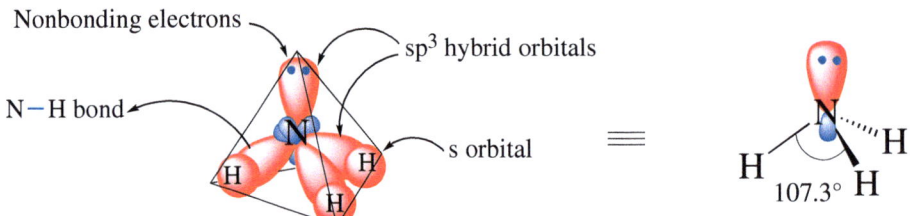

Figure 1.7 Hybridization and bonding in ammonia (NH_3).

The bonds made by nitrogen with hydrogen atoms are equivalent. While three of the four sp³ hybrid orbitals of the nitrogen atom are used for N—H bonds, the remaining fourth sp³ orbital contains the lone pair of electrons. Because the four hybrid orbitals in the ammonia molecule are not used equally as they are used in methane, a small deviation from the pyramidal structure occurs. One would expect this angle (107.3°) to be larger than in methane (109.5°) because of the repulsion between hydrogen atoms. As these nonbonding electrons are not shared, they are relatively close to the nitrogen atom. Therefore, they exert increased repulsion on the N—H bond electrons, thereby leading to bond-angle compression. Such a push brings the hydrogen atoms closer together.

When ammonia is treated with an acid (HCl), ammonia will be protonated and an ammonium salt will be formed. Hybridization of the nitrogen atom does not change. It remains as sp³. Because four hydrogen atoms are attached to the nitrogen atom, they form a smooth tetrahedron and the angle between the hydrogen atoms changes from 107.3° to 109.5°. When determining hybridization, we need to think of nonbonding electrons like a substituent.

We are faced with a similar situation in the H_2O molecule. It may be thought that this molecule is linear in the first stage. However, we know from the experimental measurements that the bond angle in water is 104.5°. Let us examine the basic electronic configuration of oxygen first. Two of the six electrons in the outer shell are located in the 2s orbital. The other four electrons are in the three 2p orbitals. Two electrons populate the $2p_x$ orbital. Therefore, oxygen has two unpaired electrons in two $2p_y$ and $2p_z$ orbitals, which reveals why oxygen forms two covalent bonds with other atoms (Figure 1.8).

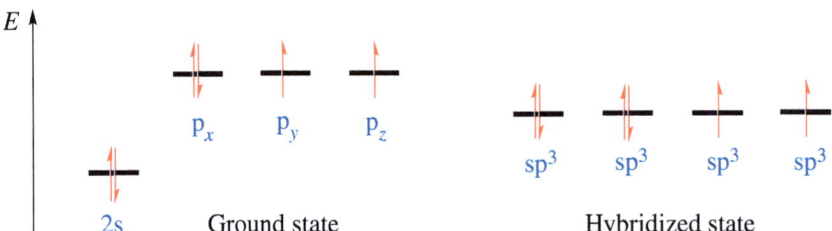

Figure 1.8 The ground-state electronic configuration of the atomic oxygen atom and the hybridized oxygen atom.

Using the atomic orbitals (p orbitals) directly for bonding is not a good model for describing bonding in water, as the angle between the bonds would be expected to be 90°. As mentioned, we know that is not the case. Again, the best solution is sp³ hybridization. During the formation of water, one 2s orbital and three 2p orbitals of oxygen combine to form four hybrid orbitals, which is then considered an sp³ type of hybridization as for the carbon atom. Two sp³ orbitals are used for overlapping with the s orbitals of hydrogen atoms, leaving the two pairs of nonbonding electrons in the remaining 2 sp³ hybrid orbitals (Figure 1.8). Because the sp³ hybrid orbitals form a pyramidal structure, the hydrogen atoms attached to the oxygen atom with the nonbonding electrons also form a pyramidal structure. The H—O—H bond angle is 104.5° in water and it has an almost tetrahedral structure. Now, we can understand why water does not have a linear structure (Figure 1.9). The H_2O molecule is not linear but angular. The repulsions between lone pairs and bonding pairs is expected to be greater, causing the H—O—H bond angle to be smaller than the ideal 109.5°.

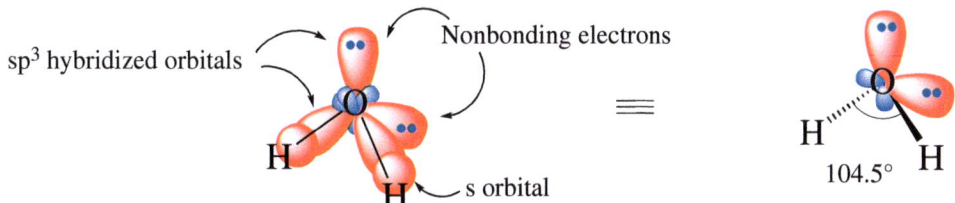

Figure 1.9 Hybridization and bonding in the H_2O molecule.

1.2.2 sp²-Hybridization of Carbon

Before examining sp²-hybridization, let us briefly consider the overlap of the orbitals and how double bonds are formed. Orbitals can overlap in two ways:

1. Head-to-head overlap
2. Side-to-side overlap

So far, we have seen the head-to-head overlap of different orbitals. Head-to-head overlap of atomic orbitals or hybridized orbitals generates σ bonds where the electron density is centered along the internuclear axis. As one can see from Figure 1.10, pure orbitals as well as hybridized orbitals can interfere and form σ bonds. The electrons in these orbitals are called σ electrons.

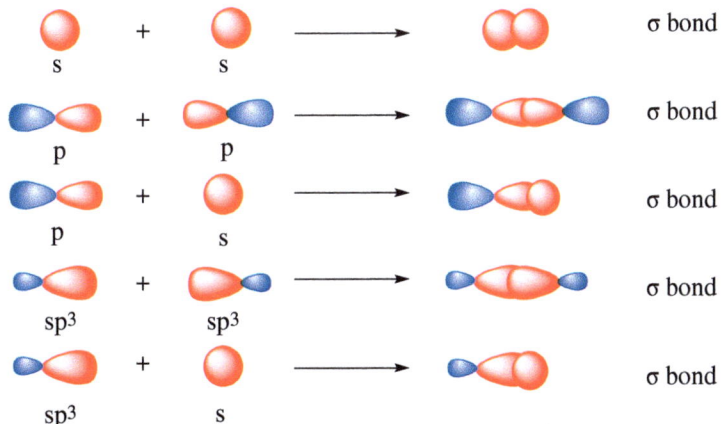

Figure 1.10 Formation of σ bonds by head-to-head overlap of different orbitals.

We have discussed the formation of a σ bond between different elements. The question is: *Is it possible to make a double bond between the carbon atoms and how can we do it?* The two carbon atoms are most easily connected along the axis via σ bonds. Because a σ bond fills the space between the two carbon atoms with electrons, it is not possible to connect these atoms with a second σ bond by head-to-head overlap. Therefore, the two carbon–carbon bonds in a double bond cannot be identical and so the second or third bonds must be different from the σ bond. The second bond of the double bond in ethylene can be constructed from side-to-side overlap of pure p orbitals of each carbon atom. Some elements use their d orbitals to form a bond.

Since there is no d orbital in the carbon atom, we will not examine the overlap with d orbitals here. However, these overlaps are important in inorganic elements having d orbitals such as sulfur and phosphorus. As can be seen in Figure 1.11, p orbitals can overlap side-to-side to form a new bond above and below the axis that connects atoms. This bond is called a *π bond*. π Bonds are formed by the overlap of pure p orbitals (unhybridized orbitals). The electrons that make up π bonds are called π electrons.

Figure 1.11 The formation of the second C—C bond by side-to-side overlap of pure p orbitals.

Now, let us try to form a double bond between the two carbon atoms. In organic chemistry, two elements can be connected by single, double, or triple bonds. For these bonds, two, four, and six electrons are needed, respectively. Regardless of the number of bonds, one of these bonds is definitely a σ bond. Because σ bonds are axially connecting the atoms, there cannot be more than one bond on this line. The second and third bonds are formed only through pure p orbitals, and these bonds are π bonds. Thus, we can make the following generalization:

A—B	Single bond	σ bond
A=B	Double bond	σ bond + π bond
A≡B	Triple bond	σ bond + 2 π bond

If there is a single bond between the two atoms, this bond is always a σ bond. When there are two bonds, one is a σ bond and the other is a π bond. In triple bonds, the first bond is a σ bond and the other two are π bonds. This rule always holds. Before we create a double bond between two carbon atoms, let us first examine the ethylene molecule, the smallest member of carbon compounds containing a double bond. The angle between the hydrogen atoms is approximately 119° and the angle between the H—C=C atoms is 121°.

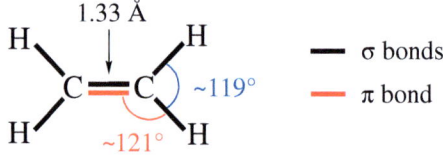

Each carbon is bonded to three atoms. As one of the pure p orbitals will be used for the π bond, the remaining orbitals, an s orbital and two of the p orbitals, will hybridize (Figure 1.12).

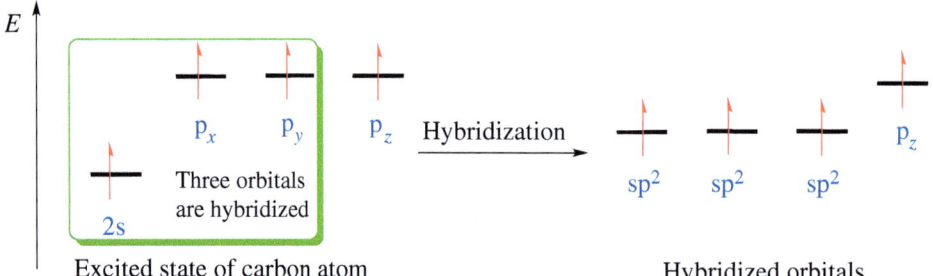

Figure 1.12 sp^2-Hybridization.

Let us examine the orbitals in the second shell of the carbon atom again. They are an s orbital and three p orbitals oriented in the *x*, *y*, and *z* directions. Now, we remove any of these p orbitals, which will be used for making the double bond, and examine the position of the remaining orbitals. The remaining orbitals (1 s and 2 p orbitals) are located in a plane (Figure 1.13). The s orbital is also in this plane because of its spherical structure. In the ethylene molecule, carbon atoms make three σ bonds apart from the π bond. If carbon uses an s and two p orbitals to form the bonds with carbon and hydrogen atoms, the bond formed with p orbitals will be different from the bond formed with the s orbital. We would expect to have an angle of at least 90° between the two groups bound to the carbon atom via p orbitals. However, it is known that the angles between the groups bound to carbon in the ethylene molecule are approximately 120°. This observation reveals that the carbon atom does not use pure orbitals when forming the σ bonds.

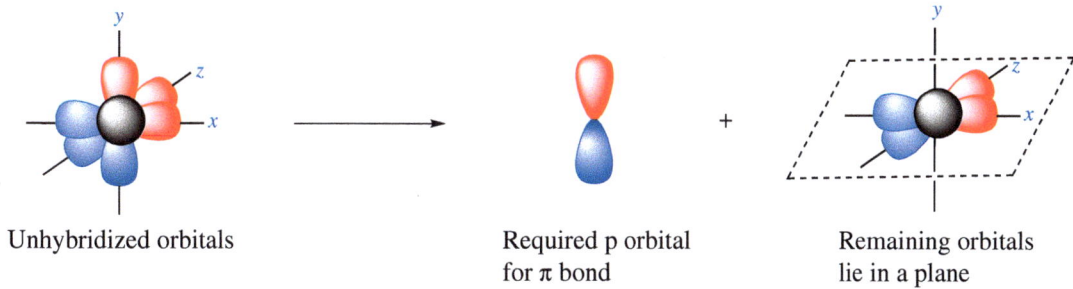

Figure 1.13 Unhybridized carbon orbitals.

This can only be explained in the ethylene molecule by hybridization. Now, these three orbitals, a single 2s orbital and two 2p orbitals, undergo hybridization and three new hybrid orbitals are formed. These orbitals are called sp^2 *hybrid orbitals*. These orbitals have the same shape and the electrons in each orbital have the same energy and they lie in a plane. To minimize electron repulsion, the three sp^2 orbitals need to get as far from each other as possible. These hybrid orbitals are separated by 120° and they are directed toward the corners of a triangle (Figure 1.14). This type of hybridized orbital is called an sp^2 (pronounced "s-p-two" not "s-p-squared") orbital.

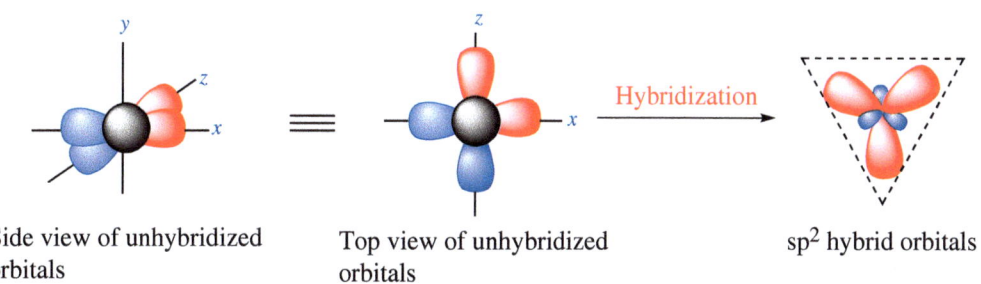

Figure 1.14 Orbitals before and after sp^2 hybridization.

The carbon atoms in ethylene form two bonds with each other. Two bonds connecting two carbon atoms are called a *double bond*. Let us try to explain how this double bond is formed. One of them, a σ bond, results from the overlap of an sp^2-hybrid orbital of one carbon with an sp^2-hybrid orbital of the other carbon (sp^2 + sp^2). This bond is formed by head-to-head overlap. The second C—C bond, a π bond, results from side-to-side overlap of the two unhybridized p orbitals, which are perpendicular to the plane formed by sp^2 hybrid orbitals. These two p orbitals must align parallel to each other so that a maximum overlap can occur. Each carbon uses the remaining sp^2 orbitals to overlap with an s orbital of a hydrogen to form the C—H bonds (Figure 1.15). As the π bond is perpendicular to the plane, π electrons are located above and below this plane.

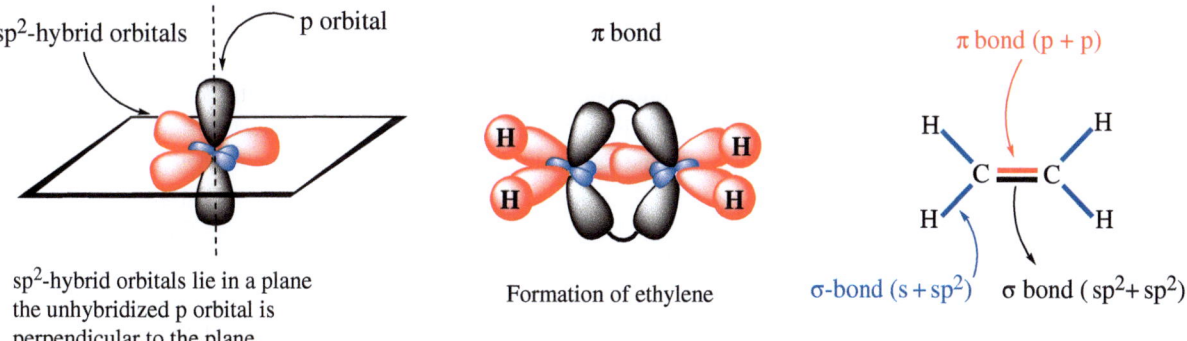

Figure 1.15 Formation of an ethylene molecule.

Because the three new sp² orbitals formed are equal, the bond angles should be 120°. However, because two different atoms (one carbon, two hydrogens) are connected to the double-bond carbon atom in ethylene, the σ bonds are not equal and a slight deviation from the ideal geometry is observed.

After discussion of sp²-hybridization, we can make a generalization. Atoms bonded to a double-bond carbon with a σ bond (whatever atom) including the double-bond carbon atoms are located in the same plane.

In ethylene, the carbon atoms are held by four electrons. However, in the case of a C—C single bond, only two electrons hold the carbons together. Therefore, a C=C double bond is stronger (174 kcal/mol or 728 kJ/mol) and shorter (1.33 Å) than a C—C single bond (90 kcal/mol or 377 kJ/mol, and 1.54 Å).

Rotation around a single bond occurs readily at room temperature, while rotation around a double bond is restricted. Restricted rotation of groups attached to the double bond causes a new type of isomerism as shown below. *cis*- and *trans*-But-2-ene are isomeric compounds, and they have different physical and chemical properties.

$$\begin{array}{cc} \text{H}_3\text{C} \quad \text{CH}_3 & \text{H}_3\text{C} \quad \text{H} \\ \text{C}=\text{C} & \text{C}=\text{C} \\ \text{H} \quad \text{H} & \text{H} \quad \text{CH}_3 \\ \textit{cis}\text{-but-2-ene} & \textit{trans}\text{-but-2-ene} \end{array}$$

Just like π bonds formed between carbon–carbon atoms, π bonds can also be formed between carbon and heteroatoms. For example, let us consider a double bond in a carbonyl group. We are going to look at the bonding in acetone, but it could equally apply to any other compound containing C=O. The interesting thing is the nature of the carbon–oxygen double bond, not what it is attached to. The hybridization of the carbon atom will be of course sp². Oxygen will form a double bond with the carbon atom, which indicates that the oxygen atom will use one of the p orbitals for making the double bond. The remaining one s and two p orbitals undergo sp²-hybridization as well. While one of the sp²-hybrid orbitals forms the σ bond with the sp² hybrid orbital of the carbon atom, the four nonbonding electrons settle in the empty sp² hybrid orbitals (Figure 1.16).

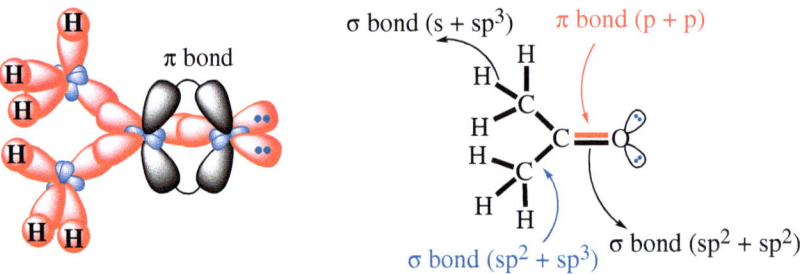

Figure 1.16 Hybrid orbitals of acetone.

The distribution of electrons in the π bond is heavily distorted toward the oxygen end of the bond because oxygen is much more electronegative than carbon. Because the hybridization of the carbon and oxygen atom in acetone is sp², all the carbon atoms lie in a plane. For example, starting from the structure of acetone, it should not be difficult to guess that formaldehyde (CH_2=O) has a planar structure.

1.2.3 sp-Hybridization of Carbon

As we mentioned above, some molecules have triple bonds between two atoms. For example, nitrogen atoms in the nitrogen molecule are connected by a triple bond. In organic chemistry, carbon atoms can also be connected to each other via a triple bond. As the simplest example, we can give the acetylene molecule. A triple bond can also be formed between carbon and nitrogen atoms (as in nitrile compounds).

$$:\text{N}\equiv\text{N}: \qquad \text{HC}\equiv\text{CH} \qquad \text{CH}_3-\text{C}\equiv\text{N}:$$

How is a triple bond formed? As we have discussed above, the two atoms can form only one σ bond along the axis that connects them. Furthermore, we explained that the second bond is a π bond. Therefore, the third bond must also be a π bond. In acetylene, carbon atoms form a triple bond using two pure p orbitals. The remaining s and p orbitals are hybridized to form two sp hybrid orbitals (Figure 1.17).

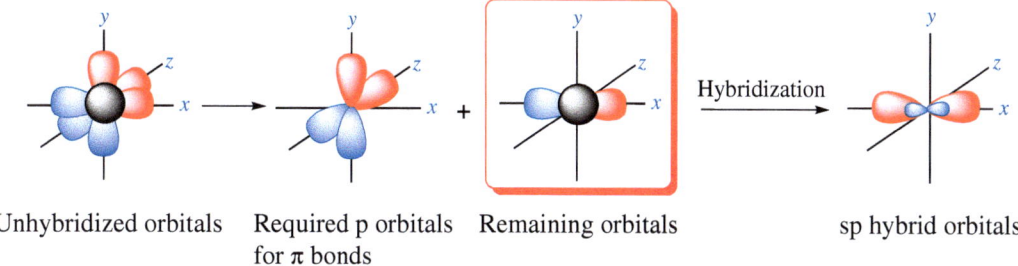

Unhybridized orbitals   Required p orbitals   Remaining orbitals   sp hybrid orbitals
                        for π bonds

Figure 1.17 sp Hybrid orbitals.

The remaining p orbitals are not hybridized. The sp orbitals have the same shape but they point in the opposite direction, and the bond angles between them are 180° again to provide maximum separation of electrons. The two p orbitals that are not hybridized are perpendicular to the axis that passes through the center of the carbon atoms. They will form the two π bonds. The two p orbitals on each carbon atom overlap side-to-side to form the π bonds. One of the sp orbitals of one carbon in acetylene overlaps with an sp orbital of the other carbon to form a C—C σ bond. The remaining sp orbital overlaps with the s orbital of the hydrogen atom to form a C—H σ bond (Figure 1.18). Because the angle between the sp hybrid orbitals is 180°, the acetylene molecule has a linear structure.

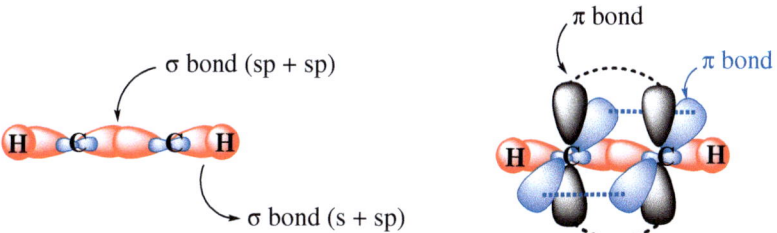

Figure 1.18 Hybrid orbitals of acetylene.

Nitrogenous compounds can also form triple bonds. The nitrogen molecule (N_2) is an "unusually stable" compound, particularly because nitrogen forms a triple bond with itself. The nitrogen atom has five valence electrons, and so it needs three more electrons to complete its octet. It is possible either to bond three atoms forming three covalent bonds or to share three electrons with another nitrogen atom, forming a triple bond. In fact, the triple bond in nitrogen is one of the strongest bonds known. Of course, nitrogen can also form a triple bond with the carbon atom. For example, acetonitrile (CH_3-C≡N) has a triple bond between nitrogen and carbon atoms. The hybridization of both nitrogen and carbon atoms is sp. The nonbonding electrons of nitrogen are in the sp hybrid orbital. The σ bond between the two carbons is formed by overlap of an sp orbital and sp^3 hybrid orbitals, and the two carbon atoms and nitrogen atom are located on a line.

In order for hybridization to be sp in a carbon atom, it is not necessary for the carbon atom to form a triple bond. For example, let us examine a compound of the cumulene type, compounds with two double bonds attached to a carbon atom such as allenes, isocyanates, and ketenes.

H_2C=C=CH_2 R—N=C=O   R—CH=C=O
Allene   Isocyanate   Ketene

The cumulated double bonds give allenes an unusual geometry because of the sp hybridization of the central carbon atom. In allenes, the two terminal carbons are sp^2-hybridized. One of the unhybridized p orbitals of the central carbon atom overlaps with a p orbital of an adjacent sp^2 carbon atom. The second p orbital of the central carbon atom overlaps with a p orbital of the other sp^2-hybridized carbon atom so that the central carbon atom produces two π bonds. These π bonds are perpendicular and so the molecule is twisted (Figure 1.19). An interesting consequence of this configuration is that allenes having two different substituents on each of the terminal carbon atoms are chiral.

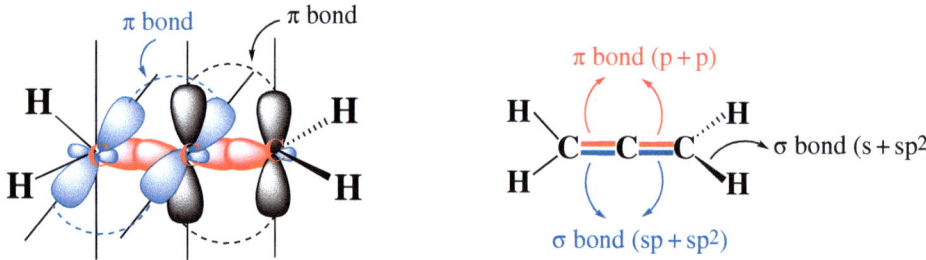

Figure 1.19 The structure of allenes and hybridization of the bonds.

Cumulenes may also have more than two double bonds. For example, 1,2,3-butatriene (CH_2=C=C=CH_2) and 1,2,3,4-pentatetraene (CH_2=C=C=C=CH_2) also belong to the cumulene class. The carbon atoms in such compounds all have a linear structure. However, the configuration of the terminal substituents depends on the number of double bonds. For an even number of double bonds, as in allene, an orthogonal configuration of terminal substituents will be seen. For an odd number of double bonds, the terminal substituents and all the carbons will lie in a plane.

After examining the hybridization of neutral compounds, let us briefly discuss the hybridization of ionic compounds (carbocations and carbanions) and radicals. Carbocations are carbon compounds that carry a positive charge on a carbon atom, which is attached to three groups. For example, positively charged carbon in the methyl cation is bonded to three hydrogen atoms and the hybridization of the carbon atom is sp^2. Therefore, the carbon atom forms three covalent bonds by using the three sp^2 hybrid orbitals. The remaining p orbital is empty and perpendicular to the plane formed by hydrogen and carbon atoms (Figure 1.20).

Radicals are molecules that contain at least one unpaired electron. The hybridization of the carbon atom in the methyl radical is also sp^2. The methyl radical differs from the methyl cation by having one unpaired electron, which is located in the p orbital. The p orbital is perpendicular to the plane formed by methyl protons and the carbon atom.

The situation is different in carbanions. The structure of the methyl anion can be compared with the electronic structure of ammonia. The hybridization of the carbon atom is sp^3. Three of the sp^3 hybrid orbitals overlap with the s orbital of hydrogen, forming σ-bonds. The remaining sp^3 orbital holds the electrons (Figure 1.20).

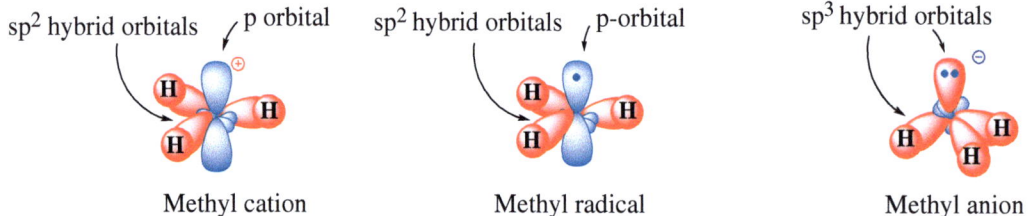

Figure 1.20 Hybridization in methyl cation, methyl radical, and methyl carbanion.

We can summarize what we have seen so far as follows. Single bonds are always made from σ bonds. One of all double bonds is a σ bond, and the other is a π bond. Triple bonds consist of two π bonds and one σ bond. The simplest method to determine the hybridization of a carbon atom or any heteroatom is to look at the number of π bonds that the atom has formed. If the atom does not form any π bonds, the hybridization is sp^3, and if there is one π bond, then the hybridization is sp^2. In the case of having two π bonds, the hybridization is sp.

1.2.4 Bond Lengths

After examining bond structures, let us take a brief look at bond lengths and bond energies. Bond energies are directly related to the reactivity of molecules. As the number of bonds connecting the two atoms increases, the distance between the atoms becomes shorter. Triple bonds are shorter than double bonds and double bonds are shorter than single bonds. The hybridization affects the C—C and C—H bond lengths (Table 1.1).

Table 1.1 The bond distances between carbon and heteroatoms in Å.

	C	N	O	Type of bonding
C	1.54	1.47	1.43	Single bond
	1.33	1.28	1.23	Double bond
	1.20	1.21	—	Triple bond
N	1.47	1.46	1.41	Single bond
	1.28	1.25	1.14	Double bond
	1.16	1.10	—	Triple bond
O	1.43	1.41	1.49	Single bond
	1.23	1.14	1.21	Double bond

The carbon–carbon single bond is 154 Å in length, the double bond is 1.33 Å, and the triple bond is around 1.20 Å. Interestingly, C—H bonds are generally much shorter and stronger than C—C single bonds. This is because the s orbital of hydrogen is closer to the nucleus than is the sp^3 orbital of carbon. The length and strength of a C—H bond depend on the hybridization of the carbon atom to which the hydrogens are connected. The more s characteristic used in the orbital of carbon to make a bond, the shorter and stronger the bond is. Therefore, a C—H bond in ethane formed by overlap of a sp^3-hybridized orbital (25% s ratio) is longer and weaker than a C—H bond formed by overlap of an sp^2 hybridized orbital (33.3% s ratio). On the other hand, a C—H bond formed by overlap of an sp hybrid orbital is shorter and stronger than the other bonds. For example, the C—H bond in acetylene (1.06 Å) is shorter than the C—H bond in ethylene (1.08 Å). The C—H bond in ethane (1.10 Å) is longer. Generally, s orbitals are spherical (more compact) and closer to the atomic nucleus. When the distance between two atoms is short, the electron density is high and the bond is stronger.

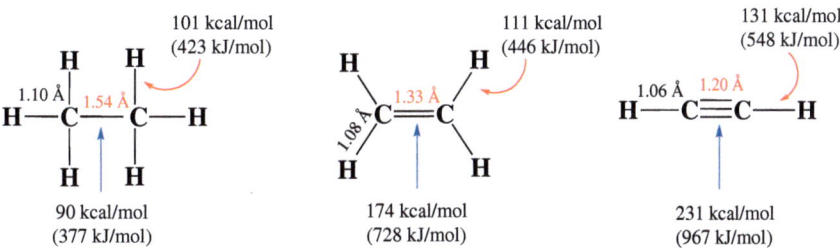

Let us compare the energies of carbon–carbon single bonds with those of carbon–carbon double bonds. A C=C double bond (174 kcal/mol) in ethylene is stronger than a C—C bond (90 kcal/mol) in ethane. As one can see, the C=C double bond of ethylene is not twice as strong. Let us calculate how strong the π bond is compared to a σ bond. The strength of the σ bond in ethane formed by sp^3–sp^3 overlap is 90 kcal/mol. The strength of the σ bond in ethylene cannot have this value because that bond is formed by sp^2–sp^2 overlap. Therefore, it must be stronger because of the increased s ratio in the overlapping orbitals. Experiments reveal that the energy of the σ bond in ethylene is approximately 112 kcal/mol. Now, we can calculate the strength of the π bond as 174−112 = 62 kcal/mol. As expected, the π bond is weaker than the σ bond because of the side-to-side overlap. As a consequence of this bond weakness, π bonds undergo chemical reactions more easily than σ bonds do.

1.3 Electrophiles and Nucleophiles

There are some basic concepts in organic chemistry that need to be learned before moving on to the reaction mechanisms. Once the basic concepts have been firmly grasped, the reaction mechanisms will be better understood. Therefore, in this section, we will primarily focus on these basic concepts.

1.3.1 Electrophiles (Electrophilic Compounds)

Chemical reactions are those that are based on electron exchange. During a chemical reaction, one compound donates electrons and the other accepts them. For a compound to accept electrons, it must be electron deficient. Therefore, to understand how the functional groups react, we must first learn to determine electron-deficient and electron-rich atoms and molecules. Electron-deficient atoms and molecules, which can take electrons, are called electrophilic compounds or *electrophiles*. Electrophile literally indicates electron loving. Electrophiles react with electron-rich compounds and try to achieve a stable shell of electrons like that of a noble gas. We can categorize electrophilic compounds in various groups.

- Cations (compounds that contain a positive charge)
- Lewis acids
- Carbonyl compounds
- Compounds containing polarized bonds

Cations: All atoms or molecules such as carbon, nitrogen, and halogens that have lost one or more valence electrons form an ion with a positive charge. For example, all carbocations (a more detailed description of carbocations is given in the carbocation section) with sp^2-hybridization and bound to three substituents are electrophiles. They are electron-deficient compounds. By reacting with nucleophiles, they accept electrons to neutralize the positive (+) charge and form a stable electronic configuration. Some examples of electrophiles having a positive charge are given below.

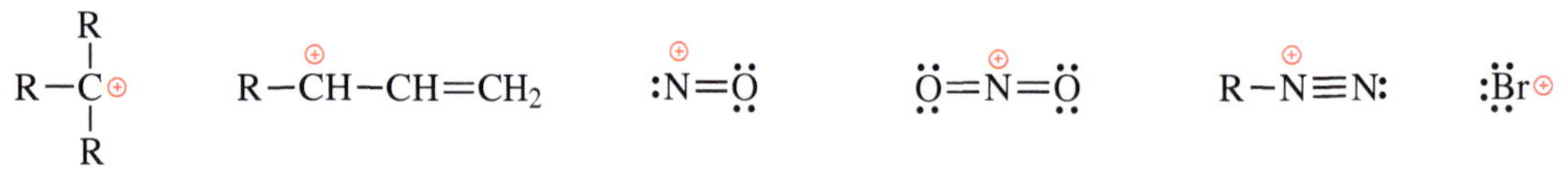

Lewis acids cannot deliver protons like classical acids. Compounds that are electron deficient are included in this group. Molecules with an incomplete octet of electrons such as AlCl$_3$, BF$_3$, and FeCl$_3$ can act as a Lewis acid. Generally, the number of electrons in the outer shell of these compounds is six. These compounds can easily react with electron-rich compounds and accept an electron pair to fill its valence shell.

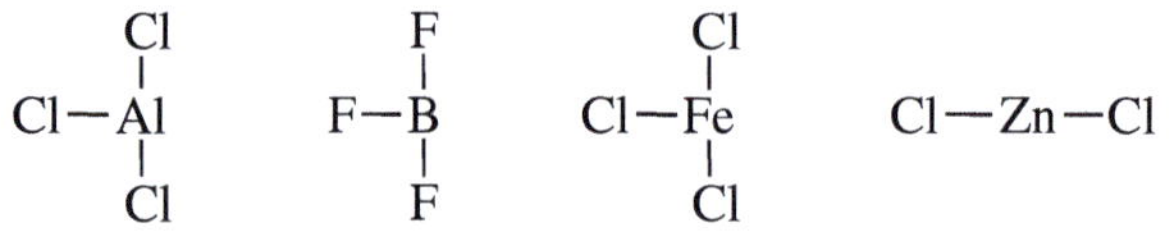

Carbonyl compounds are also potential electrophilic compounds. Double-bond electrons between carbon and oxygen are polarized because of the fact that oxygen is more electronegative than carbon. There is a partial positive charge on the carbon and a partial negative charge on the oxygen. The resonance structure of a carbonyl group is shown below to emphasize the charge separation in the carbonyl group. The structure has only one bond between the carbon and oxygen. In this structure, oxygen has an octet but carbon does not. Consequently, carbonyl carbon shows electrophilic properties and reacts easily with nucleophiles. The double bond between sulfur and oxygen is also polarized and sulfur shows electrophilic properties.

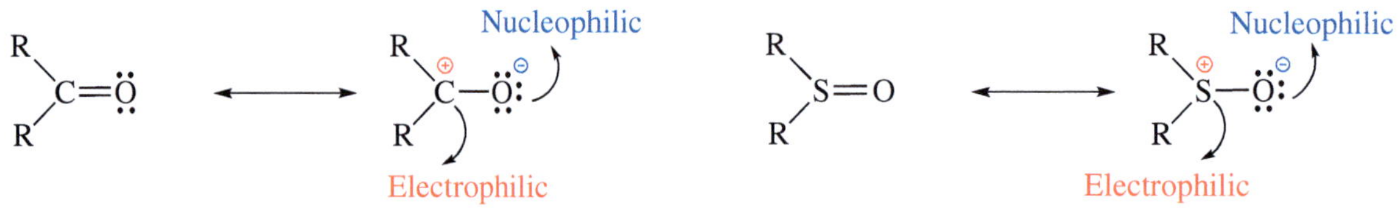

Apart from these compounds, any compound with bond polarization shows electrophilic properties. For example, alkyl halides contain a polarized σ bond between carbon and halogen atoms because halogens are more electronegative than carbon atoms. The halogens attract carbon–halogen bond electrons toward themselves. Consequently, the carbon atoms are partially positively (δ+) charged, while the halogens are negatively (δ−) charged. For this reason, the carbons bound to the halogen atom have electrophilic properties and react easily with nucleophiles. Strong electron-withdrawing groups such as tosyl and acetoxyl groups also make the carbon atom to which they are attached electrophilic as they also attract the bonding electrons toward themselves.

1.3.2 Nucleophiles (Nucleophilic Compounds)

Nucleophiles are chemical species that donate electrons to form a chemical bond and they attack electron-deficient regions in chemical reactions. Nucleophiles are "nucleus loving" compounds. We can classify nucleophiles in different groups.

- Compounds having lone pairs
- Olefinic and aromatic compounds
- Anions and Lewis bases

All elements in the fifth and sixth groups of the periodic table contain nonbonding electrons. All of these fall into the nucleophilic compound class because there are electrons in their outer shell and they react easily with electrophiles. For example, compounds such as amines, ethers, alcohols, and mercaptans are the most important representatives of this group. These compounds belong to the Lewis base group as well.

Anions also fall into the nucleophile group because they are negatively (−) charged and react easily with electrophiles.

Carbon–carbon double bonds are compounds in which the electron density is rich. They donate a pair of electrons as well. This not only covers double bonds but also triple bonds (alkynes) and even enols and enolates. Aromatic compounds are electron-rich compounds because they have π electrons that are freely delocalized. As these compounds react easily with electrophiles, they are also considered nucleophiles.

1.4 Inductive and Mesomeric Effects

1.4.1 Inductive Effect

In the introductory section, we emphasized that the inductive and mesomeric effects should be learned very well in order to understand the reaction mechanisms and the estimation of possible reaction products, etc., because these are the factors that directly determine and affect the electron density in various regions of the molecule. In symmetrical σ bonds (bonds formed between the same atoms), the electron density is equally distributed between the two atoms because they have the same tendency to attract the bonding electrons. For example, in compounds such as hydrogen, chlorine, bromine, and ethane, the bond electrons are not polarized; they are homogeneously distributed between the two atoms.

$$H-H \quad Cl-Cl \quad Br-Br \quad O=O \quad N\equiv N \quad H_3C-CH_3$$

If atoms are different from each other, bond electrons are attracted more strongly by one of the atoms and the distribution of the electron density varies. Let us consider a σ bond that binds A and B atoms. If atom A is more electronegative than atom B, A will attract electrons toward itself and will decrease the electron density around atom B, while the electron density around A will increase. A polar bond is a covalent bond in which there is a separation of charge between one end and the other. One end of the bond is slightly positively polarized and the other slightly negatively polarized. Partial charges are indicated by "δ" (delta). The signs $\delta+$ and $\delta-$ indicate partial charges.

$$A—B \qquad A\text{-}\text{-}\text{-}B \qquad A\bullet B \qquad \overset{\delta-\quad\delta+}{A—B}$$

While atom B becomes electrophilic because of bond polarization, atom A is nucleophilic. The presence of the partial negative and positive charges on these atomic centers makes them liable to be attacked by other charged groups or atoms. This bond polarization determines where electrophiles and nucleophiles attack. To determine how a bond between the two atoms is polarized, it is sufficient to look at a table showing electronegativities. With simple rules, it is easy to estimate which element is more electronegative relative to the location of the elements in the periodic table. The electronegativities of some elements that are important in organic chemistry are given in Table 1.2. As is known from general chemistry classes, electronegativity increases from left to right within a row and decreases in a group from top to bottom in the periodic table.

Table 1.2 Electronegativities of selected elements.

H 2.1							
Li 1.0	Be 1.6	B 2.0	C 2.5	N 3.0	O 3.4	F 4.0	
Na 0.9	Mg 1.3	Al 1.6	Si 1.9	P 2.2	S 2.6	Cl 3.2	Increasing electronegativity ↑
K 0.8	Ca 1.0				Se 2.4	Br 3.0	
Rb 0.8	Sr 1.0					I 2.7	

Increasing electronegativity →

One of the best examples we can give for bond polarization is alkyl halogen bonds. As can be seen from Table 1.2, halogens (F, Cl, Br, and I) are more electronegative than the carbon atom; they attract bonding electrons and polarize the C-halogen bonds. The carbon atom is then partially positively charged and halogens are negatively charged. This charge distribution (polarization) is responsible for all kinds of chemistry of alkyl halides.

If an element is more electronegative than hydrogen, this indicates that it attracts more electrons than hydrogen does, the inductive effect of that element is (−), and the effect is shown as −I. If the element is less electronegative than hydrogen, that is, it repels electrons, the inductive effect of the element is (+) and is indicated as +I. In general, the inductive effects of various substituents are (−) because the electronegativity of almost all the elements bound to the carbon atom is higher than that of hydrogen. However, the inductive effects of the alkyl groups are (+). The electron donor feature of the alkyl groups can be explained as follows. Hydrogen atoms attached to the methyl carbon atom are electron donors relative to the carbon atom. Therefore, the hydrogen–carbon bond polarizes toward the carbon atom and the electron density increases around the carbon atom. When the electron density on carbon increases, this carbon atom repels electrons toward the neighboring carbon atom to reduce the electron density.

1.4 Inductive and Mesomeric Effects

The inductive effect is transmitted only through σ electrons. Let us try to explain the inductive effect in more detail with a fluorine atom attached to the end of a saturated carbon chain. Because the fluorine atom is more electronegative than the carbon atom, it is partially negatively ($\delta-$) charged by attracting electrons toward itself. The carbon atom C1, whose electrons are drawn, is partially positively ($\delta+$) charged. We describe this situation as bond polarization.

$$-\overset{5}{CH_2}-\overset{4}{CH_2}-\overset{3}{CH_2}\rightarrow\overset{2}{CH_2}\Rrightarrow\overset{1}{CH_2}\Rrightarrow F$$

The partially positively ($\delta+$) charged carbon atom C1 withdraws bond electrons from the neighboring carbon atom, C2, to remove the reduced electron density around itself. However, this withdrawal is never at a level to neutralize the (+) charge on the carbon atom C1. When C1 pulls bond electrons from C2, C-2 will also be partially positively ($\delta+$) charged, and the C1—C2 bond will also be polarized. However, the positive ($\delta+$) charge formed on C2 is never as much as that on C1. C-2 behaves similarly, drawing electrons from C3. C3 is also partially positively ($\delta+$) charged. *For how many atoms does this last?* The inductive effect usually loses its effect after the third bond. The effect is not observed after the fourth and fifth bonds. The inductive effect propagates through the σ skeleton and drops exponentially from one carbon to another.

It is possible to show through some experiments that the inductive effect decreases exponentially on a chain. As we will discuss in the acid–base part in detail, the substituents significantly affect the pK_a values of organic acids. An organic acid turns into a carboxylate ion by removing the proton, as shown below. The farther this balance shifts to the right, the greater the acid strength is. The greater acidity of acetic acid is the result of resonance stabilization of the acetate anion.

$$H_3C-COOH \rightleftharpoons H_3C-COO^{\ominus} + H^{\oplus}$$

There is a negative charge (−) on the carboxylate ion. If this charge can be distributed over groups bound to the acid functional group, the anion can be stabilized. In general, there are two ways to stabilize the positive (+) or negative (−) charges: either neutralize these charges or distribute them as much as possible over other groups. In order to distribute the negative (−) charge over other groups, the groups must have electron-withdrawing substituents.

For example, the pK_a value of acetic acid is 4.86. When one of the methyl protons in acetic acid is replaced by a halogen, the carboxylate anion formed will be stabilized because of the inductive effect of halogen. The strength of the acid will increase, and the corresponding pK_a value of the acid will decrease. The change in pK_a values is directly proportional to the inductive effect of halogen. It can be seen from Table 1.3 that the strongest acid is fluoroacetic acid because fluorine is the most electronegative element. In the second group of acids, the pK_a value of butanoic acid is compared with the pK_a values of chlorobutanoic acids. If the chlorine atom is bound to the carbon atom C2, the pK_a value drops from 4.86 to 2.80. This is a very important change. It shows that the inductive effect of chlorine is very important; it can stabilize the anion. When the chlorine atom is attached to the carbon atom C3, a significant change in the pK_a value is observed. This time, the pK_a value rises from 2.86 to 4.06, which indicates that the strength of the acid decreases. This value shows that the inductive effect of the chlorine atom decreases significantly as the carbon chain is extended. The change in pK_a value is quite small when the chlorine atom is bound to the carbon atom C4. If we extend the carbon chain by one more carbon and connect the chlorine atom to the carbon atom C5, this time we will see that the chlorine atom has no effect on the pK_a value. Finally, observe that the effect falls off quite quickly as the attached halogen gets further away from the –COO– end. This table clearly shows that the inductive effect is transmitted through the σ bonds and the effect drops exponentially within the three bonds.

Table 1.3 pK_a values of some carboxylic acids.

Acid	pK_a	Acid	pK_a
CH_3COOH	4.86	$CH_3CH_2CH_2COOH$	4.82
FCH_2COOH	2.59	$CH_2CH_2CHCOOH$ $\mid$ Cl	2.84
$ClCH_2COOH$	2.87	CH_3CHCH_2COOH $\mid$ Cl	4.06
$BrCH_2COOH$	2.90		
ICH_2COOH	3.18	$CH_2CH_2CH_2COOH$ $\mid$ Cl	4.56

Another important example that experimentally shows how the inductive effect changes along the σ skeleton is the chemical shift values observed in the ¹H-NMR spectra of the compounds. The ¹H-NMR chemical shifts of compounds reflect a linear correlation between the electron density and the chemical shifts. For example, the terminal methyl protons of an alkyl chain resonate at 0.9 ppm. When the carbon atom C2 is replaced by an oxygen atom, the oxygen atom inductively attracts electrons and the electron density around the methyl group decreases. Thus, the resonance signal of the terminal methyl protons shifts from 0.9 to 3.3 ppm. The change observed in the chemical shift is 2.4 ppm. This value is a significant change in NMR spectroscopy. When the carbon atom C3 is replaced with an oxygen atom, the chemical shift of the terminal methyl group appears at 1.3 ppm. The chemical shift's change is now 0.4 ppm. This clearly indicates that the inductive effect of the oxygen atom on the methyl protons decreases significantly in the C3 position. In the C4 position, the effect drops to 0.1 ppm and then to zero. This example again shows that the inductive effect changes exponentially from one carbon to another.

$$\begin{array}{ccccc} 5 & 4 & 3 & 2 & 1 \\ -CH_2-CH_2-CH_2-CH_2-CH_3 & & & & 0.9\ \text{ppm} \\ -CH_2-CH_2-CH_2-O-CH_3 & & & & 3.3\ \text{ppm} \\ -CH_2-CH_2-O-CH_2-CH_3 & & & & 1.3\ \text{ppm} \\ -CH_2-O-CH_2-CH_2-CH_3 & & & & 1.0\ \text{ppm} \\ -O-CH_2-CH_2-CH_2-CH_3 & & & & 0.9\ \text{ppm} \end{array}$$

Table 1.4 The types of inductive effect: functional groups.

−I Groups			+I Groups	
$-NH_3^+$	$-CHO$	$-SH$	$-CH_3$	$-COO^-$
$-NR_3^+$	$-COR$	$-SR$	$-CH_2R$	$-O^-$
$-NO_2$	$-F$	$-CR=CH_2$	$-CHR_2$	$-NH^-$
$-C\equiv N$	$-Cl$	$-CR=CR_2$	$-CR_3$	
$-COOH$	$-Br$	$-C\equiv CH$		
$-COOR$	$-OR$			

As can be seen in the Table 1.4, almost all other substituents, apart from the alkyl groups, are inductively electron-attracting groups (−I effect) because they are more electronegative than the hydrogen atom is. Carbon–carbon double bonds and carbon–carbon triple bonds are also inductively electron-attracting groups. The σ bonds of these groups are formed by overlap of sp^2 and sp hybrid orbitals. Because the s ratio in these orbitals is higher than the ratio in sp^3 hybrid orbitals, they attract the electrons more. Alkyl groups are generally electron-donating groups (+I effect). In the carboxylate anion, the carbonyl group is not an electron-attracting group because the electron density on the carbon atom is increased. Negatively charged oxygen and nitrogen atoms also lose their electron-withdrawing properties and become electron donors.

Functional groups such as aldehyde and ketones are inductively electron-attracting groups. When the resonance structures of these groups are drawn, it is seen that a positive charge is created on the carbon atom. Because carbonyl carbon is positively charged, it attracts the neighboring σ bond electrons. This behavior of the carbonyl group makes the carbonyl groups inductively electron attracting.

1.4.2 Mesomeric Effect (Resonance Structures)

Before describing the mesomeric effect, let us introduce the resonance structures of compounds. When a compound can be represented by more than one Lewis structure, the compound is said to possess resonance. If there are multiple double bonds in organic compounds and they are conjugated, double-bond electrons can change their position within the molecule. If there are nonbonded electrons adjacent to double-bond electrons, they also participate in conjugation. To clarify the issue, consider the Lewis structure of the carbonate ion, CO_3^{2-}, and acetate ion.

1.4 Inductive and Mesomeric Effects | **19**

The carbonate anion has three oxygen atoms attached to the carbon atom. The Lewis structure for this ion has a carbon–oxygen double bond (C=O) and two carbon–oxygen single bonds (C—O). Each of the single bonded oxygen atoms bears a formal charge of −1. We know that single and double bonds have different bond lengths and a double bond is shorter than a single bond. All three carbon–oxygen bond lengths in the carbonate ion have been shown experimentally to be equal. Moreover, all three oxygen atoms bear equal amounts of negative charge. However, we cannot write a Lewis structure that shows that these three bonds are equal. Under these circumstances, the concept of resonance has to be used. This indicates that the electrons are delocalized over four atoms and the molecule is resonance stabilized if two or more resonance structures can be drawn. The carbonate anion is actually a hybrid structure consisting of three resonance structures.

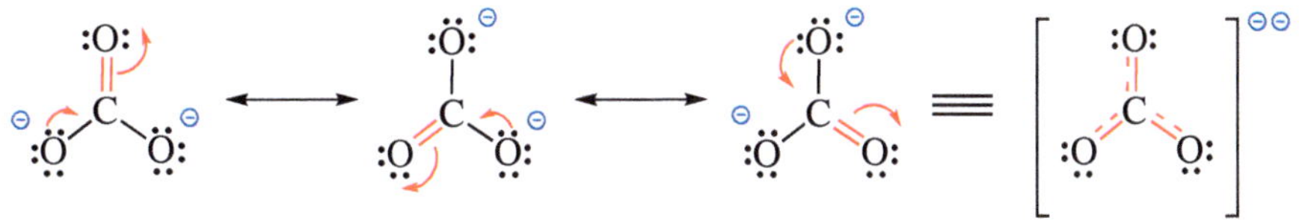

There is a similar situation in organic acids. Accordingly, one of the oxygen atoms is bonded with the double bond to the carbon atom and the other with a single bond. The distance between carbon and oxygen atoms should be different. Experimental and spectroscopic findings show that these two oxygen atoms are equivalent. Therefore, structures A and B do not show the true structure of the acetate ion. The true structure of the acetate ion is something between these two structures. The negative charge (−) and the π electrons are evenly distributed on the two oxygen atoms. This is also called resonance and the two individual line-bond structures for the acetate ion are called resonance forms. The actual structure is shown by C. In similar cases, if there are two or more resonance structures, the resonance relationship is indicated by a double-headed arrow (↔) between them. It is wrong to place two arrows (⇌) between the resonance structures. Two arrows indicate balance.

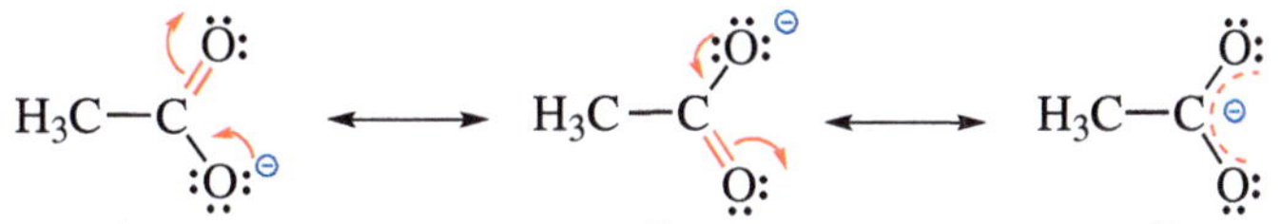

Electrons can delocalize over many atoms. Let us examine the structure of benzene. For example, structures A and B of benzene shown below do not reveal its true structure.

If the benzene structure with alternating single and double bonds had been correct, there would have been two different isomers of 1,2-disubstituted benzene derivatives. For example, the isomers of 1,2-dibromobenzene should be different according to the carbon-carbon bonds between the bromine atoms.

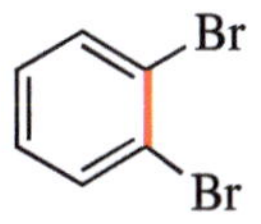

— between bromine atoms

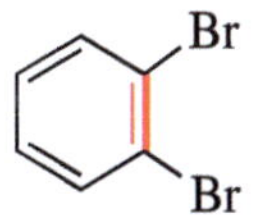

= between bromine atoms

However, spectroscopic investigations at very low temperatures showed that there are not two different structures. Experimental studies, especially those employing X-ray diffraction, showed that benzene has a planar structure with equal C—C distances of 1.39 Å, which is shorter than that of a carbon–carbon single bond (1.54 Å) but longer than that of a C=C double bond (1.34 Å). In other words, benzene does not have alternating single and double bonds. These intermediate distances are explained by electron delocalization. The carbon atoms in benzene are sp²-hybridized. The p orbitals from each carbon atom overlap to form a delocalized molecular orbital that extends around the ring, giving stability and decreased reactivity to the benzene ring. To reflect the delocalized nature of the electrons, benzene is often depicted by a circle inside a hexagonal arrangement of carbon atoms as shown above. π Electrons are not localized in benzene but are distributed evenly over six carbon atoms. The distribution of electrons in the benzene molecule is a resonance (mesomer) event, not a valence isomerization. The real structure of benzene should be shown by C.

From what we have seen so far, we come to the following conclusion. π Electrons and nonbonding electrons in conjugated systems can be distributed through p orbitals over the system. As a result of the distribution of electrons on the molecule, charge polarization can occur, which is extremely important for the reactions.

Let us examine the resonance process in another system. In the α,β-unsaturated carbonyl compounds, the C=C double bond, and the C=O double bond are separated just by one C—C single bond. This indicates that the double bonds are conjugated. Because oxygen is more electronegative in the C=O bond, electrons are withdrawn toward the oxygen atom and oxygen is negatively (−) charged. Because carbonyl carbon is positively (+) charged, it attracts the neighboring π electrons. Thus, the double bond between the carbon–carbon bond changes place. The structures shown below are resonance structures. Conjugation of a double bond to a carbonyl group transfers the electrophilic characteristic of the carbonyl carbon to the β-carbon atom of the double bond. According to this polarization, while the β-carbon atom becomes electrophilic, the oxygen atom becomes nucleophilic. The polarized structure thus formed determines the chemistry of the α,β-unsaturated carbonyl compounds. For example, nucleophiles attack the C4 carbon atom, while electrophiles bind to oxygen. Therefore, in order to decide where reagents will attack in a chemical reaction, we need to know the bond polarization within the molecule, in other words, the charge distribution.

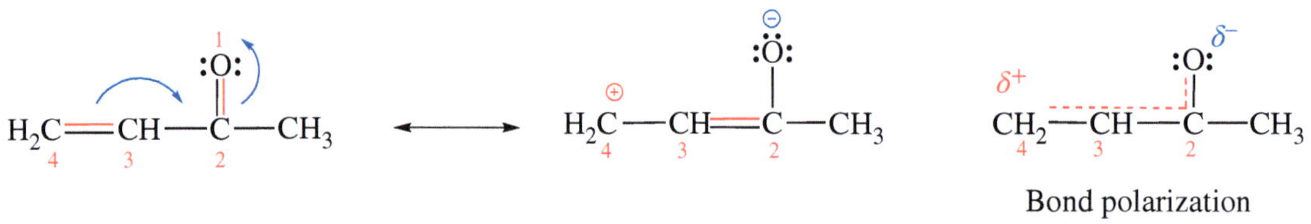

Bond polarization

The most important difference between the mesomeric effect and the inductive effect is that with the latter, σ electrons are pulled or pushed, so that the electron density is shifted toward the more electronegative atom and it is a permanent effect. With the mesomeric effect, π electrons are pulled or pushed through p orbitals. In the inductive effect, bond electrons do not change their position; they are only polarized. In the mesomeric effect, π electrons change their positions. Of course, the system must be conjugated. Nonbonding electrons can also be involved in the mesomery. For this, the nonbonding electrons must be conjugated with the double bonds. The mesomeric effect can be categorized as "negative" and "positive" based on the properties of the substituent. The mesomeric effect is positive (+M) when the substituent is an electron-donating group and negative (−M) when the substituent is an electron-withdrawing group.

Now let us return to the α,β-unsaturated carbonyl compound we reviewed above. With carbonyl group polarization, the π-electrons of the double bond are attracted and the electrons change position. The double-bond electrons are shifted between the C2 and the C3 bond. Here, the mesomeric effect of the carbonyl group is negative (−).

It is extremely simple to determine whether the mesomeric effect of any functional group is +M or −M. Suppose that there is a double bond or triple bond between X and Y (−X=Y or −X≡Y) that is conjugated with a carbon–carbon double bond or an aromatic system. If atom Y is more electronegative than atom X, double-bond electrons will be withdrawn by the electronegative element Y. As a result, element X will be positively (+) charged and it will withdraw the conjugated double-bond electrons mesomerically toward it. In this case, the mesomeric effect of this functional group, −X=Y, will be −M. Almost all of the double bonds in organic chemistry have a −M effect.

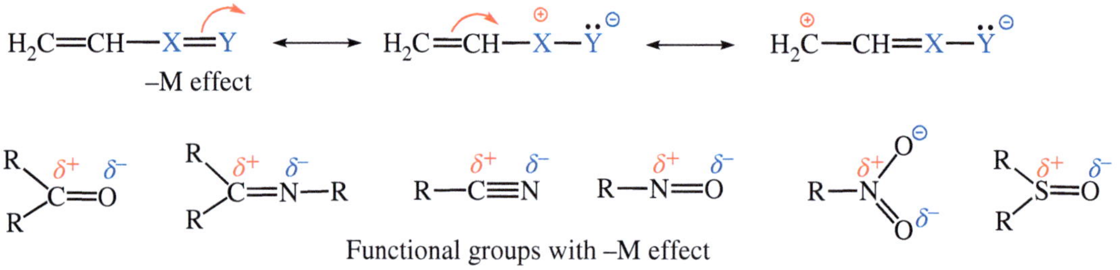

−M effect

Functional groups with −M effect

1.4 Inductive and Mesomeric Effects

Functional groups with +M mesomeric effects are usually heteroatoms with nonbonding electrons. When these groups are directly attached to a double bond or aromatic ring, they act as electron-donating groups based on the resonance structures. Let us examine the behavior of an oxygen atom attached to the aromatic ring and how it affects the reactivity of the group to which it is attached.

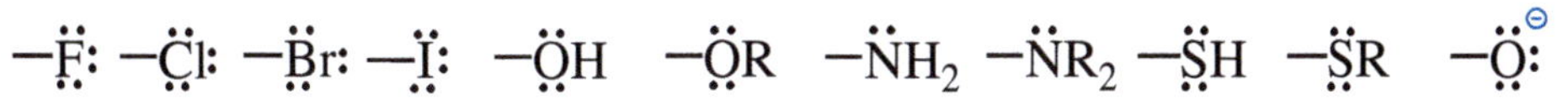

$$-\ddot{F}: \quad -\ddot{C}l: \quad -\ddot{B}r: \quad -\ddot{I}: \quad -\ddot{O}H \quad -\ddot{O}R \quad -\ddot{N}H_2 \quad -\ddot{N}R_2 \quad -\ddot{S}H \quad -\ddot{S}R \quad -\ddot{O}:^{\ominus}$$

Functional groups with +M effect

The resonance structures of phenol are studied in detail in basic organic chemistry courses. There are very important points that we need to emphasize here. As is known, oxygen is an electronegative atom that attracts electrons inductively (through σ bonds).

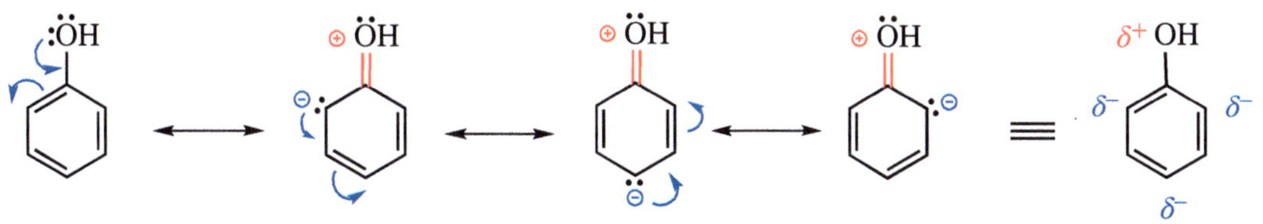

Resonance structures for phenol

In contrast to this behavior, oxygen donates nonbonding electrons to the aromatic ring. These electrons are delocalized over the benzene ring. Oxygen has performed an electron-donating task here. One of the most serious mistakes made by students is that they always perceive oxygen as an electron-withdrawing atom because of the electronegativity of oxygen. As can be seen here, oxygen not only attracts electrons but it also donates them to the conjugated systems. In that case, *the oxygen atom is an inductively electron-attracting (−I) mesomeric electron donor (+M). In this system, while the oxygen atom donates its nonbonding electrons to the ring, it draws electrons from the ring through the σ electrons. What happens if an atom draws electrons inductively at the same time as it does here and donates the electrons mesomerically? The mesomeric effect, in most cases, dominates.* We have seen that the inductive effect is observed up to a maximum of three to four bonds. With the mesomeric effect, because electrons move and change their location, it takes place at longer distances. There are four bonds between the *para* position of phenol and oxygen. The electron density is high at the *para* position.

It is necessary to pay particular attention to one point. In order for the electron donor feature of oxygen to come to the fore, it must be connected to a conjugated system. The oxygen atom in phenol increases the electron density at the ring by donating its electrons to the benzene ring. The resonance structures of phenol show that the electron density at the *ortho* and *para* positions is increased. This does not mean that the electron density at the *meta* position is decreased. The electron density is increased around the entire benzene ring. However, the increase in density at the *meta* position is not as much as that at the other positions. In that connection, we can understand why the electrophiles exclusively attack the *ortho* and *para* positions of phenol. For more information, see Chapter 6. To illustrate this situation on the figure, the structure given below on the left is incorrect. The drawing on the right expresses the true structure.

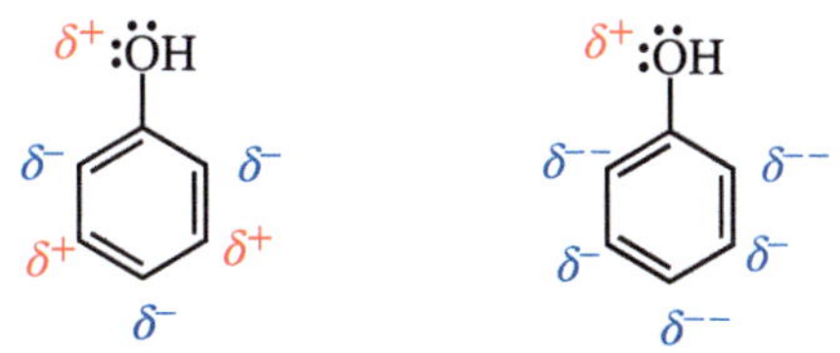

Wrong   Correct

Phenol is potentially more reactive toward electrophilic substitution reactions than benzene is because the hydroxyl group strongly activates the benzene ring by electron donation. Then, the electrophiles attack the region with higher electron density.

NMR spectroscopy gives accurate information about the electron density within a molecule. The chemical shifts are mainly affected by electron density. In the ^{1}H-NMR spectrum of phenol, the resonances of all aromatic protons are shifted to a higher field compared to benzene. This is due to the increased electron density in the aromatic ring. The *ortho* and *para* protons are more affected than the *meta* protons are as expected from the resonance structures of phenol. Thus, we can draw the following conclusion: ^{1}H-NMR chemical shifts of substituted benzenes largely depend on the mesomeric interaction between the substituent and the benzene ring.

Let us examine the relationship between the mesomeric effect and the inductive effect on benzyl alcohol. In benzyl alcohol, a methylene group (–CH$_2$–) separates the hydroxyl group (–OH) from the benzene ring. *How does oxygen act here and how does it affect the electron density on the benzene ring?*

Phenol   Benzyl alcohol — Inductively electron-withdrawing   2-Phenylethan-1-ol

The nonbonding electrons of the oxygen atom cannot conjugate with the benzene ring because the conjugation is interrupted by the methylene (–CH$_2$–) group. Thus, the oxygen atom cannot donate electrons to the ring by resonance. However, oxygen inductively attracts the σ electrons and to a lesser extent reduces the electron density on the benzene ring. If an additional methylene group (–CH$_2$–) is added to benzyl alcohol, this time the inductive effect of oxygen will not be observed on the benzene ring. However, if the ethylene group is replaced by a C=C double bond, now oxygen can deliver electrons to the benzene ring via resonance because of the uninterrupted conjugation.

No conjugation   Conjugation

Let us examine the inductive and mesomeric effects of the oxygen atom on another system. The secondary butyl carbocation is more stable than the primary butyl carbocation. *How will the stability of this carbocation be affected if we replace the methyl group far from the carbocation center with an OCH$_3$ group?* Because the oxygen atom is not directly connected to the carbocation center, the nonbonding electrons on the oxygen atom cannot conjugate with the carbocation because the conjugation is interrupted by the methylene group. In this case, only the inductive effect of oxygen will occur, so the stability of the carbocation will decrease, because of the electron-withdrawing effect of the oxygen atom. If the oxygen atom is directly attached to the carbocation center, the stability of the carbocation will increase. Actually, oxygen will decrease the stability by attracting σ electrons more strongly. On the other hand, the nonbonding electrons on oxygen will conjugate with the empty p orbital located in the carbocation center. In other words, the oxygen atom releases its electrons mesomerically and withdrawing inductively. In cases where both effects occur simultaneously, the oxygen atom will make the carbocation more stable in the α-position, as the mesomeric effect is the dominant one.

Butan-2-ylium   1-Methoxypropan-2-ylium   **Less stable**

1-Methoxypropan-1-ylium   **More stable**

For a better understanding of the subject, let us examine enol ethers, in which oxygen is directly attached to a double bond. First, it might be thought that oxygen will attract electrons from the double bond. This is true. However, while the oxygen atom inductively withdraws electrons, it releases its nonbonding electrons by resonance as we discussed in the case of phenol. Because the mesomeric effect is dominant, the electron density in the double bond increases. However, this increase is concentrated on the β-carbon atom. The positive charge is located on the oxygen and α-carbon atom. The ^{1}H-NMR spectra of enol ethers show this charge distribution well. The chemical shift value of the α-proton shifts approximately 1 ppm to a lower field, while the chemical shift value of the β-proton shifts 1 ppm to a higher field. The difference in the chemical shift of these protons is about $\Delta\delta = 2$ ppm, revealing serious electron polarization of the double bond. Again, NMR shows the bond polarization clearly. If this polarization is known, it is extremely easy to understand the chemistry of enol ethers. Enol ethers react easily with electrophiles. Because the electron density on the β-carbon atom is increased, electrophiles attack the β-carbon atom and open the carbon–carbon double bond. Therefore, enol ethers are extremely sensitive to acids.

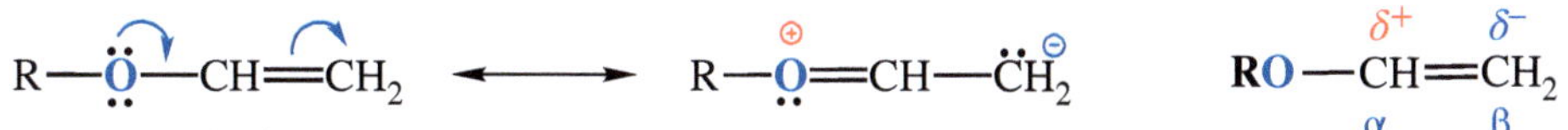

$$\text{R}-\overset{..}{\underset{..}{\text{O}}}-\text{CH}=\text{CH}_2 \quad \longleftrightarrow \quad \text{R}-\overset{\oplus}{\underset{..}{\text{O}}}=\text{CH}-\overset{..}{\underset{..}{\text{C}}}\text{H}_2^{\ominus} \qquad \overset{\delta+}{\text{RO}}-\underset{\alpha}{\text{CH}}=\underset{\beta}{\overset{\delta-}{\text{CH}_2}}$$

Enol ether

Let us analyze the effect of an electron-withdrawing group attached to benzene, for example, benzaldehyde. Because oxygen withdraws the electrons toward itself, the carbonyl group is polarized. Thus, the aldehyde is a group with a −M effect. The polarized carbonyl carbon will withdraw the π electrons from the aromatic ring. Thus, electron delocalization occurs in the ring. The resonance structures of benzaldehyde are shown below. These structures show that the electron density at the *ortho* and *para* positions is decreased. This does not mean though that the electron density at the *meta* position is increased. The electron density is decreased around the entire benzene ring. However, the decrease in electron density at the *meta* position is not as great as that at the other positions. The ^{1}H-NMR spectrum of benzaldehyde also demonstrates the electron density distribution on the benzene ring. Comparison of the chemical shifts of the aromatic protons of benzaldehyde with those of benzene clearly indicates that the resonances of all protons are shifted to a lower field because of the decreased electron density at the benzene ring. The *meta* positions are less affected.

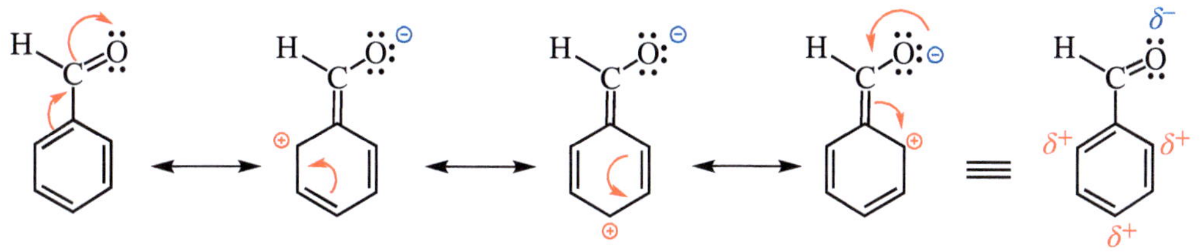

Resonance structures of benzaldehyde

Aromatic compounds, having substituents with a −M effect, undergo electrophilic substitution reaction at the *meta* position. Such compounds generally undergo more difficult electrophilic substitution reactions than benzene does. The reaction conditions must be changed relative to benzene (high temperature, strong catalyst, etc.). Therefore, electron-withdrawing groups direct the electrophiles to the *meta* position because the *meta* positions have less positive charge as shown below.

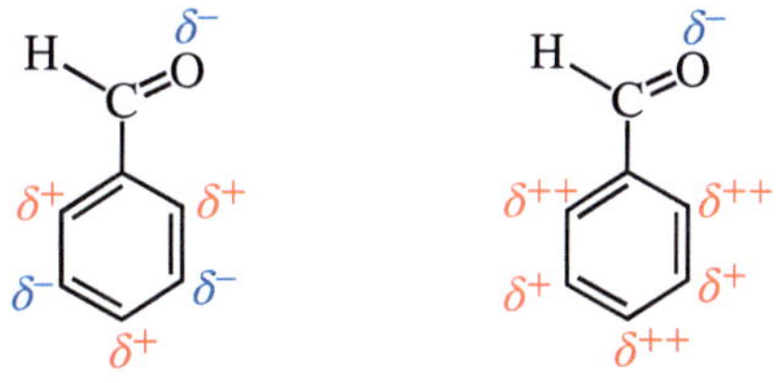

Wrong   Correct

In the examples provided above, we have examined oxygen-containing functional groups. For example, aniline behaves exactly like phenol. Although the amine group is an inductively electron-attracting group, it is mesomerically electron donating, so that the nonbonding electrons on the nitrogen atom can be involved in the resonance with the π electrons of the ring. Aromatic amines increase the electron density in the benzene ring. If sulfur atoms are directly attached to the ring such as −SH or −SR, they behave similarly because they can easily donate the nonbonding electrons to the benzene ring. If nitrogen or sulfur, directly attached to the ring, contains a double bond with oxygen, the situation changes. Now, functional groups such as −NO or −SO− will attract the ring electrons inductively as well as mesomerically as shown below.

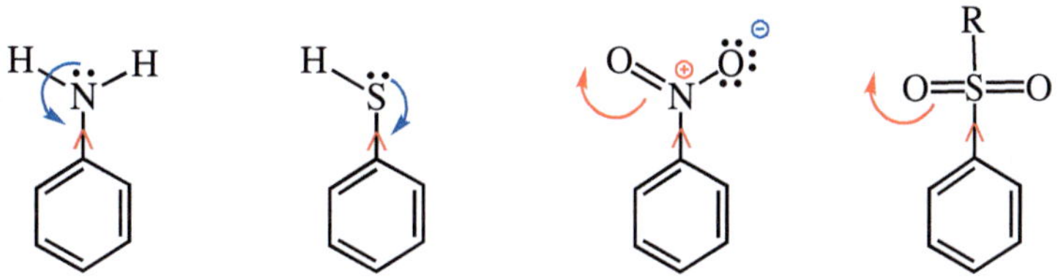

−I, +M   −I, +M   −I, −M   −I, −M

Inductive and mesomeric effects should always be considered jointly. It is now an easy process to predict how electron polarization will be in a molecule through these two effects. Once polarization is known, it is much easier to predict reactions because in all reactions, electron-rich functional groups attack the electron-deficient positions. Thus, knowing polarization is often enough to predict products that can occur as a result of a reaction. This subject will be frequently touched upon in the following chapters.

1.5 Formal Charge and Oxidation Number

1.5.1 Formal Charge

Most organic compounds are represented by Lewis structures having the normal number of bonds. On the other hand, some organic ions or molecules contain less than the customary number. For example, the carbon atom in methane contains four bonds (eight electrons) and methane does not have a charge. However, the carbon atom in the methyl cation has fewer bonding electrons (six electrons) and has a charge. The H_2O molecule forms a hydronium ion by treatment with acids. On the oxygen atom, there is a positive charge. If we remove a proton from water, a hydroxide ion will be formed, which has a negative charge on the oxygen atom. As one can see, in organic compounds or ions, there are some charges that are assigned to an atom in a molecule. These positive (+) and negative (−) charges do not show the oxidation number like they do in inorganic chemistry.

Ammonia is a neutral molecule. If it reacts with a mineral acid, an ammonium salt is formed. The molecule gains a positive (+) charge, which is assigned to the nitrogen atom. All these charges do not show the oxidation number of the corresponding atoms.

As we know, the oxidation numbers of oxygen in H_2O and nitrogen in NH_3 molecules are −2 and −3, respectively. Furthermore, we know that protonation or proton abstraction of these compounds does not change the oxidation numbers. Then, we can ask: *What are these charges assigned to some atoms? How can we predict the charges?* These charges are called *formal charges. How is the formal charge of an atom calculated?* When calculating the formal charge of an atom, the electronegativities of other atoms attached to that atom are never taken into account. Formal charge is a hypothetical charge given to every atom in a molecule considering that the bonding electrons are shared equally between the atoms. Before starting to count the formal charge, first we determine how many valence electrons the corresponding atom has in the case where it is not bonded to any other atoms. In other words, we determine the group number of the atom in the periodic table. Half of the bonded electrons are assigned to each atom, irrespective of the electronegativities. Unshared electrons belong exclusively to the parent atom. Keeping this information in mind, let us calculate the formal charge of the oxygen and nitrogen atoms in H_2O and NH_3 molecules. As we see below, the oxygen atom in water has two pairs of nonbonding electrons. We divide all of the electrons in bonds equally between the atoms (hydrogen and oxygen) that share them. Thus, each hydrogen is assigned one electron. We subtract this number (one) from 1, which is the number of valance electrons in a hydrogen atom. The result is (1−1 = 0) zero, which is the formal charge of hydrogen atoms in water. The oxygen atom is assigned six electrons (four nonbonding electrons + two shared electrons). We now subtract 6 from 6, which is the number of valence electrons in an oxygen atom to give the oxygen a formal charge of zero. The nitrogen atom in ammonia is assigned five electrons (two nonbonding electrons + three shared electrons). We subtract 5 from 5, which is the number of valence electrons of a nitrogen atom to give the nitrogen a formal charge of zero.

1.5 Formal Charge and Oxidation Number

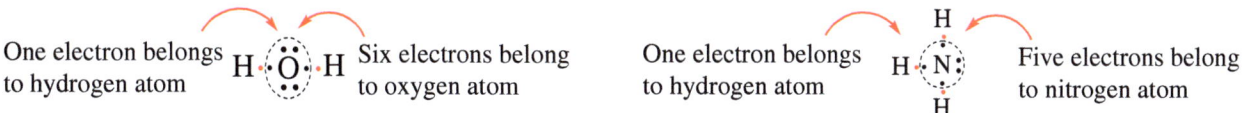

One electron belongs to hydrogen atom — H·Ö·H — Six electrons belong to oxygen atom

One electron belongs to hydrogen atom — H·N·H (with H above and H below) — Five electrons belong to nitrogen atom

The formal charges of oxygen and nitrogen atoms in protonated H_2O and NH_3 molecules are shown below.

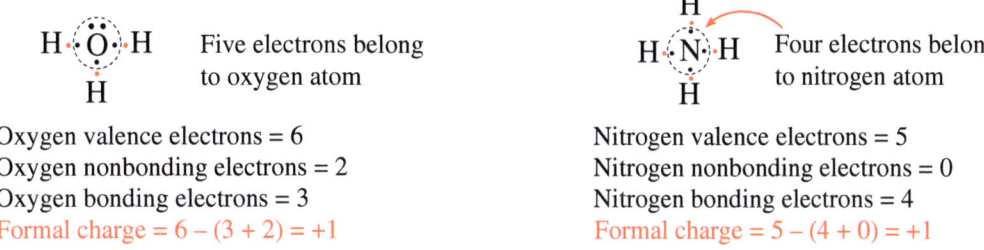

H·Ö·H with H below — Five electrons belong to oxygen atom

Oxygen valence electrons = 6
Oxygen nonbonding electrons = 2
Oxygen bonding electrons = 3
Formal charge = 6 − (3 + 2) = +1

H·N·H with H above and H below — Four electrons belong to nitrogen atom

Nitrogen valence electrons = 5
Nitrogen nonbonding electrons = 0
Nitrogen bonding electrons = 4
Formal charge = 5 − (4 + 0) = +1

The acid-catalyzed reaction of carbonyl groups occurs by initial protonation of the nonbonding electrons of the carbonyl oxygen atom to produce resonance-stabilized intermediates. The carbonyl group contains a double bond between oxygen and carbon. Two of the four electrons belong to oxygen. We assign nonbonding electrons to oxygen. The oxygen atom now has six electrons, so that the formal charge of oxygen is zero. In the case of protonation, the oxygen atom has only two nonbonding electrons and the number of bonds made by oxygen becomes three. The number of electrons belonging to oxygen is now five. The formal charge of oxygen is then +1. In the resonance structure of the carbonyl group, the formal charge of oxygen is zero; however, the formal charge of the carbon atom changes from zero to +1.

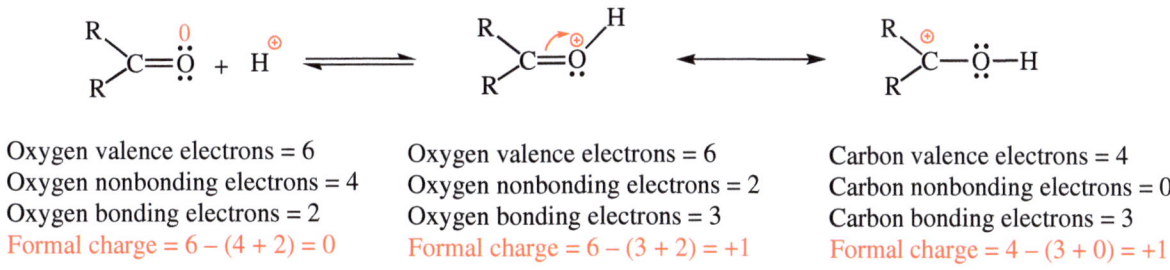

Oxygen valence electrons = 6
Oxygen nonbonding electrons = 4
Oxygen bonding electrons = 2
Formal charge = 6 − (4 + 2) = 0

Oxygen valence electrons = 6
Oxygen nonbonding electrons = 2
Oxygen bonding electrons = 3
Formal charge = 6 − (3 + 2) = +1

Carbon valence electrons = 4
Carbon nonbonding electrons = 0
Carbon bonding electrons = 3
Formal charge = 4 − (3 + 0) = +1

1.5.2 Oxidation Number

The oxidation number is another way of characterizing atoms in molecules. Let us try to determine the oxidation numbers after the formal charge calculation. The process is very similar. In contrast to the formal charge, in which the electrons in a bond are assumed to be shared equally, the oxidation number is a charge that an atom would have if the bonding electrons were assigned exclusively to the more electronegative atom.

The following diagram compares the formal charge and oxidation numbers in carbon monoxide. The formal charges of the carbon atom and oxygen atom are calculated as −1 and +1, respectively, because the bonding electrons are equally shared between the carbon atom and oxygen atom. In the case of calculating the oxidation numbers, the bonding electrons are assigned to the more electronegative atom, which is the oxygen atom. Then, the oxygen atom has eight electrons in the outer shell and its oxidation number is −2. The oxidation number of the carbon atom is calculated as +2.

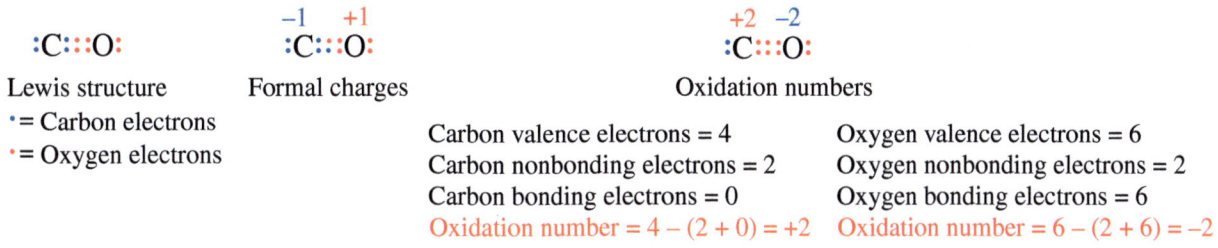

:C:::O:
Lewis structure
• = Carbon electrons
• = Oxygen electrons

−1 +1
:C:::O:
Formal charges

Carbon valence electrons = 4
Carbon nonbonding electrons = 2
Carbon bonding electrons = 0
Oxidation number = 4 − (2 + 0) = +2

+2 −2
:C:::O:
Oxidation numbers

Oxygen valence electrons = 6
Oxygen nonbonding electrons = 2
Oxygen bonding electrons = 6
Oxidation number = 6 − (2 + 6) = −2

26 | *1 Basic Concepts*

Various functional groups such as alcohol, ketone, aldehyde, and carboxylic acid are formed as a result of bonding between the carbon and oxygen atoms. Let us calculate the oxidation numbers of carbon and oxygen atoms in these functional groups. As seen below, the oxidation number of the oxygen atom in all of these groups is −2. However, the oxidation number of the carbon atom varies according to the number of bonds between the carbon and the oxygen atom.

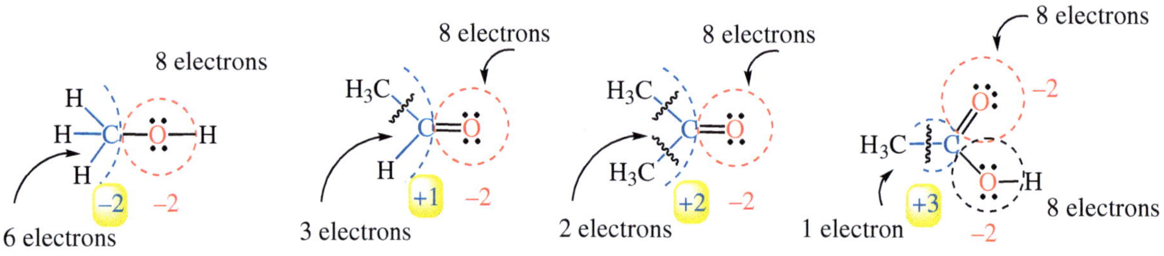

If there are one or more bonds between two carbons, due to the same electronegativity, half of the bond electrons belong to one carbon atom and half to the other. We can easily find that the oxidation number of one of the carbon atoms of ethanol is −3 and that of the other is −1.

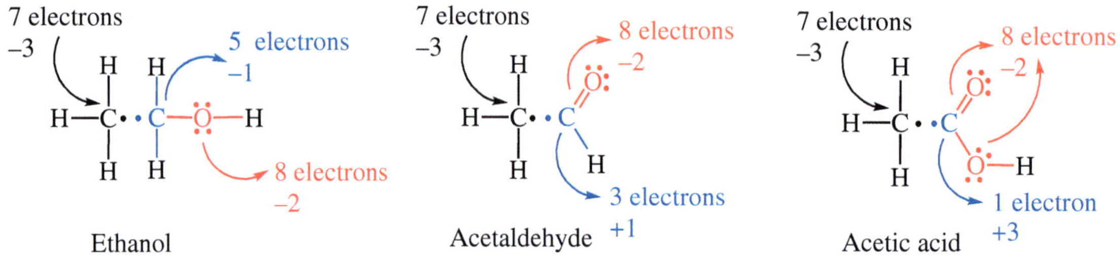

The oxidation numbers of carbon atoms in an unsaturated carbon chain can vary. For example, the oxidation numbers of carbon atoms in different compounds are given below.

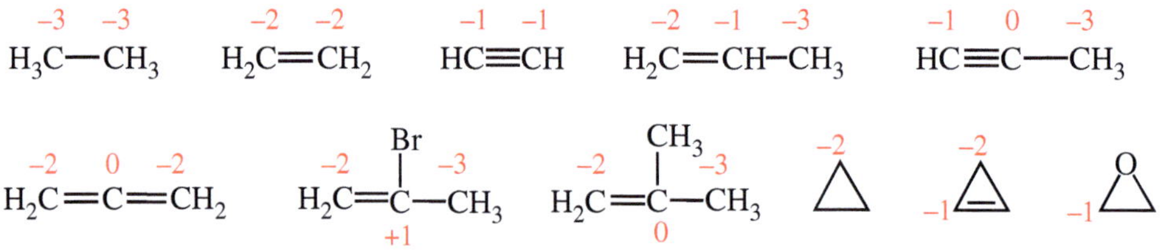

After examining the oxidation numbers of carbon atoms at various positions, let us briefly examine the oxidation and reduction phenomena in organic chemistry. In inorganic chemistry, it is easy to determine whether there is oxidation or reduction during a reaction as the oxidation numbers of the elements are always given. Because formal charges are generally used in organic chemistry, students often have difficulty determining the reduction and oxidation reactions. Occasionally, these reactions are confused with hydrolysis reactions. This confusion must be resolved before starting with the mechanism. Let us start with the simple examples known.

Conversion of ethanol to acetaldehyde is an oxidation reaction as presented. The oxidation number of the carbon atom bearing the hydroxyl group increases from −1 to +1. Conversion of aldehyde to a carboxylic acid is also an oxidation reaction. Here, the oxidation number of the carbon atom increases from +1 to +3. The reverse reactions, of course, are reduction reactions. The conversion of a secondary alcohol to a ketone is an oxidation reaction because the oxidation number of the carbon atom changes from 0 to +2. Therefore, we have to use oxidizing reagents such as permanganate and chromate to perform this reaction.

1.5 Formal Charge and Oxidation Number | 27

$$H_3C-\underset{-1}{CH_2}-OH \underset{\text{Reduction}}{\overset{\text{Oxidation}}{\rightleftarrows}} H_3C-\underset{+1}{\overset{O}{\underset{H}{C}}} \underset{\text{Reduction}}{\overset{\text{Oxidation}}{\rightleftarrows}} H_3C-\underset{+3}{\overset{O}{\underset{OH}{C}}}$$

$$H_3C-\underset{0}{\underset{|}{\overset{OH}{CH}}}-CH_3 \underset{\text{Reduction}}{\overset{\text{Oxidation}}{\rightleftarrows}} H_3C-\underset{+2}{\overset{O}{\underset{\parallel}{C}}}-CH_3$$

The Baeyer–Villiger oxidation, the reaction of ketones with *m*-chloroperbenzoic acid, is an oxidative cleavage of a carbon–carbon bond adjacent to a carbonyl group, which converts ketones to esters and cyclic ketones to lactones. To understand whether these reactions are redox reactions, it will be sufficient to examine the oxidation numbers of the carbon atoms reacted. In straight-chain ketones, the carbonyl group is oxidized from +2 to +3 and the carbon atom to which oxygen bonded is oxidized from −3 to −2. In cyclic ketones, carbonyl carbon is oxidized from +2 to +3 and the other carbon atom from −2 to −1.

Alkenes usually undergo electrophilic addition reactions at carbon–carbon double bonds and the π bond is removed and two new σ bonds are formed. Let us examine these reactions in terms of oxidation and reduction. Epoxidation of double bonds is a very common reaction applied in organic chemistry. As you can see, the oxidation numbers of the double-bond carbon atoms both change from −1 to 0, which is an oxidation reaction. The hydroxylation reaction of the double bonds is also an oxidation reaction. Here, double-bond carbon atoms are also oxidized from −1 to 0, while OsO$_4$ is reduced. It may not be possible to see in the first stage that the bromination reaction, which is one of the most common and classic reactions in organic chemistry, is also an oxidation reaction. However, when we look at the oxidation numbers of the carbon atoms to which bromine is added, it is immediately understood that they are oxidized from −1 to 0, as in epoxidation and hydroxylation reactions.

Cleavage of carbon–carbon bonds is usually an oxidation reaction. Of course, the electronegativity of the element that will be attached to the carbon atom after the cleavage will determine this. Generally, in organic chemistry, as elements other than hydrogen are more electronegative than carbon, the carbon–carbon cleavages are oxidation reactions.

$$\underset{-3}{H_3C}-\underset{-3}{CH_3} \longrightarrow 2\underset{-2}{H_3C}-OH \quad \text{Oxidation}$$

Are Diels–Alder reactions redox systems? To answer this question, we need to look at the oxidation numbers of the reacting carbon atoms. For example, the cycloaddition reaction of butadiene with maleic anhydride is neither oxidation nor reduction because the oxidation numbers of the starting materials are not changed. However, if the dienophile is an oxidizer, the diene can be oxidized. For example, the reaction of singlet oxygen with butadiene results in the formation of monocyclic endoperoxide. The carbon atom to which oxygen is bonded is oxidized and oxygen itself is reduced.

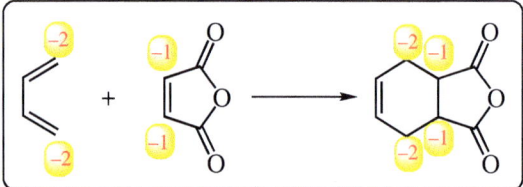

Elimination of H₂O from alcohols is a simple way to generate a double bond. The oxidation number of the carbon atom to which the hydroxyl group is attached is 0, and the oxidation number of the carbon atom from which the proton is removed is −2. The oxidation numbers of double-bond carbons formed as a result of elimination are −1. Here, while one of the carbon atoms is oxidized, the other is reduced. Substitution of the hydroxyl group with a bromide ion forms the corresponding brominated compound. As bromine and oxygen atoms are more electronegative than carbon atoms, there is no change in the oxidation number of carbon atoms. However, if the bromide atom is substituted with an element with a smaller electronegativity than the carbon atom, such as the hydride (H⁻) ion, then the carbon atom is reduced. Here, the oxidation number of the carbon atom changes from 0 to −2.

After examining the oxidation and reduction in many different reactions, let us take a brief look at the hydrolysis reactions where students become confused. Hydrolysis of esters in an acidic or basic medium is the conversion of esters to carboxylic acid and alcohol by addition of water to the medium. There is no change in the oxidation numbers of carbon atoms in ester hydrolysis. Thus, hydrolysis of a ketal to a ketone is also a reaction without oxidation or reduction.

Cyclization reactions such as conversion of diols to cyclic ethers or ring opening of an epoxide ring to give diols are not oxidation or reduction reactions. The oxidation numbers of the carbon atoms do not change. However, Wurtz-type reactions are reduction reactions.

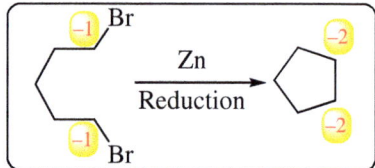

1.6 Acids and Bases

All chemical reactions are acid–base reactions. In order to predict what kind of product will be formed as a result of a chemical reaction, we have to know the acid and base properties of the functional groups in the reactants. For this reason, the concepts of acids and bases must be understood well. There are different descriptions for acids and bases. We will examine acids and bases in four different ways.

- Arrhenius theory
- Brønsted–Lowry theory
- Lewis theory
- Pearson hard–soft acid–base (HSAB) theory

1.6.1 Arrhenius Acid–Base Theory

The Arrhenius acid–base theory [1, 2] is the most classical theory. In 1884, Arrhenius suggested that NaCl dissociate when they dissolve in water to give particles and he called these particles ions.

$$\text{NaCl} \xrightarrow{\text{H}_2\text{O}} \text{Na}^+ + \text{Cl}^-$$

Shortly after this, he extended his theory to acids and bases. According to his theory, an Arrhenius acid is a substance that dissociates in water to form hydrogen ions (H$^+$) and protonation of water yields the hydronium ion, H$_3$O$^+$. An Arrhenius base is a substance that dissociates in water to form hydroxide (OH$^-$) ions. In 1903, Svante August Arrhenius (1859–1927) received the Nobel Prize in Chemistry for his electrolytic dissociation theory of acids, bases, and salts.

$$\text{HCl} + \text{H}_2\text{O} \longrightarrow \text{H}_3\text{O}^+ + \text{Cl}^-$$
Acid

$$\text{NaOH} \longrightarrow \text{Na}^+ + \text{OH}^-$$
Base

According to the International Union of Pure and Applied Chemistry (IUPAC) definition, H$_3$O$^+$ ions are called oxonium ions. Later, the IUPAC proposed calling these ions *hydronium ions*. The OH$^-$ ions formed by bases are called *hydroxide ions*. The definition of Arrhenius acids and bases is a very classic definition but it has some limitations. For example, it is only applicable to compounds having the formula HX for acids and MOH (M = metal) for bases. This theory cannot be applied to explain the acidic properties of compounds such as AlCl$_3$, CO$_2$, and SO$_2$ as they cannot dissociate into H$_3$O$^+$ ions. In order for a molecule to be a base, it must be decomposed into hydroxide ions. This can only be created with metal hydroxides. Similarly, this theory cannot explain the basic properties of NH$_3$, pyridine, Na$_2$CO$_3$, etc. It also cannot explain the acidic and basic properties of HCl and NaOH in benzene or acetone or in the gaseous state. Therefore, we will not generally use the Arrhenius acid–base theory when studying organic reactions. This theory was later expanded by Brønsted and Lowry and interpreted to include a larger group.

1.6.2 Brønsted–Lowry Acid–Base Theory

According to the Brønsted–Lowry acid–base theory [3], an acid is a species that can donate a proton, H$^+$ (in accordance with Arrhenius theory), and a base is a species that can accept a proton, H$^+$ [4, 5]. For example, when hydrogen chloride dissolves in water, almost all of the molecules dissociate into the corresponding ions, which indicates that the products are favored at equilibrium and the equilibrium lies to the right. HCl is an acid because it donates a proton, H$^+$. Water is a base because it accepts a proton from HCl. As water has two pairs of nonbonding electrons, it can form a covalent bond with a proton as shown below. When we examine this reaction from right to left, the H$_3$O$^+$ ion is an acid because the oxonium ion donates a proton, H$^+$. The chloride ion, Cl$^-$, is a base because it accepts a proton.

$$\text{HCl} + \text{H}_2\ddot{\text{O}}\text{:} \rightleftharpoons \text{H}-\overset{+}{\underset{..}{\text{O}}}(\text{H})-\text{H} + \text{Cl}^-$$
Acid   Base   Acid   Base

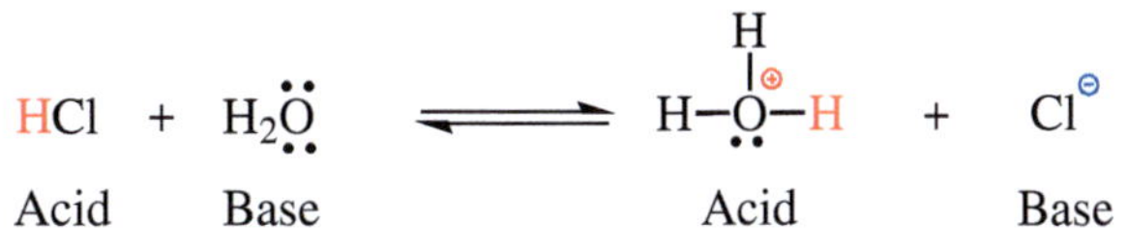

Let us examine the reaction between HCl and NH$_3$. Here, ammonia (NH$_3$) is a base because it accepts a proton from HCl. The proton forms a covalent bond with the nonbonding electrons of the nitrogen atom in ammonia. When we examine the reaction from right to left, we can see that the ammonium ion (NH$_4^+$) is an acid and the Cl$^-$ ion is a base. We can generalize the Brønsted–Lowry acid–base theory as follows. Species that have a hydrogen atom can potentially act as an acid and those possessing nonbonding electrons can potentially act as a base.

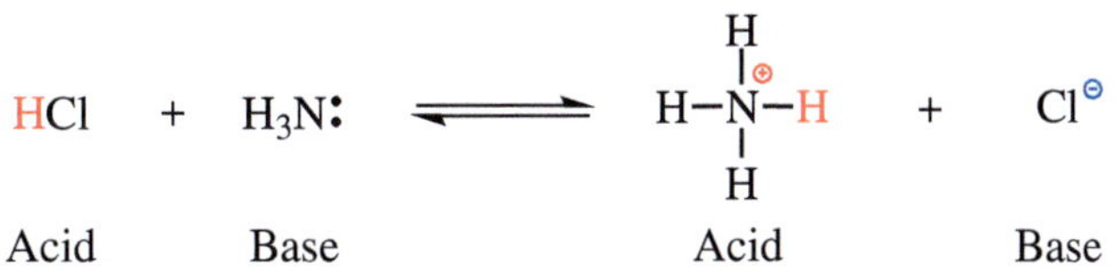

$$\text{HCl} + \text{H}_3\text{N:} \rightleftharpoons \text{H}-\overset{+}{\underset{\text{H}}{\text{N}}}(\text{H})-\text{H} + \text{Cl}^-$$
Acid   Base   Acid   Base

The concept of *conjugate pairs* is useful in describing Brønsted–Lowry acid–base reactions. When an acid donates H$^+$, the remaining species is called its *conjugate base*. Thus, Cl$^-$ is the conjugate base of HCl and H$_2$O is the conjugate base of H$_3$O$^+$. Likewise, when a base accepts H$^+$, the resulting species is called its *conjugate acid*.

In a reaction between water and ammonia, water is an acid because it donates a proton. Ammonia accepts this proton and therefore ammonia is a base. Thus, the NH$_4^+$ ion is the conjugate acid of NH$_3$ and the OH$^-$ ion is the conjugate base of H$_2$O. The reaction between water and ammonia illustrates this idea. In the forward direction, water acts as an acid by donating a proton to ammonia and subsequently becoming a hydroxide ion, OH$^-$, the conjugate base of water. In the reverse reaction, a OH$^-$ ion acts as a base in accepting a proton from the NH$_4^+$ ion, which acts as an acid.

$$\text{HOH} + \text{H}_3\text{N}: \rightleftharpoons \text{H}_4\text{N}^+ + \text{OH}^-$$

Acid	Base	Acid	Base
		Conjugate acid	Conjugate base

It is possible to draw the following conclusion from the simple examples we have given above. Brønsted–Lowry acids turn into Brønsted–Lowry bases by giving protons, and Brønsted–Lowry bases turn into Brønsted–Lowry acids by bonding a proton. Important conjugate acid–base pairs are given below.

Acid	Conjugate bases	Bases	Conjugate acids
H$_2$SO$_4$	HSO$_4^-$	NH$_3$	NH$_4^+$
HCl	Cl$^-$	H$_2$N$^-$	NH$_3$
H$_2$O	HO$^-$	SO$_4^{--}$	HSO$_4^-$
H$_2$S	HS$^-$	H$_2$O	H$_3$O$^+$
NH$_4^+$	NH$_3$	NO$_3^-$	HNO$_3$
H$_3$O$^+$	H$_2$O	HPO$_4^{--}$	H$_2$PO$_4^-$

The Brønsted–Lowry acid–base theory has occasionally sparked controversy within the scientific community. Ideas have also been proposed to change the definition as follows. According to this theory, compounds that are capable of detaching protons are called acids, while compounds that can abstract a proton from another compound are called bases. Not every compound can easily donate protons, but if there are strong bases, then a proton can be abstracted.

For example, acetylacetone is not a compound that can donate a proton like HCl can. However, a strong base can easily abstract a proton from the methylene group between the two carbonyl groups to generate an anion. Therefore, we say that the methylene protons are acidic protons.

$$\text{H}_3\text{C}-\overset{\overset{\text{O}}{\|}}{\text{C}}-\text{CH}_2-\overset{\overset{\text{O}}{\|}}{\text{C}}-\text{CH}_3 \xrightarrow{\text{Base}} \text{H}_3\text{C}-\overset{\overset{\text{O}}{\|}}{\text{C}}-\overset{\ominus}{\text{CH}}-\overset{\overset{\text{O}}{\|}}{\text{C}}-\text{CH}_3 + \text{Base}\text{H}^+$$

1.6.3 Lewis Acid–Base Theory

The Arrhenius acid–base theory [4] does not take into account the acid–base phenomenon in anhydrous environments. The Brønsted acid–base theory is very restrictive and focuses primarily on acids and bases acting as proton donors and acceptors. Furthermore, it completely eliminates compounds that do not have a proton. In 1923, G. N. Lewis (1875–1946) at UC Berkeley proposed an alternate theory: a *Lewis acid* is a species that accepts an electron pair and a *Lewis base* is a species that donates an electron pair [6]. According to this definition, a proton is a Lewis acid. This definition includes many compounds that do not contain protons. Through the use of Lewis' definition of acids and bases, chemists can now predict a wider variety of acid–base reactions.

Let us examine the reaction between a proton (H$^+$) and H$_2$O within the framework of the Lewis acid–base theory. Because proton is an electron-deficient species (Lewis acid), it can easily accept a pair of electrons. Oxygen (Lewis base), on the other hand, can form a new σ bond by transferring a pair of nonbonding electrons to the proton.

$$H^+ + H_2\ddot{O} \rightleftharpoons H-\overset{\overset{H}{|}}{\underset{..}{\overset{\oplus}{O}}}-H$$

Acid   Base

The Lewis acid–base theory explains why BF_3 reacts with NH_3 to form a complex. Boron trifluoride has a trigonal plane structure and its hybridization is sp^2 and it has an empty p orbital, which can accept electrons. According to the Lewis acid–base theory, boron trifluoride with six electrons in the outer shell is a Lewis acid. Ammonia consists of a nitrogen atom as the central atom with a pair of nonbonding electrons and so it is a Lewis base. The nonbonding electron pair from NH_3 is donated to the empty orbital of boron, forming a σ bond between the nitrogen and boron. As no proton (H^+) transfer is involved in this reaction, it qualifies as an acid–base reaction only according to Lewis' definition.

$$H_3N: + BF_3 \rightleftharpoons H_3\overset{\oplus}{N}-\overset{\ominus}{B}F_3$$

Base   Acid

In the same way, $AlCl_3$, a compound of third row elements, is a Lewis acid, because aluminum has a sextet of electrons in its outer shell. Similarly, many other compounds, such as $ZnCl_2$, $FeCl_3$, $SnCl_4$, and $TiCl_4$, are also Lewis acids.

Carbocations also fall into the Lewis acid class. A carbocation, as seen earlier, has an empty p orbital. Carbocations are highly prone to bonding by accepting electrons from Lewis bases. Double bonds with low electron density, for example, enon-type compounds, accept electrons in the β-position and form a covalent bond. The double-bond carbon atoms of tetracyanoethylene also act like Lewis acids. Cyano groups reduce the electron density of the double bond because of their strong electron-withdrawing properties. The double bond can accept an electron pair from another compound and therefore such double bonds with electron deficiency are Lewis acids.

Lewis acid ; $H_2C=CH-C(=O)R$ ⟷ $H_2\overset{}{C}-CH=C(-\overset{\ominus}{\ddot{O}:})R$ Lewis acid ; $(NC)_2C=C(CN)_2$ Lewis acid

The hydride ion (H^-) is a typical Lewis base. It reacts easily with Lewis acids and forms a covalent bond. Nonpolarized double-bond electrons are also Lewis bases because of their high electron density. They can easily react with Lewis acids.

Polarized bonds form dipoles. One end of these bonds is a Lewis acid while the other end is a Lewis base. For example, a carbon–halogen bond is a polarized bond. Because the halogen atom attracts σ electrons between carbon and halogen more strongly and has nonbonding electrons, the halogen is a Lewis base. On the other hand, the carbon atom, whose electron density is reduced, acts as a Lewis acid. The carbonyl groups are polarized. Therefore, the carbonyl carbon atom is a Lewis acid and the carbonyl oxygen is a Lewis base.

$H_3C^{\delta+}\rightarrow\ddot{\underset{..}{Cl}}:^{\delta-}$ — Lewis base / Lewis acid

$R_2C^{\delta+}=\ddot{O}^{\delta-}$ — Lewis base / Lewis acid

1.6.4 Pearson Hard and Soft Acid–Base Theory

In 1963, Ralph Pearson first introduced the qualitative *theory of hard and soft acids and bases (HSAB)*, which initially included inorganic compounds and later organic compounds [5, 6]. According to this theory, Lewis acids and Lewis bases are divided into two classes: hard and soft.

Hard acids and hard bases tend to the following:

- Be small
- Have high oxidation state (the charge criterion applies mainly to acids and to a lesser extent to bases)
- Be difficult to polarize
- Have high electronegativity
- Have low-lying highest occupied molecular orbital (HOMO) energy (bases) and high-lying lowest unoccupied molecular orbital (LUMO) energy (acids).

Soft acids and soft bases tend to:

- Be large
- Have low or zero oxidation state
- Have high polarizability
- Have low electronegativity
- Have high-lying HOMO energy (bases) and low-lying LUMO energy (acids).

According to this theory, hard acids prefer to bind to hard bases, whereas soft acids prefer to bind to soft bases.

1.6.4.1 Hard Acids and Bases

Hard bases are characterized by low-energy HOMOs. Generally, the compounds derived from strong electronegative elements and their anions are hard bases. For example, because the central atoms of elements such as HO^-, F^-, Cl^-, and NH_3 are strongly electron withdrawing, HOMO energy levels are generally low. In the case of hard acids, LUMO energy levels are very high. Alkali metal cations (Li^+ to Cs^+), alkaline earth metal cations (Be^{2+} to Ba^{2+}), transition metal cations in higher oxidation states (Ti^{4+}, Cr^{3+}, and Fe^{3+}), and the proton (H^+) are hard acids. When hard acids react with hard bases, the strong bond formed between the acid and the base is usually an ionic bond. The large electronegativity differences between hard acids and hard bases give rise to strong ionic interactions.

When the energy gap between the orbitals participating in the reaction is very high, the probability of forming a covalent bond decreases. Because of the large energy gap between the atomic orbitals that will form molecular orbitals, the energy gain by forming molecular orbitals will be less, as depicted in Figure 1.21.

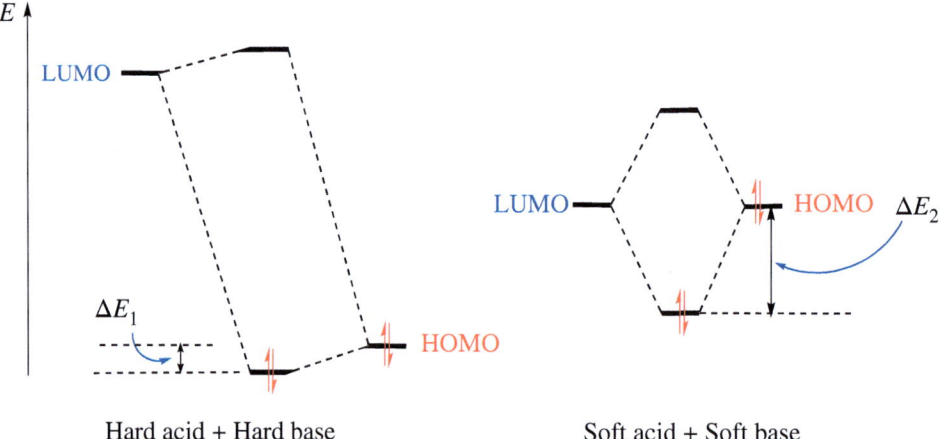

Figure 1.21 Formation of molecular orbitals by interaction of HOMOs and LUMOs with different energy levels and energy gain.

For example, let us examine the formation of a bond between Na and F atoms. In this reaction, the sodium atom loses its single valence electron to the fluorine atom, which accepts it. The ions produced are oppositely charged and are attracted to one another because of the electrostatic forces. If the two elements form a covalent bond, the energy level of the molecular orbital will not drop so much because the energy difference between the HOMOs and LUMOs is quite large (Figure 1.21). Therefore, they cannot form a covalent bond.

1.6.4.2 Soft Acids and Bases

While the HOMO levels of soft bases are quite high, the LUMO levels of soft acids are low. When soft acids and soft bases react, they form a covalent bond. Because the energy levels of the orbitals (HOMO and LUMO) are close to each other, the energy levels of the formed molecular orbitals will drop. When the electrons settle into the bonding orbital, the energy released is high and the bond becomes strong (Figure 1.21). These compounds are easily polarized and their electronegativity is very weak. Some examples are given in Table 1.5 for HSAB.

Table 1.5 Examples of hard and soft acids and bases.

Type of acids and bases	Examples
Hard bases	H_2O, HO^-, F^-, Cl^-, CH_3COO^-, PO_4^{3-}, SO_4^{2-}, NO_3^-, CO_3^{2-}, ROH, RO^-, R_2O, NH_3, RNH_2, N_2H_4
Hard acids	H^+, Li^+, Na^+, K^+, Be^{2+}, Mg^{2+}, Ca^{2+}, Al^{3+}, Fe^{3+}, Cr^{3+}, Ti^{4+}, Sn^{4+}, Si^{4+}, BF_3, $B(OR)_3$, $AlCl_3$, SO_3, RCO^+, CO_2, R_3C^+
Soft bases	R_2S, RSH, RS^-, I^-, SCN^-, PR_3, $P(OR)_3$ CN^-, CO, $CH_2=CH_2$, C_6H_6, H^-, R^-
Soft acids	Cu^+, Ag^+, Hg^+, Pd^{2+}, Hg^{2+}, Pt^{2+}, I^+, Br^+, RO^+, I_2, Br_2, $(CN)_2C=C(CN)_2$, carbenes

In 1983, Ralph Pearson and Robert Parr [7] converted the qualitative definition of HSAB to a quantitative one by using the following equation. The chemical hardness is half the difference between the ionization potential (I) and the electron affinity A.

$$\text{Hardness} = \eta = \tfrac{1}{2}(I - A) \quad \text{Softness} = \sigma = 1/\eta$$

Ionization energy (I) is the minimum amount of energy required to remove the most loosely bound valence electron. Electron affinity (A) is the amount of energy released when an electron is attached to a neutral atom or molecule to form a negative ion.

Table 1.6 Chemical hardness of some atoms, acids, and bases.

Atom	η	Cations	η	Anions	η
H	6.4	H^+	∞	H^-	6.8
Li	2.4	Li^+	35.1	F^-	7.0
C	5.0	Mg^{2+}	32.5	Cl^-	4.7
N	7.3	Na^+	21.1	Br^-	4.2
O	6.1	Ca^{2+}	19.7	I^-	3.7
F	7.0	Al^{3+}	45.8	CH_3^-	4.0
Na	2.3	Cu^+	6.3	NH_2^-	5.3
Si	3.4	Cu^{2+}	8.3	OH^-	5.6
P	4.9	Fe^{2+}	7.3	SH^-	4.1
S	4.1	Fe^{3+}	13.1	CN^-	5.3
Cl	4.7	Hg^{2+}	7.7		

Table 1.6 shows the chemical hardness of some selected atoms, acids, and bases. Hardness increases with electronegativity. The order for the halogens is $F^- > Cl^- > Br^- > I^-$ and for the second-row anions is $F^- > OH^- > NH_2^- > CH_3^-$. For cations, hardness increases with positive charge and decreases with size, so that $H^+ > Al^{3+} > Li^+ > Mg^{2+} > Na^+$. The hardness of the proton is infinite as it does not have an electron. The Na^+ ion is harder than the Ag^+ ion because sodium has a smaller volume. The alkoxide ion RO^- is harder than the RS^- ion; because the oxygen atom is more electronegative, it increases the hardness. The Cu^{2+} ion is harder than Cu^+. The increased oxidation number also increases the hardness.

Let us try to apply the HSAB concept to some chemical reactions. Hard bases and hard acids generally form ionic compounds. There are always some exceptions. For example, HO^-, which is a hard base, reacts with a hard acid, H^+, to form

H₂O, where the bond between the hydrogen atom and the oxygen atom is a covalent bond. C=C double bonds are generally classified as soft bases. Because H⁺ is a hard acid, it does not react easily with double bonds. On the other hand, the bromonium ion, Br⁺, which is very soft, reacts quickly with double bonds to give addition products.

An ambident nucleophile is the one that can attack from two or more places, resulting in two or more products. An example of an ambident nucleophile is the nitrile ion, which has the chemical formula CN⁻. This ion can execute nucleophilic attacks from either the nitrogen atom or the carbon atom. The nucleophilic substitution reactions of alkyl halides involving this ion often result in the formation of a mixture of products. Therefore, an ambident nucleophile can be thought of as an anionic nucleophile in which the negative charge of the ion is delocalized by resonance over two different atoms. When the negative charge is on the nitrogen atom, the nitrile ion is much harder. The product distribution in the reaction of the CN⁻ ion with alkyl halides varies depending on the reaction conditions. When NaCN is used, the nitrile group is attached via carbon, and when AgCN is used, the nitrile group is attached to the alkyl group on nitrogen and isonitrile is formed (for a detailed discussion, see Section 2.1.6).

Enolate anions are nucleophiles and the negative charge on an enolate ion is delocalized over two atoms. Because of the nucleophilic characteristic of enolate anions, they can undergo alkylation reactions. Because of the enolate's ambident characteristic, *O*-alkylation or *C*-alkylation may take place. The question is which position in the enolate ion is a harder base. In alkylation with alkyl halides, if the electrophile is a soft acid, *C*-alkylation of the enolate is favored, because of carbon's lower hardness. Hard acids prefer the reaction of the enolate anion with the oxygen atom.

Oxetan-2-one can undergo ring-opening reactions with nucleophiles in two different ways. Nucleophiles can attack both the carbonyl group and the alkoxy carbon atom. Because the carbonyl group is more polarized and positively ($\delta+$) charged, it is a hard acid. The alkoxide anion, which is a hard base, attacks preferentially the carbonyl group and opens the β-lactone ring. On the other hand, thiolate is a soft base because of the sulfur atom and it attacks the alkoxy carbon atom, which has a soft acid property. Similar examples will be examined in more detail in the nucleophilic substitution section.

After examining different acid and base theories, in the next section, we will examine the pK values of acids and bases and the factors affecting the acid and base strength.

1.6.5 pKa Values of Acids

Many organic reactions involve the transfer of a proton (H⁺) by an acid–base reaction. Water is the reference solvent commonly used to compare the strengths of acids. When a strong acid H–X is dissolved in water, a proton is transferred from H-X to H₂O to form H₃O⁺ and X⁻. Almost all the molecules dissociate, which indicates that the products are favored. However, if a weak acid, such as acetic acid, is dissolved in water, very few molecules dissociate and so the reactants are favored. In both cases, an equilibrium will be established. A quantitative measure of the acidity is given by the equilibrium constant for ionization, which is obtained from the following equation. Brackets are used to indicate the concentrations of all species, which are in equilibrium.

$$H-X + H_2O \rightleftharpoons H_3O^+ + X^- \quad K_a = \frac{[H_3O^+][X^-]}{[HX]} \; ; \; pK_a = -\log K_a$$

It is not appropriate to express the strengths of acids with K_a values. For convenience, the strength of an acid is generally expressed by its pK_a values rather than its K_a values. The pK_a values are the negative logarithm of K_a. The pK_a values of some selected compounds are listed in Table 1.7. Acids with pK_a values lower than 1 are strong acid, those with pK_a values between 1 and 3 are medium strength acids, and those with pK_a values higher than 4 are weak acids. The pK_a values of extremely weak acids are generally greater than 15.

Table 1.7 pK_a values of some selected acids.

Acids	pK_a	Acids	pK_a
H_2SO_4	−10	HCN	9.1
HI	−9	NH_4^+	9.2
HBr	−8	Phenol	10.0
HCl	−7	CH_3OH	15.3
H_3PO_4	2.2	H_2O	15.7
HF	3.2	NH_3	33
HCOOH	3.75	H_2	38
CH_3COOH	4.76	CH_4	50
H_2S	7.0		

1.6.5.1 Factors Affecting the Acidity Strength of Organic Compounds

There are many factors that affect the acidity strength of organic compounds. The three most important factors are as follows:

- Strength energy of the H–X bond
- Electronegativity of X
- Stability of X after dissociation

Because the effect of the H–X bond is weak compared with the other factors, we will focus on those other factors. When the pK_a value of methanol (15.3) is compared with that of methane (50), the difference is very large. This difference can be explained by electronegativity. When we examine the pK_a values of H–X compounds within a period in the periodic system, it is immediately recognized that the change is directly proportional to electronegativity. First, we will focus on individual atoms and we will use the simple organic compounds methane, ammonia, water, and hydrogen fluoride.

$$\begin{array}{ccccc} & H_3C-H & H_2N-H & HO-H & F-H \\ pK_a & 50 & 33 & 15.7 & 3.2 \end{array}$$
$$\xrightarrow{\text{Acidity increases}}$$

A clear trend can be seen by moving from left to right along the second row of the periodic table from carbon to fluorine. We have to consider the stability of the conjugate base in all cases. The negative charge ends up in each conjugate base. The electronegativity also increases by moving from left to right along a row of the periodic table. As the fluorine atom is the most electronegative of the four atoms and carbon the least, the strongest acid is HF. The bond electrons are withdrawn strongly and it becomes easier to separate the proton. The fluorine atom can stabilize the negative (−) charge better than the carbon atom.

It turns out that when moving from top to bottom within a given column of the periodic table, we again observe a clear increasing trend in acidity. We cannot explain this trend by the electronegativity of the elements. Because the most electronegative element is fluorine, one would expect that HF would be the strongest acid. However, this is not the case; the strongest acid is HI. The stability of the conjugated base determines directly the strength of acids. As we go from top to bottom in a group, the radius of the elements increases because of the increasing number of shells. The atomic radius of iodine is approximately twice that of fluorine, and so in an iodide ion, the negative charge is spread out over a significantly larger volume. The larger the area where the charge is distributed, the more stable the ion is. As the iodide anion is more stable than the fluoride ion, HI can release the proton much more easily than HF.

	H—F	H—Cl	H—Br	H—I
pK_a	3.2	−7	−8	−9

Acidity increases →

After the proton has been removed from an acid molecule, the charge formed can be stabilized via both inductive and mesomeric effects. For example, let us examine the acid strengths in two different molecules, methanol and formic acid. In both cases, the removable proton is attached to an oxygen atom. *Which O-H proton is more acidic, methanol or acetic acid?* While the pK_a value of methanol is 15.3, the pK_a value of formic acid is 3.75. Deprotonation of methanol affords a methoxide ion, which has no resonance (only one Lewis structure can be drawn). The negative charge (−) on the oxygen atom is localized and can be only stabilized by the inductive effect of the oxygen atom. However, after the deprotonation of formic acid, the negative charge will be evenly distributed on both oxygen atoms through resonance. The negative charge will be stabilized by inductive as well as by mesomeric effects. Then, the acid strength increases by a substantial amount. As a result, formic acid is a much stronger acid than methanol. Furthermore, the formate ion is a weaker base than the ethoxide ion. Remember that weaker bases have stronger conjugate acids.

H$_3$C—O—H   ⇌   H$_3$C—Ö:$^{\ominus}$ + H$^{\oplus}$
pK_a = 15.3   Inductive effect

pK_a = 3.75   Inductive effect and mesomeric effect

We have already discussed the influence of the inductive effect on the strength of acidity in the section. Here, we will give more examples. Because the carbon atom in a methyl group attracts bond electrons, the electron density around the carbon atom increases. Therefore, the methyl group donates bond electrons toward the adjacent carbon atom. Let us compare the acidity of formic acid with that of acetic acid. The pK_a value of formic acid is 3.75 while that of acetic acid is 4.76. The methyl group donates electrons toward the carbonyl group and this increases the electron density; as a consequence, the strength of acetic acid decreases. However, the argument here has nothing to do with resonance because no additional resonance contributors can be drawn for acetic acid. If one of the methyl protons in acetic acid is replaced with a methyl group (propionic acid), the strength of acidity is decreased from 4.76 to 4.86. However, this change is smaller than the change seen with formic acid because the number of bonds separating the methyl group and oxygen atom increases. Remember, the inductive effect decreases exponentially on a chain. As the number of methyl groups increases, the acid intensity decreases in parallel with the number of methyl groups.

H—COOH	H$_3$C—COOH	H$_3$C—CH$_2$—COOH	H$_3$C—CH(CH$_3$)—COOH	H$_3$C—C(CH$_3$)$_2$—COOH
3.75	4.76	4.86	4.88	5.05

When the hydrogen atoms in the methyl group in acetic acid are replaced with halogens, the strength of the acids increases in parallel with the electronegativity of the halogen. The presence of chlorines clearly increases the acidity of the carboxylic acids. As chlorine is more electronegative than carbon, the chlorine withdraws the bonding electrons from the adjacent carbon atom toward itself and away from the carboxylate group. As a result, the electrons of the oxygen atom are drawn away from the O—H bond. Therefore, the proton can ionize easily. When more than one halogen atom is attached to the carbon atom, the strengths of acids increase according to the number of chlorine atoms (see also Table 1.3). Trichloroacetic acid falls into the strong acid class. Trifluoroacetic acid is stronger than trichloroacetic acid.

$$H_3C-COOH \qquad ClH_2C-COOH \qquad Cl_2HC-COOH \qquad Cl_3C-COOH$$
$$4.76 \qquad\qquad 2.86 \qquad\qquad 1.29 \qquad\qquad 0.65$$

After examining how the inductive effect affects the pK_a values of acids, let us now examine how the mesomeric effect changes their pK_a values.

$$p K_a = 16.0 \qquad p K_a = 9.9 \qquad p K_a = 4.8$$

As is known, phenols are more acidic compounds than alcohols are. Phenol, for example, is about a million times stronger acid than cyclohexanol is but is weaker than organic acids. The alkoxide anion formed after the removal of a proton from cyclohexanol is stabilized only by oxygen's inductive effect because the conjugate base of cyclohexanol has a single resonance structure. However, the conjugate base of phenol is stabilized by distribution of the negative charge throughout the benzene ring by four significant resonance structures. This provides the stability of the phenolate, in other words, easier separation of the proton from oxygen.

Resonance structures of phenolate anion

To increase the strength of acids, we have always emphasized that the remaining anion, formed after removal of the proton, should be stabilized as much as possible. This phenolic acidity can be further enhanced by attaching electron-withdrawing substituents to the ring. The positions of the groups are also very important. In the phenolate anion, the negative charge concentrates more on the *ortho* and *para* positions, and so it is more effective to connect the electron-withdrawing groups to these positions. It is noteworthy that the influence of a nitro group is over ten times stronger in the *para* position than it is in the *meta* position. *para*-Nitrophenol is more acidic than *ortho*-nitrophenol because of the strong intramolecular hydrogen bonding between the OH hydrogen and the NO$_2$ group.

$$p K_a = 9.9 \qquad p K_a = 7.2 \qquad p K_a = 8.35 \qquad p K_a = 7.14$$

Furthermore, additional nitro groups have an additive influence if they are positioned in *ortho* or *para* locations. The trinitro phenol is a very strong acid called picric acid.

OH
pKa = 9.9

OH, NO2
pKa = 7.2

OH, NO2, NO2
pKa = 4.01

O2N, OH, NO2, NO2
pKa = 1.02   Picric acid

Connecting electron-withdrawing substituents to the phenol ring increases acidity, while connecting electron-donating substituents reduces it. These groups increase the electron density on the ring, as well as on the phenol oxygen so that the proton cannot be easily released. Inductively, the electron-donating methyl group reduces the acidity of the phenol, albeit slightly, depending on its location.

OH
pKa = 9.9

OH, CH3
pKa = 10.28

OH, CH3
pKa = 10.08

OH, CH3
pKa = 10.19

1.6.6 pK_b Values of Bases

There is a close relationship between the acidity of acids and the basicity of bases. An easy way to look at basicity is based on electron pair availability. The more available the electrons, the more readily they can donate electrons to form a new bond to the proton, and therefore the stronger the base. The strength of a base is expressed by the pK_b value. In this section, we will focus on amine bases and show how base strengths are affected.

Weak bases do not have a large attraction for the proton of an acid. For example, ammonia is a weak base, and when it dissolves in water, a low concentration of ammonium hydroxide is formed. The position of equilibrium lies well to the left. When we take the H$_2$O concentration constant, the equilibrium is shown as follows. According to this equation, the smaller the pK_b value of a base, the stronger the base.

$$H_3N: + HOH \rightleftharpoons H_3NH^+ + HO^- \qquad K_b = \frac{[NH_4^+][OH^-]}{[NH_3]} \quad ; \quad pK_b = -\log K_b$$

1.6.7 Factors Affecting the Strengths of Bases in Nitrogen-Containing Compounds

Neutral nitrogen compounds contain a pair of nonbonding electrons. Because this electron pair can bond the proton, nitrogen compounds fall into the base class. The groups that are attached to the nitrogen atom significantly affect the basicity depending on their electronic properties. Electron-donating groups have a tendency to push electrons toward the nitrogen atom and increase the electron density around the nitrogen atom, and this makes the lone pair even more attractive to hydrogen ions.

Ammonia has a pK_b value of 4.75. In general, primary alkyl amines are stronger bases than ammonia. A methyl group in methyamine decreases the pK_b value from 4.75 to 3.36 and a second methyl group to 3.23 as expected. Attachment of a third methyl group, contrary to expectations, increases the pK_b value. These values given below are observed in the aqueous environment.

H—N(H)—H
pK_b = 4.75

H$_3$C→N(H)—H
pK_b = 3.36

H$_3$C→N(H)←CH$_3$
pK_b = 3.23

H$_3$C→N(↓CH$_3$)←CH$_3$
pK_b = 4.20

Two factors influence the strength of a base: how easily a lone pair picks up a proton and the stability of the ions being formed. The nitrogen atom turns into ammonium salt by abstracting a proton. We also need to discuss the stability of the ammonium ion formed by bonding a proton. In the aqueous environment, the ammonium ion can be stabilized by solvation of the water molecules. However, when three methyl groups are bonded to the nitrogen, the stability of ammonium salts decreases because of the hindered solvation. For this reason, the base strength decreases. Trimethylamine is only a slightly stronger base than ammonia despite the presence of three electron-donating alkyl groups. These three methyl groups take up quite a large space. As the nitrogen atom is crowded by the three alkyl groups, it makes it difficult for the nitrogen atom to approach another atom. To sum up, the inductive effect of the alkyl groups increases the base strength, while the reduced solvation and steric effects decrease it.

If one of the hydrogen atoms in ammonia is replaced by a benzene ring, the base strength of the compound (aniline) decreases. In order for a base to be strong, it is crucial that the nonbonding electrons be localized as much as possible. Aniline is a weak base because the lone pair on nitrogen is delocalized into the aromatic ring. This indicates that the nonbonding electrons are no longer fully available to bond a proton. In other words, electrons are delocalized, not localized. The electron density on the nitrogen atom decreases and it is not as much as that on the nitrogen atom in ammonia. Indeed, aniline is a weaker base than ammonia.

NH_3 $pK_b = 4.75$ Ph–NH_2 $pK_b = 9.38$ Aniline

Resonance structures of aniline

Substituted anilines of course have different basicities depending on the electronic nature of the substituent. For example, if an electron-withdrawing group such as a nitro group is attached to the *para* position (*p*-nitroaniline), the nonbonding electrons on the nitrogen atom will delocalize more in the ring and the base intensity will decrease because the nitro group will attract electrons more strongly. The measured pK_b value for nitroaniline (13.02) clearly indicates this finding. If the substituent attached to the ring is an electron-donating group because of the increased electron density on the ring, the delocalization of the free electrons on the nitrogen atom will be hindered. The stronger the electron donor on the ring, the higher the basicity of the aniline.

O_2N–C₆H₄–NH_2 $pK_b = 13.02$ C₆H₅–NH_2 $pK_b = 9.38$ H_3CO–C₆H₄–NH_2 $pK_b = 8.71$

The effect of the substituents in the *ortho* position on the base strength can be either electronic or steric. Let us try to explain how the steric effect dominates over the electronic effect. When we compare the base strength of 2,4,6-trinitroaniline with that of *N,N*-dimethyl-2,4,6-trinitroaniline, the latter is known to be 40 000 times more basic. Of course, the methyl groups will increase the electron density on the nitrogen atom through their +I effects. However, this big difference cannot be solely attributed to the electron-donating effect of the methyl groups. This difference can only be explained by the steric effect. Of course, the three nitro groups will attract the nonbonding electrons on the nitrogen atom as we discussed in the case of aniline. When two methyl groups are attached to the nitrogen atom and if the base strength increases, the first thing that comes to mind is that the electrons cannot delocalize in the ring and they are localized on the nitrogen. *How can such a situation occur?*

The dimethylamino group is sufficiently large to interfere sterically with the very large nitro groups in both o-positions. Because of this steric interaction, a bond rotation about ring carbon to nitrogen occurs and the O atoms of NO₂ and the Me groups of NMe₂ move out of each other's way, so that the p orbital on the N atom is now no longer parallel to the p orbitals of the ring carbon atoms. As a consequence, the nonbonding electrons on the nitrogen cannot delocalize in the aromatic ring. For an ideal conjugation, parallel alignment of the orbitals is required. Localization of the bonding electrons on the nitrogen atom will increase the base strength. The presence of two hydrogen atoms on nitrogen does not lead to a steric interaction between the O atoms of NO and the protons of the NH₂ group.

When the nonbonding electrons of a nitrogen atom conjugate with a carbonyl group, the base strength of the nitrogen atom is also reduced. For example, the nonbonding electrons on nitrogen in amides conjugate with the carbonyl group and the base strength decreases. Generally speaking, when a carbonyl group is attached to the nitrogen atom, the ability of the nitrogen atom to bond a proton decreases because of both the −I and −M effects of that group.

1.6.8 Heterocyclic Bases

Heterocyclic bases containing nitrogen atoms are the most commonly used bases in organic chemistry. Among these, pyridine has the widest area of use. The pK_b value of pyridine is 8.96. Pyridine is a weaker base than saturated amines. As can be seen from the drawing below, the sp² hybrid orbital bearing the nonbonding electrons is perpendicular to the p orbitals in the ring. They are not part of the aromatic sextet. In other words, the lone pair of electrons is not delocalized in the ring. Then, why is the base strength of pyridine weaker than that of aliphatic amines? We emphasized that the higher the s ratio in the hybridized orbitals, the more the electrons are attracted by the nucleus. In amines, hybridization of the nitrogen atom is sp³ and the ratio of s is 25%. In pyridine, because of sp² hybridization, the ratio of s in hybrid orbitals is 33.3%. This indicates that the electrons in the sp² orbitals will be tightly bound by the nucleus. Therefore, the tendency of pyridine to bind a proton is weaker than that of aliphatic amines.

When pyridine and pyrrole are compared, it appears that pyrrole is a weaker base. In pyrrole, the nonbonding electrons of the nitrogen atom are involved in the conjugation with the two double-bond electrons of pyrrole to form a 6π electron system. Therefore, the nonbonding electrons in pyrrole are much less available for bonding to a proton. For these reasons, pyrrole nitrogen is not a base. Pyridine, on the other hand, already has a stable conjugated 6π electron system. Hence, the nonbonding electrons on the nitrogen atom in pyridine can be easily donated to a proton. On the other hand, pyrrolidine, a reduced form of pyrrole, is a strong base. There is no electron delocalization in this system; the electrons are localized on the nitrogen atom. The pK_b value of pyrrolidine is 2.73.

Aniline Pyridine Pyrrole Pyrrolidine Diethylamine
$pK_b = 9.38$ $pK_b = 8.96$ $pK_b = 13.6$ $pK_b = 2.73$ $pK_b = 3.07$

1.7 Reaction Kinetics and Energy Diagrams

Reaction kinetics is a field that investigates the rate of chemical reactions. *What is the mechanism of the reaction? What intermediates are occurring during a chemical reaction?* These questions should be addressed to understand how a reaction occurs. The energy change that occurs during the reaction is also very important in terms of the reaction mechanism. It is not sufficient to know the products formed to explain the mechanism of the occurrence of a chemical reaction. It is also necessary to know the intermediates formed between the reactant and the product. There are many experiments designed to investigate how reactions happen. One of these methods used is chemical kinetics, in which the rate of a reaction is measured. In most chemical reactions, compounds A and B do not directly form products. Intermediates often occur, and these intermediate products can form the product as well as return to the reactants.

A + B ⇌ [Intermediates] ⟶ C + D
Reactants   Products

What is the effect of energy on the chemical reaction? The effect of energy on reaction rates is indisputably important. Collision between the reactants is essential in order to react or change chemically. We should not expect all collisions between molecules to result in chemical reactions. For example, nitrogen and oxygen molecules are present in the air and they constantly collide. However, there is no chemical reaction between them as a result of these collisions. If collisions alone cannot explain reactivity, there must be other factors responsible for chemical reactions. The kinetic energy of molecules is important here. Slow moving molecules with lower kinetic energy do not take part in chemical reactions. Collisions between molecules with higher energy that lead to a chemical reaction are called *effective collisions*. We should then answer the question: *Why do reactions require collisions of molecules with high energy?* We know that during a chemical reaction, some bonds are broken and some bonds are newly formed. We know that breaking bonds requires energy and the formation of new bonds releases energy. The minimum energy required for an effective collision to break some bonds and start the reaction is called the *activation energy* (Figure 1.22).

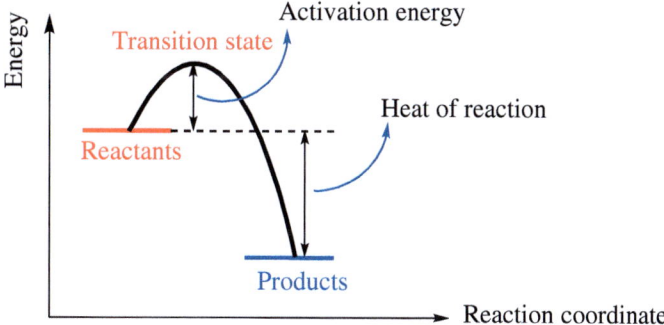

Figure 1.22 Reaction coordinates for an exothermic reaction and activation energy.

This diagram (Figure 1.22) plotting energy as a function of reaction progress is called a potential energy diagram. According to this diagram, reactants must reach a certain energy level. The energy rises to a maximum value called the *transition state*. At the transition state, the reactants have sufficient energy so that the bonds in the reactant start to break. The bonds to be broken in the transition complex are partially weakened and new bonds to be formed begin to form partially. Let us try to explain this phenomenon with a chemical reaction. In the transition state for the nucleophilic substitution reaction of methyl bromide by a hydroxide ion, a bond between the oxygen and carbon is partially formed and the bond between carbon and bromide is partially broken. The structure of the transition state lies somewhere between that of the product and that of the reactants. The dashed lines show partially broken and partially formed bonds.

$$HO^{\ominus} + H_3C-Br \longrightarrow \overbrace{HO\text{----}CH_3\text{----}Br}^{\ominus} \longrightarrow HO-CH_3 + Br^{\ominus}$$

Transition state complex

The half-life of the transition state complex is very short. When moving to the right on the reaction coordinate, the system starts to release energy and begins to form products. The higher the activation energy, the slower the reaction; the lower the activation energy, the faster the reaction. For example, if the activation energy of a chemical reaction is 40 kcal/mol, the reaction rate will be indeterminably slow at room temperature (25 °C). This indicates that the reaction will not take place at room temperature. If the activation energy of a reaction is 15 kcal/mol, this reaction takes place quickly at room temperature.

To increase the rate of a reaction, the amount of energy required for the formation of the transition state must be reduced. Compounds that reduce the activation energy required for the formation of the transition state and increase the rate of a reaction are called *catalysts*. Catalysts are not consumed during the reaction, they actively participate in the formation of intermediates, and they do not appear in the net reaction equation. They are recovered after the reaction and certainly do not affect the energies of the reactants or products (Figure 1.23).

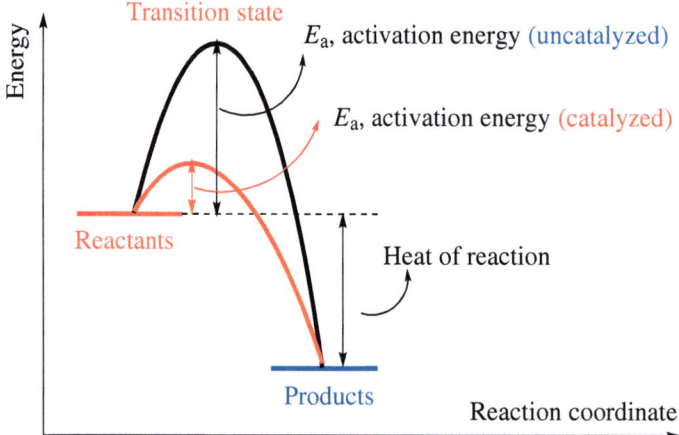

Figure 1.23 Potential energy diagram comparing catalyzed and uncatalyzed processes.

Chemical reactions cannot be limited to a single transition state. Some reactions occur in two or more steps as in the case of addition of HBr to ethylene. In the first transition state, double-bond electrons attack the proton of HBr and begin to bond it using the π electrons to the carbon atom. The energy then decreases until an *intermediate*, a carbocation, is formed. In the next step, which has its own activation energy, the carbocation starts to react with the bromide ion to form the final addition product (Figure 1.24).

The carbocation formed is more stable than either of the two transition states. In most cases, the intermediates are not stable and they cannot be isolated. One should not confuse intermediates with transition states. *Intermediates always have fully formed bonds, while transition states always have partially formed bonds.*

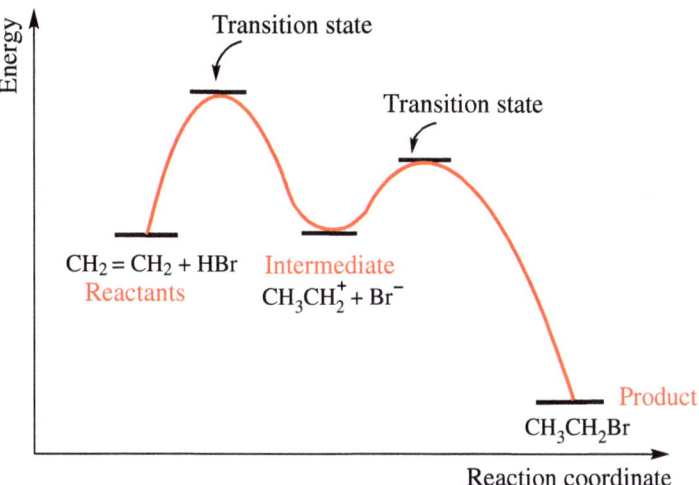

Figure 1.24 Reaction coordinate diagram for the addition HBr to ethylene.

1.7.1 Thermodynamic vs. Kinetic Control of Reactions

Let us discuss a chemical reaction in which two isomeric products are formed. The energies of these products will most likely be different. For example, conjugated dienes undergo electrophilic addition reactions to form a mixture of 1,2-adduct and 1,4-adduct. The reaction of hydrogen bromide with 1,3-butadiene gives different product distributions under different conditions. When the reaction is carried out at 0 °C, the major product is the 1,2-addition product. However, when the same reaction is carried out at 40 °C, the major product is the 1,4-addition product. When the structures of these molecules are compared, it is easy to recognize that the 1,4-addition product is more stable than the 1,2-addition product is. The 1,4-addition product is the thermodynamic product with the lowest energy. The 1,2-addition product is a less stable product called a kinetic product.

$$H_2C=CH-CH=CH_2 \xrightarrow{HBr} HH_2C-\overset{\oplus}{C}H-CH=CH_2 \longleftrightarrow HH_2C-CH=CH-\overset{\oplus}{C}H_2$$

Butenyl cation

	1,2-Addition product	1,4-Addition product
	Kinetically controlled product	Thermodynamically controlled product
At 0°C	71%	29%
At 40°C	15%	85%

The first step is the protonation of one of the C=C double bonds in butadiene. The protonation occurs regioselectively and gives the more stable carbocation because of conjugation with the adjacent double bond. The positive charge (+) is distributed over two carbon atoms. The bromide ion can attack either carbon. Attacking the C2 carbon leads to the kinetic product; attacking the terminal carbon, C4, leads to the thermodynamic product. The activation energy leading to the 1,2-addition product is less than that leading to the 1,4-addition product, even though the 1,4-addition product is more stable. Therefore, at low temperature, 1,2-adduct, the kinetically controlled product, is formed as the major product. At low temperatures, formation of these products is irreversible. The energy is not enough to cross back over the barrier to form a butenyl cation. However, at higher temperatures, there is sufficient energy for both adducts to revert to the butenyl cation. Because the activation barrier for 1,2-adduct is smaller for conversion back to the butenyl cation than that for 1,4-adduct, more 1,2-adduct will revert to the butenyl cation, which will again form both adducts. Finally, the equilibrium will shift to the side of the thermodynamically more stable product, the 1,4-addition product. In summary, the 1,2-adduct is the *kinetically controlled product* with the lower activation barrier, while the 1,4-adduct is the *thermodynamically controlled product* with the higher activation barrier (Figure 1.25).

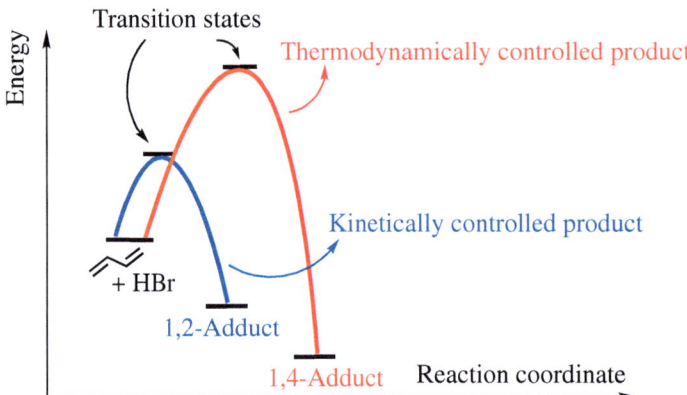

Figure 1.25 An energy diagram for the formation of 1,2-adduct and 1,4-adduct.

In contrast to many other electrophilic aromatic substitution reactions, aromatic sulfonation is reversible; in other words, it is an equilibrium. Two products are formed in the sulfonation reaction of naphthalene. The α-substitution product is less stable and occurs at 80 °C. The β-substitution product is thermodynamically more stable and occurs at 160 °C. Therefore, the transition state leading to the α-substitution product will be lower in energy and it will be formed fastest. Therefore, the α-substitution product is the kinetically favored product. However, the β-substitution product is more stable than its α-isomer is. This is due to the 1,8 steric repulsion present in the α-isomer. Again, here is also an equilibrium between naphthalene and the products. The formation of such a balance at high temperature is demonstrated by the conversion of the α-substitution product into a β-substitution product when heated to 160 °C.

1.7.2 Reaction rate

Knowing whether a chemical reaction is endothermic or exothermic does not give us any information about the reaction rate. The activation barrier must be overcome first. Otherwise, a chemical reaction cannot proceed. Therefore, there are several factors that affect the reaction rate. What are these factors?

- Number of collisions between molecules in a given unit of time and volume,
- Number of colliding molecules with sufficient energy,
- Number of molecules colliding in the appropriate position.

These three factors determine the speed of chemical reactions. The first is called the collision frequency, the second the energy factor, and the third the orientation factor. The reaction rate directly depends on the concentration of the reactants. The higher the reactant concentration, the greater the probability of molecules colliding. The speed of each reaction is different and the reaction rate is determined by experiments.

A chemical reaction takes place when particles collide in the correct orientation and with sufficient energy. If they do not collide, there can be no reaction. These parameters can either speed up or slow down the rate of the reaction.

Some reactions contain only one reactant. For example, when sulfolane is heated, SO_2 is released and it turns into butadiene. Because there is only one reactant in this reaction, the reaction rate directly depends on the concentration of sulfolane.

If its concentration is doubled, the reaction rate is doubled. The reaction rate formula is given below. The rate is directly proportional to the concentration of the reactant. Such reactions are called *first-order reactions*.

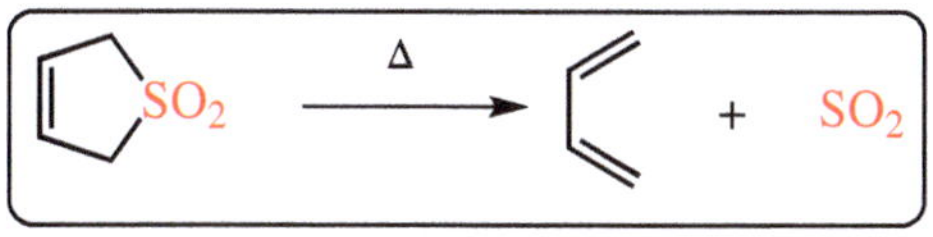

$$\text{Reaction rate} = k_a [A]$$

A = concentration of sulfolane

k_a = rate constant

Let us look at the kinetics of a simple nucleophilic substitution reaction, the reaction of ethyl mesylate with Br⁻. If we double the concentration of mesylate, we find that the rate of the reaction also doubles. Similarly, if we double the concentration of the Br⁻ ion, the reaction rate will again double. In this case, the rate of the reaction linearly depends on the concentration of two species. This reaction is called a *second-order reaction*.

$$\text{Br}^- + \text{CH}_3\text{CH}_2\text{-OSO}_2\text{Me} \longrightarrow \text{H}_3\text{CH}_2\text{C-Br} + {}^-\text{OSO}_2\text{Me} \qquad \text{Reaction rate} = k_a [\text{Et-OSO}_2\text{Me}][\text{Br}^-]$$

Most nucleophilic substitution reactions take place by this mechanism discussed above. However, when a substitution reaction is carried out with an unhindered substrate, such as methyl bromide with a neutral nucleophile (H_2O) in a protic solvent, the reaction is one of the slowest substitution reactions. When the same reaction is carried out with *t*-butyl bromide and water, the reaction is one million times faster to give *t*-butanol. The rate of this reaction depends only on the concentration of *t*-butyl bromide. The mechanism of this reaction differs from the mechanism shown above. In the first step, the leaving group (Br⁻) leaves and then a carbocation is formed. This step is the rate-determining step. The formation of a bond between the nucleophile and the carbocation in the second step occurs rapidly. In the rate-determining step, only one substrate is involved. Therefore, the rate of the reaction depends only on the concentration of *t*-butyl bromide. Hydrolysis of CH_3Br is a second-order reaction. The rate of the reaction depends on the concentration of CH_3Br and H_2O. However, if water is used as a solvent (large excess), only a small fraction of water in excess will be consumed, and its concentration can be considered to stay constant. The rate of the reaction will depend only on the concentration of CH_3Br. This is also a *first-order reaction*, although two different substrates are used as the starting materials.

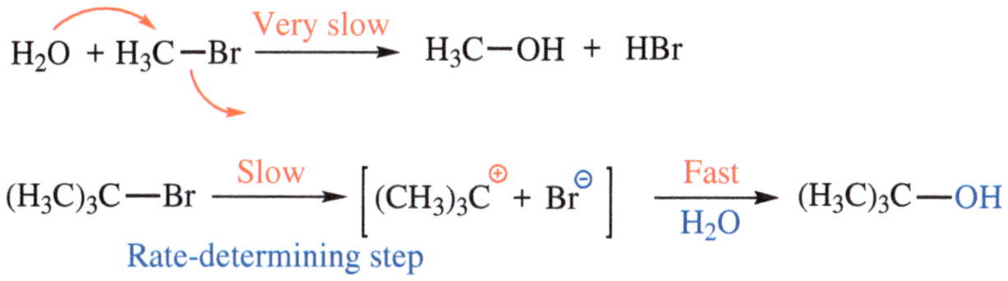

	rate
$H_2O + H_3C\text{-Br} \xrightarrow{\text{Very slow}} H_3C\text{-OH} + HBr$	1
$(H_3C)_3C\text{-Br} \xrightarrow{\text{Slow}} [(CH_3)_3C^+ + Br^-] \xrightarrow[H_2O]{\text{Fast}} (H_3C)_3C\text{-OH}$ Rate-determining step	1.000.000

$$\text{Reaction rate} = k_a [(CH_3)_3C\text{-Br}]$$

A molecule can undergo a reaction with itself. Such reactions are also second-order reactions. In such a reaction, if the concentration of molecule A is doubled, the reaction rate is quadrupled.

$$\text{Reaction rate} = k_a [A][A] = k_a [A]^2$$

The reaction rate should not be confused with the rate constant. The reaction rate depends on concentration and the rate constant. However, the rate constant k_a is not a concentration-dependent parameter. The value of the rate constant depends on the temperature and the activation energy E_a and is expressed by the Arrhenius equation.

$$\text{Arrhenius equation } k = A e^{-\frac{E_a}{RT}}$$

k = rate constant
E_a = activation energy
T = temperature (K)
A = a constant for the frequency of particle collisions (frequency factor).

Here, E_a is the activation energy. As E_a increases, k will be smaller, and a smaller rate constant corresponds to a slower chemical reaction. Temperature appears in the exponent and therefore the rate constant will be very sensitive to this parameter. Small changes in temperature can have a drastic effect on the rate of the reaction. Frequency factor A is a measure of the number of molecules colliding at the appropriate position. It varies slightly with temperature. During the reaction, only a small part of the colliding molecules collide into the product (in the appropriate position), and the energy of a small proportion of the colliding molecules is sufficient to form the product.

The Arrhenius equation indicates that the reaction rate depends on the number of molecules with kinetic energy of at least E_a. Not all molecules at a given temperature have the same kinetic energy. The energies and numbers of molecules in a gas phase are shown in the graphic in Figure 1.26. The red curve shows the energy distribution of molecules at room temperature, while the blue curve shows the energy distribution of molecules at a higher temperature. Suppose the activation energy required for a reaction is 20 kcal/mol. The activation energy is shown with a dashed black line. The shaded area on the right side of the line where this line crosses the red curve shows the amount (fraction) of molecules that have the activation energy necessary for the reaction to occur. If the activation energy is 40 kcal/mol, the number of molecules with this energy appears to be very low. When the reaction temperature is increased (blue curve), the number of molecules with the activation energy required for the reaction increases (the entire shaded area). This indicates that the reaction rate increases with increasing temperature.

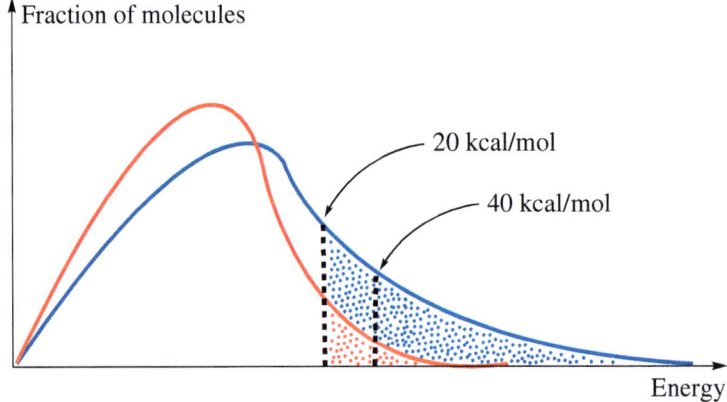

Figure 1.26 Variation in the kinetic energy of the molecules with temperature. Red curve: room temperature, blue curve: a higher temperature.

Problems

1.1 What is the hybridization of each bond (10 bonds) in propane ($CH_3CH_2CH_3$).

1.2 Find out from which orbitals the bonds are formed in the following molecules.

$$CH_3-CH_2-O-H \quad CH_3-O-CH_2-NH_2 \quad Br-CH_2-O-N(CH_3)_2 \quad H_2N-NH_2$$

1.3 Determine the hybridization of all the bonds of the following compounds.

$$H_2C=CH-CH_3 \quad H_2C=CH-CH=CH_2 \quad H_3C-CH=CH-C\equiv CH \quad H_3C-N=C=O$$

1.4 Determine the hybridization of the carbon atoms in the following compounds. Draw the molecules with their orbitals and discuss the positions of the hydrogen atoms.

$$H_2C=C=C=CH_2 \quad H_2C=C=C=C=CH_2$$

1.5 Determine the hybridization of the carbon atoms in the molecules whose structures are given.

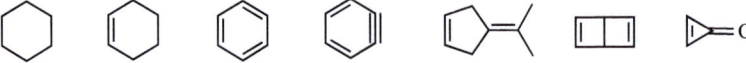

1.6 Determine the hybridization of the carbon atoms of the following molecules.

1.7 In the molecule below, determine the five-membered rings' position relative to each other, taking into account the hybridization.

1.8 Discuss the conformation of butadiene, taking into account the hybridization of butadiene carbon atoms.

1.9 Discuss the geometric structures of the molecules below, taking hybridization into account.

$$H_3C-C\equiv C-C\equiv CH \qquad H_2C=CH-C\equiv C-CH=CH_2$$

1.10 Discuss the stability of the molecules below, taking hybridization into account.

1.11 Find the formal charges of the following compounds.

$$H_3C-\underset{..}{\overset{CH_3}{\underset{|}{O}}}-CH_3 \qquad H_3C-\underset{H}{\overset{CH_3}{\underset{|}{N}}}-CH_3 \qquad H_3C-\underset{..}{\overset{CH_3}{\underset{|}{C}}}-CH_3 \qquad H-\underset{H}{\overset{CH_3}{\underset{|}{N}}}-\underset{H}{\overset{H}{\underset{|}{B}}}-H$$

1.12 Find the formal charges of the elements given below.

$$H_3C-\overset{CH_3}{\underset{|}{C}}-CH_3 \qquad R-\ddot{N}=\ddot{N}: \qquad R-N\equiv N: \qquad \underset{R}{\overset{R}{>}}C: \qquad R-C=O \qquad R-\underset{H}{\overset{R}{\underset{|}{C}}}=N\underset{R}{\overset{R}{<}}$$

$$R-N\underset{\ddot{O}:}{\overset{\ddot{O}:}{<}} \qquad R-\underset{:O:}{\overset{:O:}{\underset{||}{\overset{||}{S}}}}-Cl \qquad R-N\equiv C: \qquad \underset{R}{\overset{R}{>}}\dot{C}-R$$

1.13 Write the resonance structures of the following compounds and the formal charges of the atoms involved.

$$CH_2N_2 \qquad RN_3$$

1.14 Find the oxidation numbers of the central atoms reacting in the reactions given below and discuss the reactions in terms of oxidation and reduction.

(a) [reaction showing HO-C(=O)-CH=CH- with I_2 giving a cyclic product with O and I]

(b) [reaction of cyclohexene with pendant OH with Br_2 giving bicyclic product with O and Br]

1.15 Unsaturated bicyclic endoperoxides rearrange into unsaturated hydroxy ketones in the presence of a base or an acid. Discuss this transformation in terms of oxidation and reduction.

[bicyclic endoperoxide] —Acid or base→ [cyclohexenone with OH]

1.16 Let us examine the reactions given below in terms of oxidation and reduction.

(a) [bicyclic azo compound] —Heat→ [cyclopropene] + :N≡N:

(b) Cl₃C–C(=O)Cl (hexachloro) —Zn→ Cl₂C=C=O

1.17 Which of the following compounds is the stronger acid?

1. (a) CH₃OCH₂CH₂OH   2. (a) H₃C-C(=O)-CH₂OH   3. (a) CH₃OCH₂CH₂OH

   (b) CH₃CH₂CH₂CH₂OH   (b) CH₃CH₂-C(=O)-OH   (b) CH₃CH₂OCH₂OH

1.18 Sort the following molecules according to their acidity strength.

(a) CH₃CH₂CH₂OH   (b) CH₃CHFCH₂OH   (c) CH₃CHClCH₂OH   (d) ClCH₂CH₂CH₂OH

1.19 Sort the following molecules according to their acidity strength.

(a) ClCH₂CH₂COOH   (b) CH₃CHFCOOH   (c) BrCH₂CH₂COOH   (d) ClCH₂CHFCOOH

1.20 Which of the following compounds is more acidic?

H₃C-C(=O)-OH   H₃C-S(=O)(=O)-OH

1.21 HCl is a weaker acid than HBr. ClCH₂COOH is a stronger acid than BrCH₂COOH. Give a reasonable explanation.

1.22 Sort the following molecules according to their acidity strength.

H₃C-C(=O)-CH₃   H₃C-C(=O)-CH₂-C(=O)-CH₃   C₆H₅-OH   H₃C-C(=O)-OH

(a) pK_a = 20   (b) pK_a = 9   (c) pK_a = 10   (d) pK_a = 4.7

1.23 Which of the above compounds reacts with NaOH? (pK_a value of H₂O = 15.7).

1.24 Does the *t*-butoxide anion (KOC(CH₃)₃) react with H₂O? Can an aqueous solution of this anion be obtained?

1.25 We have two bottles. In one bottle, we have phenol (pK_a = 9) and in the other we have an organic acid, which is not soluble in water (pK_a = 5.0). How can we distinguish between the two compounds using sodium bicarbonate (NaHCO₃) (pK_a value of H₂CO₃ = 6.4)?

References

1 Finston, H.L. and Rychtman, A.C. (1982). *A New View of Current Acid-Base Theories*, vol. 140. New York: Wiley.
2 Hawkes, S.J. (1992). *J. Chem. Edu.* 69: 542.

3 Bronsted, J.N. (1923). *Recl. Trav. Chim. Pays-Bas* 42: 718.
4 Jensen, W.B. (1980). *The Lewis Acid-Base Concepts: An Overview*. New York: Wiley-Interscience.
5 Pearson, R.G. (1963). *J. Am. Chem. Soc.* 85: 3533.
6 Pearson, R.G. (1997). *Chemical Hardness – Applications From Molecules to Solids*, vol. 198. Weinheim: Wiley-VCH.
7 Parr, R.G. and Pearson, R.G. (1983). *J. Am. Chem. Soc.* 105: 7512.

2

Nucleophilic Substitution Reaction

2.1 Types of Chemical Reactions

During chemical reactions, the original chemical bonds between the atoms are broken and new bonds are formed. When we examine the reactions in terms of breaking and forming new bonds, we encounter three groups.

2.1.1 Polar Reactions

Bond breaking in which the covalent bond between two chemical species is broken unequally so that the bonding electrons are retained by one of the chemical species. This kind of bond breaking is called *heterolytic bond cleavage*. When a neutral molecule undergoes heterolytic bond cleavage, one of the products will have a positive charge, while the other will have a negative charge. The charge distribution depends on the electronegativity difference between the atoms.

2.1.2 Radical Reactions

Bond breaking in which the bonding electron pair is split evenly between the products so that one electron is retained by each of the original fragments of the molecule. This kind of bond breaking is called *homolytic bond cleavage*. When a neutral molecule undergoes homolytic bond cleavage, two free radicals are formed. Because radicals are very unstable, they react in a variety of ways (see Section 9.1).

2.1.3 Pericyclic Reactions

Although most organic reactions take place by way of ionic or radical intermediates, many useful reactions occur in one-step processes that do not form reactive intermediates. They are not affected by solvent changes. Bond breaking and bond formation take place simultaneously in a concerted manner. For example, the Diels–Alder reactions fall into this group.

Polar reactions constitute a significant part of organic chemistry. Polar reactions should also be classified.

In nucleophilic substitution reactions, an electronegative atom or an electron-attracting group is separated from the molecule and replaced by another atom or group. If a halogen is bonded to a carbon atom, the two atoms do not share the bonding electrons equally because halogens are more electronegative than carbon. The bond electrons are more attracted by halogen and the carbon–halogen bond is polarized. The carbon atom carries a partial positive charge ($\delta+$), while halogen carries a partial negative charge ($\delta-$). Therefore, nucleophiles attack the carbon atom and remove the halogen. A new bond is formed between the nucleophile and the carbon atom. The halogen that is substituted in this reaction is called *the leaving group*. This displacement reaction is called a *nucleophilic substitution reaction* because the atom or group that replaces the leaving group is a nucleophile. The reaction occurs on saturated, sp³-hybridized carbon atoms. The equation below shows the nucleophilic substitution reaction of an alkyl halide.

$$\text{Nu}:^{\ominus} + \text{R—H}_2\text{C}^{+\delta}\text{—X}^{-\delta} \xrightarrow{\text{Solvent}} \text{R—H}_2\text{C—Nu} + :\text{X}^{\ominus}$$

Nucleophile   Alkyl halide   Product   Leaving group

In nucleophilic substitution reactions, the bond between the leaving group and the substrate carbon undergoes heterolytic cleavage so that the leaving group is separated from the molecule with the bonding electrons. The new bond formed between the carbon atom and the nucleophile is created by the unshared electron pair of the nucleophile. In order to understand the mechanism of nucleophilic substitution reactions, the function of the nucleophile, base strength, reactant, leaving group, product, and solvent should be discussed in detail. Let us examine nucleophilicity first.

2.1.3.1 Relationship Between Nucleophilicity and Basicity

Basicity is a measure of how well a compound shares its lone pair of electrons. In other words, basicity refers to the ability of a base to accept a proton. Nucleophilicity is a measure of how fast a compound can attack a partially positively charged carbon atom. In the reaction shown above, there is an electron pair on the nucleophile and the nucleophile attacks the carbon atom with this electron pair. Hence, there must be a relationship between nucleophilicity and basicity. Nucleophilicity affects the rate of the substitution reaction (rate constant, k_a), whereas basicity affects the equilibrium constant (K_a). With some examples, let us try to explain the points in common and not in common between basicity and nucleophilicity. In a more general sense, we can separate these two concepts from each other as follows. Nucleophilicity refers to the rate of substitution reactions at the halogen-bearing carbon atom, while basicity refers to the ability of a base to accept a proton.

$$\text{Base} + \text{H—X} \xrightleftharpoons{K} \text{Base—H}^{\oplus} + \text{X}^{\ominus} \quad \text{Basicity}$$

$$\text{Nu} + \text{R—X} \xrightleftharpoons{k} \text{R—Nu}^{\oplus} + \text{X}^{\ominus} \quad \text{Nucleophilicity}$$

Let us discuss these two concepts with some examples. We can compare the basicity and nucleophilicity of the halides. The strongest acid among HI, HBr, HCl, and HF is HI. HF is the weakest acid, and the F⁻ ion is the strongest base and the order of basicity is as follows: F⁻ > Cl⁻ > Br⁻ > I⁻. The stability of the conjugate base directly determines the strength of acids. As the atomic radius of iodine is approximately twice that of fluorine, iodine can distribute the negative charge over a significantly larger volume. Therefore, iodide is more stable and the least basic. The least basic one, iodide, is the most nucleophilic; on the other hand, the most basic one, fluoride, is the least nucleophilic. Within a group in the periodic table, because of the increasing polarization, nucleophilicity decreases in the following order: I⁻ > Br⁻ > Cl⁻ > F⁻. Polarizability is a measure of how easily an electron cloud of an atom can be distorted. In the case of a larger atom such as iodine, the outer electrons experience a weaker force of attraction from the nucleus (although the nuclear charge is higher). The weaker attraction of the nucleus on the electrons indicates that the electron cloud can more easily be distorted (Figure 2.1).

   Orbital polarization

Figure 2.1 Approximate representation of an orbital polarization.

The solvent has a drastic effect on the nucleophilicity of halides. Iodide (I⁻) is a better nucleophile than fluoride (F⁻) in *polar protic* solvents, whereas fluoride is a better nucleophile in *polar aprotic* solvents. Protic solvents, such as H₂O, MeOH, and EtOH, are hydrogen donors. If ions are dissolved in such solvents, they are surrounded by solvent molecules; they are solvated. Solvation weakens the nucleophilicity of anions by forming a shield of solvent molecules around the anions and impeding attack of the electrophiles. This interaction between the ion and dipole of the protic solvent is called an *ion–dipole interaction*. Of course, smaller ions, such as fluoride, are more tightly solvated than the larger ones, such as iodide, because the charge of smaller ions is more concentrated. A solvated nucleophile has to push this shield of solvent molecules out of the way to attack the carbon bearing the leaving group. Weak bases, such as iodide, with a low charge density are loosely solvated. There are only a few solvent molecules to push out of the way. Therefore, iodide is more nucleophilic than fluoride in polar protic solvents (Figure 2.2).

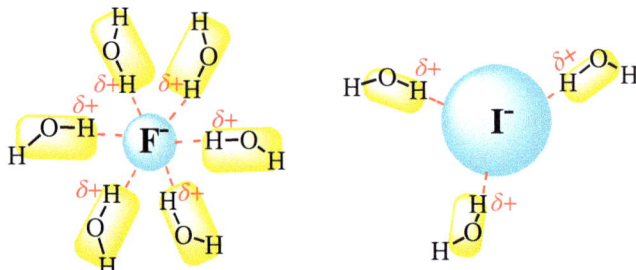

Figure 2.2 Solvation of a small anion (F⁻) and a larger anion (I⁻).

However, the situation is entirely different in polar aprotic solvents. Among the halides, the most nucleophilic ion is fluoride (F⁻). Polar aprotic solvents such as dimethylformamide (DMF), dimethyl sulfoxide (DMSO), and acetone can dissolve ionic compounds. However, because they do not have polar groups such as –OH and –NH$_2$, they cannot form hydrogen bonds with the anions. Polar aprotic solvents have a partial negative charge on their surface that can solvate only cations. Of course, they also have a partial positive charge, which is located inside the molecule. The solvation of cations leaves anions freer to act as nucleophiles (Figure 2.3). For example, bromomethane reacts with potassium iodide (KI) faster in acetone than in methanol.

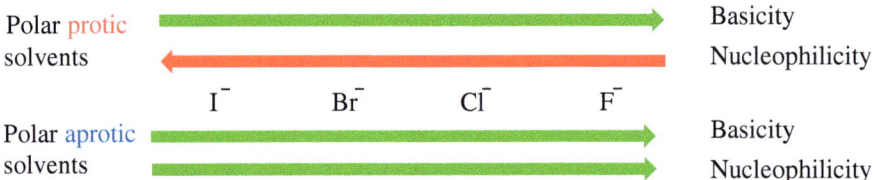

Figure 2.3 Solvation of a cation (K⁺) by DMF.

In summary, in polar protic solvents, within a group of the periodic table, the order of nucleophilicity increases from top to bottom, and the basicity runs opposite to the order of nucleophilicity. In polar aprotic solvents, nucleophilicity increases from bottom to top, parallel to basicity.

The same effects are seen when we compare H$_2$S with H$_2$O. Sulfur and oxygen are in the same group. Sulfur is a stronger nucleophile because the sulfur atom is larger than the oxygen atom and sulfur cannot be well solvated in a polar solvent.

How can we measure the nucleophilicity of a compound (nucleophile)? The two nucleophiles can be mixed and reacted with the same alkyl halide (R-X) under the same conditions. No matter which nucleophile has formed the main product, that nucleophile is a stronger nucleophile than the other [1]. Experiments show that there is a parallelism between the nucleophilic power and the base strength of elements in the same period. There will not be much difference between the sizes of the elements in that period. For example, an amide is a stronger base than the hydroxide ion. The hydroxide ion is a stronger base than the fluoride ion. The methoxide ion is more nucleophilic than the fluoride anion. The oxygen atom is less electronegative than the fluoride atom. Therefore, oxygen can donate the nonbonding electrons easily to carbon in nucleophilic substitution reactions. On the other hand, the methoxide ion is more basic than the fluoride anion. Within a period, the basicity and nucleophilicity increase in parallel from right to left.

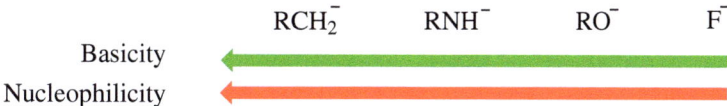

Consider a series of nucleophiles that attack with the same atom shown below. Base strength increases from right to left in the following compounds. Nucleophilicity also increases from right to left in parallel to the base strength.

$$RO^\ominus > OH^\ominus > PhO^\ominus > R-C(=O)O^\ominus > ROH > H_2O$$

Basicity ←
Nucleophilicity ←

The steric effect on base strength is not very great; however, it affects nucleophilicity significantly. For example, although bulky *t*-butoxide is a stronger base than the ethoxide ion, it is a poorer nucleophile than ethoxide. The bulky nucleophile *t*-butoxide cannot approach the carbon atom as easily as the less sterically hindered ethoxide ion. Therefore, ethoxide is a stronger nucleophile and less basic than the *t*-butoxide ion.

$$H_3C-CH_2-O^\ominus \quad H_3C-CH(CH_3)-O^\ominus \quad H_3C-C(CH_3)_2-O^\ominus$$

Basicity →
Nucleophilicity ←

We observe similar situations with compounds whose central atom is nitrogen. When the steric effect increases, the nucleophilicity of the relevant ion or molecule decreases. Some amine compounds are listed below according to their nucleophilicity.

[LDA-type diethylamide Li⁺ ≫ diisopropylamide Li⁺]   [pyridine ≫ 2,6-di-*t*-butylpyridine]

When a nucleophile is attached to an atom with nonbonding electrons, its nucleophilicity increases; on the other hand, its basicity decreases. Hydrazine (H₂N-NH₂) is considerably more nucleophilic than ammonia (NH₃), although it is less basic. Basicity is decreased because of the electron-withdrawing effect of the second nitrogen atom.

Nucleophilicity: $H_2N-NH_2 \gg NH_3$
$HOO^\ominus \gg HO^\ominus$

Molecular orbital (MO) theory can explain the change in nucleophilicity. The nonbonding electrons on nitrogen or oxygen atoms interact with each other and form new MOs. Now, there are four electrons to populate these two orbitals. Two of the electrons will occupy the antibonding orbital. This interaction raises the energy of two nonbonding electrons relative to the energy in the absence of other nonbonding electrons. Raising the highest occupied molecular orbital (HOMO) energy causes increased nucleophilicity (Figure 2.4). This effect of the adjacent heteroatom is called the *α-effect* [2]. The hydroxide (HO⁻) ion is 16 000 times more basic than the peroxide (HOO⁻) ion. However, the peroxide ion attacks the carbonyl group 200 times faster than the nucleophilic hydroxide ion does.

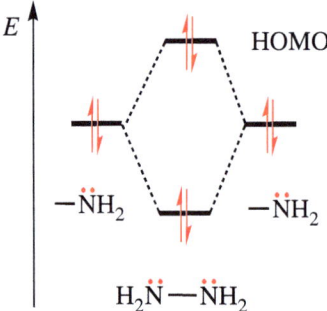

Figure 2.4 Molecular orbital diagram of hydrazine.

In gas-phase experiments, inverse results were obtained. The fact that hydrazine and hydroperoxide are stronger nucleophiles in the liquid phase can be attributed to weak solvation of these compounds [3].

Sodium hydride (NaH) in a small volume is a strong base and it is not easily polarizable. The nucleophilicity of NaH is very weak and therefore it does not undergo substitution reactions.

2.1.3.2 Leaving Group

The nature of the groups substituted by the incoming nucleophile is important in terms of reaction rate and reaction mechanism. In most cases, the leaving groups are expelled with a negative charge. The best leaving groups are those that best stabilize the negative charge. Otherwise, it would not be easy to separate the group from the molecule. Groups that exhibit weak base properties after leaving the carbon atom are generally good leaving groups. For example, it is difficult to remove a hydroxide ion (HO^-) from alcohol. However, when the hydroxyl group is first protonated, it can then be easily removed as water because H_2O is a weaker base than the hydroxide anion. Among the halides, the iodide ion is the best leaving group, whereas the fluoride ion is the poorest. Within a group, leaving-group ability increases as you go down the column in accordance with decreased basicity ($I^- > Br^- > Cl^- > F^-$). Iodine can stabilize the negative charge much better than the other halides. The most exciting feature of iodine is that it is both a good nucleophile and a good leaving group.

A good leaving group can distribute the negative charge on as many atoms as possible. Alkyl triflate, alkyl tosylate, and alkyl mesylate are good leaving groups.

If there is a –OH group in a molecule and it is desired to substitute this group by nucleophilic substitution reaction, it is more convenient to first convert the hydroxyl group into a better leaving group, sulfonate, and then perform the substitution reaction. Ethers do not undergo substitution reactions because alkoxide ions are bad leaving groups. However, an epoxide, three-membered cyclic ether, can undergo a ring-opening reaction by nucleophilic substitution reactions because of the angle strain in the three-membered ring. One of the epoxide bonds opens, the resulting group remains in the molecule, and 26 kcal/mol energy is released by ring opening.

Nucleophilic substitution reactions are divided into four groups according to the charges of the reactants. Let us explain these with examples [4].

1. *Neutral Reactant and Neutral Nucleophile*: Nucleophilic substitution takes place between two neutral compounds. The resulting products can be neutral or charged.

$$H_3C-Br + \overset{..}{N}H_3 \longrightarrow H_3C-\overset{\oplus}{N}H_3 + Br^{\ominus}$$

$$\underset{CH_3}{\overset{CH_3}{H_3C-\underset{|}{\overset{|}{C}}-Cl}} + H_2\overset{..}{\underset{..}{O}} \longrightarrow \underset{CH_3}{\overset{CH_3}{H_3C-\underset{|}{\overset{|}{C}}-OH}} + HCl$$

2. *Neutral Reactant and Negatively (−) Charged Nucleophile*: This is the most frequently encountered nucleophilic substitution reaction.

$$H_3C-Br + {}^{\ominus}OH \longrightarrow H_3C-OH + Br^{\ominus}$$

$$C_3H_7-OTs + {}^{\ominus}OCH_2H_5 \longrightarrow C_3H_7-OC_2H_5 + TsO^{\ominus}$$

3. *Positively (+) Charged Reactant and Neutral Nucleophile*: This kind of substitution reaction is observed in charged nitrogen and sulfur compounds.

$$H_3C-\overset{\oplus}{\underset{CH_3}{\overset{CH_3}{S}}} + :N(CH_3)_3 \longrightarrow H_3C-\overset{\oplus}{\underset{CH_3}{\overset{CH_3}{N}}}-CH_3 + :S\overset{CH_3}{\underset{CH_3}{}}$$

4. *Positively (+) Charged Reactant and Negatively (−) Charged Nucleophile*:

$$C_2H_5-\overset{\oplus}{\underset{C_2H_5}{\overset{C_2H_5}{O}}} + Br^{\ominus} \longrightarrow C_2H_5-Br + :\overset{C_2H_5}{\underset{C_2H_5}{O}}:$$

Nucleophilic substitution can occur at a saturated carbon atom. As a result of detailed research, it has been determined that there are two different mechanisms for nucleophilic substitution reactions.

1. Unimolecular nucleophilic substitution reaction, S_N1
2. Bimolecular substitution reaction, S_N2

We will examine these two mechanisms in detail. There are important factors that affect reaction mechanisms. These are as follows:

- Structure of the reactant
- Nucleophilic power
- Concentration of the nucleophile
- Solvent

2.2 Unimolecular Nucleophilic Substitution Reactions, S_N1

The unimolecular nucleophilic substitution reaction, S_N1, occurs in two steps. In the first step, the bond between the leaving group and the carbon atom breaks to produce a carbocation. Most commonly, an anionic leaving group is removed from the molecule by taking the previously shared pair of bond electrons. The concentration of the nucleophile does not affect the rate of the reaction. The carbocation intermediate formed has a positive charge on the carbon atom with three substituents. In the second step, the carbocation reacts with nucleophiles to give the product.

$$\underset{CH_3}{\overset{CH_3}{H_3C-\underset{|}{\overset{|}{C}}-Br}} \underset{Slow}{\rightleftharpoons} \underset{CH_3}{\overset{CH_3}{H_3C-\underset{|}{\overset{|}{C}}{}^{\oplus}}} + H_2\overset{..}{\underset{..}{O}} \xrightarrow{Fast} \underset{CH_3}{\overset{CH_3}{H_3C-\underset{|}{\overset{|}{C}}-\overset{\oplus}{\underset{H}{\overset{H}{O}}}}} \underset{Fast}{\overset{-H^+}{\rightleftharpoons}} \underset{CH_3}{\overset{CH_3}{H_3C-\underset{|}{\overset{|}{C}}-OH}}$$

Carbocation

This mechanism is called an S_N1 mechanism, with "S" for substitution, "N" for nucleophilic, and "1" for unimolecular. Unimolecular indicates that one molecule is involved in the transition state of the rate-determining step. The order of the reaction (first-order, second-order, etc.) is defined as the number of compounds whose concentrations affect the rate of the reaction. The molecularity of a reaction refers to the number of molecules or ions that participate in the transition state of the rate-determining step. As we will see when discussing S_N2 reactions, the reaction molecularity and the order of a reaction may vary. In the reaction shown above, the formation of the carbocation is the slow and rate-determining step. The bond formation between the carbocation and nucleophile is a very fast process. Because the rate-determining step involves only one compound (substrate), the reaction is unimolecular. An energy diagram showing the progress of an S_N1 reaction is given in Figure 2.5.

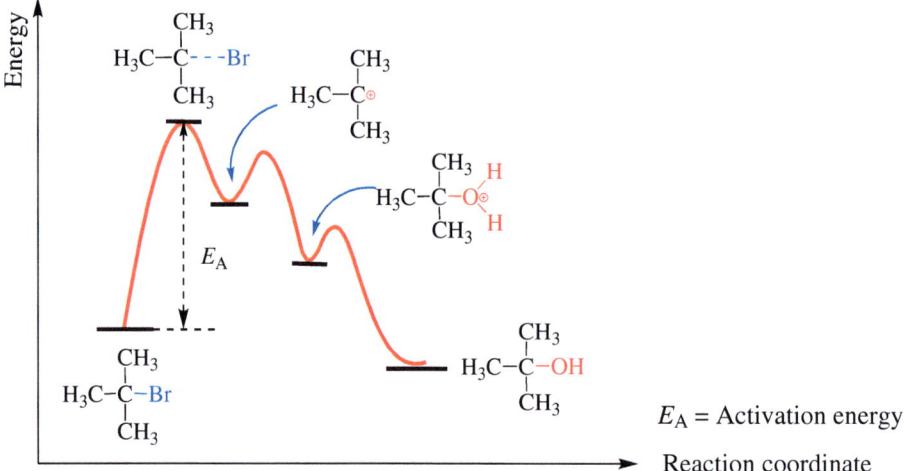

Figure 2.5 Energy diagram for an S_N1 reaction.

In unimolecular nucleophilic substitution reactions, the first step is the dissociation of the leaving group to form the intermediate, a carbocation. The stability of the carbocation determines the mode of the reaction. We have shown that a tertiary carbocation is more stable than a secondary carbocation. Therefore, the former is formed more easily. Generally, carbocation stability decreases from tertiary carbocations to the primary ones in saturated systems. Secondary alkyl halides can react either by the S_N1 mechanism or by the S_N2 mechanism (next chapter). Primary alkyl halides do not react according to the S_N1 mechanism because of the instability of the methyl carbocation. Ethyl bromide reacts very slowly. Generally, secondary and tertiary carbon compounds react according to the S_N1 mechanism.

$$\underset{\underset{CH_3}{|}}{\overset{\overset{CH_3}{|}}{H_3C-C-Br}} \gg \underset{\underset{CH_3}{|}}{H_3C-CH-Br} \gg H_3C-CH_2-Br \gg H_3C-Br$$

⟵ S_N1 reactivity

Resonance stabilization of a carbocation can also promote S_N1 reactions. For example, allyl bromide undergoes an S_N1 reaction easily as fast as a secondary halide, although it is a primary halide. The allylic system stabilizes the carbocation through overlap of the vacant p orbital with the double bond electrons. Benzyl halides form resonance-stabilized benzyl carbocations. Therefore, they can easily undergo a substitution reaction by S_N1.

$$H_2C=CH-CH_2-Br \longrightarrow H_2C=CH-\overset{\oplus}{C}H_2 \longleftrightarrow \overset{\oplus}{H_2C}-CH=CH_2 \longrightarrow H_2C=CH-CH_2-OH$$

$$Ph-CH_2Br \xrightarrow{S_N1} \left[Ph-\overset{\oplus}{C}H_2 \right] \xrightarrow{H_2O} Ph-CH_2OH$$

Resonance stabilized

Because of resonance stabilization, primary allylic and benzylic carbocations have about the same stability as secondary alkyl carbocations. The stability order of some carbocations by S_N1 reactions is given below.

$$CH_3-\underset{\underset{CH_3}{|}}{\overset{\overset{CH_3}{|}}{C}}{}^{\oplus} \gg CH_3-\underset{\underset{CH_3}{|}}{\overset{\oplus}{CH}} \approx C_6H_5-\overset{\oplus}{C}H_2 \approx H_2C=CH-\overset{\oplus}{C}H_2 \gg CH_3-\overset{\oplus}{C}H_2 \gg \overset{\oplus}{C}H_3 \gg H_2C=\overset{\oplus}{C}H$$

⟵ Carbocation stability

Because the first step in S_N1 reactions involves removal of the leaving group from the molecule, the stability of the carbocation formed, as well as the electronic structure of the leaving group, has an essential effect on the reaction. We have already discussed the stability of the carbocation. The stability of the leaving group also has a significant effect on the mechanism. It is preferred that the bond between the leaving group and the carbon atom be weak. The best leaving groups are those that are more stable anions or neutral compounds when they depart. The most stable anions and the best leaving groups are the conjugate bases of strong acids. Reactions with a series of halogens bound to the same carbon atom show that reactivity decreases from the iodine atom to the fluorine atom. The halides are good leaving groups as they can stabilize the negative charge because of their size and electronegativity. Triflate, tosylate, and mesylate are also good leaving groups because they can also stabilize the negative charge.

$$H_3C-\underset{\underset{CH_3}{|}}{CH}-I \gg H_3C-\underset{\underset{CH_3}{|}}{CH}-Br \gg H_3C-\underset{\underset{CH_3}{|}}{CH}-Cl \gg H_3C-\underset{\underset{CH_3}{|}}{CH}-F$$

⟵ Reactivity of alkyl halides in S_N1 reactions

The hydroxide anion (OH$^-$), methoxide anion (CH$_3$O$^-$), and amide (NH$_2^-$) are poor leaving groups because they are strong bases. However, the hydroxyl group can be turned into an excellent leaving group either by protonation or by conversion into tosylate or mesylate.

The electronic nature of the nucleophile does not affect the S_N1 reaction because the rate-determining step in the S_N1 reaction is the dissociation of the leaving group in which the added nucleophile has no part. Therefore, the nucleophile cannot affect the reaction rate. For example, the reaction of *t*-butanol with HX occurs at the same rate regardless of the electronic nature of halides.

$$CH_3-\underset{\underset{CH_3}{|}}{\overset{\overset{CH_3}{|}}{C}}-OH \quad \xrightarrow{H-X} \quad CH_3-\underset{\underset{CH_3}{|}}{\overset{\overset{CH_3}{|}}{C}}-X \quad + \quad H_2O$$

X = I, Br, Cl same rate

The polarity of the solvent also plays an important role. A polar solvent stabilizes the transition state and carbocation intermediate as well as an anionic leaving group. Therefore, S_N1 reactions occur more readily in polar solvents. For example, secondary alkyl halides can react by either S_N1 or S_N2 reaction mechanism. Polar solvents tend to shift the mechanism toward an S_N1 reaction.

In S_N1 reactions, the first step is the formation of a carbocation. Once a carbocation is formed, there are several options for further reactions.

- A nucleophilic substitution reaction (S_N1) occurs when a carbocation reacts with a nucleophile.
- Elimination (E1) can occur by loss of a proton (see Chapter 3) to form an alkene.
- The carbocation can undergo rearrangement. The driving force for the rearrangement is the formation of a more stable carbocation. Once the carbocation is rearranged, the new carbocation formed can undergo either a substitution reaction or elimination or both.

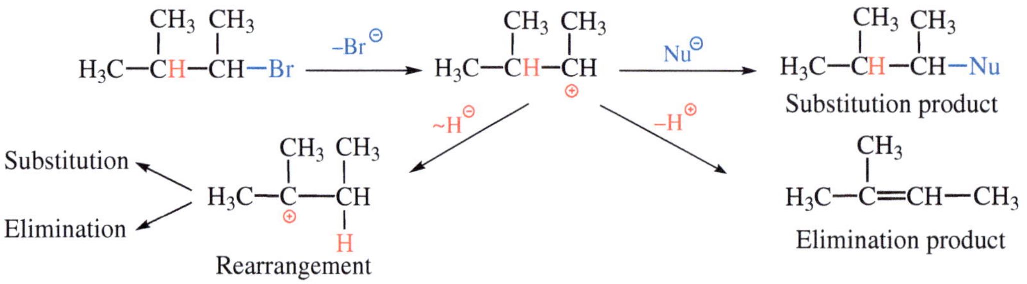

2.2.1 Stereochemistry in S$_N$1 Reactions

The first step of an S$_N$1 reaction is the bond breaking between the carbon and the leaving group, and the bonding electrons stay with the leaving group. The result is the formation of a carbocation intermediate. To minimize electron repulsion, the positively charged carbon atom changes the hybridization from sp^3 to sp^2. Three substituents, including the carbon atom, form a planar structure with an empty unhybridized p orbital perpendicular to that plane. The nucleophile can approach the carbocation from either side of the plane (Figure 2.6).

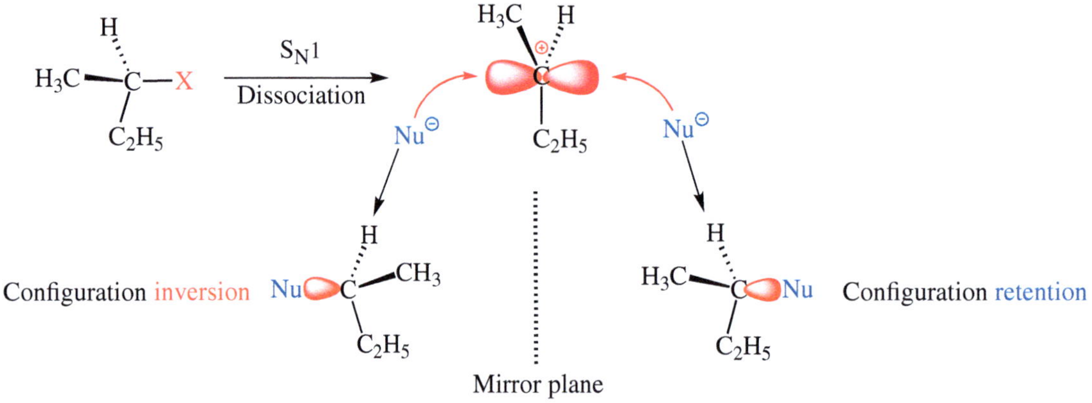

Figure 2.6 Stereochemistry of an S$_N$1 reaction.

Any tetrahedral carbon atom with four different substituents is called *chiral*. These compounds are optically active and they have a chiral center. If they form a carbocation, because of the carbocation's trigonal planar shape, both sides are susceptible to attack by the nucleophile with the same probability to form different stereoisomers (Figure 2.6). These isomers are mirror images of each other. They are different and non-superimposable and they are called *enantiomers*.

When the configurations of the isomers are compared with that of the reactant, it is seen that the configuration in the starting compound is maintained (*configuration retention*) in the product on the right side (Figure 2.6) and the configuration is inverted (*configuration inversion*) in the product on the left side. The configuration inversion is comparable to the flipping of an umbrella in windy weather. A mixture of equal amounts of the two enantiomers is called a *racemic mixture* and the process is called *racemization*. These enantiomers can be represented by the right hand and left hand, which are also mirror images of each other. Just as the right hand is different from the left hand, the enantiomers are also different from each other. Let us try to answer the question *What is the difference between the right hand and the left hand?* Let us take a right-hand glove and try to put on the left hand. While the right hand easily enters the right-hand glove, the left hand does not enter the right-hand glove.

It is expected that equal amounts of both enantiomers should be formed in an S$_N$1 reaction. If the reaction leads to equal amounts of the two enantiomers, the reaction is called *complete racemization*. However, in most cases, a higher amount of the product with an inverted configuration is obtained. The nucleophile will approach the carbocation predominately from the side opposite the leaving group because the leaving group is blocking one face of the carbocation. The environment around the carbocation is called *an asymmetric environment*. The asymmetric nature of the environment prevents the nucleophile from attacking the carbocation equally in both directions, even though the carbocation has a symmetrical structure (Figure 2.7). When one of the enantiomers is in excess, the optical activity will not be lost entirely. This process is called *partial racemization*. Winstein has explained why the enantiomer with the inverted configuration is formed in a higher amount. He postulated the formation of an *intimate ion pair*, which indicates that the bond between the carbon atom and the leaving group is broken, but they remain close to each other. In this case, one face is blocked. This result is particularly often observed when a reactant with a relatively poor leaving group (e.g. chloride) is used. Hydrolysis of 1-chloroethylbenzene is an example of incomplete racemization. However, in the hydrolysis of 1-bromoethylbenzene, complete racemization is obtained because the bromide anion is a very good leaving group in contrast to the chloride anion.

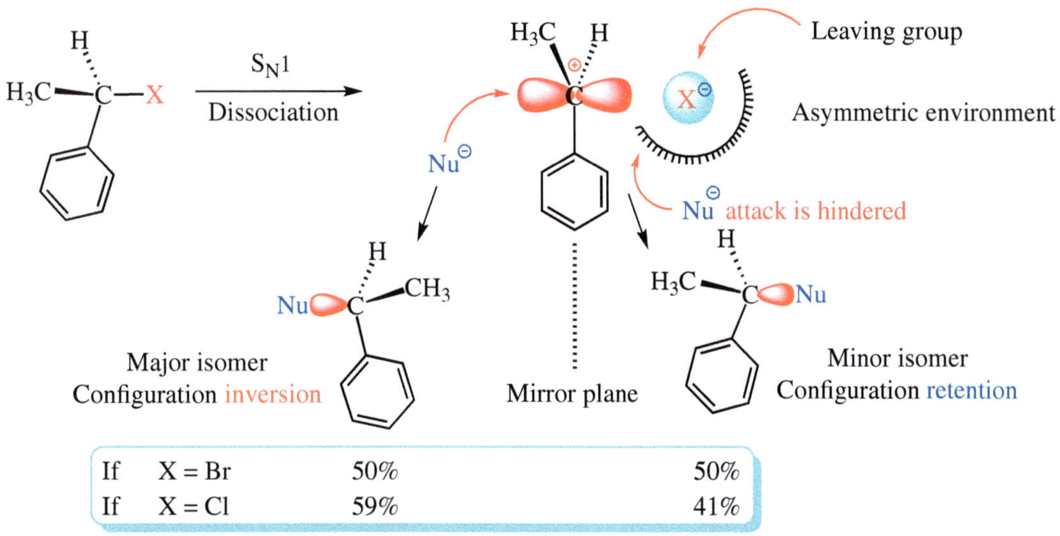

Figure 2.7 Presentation of partial racemization in S_N1 reactions.

The enantiomers of a chiral compound have identical properties such as melting point, boiling point, density, solubility, and spectroscopic properties (NMR, IR, and UV). However, they differ in the way they affect polarized light. When polarized light passes through a pure enantiomer, one enantiomer rotates the plane of polarized light to the right while another rotates the polarized plane an equal amount to the left.

Enantiomers also have the same chemical properties in an *achiral environment*. However, they have substantially different biological activities because they bind to receptors in the body that are also chiral. For example, (+)-glutamate, which has an asymmetric carbon atom, is used as a sweetening and flavoring additive in the food industry, while (−)glutamate has no taste [5]. D-Asparagine has a sweet taste, while the natural L-asparagine is tasteless. Carvone has two enantiomeric forms. One of these forms, (−)-carvone, is found in mint leaves and it has a distinctive odor of mint. The other form, (+)-carvone, is found in caraway seeds. This form has a very different smell and is typically used to flavor rye bread [6]. (+)-Thalidomide was used as a sedative for many years. However, when this compound was used as a racemic mixture, it was observed after the birth of 10 000 children that the (−)-thalidomide isomer caused pregnant women to give birth to disabled babies [7]. As seen, the biological properties of enantiomers can be very different.

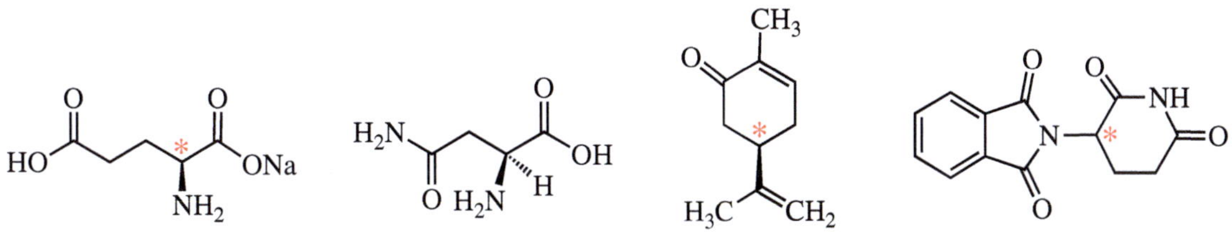

(+)-Sodium glutamate   D-Asparagine   (−)-Carvone   (+)-Thalidomide

2.2.2 Optical Activity

An asymmetric carbon atom is defined as a carbon atom within an organic compound that contains four different substituents bonded to it. These compounds are optically active. Some examples of optically active compounds are given below. There are compounds with a single optically active carbon atom in organic molecules, as well as compounds containing more than one optically active carbon atom.

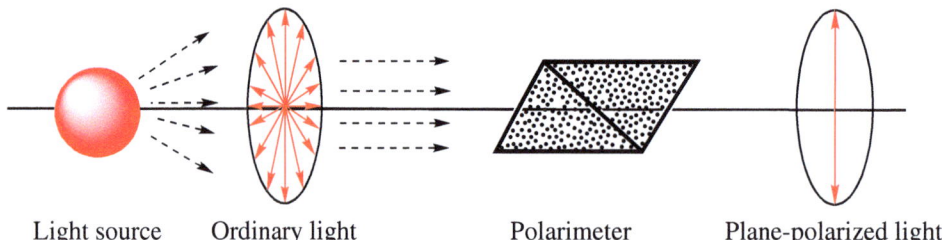

When plane-polarized light is passed through a solution of an enantiomer, an optically active compound, the plane of polarization is rotated in one direction to the right or left. If the same experiment were carried out with another enantiomer, the plane of polarization would be rotated by precisely the same amount but in the opposite direction.

What is Plane-Polarized Light? Ordinary light is a bundle of electromagnetic waves that oscillate simultaneously in an infinite number of planes. When an ordinary light beam is passed through a device called a *polarizer* (also called polarimeter), it will be converted to a beam of plane-polarized light that oscillates in a single plane. Light waves in all other planes are blocked out (Figure 2.8).

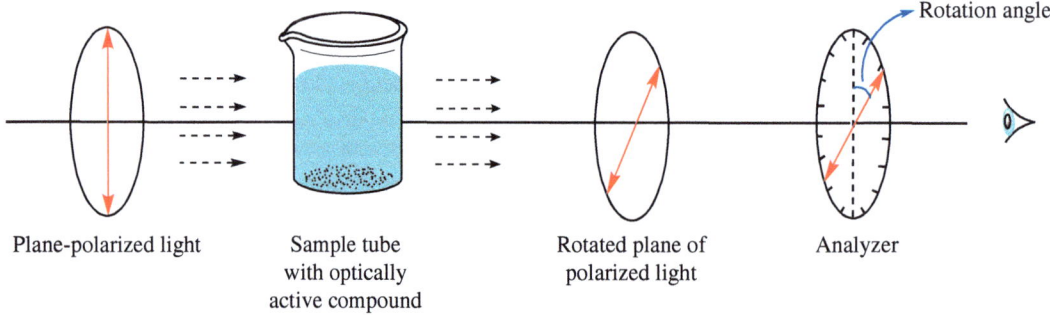

Figure 2.8 Representation of a polarimeter.

If the plane-polarized light passes through a solution containing an optically active compound, the plane of the polarized light will be rotated (optical rotation) by an interaction with the chiral compound either clockwise or counterclockwise (Figure 2.9). The size of the rotation angle depends on the structure of the molecule as well as on the concentration of the optically active compound. Chiral molecules rotate plane-polarized light; they are called *optically active*. The measured rotation in degrees is the *specific optical rotation*.

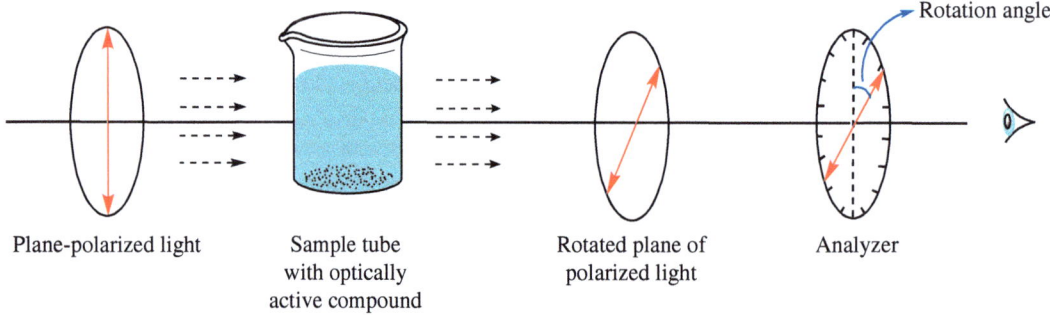

Figure 2.9 Schematic diagram of a polarimeter with a sample tube containing a solution of an optically active compound.

The most crucial point is that two enantiomers will have the same specific rotation but in opposite directions. If both enantiomers are present in a solution in equal amounts (50 : 50 mixture), the plane of the polarized light will not be rotated because the effects of the two stereoisomers will cancel each other out. The specific rotation of this mixture will be zero. These compounds are called *optically inactive* compounds.

If an enantiomer rotates the plane of the polarized light in a clockwise direction, the compound is *dextrorotatory* (*dexter*, Latin, right), and the compound is referred to as the (+) enantiomer. This enantiomer may also be called the *d form*, where *d* refers to dextrorotatory such as (+)-pinene or *d*-pinene. If an enantiomer rotates the plane of the polarized light in a counterclockwise direction, the compound is *levorotatory* (*leavus*, Latin, left), and the compound is referred to as the (−) enantiomer. This enantiomer may also be called the *l form*, where *l* refers to levorotatory. However, these should not be confused with *R* and *S*. If the configuration of a compound is *R*, it is possible that the *R*-configurated compound can rotate the plane of the light to the right or left depending on the chemical structure.

The specific rotation, $[\alpha]_D$, is a physical constant characteristic of an optically active compound, and the degree of the rotation depends on the wavelength of the light, the length of the cell through which the light passes, the concentration of the optically active compound in the solution, and the observed rotation of the sample.

$$[\alpha]_D^T = \frac{\alpha}{l \times c}$$

$[\alpha]$ = Specific rotation
α = Observed rotation
T = Temperature (20 °C)
D = Wavelength of incident light (589 nm, Na-D)
l = Length of the cell in decimeters (1 dm = 10 cm)
c = Concentration g/100 ml

The organic molecules in living systems, both plants and animals, are chiral. Although the compounds can form different stereoisomers, almost invariably only one stereoisomer is found in nature. In some cases, more than one isomer is found, but these rarely exist together in the same biological system. In experiments carried out under laboratory conditions, a racemic mixture is always obtained unless special methods are applied. It is possible to obtain one of the enantiomers at a purity close to 100% in enzyme catalyst experiments. Racemic mixtures can also be easily separated into enantiomers by various techniques.

2.2.3 Other Factors Affecting S_N1 Reactions: Steric Factors

We have already discussed the effect of the molecular structure on the reaction rate in S_N1 reactions. We have mainly focused on the stability of the carbocations. Other structural effects influence the rate of S_N1 reactions. Steric crowding also significantly controls the reaction rate. The hybridization of the carbon atom on which the substitution reactions take place is sp^3 and the angle between the substituents is approximately 109.5°. If the leaving group departs from the molecule, the hybridization of the carbon atom to which the substituents are bonded changes from sp^3 to sp^2. As the carbon atom formed has a planar structure, the angle between the substituents changes from 109.5° to 120.0°. If the alkyl groups attached to the carbon atom are sterically more crowded in the substrate than they are in the carbocation, the steric relief will accelerate the rate of the formation of the carbocation so that the steric interaction between the substituent in carbocation will be reduced. The examples given in Table 2.1 show that the rate of nucleophilic substitution reaction increases with increasing branching in R groups [8].

Table 2.1 Relative rate constants for solvolysis of R(CH$_3$)$_2$C–Cl in aqueous ethanol at 25 °C.

R	Relative rate
CH$_3$	1.0
CH$_2$CH$_3$	1.67
CH$_2$CH$_2$CH$_3$	1.58
CH(CH$_3$)$_2$	0.88
C(CH$_3$)$_3$	1.21
CH$_2$C(CH$_3$)$_3$	22.4

In contrast to the examples discussed above, in some structures, the formation of a carbocation leads to a more significant steric strain than in the reactant. This effect is particularly observed in bridgehead substituted bicyclic systems in which molecular geometry forces the carbocation to adapt a planar structure. This increases the strain in the molecule. Therefore, systems substituted at the bridgehead are generally resistant to S_N1 reactions.

For the formation of a carbocation at a bridgehead, it is necessary to minimize the strain that occurs after carbocation formation. This is only possible by extending the carbon chains (a, b, c) connecting the bridge carbons (a similar situation is observed in elimination reactions at the bridge; see Bredt's rule in Section 3.4.4). The relative rates of solvolysis reactions of some bicyclic systems substituted at the bridgehead carbon are given below [9, 10].

Relative reactivities in solvolysis reactions.

2.3 Bimolecular Substitution Reactions, S$_N$2

S$_N$1 reactions proceed in two steps. In the first step, the bond between the leaving group and the carbon atom breaks to form a carbocation. In the second step, the carbocation reacts quickly with the nucleophile to form the product. The S$_N$2 reaction proceeds in a single step. In the S$_N$2 reaction, the attack of the nucleophile on the carbon atom (for the formation of a new bond between the carbon atom and the nucleophile) and departure of the leaving group occur simultaneously. The reaction is concerted and the substrate and the nucleophile are both present in the transition state.

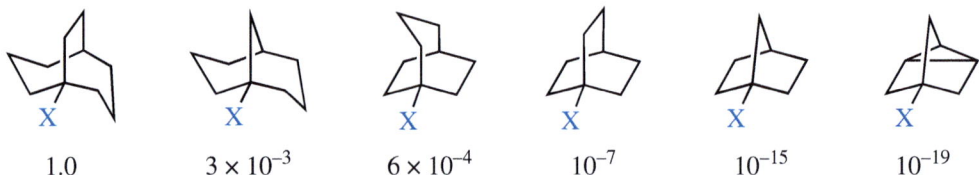

According to this reaction mechanism, a carbocation is not formed. A transition state occurs. In the transition state, the bond between the nucleophile and the carbon atom is partially formed, and the bond between the carbon atom and the leaving group is partially broken. As the substrate (RX) and the nucleophile (Nu) are both involved in the formation of the transition state, this reaction is called a *bimolecular nucleophilic substitution reaction* (S$_N$2) or a *second-order reaction*. In the S$_N$1 reaction, the nucleophile did not contribute to the formation of the intermediate, a carbocation. In S$_N$2 reactions, no intermediates are formed. The rate of S$_N$2 reactions depends on the concentration of both reactants (R-X and Nu). If the concentration of the nucleophile is doubled, the reaction rate is also doubled. When the concentration of both reactants is increased twofold, the reaction rate increases fourfold. The rate of the reaction is expressed by the equation below.

Reaction rate = k_a [R − X] [Nu]

The reaction rate in S$_N$2 reactions may not always depend on the concentration of both reactants. Let us examine a solvolysis reaction. *Solvolysis* is a chemical reaction in which the solvent, such as water, alcohol, or acid, is one of the reagents and becomes part of the reaction product. When the solvent is water, the reaction is called *hydrolysis*. The reaction of ethyl bromide with methanol forms ethyl methyl ether in a slow reaction. This is a solvolysis reaction.

If the nucleophile is used as a solvent in excessive amounts, the concentration of the solvent does not vary significantly during the reaction. Therefore, the rate of the above reaction depends only on the concentration of ethyl bromide. Then, the reaction is a first-order reaction because its rate varies only depending on the concentration of one reactant. The concentration of the nucleophile, methanol, does not have any effect on the rate of the reaction. However, to form a transition state, ethyl bromide and methanol have to come together. Therefore, the reaction molecularity is bimolecular. The reaction rate is first order. Such a situation is known as a *pseudo first-order reaction*. In the S_N2 expression, the number "2" on N indicates not the reaction rate but the molecularity of the reaction, i.e. how many molecules must come together to form a transition complex. The reaction rate and reaction molecularity should not be confused. S_N2 reactions are bimolecular reactions. The reaction rate can be first order as well as second order. Thus, the reaction rate depends on the concentration of a single compound and the rate of the reaction is

Reaction rate = k_a [R – X]

In an S_N2 reaction, a nucleophile approaches from the back of the carbon-leaving group bond (the so-called backside attack). The formation of a bond between the carbon atom and the nucleophile and breaking of the bond between the carbon atom and the leaving group occur simultaneously. Therefore, the transition complex formed turns directly into the product. According to the S_N2 mechanism, there is a single transition state with no intermediates (Figure 2.10). In the transition state, the nucleophile, the leaving group, and three substituents are connected to the carbon atom. Larger substituents can result in increased strain, leading to an increased energy level of the transition complex and slower reaction rates.

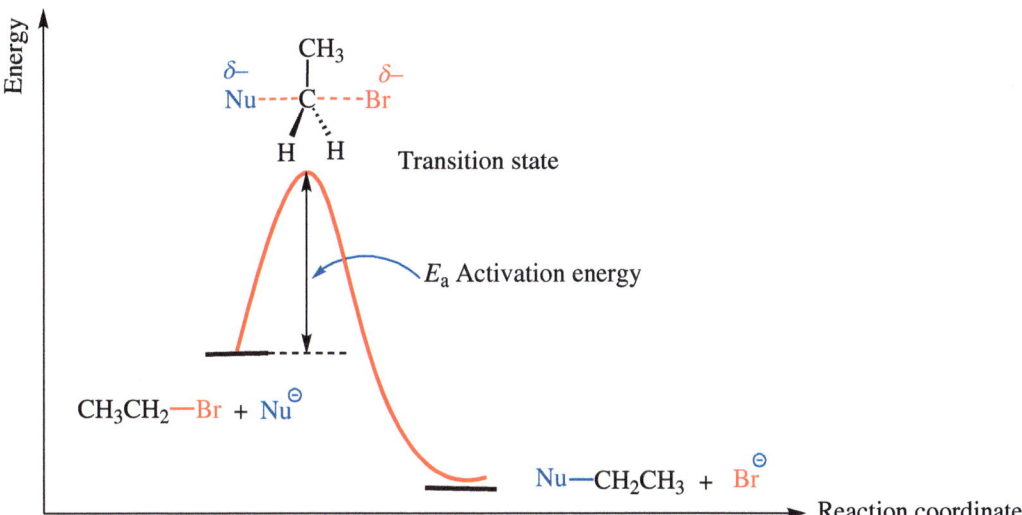

Figure 2.10 Energy diagram for an S_N2 reaction.

S_N2 reactions are particularly sensitive to steric factors because they are significantly retarded by steric crowding at the carbon atom. If we successively replace each of the hydrogens in methyl bromide with methyl groups, the rate of the S_N2 reaction will significantly diminish because the addition of methyl groups will hinder the backside attack to the electrophilic carbon. The examples given below illustrate this concept, showing that electrophilic carbon attached to three hydrogen atoms results in faster nucleophilic substitution reactions compared to primary and secondary haloalkanes. The addition of a third methyl group to this molecule creates a carbon that is entirely blocked. Tertiary haloalkane, with three methyl groups, does not undergo S_N2-type nucleophilic substitution.

	CH_3—Br	>>	H_3C—CH_2—Br	>>	H_3C—CH(CH$_3$)—Br	>>	H_3C—C(CH$_3$)$_2$—Br
Relative rates	221 000		1350		1		0

Reactivity of alkyl halides in S_N2 reactions

2.3.1 Stereochemistry of S$_N$2 Reactions

Let us examine the nucleophilic substitution reaction of an optically active reactant. In S$_N$2 reactions, the nucleophile attacks the back of the carbon atom that is bonded to the leaving group. The lone pairs of the leaving group create a region of high electron density that effectively blocks the front of the molecule so that the nucleophile can only attack from the back. In the transition state, the central carbon atom is partially bonded to the nucleophile and the leaving group. The three substituents and the carbon atom form a plane. As the nucleophile gets closer to the carbon atom, the leaving group moves farther away from it, and the three substituents also move in the same direction. Finally, the bond between the nucleophile and the carbon atom is fully formed, while the bond between the leaving group and the carbon atom is completely broken. During this process, the carbon atom changes the configuration. As a result, the reaction will proceed with an inversion of configuration (Figure 2.11). Paul Walden was the first to observe such an inversion, in 1896; therefore, it is called *Walden inversion*.

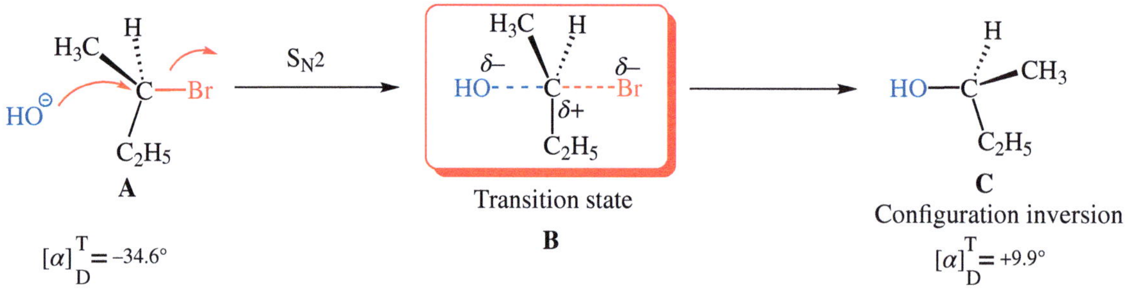

Figure 2.11 Configuration inversion in an S$_N$2 reaction.

Generally, configuration inversion is observed in S$_N$2 reactions. However, there are also cases in which configuration retention occurs under S$_N$2 conditions. This will be discussed in Section 2.3.5 under "Neighboring Group Participation in Nucleophilic Substitution Reactions."

Let us compare the angle of rotation obtained before and after the reaction in the example above. The angle of rotation of the reactant is $[\alpha] = -34.6°$. The configuration of the alcohol formed after the reaction is inverted. If the reactant rotates the plane of the polarized light to the left, it does not mean that the alcohol with inverted configuration will rotate to the right. However, we should keep in mind that the product has a structure different from that of the reactant and it is not a mirror plane isomer of the reactant. In this particular case, the product formed rotates the plane of the polarized light to the right. The specific rotation values of the reactant and product are different. In summary, we can say that even if we know the specific rotation value of the reactant, we can never predict which direction the product will rotate the plane of the light. The nature of the substituents will determine the direction of the rotation.

Let us take another example to analyze the changes observed in configurations in S$_N$2 reactions. In the reaction sequence shown below, the first step is the conversion of optically active 1-phenylpropan-2-ol to the corresponding tosylate. Because the hydroxyl group is a bad leaving group in nucleophilic substitution reactions, this group must first be transformed into a good leaving group. The reaction of an alcohol with tosyl chloride in pyridine results in the formation of the tosylate. As this reaction occurs on the oxygen atom, the configuration of the carbon atom cannot be changed. The tosylate must also be optically active. The specific rotation of the tosylate is changed because of the change in the substituents.

In the second step, the acetate anion attacks the optically active carbon center from the back of the molecule, causing inversion in the configuration because of the S$_N$2 reaction mechanism. The specific rotation of the resulting acetate is $[\alpha] = -70.6°$. Based on this value, it is not possible to estimate whether there is a configuration change in the molecule or not. As a new group is bonded to the carbon atom, it is expected that the specific rotation will change by a certain value.

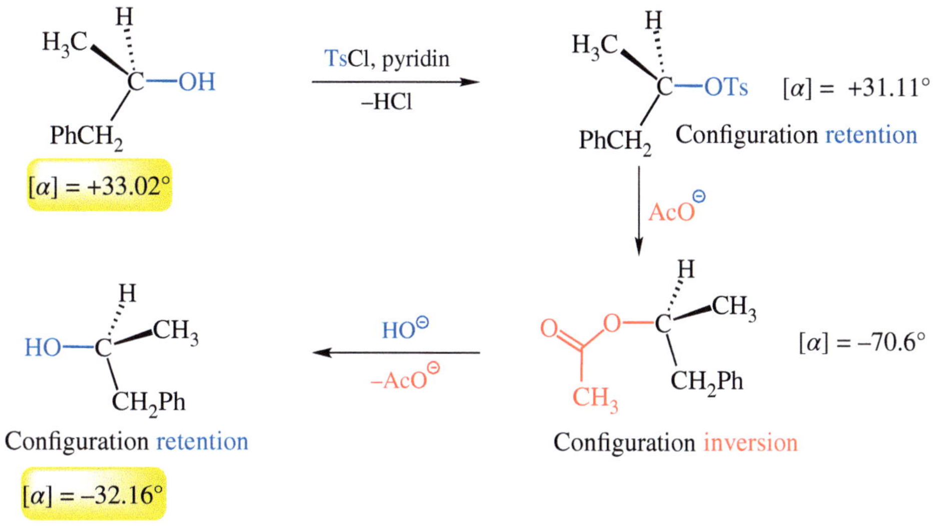

The last step is a hydrolysis reaction. According to the basic ester hydrolysis mechanism, the base attacks the carbonyl carbon and ester hydrolysis takes place, as shown below. As the reaction proceeds on the carbonyl carbon, there cannot be any change in the configuration of the asymmetric carbon atom. The specific rotation of the alcohol obtained as a result of hydrolysis is $[\alpha] = -32.16°$, while a specific rotation of the reactant is $[\alpha] = +33.02°$. We can see that there is not much change between the absolute values of the specific rotations. However, the reactant rotates the plane of the polarized light to the right, while the resulting product rotates the plane of the polarized light to the left. This observation indicates that the configuration of the reactant is changed in one of the reaction steps. In the reaction sequence given above, there is only one step in which the substitution occurs at the optically active carbon atom. If the mechanism of this reaction had been S_N1, complete racemization of the configuration would have been observed in the final product.

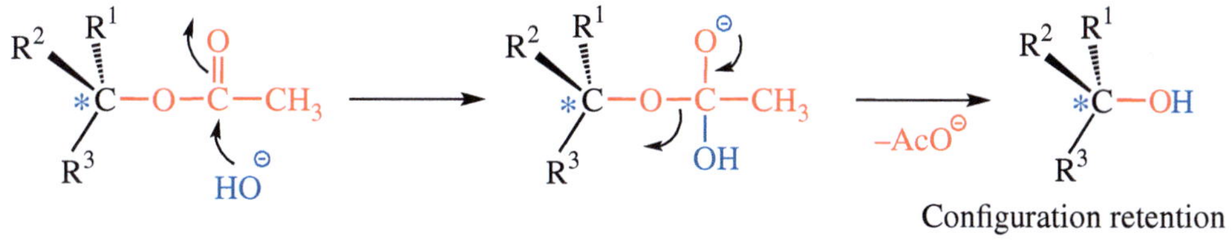

So far, we have discussed the S_N2 reaction mechanism and showed that nucleophiles attack the carbon atom from the back. *Can the nucleophiles attack the carbon atom from the front, from the side of the leaving group?* We will try to answer this question. We will benefit from MO theory when addressing this question. According to that theory, the electron density flows from the HOMO of the nucleophile into the lowest unoccupied molecular orbital (LUMO) of the electrophile. When a nucleophile substitutes a leaving group, the nucleophile HOMO interacts with the LUMO of the carbon atom bearing the leaving group. The overlapping MOs must be in phase to generate a new bond. Figure 2.12 shows that in the case of a backside attack, a bonding interaction occurs between the relevant orbitals; however, in the case of a front side attack, there is an antibonding interaction between the orbital of the nucleophile and the larger loop of the σ* orbital of the carbon atom.

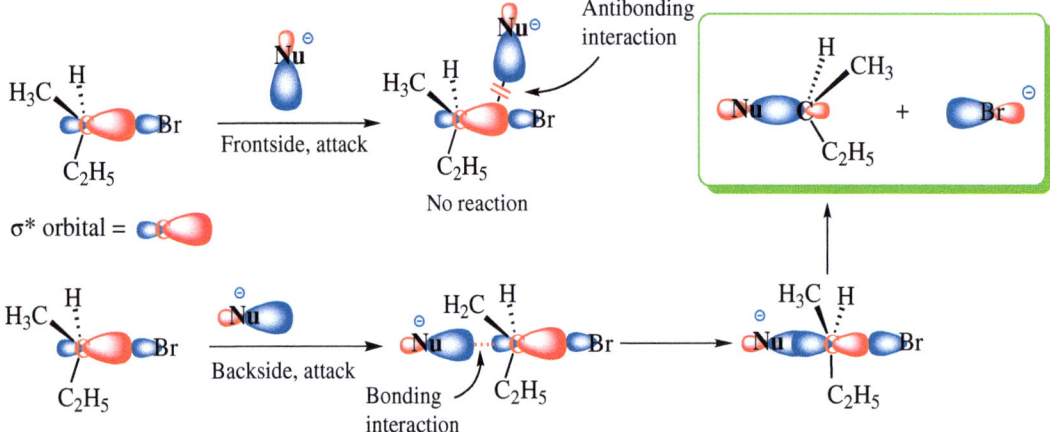

Figure 2.12 Interaction of the orbitals during a backside attack and a front side attack.

2.3.2 Factors Affecting the S_N2 Reaction Mechanism

We have seen that nucleophilic substitution reactions follow two different mechanisms, S_N1 and S_N2. In this section, we will examine under what conditions these reactions occur. The factors affecting the nucleophilic substitution reaction mechanisms are as follows:

- Structure of the substrate
  1. Electronic effect (inductive effect and mesomeric effect)
  2. Steric effect
- Solvent effect
- Structure of the leaving group
- Structure of the nucleophile

2.3.2.1 The Structure of the Substrate

We have shown that S_N1 reactions are two-step reactions in which the first step is the ionization of alkyl halides to form a carbocation. Then, the carbocation reacts with a nucleophile to form the product. The most critical factor that determines the reactivity of substrates is the relative stability of the carbocation that is formed. Of course, the stability of the carbocation depends on the molecular structure. Tertiary carbocations are stabilized by hyperconjugation (see Section 7.1.1), and they are generally stable and react according to the S_N1 mechanism. As the methyl and primary carbocations are not as stable as the tertiary carbocations, they prefer to undergo a substitution reaction according to the S_N2 mechanism. However, the secondary alkyl halides can undergo both S_N1 and S_N2 reactions, depending on the nucleophile and the solvent. Because other factors besides the molecular structure also affect the reaction mechanism, it is possible to direct the reaction mechanisms of the secondary alkyl compounds according to the reaction conditions (solvent, nucleophile, etc.). Vinylic and aryl halides do not undergo either S_N1 or S_N2 reactions.

In addition to carbocations' stabilization by inductive effect, there are also compounds stabilized by conjugation, which react according to the S_N1 or S_N2 mechanism. For example, a primary benzylic or allylic carbocation is about as stable as a secondary carbocation. Therefore, it can undergo S_N1 as well as S_N2 reactions. On the other hand, a secondary benzylic or secondary allylic carbocation is about as stable as a tertiary carbocation.

In S_N1 reactions, if alkoxy groups and halogens are attached to the central carbon atom, both increase the reaction rate and allow the reaction to proceed according to the S_N1 mechanism. However, if the halogen and alkoxy groups are bonded to the adjacent carbon atoms, they inhibit carbocation formation with their inductive effects (−I) so that the rate of the reaction decreases. In the β-position, they cannot stabilize the carbocation by the mesomeric effect. The nucleophilic substitution reaction in allylic systems will be discussed in more detail in Section 2.3.3.

Inductively electron-withdrawing groups attached to the carbon atom bearing the leaving group reduce the electron density on the carbon atom, making it more electropositive. The possibility of the formation of a carbocation is then reduced. A highly polarized carbon atom is susceptible to attack by nucleophiles according to the S_N2 reaction mechanism. When functional groups such as carbonyl, ester, or nitrile are attached to the carbon atom, substitution reactions generally proceed according to the S_N2 mechanism. The reaction rate depends on the electron-withdrawing ability of the substituent.

Relative rates: $k = 1$; 1.600; 2.800; 33.000; 100.000

The *Finkelstein reaction* is an S_N2 reaction in which alkyl iodides are prepared by reaction of alkyl bromide or chlorides with potassium or sodium iodide in acetone [11, 12]. Iodide is a stronger nucleophile than bromide or chloride. In this reaction, the relatively strong carbon–chloride bond (84 kcal/mol; 350 kJ/mol) is replaced by a rather weak carbon–iodine bond (57 kcal/mol; 239 kJ/mol). *What is the driving force behind this reaction?* As potassium iodide is soluble in acetone but sodium bromide and sodium chloride are not soluble in acetone, the equilibrium shifts to the side of the product according to Le Chatelier's principle. When electron-withdrawing groups are attached to the carbon atom, the reaction rate increases. In particular, the carbonyl group has an enormous effect on the reaction rate.

$$R-Cl \xrightarrow[\text{Acetone}]{KI} R-I \quad \text{Finkelstein reaction}$$

The structures of some compounds directly affect both the reaction rate and reaction molecularity. Compounds having a leaving group at the bridgehead are extremely reluctant to undergo substitution reactions. Bridged compounds prevent a substitution reaction by the S_N2 mechanism. In an S_N2 reaction, the nucleophile and the reactant must come together to form the transition state. Therefore, the nucleophile must approach the carbon atom from the back. It is impossible for a nucleophile to approach the carbon atom bearing the leaving group because of the cage structure.

For this reason, substitution reactions at bridgeheads are not possible according to the S_N2 reaction mechanism. For an S_N1 reaction, steric hindrance is not very significant because the nucleophile is not involved in the formation of the carbocation. Even this is not possible because carbocation formation requires a planar structure, which would cause remarkable ring strain. Therefore, bridgehead compounds generally fail to react by either an S_N1 or an S_N2 mechanism. 1-Bromobicyclo[2.2.2]octane undergoes a substitution reaction with AgNO$_3$ under harsh conditions (160 °C) by the S_N1 mechanism. AgNO$_3$ is generally an excellent initiator for S_N1 reactions. Triptycyl bromide on boiling with 30% NaOH in ethanol gives no substitution product.

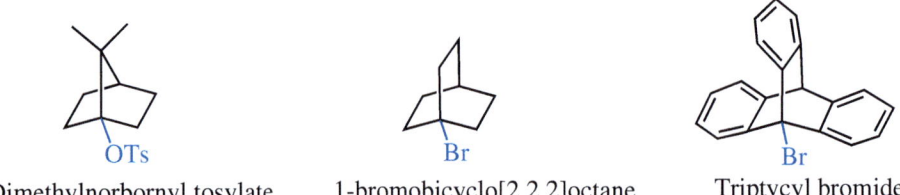

Dimethylnorbornyl tosylate   1-bromobicyclo[2.2.2]octane   Triptycyl bromide

2.3.2.2 Solvent Effect

In S_N1 reactions, a carbocation is formed as an intermediate. Polar solvents can stabilize charged species better than the nonpolar ones can. Therefore, polar solvents increase the rate of S_N1 reactions, whereas they decrease the rate of S_N2 reactions. Polar protic solvents containing –OH or –NH groups are generally worst for S_N2 reactions. Polar solvents have a higher dielectric constant than the nonpolar ones. A polar solvent with a higher dielectric constant can solvate a polar nucleophile and polar transition state better. The dielectric constants of some solvents frequently used in organic chemistry are given in Table 2.2.

Table 2.2 The dielectric constants of some common solvents (ε, at 25 °C).

Protic solvents	Dielectric constant	Aprotic solvents	Dielectric constant
HCN	123	CH$_3$SOCH$_3$ (DMSO)	47
HCONH$_2$	110	MeCN	38
H$_2$SO$_4$	110	(CH$_3$)$_2$NCHO (DMF)	37
H$_2$O	79	Acetone	22
HCOOH	50	Tetrahydrofuran (THF)	7.6
MeOH	33	CH$_3$COOEt	6
EtOH	25	Diethylether	4.3
t-Butanol	11	Benzene	2.3
Acetic acid	6	Pentane	1.8

Solvents such as H$_2$O, formic acid, and formamide are polar solvents having high dielectric constants. These solvents stabilize the ions and facilitate the formation of carbocations in S_N1 reactions. The situation is somewhat different in S_N2 reactions. For example, let us examine the type of reaction given below.

$$HO^{\ominus} + H_3C-Br \longrightarrow [\overset{\delta-}{HO}---CH_3---\overset{\delta-}{Br}] \longrightarrow HO-CH_3 + Br^{\ominus}$$

Reactants   Transition state   Products

In this reaction, the nucleophile hydroxide ion is negatively (−) charged and this charge is localized on the oxygen atom. In the transition state, this charge is distributed on the oxygen atom and the leaving group. Suppose the charge on the reactant is higher than the charge on the transition state complex; in this case, increasing the polarity of the solvent will decrease the rate of the S_N2 reaction. The polar solvents will stabilize the reactants more than the transition state. It is recommended to use apolar solvents for this reaction. We can generalize this. In polar solvents, the mechanism of nucleophilic substitution reactions shifts mostly toward S_N1, while in low-polarity solvents, it shifts toward S_N2. This generalization does not always apply to every system. For example, the nucleophiles in nucleophilic substitution reactions can also be neutral species and the product formed can be ionic. An example of the reaction between a neutral nucleophile and a neutral substrate, such as the reaction of methyl bromide with trimethylamine, is given below.

$$H_3C-Br + :N(CH_3)_3 \longrightarrow H_3C-\overset{\oplus}{N}(CH_3)_3 + Br^{\ominus}$$

Neutral reactant Neutral nucleophile Product Leaving group

If the charge on the product is greater than the charge on the reactant, a polar solvent will stabilize the product more than it will stabilize the reactant. In this case, increasing the polarity of the solvent will also increase the rate of the S_N2 reaction.

In the part on S_N1 reactions, we discussed the effect of polar aprotic solvents on substitution reactions. These solvents can dissolve many salts because of their high dielectric constant, and they tend to solvate cations rather than anions. As a result, unsolvated anions gain a greater nucleophilicity in S_N2 reactions. For example, an increase in the rate of 200 000 has been observed by changing the solvent from methanol to hexamethyl phosphoric triamide (HMPA) for the reaction of *n*-butylbromide with azide ion [13].

$$CH_3CH_2CH_2CH_2-Br + N_3^{\ominus} \xrightarrow{\text{Solvent}} CH_3CH_2CH_2CH_2-N_3 + Br^{\ominus}$$

Solvent	CH_3OH	H_2O	DMSO	DMF	CH_3CN	HMPA
Relative rate	1	7	1300	2800	5000	200 000

2.3.2.3 Leaving Group Effect

In both unimolecular and bimolecular substitution reactions, the bond between the carbon atom and the leaving group is broken. Because the leaving group is removed from the substrate, the best leaving groups are those that best stabilize the charge in the transition state. The weakest bases can stabilize the negative charge well. Let us examine halogens. The best leaving group among the halogens is iodine and the poorest one is fluorine. The bond between the carbon atom and the halogen atom weakens when the transition complex is formed. The weaker the energy of the bond between carbon and halogen, the easier it is to remove the halogen. Carbon–halogen bond energies are given below.

R—I	57 kcal/mol
R—Br	70 kcal/mol
R—Cl	84 kcal/mol
R—F	108 kcal/mol

When alkyl halides having the same alkyl groups were allowed to react with the same nucleophile under the same reaction conditions, alkyl iodide was the most reactive one and alkyl fluoride the least reactive [14]. The difference between these reactions is the nature of the leaving group. The relative rates of the reactions show that iodide is the best leaving group and fluoride is the worst. Iodide is a weak base and not bonded strongly to the carbon atom; therefore, the weaker bond is more easily broken. In summary, the conjugate bases of strong acids are good leaving groups. A poor leaving group can often be converted into an excellent one by protonation because protonated species are much weaker bases.

Reaction	Relative reaction rates
$HO^{\ominus}$ + R—F ⟶ HO—R + $F^{\ominus}$	1
$HO^{\ominus}$ + R—Cl ⟶ HO—R + $Cl^{\ominus}$	200
$HO^{\ominus}$ + R—Br ⟶ HO—R + $Br^{\ominus}$	10 000
$HO^{\ominus}$ + R—I ⟶ HO—R + $I^{\ominus}$	30 000

Because the leaving groups are generally anions, it is essential to stabilize them. Resonance-stabilized structures are weak bases. Therefore, leaving groups that form resonance structures upon leaving are excellent leaving groups. The following diagram shows that it becomes more difficult to separate the leaving groups from the molecule when going from left to right.

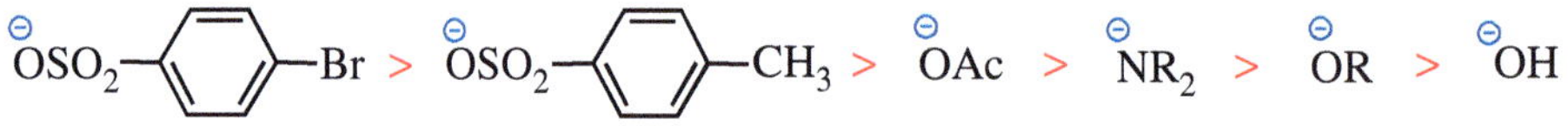

$$^{\ominus}OSO_2\text{-}C_6H_4\text{-}Br > {}^{\ominus}OSO_2\text{-}C_6H_4\text{-}CH_3 > {}^{\ominus}OAc > {}^{\ominus}NR_2 > {}^{\ominus}OR > {}^{\ominus}OH$$

Alkyl sulfates and sulfonates are excellent leaving groups. This is due to the formation of a resonance-stabilized structure upon leaving. Nucleophiles cannot substitute anions such as F^-, OH^-, and OR^-. To carry out an S_N2 reaction with a –OH group, it must be first converted into a better leaving group such as a tosylate or mesylate. When nucleophilic substitution reactions are carried out in an acidic environment, these groups can be easily removed from the molecule as H_2O or ROH by protonation of the nonbonding electrons on oxygen.

Likewise, NH_2^- functions seldom serve as leaving groups in nucleophilic substitution. Heating an amine with HI or HBr does not usually convert it into the corresponding alkyl halide, as in the case of alcohols and ethers. One group of amine derivatives that have proven useful in S_N2 reactions is that composed of tetraalkylammonium salts, which are separated from the molecule as a neutral leaving group.

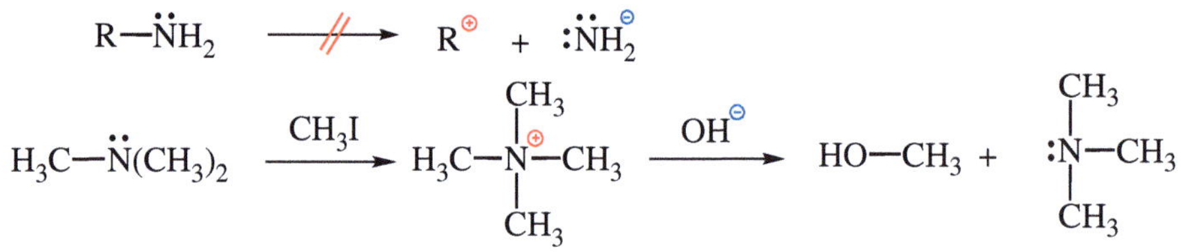

$$R\text{-}\ddot{N}H_2 \xrightarrow{\,/\!/\,} R^{\oplus} + :\ddot{N}H_2^{\ominus}$$

$$H_3C\text{-}\ddot{N}(CH_3)_2 \xrightarrow{CH_3I} H_3C\text{-}\overset{\oplus}{N}(CH_3)_3 \xrightarrow{OH^{\ominus}} HO\text{-}CH_3 + :N(CH_3)_3$$

The reaction rate in S_N2 reactions depends on the electronic structure of the leaving group. For example, the reaction rates of alkyl fluoride, alkyl bromide, and alkyl tosylate with the same nucleophile are different. In Table 2.3, the relative rates with the different leaving groups are given. If an alkyl bromide is reacted at the same time with more than one nucleophile, different substitution products will be formed. The ratio of these products will be affected by the electronic nature of the leaving group. However, if a similar reaction is carried out under S_N1 conditions, the ratio of the products formed will not be affected by the nature of the leaving group because the nucleophiles attack the carbon atom after dissociation of the leaving group.

Table 2.3 Relative rates of reaction depending on the leaving group.

R–X	X	Relative rate
	$CF_3SO_3^-$ (triflate)	1.4×10^8
	Nitrobenzenesulfonate	4.4×10^5
	p-Toluenesulfonate (tosylate)	3.7×10^4
	$CH_3SO_3^-$ (mesylate)	3.0×10^4
	I^-	91
	Br^-	14
	CF_3COO^-	2.1
	Cl^-	1.0
	F^-	9×10^{-6}
	CH_3COO^-	1.4×10^{-6}

2.3.2.4 Structure of the Nucleophile

In nucleophilic substitution reactions, the structure of the nucleophile attacking the polarized carbon atom can be complex. After completion of the reaction, a bond between the nucleophile and reactant will be formed. Strong nucleophiles can easily share their electrons to form a bond. The same nucleophilic atom with a negative charge is always a stronger nucleophile than a neutral species. For example, RO^- is always a stronger nucleophile than ROH. Electronegative elements strongly withdraw the valence electrons and the electrons are held tightly to the atom, which makes these elements weaker nucleophiles. In the periodic table, within a period nucleophilicity decreases from left to right. This is also a criterion for base strength. Compounds that can easily release their electrons for bond formation are also strong bases. Therefore, nucleophilicity roughly parallels basicity. This topic is covered in detail in the introduction of this chapter. Within a group, from top to bottom, nucleophilicity increases. Furthermore, as larger atoms are more polarizable, they are stronger nucleophiles.

There are numerous factors that affect the nucleophilic power of a compound. These factors are reduced to five by Streitwieser:

- Solvation energy of the nucleophile,
- Bond strength created between the nucleophile and the reactant,
- Size of the nucleophile,
- Electronegativity of the nucleophile,
- Polarizability of the nucleophile.

Relative rates of a substitution reaction with different substituents are given in Table 2.4.

Table 2.4 The effect of the substituents on the rate of the reaction in S_N2 reactions.

$$Nu^{\ominus} + R-X \longrightarrow Nu-R + X^{\ominus}$$

Nucleophile	Relative rate
NO_3^-	1
CH_3COO^-	20
Cl^-	80
$C_2H_5O^-$	1000
I^-	3700
PhS^-	47 000

2.3.3 Nucleophilic Substitution in Allylic Systems: Allylic Rearrangement

In allylic systems, nucleophilic substitution reactions proceed according to both S_N1 and S_N2 reaction mechanisms. First, let us examine the behavior of the allylic system according to the S_N1 mechanism. The unimolecular mechanism of substitution in allylic compounds is similar to the well-known S_N1 mechanism of nucleophilic aliphatic substitution. When an allyl bromide or allyl chloride is heated in a polar solvent, the allyl cation will be first formed. Allylic systems undergo ionization more easily than the alkyl groups because the allyl cation can be stabilized by resonance with the adjacent double-bond electrons.

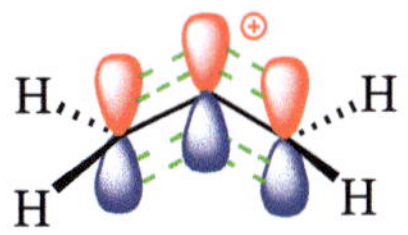

Unsubstituted allyl cation

The positive charge is distributed over two carbon atoms. The resonance structures of allyl cations are shown below. Because of resonance stabilization, primary allyl cations are as stable as secondary carbocations. On the other hand, secondary allyl cations are about as stable as tertiary carbocations.

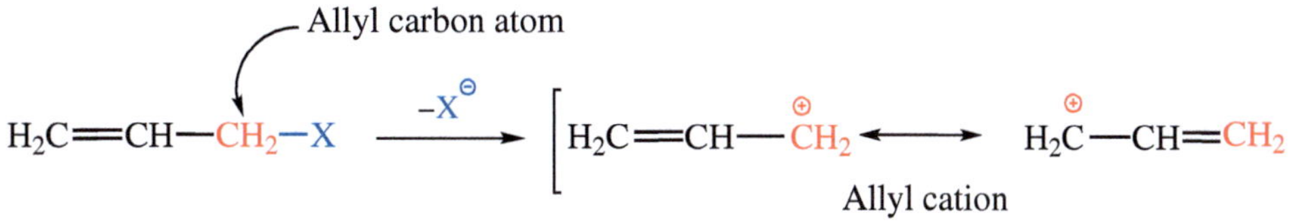

$$H_2C=CH-\underset{\text{Allyl carbon atom}}{CH_2}-X \xrightarrow{-X^{\ominus}} \left[H_2C=CH-\overset{\oplus}{C}H_2 \longleftrightarrow H_2\overset{\oplus}{C}-CH=CH_2 \right]$$

Allyl cation

In alkyl systems, the positive (+) charge is localized on a single carbon atom, whereas in allyl cations, the charge is distributed. Unlike in alkyl systems, nucleophiles can attack either of the electron-deficient sites, forming rearranged products.

This kind of rearrangement is called *allylic rearrangement*. Because the unsubstituted allyl cation has a symmetrical structure, an allylic rearrangement is undetectable in this case. However, if other groups are attached to the allylic system and these groups disrupt symmetry in the allyl carbocation, different substitution products will be formed. For example, let us examine an allylic system in which a methyl group is attached to the terminal double-bond carbon atom.

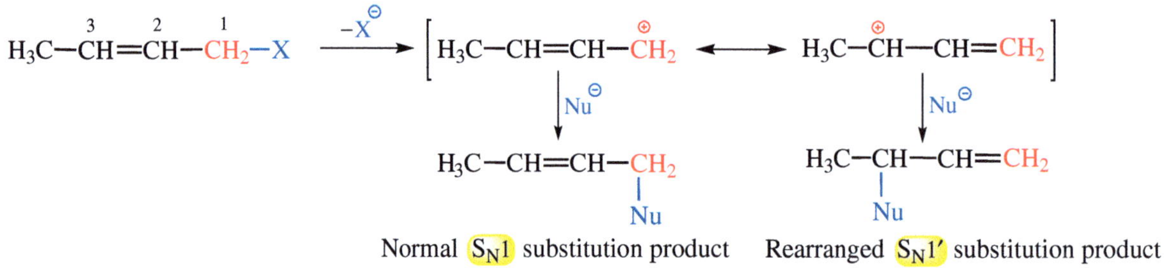

$$H_3C\overset{3}{-}CH=\overset{2}{CH}-\overset{1}{CH_2}-X \xrightarrow{-X^{\ominus}} \left[H_3C-CH=CH-\overset{\oplus}{CH_2} \longleftrightarrow H_3C-\overset{\oplus}{CH}-CH=CH_2 \right]$$

$$\downarrow Nu^{\ominus} \qquad\qquad\qquad \downarrow Nu^{\ominus}$$

$$H_3C-CH=CH-\underset{Nu}{CH_2} \qquad H_3C-\underset{Nu}{CH}-CH=CH_2$$

Normal S_N1 substitution product   Rearranged S_N1' substitution product

When this compound is treated with a nucleophile under S_N1 conditions, two products are usually formed, the normal one and the rearranged one, because of the unsymmetrical nature of the allyl cation. The allyl cation is stabilized by delocalization of the p electrons between two resonance structures. The partial positive charge will be on both C1 and C3. The nucleophile can attack either of the carbons bearing the positive charge. If the nucleophile attacks the carbon atom to which the leaving group is attached, the *normal substitution product* will be formed. However, if the nucleophile attacks the other carbon atom, a *rearranged product* will be formed. As part of this rearrangement, the double bond will change place in the molecule. The normal product is called the S_N1 product and the second product is the S_N1' product.

A nucleophilic substitution reaction in an allylic system can also take place by an S_N2 mechanism, in which case, no allylic rearrangement is expected. As in S_N2 reactions, a strong nucleophile can attack the allylic carbon atom and displace the leaving group without forming a carbocation as an intermediate. In this case, a normal substitution product will be formed. However, allylic rearrangement can also take place in S_N2 reactions when the nucleophiles attack the unsaturated γ-carbon atom rather than the carbon atom bearing the leaving group to yield the rearranged product. This process is called *abnormal bimolecular substitution* and is represented by S_N2'. The S_N2' rearrangement occurs when the carbon atom bearing the leaving group is sterically crowded. With this attack on the double bond electrons, the electrons move to the next carbon atom and substitute the leaving group, as shown below.

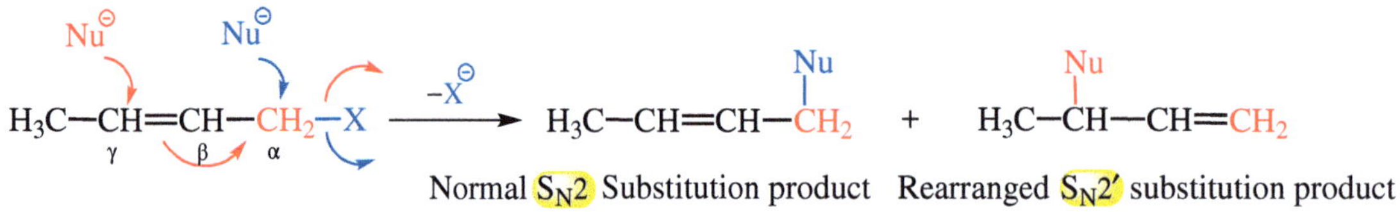

$$H_3C\underset{\gamma}{-}CH=\underset{\beta}{CH}-\underset{\alpha}{CH_2}-X \xrightarrow{-X^{\ominus}} H_3C-CH=CH-\underset{Nu}{CH_2} + H_3C-\underset{Nu}{CH}-CH=CH_2$$

Normal S_N2 Substitution product   Rearranged S_N2' substitution product

Hydrolysis of crotyl chloride (1-chlorobut-2-ene) in aqueous acetone yielded a mixture of two isomeric alcohols [15]. The normal substitution product was formed in a yield of 56%, while the rearranged product was formed in 44% yield. An ionic mechanism, S_N1, was proposed for this reaction. It is possible to change the ratio of the products by changing the reaction conditions applied.

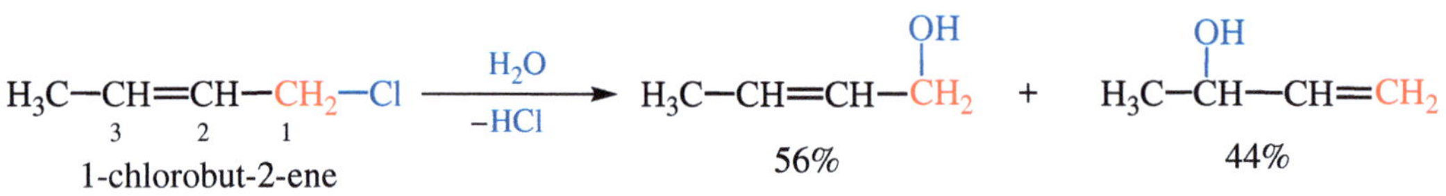

$$H_3C-\overset{3}{CH}=\overset{2}{CH}-\overset{1}{CH_2}-Cl \xrightarrow[-HCl]{H_2O} H_3C-CH=CH-\underset{OH}{CH_2} + H_3C-\underset{OH}{CH}-CH=CH_2$$
1-chlorobut-2-ene   56%   44%

When crotyl chloride was refluxed in ethanol, two different ethers were formed as expected. The normal substitution product was formed in a yield 91%, while the rearranged product was formed in a yield of 9%. When 3-chlorobut-1-ene was refluxed in ethanol under the same reaction conditions, the same products were formed in yields of 47% and 53%, respectively. When we compare the results of these two reactions, we can get some information about the reaction mechanism. If the reaction mechanism in both cases is according to S_N1, the same mixture of ethers should be formed because these two reactants would generate the same allyl cation as the intermediate, although the starting compounds are different.

$$H_3C-CH=CH-CH_2-Cl \xrightarrow[-HCl]{C_2H_5OH} H_3C-CH=CH-\underset{\underset{91\%}{|}}{CH_2} + H_3C-\underset{\underset{9\%}{|}}{CH}-CH=CH_2$$

$$H_2C=CH-\underset{\underset{Cl}{|}}{CH}-CH_3 \xrightarrow[-HCl]{C_2H_5OH} H_2C=CH-\underset{\underset{\text{47\%}}{\underset{\text{Normal substitution product}}{|}}}{CH}-CH_3 + H_2\underset{\underset{\text{53\%}}{\underset{\text{Rearranged product}}{|}}}{C}-CH=CH-CH_3$$

3-chlorobut-1-ene

From these results, we can first conclude that the mechanisms of these reactions should be different. We should keep in mind that both reaction mechanisms, S_N1 and S_N2, are ideal boundary mechanisms. Most of the time, reactions can proceed according to both mechanisms. Some of the products can be formed according to the S_N1 mechanism and the other part according to the S_N2 mechanism [16]. When these two reactants, crotyl chloride and 3-chlorobut-1-ene, were submitted to a silver-ion-catalyzed hydrolysis reaction, the same mixture of alcohols was formed. This indicates that both reactions generate the same allyl cation. Remember that silver ions can easily remove the chloride and force the reaction mechanism toward S_N1.

Generally, an S_N2' reaction does not compete with an S_N2 reaction; the latter is usually predominant. Under the steric conditions in which S_N2 is inhibited, a S_N2' reaction can be a dominant one or it can be an exclusive reaction. The presence of alkyl groups at the α-carbon atom will hinder the S_N2 reaction, but S_N2' can still take place. Larger nucleophiles can also increase the amount of S_N2' reaction at the expense of the S_N2 reaction.

The reaction of 3-chloro-3-methylbut-1-ene with benzene thiolate exclusively forms the S_N2' product because the attack is sterically hindered because of the crowding at the α-carbon atom.

3-chloro-3-methylbut-1-ene

In recent years, allylic substitution reactions have been successfully applied to form a carbon–carbon bond using various copper catalysts [17, 18]. Copper-catalyzed allylic substitution typically occurs to form products from the addition of copper alkyl compounds (hard nucleophiles) at the γ-position to the leaving group to form the rearranged product exclusively.

2.3.3.1 Stereochemistry in Allylic Substitution Reactions

The stereochemistry of the normal products formed by allylic substitution reactions is similar to that of the products formed by S_N1 and S_N2 reactions. If the reaction proceeds by the S_N1 mechanism, the bond between the carbon and the leaving group breaks to form a carbocation as an intermediate. Because of the planar structure, the carbocation will be attacked at either side to form a racemic mixture. In S_N2-type reactions, the nucleophile attacks the back of the carbon atom that is bonded to the leaving group, forming a product with configuration inversion. However, the situation is somewhat different in S_N2' reactions. The nucleophile approaches the double bond on the side from which the leaving group departs, which is called a *syn-attack*. In such an attack, double-bond electrons open backward to remove the X-group. Thus, the electron density is increased at the back of the leaving group. Substitution occurs by the attack of electrons on the carbon atom to which the leaving group is bonded. If the nucleophile attacks the double bond from the opposite direction (*anti-attack*), the electron density will increase at the side of the leaving group, which will not be suitable for a substitution reaction.

syn-attack

The *syn*-attack can be nicely demonstrated in cyclic structures. In the example given below, the nucleophile attacks the molecule from the side of the leaving group and removes benzoate [19].

2.3.4 Internal Nucleophilic Substitution Reaction, S_Ni

In the previous sections, we have discussed S_N1 and S_N2 mechanisms. S_N2 reactions proceed through the formation of a transition state, resulting in configuration inversion of the product. There are still other reactions whose stereochemical outcome cannot be explained by S_N1 or S_N2 mechanisms. In some nucleophilic substitution reactions, although the reaction molecularity is bimolecular, retention of the configuration is observed instead of inversion. These and similar reactions are often observed by the reaction of chiral alcohol with thionyl chloride to give the corresponding alkyl halide. In the first step, the oxygen atom of alcohol attacks the sulfur atom of thionyl chloride and removes one of the chlorine atoms attached to the sulfur atom to form alkyl chlorosulfite, which can be isolated. At this stage, there is no configurational change at the chiral carbon atom as the nucleophilic substitution reaction takes place on the sulfur atom. In the second step, the chlorine atom attached to the sulfur atom attacks the carbon atom from the front. This attack results in the retention of configuration. This reaction is called *an internal nucleophilic substitution reaction* and is represented by S_Ni. It has been found that the rate of dissociation of the chlorosulfite to the product increases with the increased solvent polarity and the stability of the carbocation formed. Therefore, the formation of a close intimate ion pair of the type R^+ and ^-OSOCl within the solvent cage has been suggested. These ions are in very close association. The collapse of the intimate ion pair rapidly forms a product with configuration retention.

When the reaction is carried out in the presence of pyridine, the alkyl halide is formed with an inverted configuration. Pyridine reacts with alkyl sulfite to give 1-(methoxysulfinyl)pyridin-1-ium salt. The HCl gas formed as the by-product reacts with pyridine to form a salt. The free chloride ion attacks the chiral carbon atom from the back, causing a configuration inversion.

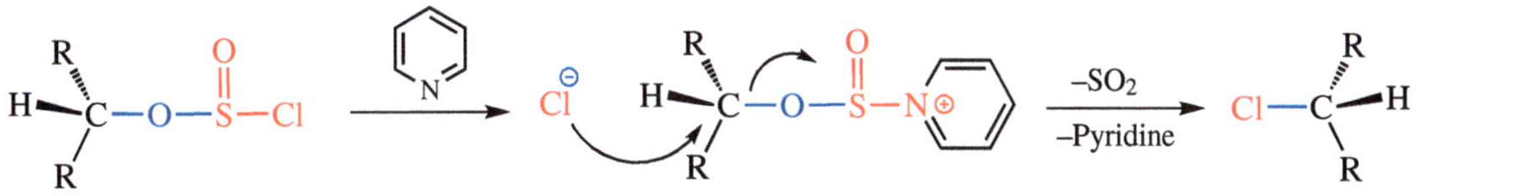

When the reaction is carried out in the presence of dioxane as the solvent, a product with retained configuration is formed. Probably, the retention here arises because of the two consecutive S_N2 reactions, as shown below.

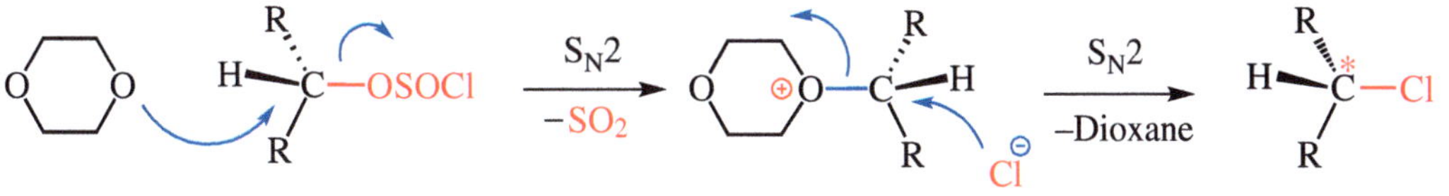

Inversion + Inversion = Retention

A new mechanism for the classic internal nucleophilic substitution reactions (S_Ni) has recently been suggested through computational studies in the gas phase that fits the experimental observations better [20]. According to this mechanism, alkanesulfonyl chloride produces an olefin by the simultaneous elimination of HCl and SO_2. The final *syn*-addition of HCl to the olefin leads to alkyl chloride with the retention of configuration.

Internal substitution reactions are also observed in allylic systems. Allyl alcohols can form abnormal substitution products as well as normal substitution products as a result of internal substitution. Such reactions are expressed as S_Ni' reactions.

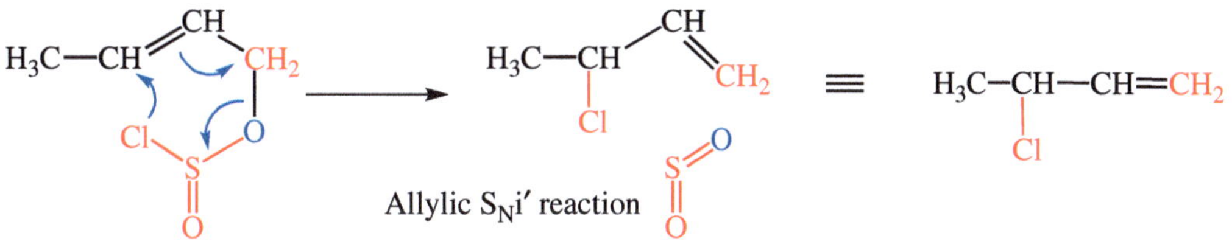

Allylic S_Ni' reaction

2.3.5 Neighboring Group Participation in Nucleophilic Substitution Reactions

Nucleophilic substitution reactions are not only those initiated by a nucleophile added to the reaction medium. The nucleophile and the leaving group can be bonded to the same molecule. The nucleophile can interact with the reaction center with a lone pair of electrons in an atom or the electrons present in a σ bond or a π bond. Such groups are called *neighboring groups*. Neighboring groups can increase the reaction rate or affect the stereochemistry of the reaction, making it abnormal compared with a normal reaction. The involvement of a neighboring group in a nucleophilic substitution reaction is called *neighboring group participation*. A neighboring group can displace the leaving group stereoselectively through a backside attack and a bond is formed between the neighboring group and the central carbon atom. A configuration inversion takes place at the central carbon atom during this attack. Now, this group bonded to the central carbon atom is displaced in a second step by the external nucleophile through another backside attack, causing a configuration inversion again. Both reactions are S_N2 reactions. The neighboring group itself does not undergo any evident change during the reaction. This event is called *anchimeric assistance*. There are two different nucleophiles in the reaction medium. The first one is the nucleophile bonded to the substrate, while the second one is the free nucleophile added to the reaction medium. In order for the neighboring group effect to occur, it is crucial that the nucleophile on the molecule attacks the molecule center faster than the free nucleophile in the reaction medium. Otherwise, a normal substitution reaction takes place and the neighboring group effect will not be observed.

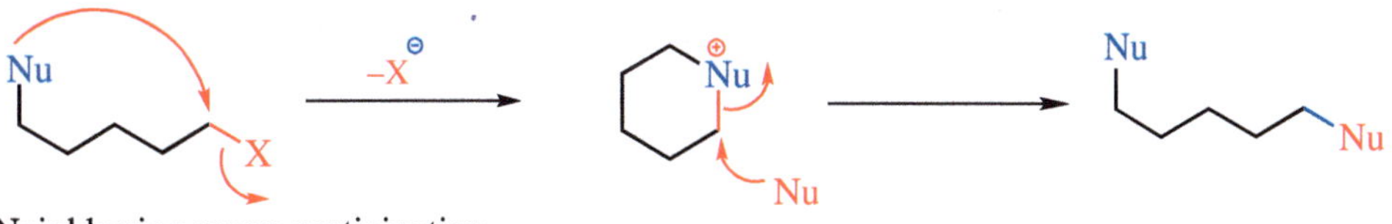

Neighboring group participation

In summary, a neighboring group behaves like a nucleophile and attacks the carbon atom to which the leaving group is attached. Now, the neighboring group bonded to the carbon atom acts as a new leaving group with the attack of another nucleophile. The bond between the neighboring group and carbon atom is breaking, and the neighboring group is separated from the central carbon atom and returns to its former position. *What effect would such behavior have on nucleophilic substitution reactions?*

- Rate increase
- Configuration retention
- Rearrangement.

2.3.5.1 Reaction Rate

In 1950, Winstein observed that certain alkyl halides with good leaving groups underwent substitution reactions several hundred times more rapidly compared with structurally similar compounds. The one common structural feature of these compounds was the presence of a neighboring group with a pair of nonbonding electrons. This observation strongly supported neighboring group participation. For example, the hydrolysis rate of β-chlorothioether is 10^4 times faster than that of β-chloroether [21].

Rate : 1   Rate : 10 000

The difference observed between the reaction rates of these two reactions cannot be explained by the formation of any intermediates observed in S_N1 or S_N2 reactions. Oxygen is more electronegative than sulfur. Consequently, it is expected that oxygen will strongly withdraw the electrons through σ bonds and make the carbon atom bearing the chlorine more electropositive. In such a case, the oxygenated molecule is expected to react faster. However, this is not the case. Because sulfur is a stronger nucleophile than oxygen, the sulfur lone pair displaces chloride ion and forms a cyclic sulfonium ion. This step results in the inversion of configuration at the carbon that was bonded to the leaving group. The intermediate product formed is highly strained because of the three-membered ring. In a second reaction, H_2O attacks the back of the carbon–sulfur bond, resulting in the second inversion of configuration at this carbon atom.

A dramatic example of neighboring group participation is provided by β,β-dichlorodiethyl sulfur (mustard gas). Although the boiling point (218 °C) of this compound is very high, it hydrolyzes readily and releases HCl. This compound was used in World War I as well as in 1988 as a chemical warfare agent [22]. This compound reacts rapidly with nucleophiles; the rate of the reaction is first order. The added nucleophiles do not affect the reaction rate. Mustard gas releases HCl efficiently into the lungs of soldiers and death is very painful. The cyclic sulfonium ion can easily undergo a reaction with the amino groups of deoxyribonucleic acid (DNA) and block DNA molecules.

An analogous work with nitrogen compounds showed that three-membered ring formation occurs at a greater rate. However, the rate for the formation of a five-membered ring was much faster and the product formed was stable.

In the examples we have given so far, we have examined the neighboring group participation of nonbonding electrons. In the example below, we will show that π-electrons are also involved in neighboring group participation. Phenylethyl tosylate undergoes solvolysis 3040 times faster than ethyl tosylate [23]. It was suggested that a phenonium ion is formed as an intermediate by the intervention of π-electrons. The formation of phenonium ion was proven by isotope labeling. The deuterium atoms were located on the carbon atom bearing the leaving group. However, the deuterium atoms were scrambled in the reaction product in a ratio of 1 : 1. This outcome can only be explained by the formation of a phenonium ion as an intermediate.

2.3.5.2 Configuration Retention

Besides the pronounced difference in rate, some compounds reveal a striking difference in stereochemistry. In S_N2 reactions, the carbon atom bearing the leaving group is inverted. In some reactions, although the substitution proceeds according to the S_N2 mechanism, retention is observed at the configuration of the central carbon atom instead of inversion. For example, when bromopropionic acid methyl ester is treated with sodium methoxide, it reacts according to the S_N2 mechanism, resulting in the formation of a product with an inverted configuration as expected. However, when the same reaction is performed with bromopropionic acid, the configuration retention is observed. Only the effect of neighboring groups can explain this situation.

The reaction of the ester with the nucleophile proceeds with inversion of configuration because of the normal S_N2 reaction. However, the reaction of a carboxylic acid is different. The stereochemical outcome is a result of two consecutive inversions. The base first abstracts the acidic proton from the acid to form a carboxylate anion and so a second nucleophilic center occurs. The carboxylate oxygen atom attacks the α-carbon atom from the back, forming α-lactone with inverted configuration. Because the α-lactone has a strained structure, the methoxide anion attacks the lactone carbon atom from the back to generate a product with the same configuration as the reactant [24].

Inversion + Inversion = Configuration retention

In the nucleophilic substitution reactions of the 3-chlorotropane system, no configuration isomerization was observed. These results can be explained by the participation of the nonbonding electrons on the nitrogen atom. The nitrogen atom attacks the carbon atom bearing the chlorine from the back and displaces the leaving group via inversion of the configuration. In the second step, the bond between the nitrogen atom and the carbon atom is broken by the attack of the nucleophile via a second inversion of the configuration [25].

Nu = N_3^-, CN^-, NH_2NHPh Inversion + Inversion = Configuration retention

2.3.5.3 Molecular Rearrangement

Another effect of the neighboring groups observed in nucleophilic substitution reactions is the molecular rearrangement of the compounds. For example, when both 2-(ethylthio)propan-1-ol and 1-(ethylthio)propan-2-ol are reacted with HCl, they are converted to the same compound, although they have different structures.

A 1-(ethylthio)propan-2-ol   (2-chloropropyl)(ethyl)sulfane   **B** 2-(ethylthio)propan-1-ol

When the structure of the resulting product is compared with those of the starting compounds, this product formed is a substitution product derived from the isomer **A**. However, this structure is not in agreement with the structure expected from the substitution of the isomer **B**. The compound is rearranged. This constitutional change can only be explained by neighboring group participation. In both cases, the nonbonding electron pairs on sulfur displace the protonated leaving group, forming the same intermediate. The sulfonium salt formed is now open to any nucleophilic attack. The chloride ion exclusively attacks the carbon atom to which the methyl group is bonded, forming the rearranged product. For further reactions, see "The carbocation rearrangement" in Section 7.4.

Normal substitution product for **A** →   ← Rearranged product for **B**

2.3.6 Ambident Nucleophiles

Some nucleophiles have a pair of electrons that can delocalize over two or more atoms. Such nucleophiles can attack a polarized carbon atom with two different atoms to give different products [26–28]. Such compounds are called *ambident nucleophiles* (Latin, ambi = two, dent = tooth). For example, the cyanide and enolate anions are the simplest ambident nucleophiles in this group. As can be seen from the resonance structures, both ions can attack with either of two atoms depending on the structures of the reactants and the reaction conditions.

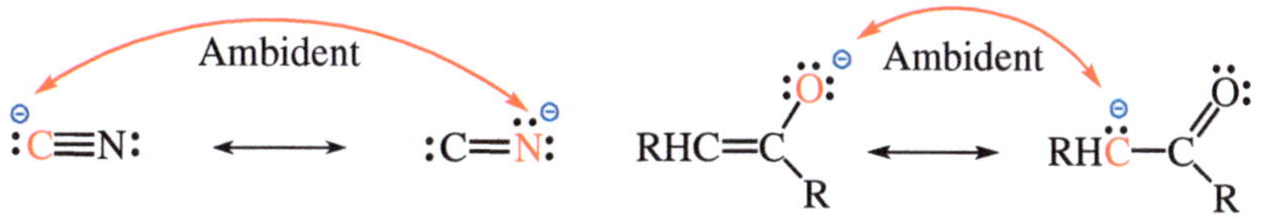

It might be expected that the more electronegative atom of ambident nucleophiles would attack; however, this is not the case. Remember hard and soft acids and bases (HSAB) theory: Hard acids prefer to react with hard bases and soft acids with soft bases. The cyanide ion can react with electrophiles either at the heteroatom to form isonitriles or at the carbon atom to form nitriles.

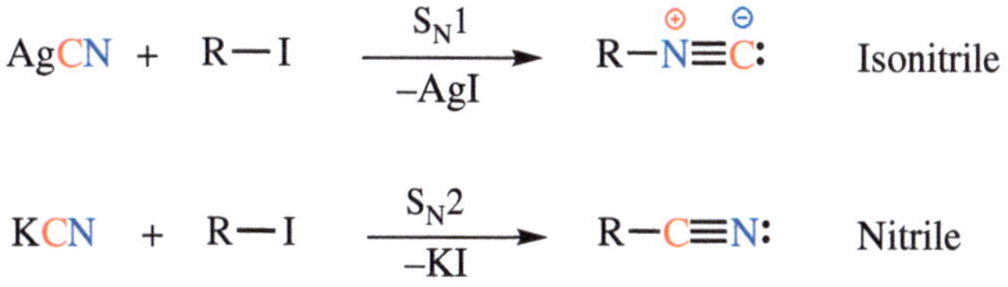

$$AgCN + R-I \xrightarrow[-AgI]{S_N1} R-N\equiv C: \quad \text{Isonitrile}$$

$$KCN + R-I \xrightarrow[-KI]{S_N2} R-C\equiv N: \quad \text{Nitrile}$$

When the reaction conditions are selected appropriately, reactions with ambident nucleophiles can be controlled so that only one product is formed. If the reaction proceeds by the S_N1 mechanism, the carbocation will be formed as the intermediate, which is a hard acid. Because the more electronegative atom of the cyanide ion is the nitrogen atom, which is a harder base, the attack will occur via the nitrogen atom. This reaction should be carried out in polar solvents to stabilize the carbocation, and it should be used with AgCN, which can easily remove the halogen. When the mechanism of a given reaction is changed from S_N1 to S_N2, an ambident nucleophile will attack with the less electronegative atom (a weak base). The solvent can also influence the direction of the reaction. In protic solvents, the more electronegative atom is better solvated than the less electronegative atom. Therefore, in a polar solvent, a less electronegative atom will attack the carbocation. On the other hand, in a polar aprotic solvent, the more electronegative end of the ion is freer as polar aprotic solvents solvate the cations. Therefore, in polar aprotic solvents, the more electronegative atom will attack.

We are faced with a similar situation in the nitrite ion. There are two different atoms in the nitrite ion that can act as nucleophiles. In the nitrite ion, a significant part of the negative (−) charge is concentrated on oxygen atoms. Therefore, solvation around oxygen is stronger because of the greater electronegativity. Under S_N2 conditions, the nucleophilic attack is carried out by the nitrogen atom and nitroalkanes are formed. We can also apply the HSAB theory here. Because oxygen is a harder base than nitrogen, alkyl nitrites are formed under S_N1 conditions.

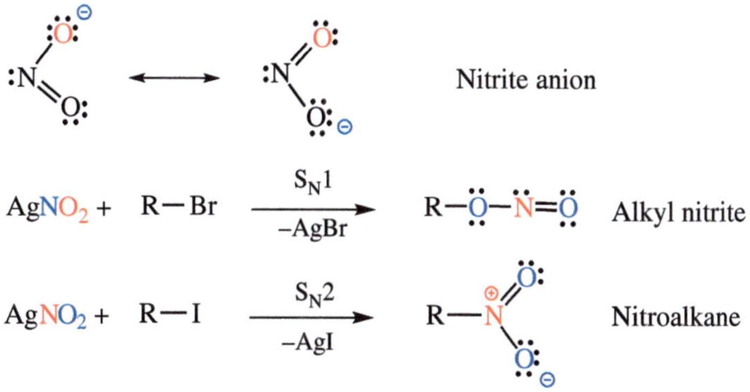

$$AgNO_2 + R-Br \xrightarrow[-AgBr]{S_N1} R-\ddot{O}-N=\ddot{O} \quad \text{Alkyl nitrite}$$

$$AgNO_2 + R-I \xrightarrow[-AgI]{S_N2} R-N\overset{O}{\underset{O}{\diagdown}} \quad \text{Nitroalkane}$$

The most common ambident nucleophiles are enolate anions. The enolate ion reacts with electrophiles over both the oxygen and the carbon atom. Synthetic application is also significant if the direction of the reaction can be controlled. Electrophiles generally add to carbon rather than to oxygen. Some very powerful electrophiles such as silyl chlorides and oxophilic electrophiles generally add to oxygen. For example, the proton (H⁺), which is a hard acid, prefers to react with oxygen because it is a harder base compared to the carbon atom. In alkylation reactions that proceed according to the S_N2 reaction mechanism, the alkyl groups are usually bonded to the carbon atom.

2.3.7 Various Nucleophilic Substitution Reactions

2.3.7.1 Williamson Ether Synthesis

One of the most common and classical methods for making ethers is Williamson ether synthesis, named after British chemist Alexander Williamson (1824–1904), who developed this method in the nineteenth century. This method is used to convert an alcohol and an alkyl halide or alkyl sulfonate to an ether. The alcohol must be first converted to alkoxide by the reaction with sodium metal or sodium hydride. Then, the alkyl halide is introduced into the reaction medium. An alkoxide ion attacks an alkyl halide, substituting the leaving group to form the corresponding ether.

The reaction is carried out under typical S_N2 conditions. The alcohol from which the alkoxide is derived can be used as a solvent as well as DMSO or HMPA can be used. It is not possible to perform all kinds of ether synthesis with this method. In some cases, elimination reactions may occur instead of substitution. For example, two different ways can be applied for the synthesis of sec-butyl methyl ether. The reaction of methoxide with sec-butyl bromide or reaction of sec-butoxide with methyl bromide should result in the formation of sec-butyl methyl ether. The reaction with sodium methoxide results in the formation of an alkene, whereas the second reaction leads to the formation of the desired product. Primary halides or tosylates work much better than the hindered bromide or tosylates. In the case of hindered bromides, E2 elimination occurs to form an alkene. Therefore, unsymmetrical ether should be synthesized by the reaction of less hindered halides or tosylates with more hindered alkoxides.

Cyclic ethers can be synthesized by treatment of halohydrins with a base such as sodium hydroxide. The nucleophile approaches the electrophilic carbon atom from the back of the leaving group. In the case of epoxide formation, only one conformation of the haloalkoxide can lead to the product. This method is used very often in organic chemistry, especially for epoxide synthesis.

Epoxides are formed as a result of an intramolecular nucleophilic substitution reaction if the alkoxide and the leaving group (halogen, tosylate, etc.) are bonded to the adjacent carbon atoms. The nucleophile approaches the electrophilic carbon atom from the back of the leaving group. For epoxide formation, only one conformation of the haloalkoxide, antiperiplanar conformation, can lead to the product.

Three-membered rings along with five-membered rings are formed fastest, followed by six-, four-, seven-, and lastly eight-membered rings. Enthalpic and entropic factors influence the relative rate of the reaction. The most strained compound is epoxide. The most obvious enthalpy effect is the ring strain. If enthalpy plays the most crucial role, then epoxide should be formed at the lowest rate. The proximity effect operates in this case. Smaller rings have less entropy. Therefore, they are more favorable for ring closure. As the ring size increases, the strong reduction in ring strain outweighs the worsening entropy factor (no proximity factor).

2.3.7.2 Ether Cleavage

Ethers are usually rather inert compounds. However, they undergo oxidation with air oxygen by a radical mechanism to form hydroperoxide, which can decompose explosively. Therefore, extreme care should be taken. Here, we will mainly discuss the decomposition of ethers under ionic conditions. As the ethers are formed by nucleophilic substitution reactions, they can also be cleaved by a nucleophilic substitution reaction. As mentioned in the previous sections, alkoxides are generally bad leaving groups. They must be turned into easily leaving groups before they leave the molecule. This is achieved in an acidic environment. When ethers are treated with a strong acid (HBr, HI) whose conjugate bases are good nucleophiles, they can be cleaved to give alkyl halides and alcohols. In the first step, acid protonates the ether oxygen, which turns it into a better leaving group. In the next step, halide attacks the α-carbon atom according to the S_N2 reaction mechanism to form alkyl halide and the alcohol.

Ethers can be cleaved by either the S_N1 or S_N2 reaction mechanism. For an S_N1 reaction, it is expected that the leaving group will form a stable carbocation. Methyl, vinyl, aryl, and primary carbocations cannot act as leaving groups. For example, an ether having a *t*-butyl group can be cleaved according to the S_N1 mechanism. The S_N2 reaction cannot take place as the tertiary carbons are much too hindered for a backside attack. Furthermore, they can form relative stable carbocations, which then can be attacked by the iodide to give alcohol and R-I.

In S_N2 reactions, halide ion attacks the less substituted carbon atom of the two alkyl groups, and an alkyl halide is formed. The cleavage reaction cannot be performed with the halide salt, a proton is required to protonate the oxygen atom to form alcohol as the leaving group. Iodide can attack the α- or α′-carbon atoms. The methyl group is more electropositive because of the positive (+) charge on the oxygen atom. Furthermore, the methyl group is less branched. Therefore, the iodide ion preferentially attacks the methyl group to form ethanol and methyl iodide.

If one of the ether components contains an optically active carbon attached directly to the oxygen atom, the reaction mechanism is more easily determined. For example, when *sec*-butyl methyl ether is cleaved with HI and optical activity is not lost in the resulting product, this reveals that the cleavage proceeded according to the S_N2 mechanism.

2.3.7.3 Reactions of Epoxides

Epoxides are more reactive than dialkyl ethers because of the large strain energy of the three-membered ring. Epoxides can undergo ring-opening reactions under both acidic and basic conditions. Epoxides react with strong acids. The first step is the protonation of the oxygen atom. The protonated oxygen acts as a leaving group. Therefore, it can easily be attacked by nucleophiles. In the case of 2,2-dimethyloxirane, the more substituted carbon atom is more likely to be attacked because the C—O bond begins to break, and the positive charge begins to build up on the more substituted carbon atom. The nucleophile, methanol, attacks the electrophilic carbon atom before a complete carbocation intermediate is formed. Probably, the best way to describe this acid-catalyzed epoxide ring-opening reaction is to say that it proceeds by something between an S_N2 and S_N1 mechanism. It is not a pure S_N1 reaction because the expected carbocation is not formed as an intermediate. It is also not a pure S_N2 reaction because the nucleophiles in an S_N2 reaction attack the carbon that is *least* hindered.

Ethers do not generally react with nucleophiles. However, epoxides can undergo ring-opening reactions by various nucleophiles because of the strain in the three-membered ring. The leaving group will be an alkoxide anion, which is a poor leaving group. Therefore, without a nucleophilic attack, the epoxide ring cannot be opened. Unsymmetrical epoxides can form different products under base-catalyzed reactions. Propylene epoxide, which has an asymmetrical structure, is opened in basic conditions. There are two different carbon atoms that the base can attack. The base generally attacks the less branched carbon, which is more accessible to attack. The resulting alkoxide ion then picks up a proton from the solvent and the methoxide ion is regenerated.

The aromatic hydrocarbon can be enzymatically oxidized into an arene oxide by an enzyme called cytochrome *P*-450 in the liver. The arene oxides formed can undergo ring-opening reactions. Arene oxides react with nucleophiles present in enzymes, ribonucleic acid (RNA), and DNA. A particularly dangerous arene oxide is the epoxide of benzo[*a*]pyrene. Epoxidation with *P*-450 followed by a ring-opening reaction gives a diol and further epoxidation of this diol results in the formation of an epoxydiol. The reaction of this epoxydiol with an amino group in DNA causes cancer because the structure of DNA is altered.

2.3.7.4 Substitution in Unsaturated Systems

We have seen so far substitution reactions in which the leaving group was bonded to an sp³-hybridized carbon atom. *Is a nucleophilic substitution reaction possible, when the leaving group is bonded to a double bond or triple bond?* When a group departs from a vinyl group with electrons, a vinyl cation occurs. The hybridization of the vinyl carbon is sp². The sp² hybrid orbital withdraws the electrons stronger than the sp³ hybrid orbital because of the increased s ratio. It is well established that Ag⁺ ions promote S_N1 reactions. However, vinylic halides are unreactive in the presence of silver nitrate, which does not precipitate silver halides because of the instability of vinyl cations. According to the S_N1 mechanism, a vinyl carbocation will be formed after separation of the leaving group. Studies show that vinyl cations are more unstable than saturated carbocations. For example, a secondary vinyl carbocation is less stable than a secondary alkyl carbocation. In order to have a reaction according to the S_N1 mechanism in vinyl systems, a very good leaving group such as triflate (-OSO₂CF₃) should be bonded to the vinyl system. An aryl substituent with an electron-donating substituent can stabilize the vinyl cation. Because the vinyl carbocation has a linear structure, the configuration of the substitution product will change and a mixture of *cis* and *trans* products will be formed after substitution.

The same argument is also valid for alkynyl systems. Because the hybridization of alkyne carbon is sp, the bonding electrons will be attracted more than in vinyl systems. Thus, the substitution will become more difficult in alkyne systems.

Can nucleophilic substitution reactions occur in vinyl or aryl systems according to the S_N2 mechanism? Such reactions will be even more difficult because the nucleophile must approach the molecule from the backside, which will lead to strong repulsion between the π bond electrons and the nucleophile electrons. However, a nucleophilic substitution reaction can occur in aromatic systems when strong electron-attracting groups are attached to aromatic compounds. In aromatic systems without electron-withdrawing groups, nucleophilic substitution reactions proceed according to the addition–elimination mechanism (for details, see Section 6.8.1).

If the carbon-carbon double bond is conjugated with a carbonyl group, the nucleophilic substitution of a group attached to the β-carbon atom of the double bond may be possible. However, the reaction is not a direct substitution reaction. Substitution reactions take place as a result of addition and elimination.

Problems

2.1 Write the products of each of the following bimolecular (S_N2) reactions.

(a) $H_3C-\underset{\underset{Br}{|}}{\overset{\overset{CH_2OCH_3}{|}}{C}}-CH_2CH_3 + HO^-$

(b) $H_3C-\underset{\underset{Br}{|}}{\overset{\overset{CH_2OCH_3}{|}}{C}}-CH_2CH_3 + HO^-$
(*S*)

2 Nucleophilic Substitution Reaction

2.2 Rank each of the following sets of compounds in order for increasing S_N2 reactivity.

(a) $H_3CCH_2-Br + HO^-$ or $H_3CCH_2-Br + H_2O$

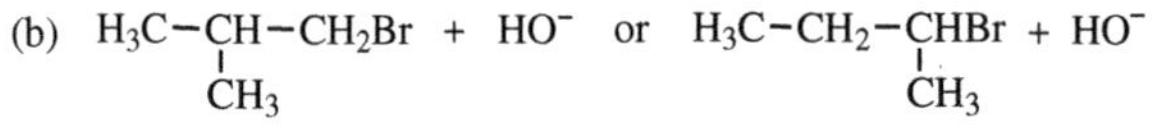

(b) $H_3C-\underset{CH_3}{CH}-CH_2Br + HO^-$ or $H_3C-CH_2-\underset{CH_3}{CHBr} + HO^-$

(c) $H_3CCH_2-Br + CH_3O^-$ or $H_3CCH_2-Br + {}^-SCH_3$

2.3 Rank the following compounds in order for increasing S_N1 reactivity.

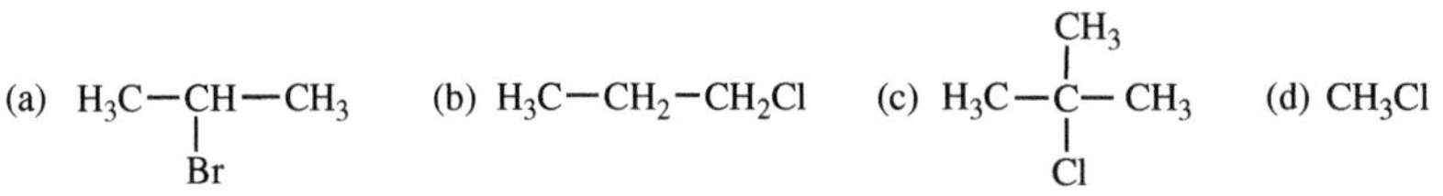

(a) $H_3C-\underset{Br}{CH}-CH_3$ (b) $H_3C-CH_2-CH_2Cl$ (c) $H_3C-\underset{Cl}{\overset{CH_3}{C}}-CH_3$ (d) CH_3Cl

2.4 The following compounds react with HO^- and HS^- in a polar solvent according to S_N1 conditions. How do the rate of the reactions change depending on the nucleophile?

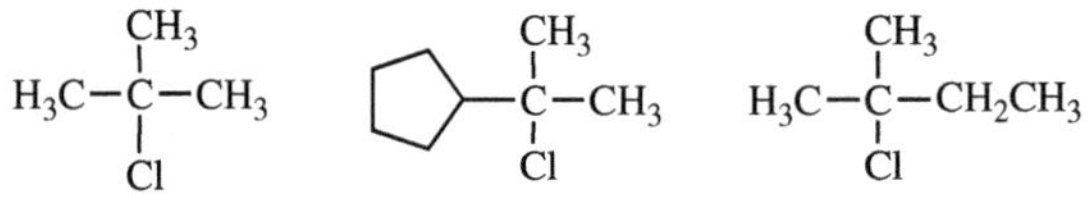

$H_3C-\underset{Cl}{\overset{CH_3}{C}}-CH_3$ cyclopentyl-$\underset{Cl}{\overset{CH_3}{C}}-CH_3$ $H_3C-\underset{Cl}{\overset{CH_3}{C}}-CH_2CH_3$

2.5 Which of the following compounds is more reactive under S_N1 conditions?

(a) $H_3CO-CH=CH-CH_2Br$ (b) $H_3CO-CH_2-CH=CH-CH_2Br$ (c) $H_2C=CH-CH_2Br$

2.6 Which of the following compounds is more reactive under S_N1 conditions?

(a) $H_3CH_2C-\underset{Br}{CH}-\overset{H}{\underset{H}{C}}=C\overset{CH_2CH_3}{\underset{H}{}}$ (b) $H_3C-\underset{Br}{CH}-CH_2-\overset{H}{\underset{H}{C}}=C\overset{CH_2CH_3}{\underset{H}{}}$

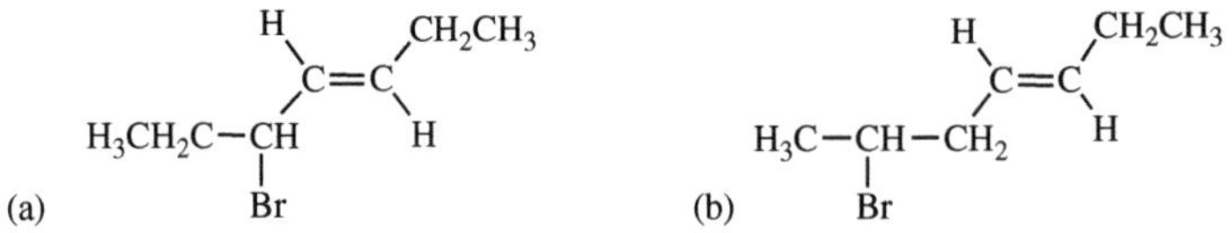

2.7 Give the potential product of each of the following reactions. Which reaction rate increases with increasing nucleophile concentration?

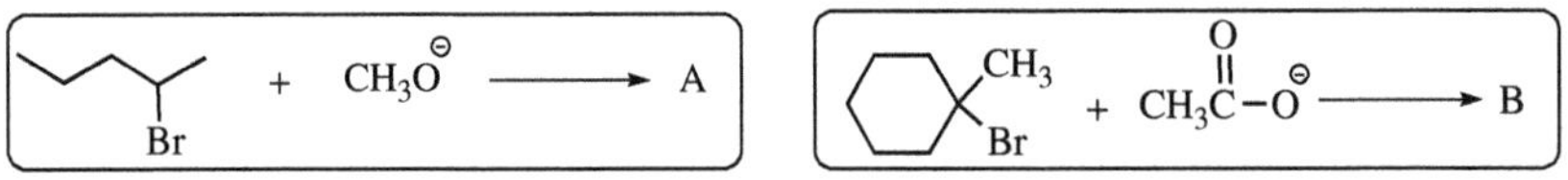

$\underset{Br}{\text{sec-butyl}} + CH_3O^{\ominus} \longrightarrow A$

$\underset{Br}{\text{1-methylcyclohexyl}} + CH_3C(=O)-O^{\ominus} \longrightarrow B$

2.8 When the dibromine reacts with methanol, only one bromoether is formed. Write down the structures of the compound and explain the exclusive formation of that product.

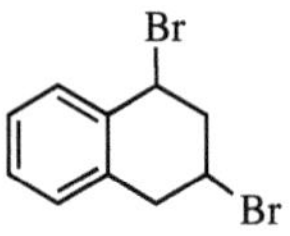

(1,3-dibromotetralin structure with Br at position 1 and Br at position 3)

2.9 Which of the following pairs of compounds react faster with $AgOCOCH_3$ in acetic acid?

(a) $n\text{-}C_3H_7Br$ or $i\text{-}C_3H_7Br$ (b) $H_3C\text{-}O\text{-}CH_2Cl$ or $H_3C\text{-}O\text{-}CH_2\text{-}CH_2Cl$

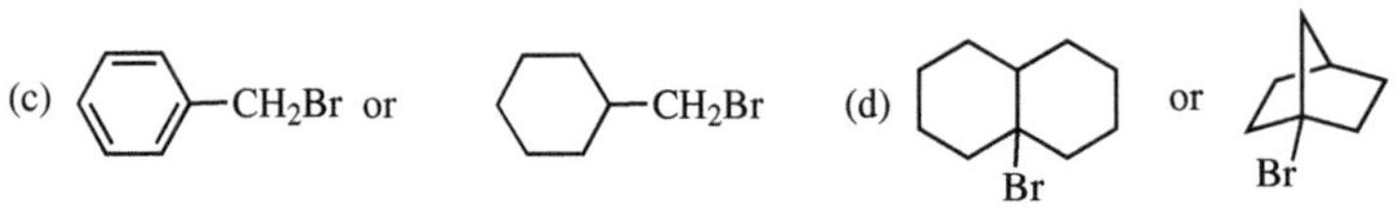

(c) $\text{C}_6\text{H}_5\text{-CH}_2\text{Br}$ or cyclohexyl-CH_2Br (d) decalin-Br or norbornyl-Br

2.10 Write down the mechanism of the reactions proceeding in the presence of Ag⁺ ions.

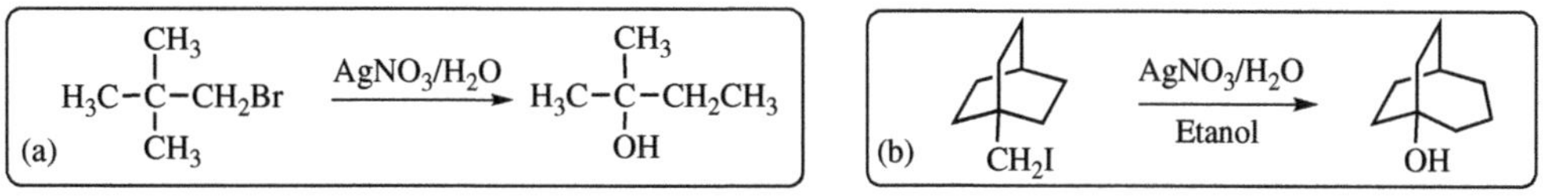

(a) $H_3C-C(CH_3)_2-CH_2Br \xrightarrow{AgNO_3/H_2O} H_3C-C(CH_3)_2-CH_2CH_3$ with OH

(b) bicyclic-CH₂I $\xrightarrow{AgNO_3/H_2O, \text{Etanol}}$ bicyclic-OH

2.11 Show how the following transformation occurs.

$$H_2C=CH-CHBr-CH_3 \xrightarrow[\text{Acetone}]{LiBr} H_3C-CH=CH-CH_2-Br$$

2.12 Write the products of each of the following reactions.

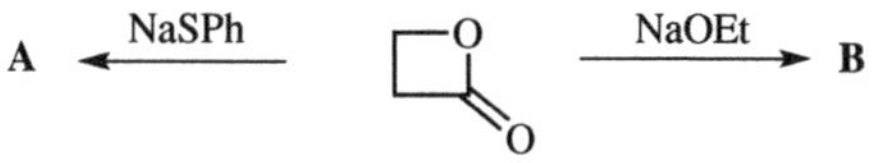

A $\xleftarrow{\text{NaSPh}}$ β-propiolactone $\xrightarrow{\text{NaOEt}}$ B

2.13

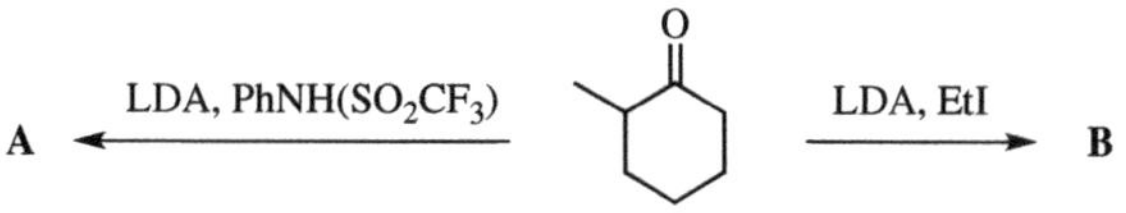

A $\xleftarrow{\text{LDA, PhNH(SO}_2\text{CF}_3\text{)}}$ 2-methylcyclohexanone $\xrightarrow{\text{LDA, EtI}}$ B

2.14

A $\xleftarrow{\text{AgCN, DMF}}$ prenyl bromide $\xrightarrow{\text{NaSPh, THF}}$ B

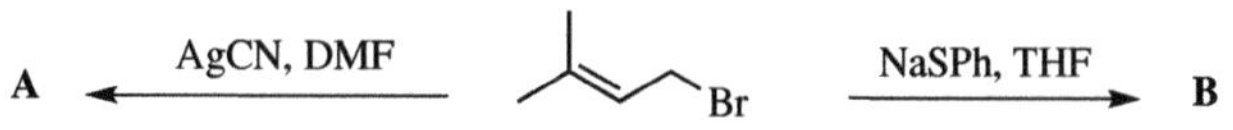

2.15 Which alkyl halides and alkoxides are used in the synthesis of the ethers given below?

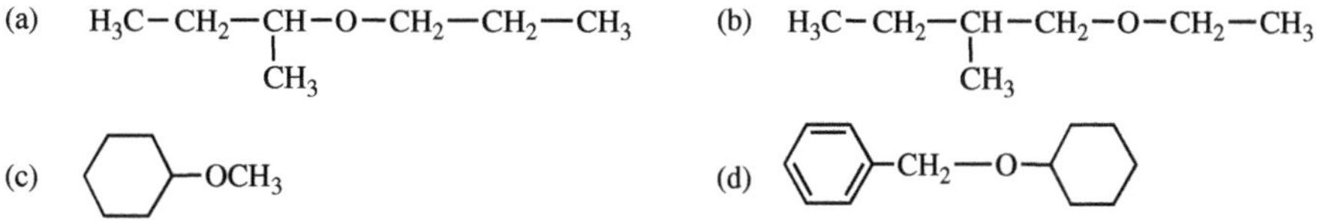

(a) $H_3C-CH_2-CH(CH_3)-O-CH_2-CH_2-CH_3$
(b) $H_3C-CH_2-CH(CH_3)-CH_2-O-CH_2-CH_3$
(c) cyclohexyl-OCH₃
(d) Ph-CH₂-O-cyclohexyl

References

1 Carroll, F.A. (1998). *Perspective on Structure and Mechanism in Organic Chemistry*, vol. 498. An International Thomson Publishing Company.
2 Edwards, J.O. and Pearson, R.G. (1962). *J. Am. Chem. Soc.* 84: 16.
3 Evanseck, J.D., Blake, J.F., and Jorgensen, W.L. (1987). *J. Am. Chem. Soc.* 109: 2349.
4 Katritzky, A.R. and Brycki, B.E. (1990). *Chem. Soc. Rev.* 19: 83.
5 Kimber, L., Rundlett, D., and Daniel, W.A. (1994). *Chirality* 6: 277.
6 Leitereg, T.J., Guadagni, D.G., Harris, J. et al. (1971). *J. Agric. Food. Chem.* 19: 785.
7 Smithells, D. (1998). *Drug Safety* 19: 339.
8 Brown, H.C. and Fletcher, R.S. (1949). *J. Am. Chem. Soc.* 71: 1845.
9 Bingham, R.C. and Schleyer, P.v.R. (1971). *J. Am. Chem. Soc.* 93: 3189.
10 Caroll, F.A. (1998). *Perspective on Structure and Mechanism in Organic Chemistry*, 466. Brooks/Cole Publishing Company.
11 Finkelstein, H. (1910). *Ber. Dtsch. Chem. Ges.* 43: 1528.
12 Lowe, P.T., Cobb, S.L., and O'Hagan, D. (2019). *Org. Biomol. Chem.* 17: 7493.
13 McMurry, J.E. *Organic Chemistry*, 8e, vol. 2012, 384. BROOKS/COLE CENGAGE Learning.
14 Bruice, P.Y. *Organic Chemistry*, 5e, vol. 2007, 352. Person International.
15 DeWolfe, R.H. and Young, W.G. (1956). *Chem. Rev.* 56: 753.
16 Young, W.G., Winstein, S., and Goering, H.L. (1951). *J. Am. Chem. Soc.* 73: 1958.

17 Trost, B.M., Van Vranken, D.L., and Crawley, L. (1996). *Chem. Rev.* 96: 395.
18 Yorimitsu, H. and Oshima, K. (2005). *Angew. Chem. Int. Ed.* 44: 4435.
19 Stork, G. and White, W.N. (1956). *J. Am. Chem. Soc.* 78: 4609.
20 Aurell, M.J., González-Cardenete, M.A., and Zaragozá, R.J. (2018). *Org. Biomol. Chem.* 16: 1101.
21 Böhme, H. and Sell, K. (1948). *Chem. Ber.* 81: 123.
22 Duchovic, R.J. and Vilensky, J.A. (2007). *J. Chem. Educ.* 84: 944.
23 Winstein, S., Lindegren, C.R., Marshall, H., and Ingraham, L.L. (1953). *J. Am. Chem. Soc.* 75: 147.
24 Rodriquez, C.F. and Williams, I.H. (1997). *J. Chem. Soc. Perkin Trans.* 2: 953.
25 Archer, S., Bell, M.R., Lewis, T.R. et al. (1958). *J. Am. Chem. Soc.* 80: 4677.
26 Shevelev, S.A. (1970). *Russ. Chem. Rev.* 39: 844.
27 Gompper, R. and Wagner, H. (1976). *Angew. Chem. Int. Ed.* 15: 321.
28 Edenborough, M. (1998). *Organic Reaction Mechanism*. Taylor Francis.

3

Elimination Reactions

Removal of two atoms or groups from a molecule is called *elimination*. In a substitution reaction, the leaving group departs from the molecule with the bonding electrons and the nucleophile attacks a carbon atom in the substrate. In elimination reactions, one of the groups leaves the molecule with the bonding electrons, while the nucleophile acts as a base and removes a proton from the adjacent carbon atom. A double bond is formed between two carbon atoms, from which the leaving group and proton are eliminated. We can divide elimination reactions into three subgroups depending on the location of the leaving groups.

α-Elimination: When two atoms or groups are eliminated from a single atom of the substrate, such type of elimination reaction is called *α-elimination*, *1,1-elimination*, or *geminal elimination*.

Because both groups leaving the molecule are bonded to the same carbon atom, a new bond will not be formed as a result of α-elimination. An electron pair remains on the atom from which the groups are eliminated. The product is a carbene when the groups are eliminated from a single carbon atom. The carbon atom is a divalent carbon.

β-Elimination: When the atoms or groups are eliminated from the adjacent atoms, it is called *β-elimination*, *1,2-elimination*, or *vicinal elimination*. A double bond is formed between the atoms from which the groups or atoms are eliminated.

γ- and Higher eliminations: There is a third type of elimination, which is called *γ-elimination* or *1,3-elimination*, in which a three-membered ring is formed. Of course, higher cyclic systems can also be created depending on the position of the leaving groups in the molecule.

In this section, we will only discuss 1,2-elimination (β-elimination) reactions because they are one of the most important methods applied to generate a double bond as alkenes are essential industrial compounds. The application area of these reactions is extensive and they are also mechanistically very important.

3 Elimination Reactions

Elimination can be described by the following three model mechanisms:

1. E1 Elimination: unimolecular elimination
2. E2 Elimination: bimolecular elimination
3. E1cb Elimination: conjugate base elimination

3.1 Unimolecular Elimination Reactions, E1

Nucleophilic substitution reactions of alkyl halides, tosylates, and similar groups were examined in detail in the previous section. In the S_N1 reaction, we have seen that nucleophiles readily trap a carbocation formed in the first step through an attack at the positively charged carbon atom. However, this is not the only mode of the reaction. If there is a base in the reaction medium, following the formation of the carbocation, the base can remove a proton from the adjacent carbon atom (β-carbon atom) to form a double bond.

A base is needed for the elimination reaction given in the above equation. The base can also be weak. *How can a weak base such as H_2O remove a proton from an sp^3-hybridized carbon atom ($pK_a > 50$)?* When a carbocation is formed, the acidity of the proton attached to the adjacent carbon atom is increased because of the positive charge on the carbon atom and the pK_a value is significantly reduced. Then, even a weak base can take the proton, forming a double bond between the carbon atoms bearing the departed groups. As seen from the example above, a common intermediate, a carbocation, is formed in both unimolecular substitution and unimolecular elimination reactions. This cation undergoes substitution or elimination reactions. At this stage, it is important to mention that elimination always competes with substitution.

3.1.1 E1 Reaction Mechanism

Remember that the S_N1 reaction proceeds in two steps. The formation of a carbocation is the rate-determining step. Similarly, an E1 reaction also proceeds in two steps, and the rate-determining step is the formation of a carbocation. "E" stands for elimination, while "1" stands for the unimolecular reaction. The carbocation formed must be stable. Furthermore, it is desired to use polar solvents to stabilize the carbocation and to have good leaving groups. Strong nucleophiles attack the carbocation, forming substitution products, while weak nucleophiles form elimination products. Often, both products can occur simultaneously. For example, let us examine the solvolysis of *t*-butyl bromide in ethanol. Because ethanol is a polar solvent, it both facilitates the removal of bromide and stabilizes the carbocation formed. The *t*-butyl cation is a common intermediate in E1 and S_N1 reactions. Ethanol attacks the carbocation as a nucleophile and forms the *t*-butyl ethyl ether. On the other hand, ethanol acts as a base and abstracts a proton from the β-carbon to form an alkene. The nucleophilic substitution product is formed in 81% yield, while the elimination product yield is about 19%.

The solvolysis of *t*-butyl bromide proceeds by the E1 mechanism. The first step is the formation of a carbocation by heterolytic cleavage of the C—Br bond. In the second, a more rapid deprotonation step, ethanol removes a proton from the adjacent β-carbon atom. If there is no proton in the β-position, then the carbocation can either undergo rearrangement or react with a nucleophile. In the case of *t*-butyl cation, there are nine hydrogens in the β-position, and all of these hydrogens are equal. Elimination of one of these hydrogens results in the formation of the alkene.

In contrast to *t*-butyl bromide, 2-bromo-2-methylbutane has two structurally different β-carbons from which a proton can be eliminated. Therefore, when 2-bromo-2-methylbutane undergoes solvolysis with ethanol, two elimination products are formed. When more than one alkene can be formed, the more substituted alkene will be formed as the major product. This more stable alkene is called the *Zaitsev product*. Alexander M. Zaitsev, a Russian chemist (1841–1910), explained that the more substituted alkene would be obtained when a proton is removed from a β-carbon atom that is bonded to the fewest hydrogens.

In most E1 elimination reactions in which there are two or more different β-hydrogens, a mixture of elimination products will be formed. The product with the most substituted double bond will be the major product. In the example shown above, the major product is a trisubstituted double bond with three alkyl groups. The second product is a disubstituted alkene with two alkyl groups. The stability of a double bond increases with the number of alkyl groups bonded to it. The stability increase can be explained by hyperconjugation. As the more substituted alkene is the more stable one formed in this reaction, it has a more stable transition state, leading to the formation of the Zaitsev product (Figure 3.1). However, it is not always observed that the Zaitsev product is the major product in elimination reactions. Systems in which Zaitsev products are not preferentially formed will be discussed in the following chapters.

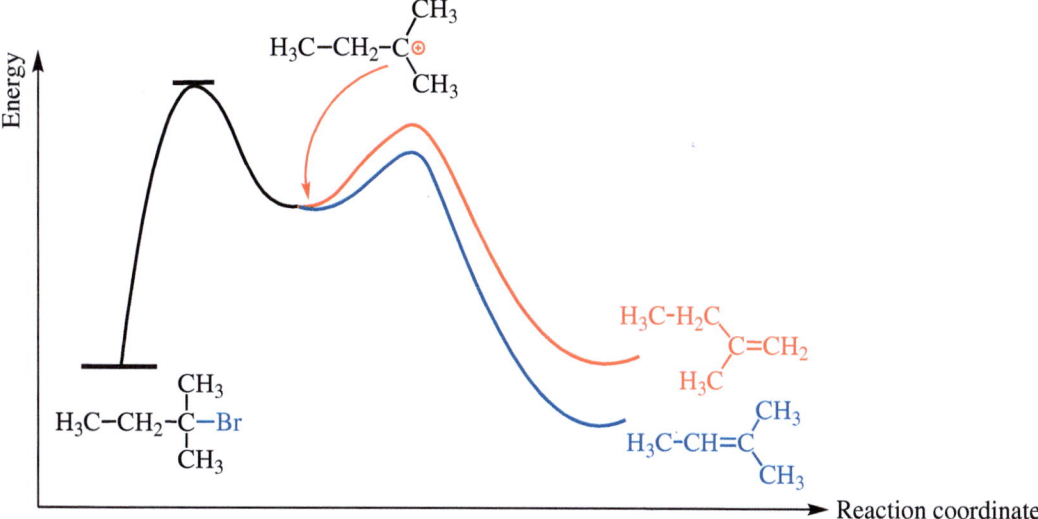

Figure 3.1 Reaction coordinate diagram for the E1 reaction and formation of the major and minor products.

Methanol is a weak base. *How can methanol remove a proton from an sp³-hybridized carbon atom?* An electron pair on the oxygen atom in methanol attacks a hydrogen atom on a carbon adjacent to the carbon atom bearing the positive charge. The proton is transferred to methanol, leaving the bonding electrons on the carbon atom. The hybridization of the carbon atom changes from sp³ to sp², and a double bond is formed between the carbon atoms (Figure 3.2) [1].

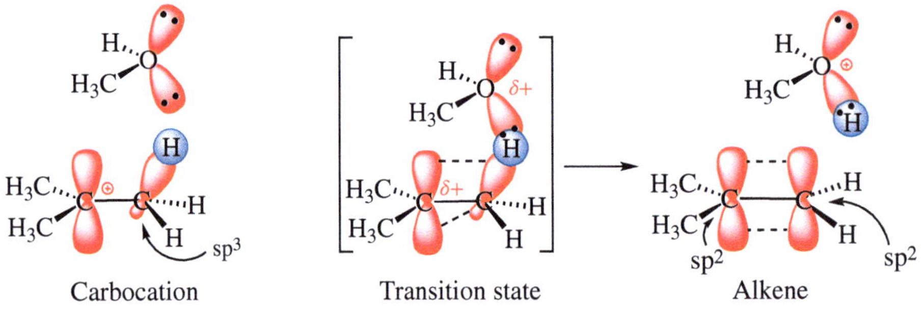

Figure 3.2 The alkene-forming step in the E1 reaction. Deprotonation by the solvent methanol.

Menthol undergoes dehydrohalogenation in aqueous ethanol to form alkenes. As there are two different β-protons in the molecule, two products are formed as a result of two different eliminations. As expected, the major product, 1-isopropyl-4-methylcyclohex-1-ene, is formed in a yield of 68%, whereas the minor product, 3-isopropyl-6-methylcyclohex-1-ene is formed in 32% yield. The major product, the Zaitsev product, has a trisubstituted double bond, while the minor product has a disubstituted double bond.

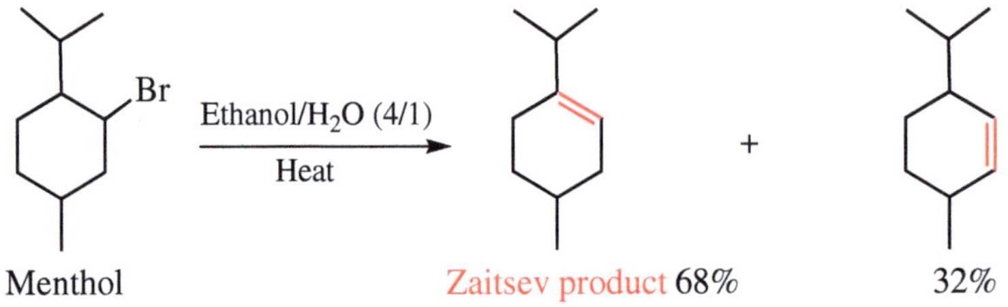

The E1 reaction involves a carbocation intermediate. The stability of the resulting carbocation determines the mode of the reaction. Carbocation can be transformed into alkene by the removal of a proton. If a carbocation can generate a more stable carbocation, it will undergo rearrangement before undergoing an elimination or substitution reaction. Consider the reaction of 3-chloro-2-methyl-2-phenylbutane in methanol. The first step is the removal of chloride and the formation of a secondary carbocation in a methanolic medium. Now, the rearrangement occurs. The driving force of this rearrangement is the formation of a more stable carbocation. The less stable, secondary carbocation rearranges to a more stable carbocation. The rearrangement occurs through the migration of one of the methyl groups from the carbon atom adjacent to the carbon atom bearing the positive (+) charge. The methyl group migrates with the bonding electrons. The migration of the methyl group from one carbon to an adjacent one is called a 1,2-shift. Generation of a tertiary carbocation and additionally conjugation with the phenyl group make the system even more stable. The Zaitsev product occurs as a result of proton elimination.

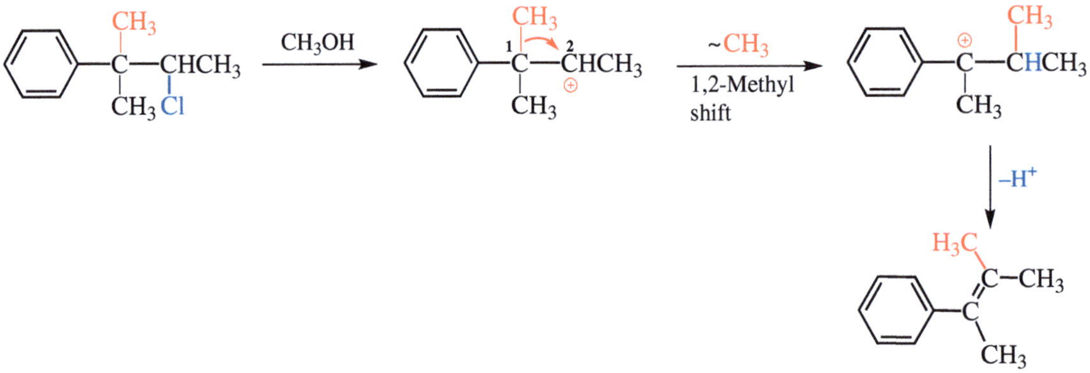

We have shown that constitutional isomers can be formed by elimination of different β-protons bonded to different carbon atoms. On the other hand, if the β-carbon from which a proton is to be removed is bonded to two hydrogens, two configurational isomers of an alkene can be formed. The carbocation created in the first step has a planar structure. Now, elimination can proceed with one of these two β-protons forming both the E and Z products. For example, (2-chloropentan-2-yl)benzene reacts with the base to eliminate hydrogen chloride to create two different olefins. Elimination in which *cis*- and a *trans* double bonds are formed is called *stereogenic elimination*.

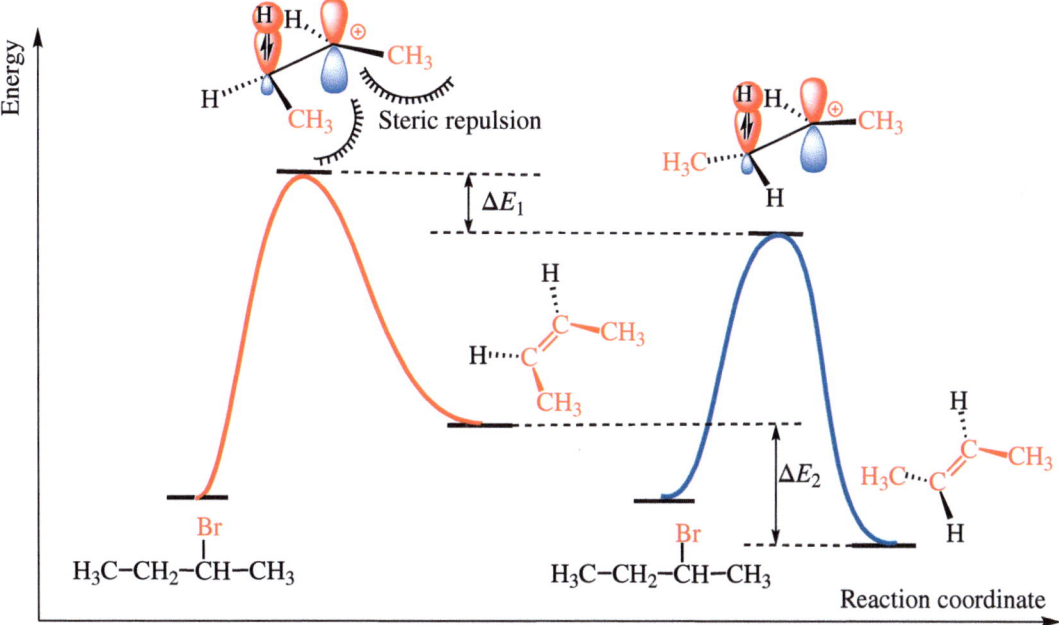

trans-Alkenes are more stable than their *cis*-isomers because of the nonbonded interaction strain between alkyl groups on the same side of the double bond in the *cis*-isomers. The ratio of *cis*- and *trans*-products depends on the reaction mechanism. In E1 reactions, this ratio is determined by the transition state. The first step is the formation of a carbocation. For elimination of the β-proton, the C—H bond from which the proton will be eliminated and the empty p orbital must align parallel to form the C=C double bond. The *cis*-transition state always has higher energy than the *trans*-transition state because of steric repulsion between the substituents. The double bond formation takes place "preferentially via the lower energy pathway," which leads to the formation of the *trans*-isomer (Figure 3.3).

Figure 3.3 Energy profile of *cis*- and *trans*-transition states in E1 elimination reactions. ΔE_1 is the energy difference between two transition states and ΔE_2 is the energy difference between the olefins.

3.1.1.1 Dehydration of Alcohols

Elimination of a water molecule from alcohols is a dehydration reaction. This reaction requires an acid as a catalyst to convert the hydroxyl groups into a good leaving group. Mineral acids, as well as Lewis acids, are used for H_2O elimination. *t*-Alcohols easily undergo elimination and form alkenes. If there is more than one adjacent carbon atom, a mixture of alkenes is formed. Regioselectivity is observed in the products. As seen from the example below, thermodynamically stable olefin is produced as the major product in 95% yield.

Alkenes formed under E1 conditions can further rearrange under the reaction conditions. For example, when 2-cyclohexyl-2-propanol is treated with $BF_3 \cdot OEt_2$, the endocyclic product is formed in 90% yield as the major product, whereas the expected Zaitsev product, an exocyclic product, is formed in a yield of 10% [2].

If both the double-bond carbons are within the ring, the bond is called *endocyclic*, while if only one of them is within a ring, the bond is said to be *exocyclic*. It is known that 1-methylcyclohexene is 2 kcal/mol more stable than methylene cyclohexene.

The major product in this reaction cannot arise from a β-elimination process. Furthermore, the carbon atom to which the leaving group was bonded does not appear in the double bond. When any of these compounds formed is separately reacted under the same reaction conditions, equal product distribution is observed irrespective of the starting product. Based on these results, we can conclude that the main product is thermodynamically the most stable one. However, if the double bond substitution of both products is compared, the main product contains a trisubstituted double bond, while the minor product contains a tetrasubstituted double bond. The *exo* position of the double bond in the cyclohexene ring increases the strain in the molecule. For this reason, the tetrasubstituted product, which is the Zaitsev product, is formed here as the side product.

3.1.2 Factors Affecting the Ratio of E1 and S_N1

Because S_N1 and E1 reactions proceed through the formation of a common intermediate, a carbocation, they compete with each other. Often, it is possible to find a mixture of elimination and substitution products. As the first step in both reactions is the removal of the leaving group and the formation of the carbocation, the electronic nature of the leaving group cannot affect the ratio of the product distribution. The critical point is the formation of a carbocation; it does not matter how the carbocation ion is formed. For example, *t*-butyl iodide and *t*-butyl chloride in 80% aqueous ethanol give the same mixture of elimination and substitution products. However, as iodide is a better leaving group, *t*-butyl iodide reacts 100 times faster. The product distribution is the same.

3.1 Unimolecular Elimination Reactions, E1

The structure of the alkyl group directly affects the ratio of substitution/elimination products. Elimination becomes more competitive with substitution as the branching in the β-position increases. The carbocation formed in the first step has a planar structure. With the removal of the leaving group, the steric repulsion between the groups in the starting material is partially reduced. If the carbocation reacts with a nucleophile, which will be attached to the carbocation, the steric repulsion between the substituents will again increase. However, if elimination occurs, the steric effect will be further reduced because another group will depart from the molecule. The reaction of various branched alkyl chlorides with 80% ethanol at 65 °C shows that the amounts of elimination products increase with the branching, while the amounts of substitution products decrease (Table 3.1) [3].

Table 3.1 The percentages of alkenes formed by solvolysis of branched alkyl chlorides in 80% ethanol.

(CH₃)₃C–Cl	(CH₃)₂(C₂H₅)C–Cl	(CH₃)(C₂H₅)₂C–Cl	(C₂H₅)₃C–Cl =0
16%	34%	41%	40%
(CH₃)₂CH–C(CH₃)₂–Cl	(CH₃)₃C–C(CH₃)₂–Cl	(CH₃)₂CH–C(CH₃)(CH(CH₃)₂)–Cl	(CH₃)₃C–C(C₂H₅)₂–Cl
62%	61%	78%	90%

The slow and rate-determining step in E1 reactions is the removal of the leaving group from the molecule. Therefore, it is desired to have a good leaving group. We have presented examples where the leaving groups were halides and a hydroxyl group. Elimination reactions can also be performed with sulfonium and ammonium salts. Thioethers can also undergo elimination reactions. However, the thioether functionality first must be converted into a good leaving group, such as sulfonium salts. If the substrate can form a carbocation after removal of the thioether group, an elimination product will be formed [4].

Zaitsev product 87%   Minor product 13%

In general, the following conclusions can be drawn from the E1 elimination mechanism described above:

1. Reaction kinetics is first order,
2. The rate-determining step is the formation of a carbocation, as in the case of S_N1 reactions, E1 and S_N1 reactions compete,
3. Rearranged products can be formed,
4. If there is the possibility of forming more than one elimination product, the double bond to which the most substituent is attached is formed as the major product (Zaitsev product).

Rearrangement of the carbocation and its detailed mechanism will be discussed in Section 7.1.

3.2 Bimolecular Elimination Reactions, E2

Elimination reactions can also proceed under second-order conditions with a strong base. In S_N2 reactions, the transition state is formed between the substrate and the nucleophile. Therefore, the reaction is a second-order reaction. E2 reactions are also second-order reactions and the transition state is formed between the substrate and the base. Like the S_N2 reaction, the E2 reaction also proceeds in one step with one transition state and without the formation of any intermediates. The bond-breaking and bond forming that occur during E2 reactions are shown by the curved arrows.

E2 elimination mechanism

In a bimolecular elimination reaction, the abstraction of a proton from a carbon atom next to the carbon bearing the leaving group, cleavage of the C—X bond, and the formation of a double bond occur in a concerted manner in a single step. The E2 eliminations are observed in systems that cannot produce stable carbocations. Primary alkyl compounds are transformed into alkenes via E2 mechanism. There are remarkable similarities between E2 and S_N2 reactions, as between E1 and S_N1 reactions. Because the first attack comes from the base in E2 reactions, it is essential to have a strong base to abstract the proton. While the base abstracts the proton, the bonding electrons that it shared with carbon move toward the adjacent carbon atom bearing the leaving group. The electrons attack the carbon atom from the back as a nucleophile and depart the leaving group. When the reaction is completed, the electrons that were originally bonded to the hydrogen atom form the π bond.

Like the S_N2 reaction, an E2 reaction requires a precise geometry. According to the mechanism given above, the four reacting atoms, i.e. the hydrogen to be eliminated, two carbon atoms to which the hydrogen and the leaving group are bonded, and the leaving group, must lie in a plane. There are two conformations in which the C—H bond and the C—X bond can be in the same plane. In an arrangement like that, the orbitals align appropriately for the formation of a π bond. The proton and the leaving group can be in either an *anti-periplanar* or *syn-periplanar* conformation. If the elimination proceeds from the *anti*-periplanar conformation, it is called *anti-* or *trans-elimination*. If it proceeds from the *syn*-periplanar conformation, it is called *cis-* or *syn-elimination*. Of these possible two conformations, the *anti*-periplanar conformation is the more commonly encountered in E2 reactions. In the transition state, the base is far away from the leaving group. The transition state for *syn*-periplanar elimination has higher energy resulting from the eclipsed interaction between the substituents. Furthermore, the base must approach close to the leaving group, and there will be strong repulsion between the leaving group and the base. However, some cyclic compounds have rigid geometry. Such compounds can undergo an E2 elimination reaction by a concerted *syn*-periplanar mechanism. In acyclic systems, because the carbon–carbon bonds rotate freely, the molecule can always adapt the necessary conformation. However, it is not still possible to have an ideal *trans*-conformation in cyclic systems because free bond rotation is hindered.

3.2 Bimolecular Elimination Reactions, E2

anti-Periplanar conformation → Base → *trans*- or *anti*-Elimination

syn-Periplanar conformation → Base → *cis*- or *syn*-Elimination

Let us examine an E2 elimination reaction and product distribution. *i*-Propylbromide reacts with sodium ethoxide (NaOC$_2$H$_5$) in ethanol to form 80% elimination product and 20% substitution product. Often, there is a parallelism between basicity and nucleophilicity. Thus, a base used for an elimination reaction is also a nucleophile. The base can abstract a proton from the β-carbon atom and can also attack as a nucleophile the carbon atom to which the leaving group is attached. If a stronger base is used, the rate of the E2 reaction will be greater than the rate of S$_N$2 reaction. Then, the elimination reaction will be the dominant one. The results of the reaction (*i*-propyl bromide + sodium ethoxide) are given below.

$$\text{H-C(CH}_3\text{)(CH}_3\text{)-Cl} + \text{CH}_3\text{CH}_2\text{O}^\ominus \longrightarrow \text{CH}_2=\text{C(CH}_3\text{)H} + \text{CH}_3\text{CH}_2\text{O-C(CH}_3\text{)(CH}_3\text{)-H}$$

Elimination product 80%   Substitution product 20%

For the formation of the transition state, the compound must be in the *anti*-periplanar conformation. In the transition state, the C—H bond and C—Cl bonds are partially cleaved and the double bond is partially formed. The transition state has an alkene-like structure. The partially broken and partially formed bonds are shown by dashed lines in Figure 3.4.

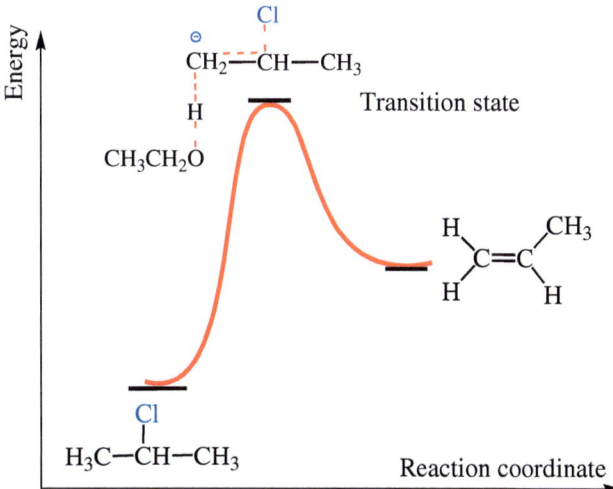

Figure 3.4 Reaction coordinate diagram for the E2 reaction for *i*-propyl chloride and sodium ethoxide

The product distribution is sensitive to the nucleophile and the base. When 2-bromopropane is reacted with sodium ethanethiolate, which is a strong nucleophile, only a substitution product is formed because the reaction proceeds through an S_N2 mechanism. However, when the same compound is treated with sodium ethoxide, which is a strong base, elimination is favored. With a weak nucleophile such as ethanol, the substitution product is formed as the major product [5].

As we discussed in E1 elimination reactions, isomeric alkenes can also be formed in E2 reactions, if there are different β-hydrogen atoms. The reaction of 2-butylbromide with sodium ethoxide in ethanol results in the formation of two olefins. The Zaitsev product, *trans*-2-butene, which is a thermodynamically more stable product, is formed as the major product in 81% yield. 1-Butene is formed as a side product in 19% yield. When the elimination reaction is performed with DBU (1,8-diazabicyclo-[5.4.0]undec-7-en), only elimination products (no substitution products are found) occur [6]. Because DBU is a strong base and a weak nucleophile, it attacks only the proton. The total yield in the reaction is 95% and the product distribution is 71 : 29. These values are close to the results of the experiment with NaOEt.

Branching in the molecular structure affects the amount of the Zaitsev product. For example, ethoxide-promoted elimination of HBr from 2-bromo-2-methylbutane produced 70% 2-methylbut-2-ene (Zaitsev product) and 30% 2-methylbut-1-ene. The introduction of a methyl group in the β-position decreases the amount of Zaitsev product to 62%. Further branching decreases the yield of the Zaitsev product to 14%. Branching at the β-position prevents the base from attacking the β-protons, which would form the Zaitsev product [7]. The crowding around the β-proton, which would form the Zaitsev product, hinders the attack of the base to abstract this proton and the formation of the thermodynamically more stable compound as the major product. Then, the base preferentially removes the most accessible proton.

3.2 Bimolecular Elimination Reactions, E2 | 97

$$H_3C-CH_2-\underset{CH_3}{\underset{|}{\overset{Br}{\overset{|}{C}}}}-CH_3 \xrightarrow[\text{EtOH}]{\text{NaOEt}} \underset{H}{\overset{H_3C}{>}}C=C\underset{CH_3}{\overset{CH_3}{<}} + \underset{H_3C}{\overset{H_3CH_2C}{>}}C=C\underset{H}{\overset{H}{<}}$$

Zaitsev product 70%   Minor product 30%

$$H_3C-\underset{CH_3}{\underset{|}{CH}}-\underset{CH_3}{\underset{|}{\overset{Br}{\overset{|}{C}}}}-CH_3 \xrightarrow[\text{EtOH}]{\text{NaOEt}} \underset{H_3C}{\overset{H_3C}{>}}C=C\underset{CH_3}{\overset{CH_3}{<}} + H_3C-\underset{H_3C}{\underset{|}{CH}}\overset{CH_3}{\overset{|}{}}\underset{H_3C}{C}=C\underset{H}{\overset{H}{<}}$$

Zaitsev product 62%   Minor product 38%

$$H_3C-\underset{CH_3\;CH_3}{\underset{|\quad|}{\overset{CH_3}{\overset{|}{C}}}}-\underset{CH_3}{\underset{|}{CH}}-\underset{}{\overset{Br}{\overset{|}{C}}}-CH_3 \xrightarrow[\text{EtOH}]{\text{NaOEt}} H_3C-\underset{CH_3}{\underset{|}{\overset{CH_3}{\overset{|}{C}}}}-CH=C\underset{CH_3}{\overset{CH_3}{<}} + H_3C-\underset{CH_3}{\underset{|}{\overset{CH_3}{\overset{|}{C}}}}-CH_2-\underset{CH_3}{\overset{CH_2}{\overset{\|}{C}}}$$

Zaitsev product 14%   Major product 86%

With the examples given so far, we have shown how branching at the α- and β-positions of the molecule affects the product distribution. The amount of Zaitsev product gradually decreases as the branching increases. The reason for this behavior is related to the steric effect. The ratio of the Zaitsev product is also affected by branching in the structure of the base. 2-Bromo-2-methylbutane is reacted with different bulky bases under the same reaction conditions. Reaction with sodium ethoxide provides the Zaitsev product as the major product. However, when the base is changed to potassium *t*-butoxide, the Zaitsev product is formed in 28% yield as the minor product. The base, potassium *t*-butoxide, is an extremely bulky base. When the base approaches the proton from which the Zaitsev product will be formed, a steric interaction occurs. The repulsion between the electron clouds of methyl groups will raise the energy of the transition state. This will slow down the reaction. The greater crowding in the structure of the reactant, as well as in the structure of the base, forces the base to abstract a proton from the methyl group.

$$H_3C-CH_2-\underset{CH_3}{\underset{|}{\overset{Br}{\overset{|}{C}}}}-CH_3 + NaO-CH_2-CH_3 \longrightarrow \underset{H}{\overset{H_3C}{>}}C=C\underset{CH_3}{\overset{CH_3}{<}} + \underset{H_3C}{\overset{H_3CH_2C}{>}}C=C\underset{H}{\overset{H}{<}}$$

Zaitsev product 70%   30%

$$H_3C-CH_2-\underset{CH_3}{\underset{|}{\overset{Br}{\overset{|}{C}}}}-CH_3 + NaO-\underset{CH_3}{\underset{|}{\overset{CH_3}{\overset{|}{C}}}}-CH_3 \longrightarrow \text{Zaitsev product 28\%} \qquad 72\%$$

$$H_3C-CH_2-\underset{CH_3}{\underset{|}{\overset{Br}{\overset{|}{C}}}}-CH_3 + NaO-\underset{CH_2CH_3}{\underset{|}{\overset{CH_2CH_3}{\overset{|}{C}}}}-CH_2CH_3 \longrightarrow \text{Zaitsev product 11\%} \qquad 89\%$$

Although the E2 dehydrohalogenation of alkyl iodides, alkyl bromides, and alkyl chlorides results in the formation of more substituted alkenes, the E2 elimination of alkyl fluorides forms the less substituted alkene as the major product. The removal of hydrogen chloride, hydrogen bromide, and hydrogen iodide is a synchronous process. Breaking of the C—H bond, formation of the double bond, and removal of the leaving group take place at the same time. The negative charge does not build up on the carbon that is losing a proton. Therefore, the transition state resembles an alkene. Fluoride is a stronger base than the other halides, and it is a bad leaving group. When the base starts to abstract the proton, the fluoride ion does not depart at the same time. As a result, a negative charge develops on the carbon atom. Consequently, the transition state resembles a carbanion rather than a double bond. Then, we have to look at the stabilities of carbanions. Alkyl groups destabilize the carbanions by their electron-donating abilities. Therefore, a carbanion built up on a primary carbon is more stable than the one on secondary or tertiary carbons. Finally, the transition state leading to the formation of 1-pentene is more stable and it is formed as the major product [8].

$$\underset{\text{CH}_3-\text{CH}-\text{CH}_2\text{CH}_2\text{CH}_3}{\overset{\text{F}}{|}} \xrightarrow{\text{NaOCH}_3} \underset{\substack{\text{1-Pentene} \\ 70\%}}{\text{CH}_2=\text{CH}-\text{CH}_2\text{CH}_2\text{CH}_3} + \underset{\substack{\text{2-Pentene (mixture of } E \text{ and } Z) \\ 30\% \text{ Zaitsev product}}}{\text{CH}_3-\text{CH}=\text{CH}-\text{CH}_2\text{CH}_3}$$

3.2.1 Kinetic Isotope Effect

In a chemical reaction, if an atom is replaced by one of its isotopes, relevant information can be obtained about the reaction mechanism. In many cases, reactions involve more than one step. Even if we have enough information about the reaction mechanism, we may not know which step is the rate-determining step. Consider an elimination reaction of an alkyl halide. We suppose that the bond-breaking step is the rate-determining step. If we replace the hydrogen with a deuterium atom, the rate of the reaction will be affected. This process is called *a primary isotope effect* [9]. If other steps are rate determining, no change in the rate of the reaction will be observed. A widespread isotope substitution takes place when hydrogen is replaced by deuterium. For example, if the hydrogen atom is replaced by deuterium, the mass of the hydrogen atom increases by 100%. In a significant proportion of the E2 reactions, the first step is the breaking of the C—H bond. The C—H bond is much more easily broken than the C—D bond because the C—D bond is about 1.2 kcal/mol stronger than the C—H bond. Therefore, reactions carried out with the C—H bond are six to seven times faster than those carried out with the C—D bond. Significant effects are seen because the percentage mass change between hydrogen and deuterium is great. Heavy atom isotope effects are much smaller. For example, if a C-13 isotope replaces a C-12 carbon atom, the increase in carbon mass is only around 8%.

$$\text{Deuterium kinetic isotope effect} = \frac{k_\text{H}}{k_\text{D}} = \frac{\text{Rate constant for H-containing compound}}{\text{Rate constant for D-containing compound}}$$

In E2 reactions, the C—H bond is partially broken in the transition state. As seen in Figure 3.4, specific activation energy is needed for this process. If the reaction is carried out with a compound having a C—D bond instead of a C—H bond, the activation energy required to break the C—D bond is greater. Excess activation energy indicates a decreased rate of the reaction. In an elimination reaction, if a β-proton is replaced by deuterium and the reaction rate is measured and found to be reduced, it can be concluded that the rate-determining step is the bond-breaking step. In other words, the reaction proceeds according to the E2 mechanism. The change in the reaction rate because of isotopes is called the *kinetic isotope effect* [9].

Consider the elimination reaction of (2-bromoethyl)benzene. When the rate constant for the elimination of HBr from (2-bromoethyl)benzene is compared with the rate constant for the elimination of DBr, k_H is found to be 7.1 times greater than k_D. This value indicates that the rate-determining step is the bond-breaking step of the C—H (C—D) bond, which is also consistent with the proposed E2 reaction mechanism.

3.2.2 Stereochemistry in E2 Elimination Reactions

We mentioned that the E2 reaction is a synchronous reaction. When the bond between the hydrogen and carbon atom breaks, a double bond is formed and the leaving group departs. As discussed before, an *anti*-periplanar geometry is required

3.2 Bimolecular Elimination Reactions, E2

for the elimination reaction. We have seen that elimination reactions occur through a backside attack; in that way, they achieve the best overlap of the corresponding orbitals. This overlap must occur partially in the transition complex. For this reason, the orbitals in question must be in the same plane (Figure 3.5). There are two possible conformations in which the departing hydrogen, the leaving group, and two carbon atoms can lie in a plane: *anti*-periplanar conformation and *syn*-periplanar conformation, as depicted below.

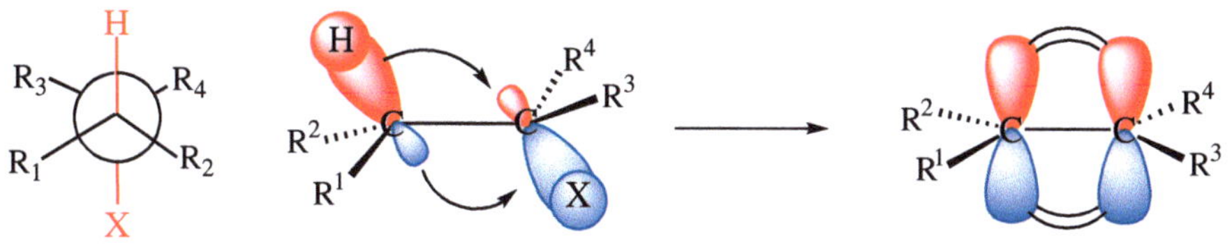

Figure 3.5 The concerted transition state of the E2 reaction. The proton, the leaving group, and the two carbons must lie in a plane.

To more easily follow and determine the stereochemistry of the products formed in elimination reactions, we will try to show the molecules in *syn*-periplanar and *anti*-periplanar conformation, as seen below. A Newman projection is a representation of the molecule looking through a C—C single bond. The front carbon is represented by three lines in a "Y" shape and the bonds attached to the back carbon are represented by lines to the edge of the circle. The drawings on the left side of the Newman projections show the side view of the molecules. Let us examine the various conformations of a molecule with the Newman projection.

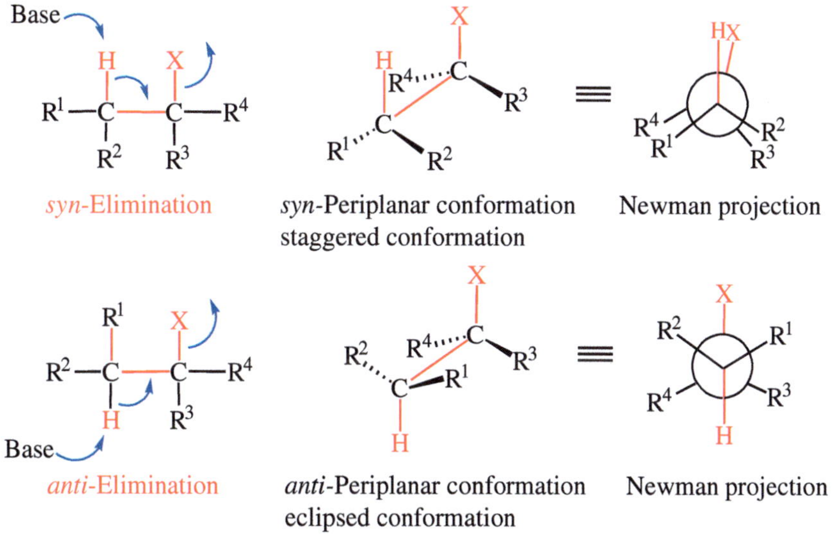

syn-Elimination   *syn*-Periplanar conformation   Newman projection
                         staggered conformation

anti-Elimination   *anti*-Periplanar conformation   Newman projection
                         eclipsed conformation

In straight-chain systems, molecules are found in various conformations by rotation about the carbon–carbon σ bonds. Let us examine the potential energy of the multiple conformations that 2,3-dibromo-2,3-dimethylbutane is found in during bond rotation. Because the bulkiest substituents in this molecule are bromine atoms, the most stable conformation is the *staggered conformation* (see conformer 1) in which two bromine atoms are as far away from each other as possible (Figure 3.6). Bromine atoms are in an *anti*-periplanar position. Steric hindrance between the substituents is minimized. Rotation of the back carbon in the Newman projections about 60° clockwise produces an *eclipsed conformation* (conformer 2) with two CH_3–Br interactions and one CH_3–CH_3 interaction. This conformer is higher in energy than the *anti*-conformer. Further rotation produces a new staggered conformation (conformer 3) in which the two bromine atoms are closer than they were in the *anti*-conformer. Even further rotation forms a new eclipsed conformation in which two bromine atoms and the methyl groups are superposed. Because the two bulky bromines are eclipsed in this conformer, it is energetically highest. It is not possible to give a general answer to the question *What is the energy difference between the staggered conformation and the eclipsed conformation?* because this energy difference depends entirely on the volumes and electronic structures of the substituents. For example, while the energy difference between the two conformations in the ethane molecule is 2.9 kcal/mol, this difference in butane increases to 4 kcal/mol. This is not a large amount of energy;

at room temperature, most molecules have enough kinetic energy to overcome this rotational barrier. In E2 elimination reactions, because of the free rotation, the molecules can always provide the desired conformation for elimination.

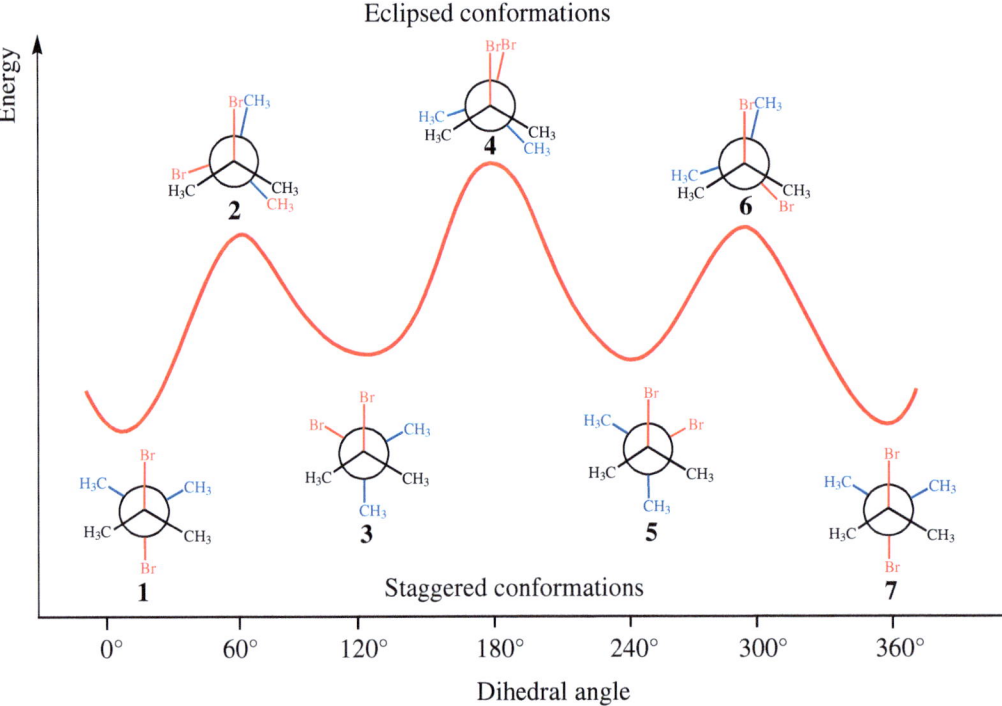

Figure 3.6 Potential energy diagram for the rotation about the C2–C3 bond in 2,3-dibromo-2,3-dimethylbutane.

As discussed above, in *syn*-elimination, the base approaches the carbon atom from the same side of the leaving group. There will be strong electrostatic repulsions between the electron-rich base and the leaving group. The transition state for *syn*-periplanar elimination will have higher energy resulting from the eclipsed interaction between the substituents. Such electrostatic repulsions do not occur when the base approaches the molecule from the back. This is the most critical reason why *anti*-elimination is generally preferred over *syn*-elimination. Moreover, in *syn*-elimination, the proton and the leaving group must be present in *syn*-periplanar conformation. *syn*-Elimination is not observed in straight-chain systems. Before examining the stereochemistry of E2 reactions in detail, we need to present the configuration and conformation of the reactants.

3.2.2.1 Erythro- and Threo-Configurations

Numerous natural products such as steroids, terpenes, and carbohydrates have more than one asymmetric center. Now, let us consider molecules with two asymmetric centers. For example, 2,3-dibromopentane has two asymmetric centers. Two groups are common (Br and H) to each carbon, while the third one (CH_3 and CH_2CH_3) is different. The configuration of those compounds is designated as *erythro* or *threo*.

$$H_3C-\overset{*}{C}H-\overset{*}{C}H-CH_2CH_3$$
$$\quad\quad\quad\;\; |\quad\;\; |$$
$$\quad\quad\quad\; Br\quad Br$$

2,3-Dibromopentane

On drawing from the Newman projection directly, if the similar groups are on the same side in the eclipsed form, the configuration is called *erythro*; if they are on the opposite side, the configuration is called *threo*. Two *threo* isomers constitute a pair of enantiomers that, when mixed, give a racemate. The *erythro* isomers are also a pair of enantiomers, which also form a racemate on mixing. All physical (melting point, boiling point, refractive indices, spectroscopic data, etc.) and chemical properties of *erythro* and *threo* compounds are different. These compounds (*erythro* and *threo*) are called *diastereomers*. They do not turn into each other by the rotation of carbon–carbon bonds.

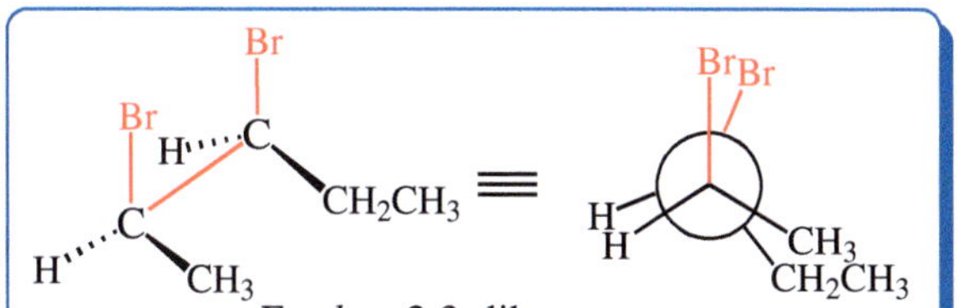

3.2 Bimolecular Elimination Reactions, E2

Erythro-2,3-dibromopentane

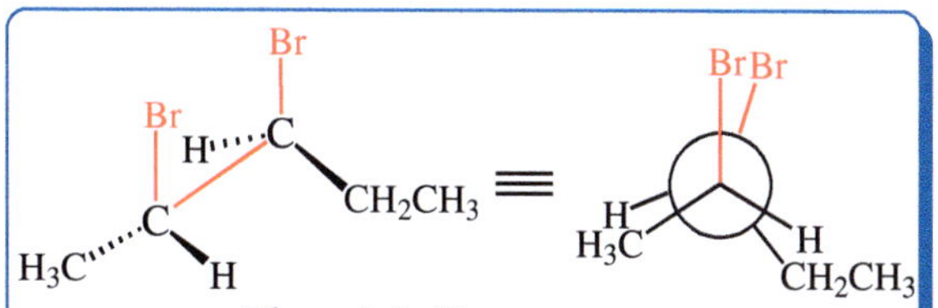

Threo-2,3-dibromopentane

Two different examples of *erythro* and *threo* compounds are given below. Chemists prefer to use perspective formulas, which show the molecules in a three-dimensional structure. For those molecules, we can draw various conformations. The staggered and eclipsed conformations are given below.

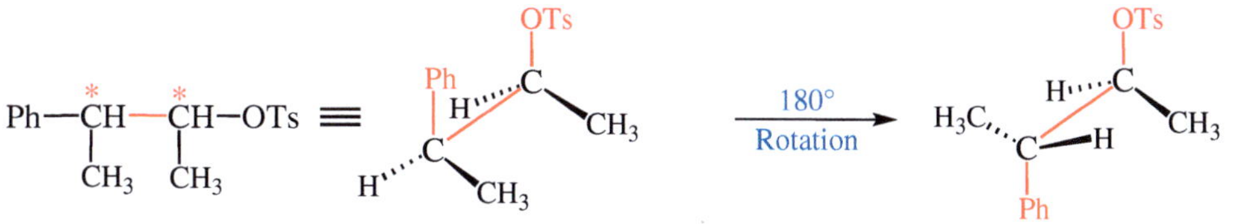

Erythro

syn-Periplanar conformation
eclipsed conformation

anti-Periplanar conformation
staggered conformation

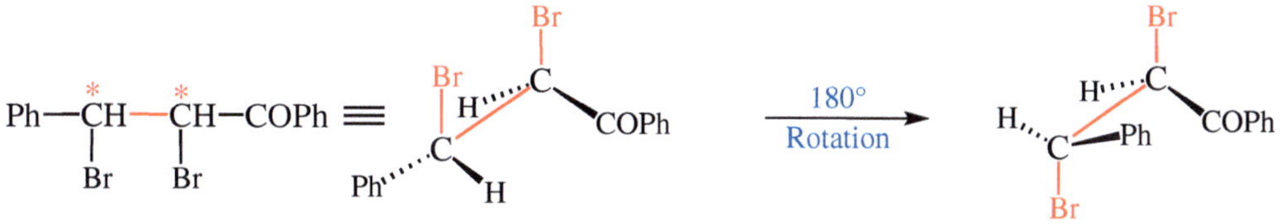

Threo

syn-Periplanar conformation
eclipsed conformation

anti-Periplanar conformation
staggered conformation

Let us examine a compound with two asymmetric centers, where all substituents are common. For example, 2,3-dihydroxybutane, 2,3-dibromobutane, and 1,2-dibromo-1,2-dichloroethane are compounds that fall into this group.

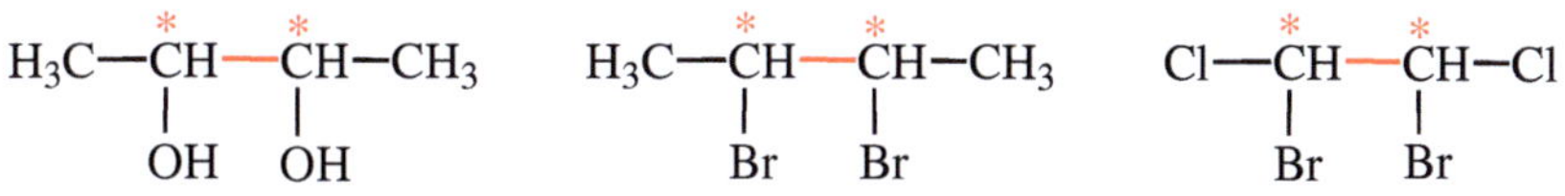

$$H_3C-CH-CH-CH_3 \quad H_3C-CH-CH-CH_3 \quad Cl-CH-CH-Cl$$
$$\quad\;\;| \;\;\;\; | \quad\quad\quad\quad\;\; | \;\;\;\; | \quad\quad\quad\quad\;\; | \;\;\;\; |$$
$$\quad\;\;OH \; OH \quad\quad\quad\;\;\; Br \; Br \quad\quad\quad\quad\;\; Br \; Br$$

In the above examples, the groups bonded to both carbon atoms are equivalent. When we rotate these compounds about the carbon–carbon bond, we encounter two possibilities: these groups will overlap mutually or not. In the case of overlapping, the compound is called *meso*. The two stereocenters are superimposable. A characteristic feature of a *meso* compound is the presence of an internal mirror plane that bisects the molecule in such a way that the top half is the reflection of the bottom half. If the configuration of one of the carbon atoms is "*R*," the configuration of the other carbon atom is "*S*" because it is a mirror plane isomer. This reveals that the molecule cannot be optically active and all *meso* compounds are *achiral*. Generally, if a compound has a plane of symmetry, it will not be optically active. Furthermore, the *meso* compounds will not have an enantiomer, even though it has asymmetric centers.

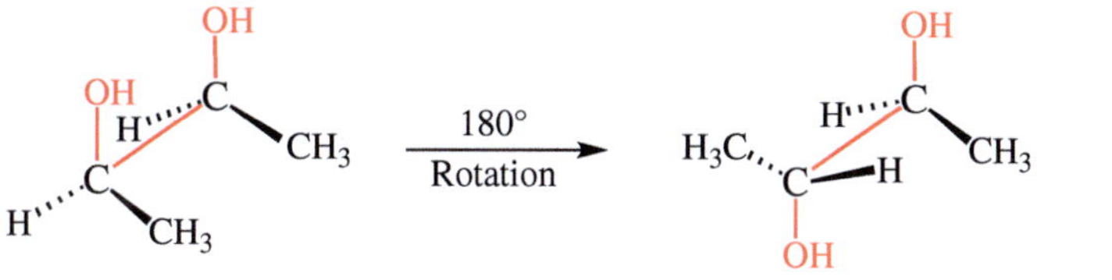

meso
syn-Periplanar conformation

meso
anti-Periplanar conformation

meso
Mirror plane
Optically inactive

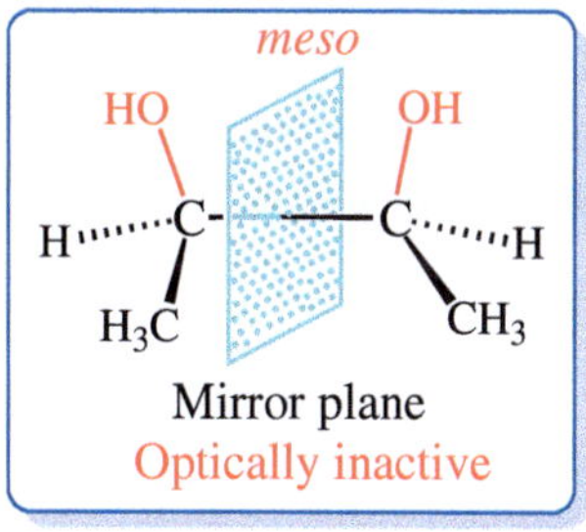

If three identical substituents bonded to the carbon atoms do not come face to face by rotation about the C—C bond, those compounds are chiral. They do not have a plane of symmetry. If we draw a mirror image of this compound, we can see that it is not superimposable on the other structure. These compounds are called DL-*pairs*. In summary, there are two molecules in the DL-configuration. Their configurations are *R,R* or *S,S*. Both compounds are optically active and are also enantiomers of each other. The relationship between the DL-pair and *meso* compounds is different. Both DL-pairs are diastereomers relative to *meso* compounds.

Compounds having three different substituents attached to two adjacent carbon atoms are also found in cyclic structures. If the groups are face to face, they are called *meso* compounds. For example, *cis*-1,2-dichlorocyclobutane has two asymmetric carbon atoms, yet it is achiral because it and its mirror image are superimposable. This compound is called *meso*. Furthermore, the *cis*-isomer has a plane of symmetry that bisects the molecule into mirror image halves. On the other hand, the *trans*-1,2-dichlorocyclobutane is a chiral compound and exists as a pair of enantiomers. The enantiomers are called DL-pairs.

Because E2 reactions are stereospecific, the structures of the products formed depend on the configurations of the starting molecules. Because a carbocation with a planar structure is formed in the first step of E1 reactions, the configuration of the reactants (*R* or *S*) does not affect the configuration of the product. The situation is very different in E2 reactions. In E2 reactions, it is possible to estimate the configuration of the product that will occur when the configuration of the reactant is known. Let us try to explain the elimination of E2 with some examples. Elimination of HBr from 1-bromo-1,2-diphenylpropane in a basic medium results in the formation of an olefin [10]. The stereochemistry of this reaction can be investigated if we know the configuration of the starting compound.

$$C_6H_5-\overset{1}{C}H-\overset{2}{C}H-\overset{3}{C}H_3 \xrightarrow[-HBr]{HO^{\ominus}} C_6H_5-CH=C\overset{CH_3}{\underset{C_6H_5}{}}$$
$$||$$
$$BrC_6H_5$$

This starting molecule, 1-bromo-1,2-diphenylpropane, has two asymmetric carbon atoms. Two of the substituents attached to these carbon atoms, phenyl and hydrogen, are present in both carbon atoms, and the third substituents (methyl and bromine) are different. Thus, the configuration of the molecule can be either *erythro* or *threo*. We have stated that all physical and chemical properties of *erythro*- and *threo*-configurated isomers are different. To reveal this difference, let us examine the elimination reaction of both isomers with a strong base. We assume that the reaction proceeds according to the E2 mechanism. The rate of reaction depends on the concentration of the base and the reactant and the reaction takes place in one step. The substituents to be eliminated are hydrogen and bromine atoms. When the hydrogen and bromine are *anti*-periplanar to each other and the attacking base is also in the same plane, then elimination can occur. When the desired conformation cannot be achieved for any reason, elimination will not occur or it will be difficult to achieve.

Let us first examine the reaction of *erythro*-1-bromo-1,2-diphenylpropane with a base. We draw the three-dimensional structure of the molecule. In order to avoid mistakes, it is recommended to draw the molecule first in *syn*-periplanar conformation. In this conformation, we must pay attention to the fact that the equivalent groups (phenyl and hydrogen) are face to face as required for an *erythro* compound. Then, let us rotate the front carbon about the carbon–carbon bond (60°) counterclockwise to bring the groups that will be eliminated to the *anti*-periplanar conformation. Because in proton abstraction the formation of the double bond and the leaving group departure are simultaneous, the other substituents not involved in the elimination will retain their positions in the transition state. Because both phenyl groups are located on the same side of the molecule in the transition state, the *cis*-olefin will be formed. This reaction is stereospecific. The location of two large phenyl groups at the *cis*-position reveals that the reaction requires an *anti*-periplanar conformation for elimination. The *anti*-periplanar geometry for E2 elimination has specific stereochemical consequences that provide strong evidence for the proposed *anti*-elimination mechanism.

Erythro-1-bromo-1,2-diphenylpropane
syn-periplanar conformation → *anti*-periplanar conformation → *cis*-olefin

In the same way, let us examine the hydrogen bromide elimination reaction of *threo*-1-bromo-1,2-diphenylpropane with a strong base. We again draw the *syn*-periplanar conformation of the molecule. The equivalent groups cannot be face to face in this case. Then, let us rotate the front carbon about the carbon–carbon bond (60°) clockwise to bring the groups that will be eliminated into the *anti*-periplanar conformation. In this conformation, unlike in the *erythro* conformation, the phenyl groups appear to be in a *trans*-position in the transition state. The alkene isomer with the *trans*-configuration will be formed.

Threo-1-bromo-1,2-diphenylpropane
syn-periplanar conformation → *anti*-periplanar conformation → *trans*-olefin

The two discussed examples showed that the *erythro* compound gives a *cis*-isomer, whereas the *threo* compound results in a *trans*-isomer. This finding cannot be generalized. The example below shows that it is not general.

When *erythro*-2-bromo-2,3-diphenylbutane reacts under the same conditions, a *trans*-olefin occurs stereospecificly. In the previous example (HBr elimination of *erythro*-1-bromo-1,2-diphenylpropane), a *cis*-olefin was formed starting from an *erythro* compound. Therefore, it is not possible to generalize that *erythro* compounds form *cis*- or *trans*-olefins. The groups eliminated determine the configuration of the olefin to be formed. The leaving groups may be equivalent groups or not.

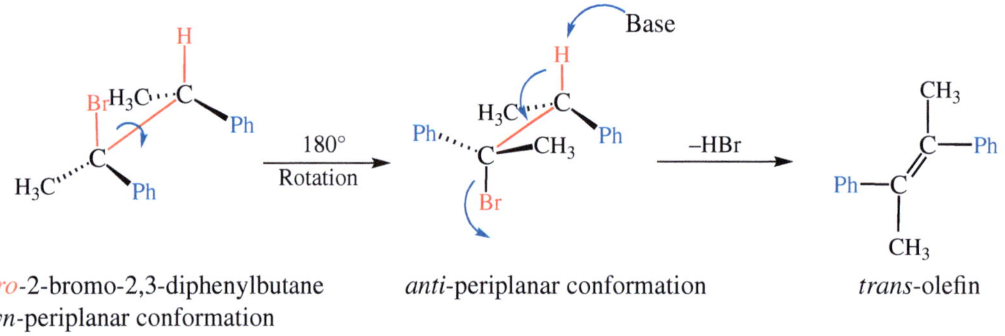

Erythro-2-bromo-2,3-diphenylbutane
syn-periplanar conformation

anti-periplanar conformation

trans-olefin

Similarly, high stereospecifity is observed in elimination reactions with *meso* and D,L compounds. The DL-pair of 1,2-dibromo-1,2-diphenylbutane produces a *cis*-olefin by elimination of HBr in the basic medium [11]. The *meso* form of this compound produces only a *trans*-isomer of the resultant olefin. It is not possible to generalize here that the *cis*-olefin is formed from the *meso* compound and the *trans*-olefin is formed from the DL-pair.

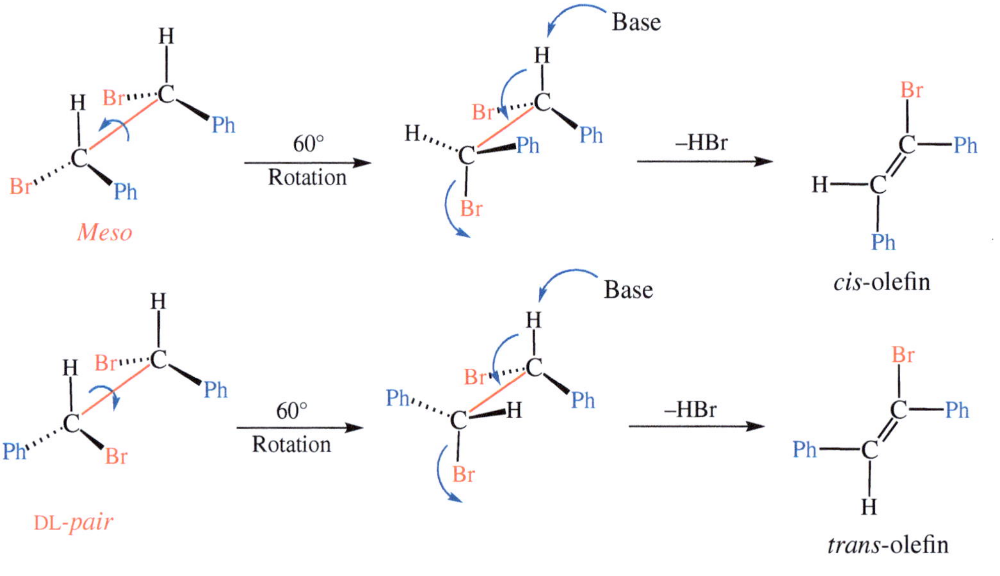

Meso

cis-olefin

DL-*pair*

trans-olefin

Instead of elimination of HBr from this compound (1,2-dibromo-1,2-diphenylethane), a bromine molecule can also be eliminated by the action of zinc. In that case, the *meso* compound forms the *trans*-olefin as a result of *anti*-elimination, while the D,L-pair produces the *cis*-olefin [12, 13]. Here, both halogen atoms are eliminated. The halogen elimination mechanism will be covered in later sections.

3.2.3 E2 Elimination in the Cyclohexane System

Elimination reactions in cyclic systems proceed according to the reaction mechanisms E1 and E2 as in acyclic systems. The molecule in which elimination is best studied in cyclic systems is the cyclohexane ring. Before examining elimination reactions in the cyclohexane system, we have to examine the conformation of the respective ring and the dynamic processes in the ring.

3.2.3.1 Conformation and Configuration in Cyclohexane

The cyclohexane ring cannot have a planar structure. If cyclohexane were planar, the internal C-C-C angles would be 120°, which is larger than the regular tetrahedral angle (109.5°). This creates a significant strain in the ring. The resulting strain prevents the cyclohexane ring from being planar. Let us draw a planar cyclohexane structure and move the carbon atom C1 up and C4 down so that the carbons 2, 3, 5, and 6 are located in a plane. This shows the framework of the six-membered ring. This conformation is the most stable one and is called the *chair conformation*. The cyclohexane ring is found in various conformations, namely, the chair conformation, boat conformation, and twisted conformation [14–16].

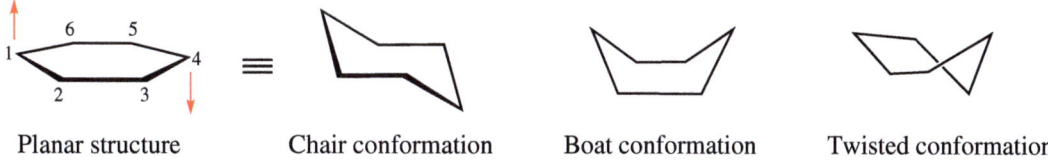

Planar structure   Chair conformation   Boat conformation   Twisted conformation

The most favored conformation is the *chair conformation*, in which the bond angles are 109.5° and the strain in the cyclohexane ring is close to zero. However, the boat conformation is not as stable. The cyclohexane ring has two chair conformations. By moving the ring carbon atoms C1, C3, and C5 in one direction (down) and the other ring carbon atoms, C2, C4, and C6, in the opposite direction (up), one chair conformation can be converted into the other. There is a fast equilibrium between these two conformations. This dynamic process is called *ring inversion*.

Chair conformation   Ring inversion   Chair conformation

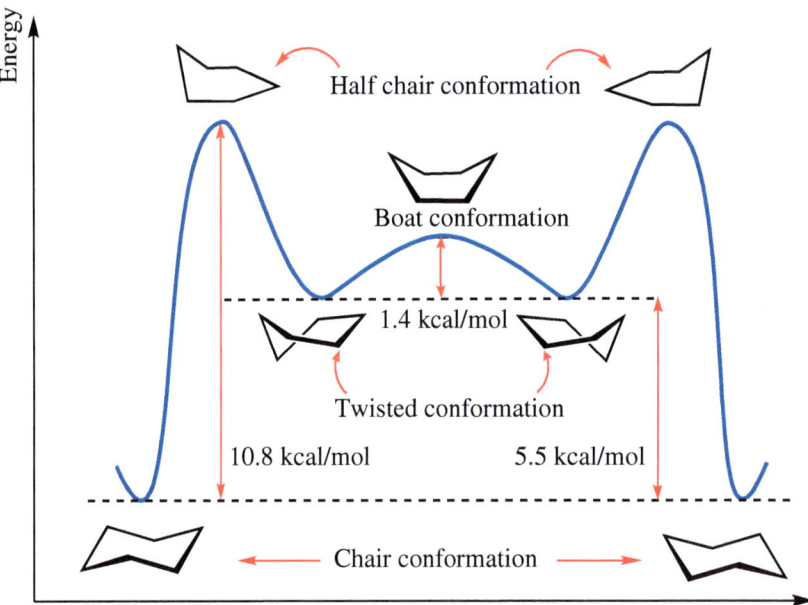

Figure 3.7 Different conformers of cyclohexane and their relative energies.

Apart from the chair conformation, cyclohexane has another conformation in which all C–C–C angles are 109.5°, which is the *boat conformation*. The boat conformation resembles the chair conformation except that one methylene group is folded upward. However, the boat conformation is not as stable because the steric strain and torsional strain make this conformation about 6.9 kcal/mol higher in energy than the chair conformation. This eclipsing of hydrogen atoms forces them to interfere with each other. These hydrogen atoms are called *flagpole hydrogens*.

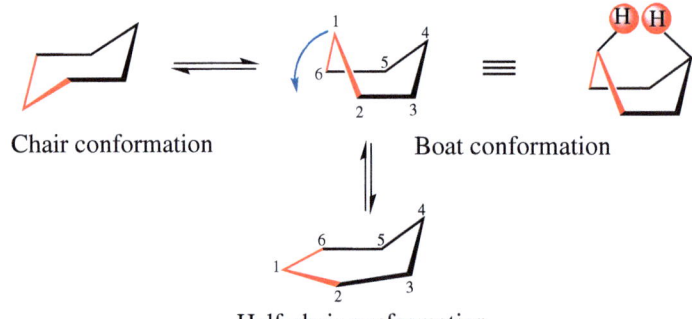

Cyclohexane passes through the intermediate conformations when being converted from one chair conformation to another (Figure 3.7). When the topmost carbons of the boat conformation are pulled down, just a slightly *twisted conformation* will be obtained. The twisted conformation is more stable than the boat conformation because the steric interaction between the flagpole hydrogens is partially reduced. When one of the topmost carbons (C1) in the boat conformation is pulled down to the point where it is in the same plane formed by C2, C3, C5, and C6, a very unstable *half-chair conformation* will be formed, which is the most unstable conformation. During inversion from one chair conformation to another, a half-chair conformation occurs first. The energy barrier for this process is about 10.8 kcal/mol. At room temperature, this energy is quickly supplied by the environment. From the half-chair conformation, the molecule passes to the twisted boat conformation. There are two half-boat conformations. The boat conformation occurs between them (Figure 3.7).

There are 12 hydrogen atoms in the cyclohexane. Two hydrogens are attached to each carbon atom. Let us assume that there is no ring inversion process. In that case, these hydrogens are not equally bonded to carbon atoms. Considering that we pass a plane through the middle of the cyclohexane ring in which all carbon atoms are located, six C—H bonds are perpendicular to this plane, three pointing up and three pointing down. These six hydrogens are called *axial hydrogens*. The other six hydrogens lie approximately in that plane. These hydrogens are called *equatorial hydrogens*. Equatorial hydrogens are usually denoted by "e" and axial hydrogens by "a."

Let us assume that the chair conformation is frozen for a moment in the cyclohexane ring. If the temperature is lowered to −70 °C, the ring inversion stops. In other words, dynamic equilibrium (ring inversion) becomes static. In that case, cyclohexane hydrogens cannot be identical because of their location. The ^{1}H-NMR measurements at lower temperatures reveal the presence of two different protons. The axial and equatorial protons resonate at individual ppm values. However, we know that cyclohexane hydrogens are all equal at room temperature.

If a rapid equilibrium occurs between the two chair conformations of the cyclohexane, all of those hydrogens that are axial become equatorial, while those that are equatorial become axial. The activation barrier of this dynamic process is about 10.8 kcal/mol [17]. Based on this value, it is calculated that ring inversion in cyclohexane occurs at room temperature approximately 200 000 times per second. The time scale of the NMR is not fast enough to distinguish between these protons. In the ^{1}H-NMR spectrum of cyclohexane obtained at room temperature, all protons resonate as a singlet, suggesting that all protons are equal. When we compare the structures of two chair conformers, it is seen that there is no difference between them. Such systems are called *degenerate systems*.

If any substituent is attached to the cyclohexane ring, the situation will be different. In methylcyclohexane, the methyl group occupies either an axial or an equatorial position because of ring inversion. These two conformers are different from each other. As we will see, the reactivity of the substituents depends on whether its position is equatorial or axial.

When the methyl group is in the equatorial position, it is located at the furthest point from other parts of the molecule. There is no steric interaction between the other hydrogens and the methyl group. However, if the methyl group is in the axial position, it is close to the other two axial hydrogens on the same side of the molecule. Steric repulsion occurs between the methyl group and the hydrogens. This repulsion is called *1,3-diaxial interaction* because of the 1.3 relation. Measurements have shown that the chair conformation with the methyl group in equatorial position is thermodynamically about 1.7 kcal/mol more stable than that of the other isomer. Because both isomers are in equilibrium, the activation barrier of this equilibrium changes according to the sizes and numbers of the substituents connected to the cyclohexane ring. This energy difference in monosubstituted systems is around 0.5–5.0 kcal/mol (2.1–21.0 kJ/mol). The energy difference between axial and equatorial in methylcyclohexane is 1.7 kcal/mol (7.1 kJ/mol) and the distribution of conformers is around 95 : 5 (equatorial : axial). It is possible to change the dynamic equilibrium into the static one by lowering the temperature. When the ^{1}H-NMR spectrum of the molecule is recorded in a case where the equilibrium is frozen, the signals arising from the methyl protons (axial and equatorial) can be separately recorded. From the integration of the signals, the ratio of the isomers can be determined.

We have shown that monosubstituted cyclohexanes are always more stable with their substituent in the equatorial position. However, the situation in disubstituted cyclohexanes is complicated because of the steric effects between the substituents. Let us examine the situation first in 1,2-dimethylcyclohexane. Suppose that both methyl groups are axially attached to the adjacent carbon atoms. As a result of ring inversion, both of the substituents switch from the axial to the equatorial position. When we examine the positions of methyl groups in the axial–axial position (bonds in blue), we can see that they are in the *trans*-position because they are on opposite faces of the ring. When the methyl groups flip to the equatorial position by ring inversion, they are still in the *trans*-position because one bond is pointing upward and the other is pointing downward. The axial methyl group connected to C1 interacts with the hydrogen atoms connected to C3 and C5, and the other axial methyl group at C2 interacts with the hydrogen atoms connected to C4 and C6. The total steric strain energy arising from these four 1.3-diaxial interactions is about 15.2 kJ/mol. In the case of the equatorial–equatorial conformation, there is no 1,3-diaxial interaction. However, there is a *gauche interaction* between the two methyl groups (3.8 kJ/mol). Therefore, the equatorial–equatorial conformation is (15.2–3.8 = 11.4 kJ/mol) more stable than the axial–axial conformation. The molecule exists in almost equatorial–equatorial conformation.

If one of the substituents is in the equatorial position and the other in the axial position, the configuration is then *cis*. In both conformations of *cis*-1,2-dimethylcyclohexane, one methyl group is in an axial position and the other is in an equatorial position. Both conformations have equal energies and they are present in equal amounts. In the *cis*-isomer, one methyl group has a 1,3-diaxial interaction with two hydrogen atoms and a gauche interaction between the two methyl groups. The total steric strain is 11.4 kJ/mol (2.7 kcal/mol).

The same kind of conformational analysis can be done for any substituted cyclohexane derivatives. Let us examine the *trans*-1,2-dihalocyclohexanes. They have been investigated by IR and NMR spectroscopy [18]. In these compounds, the axial–axial conformer is substantially populated sometimes more than the equatorial–equatorial conformer. The amount of axial–axial conformer increases across the series I > Br > Cl. In the axial–axial conformer, there is no electrostatic interaction between the halogens. However, in the equatorial–equatorial conformation, the two dipoles are oriented at a dihedral angle of 60°, and there is considerable repulsion between the halogens, which destabilizes the equatorial–equatorial conformation.

For *trans*-1,2-dibromocyclohexane, the relative population of conformers depends on the nature of solvents (Table 3.2). In nonpolar solvents such as *n*-octane, the equatorial–equatorial form is destabilized by dipole–dipole repulsion. Consequently, the axial–axial form is more stable. In polar solvents, the dipole–dipole interaction in equatorial–equatorial conformer is minimized by the intermolecular dipole–dipole interaction with the solvent molecules. Therefore, the sterically favored equatorial–equatorial conformation is more stable in methanol [19].

Table 3.2 Relative population of axial–axial and equatorial–equatorial conformers in *trans*-1,2-dibromocyclohexane.

State of aggregation	% of axial–axial conformation	% of equatorial–equatorial conformation
Liquid	65	35
Gaseous	95	5
In CCl$_4$	84	16
In benzene	52	48

Take any proton in the cyclohexane molecule. This proton can be in an axial or equatorial position. Because there will be two protons attached to the adjacent carbon atom, one of these protons will be *cis*- and the other will be *trans*.

H$_a$-H$_b$ (aa) : *trans*
H$_a$-H$_c$ (ae) : *cis*

H$_a$-H$_b$ (ee) : *trans*
H$_a$-H$_c$ (ea) : *cis*

It is not always necessary to draw three-dimensional structures in order to show the configurations of the substituents attached to adjacent carbon atoms in the cyclohexane. For example, one of the protons in the *trans*-position is indicated with a dashed line and the other with a wedged line. If a dot is placed on one of the carbon atoms, this indicates that the substituents are in the *trans*-configuration.

trans-1,2-Dibromocyclohexane

cis-1,2-Dibromocyclohexane

In cyclic systems, elimination reactions are carried out according to the E1 reaction mechanism, as in acyclic systems, as well as according to the E2 mechanism. Reactions are generally accompanied by strong bases. E1 elimination begins with dissociation of the leaving group to form a carbocation, which loses a proton to a weak base. In the E1 reaction, the base does not contribute to the formation of the carbocation. Stereoselectivity is not observed in E1 elimination.

E2 elimination in cyclohexane can take place only if the proton and the leaving group can get into a *trans*-arrangement (axial–axial conformation). This arrangement is also expected in acyclic systems. In acyclic systems, the desired *anti*-periplanar, i.e. *trans*-conformation, is always provided as the molecule can rotate freely about the carbon–carbon bond. However, free rotation of carbon–carbon bonds in cyclic systems is impossible because of molecular geometry. Therefore, the desired *anti*-periplanar conformation in cyclohexane may not be achieved. In this case, elimination according to the E2 mechanism cannot take place.

For example, the bromine atom in bromocyclohexane can be either in the axial position or in the equatorial position. When the bromine is in the equatorial position, no *anti*-periplanar conformation occurs with any of the neighboring protons. Therefore, an elimination reaction cannot take place from this conformation, although the molecule predominantly prefers this conformation. The elimination reaction can proceed only from the axial conformation.

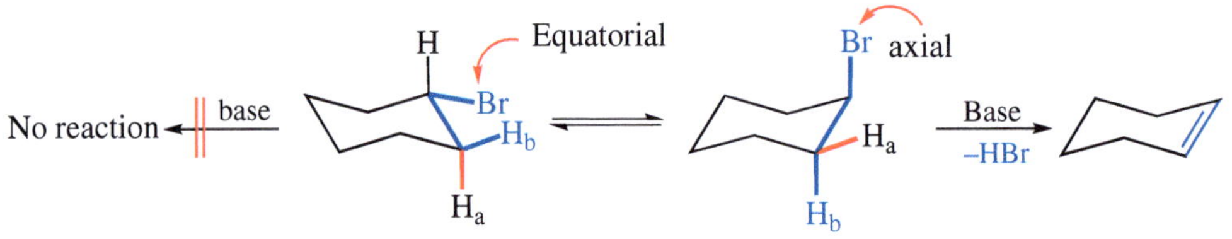

An exciting example of this is found in the E2 reaction of *trans-* and *cis-*1-chloro-2-methylcyclohexane with KOH in ethanol. The compound must adopt a conformation in which the hydrogen atom and the chlorine must be in the axial positions. As you can see from the structure, the adjacent hydrogen H$_a$ in the *trans*-isomer is in the equatorial position. The elimination of the proton H$_a$ would lead to the thermodynamically more stable Zaitsev product. However, the elimination of this hydrogen is impossible because of its equatorial orientation. The other adjacent carbon has two hydrogens (H$_b$ and H$_c$). One of them (H$_b$) is in the axial position so that elimination with this hydrogen takes place to give 3-methylcyclohexene as the sole product. In the *cis*-isomer, the chlorine atom populates an axial position in the more stable chair conformation, and there are two axial β-hydrogens (H$_a$ and H$_b$). E2 elimination can proceed with these hydrogen atoms so that the more stable 1-methylcyclohexene is therefore the predominant product (Zaitsev product).

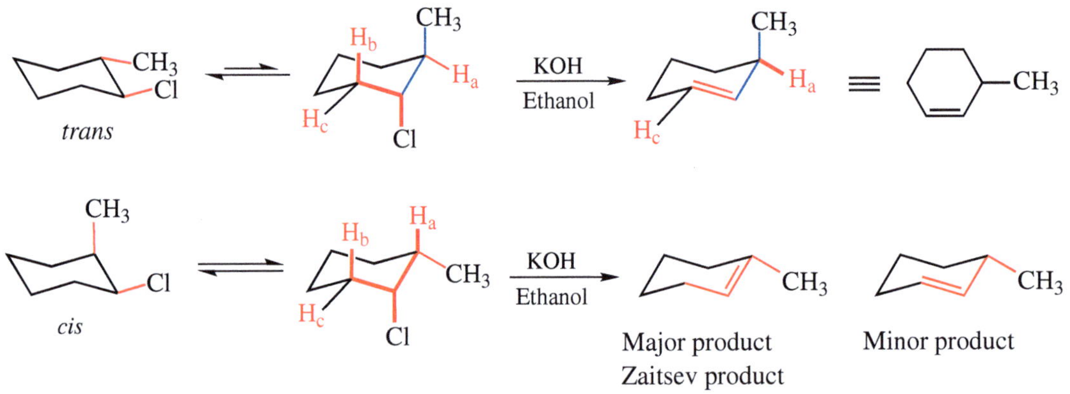

The elimination of HCl from the isomeric neomenthyl chloride and menthyl chloride is another example of axial–axial requirement for an E2 elimination reaction [20, 21]. Two conformations of neomenthyl chloride are shown below. In one of the conformations, the methyl and *i*-propyl groups are in the equatorial positions, which is the most thermodynamically stable conformation. Moreover, in this conformation, there are two hydrogen atoms (H$_a$ and H$_b$) adjacent to the chlorine atom and they are in the *anti*-periplanar positions. Sodium ethoxide can attack both hydrogen atoms to complete the E2 reaction. As a result, two products are formed. The major product is the Zaitsev product formed in 77% yield. The side product is formed in 23% yield.

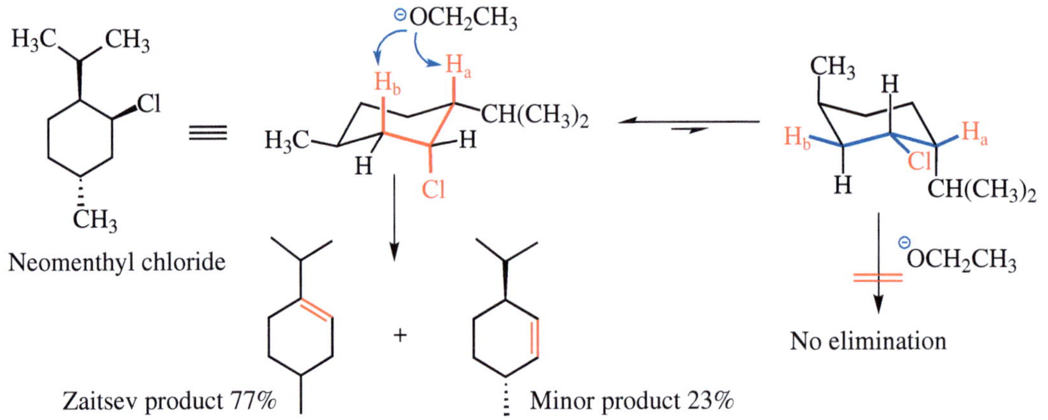

Menthyl chloride has a stable conformation in which all substituents are in the equatorial position. To achieve the desired conformation for elimination, the cyclohexane ring must undergo ring inversion to a higher energy conformation, in which now all three substituents are in axial positions. E2 elimination can occur only with the loss of H_c hydrogen, leading to the non-Zaitsev product.

Another remarkable point when these two reactions are analyzed kinetically is that neomenthyl chloride undergoes HCl elimination with sodium ethoxide 200 times as fast as menthyl chloride does. We can explain this observation as follows: while elimination in neomenthyl chloride takes place from thermodynamically the most stable conformation, elimination in menthyl chloride takes place from the less stable conformation. The different population of conformational isomers directly affects the rate of the reaction.

In cyclic systems, the presence of a hydrogen atom in the axial position may not be sufficient for E2 reactions. For example, let us examine the elimination reactions of *cis-* and *trans-*4-*t*-butylcyclohexyl tosylate in a basic medium. E2 elimination can take place only if the hydrogen and the leaving group can get into an axial–axial arrangement. In the *cis*-isomer, while the *t*-butyl group is in the equatorial position, the tosylate group is in the axial position. In this conformation, two hydrogens connected to the adjacent carbon atoms are also in the axial position. Because these hydrogens are identical, the tosyl group can be eliminated with any one of these hydrogens to form a double bond. The situation in the *trans*-isomer is somewhat different. The more stable conformer in which the tosyl and *t*-butyl groups are in the equatorial positions does not undergo an E2 reaction. Ring inversion will bring the tosyl group into the axial position. However, the *t*-butyl group will also switch to the axial position. In this case, there will be a strong repulsion between the methyl groups and the hydrogen atoms in the axial position because of the 1,3-diaxial interaction. Because of this steric hindrance, the compound will prefer the other conformation in which the tosyl and *t*-butyl groups are in the equatorial position. Therefore, elimination does not occur because of the lack of *anti*-periplanar alignment of the groups to be eliminated [22].

The reaction of *trans*-1,2-dibromocyclohexane with KOH is another example that demonstrates how important the configuration is of the groups to be eliminated. A vinyl bromide is not encountered in this reaction. The resulting product is cyclohexa-1,3-diene. In the conformation required for elimination, both bromine atoms are located in the axial position because they are *trans*. In order for vinyl bromide to form, the base must abstract the proton H_c attached to the carbon atom bearing a bromine atom. This is not possible because this hydrogen H_c is in the equatorial position. Therefore, the base abstracts the proton H_b and allylic bromide is formed. This time, for the second elimination, because it cannot remove some vinylic hydrogen, the base attacks the hydrogen H_d and HBr elimination takes place. Thus 1,3-cyclohexadiene is formed.

3.2.3.2 syn-Elimination (cis-Elimination)

In the previous section, we saw that E2 elimination requires an *anti*-periplanar arrangement of the groups to be eliminated. In certain rigid molecules, the *anti*-periplanar arrangement of the base, the leaving groups, and the carbon atoms to which they are bonded cannot be attained. In such molecules, the leaving groups may be on the same side, i.e. *syn*-periplanar, and the process is called *syn-elimination* [23]. An example is the reaction of *exo*-2-bromonorbornane, which is stereospecifically deuterated at the C3 carbon atom, with a base, and then an elimination reaction occurs to give norbornene.

In this system, elimination can take place between the bromine atom and the hydrogen or deuterium atom attached to the neighboring carbon atom. One of the exciting aspects of this reaction is that *syn*-elimination occurs and the deuterium atom is removed from the molecule. The absence of deuterium in the product indicates that the deuterium bromide has been eliminated. The dihedral angle between the hydrogen and bromine is approximately 120°. In an ideal *anti*-periplanar conformation, this angle is 180°. The dihedral angle between the bromine and deuterium atom is 0°. Deuterium is coplanar with the leaving group while hydrogen is not. In this system, *syn*-elimination is preferred [24, 25]. This example demonstrates that E2 elimination takes place if the dihedral angle between the eliminated groups is almost precisely 180°.

The results are different when the same reaction is performed with the *endo*-bromine isomer. Here, *anti*-elimination competes with *syn*-elimination and *anti*-elimination dominates. If the *syn*-elimination were the initial reaction, the product with deuterium would be the major product. The product distribution shows us that *anti*-elimination is dominant in this system. One point that should also be considered is that the approach of the base from the *exo*- and *endo*-face of the molecule is different. The *endo*-face is more hindered by the ethylene linkage, which hinders the removal of an *endo* proton that is required for *anti*-elimination.

In systems where *anti*- and *syn*-elimination are in competition, the rate of *syn*-elimination increases when solvents with low polarity are used. Ion pair promoters are suggested as capable of *syn*-elimination of the anionic leaving group. Starting from this point, a mechanism can be proposed for *syn*-elimination. When the polarity of the solvent is low, an increased amount of ion pairing occurs and a six-membered transition state can be easily formed. Elimination occurs through this transition complex. The most critical factor is whether the base is free or is in the form of an ion pair. This ion pair promotes *syn*-elimination. When the solvent is polar, an ion pair cannot be formed.

3.3 Unimolecular Conjugate Base Elimination, E1cb

There are other elimination mechanisms besides those of E1 and E2 reactions. The E1cB (elimination, unimolecular, conjugate base) mechanism is a third mechanistic pathway for elimination reactions. E2 elimination is a one-step reaction and the mechanism is a concerted process. The E1 reaction proceeds in two steps and the first step is the departure of the leaving group and the formation of a carbocation. Elimination is completed by the removal of a proton attached to the adjacent carbon atom. The E1cB reaction, like E1, involves two steps, but the order is reversed. Compounds containing an acidic proton and a poor leaving group can undergo E1cB elimination.

In contrast to the E1 reaction, which involves a carbocation intermediate, the first step in the E1cB reaction is proton abstraction to form a carbanion, which then substitutes the leaving group to form an alkene. The E1cB mechanism usually occurs with strong bases and with substrates having substituents, which can stabilize the negative charge on that carbanion

3.3 Unimolecular Conjugate Base Elimination, E1cb

center. Here, first, the proton is removed by the base and then the X-group is eliminated from the molecule. Such elimination reactions are called *unimolecular conjugate base elimination* and are expressed as E1cB. The reaction rate is controlled by a single step; base concentration does not affect the reaction rate.

Carbocation intermediate, two-step process stabilized by electron-donating groups
X = good leaving group

Concerted reaction
no intermediate

Carbanion intermediate, two-step process stabilized by electron-withdrawing groups
X = poor leaving group

It is easy to distinguish between the E2 reaction mechanism and the E1cB mechanism. The elimination reaction can be carried out in a deuterated solvent. After a certain time, the reaction can be interrupted. If the analysis of the structure of the unreacted compound shows the incorporation of a deuterium atom, this shows that the reaction proceeds according to the E1cB mechanism.

Probably, the most commonly encountered example of the E1cB mechanism in organic chemistry is the aldol condensation reaction. Aldol condensation is a reaction in which an enolate ion reacts with a carbonyl compound to form β-hydroxyaldehyde. The last step, dehydration of an aldol to give an α,β-conjugated carbonyl group, is the most common E1cB reaction. The first step is the removal of the proton with a strong base without departure of the leaving group at the same time. The carbanion formed can be stabilized by the carbonyl group by resonance. The removal of the hydroxyl group in a slow reaction generates the alkene. The poor leaving group (HO$^-$) makes the elimination step slow, which is mainly responsible for this two-step mechanism.

Aldol compound

As discussed, E1cB elimination reactions occur in systems capable of forming stable carbanions. Carbonyl compounds containing a good leaving group in the β-position, such as 4-(p-nitrophenyl)butane-2-on derivatives, also react according to the E1cB mechanism. When it reacts with a base, the β-proton is first removed to give the carbanion, which is stabilized by the carbonyl group. Then, the leaving group departs from the molecule to form an α,β-conjugated carbonyl group [26].

X = Cl, OTs

We have talked about the groups that can stabilize the carbanion formed in E1cB reactions. The carbanion can be stabilized by resonance as well as by inductive effect; for example, HF elimination from 2,2-dichloro-1,1,1-trifluoroethane proceeds by the E1cB mechanism.

3.4 Elimination Reaction in Synthesis

3.4.1 Halogen Elimination

In the elimination reactions discussed above, one of the groups eliminated was the hydrogen atom. Apart from hydrogen, other groups can be eliminated. For example, debromination of vicinal dihalogens by specific reducing agents, including the iodide ion, has been known about for many years [27]. The iodide ion is used as the dehalogenation agent in protic and aprotic solvents. The reaction is second order and is stereoselective. In order for the halogen atoms to be eliminated, they must be in the *anti*-periplanar conformation.

The reaction has been proven to be stereospecific as a result of elimination reactions with *meso*- and DL-compounds. For example, the *meso*-dibromobutane produces a *trans*-alkene by the elimination of bromine, and the DL-compound selectively forms a *cis*-alkene. We first draw the *syn*-periplanar conformation of the molecule. In the case of a *meso*-isomer, the equivalent groups must be face to face. Then, we rotate the front carbon about the carbon–carbon bond 180° to bring the groups that will be eliminated into *anti*-periplanar conformation. In this conformation, the methyl groups appear to be in a *trans*-position in the transition state; then, the alkene isomer with the *trans*-configuration will be formed. These results were regarded as evidence for a concerted and stereospecific E2 elimination reaction.

The speed of the elimination reactions with the iodide anion varies according to the location of the bromine atoms. If the bromine atoms are bonded to the terminal carbon atoms C1 and C2, the rate of the reaction increases. A different mechanism is proposed for such reactions. Nucleophilic substitution will compete with elimination. Because the steric effect at the terminal carbon atom will be less than that at other carbon atoms, iodide will prefer to substitute the bromine atom bonded to the C1 carbon rather than the elimination reaction. Then, elimination will follow the substitution reaction.

$$\text{R–CH(Br)–CH}_2\text{Br} \xrightarrow[\text{Substitution}]{I^{\ominus}, S_N2} \text{R–CH(Br)–CH}_2\text{I} \xrightarrow[\text{Elimination}]{I^{\ominus}, E2} \text{R–CH=CH}_2 + I_2 + Br^{\ominus}$$

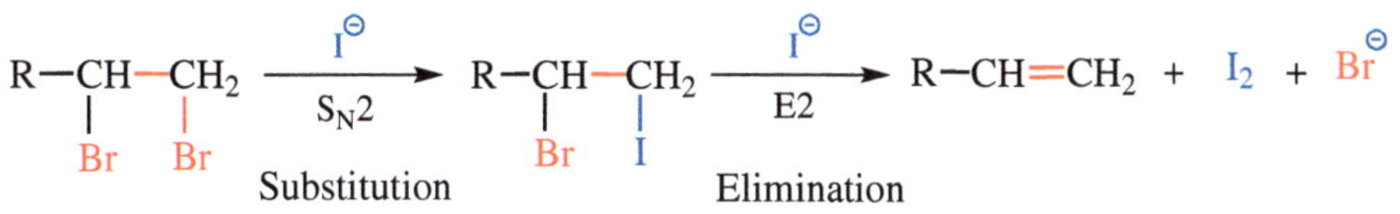

How can we prove the mechanism of the reaction? An elimination reaction with deuterium-containing compounds is necessary to determine if the iodide anion binds to the molecule and then leaves. For example, let us examine the elimination of *erythro*-1-deutero-1,2-dibromopropane with iodide. If the reaction usually proceeded according to the E2 reaction mechanism, a *trans*-olefin would be expected as a result of *anti*-elimination, as shown below.

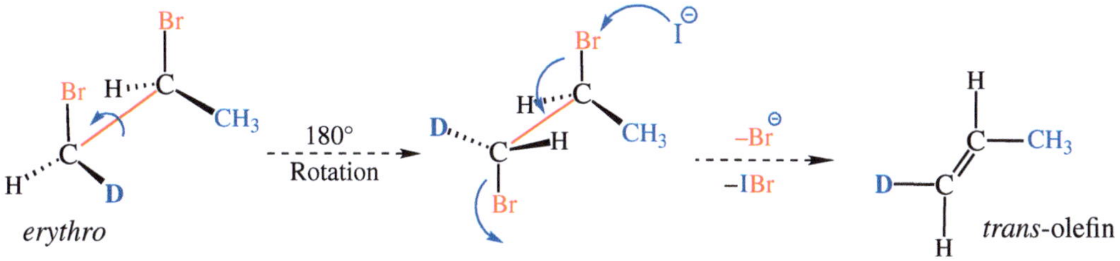

However, the configuration of the alkene formed was determined as *cis*. This outcome can only be explained by the fact that substitution and elimination reactions occur consecutively. The substitution of the bromine atom bonded to the terminal carbon atom by iodide will first change the configuration of the corresponding carbon atom by an S_N2 reaction. Then, the iodide anion will attack the iodine atom bonded to the terminal carbon atom to complete the elimination reaction to form the *cis*-olefin.

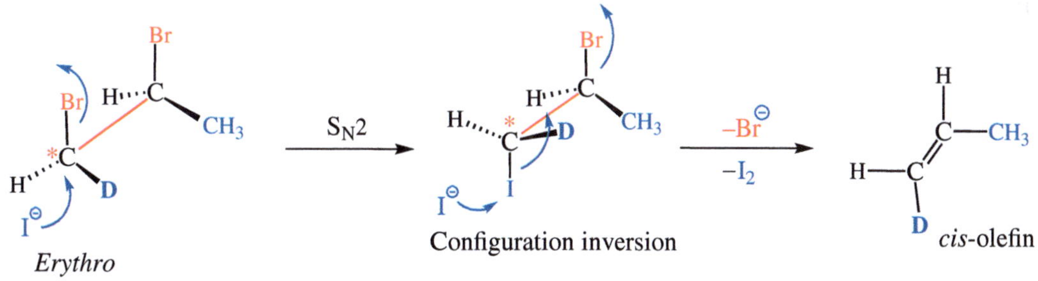

Similar reactions are observed in cyclic systems. Although *cis*-1,2-dibromocyclohexane cannot form an *anti*-periplanar conformation, it is also converted to cyclohexene upon treatment with potassium iodide, eliminating bromine at 80 °C. *trans*-1,2-Dibromocyclohexane is debrominated about 11.5 times as fast as the *cis*-isomer is. *trans*-1,2-Dibromocyclohexane also eliminates halogen under the same reaction conditions by an E2 reaction, whereas the *cis*-isomer is first converted to *trans*-1-bromo-2-iodocyclohexane by a rate-determining substitution reaction before dehalogenation [28].

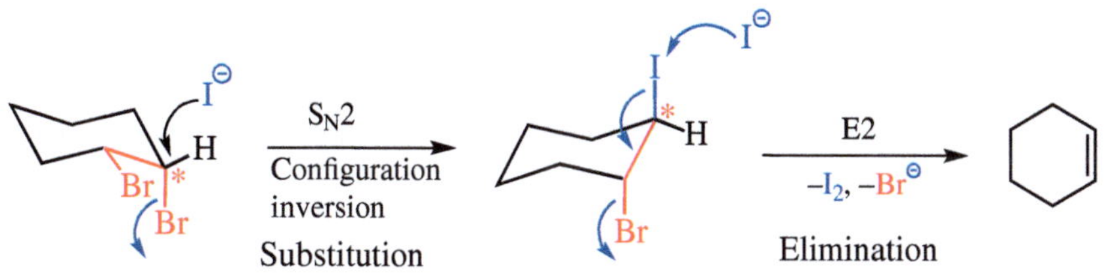

Stereospecific debromination of *syn*-configurated halogens has also been observed with bis(trimethylsilyl)mercury. The application of this methodology to 1,2-dibromoadamantane to form the highly reactive adamantene is shown below. The instability of adamantene, which can be trapped with dienes, arises from the presence of the double bond at the bridgehead (see Bredt's rule) [29].

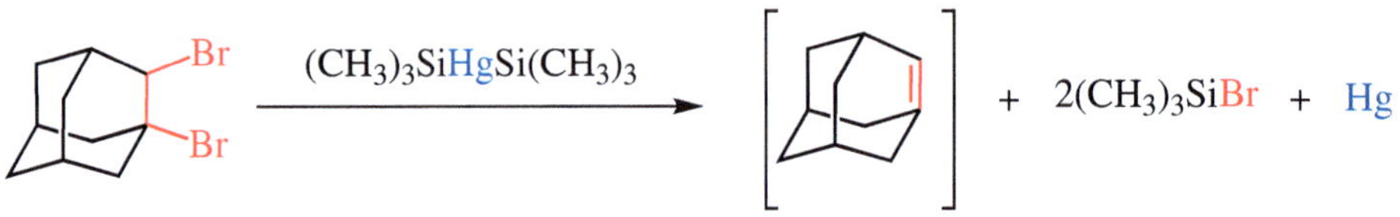

Furthermore, it is also possible to eliminate vicinal bromides with metals such as Zn and Mg.

3.4.2 Hofmann Elimination: Quaternary Ammonium Salts

In the elimination reactions of asymmetric alkyl halides, we have seen that when more than one alkene can be formed, mostly the more substituted alkene was formed as the major product. This more stable alkene was called the *Zaitsev product*. There are some elimination reactions that produce as the major product that alkene with fewer alkyl substituents connected to the double bond. Of course, the double bond formed is a thermodynamically less stable one. Elimination producing alkenes and *t*-amines from quaternary ammonium salts in the presence of a base is called the *Hofmann elimination*. Sometimes, it is referred to as the *Hofmann degradation*. The Hofmann elimination is named after its discoverer, the German chemist August Wilhelm von Hofmann (1818–1892). In the Hofmann elimination reaction, an amine is treated with excess methyl iodide to give quaternary ammonium iodide salt. The replacement of the iodine by a hydroxyl anion is carried out with silver oxide. Heating of quaternary ammonium hydroxide results in an E2 reaction and the formation of an alkene. If one of the alkyl groups is an ethyl group, ethene will be formed as an alkene. The carbon atoms directly bonded to the nitrogen atom are designated as the α-carbon atom, and so the next carbon atom from which the proton is removed is called the β-carbon atom. The base first attacks the hydrogen in the β-position and trialkyl amine departs as the leaving group [30, 31].

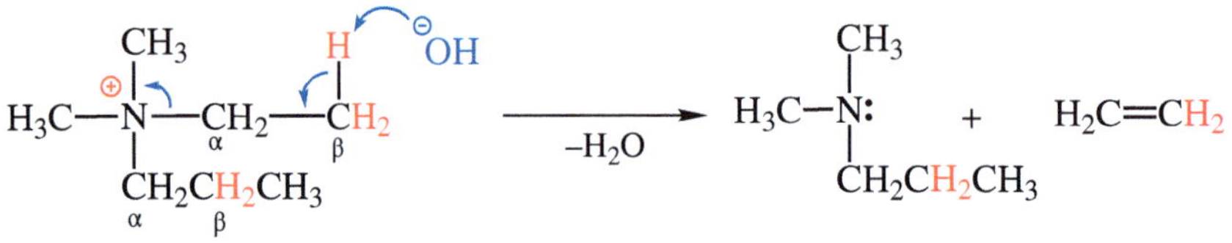

For the Hofmann reaction, four alkyl groups must be attached to the nitrogen atom. The Hofmann reaction can be carried out with primary, secondary, and tertiary amines. However, amines must be first converted into quaternary ammonium iodides by treating them with excess methyl iodide, which undergoes an S_N2 reaction repeatedly on the nitrogen atom of an amine. Methyl iodide is used as an alkylating reagent because it has no β-hydrogens and therefore cannot compete in the elimination reaction. At least one of these alkyl groups must have hydrogen at the β-position.

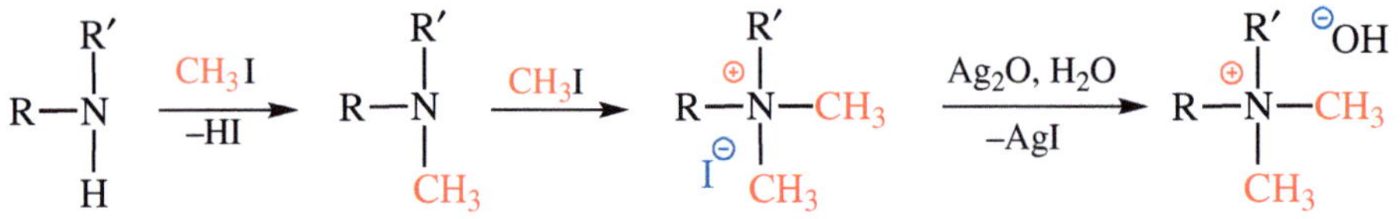

Quaternary ammonium salt

Let us try to explain why the Hofmann product is preferably formed in this elimination reaction with a concrete example. We investigate the Hofmann elimination of 2-pentyltrimethylammonium salt [32, 33]. There are two different β-hydrogen atoms in the pentyl group. If elimination occurs with the H_a proton, the more stable olefin, 2-pentene, is formed, and if the

H_b proton is removed, the less stable product, 1-pentene, is formed. Heating of 2-pentyltrimethylammonium salt gives the *Hofmann product* 1-pentene as the major product in 96% yield, whereas the Zaitsev product, the thermodynamically more stable compound, is formed only in 4% yield.

How can we explain the formation of the Hofmann product as the main product? There are several reasons to support the formation of the Hofmann product. Let us first compare the acidity of the β-protons because the first step in the Hofmann elimination is abstraction of the β-proton. The positive (+) charge on the nitrogen atom draws the bonding electrons toward the nitrogen. It indicates a positive characteristic in the nearby atoms, including the β-carbon atom. However, alkyl groups bonded to the β-carbon atom donate (+I effect) electrons, which reduces the positive characteristic of the β-hydrogen atom. Therefore, the base abstracts, preferentially, the proton from the β-carbon atom bonded to the higher number of hydrogens.

The reason for the formation of the Hofmann product can also be steric. Because of the large size of the leaving group, trialkylamine, the base probably prefers to abstract the hydrogen from the most accessible, least hindered position.

We have discussed that E2 dehydrohalogenation of alkyl iodides, alkyl bromides, and alkyl chlorides results in the formation of the more substituted alkenes, and E2 elimination of alkyl fluorides forms the less substituted alkene as the major product. Fluoride is a poor leaving group when the base starts to abstract the proton and the fluoride ion does not depart at the same time. As a result, a negative charge develops on the carbon atom (see Section 3.2). We are faced here with a similar situation. The trialkyl group is a strong base and a bad leaving group like the fluoride anion. When the base starts to abstract the β-proton, the trialkylamine group does not depart at the same time. As a result, the negative charge develops on the carbon atom. Consequently, the transition state resembles a carbanion rather than a double bond. Alkyl groups destabilize carbanions by their electron-donating abilities. Therefore, a carbanion built up on a carbon atom with fewer alkyl substituents is more stable than a carbon atom with more alkyl groups. The transition state leading to the formation of 1-pentene is more stable and it is formed as the major product.

In the example above, both β-protons were on the same alkyl group. Let us look at an example that contains β-protons on different alkyl groups. When ethyldimethylpropylammonium salt is heated, ethylene and dimethylpropylamine are formed. Because the H_a proton on the ethyl group is more acidic than the proton H_b, the base preferentially abstracts the H_a proton to form ethylene as the Hofmann product.

At the time when spectroscopic methods for analyzing the structures of complex molecules containing nitrogen atoms such as alkaloids were not available, the Hofmann elimination was applied to the structure determination of molecules by breaking them down into simpler pieces and looking for smaller and known parts among the fragments. Therefore, it is also called the Hofmann degradation. Alkaloids are natural compounds containing nitrogen atoms. For example, let us examine the Hofmann degradation to 2-methylpyrrolidine. Two consecutive Hofmann reactions break down the molecule until 1,4-pentadiene is formed [34].

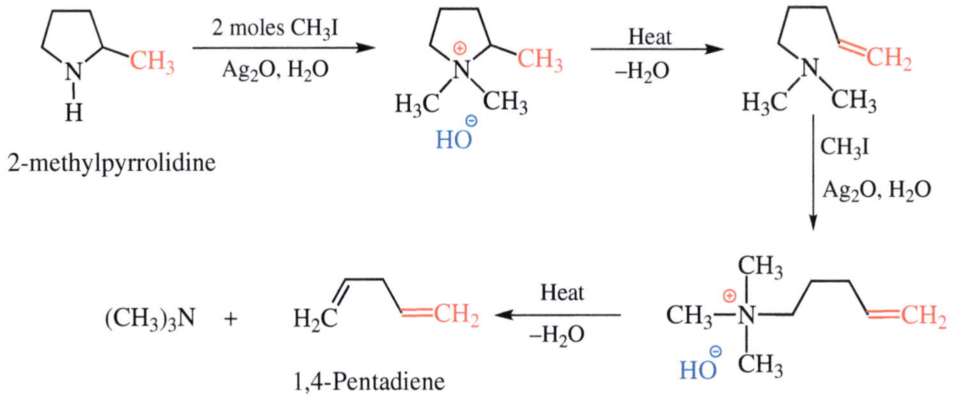

When the Hofmann elimination is applied to 3-methylprolidine, 2-methylbutadiene is obtained. When we compare the two reactions, it is seen that the alkenes formed are different. The structure of the alkenes depends on the location of the methyl group in the pyrrolidine ring. Thus, moving back from the reaction results, it is possible to find out where the methyl group was in the ring before the Hofmann elimination. This is an important clue in the structure analysis of alkaloids with a complex structure.

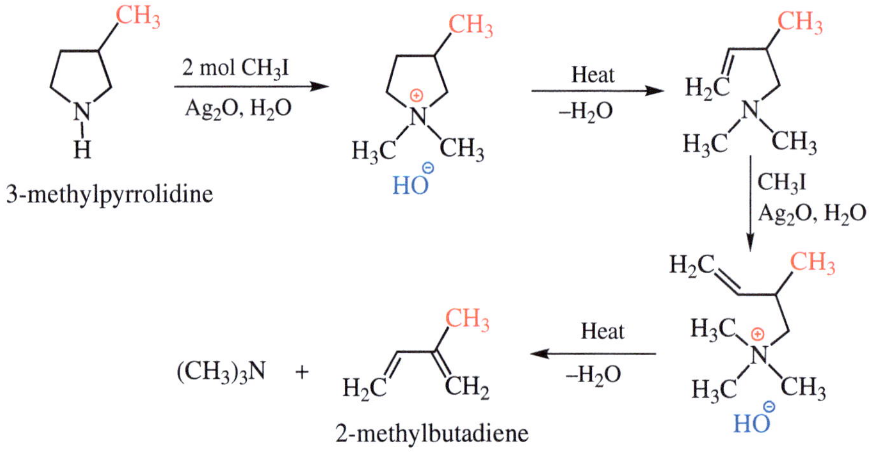

It is well established that the leaving groups in the Hofmann elimination must be in the *anti*-periplanar conformation. Otherwise, elimination cannot take place. For example, in *trans*-4-*t*-butylcyclohexyltrimethylammonium hydroxide, substituents prefer the equatorial–equatorial conformation [35, 36]. In this conformation, the molecule cannot have the required *anti*-periplanar conformation. In an *anti*-periplanar conformation, both groups (*t*-butyl and trimethylamine) will be in axial positions, which will increase the energy of the molecule. Therefore, the *trans*-isomer does not produce an elimination product. The situation is different in the *cis*-isomer. An essential part of the molecule will be in the conformation shown below. Because the hydrogen and an amine group to be eliminated are in the *anti*-periplanar conformation, elimination takes place to form a cyclohexene ring in 92% yield.

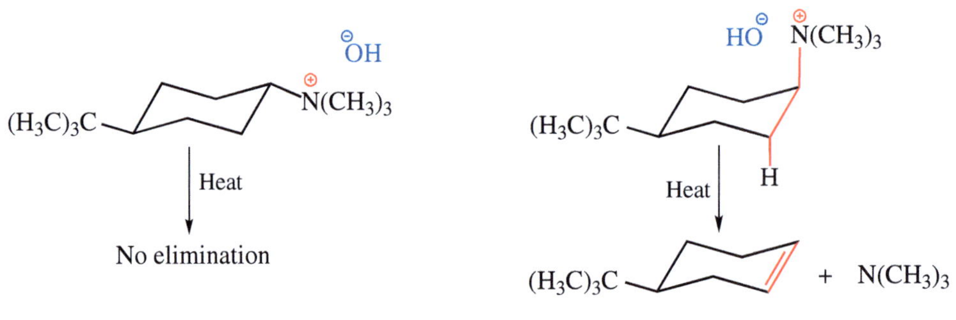

The Hofmann degradation has been successfully applied to the synthesis of some compounds. Cyclooctatetrane was first synthesized by Richard Willstaetter in 1905. He recognized its potential as a starting material for the synthesis of a carbocyclic eight-membered ring. The synthesis consists of progressive degradation of an alkaloid named pseudopelletierine obtained from the root bark of the pomegranate tree [37–39]. The synthetic steps are given below.

Hofmann eliminations are not limited to tetraalkylammonium salts. Elimination reactions take place with phosphonium and sulfonium salts of a similar structure. Base-promoted elimination of dimethyl-*sec*-butyl sulfonium salt gives 1-butene as the major product in 74% yield and 2-butenes (*cis*- and *trans*-mixture) in 26% yield. Electron withdrawal increases here the acidity of β-hydrogen atoms. The proton H_a from the methyl groups is more acidic than the proton H_b from the methylene group. Therefore, the base abstracts one of the acidic protons, H_a, and forms the thermodynamically least stable alkene.

3.4.3 Pyrolytic Elimination: Intramolecular *cis*-Elimination Reactions

We have seen several examples of reactions in which two groups bonded to the adjacent carbon atoms were eliminated. Furthermore, we emphasized that the eliminated groups should be in the *anti*-periplanar conformation in E2 reactions. We also talked about *cis*-elimination in bicyclic systems, albeit briefly. In this section, we will mainly discuss *cis*-elimination reactions. Some intramolecular elimination reactions that occur thermally are known, and many of these reactions result in *syn*-elimination. These reactions are also called *pyrolytic elimination* reactions. We will examine the reactions included in this group in three sections.

1. Ester pyrolysis: High-temperature decomposition of carboxylic acid esters
2. Elimination of xanthates: Chugaev elimination
3. Elimination of trialkylaminoxides: Cope elimination

Ester pyrolysis: Often, pyrolytic reactions are carried out in the gas phase, and they are not affected by solvents, ions, or other species. Some esters, especially acetates, when heated (thermolysis) in a solvent-free environment turn into alkenes and acids as elimination products. In order for the reaction to take place, the alkyl group bonded to the oxygen atom must have at least one hydrogen atom in the β-position.

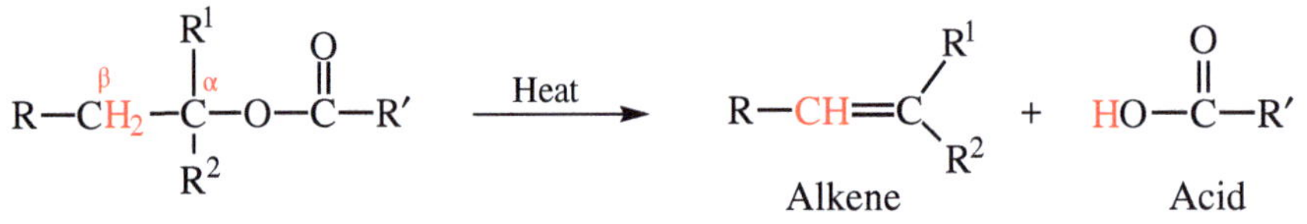

The fact that the activation entropy of the reaction is negative indicates that the groups are in a specific conformation in the transition complex. The formation of a six-membered ring in the transition complex is proposed. The leaving groups, i.e. the proton and the carboxylic acid, must be in the *cis*-position. Otherwise, the transition state cannot be formed.

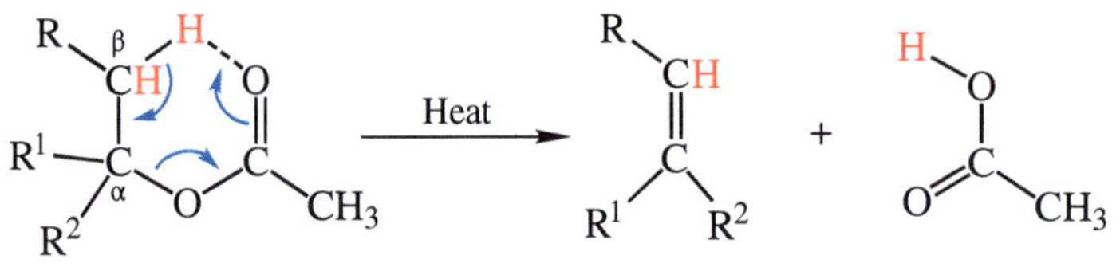

In the transition state, a hydrogen bond (an electrostatic interaction) occurs between the β-proton of the alkyl group and the carbonyl oxygen atom. This interaction weakens the bond between the hydrogen and the carbon atom, allowing hydrogen to be easily removed. The formation of a six-membered transition state is not only revealed by the measured negative entropy value. Pyrolysis experiments with labeled optically active compounds also show that the reaction is stereoselective and the pyrolysis reaction is thought to be highly concerted with little charge separation.

For example, the pyrolysis of *erythro*-2-deutero-1,2-diphenyl acetate gives the *trans*-olefin, and the entire deuterium atom remains in the alkene [40, 41]. In the *erythro*-isomer, a *syn*-periplanar conformation can be achieved with both hydrogen and deuterium. However, when a *syn*-periplanar conformation is formed with a deuterium atom, the phenyl groups will be in the *cis*-position in the transition state; they will repel each other for both steric and electronic reasons. This allows the molecule to move away from this conformation. When we rotate the front carbon atom 120° about the carbon–carbon bond, the hydrogen atom and the acetate group will be in the required conformation for elimination, and the phenyl groups will be in the *trans*-position, thereby ensuring the most stable *syn*-periplanar conformation. In elimination from this conformation, the *trans*-alkene is formed and the deuterium atom remains on the alkene in the molecule, as shown below.

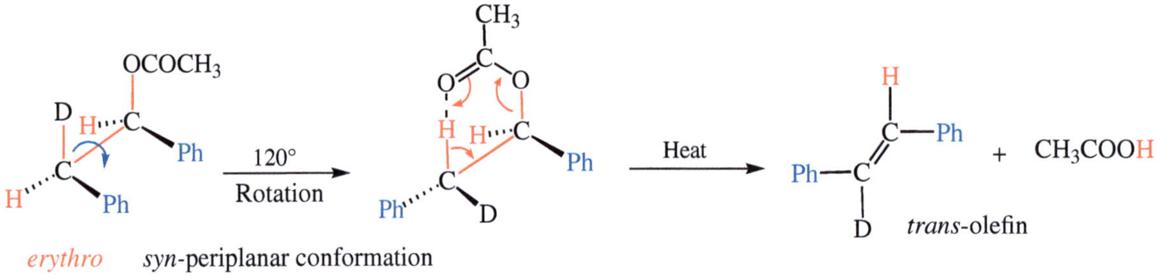

When the same experiment is carried out with the *threo*-isomer, again the *trans*-olefin occurs, but this time, the entire deuterium atom is eliminated. Here too, the most stable conformation is the one in which the phenyl groups are far apart from each other. While the phenyl groups are in a *trans*-position, the acetate can only provide *cis*-configuration with the deuterium atom. As a result of the reaction, the deuterium atom is eliminated from the molecule. In both products, the configuration is *trans*. Elimination takes place with hydrogen at one and with deuterium at the other. The stereospecificity of these two reactions makes it clear that the removal of the proton (or D) involves *syn*-elimination.

Some ketones also undergo pyrolysis by a similar mechanism. Because the pyrolysis of ketones also requires high temperatures, the reactions are not as clean as the pyrolysis of esters. For example, pyrolysis of heptane-2,6-dione, which has activated γ-hydrogen atoms at 650 °C in the gas phase, gives a mixture of acetone and methyl vinyl ketone in 22% and 17% yields, respectively [42].

Heptan-2,6-dione

β-Hydroxyalkenes decompose into alkenes and aldehydes (or ketones) when they are pyrolyzed in the gas phase at 500 °C. The example below shows that the deuterium atom bonded to the oxygen atom is transferred to the carbon atom to which the R group is attached. Such a reaction is only possible with the formation of a cyclic transition state [43]. The observation supported this conclusion that the activation energies of similar pyrolysis reactions were around 40 kcal/mol, while activation entropy was −8.8 eu. This value reveals that the transition state should be in a specific conformation.

The Chugaev elimination: One of the most studied pyrolytic eliminations is the Chugaev reaction. The Chugaev elimination is a chemical reaction that involves the elimination of water from alcohols to produce alkenes via a xanthate intermediate. The transformation of xanthates to olefins, carbon oxysulfide, and methanethiol at high temperatures is known as the *Chugaev elimination* [31, 44]. The reaction was named after its discoverer, the Russian chemist Lev Aleksandrovich Chugaev (1873–1922). Xanthates are readily synthesized by reacting potassium (sodium) xanthate, which is formed by the reaction of related alkoxides with carbon disulfide, followed by with methyl iodide.

Xanthate ester

As this elimination takes place at around 150–200 °C, it is preferred to ester pyrolysis. For the reaction to occur, the alkyl group connected to the oxygen atom must have a proton in the β-position. Otherwise, elimination does not occur. It is generally accepted that the reaction proceeds by a concerted mechanism in which the β-hydrogen atom forms a bond with sulfur in a cyclic transition state while the carbon–oxygen bond breaks. The demonstration of first-order kinetics for the reaction supports an intramolecular cyclic process and *cis*-elimination.

It has been well established that such elimination reactions in cyclic systems occur through a cyclic structure, and the configuration and conformation of the molecule significantly affect the product distribution. For example, pyrolysis experiments with xanthates derived from *cis*- and *trans*-1-phenyl-2-cyclohexanol give different results. In both compounds, the xanthate group prefers the equatorial position. In the *cis*-isomer, the phenyl group is in the axial position. In this conformation, there is only one hydrogen, H_b, in the β-position and *cis*- to the xanthate group to be eliminated. H_a hydrogen is *trans* to the xanthate group. Therefore, elimination proceeds mainly with the hydrogen H_b, and the Hofmann product is formed as the main product. The situation is different in the experiment with the *trans*-isomer. There are two hydrogens, H_a and H_b, that are *cis*- to the xanthate group. Therefore, both of these hydrogens can be involved in the elimination reaction. As expected, elimination occurs with both hydrogens and as a consequence two products are formed. The thermodynamic stability of the products and the acidity of the proton eliminated show us why the Zaitsev product is produced in 86% yield as the major product.

The stereoselectivity and *cis*-elimination observed in Chugaev reactions are not limited to cyclic systems. Experiments carried out with acyclic systems also demonstrated the stereoselectivity of the Chugaev reaction and the formation of a six-membered transition state. The thermolysis of the *erythro*-xanthate of 1,2-diphenylpropan-1-ol gave a *trans*-olefin, while the *threo*-isomer produces a *cis*-isomer. These results demonstrate the occurrence of *cis*-elimination and the formation of a cyclic transition state in cyclic as well as in acyclic systems.

The Chugaev reaction is one of the ideal methods used from time to time to eliminate H_2O from alcohol because it runs at lower temperatures than other pyrolysis reactions. When dehydration is carried out in an acidic medium, the carbocation likely formed can undergo rearrangement. Because the Chugaev elimination proceeds under neutral conditions, it is successfully applied in natural product synthesis as in the example given below to generate the exocyclic double bond at the C4 position present in lupulin C [45].

The Cope elimination: This is the conversion of *t*-amine oxides into alkene and hydroxylamine by heating in aprotic solvents. *t*-Amine oxides are synthesized in situ by oxidation of tertiary amines with peracids. The Cope elimination is mechanistically similar to the Hofmann elimination. In the Hofmann elimination, a hydroxide ion is taken as a base. In contrast, in the Cope elimination, a base is not needed, an oxide attached to the nitrogen atom acts as a base in an intramolecular elimination reaction. The reaction is stereospecific and takes place with a β-hydrogen atom in the molecule. A five-membered ring is formed in the transition state and the leaving groups depart from the molecule simultaneously. As with the Hofmann elimination, here the thermodynamically less stable product, the Hofmann product, is formed as the major product.

The reaction proceeds according to the E2 mechanism, and a *cis*-elimination takes place. To form the transition state, the groups must be in a *syn*-periplanar conformation. Experiments with *N,N*-dimethylneomentylamines reveal that the elimination is stereospecific [46]. In the compound in which the isopropyl group and the dimethylamine oxide group are *trans*, there are two hydrogen atoms at the *cis*-position with the amine oxide group. Therefore, elimination can take place with both hydrogen atoms, and the Hofmann product is formed as the main product in 64% yield. The Zaitsev product, which is thermodynamically more stable, occurs as the minor product in 36% yield. However, a single product is formed in the pyrolysis experiment with a *cis*-isomer. The resulting product is a Hofmann product. Because the hydrogen atom attached to the carbon atom bearing the isopropyl group is in the *trans*-position with amine oxide, this hydrogen cannot be eliminated. Therefore, elimination takes place with the hydrogen atom from the methylene group. These results reveal that the reaction is stereospecific and a *cis*-elimination occurs.

3.4.4 Elimination at the Bridgehead: Bredt's Rule

There are three different ways to connect two rings. Fused rings are the most common ones. They share two adjacent carbon atoms, and there is a bond between them. Bridged rings share two nonadjacent carbon atoms. However, there is no bond between these nonadjacent carbon atoms. There are one or more carbon atoms between them. If two rings share just a single carbon atom, they are called spiro compounds.

Fused compound — Adjacent carbon atoms shared by two rings

Bridge carbon / Bridgehead carbons — Nonadjacent carbon atoms shared by two rings

Spiro compound — Two rings share one carbon atom

Elimination at the bridgehead has been a field of interest for organic chemists for more than a century. A double bond cannot be placed at the bridgehead of a bridged ring system. The twisting of the p orbitals will insert strain into the molecule, which makes elimination difficult at the bridgehead. First, it was concluded that elimination could not take place at the bridgeheads as a result of the following experiments.

1-Bromocyclopentane-1,3-dicarboxylic acid can easily be converted to cyclic anhydride by eliminating H_2O. However, HBr elimination from this molecule could not be achieved despite all efforts. On the other hand, HBr elimination from dicarboxylic acid proceeded smoothly. The cyclopentene dicarboxylic acid formed could not be cyclized to anhydride. Unsuccessful reactions were explained with the thesis that double bonds cannot be formed at the bridgehead.

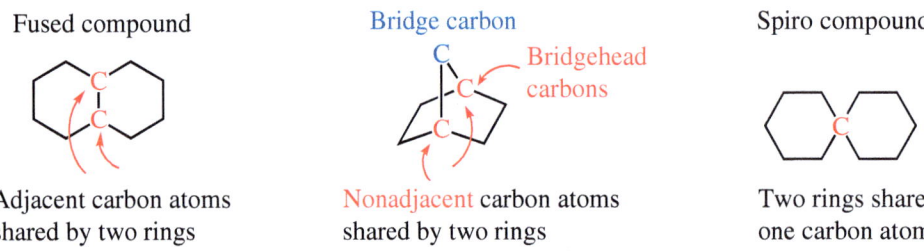

When a double bond is placed at the bridgehead, because of the geometric structure of the molecule, bending occurs between the p orbitals at the bridgehead. These orbitals must align parallel to form a strain-free double bond. Bridgehead double bonds are unstable because of poor orbital overlap. It was postulated that the strain of bridgehead alkenes is closely related to the strain of *trans*-cycloalkenes. This strain prevents the formation of a double bond. In the 1920s, German organic chemist Julius Bredt (1855–1937) was experimenting with bicyclic compounds and rings with a bridge. *Bredt's rule* was introduced in 1924 [47]. According to this rule, double bonds cannot be formed at bridgeheads. However, Bredt soon realized that stable bridgehead double bonds were possible if the bridge was large enough and later he modified his rule, but without committing himself to an exact boundary line.

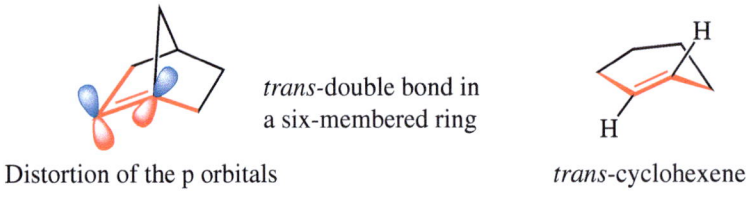

Distortion of the p orbitals

trans-double bond in a six-membered ring

trans-cyclohexene

Since the introduction of Bredt's rule, elimination reactions to create double bonds at bridgeheads have received special and increasing attention from chemists. In the studies conducted, molecules that do not comply with Bredt's rule were synthesized [48, 49]. In 1950, Prelog [50] synthesized a molecule in which the double bond was placed at the bridgehead. A diketone was subjected to a condensation reaction with a base.

Prelog showed that Bredt's rule does not apply to every system by isolating a condensation product containing a double bond at the bridgehead. The definition of the rule was changed as follows. If the number of atoms in the bridges connecting the bridgehead carbon atoms is eight or there abouts ($S = a + b + c = 8$), it is possible to place a double bond at the bridgeheads. Fawcett concluded that $S = 9$ should represent the smallest value for isolable bridgehead olefins [51].

$$S = a + b + c \geq 8$$

This value was shown to be invalid in 1967 when Wiseman [52] and Marshall [53] independently reported decarboxylation of the carboxylate anion and pyrolysis of quaternary ammonium hydroxide at 140° to produce the bridgehead alkene bicyclo[3.3.1]non-1-ene ($S = 7$). They demonstrated that the sum of the number of atoms in the bridges of a bicyclic system (S number) could also be 7. Later, Wiseman determined that the location of the double bond in a smaller or larger ring will affect the stability of the compound. Furthermore, a *trans*-configurated double bond will be more stable [54].

$$S = a + b + c = 7$$

Bredt's rule does not mean that double bonds do not occur at bridgeheads when the number "S" is below 7. The number of bridge atoms S gives us information about whether the double bond placed at the bridgehead will be stable or not. In bicyclic systems containing much smaller rings ($S < 7$), a double bond at the bridgehead can be generated; however, the product will be extremely unstable. When strong dienes are added to the reaction medium, such double bonds can be trapped by Diels–Alder reactions. For example, when *exo*-2-bromo-1-iodonorbornane is reacted with butyllithium in the presence of furan, the intermediate, norborn-1-ene, leads to stereoisomeric cycloadducts [55]. As the number of carbon atoms in the ring increases, the strain in the molecule is expected to decrease.

By similar methods, bicyclo[2.2.2]oct-1-en was synthesized and trapped with *t*-butyl-lithium and shown to be quite unstable [56]. The bridgehead olefins have been classified into three groups: [57, 58]

- isolable bridgehead olefins with olefin strain energy <17 kcal/mol
- observable bridgehead olefins with olefin strain energy between 17 and 21 kcal/mol
- unstable bridgehead olefins with strain energy >21 kcal/mol.

For example, bicyclo[3.2.1]oct-1-ene could not be isolated [59]. When the position of the double bond changes, the instability of the molecule does not change [60]. Bridgehead double bonds formed as a result of elimination reactions in [3.3.2] systems are stable at −80 °C, and they can be observed by spectroscopic methods. These compounds stable at low temperatures undergo dimerization when they are brought to room temperature. In some recent studies, it has been shown that the double bonds formed on the systems [2.2.2] and [3.2.1] can be stabilized by matrices [61].

Bicyclo[2.2.2]oct-1-ene   Bicyclo[3.2.1]oct-1-ene   Bicyclo[3.2.1]oct-5-ene

After these studies, it was revealed that Bredt's rule is not generally valid and elimination reactions may occur at the bridgeheads. Köbrich [62] established three guidelines that govern the application of Bredt's rule.

- For homologs with different S values, the ring strain varies inversely.
- For a given S, the ring strain varies inversely with the ring size of the larger of the two rings concerning which bridgehead bond is endocyclic.

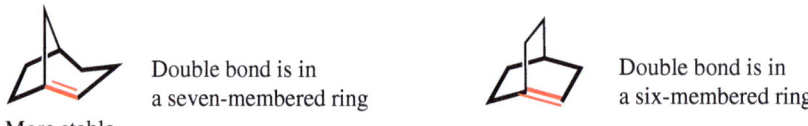

- For a given bicyclic ring system, the ring strain varies inversely with the size of the bridge containing the bridgehead double bond. These rules are called *Köbrich's rules*.

In recent years, compounds containing double bonds at the bridgehead have been synthesized by many different reactions besides elimination reactions. For example, as in the examples given below, Wittig cyclization [63] and Cope rearrangement [64] have been successfully applied and implemented to form double bonds at the bridgeheads.

3.4.5 Grob Fragmentation

According to the definition given by British chemist Cyril A. Grob (1917–2003), a heterolytic fragmentation is a reaction in which a molecule with a certain carbon–heteroatom bond breaks down into three fragments as a result of a chemical reaction [65, 66]. In β-elimination reactions, the leaving groups are attached to the adjacent carbon atoms, and a double bond occurs between these carbon atoms. In Grob fragmentation, the electrophilic and nucleophilic groups are bonded to the C1 (this atom can also be a heteroatom) and C4 carbon atoms [67, 68]. In other words, unlike β-elimination, two more saturated atoms exist between the two carbon atoms bearing the leaving groups. As a result, three fragments are formed: the electrofugal fragment, which leaves the molecule without bonding electrons, an unsaturated neutral fragment, and the nucleofuge, which leaves with the bonding electrons. This reaction is called *Grob fragmentation*. Some general examples of Grob fragmentation are presented below. Fragmentation differs from elimination. In elimination reactions, two groups are separated and a double bond is formed. In the case of fragmentation reactions, a carbon–carbon single bond is also cleaved.

The reaction mechanism can proceed via either a two-step or a single-step process, as in the case of elimination reactions. In the two-step reaction, the first step is likely the departure of the leaving group to form a carbocation as in the E1 and S_N1 reactions. The other possibility is that the electrophilic group is first separated to form a carbanion, which can then substitute the leaving group in the second step, as seen in the E1cB reaction. In the concerted process, at least five centers contribute to the transition state. Therefore, the electron pair that initiates the reaction, the leaving group, and the carbon atoms that form the double bond all must be in *anti*-periplanar conformation for maximal overlapping in the transition state. Otherwise, instead of fragmentation, reactions such as substitution or cyclization may occur.

Grob fragmentation is successfully applied to the synthesis of various natural products. For the synthesis of optically active lavandulol derivatives, Mehta and coworkers used carvone derivatives as the starting compound [69]. The carvone derivative, a diol, is selectively converted to a monomesylate and then reacted with a base (NaH). The base first abstracts the proton bonded to the oxygen atom and initiates fragmentation. In addition to the cleavage of the carbon–carbon bond, a carbonyl group and a C=C double bond are formed.

The formation of medium-sized rings in organic chemistry remains challenging because of well-known entropy and enthalpy factors. Grob fragmentation has been successfully applied to the synthesis of medium size rings by many groups. The natural product parvifoline contains an eight-membered ring. Joseph-Nathan and coworkers [70] successfully applied Grob fragmentation to the bicyclooctane system in the synthesis of (±)-parvifoline derivatives for the formation of a cyclooctane ring. The reaction of hydroxy-tosylate with sodium methoxide produced a benzocyclooctane skeleton in 80% yield.

Cyclization reactions with samarium iodide have become the focus of attention among synthetic chemists in recent years. As seen in the example given below, the intramolecular Barbier reaction can occur between the iodoalkyl chain and carbonyl group, generating a bicyclic alkoxide that can then undergo Grob fragmentation to give a nine-membered monocyclic structure [71].

The isomeric epoxy alcohols were reacted with potassium *t*-butoxide in THF at room temperature. *endo*-Epoxy alcohol underwent a substitution reaction to provide oxetane in 89% yield. However, the *exo*-alcohol reacted slowly under the same reaction conditions to provide a fragmentation product in 10% yield. In contrast to these results, *endo*-epoxide reacted within 20 minutes to provide the ring-opening product via Grob fragmentation in 100% yield [72].

Problems

3.1 3-Bromo-2,3-dimethylpentane undergoes HBr elimination under E1 conditions to form four alkenes. Write the structures and give the amounts of the alkenes qualitatively.

3.2 Which of the following compound pairs are more reactive under E2 conditions?

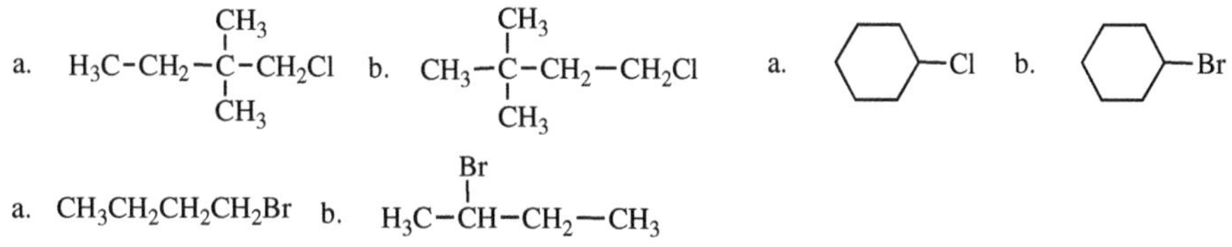

a. H₃C-CH₂-C(CH₃)(CH₃)-CH₂Cl   b. CH₃-C(CH₃)(CH₃)-CH₂-CH₂Cl   a. cyclohexyl-Cl   b. cyclohexyl-Br

a. CH₃CH₂CH₂CH₂Br   b. H₃C-CH(Br)-CH₂-CH₃

3.3 Give the structure of the major products that will occur when the following compounds are treated with NaOH.

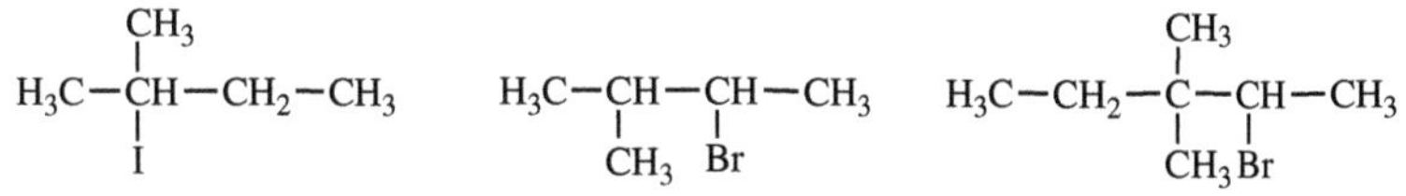

H₃C-CH(I)-CH(CH₃)-CH₂-CH₃   H₃C-CH(CH₃)-CH(Br)-CH₃   H₃C-CH₂-C(CH₃)(CH₃)-CH(Br)-CH₃

3.4 Write the products that occur when 1-chloro-1,2-diphenylethane reacts with NaOCH₃ under E2 conditions and indicate the major product.

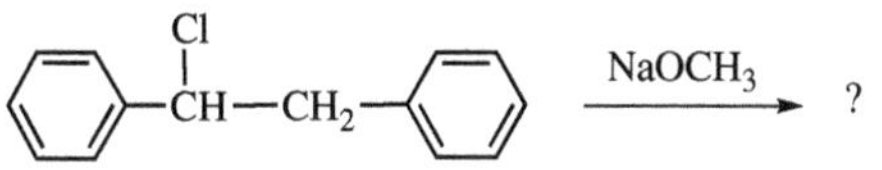

Ph-CH(Cl)-CH₂-Ph $\xrightarrow{NaOCH_3}$?

3.5 Write the mechanism for the following transformation.

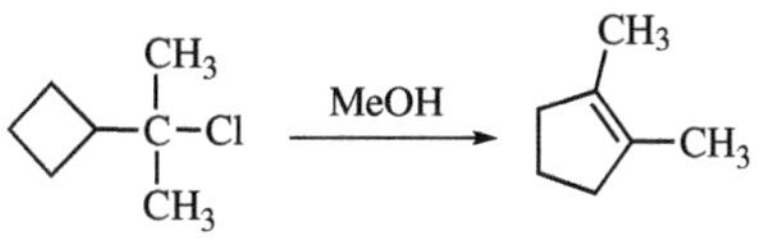

cyclobutyl-C(CH₃)₂-Cl $\xrightarrow{MeOH}$ 1,2-dimethylcyclopentene

3.6 *erythro-* and *threo-*2-Chloro-3-methylpentane are reacted with a high concentration of NaOMe. Give the structures and the configuration of the products.

3.7 Assuming the following reaction takes place under E1 conditions, write the products and formation mechanism.

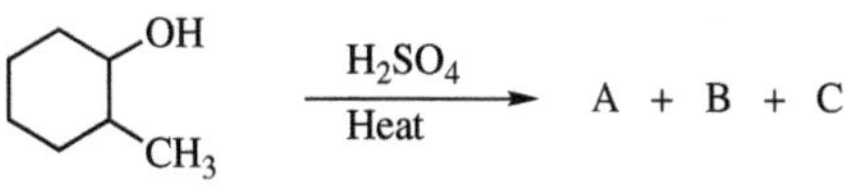

2-methylcyclohexanol (OH, CH₃) $\xrightarrow[\text{Heat}]{H_2SO_4}$ A + B + C

3.8 *erythro-* and *threo-*Tosylates react with NaOEt. Write the products and give the configurations.

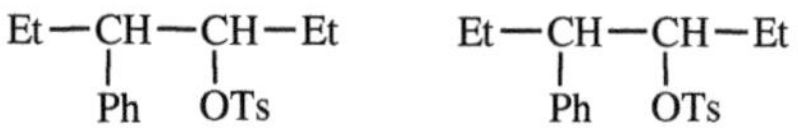

Et-CH(Ph)-CH(OTs)-Et   Et-CH(Ph)-CH(OTs)-Et
      erythro   *threo*

3.9 The *trans*-1-bromo-2-methylcyclohexane reacts with KOH to form a single elimination product. Write the structure of the product and explain the formation.

3 Elimination Reactions

3.10 Deuterobromomethylcyclohexane eliminates hydrogen bromide with NaOCH$_3$. Write the product and explain the formation.

3.11 Bromodimethylcyclohexane does not give an elimination product with sodium methoxide. Explain the reason.

3.12 Fill in the blanks in the equation below and discuss the formation of the product.

3.13 How can you synthesize the butadiene starting from pyrolidine? Discuss intermediates and the reaction mechanism.

3.14 The alkaloid obtained as an intermediate in the synthesis of Vittatine is submitted to further reaction. Give the products A and B. Write the structures and discuss the mechanism of formation.

3.15 Write down the structure and formation mechanism of product A given below.

3.16 In the reaction given below, write the products' structures and discuss the formation mechanism.

References

1 Volhardt, K.P.C. and Shore, N.E. (2007). *Organic Chemistry and Function*, 5the. New York: W. H. Freeman and Company.
2 Bruckner, R. (2002). *Advanced Organic Chemistry, Reaction Mechanisms*, 152. Harcourt Academic Press.
3 Brown, H.C. and Fletcher, R.S. (1950). *J. Am. Chem. Soc.* 72: 1223.
4 Feit, I.N. and Wright, D.G. (1975). *J. Chem. Soc. Chem. Commun.*: 776.
5 Hart, H., Craine, L.E., and Hart, D.J. (1999). *Organic Chemistry A Short Course*, 10the, 195. New York: Houghton Mifflin Company Boston.
6 Wolkoff, P. (1982). *J. Org. Chem.* 47: 1944.
7 Brown, H.C., Moritani, I., and Nakagawa, M. (1956). *J. Am. Chem. Soc.* 78: 2190.
8 Bruice, P.Y. (2007). *Organic Chemistry*, 5the, 396. Pearson.
9 Wiberg, K.B. (1955). *Chem. Rev.* 55: 713.
10 Cram, D.J., Greene, F.D., and Depuy, C.H. (1956). *J. Am. Chem. Soc.* 78: 790.
11 Lund, T., Bjoern, C., Hansen, H.S. et al. (1993). *Acta. Chem. Scan.* 47: 877.
12 Khurana, J.M., Gogia, A., and Bankhwal, R.K. (1997). *Syn. Commun.* 27: 1801.
13 Janout, V. and Cefelin, P. (1983). *Tetrahedron Lett.* 24: 3913.
14 Gill, G., Pawar, D.M., and Noe, E.A. (2005). *J. Org. Chem.* 70: 10726.
15 Leventis, N., Hanna, S.B., and Sotiriou-Leventis, C. (1997). *J. Chem. Educ.* 74: 813.
16 Sauers, R.R. (2000). Cyclohexane: Boat Form Revisited. *J. Chem. Educ.* 77: 33.
17 Lightner, D.A. and Gurst, J.E. (2000). *Organic Conformational Analysis and Stereochemistry from Circular Dichroism Spectroscopy*. Wiley.
18 Freitas, M.P., Rittner, R., Tormena, C.F., and Abraham, R.J. (2005). *Spectrochim. Acta A* 61: 1771.
19 Nasupuri, D. (1994). *Stereochemistry of Organic Compounds: Principles and Applications*, 2nde. New Age International.
20 Hughes, E.D., Ingold, C.K., and Rose, J.B. (1953). *J. Chem. Soc.*: 3839.
21 Hückel, W., Tappe, W., and Legutke, G. (1940). *Liebigs. Ann. Chem.* 543: 191.
22 Bieltrame, P., Biale, G., Lloyd, D.J. et al. (1972). *J. Am. Chem. Soc.* 94: 2240.
23 Sicher, J. (1972). *Angew. Chem. Int. Ed. Engl.* 11: 200.
24 Kwart, H., Takeshita, T., and Nyce, J.L. (1964). *J. Am. Chem. Soc.* 86: 2606.
25 Bartsch, R.A. and Lee, J.G. (1990). *J. Org. Chem.* 55: 5247.
26 Meng, Q., Du, B., and Thibblin, A. (1999). *J. Phys. Org. Chem.* 12: 116.
27 Mahai, I.M., Schug, K., and Miller, S.I. (1970). *J. Org. Chem.* 35: 1733.
28 Goering, H.L. and Espy, H.H. (1955). *J. Am. Chem. Soc.* 77: 5023.
29 Cadogan, J.I.G. and Leardini, R. (1979). *Chem. Commun.*: 783.
30 Cope, A.C. and Trumbull, E.R. (1960). *Org. React.* 11: 317.
31 DePuy, C.H. and King, R.W. (1960). *Chem. Rev.* 60: 431.
32 Feit, I.N. and Saunders, W.H. Jr. (1970). *J. Am. Chem. Soc.* 92: 5615.
33 Feit, I.N. and Saunders, W.H. Jr. (1967). *Chem. Commun.* 61.
34 Glacet, C., Hasiak, B., and Dumas, C.C.R. (1977). *Hebd. Seanc. Sci. Ser. C* 285: 93.
35 Curtin, D.Y., Stolow, R.D., and Maya, W. (1959). *J. Am. Chem. Soc.* 81: 3330.
36 Booth, H. and Gidley, G.C. (1964). *Tetrahedron Lett.* 5: 1449.
37 Willstaetter, R. and Waser, E. (1911). *Chem. Ber.* 44: 3423.
38 Willstaetter, R. and Heidelberger, M. (1913). *Chem. Ber.* 46: 517.
39 Cope, A.C. and Overberger, C.G. (1948). *J. Am. Chem. Soc.* 70: 1433.
40 Curtin, D.Y. and Kellom, D.B. (1953). *J. Am. Chem. Soc.* 75: 6011.
41 Lee, I., Cha, O.J., and Lee, B.-S. (1990). *J. Phys. Chem.* 94: 3926.
42 Bailey, W.J. and Cesare, F. (1978). *J. Org. Chem.* 43: 1421.
43 Arnold, R.T. and Smolinsky, G. (1960). *J. Org. Chem.* 25: 129.
44 (a) Alexander, E.R. and Mudrak, A. (1955). *J. Am. Chem Soc.* 72: 1810. (b) Carroll, F.A. (1998). *Perpectives on Structure and Mechanism in Organic Chemistry Brooks*, 695. Cole Publishing Company.
45 Meulemans, T.M., Stork, G.A., Macaev, F.Z. et al. (1999). *J. Org. Chem.* 64: 9178.
46 Cope, A.C. and Acton, E.M. (1958). *J. Am. Chem. Soc.* 80: 355.
47 Bredt, J., Thout, H., and Schnitz, J. (1924). *Liebigs Ann.* 437: 1.

48 Shea, K.J. (1980). *Tetrahedron* 36: 1683.
49 Wagner, P.M. (1989). *Chem. Rev.* 89: 1067.
50 Prelog, V. (1950). *J. Chem. Soc.*: 420.
51 Fawcett, F.S. (1950). *Chem. Rev.* 47: 219.
52 Wiseman, J.R. (1967). *J. Am. Chem. Soc.* 89: 5966.
53 Marshall, J.A. and Faubl, H. (1967). *J. Am. Chem. Soc.* 89: 5965.
54 Wiseman, J.R. and Pletcher, W.A. (1970). *J. Am. Chem. Soc.* 92: 956.
55 (a) Keese, R. and Krebs, E.P. (1971). *Angew. Chem. Int. Ed.* 10: 262. (b) Keese, R. (1975). *Angew. Chem. Int. Ed.* 14: 528.
56 Grootveld, H.H., Blomberg, C., and Bickelhaupt, F. (1973). *J. Chem. Soc. Chem. Commun.*: 542.
57 Maier, W.F. and Schleyer, P.v.R. (1981). *J. Am. Chem. Soc.* 103: 1891.
58 Gudipati, M.S., Radziszewski, J.G., Kaszynski, P., and Michl, J. (1993). *J. Org. Chem.* 58: 3668.
59 Chong, J.A. and Wiseman, J.R. (1972). *J. Am. Chem. Soc.* 94: 8627.
60 Dauben, W.G. and Robbins, J.D. (1975). *Tetrahedron Lett.* 16: 151.
61 Roach, P. and Warmuth, R. (2003). *Angew. Chem. Int. Ed.* 42: 3039.
62 Köbrich, G. (1973). *Angew. Chem. Int. Ed.* 12: 464.
63 Bestmann, H.J. and Schade, G. (1982). *Tetrahedron Lett.* 23: 3543.
64 Levine, S.G. and McDaniel, R.L. Jr. (1981). *J. Org. Chem.* 46: 2199.
65 Becker, K.B. and Grob, C.A. (1977). *The Chemistry of Double Bonded Functional Groups* (ed. S. Patai), 653. Chichester: Wiley.
66 (a) Grob, C.A. and Schiess, P.W. (1967). *Angew. Chem. Int. Ed.* 6: 1. (b) Kathrin Prantz, K. and Mulzer, J. (2010). *Chem. Rev.* 110: 3741.
67 Grob, C.A. and Baumann, W. (1955). *Helv. Chim. Acta* 38: 594.
68 Weyerstahl, P. and Marschall, H. (1991). *Comp. Org. Syn.* 6: 1044.
69 Mehta, G., Karmakar, S., and Chaddopadhyay, K. (2004). *Tetrahedron* 60: 5013.
70 Villagomez-Ibarra, R. and Joseph-Nathan, P. (1994). *Tetrahedron Lett.* 35: 4771.
71 Molander, G.A., Le Huerou, Y., and Brown, G.A. (2001). *J. Org. Chem.* 66: 4511.
72 Holton, R.A. and Kennedy, R.M. (1984). *Tetrahedron Lett.* 25: 4455.

4

Addition Reactions to Alkenes

An addition reaction is a chemical reaction in which two or more compounds come together to form a larger compound. However, only chemical compounds containing multiple bonds such as C=C (or C=O) double bonds or C≡C triple bonds can undergo an addition reaction. Two new σ-bonds are formed during this reaction, and a double or triple bond is usually broken to form the required single bonds. A double bond consists of a σ-bond and a π-bond. A σ-bond is more stable than a π-bond because in the former, the bonding takes place in the direction of the axis connecting the two atoms, while in the latter, the p orbitals interfere above and below the plane that forms a double bond. Therefore, this interference is not as effective as in σ-bonds. Thus, π-bonds are less stable, and the most common reactions of double bonds transform π-bonds into σ-bonds. A general addition reaction to a C=C double bond is given below.

$$\underset{R^1}{\overset{H}{\text{C}}}=\underset{R^2}{\overset{H}{\text{C}}} \xrightarrow{A-B} A-\underset{R^1}{\overset{H}{\underset{|}{\text{C}}}}-\underset{R^2}{\overset{H}{\underset{|}{\text{C}}}}-B \quad \text{Addition reaction}$$

We can classify addition reactions in two groups according to the polarity of the reactions:
Polar addition reactions, namely:
Electrophilic addition reactions
Nucleophilic addition reactions
Nonpolar addition reactions, namely:
Free radical addition reactions
Cycloadditions reactions

In this chapter, we will discuss electrophilic addition reactions to C=C double bonds and C≡C triple bonds.

The electron density between two carbon atoms of a double bond is high. Therefore, alkenes show nucleophilic properties and act as nucleophiles in addition reactions. Because of this feature of alkenes, compounds with an electron deficiency, i.e. electrophiles, can be readily attacked by double-bond electrons. In this section, we will focus mainly on the reaction of alkenes with electrophiles. The question arises here, *Can alkenes react with nucleophiles?* For alkenes to react with nucleophiles, it is essential to decrease the electron density in the double bond. *How is this achieved?* If one or two strong electron-attracting groups are attached to the double bond, the electron density decreases in the double bond, the nucleophilic double bond at this time is electrophilic, and the nucleophiles quickly attack the double bond.

4.1 Halogen Addition to Alkenes: Halogenation

In principle, most addition reactions to a double bond are exothermic. For example, the addition of hydrogen to a double bond is strongly exothermic, but the rate is very slow. A mixture of an alkene and hydrogen can remain for years without a visible reaction. A catalyst is needed to start the addition reaction. Some reagents add to the double bond without the aid of a catalyst. For example, halogens such as bromine and chlorine quickly add to double bonds, forming 1,2-dihalides [1–4]. We emphasized that alkenes react readily with electrophiles because of their nucleophilic properties. The general reaction of bromine with a double bond is shown below.

4 Addition Reactions to Alkenes

$$\underset{R^1}{\overset{H}{\underset{|}{C}}}=\underset{R^2}{\overset{H}{\underset{|}{C}}} \xrightarrow{Br-Br} Br-\underset{R^1}{\overset{H}{\underset{|}{C}}}-\underset{R^2}{\overset{H}{\underset{|}{C}}}-Br \quad \text{Electrophilic addition}$$

How does bromine or chlorine attack the nucleophilic double bond even though halogens do not have an electrophilic center? For this addition to take place, bromine must approach the double bond in the form of bromonium ion (Br^+). However, there is no polarization in the bromine molecule. Bromine atoms are equally electronegative and so there will be no charge separation. *Then, how can bromine be an electrophile?* Bromine is a very "polarizable" molecule. In other words, the electrons in the Br—Br bond are easily pushed to one end or the other. As the bromine approaches the alkene, the π-bond electrons will repel the electrons in the bromine–bromine bond, leaving the nearer bromine slightly positive and the further one slightly negative, causing polarization in the bromine molecule. Subsequently, the bromine–bromine bond is cleaved heterolytically, and an intermediate, a cyclic bromonium ion, is formed by the addition of the partially positively charged bromine atom to the double bond. Two new σ-bonds are formed between the bromine atom and the carbon atoms. Double-bond electrons form one of these σ-bonds while the other is formed by one of the nonbonding electron pairs on the bromine atom. The other bromine species, a bromide anion, attacks the bromonium ion from the opposite side to the bromine atom in the ring. The ring opens according to an S_N2 mechanism, giving a 1,2-addition product. This mode of addition is called anti-addition or *trans*-addition.

When bromine is added to ethylene, it is not possible to demonstrate that the second bromine atom is added from the opposite side, although this mechanism seems plausible. In particular, it does not explain the stereochemistry of the addition reaction. However, when the reaction is carried out on a cyclic system, it is possible to demonstrate that the addition of the second bromine takes place from the opposite side because the bond between the carbons to which the bromine atoms added cannot freely rotate.

When bromine reacts with cyclopentene, which has a symmetrical structure, a bromonium ion is formed in the first step. This ion is formed on one side of the ring; therefore, that side is blocked from any attack. The bromide anion attacks the bromonium ion in two ways, as seen in the equation below. As a result of these attacks, enantiomers occur. The fact that only *trans*-stereoisomers are formed rather than a mixture of *cis*- and *trans*-isomers indicates the formation of a bromonium ion and the anti-addition of bromine.

In bromination reactions, an alternate mechanism can be proposed as follows: the π-electrons of the double bond can attack the electropositive end of the bromine molecule; consequently, the carbon–carbon bond can break, leaving one carbon with a positive charge. The bromine–bromine bond also breaks to form a bromide anion. The bromide anion, which is free in the reaction medium, can attack the carbocation in the second step to complete the addition reaction. Before completion of the reaction, the carbon–carbon single bond can freely rotate so that the bromide can attack the carbocation from two different sides. There is no free rotation of the carbon–carbon bond in cyclic alkenes. If a carbocation was formed in these systems, *cis-* and *trans-*addition products would be expected. The high stereospecificity observed in bromination reactions indicates that a free carbocation does not occur as an intermediate. This topic will be discussed in detail in the next section.

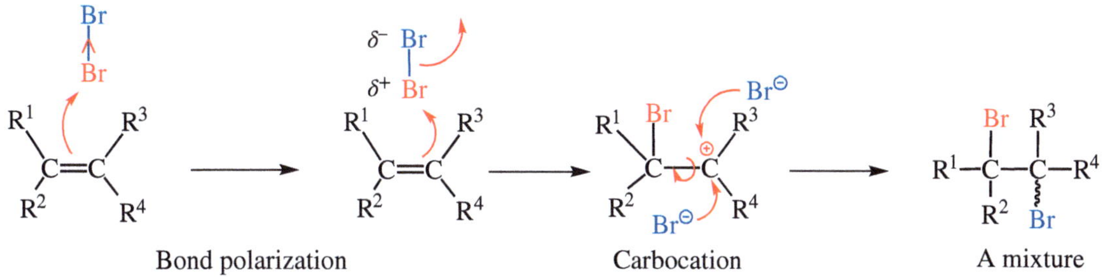

During the bromination of adamantylidenadamantane in carbon tetrachloride, Wynberg and coworkers isolated a yellow ionic compound [5]. Single-crystal X-ray analysis of the product, performed by Brown and coworkers [6], revealed it to be a stable three-membered cyclic bromonium ion, tribromide. Furthermore, the overall structure shows that there is severe crowding at the side opposite the molecule so that the bromide anion cannot attack from the back to complete the addition reaction. This unusual feature is the likely source of the stability of the ion pair. Furthermore, a bridged bromonium ion was observed on ionization of 2,3-dihalo-2,3-dimethylbutanes in antimony pentafluoride–sulfur dioxide solution at −60° directly by nuclear magnetic resonance (NMR) spectroscopy [7].

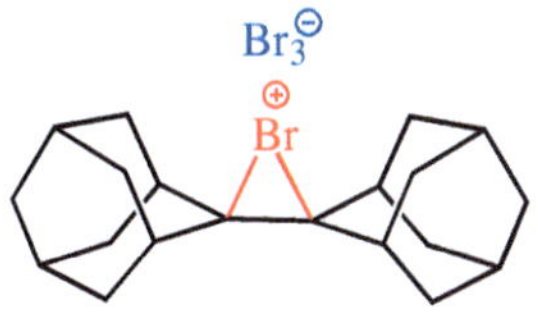

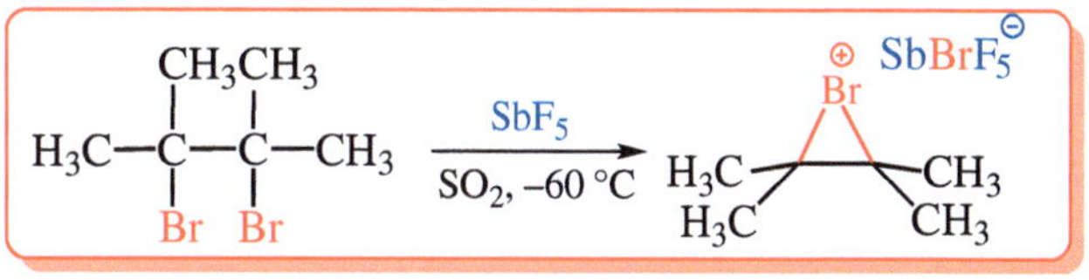

Another issue to be examined in bromination reactions is whether they take place in a single step or many steps. Simple experiments demonstrate that the bromination reaction proceeds in two steps. If it occurs in two steps and ionic intermediates are formed, it is possible to trap this intermediate in some way. For example, when bromine was added to the 1-hexene in methanol, the expected 1,2-dibromohexane, 1-bromo-2-methoxyhexane, and 2-bromo-1-methoxy-hexane were also formed [8]. The formation of these products reveals that the reaction takes place in two steps. Otherwise, only 1,2-dibromohexane would occur. The formation of 2-bromo-1-methoxyhexane indicates that a secondary carbocation is not formed as an intermediate; instead, a bromonium ion is formed.

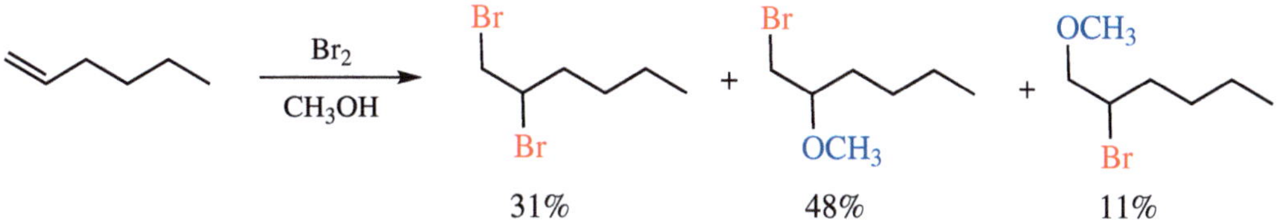

It is well established that the addition of bromine to double bonds is an electrophilic process. Electron-donating groups attached to the double bond increase the reactivity of an alkene, as evidenced by the increased rate of bromination. For example, in the bromination reaction of ethylene, propene, isobutene, 2-methylbutene, and tetramethylethylene, the reaction rate increases in parallel with the increased number of alkyl groups attached to the double bond. On the other hand, electron-withdrawing substituents attached to the double bond decrease the rate of addition reactions. Although the alkyl groups increase the reaction rate because of their electron-donor properties, they can negatively affect the reaction rates because of the steric effect they may create. Studies have revealed that electronic factors are more dominant than the steric ones. The relative rates of bromine addition to various alkenes are given below.

Br—CH=CH₂	HOOC—CH=CH₂	CH₂=CH₂	H₃C—CH=CH₂
0.03	0.03	**1.0**	2.0

CH₂=C(CH₃)CH₃	H₃C—CH=C(CH₃)CH₃	(H₃C)₂C=C(CH₃)₂	
5.0	10.0	13.0	Relative rates

When the bromination reaction is applied to a molecule containing multiple and different double bonds, electrophiles prefer to add to the double bond, where the electron density is higher. For example, bromination of 1,4,5,8-tetrahydronaphthalene with bromine gave mainly 9,10-dibromide in 60% yield as well as smaller amounts of other products formed by the addition of bromine to the other double bonds [9].

4.1.1 Stereochemistry of Halogen Addition

To demonstrate the stereospecifity of halogen addition to alkenes, it is necessary to work with alkenes of specific configurations. For example, let us examine the bromination reaction of *cis*-butene. The addition of bromine to *cis*-butene first produces an achiral bromonium ion. Subsequent attack of the bromide anion following two pathways provides a racemic mixture of the enantiomeric products. This reaction proceeds via a cyclic bromonium ion. The anti-addition of the bromide ion takes place at either of the two carbon atoms, and two enantiomers are formed in a 1 : 1 ratio. To compare the structures of these products, rotate the carbon–carbon bond about 180° to bring the bromine atoms from the anti-periplanar conformation into the syn-periplanar conformation. We see that this pair are mirror plane isomers of each other.

In contrast, the addition of bromine to *trans*-butene, as shown, produces a chiral intermediate, a bromonium ion possessing a mirror plane isomer. Regardless of whether the reaction takes pathway a or b, the same product, *meso*-2,3-dibromobutane, is always formed.

Similarly, while bromination of dimethyl fumarate gives a *meso* compound, a DL-pair is formed from the reaction of the isomeric dimethyl maleate. In both cases, the reaction proceeds via a cyclic bromonium ion and *trans*-addition takes place.

As a result of these reactions, it is well established that the bromine stereospecifically adds to the double bonds, forming anti-addition products. Stereospecific addition is explained by the formation of a cyclic bromonium ion. Once the bromonium ion is created, the configuration of the substituents cannot change. The three-membered ring formed by the bromonium ion prevents rotation about the carbon–carbon bond.

Like acyclic compounds, cyclic systems also undergo a stereospecific bromine addition reaction to give *trans*-1,2-addition products. For example, cyclohexene reacts with bromine in the same way and under the same conditions as any other alkene and forms *trans*-1,2-dibromocyclohexane. Again, the bromine is polarized by the approaching π-bond in the cyclohexene. The positively polarized bromine atom forms the bromonium ion, which is opened stereospecifically by the attack of the bromide anion to give the *DL*-pair.

There is a positive charge (+) on the bromine atom in the complex that is formed in the first step. The question arises here of whether we can write the following resonance structures for the bromonium ion. There is another question that should also be addressed. The double-bond electrons can attack the electropositive end of the bromine molecule forming carbocation **A**. The three lone pairs on the bromine atom are close in space to the empty p orbital of the carbon atom carrying the positive charge, and one of them can attack the p orbital to form the bromonium ion **B**. Intermediates **A** and **C** may also occur if one of the carbon atoms is attached to the electron-donating groups to stabilize the positive (+) charge. When a free carbocation, such as **A** and **C**, occurs, the carbon–carbon bond can freely rotate. Then, the stereochemistry of the products will change. There is another possibility: *is there a partial bond between the carbocation and bromine atom where the positive charge is mainly localized on the carbon atom?*

To determine whether a free carbocation occurs during the bromination reaction, *t*-butylethylene was reacted with bromine in methanol. If the bromonium ion is formed during the bromination reaction, it is expected that the secondary carbocation will be formed. However, this carbocation is prone to undergo rearrangement. A methyl group from the adjacent carbon atom would shift to create a more stable tertiary carbocation. The bromide anion would attack the tertiary carbon atom to yield 1,3-dibromide. The absence of 1,3-dibromide under the products reveals that a free bromine-substituted carbocation does not occur during the bromination reaction [8, 10].

4.2 Addition of Hydrogen Halides to Alkenes: Markovnikov's Rule

The addition of hydrogen halides to double bonds is an essential reaction in organic chemistry, both mechanistically and synthetically. The hydrogen halides are polarized compounds because of the difference in electronegativities of halogens and hydrogen. Therefore, all alkenes undergo addition reactions with the hydrogen halides. A hydrogen atom adds to one of the carbon atoms originally in the double bond and a halogen atom to the other. In the first step, a strong electrophile, a proton, attracts the loosely bound electrons from the π-bond of the alkene, and the electrophile forms a σ-bond to one of the carbon atoms of the double bond. Therefore, this reaction is called an electrophilic addition. This reaction is the reverse process of the deprotonation step in the E1 reaction. Then, a carbocation will be formed. In the second step, the carbocation, a strong electrophile, reacts with a nucleophile to create a second σ-bond. This last step can occur if a good nucleophile is present. The total reaction is a 1,2-addition of a hydrogen halide to a double bond.

Only one product is possible from the addition of these hydrogen halides (H-X) to symmetrical alkenes. However, if the double-bond carbon atoms are not structurally equivalent, H-X conceivably may add in two different ways. In 1869, Vladimir Vasilyevich Markovnikov (1837–1904), a Russian organic chemist at the University of Kazan, formulated in his doctoral thesis the important rule that was named after him. According to this rule: *"when an unsymmetrical alkene reacts with a hydrogen halide, the hydrogen adds to the carbon that has the higher number of hydrogen substituents, and the halogen to the carbon having the fewer number of hydrogen substituents"*. Markovnikov's rule predicted the regioselectivity of electrophilic addition reactions of alkenes and the occurrence of the most stable carbocation at the first stage [11–13].

In the bromination reaction, we have seen that a bromonium ion is formed in the first step in which the bromine atom is bonded to two carbon atoms by σ-bonds. An intermediate like that cannot be formed by the addition of a proton to the double bond. Therefore, the proton binds to one of the carbon atoms immediately, and a carbocation is formed. Then, the nucleophile attacks the carbocation to complete the addition reaction. Let us examine the addition of hydrogen chloride to propylene, the simplest asymmetric alkene.

Because the proton will add to the double bond first, it can add to two different carbon atoms. If the proton binds to the carbon atom C2, a primary carbocation occurs, which is an unstable intermediate. However, if the proton binds to the C1 carbon atom, a secondary carbocation occurs. The secondary carbocation is far more stable, and therefore, its formation is preferred over the primary carbocation. Generation of the carbocation is the rate-determining step; once it occurs, the chloride attacks the C2 carbon atom. In this reaction, a single product, 2-chloropropane, is formed. At the stage of the formation of the carbocation, the stereoselectivity will be lost because the carbon–carbon bond can rotate freely around its axis.

Similarly, the reaction of hydrogen iodide with 2-methyl-2-butene gives 2-iodo-2-methylbutane as the major product. In the formation of this compound, the proton binds to the most hydrogen-containing carbon atom, thereby forming the most stable carbocation intermediate. The product formed by the attack of iodide to the tertiary carbocation is 2-iodo-2-methylbutane, which is called the *Markovnikov product*. 2-Iodo-3-methylbutane, which occurs as a side product, is formed in a small amount. This product is called *the anti-Markovnikov product*.

4 Addition Reactions to Alkenes

$$\underset{\text{2-Methyl-2-butene}}{\underset{H_3C}{\overset{H_3C}{>}}C=CH-CH_3} \xrightarrow{H-I} \underset{\text{Markovnikov product}}{H_3C-\underset{I}{\overset{CH_3}{\underset{|}{C}}}-\underset{H}{\overset{H}{\underset{|}{C}}}-CH_3} + \underset{\textit{anti}\text{-Markovnikov product}}{H_3C-\underset{H}{\overset{CH_3}{\underset{|}{C}}}-\underset{H}{\overset{I}{\underset{|}{C}}}-CH_3}$$

In the absence of a strong nucleophile, the carbocation formed by the addition of a proton to the double bond can undergo rearrangements. For example, the counterion of trifluoroacetic acid is much less nucleophilic. The reaction of 3-methylbut-1-ene with trifluoroacetic acid, CF_3COOH, forms only about 43% Markovnikov product. The major product is formed by the attack of the nucleophile on the carbocation formed by the rearrangement of the initially formed carbocation. The driving force of this rearrangement is a conversion of the secondary carbocation into the more stable tertiary carbocation before the nucleophile attacks [14].

The trifluoromethyl group is a strong electron-attracting group, and it destabilizes an adjacent carbocation. If the trifluoromethyl group is attached to a double bond, the electron density at the double bond will decrease. Therefore, such compounds undergo electrophilic addition reactions under harsh conditions. For example, reaction of 3,3,3-trifluoroprop-1-ene with HBr and $AlBr_3$ at 100 °C produces 3-bromo-1,1,1-trifluoropropane as the sole product [15]. When we examine this reaction in terms of Markovnikov's rule, it becomes clear that the product is an *anti*-Markovnikov product. Because the proton is bonded to the *carbon having the fewer hydrogen atoms, here,* the first question that may come to mind at the first stage is whether the carbocation formed as an intermediate is a thermodynamically less stable carbocation.

Let us examine both carbocations that may occur during this reaction. If the addition occurs through a carbocation, then the regiochemistry might be attributed to the greater stability of the cation formed as an intermediate. The proton can bind to the double bond in two different ways. A secondary carbocation occurs when the proton binds to the C1 carbon atom with the most protons following Markovnikov's rule. If the proton adds to the C2 carbon atom, a primary carbocation will be formed. The direction of the reaction will depend on the formation of a carbocation intermediate with greater stability. When the strong electron-attracting trifluoromethyl group is directly attached to the carbocation center, it makes this cation unstable. The primary carbocation will be, in this case, more stable because the trifluoromethyl group is further removed from the cationic center. As a result, the primary carbocation is more stable than the secondary carbocation. Therefore, 3-bromo-1,1,1-trifluoropropane is formed as the sole product. Calling this product an *anti*-Markovnikov product can sometimes mislead the reader. The most crucial point to be considered in the addition reactions is the creation of the most stable intermediate in the first step. Therefore, Markovnikov's rule has been extended: in an electrophilic addition reaction, the electrophiles add in such a way that the most stable carbocation is formed as the intermediate.

Vinyl-trimethyl ammonium salts also act as trifluoropropene in addition reactions, forming an *anti*-Markovnikov product. If addition took place following Markovnikov's rule, the positive (+) charge on the carbon atom and the positive (+) charge on the nitrogen atom would be close to each other, and the intermediate would be unstable. However, the primary carbocation required for the formation of the *anti*-Markovnikov product is more stable than the secondary carbocation is.

Electrophilic addition reactions in cyclic systems generally take place according to Markovnikov's rule. A tertiary carbocation is preferably formed in the reaction of methylcyclohexane with HBr and the bromide ion binds to the tertiary carbon atom. The *anti*-Markovnikov product occurs as a side product. In bicyclic systems, the major product is derived again from the most stable intermediate, a tertiary carbocation.

4.2.1 Anti-Markovnikov Addition of Hydrogen Halides to Alkenes

As discussed in the previous section, alkenes usually react with hydrogen halides to give products of Markovnikov addition. However, the ratio of the Markovnikov and the *anti*-Markovnikov products changes when the same reaction is performed in the presence of peroxides and heat or light. For a long time, there was controversy surrounding this subject because the results concerning the structures of the products formed in the addition reactions varied from laboratory to laboratory. In 1933, Kharasch and Mayo found that some additions of HBr to alkenes gave products that were opposite those expected from Markovnikov's rule [16]. The structure of the product formed is completely different from that of the product formed in the ionic environment. For example, the reaction of propene with HBr produces the more substituted alkyl halide as the major product; however, in the presence of peroxides, HBr forms the less substituted alkyl bromide, an *anti*-Markovnikov product.

The formation of different addition products in the presence of peroxides indicates that peroxides play an essential role in this addition. Moreover, the rate of the reactions accompanied by peroxide is much faster than the rate of the alternate electrophilic addition reaction.

Peroxides (RO–OR) are compounds that contain an oxygen–oxygen bond and they are unstable molecules. The energy of the oxygen–oxygen bond (approximately 33 kcal/mol (138 kJ/mol)) is one of the lowest bond energies known in organic chemistry. When peroxides are heated or exposed to ultraviolet (UV) light, they undergo homolytic cleavage to generate alkoxy radicals. Radicals are electrophilic species and are unstable because they do not have an octet configuration. This alkoxy radical picks up the hydrogen from HBr to create a bromine radical, which adds to the double bond to generate an alkyl radical on the more substituted carbon atom. If the double bond attacked by the bromine radical is not symmetrical, the bromine radical has the possibility of attacking both carbon atoms of the double bond, as we saw in polar addition. The bromine radical adds to the carbon atom that is bonded to the higher number of hydrogens and a secondary alkyl radical is formed. The bromine radical will prefer to generate a more stable intermediate. If the bromine radical adds to the C2 carbon atom in propene, a primary alkyl radical is formed. As the radicals are electron-deficient species, the stability order is similar to that of carbocations. Then, the resulting alkyl radical abstracts a hydrogen atom from HBr to form the *anti*-Markovnikov product and bromine radical. The regenerated bromine radical reacts with another molecule of the alkene, continuing the chain reaction. Therefore, a catalytic amount of peroxide is needed for this reaction.

In the *anti*-Markovnikov product, the hydrogen atom is bonded *to the carbon having the fewer hydrogens*. The resulting product is called *the anti*-Markovnikov product. The opposite addition was observed in the ionic reaction. This may cause one to wonder if the *anti*-Markovnikov product created here was formed over a more unstable intermediate. The answer to this question is no. The most stable intermediate products (carbocation or carbon radical) are always formed in either radical additions or ionic additions. The difference between these two reactions is that in ionic reactions, the electrophile is *a proton*, while in radical reactions, it is *a bromine radical*. In ionic addition, first the proton adds to the double bond, followed by the addition of the bromide anion. In radical addition, first the bromine radical adds, followed by the addition of hydrogen. The most stable intermediates are always formed in both reactions. Because the order of the addition of hydrogen and bromine varies, one produces the Markovnikov product and the other provides the *anti*-Markovnikov product. However, students ignore this crucial point. They assume that the *anti*-Markovnikov addition results in the formation of the least stable intermediate. Therefore, it would be better if this formulation of *anti*-Markovnikov addition for radical addition were changed. In summary, both reactions, the addition of HBr to an alkene with and without peroxides, follow the extended statement of Markovnikov's rule: in both reactions, the electrophile adds to the double bond and generates the most stable intermediate, either a carbocation or a free radical.

A question such as *Can a normal Markovnikov product be produced during the addition of HBr in the presence of peroxides?* may arise here. Some parts of HBr in the reaction medium can undergo ionic addition and form the Markovnikov product. Kharasch [16] defined the *peroxide effect*, explaining how an *anti*-Markovnikov orientation could be achieved via free radical addition.

Only HBr can add to double bonds in the presence of peroxides. The peroxide effect is not observed with other hydrogen halides such as HF, HCl, and HI. The addition of HBr to a double bond proceeds via a chain reaction. In the first step, a bromine radical adds to the double bond and creates a new alkyl radical. Then, the resulting alkyl radical abstracts a hydrogen from HBr to complete the addition reaction. Both propagation steps in the case of HBr addition are exothermic. The addition of HI and HF in the presence of peroxides has never been observed. However, the addition of HCl in the presence of peroxides has been observed rarely because the second step, the reaction of an alkyl radical with HCl, is strongly endothermic. Similarly, the highly endothermic reaction of HF (the second step) with the double bond prevents the addition of HF. On the other hand, the reaction of iodine radical with a double bond is strongly endothermic; therefore, the peroxide effect is not observed (Table 4.1) [17, 18].

Table 4.1 Enthalpies in the propagation steps in the radical addition of HX to ethylene [17].

HX	First step (kcal/mol)	Second step (kcal/mol)
HF	−46	36
HBr	−3	−11
HCl	−17	4
HI	12	−27

Source: Data from Carroll [17].

Anti-Markovnikov additions are observed in cyclic systems as well as acyclic systems. For example, methylenecyclopentane forms an *anti*-Markovnikov product by reacting with HBr in the presence of peroxides.

4.3 Addition of Water and Alcohols to Alkenes

4.3.1 Hydration

In the elimination section, we saw that alkenes are formed by H_2O elimination from alcohols. Here, we will examine the reverse reaction. Electrophilic hydration is the reverse dehydration of alcohols. Generally, an alkene does not react with water because water does not have an electrophilic part. The O—H bond in water is too strong to allow the hydrogen to act as an electrophile.

As the water cannot protonate a double bond, the presence of a strong acidic catalyst is necessary. The acid catalyst will provide an electrophile (H^+). Acid-catalyzed addition of water to alkenes proceeds in three steps. The mechanism is the same as the reaction mechanism of the addition of hydrogen halides to alkenes. All steps of acid-catalyzed hydration of alkenes are reversible; the acid acts only as a catalyst and it is not consumed in the overall reaction. The nucleophilic double bond attacks the hydronium ion, formed in acid solution, to form a carbocation. Once the carbocation is formed, nucleophilic attack by water gives a protonated alcohol, which loses a proton to form the neutral alcohol. The first step, the rate-determining step, is slow, while the subsequent steps are fast.

Cyclic alkenes also undergo hydration reactions. Acid-catalyzed hydration of cyclohexene produces one alcohol because of the symmetrical structure of cyclohexene. There is no difference regarding to which carbon atom the hydroxyl group is bonded. The addition of water to alkenes is generally carried out in a dilute acidic solution because the use of a strong and concentrated acid would cause E1 elimination of the alcohol.

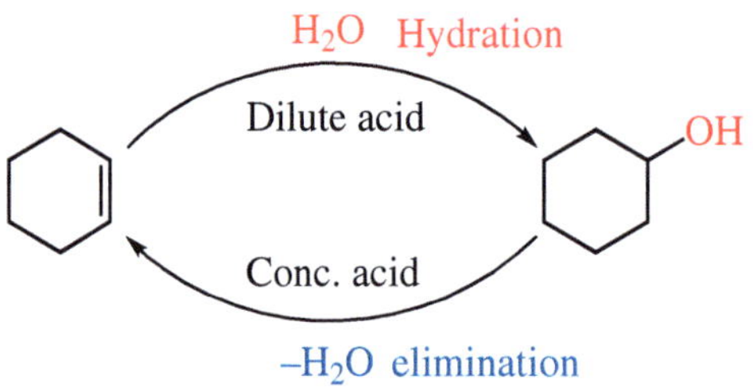

If the alkene has an asymmetrical structure, there are two kinds of addition of water. For example, if 1-methylcyclohexene is used as the reactant, there are two possibilities for addition of the hydroxyl group. The proton will prefer to add to the less substituted (with most protons) double bond to create the most stable carbocation as the intermediate. Then, water adds to the carbocation to produce alcohol with the hydroxyl group attached to the more substituted carbon atom to form the protonated alcohol. Removal of a proton gives the alcohol. It should be remembered that the more stable carbocation determines the major product of the reaction. The addition is regioselective and follows Markovnikov's rule, such as the addition of hydrogen halides to alkenes.

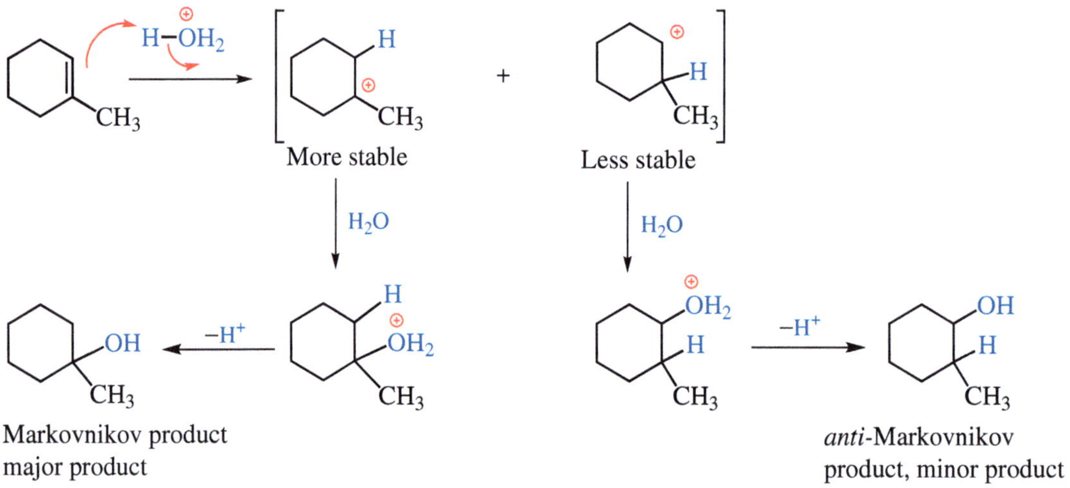

Hydrogen halides are generally used as a catalyst in hydration reactions. Alkenes react with cold and concentrated sulfuric acid to give alkyl hydrogensulfates. Alkyl hydrogensulfates are difficult to isolate because the sulfate anion is an excellent leaving group. When they are heated or hydrolyzed with water, they turn into alcohols. If the alkene has an asymmetrical structure, the regioselectivity is governed by Markovnikov's rule as shown below.

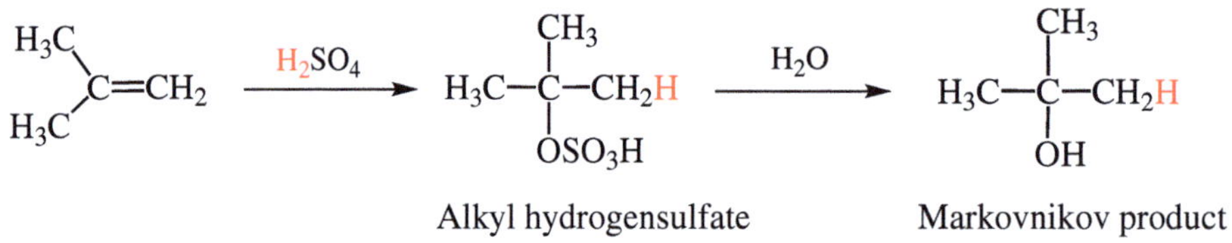

4.3.2 Alkoxylation

The acid-catalyzed alkoxylation reaction is an analogous reaction to hydration. As a result of the reaction, ethers are formed. The reaction mechanism is similar to the acid-catalyzed addition of water to alkenes, except that the nucleophile is ROH instead of H_2O. The reaction proceeds according to Markovnikov's rule. The proton adds to the less substituted carbon atom, while the alkoxy group adds to the more substituted carbon atom. There is no stereospecificity associated with this reaction because the intermediate formed is a carbocation, which can be attacked from both faces of the planar carbocation.

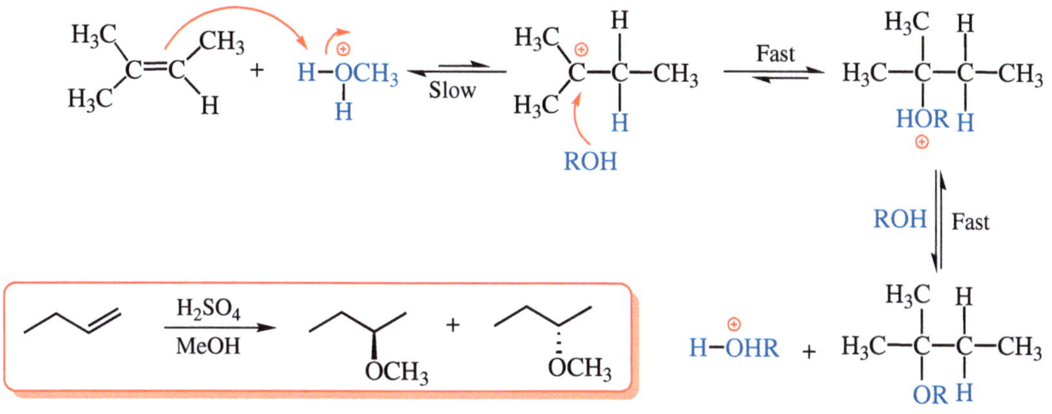

4.3.3 Formation of Halohydrins

When alkenes react with halogens in non-nucleophilic solvents such as carbon tetrachloride and chloroform, they produce vicinal 1,2-dihalides. If the reaction is performed in a nucleophilic solvent such as water, the solvent becomes the nucleophile in the second step and reacts with the halonium ion to form halohydrins.

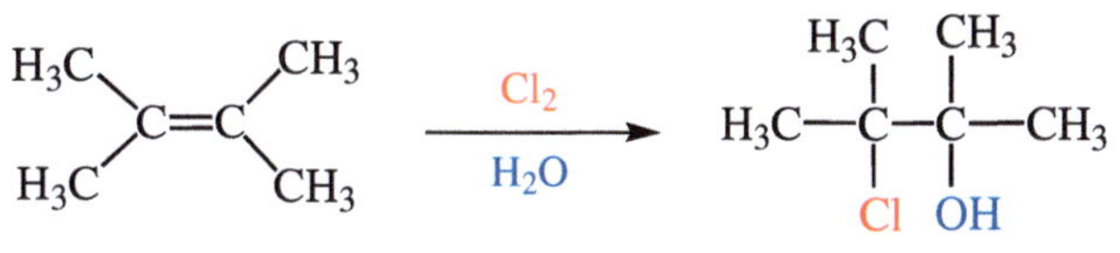

Chlorohydrin

As the halogen approaches the alkene, the π-bond electrons will repel the electrons in the halogen–halogen bond, leaving the nearer halogen slightly positive as shown below. First, a cyclic halonium ion is formed because the halogen is the only electrophile. The halogen accepts the π-bond electrons from the double bond and forms a bromonium (or chloronium) ion. The unstable cyclic halonium ion rapidly reacts with the nucleophile, water, because it is present in a much higher percentage than the halide anion. An *anti*-stereoselectivity is observed during the addition because the solvent approaches the halonium ion from the back. For example, the addition of bromine to cyclopentene in the presence of water gives *trans*-2-bromocyclopentanol.

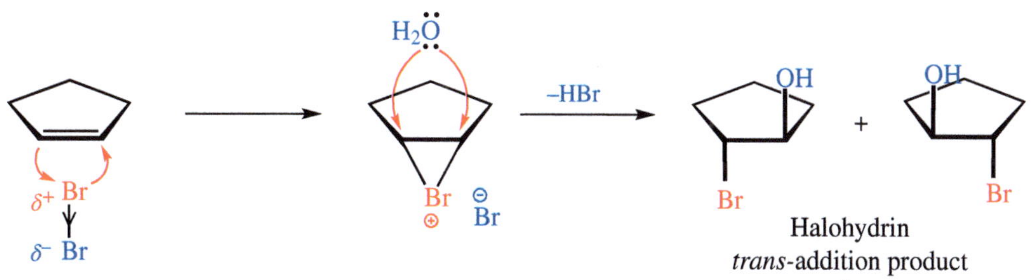

Halohydrin
trans-addition product
enantiomers

If the halonium ion formed has a symmetrical structure, the nucleophile attacks both carbon atoms with the same probability. However, if the halonium ion is not symmetrical, a regioselectivity is observed by the same mechanism we used to rationalize Markovnikov's rule. When propene reacts with bromine and water, the major product has the electrophile bonded to the less substituted carbon atom of the double bond, while the hydroxyl group is bonded to the more substituted carbon atom.

MNDO (the modified neglect of differential overlap) calculation [19, 20] shows that the bromonium ion derived from ethene has a symmetrical structure, and the distances between the carbon atoms and the bromine atom are equal. When the same calculations are made with 2-methylpropene, the results are different. If the number of alkyl groups bonded to one of the carbon atoms is higher, it is seen that the structure of the bromonium ion is quite asymmetrical and the distance between bromine and the carbon atom (C2) is extended. The two carbon atoms have partial positive charges, with a larger charge on the more substituted one [21, 22]. Therefore, the nucleophile (water) attacks this more substituted, more electrophilic carbon atom. The result is both *anti*-stereochemistry and Markovnikov orientation.

4.4 Hydration Alkenes: Oxymercuration and Demercuration

In the previous section, so far, we have seen the acid-catalyzed addition of H_2O to double bonds. This reaction has some disadvantages. Because the reaction is carried out in an acidic medium, an unwanted by-product may occur if there are acid-sensitive functional groups in the molecule. Furthermore, because a carbocation occurs as an intermediate, intramolecular rearrangements may also occur. The method that eliminates both negativities is the oxymercuration of the double bonds and the subsequent reduction reaction. Alkene hydration using the oxymercuration–demercuration [23] reaction pathway gives the Markovnikov product without carbocation rearrangement.

When 3,3-dimethyl-1-butene is subjected to a hydration reaction in an acidic medium, the rearranged addition product is formed as the sole compound because the secondary carbocation initially formed undergoes a methyl shift to create a more stable tertiary carbocation. The water molecule attacks the tertiary carbon atom to yield 1,3-addition product.

Acid-catalyzed hydration reaction

However, the application of oxymercuration–demercuration to 3,3-dimethyl-but-1-ene gives the product 3,3-dimethylbutan-2-ol in excellent yield. In contrast to the acid-catalyzed reaction, the reaction gives the Markovnikov product and without any rearrangement.

4.4 Hydration Alkenes: Oxymercuration and Demercuration

$$H_3C-\underset{\underset{CH_3}{|}}{\overset{\overset{CH_3}{|}}{C}}-CH=CH_2 \xrightarrow[H_2O]{Hg(OAc)_2} H_3C-\underset{\underset{CH_3}{|}}{\overset{\overset{CH_3}{|}}{C}}-\underset{}{\overset{\overset{OH}{|}}{CH}}-\underset{}{\overset{\overset{HgOAc}{|}}{CH_2}} \xrightarrow{NaBH_4} H_3C-\underset{\underset{CH_3}{|}}{\overset{\overset{CH_3}{|}}{C}}-\underset{}{\overset{\overset{OH}{|}}{CH}}-\underset{}{\overset{\overset{H}{|}}{CH_2}}$$

Hydration via oxymercuration and demercuration — Nonrearranged product

The oxymercuration reaction proceeds in two steps with mercuric acetate (Hg(OAc)$_2$), which is in equilibrium with the mercuric acetate cation. This electrophile, HgOAc$^+$ cation, adds to the double bond to form a three-membered ring called a *mercurinium ion*. The mercurinium ion also has a substantial amount of positive charge on both of its carbon atoms. A nucleophilic attack by water can open the resulting three-membered ring of the mercurinium ion. Attack by water occurs on this more electrophilic carbon, giving a product according to Markovnikov's rule. After completion of the addition, demercuration can be performed by reduction using sodium borohydride (NaBH$_4$) [24].

[Mechanism scheme showing propene + Hg(OAc)$_2$ → mercurinium ion → attack by H$_2$O → oxonium intermediate → loss of H$^+$ → β-hydroxy mercurial → NaBH$_4$ → 2-propanol]

In symmetrical olefins, the structure of the mercurinium ion is thought to be symmetrical, as in the bromonium ion [25]. Experimental evidence for the symmetrical structure of the mercurinium formed by the reaction of ethylene and Hg(OAc)$_2$ was obtained by NMR spectroscopy [26, 27]. If the double bond has an asymmetrical structure, the H$_2$O molecule attacks the carbon atom that carries the most alkyl groups. Experimental results [28] and theoretical calculations [29] strongly support an unsymmetrical structure of the mercurinium ion, and the positive (+) charge is concentrated on the carbon atom bearing more alkyl groups.

After completing the oxymercuration reaction, NaBH$_4$ is added to the reaction medium without isolating the product. NaBH$_4$ substitutes the mercuric acetate unit with a hydrogen atom. When the oxymercuration reaction is performed in an alcohol-containing medium, the alcohol molecules will attack the mercurinium ion. The resulting product will be an ether. This methodology can be successfully applied to the conversion of alkenes to ethers. By adding other nucleophiles to the medium, of course, it is possible to introduce different functional groups into the molecule.

[Scheme: Mercurinium ion + CH$_3$OH → H$_3$C–CH(OCH$_3$)–CH$_2$–Hg(OAc) → NaBH$_4$ → H$_3$C–CH(OCH$_3$)–CH$_2$–H (Ether)]

Although the oxymercuration step is a stereoselective reaction, stereoselectivity is generally not observed in the reduction step. For example, while stereoselectivity was observed in the products obtained in the oxymercuration of *cis*- and *trans*-2-butene, this selectivity disappeared during the reduction process, and *erythro*- and *threo*-products were formed in a 1 : 1 ratio [30].

The oxymercuration reaction has been successfully applied to the synthesis of *cis*-2-azabicyclo[3.3.0]octane derivative. Intramolecular cyclization reactions occur, especially if there is a double bond in the molecule, as well as a free amino group. Mercuric acetate adds to the double bond to create a cyclic mercurinium ion. Intramolecular attack by the amino group followed by demercuration with NaBH₄ leads to the formation of 2-azabicyclo[3.3.0]octane derivative [31].

Stereoselective one-pot double intramolecular oxymercuration has been applied to the synthesis of the mono-hydroxylated bis-tetrahydrofuran ring. This structure appears as a substructure in the natural products asimitrin and salzmanolin. Demercuration was carried out under a stream of oxygen and in the presence of NaBH₄ to give the diol in 81% yield [32].

The advantages of the oxymercuration reaction to the acid-catalyzed hydrations are that the reactions continue according to Markovnikov's rule, there is no rearrangement, and the reaction can be completed in a short time at room temperature. The only drawback is that the mercury compounds used are toxic.

4.5 Hydroboration of Alkenes: *anti*-Markovnikov Hydration

"So far, we have seen two different hydration reactions of alkenes with Markovnikov orientation: acid-catalyzed H₂O addition and oxymercuration–demercuration. *Can H₂O be added to alkenes following the anti-Markovnikov rule?* In this section, we will present an additional method that complements the other two methods: hydroboration of double bonds with *anti*-Markovnikov orientation. Such a method was not known until the discovery by Herbert Charles Brown, an American chemist (1912–2004), of the hydroboration oxidation reaction [33, 34]. H. C. Brown received the Nobel Prize in Chemistry in 1979 jointly with George Wittig "for their development of the use of boron- and phosphorus-containing compounds, respectively." Hydroboration is a three-step reaction that involves the conversion of an alkene into alcohol by *anti*-Markovnikov hydration. Brown discovered that diborane (B_2H_6) adds to alkenes with *the anti*-Markovnikov orientation to give alkyl boranes, which can be oxidized to produce to alkyl borate. The final step is the hydrolysis of borate to give the *anti*-Markovnikov alcohol [35–38].

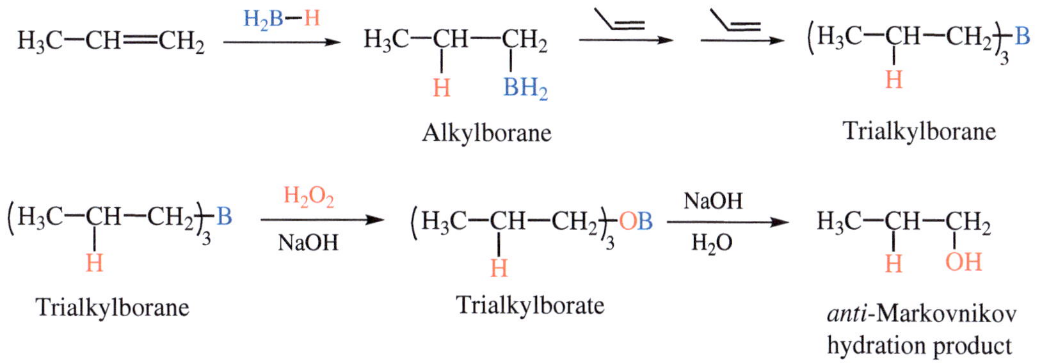

$$H_3C-CH=CH_2 \xrightarrow{H_2B-H} \underset{\underset{H}{|}\quad\underset{BH_2}{|}}{H_3C-CH-CH_2} \longrightarrow \longrightarrow (\underset{\underset{H}{|}}{H_3C-CH-CH_2})_3 B$$

Alkylborane   Trialkylborane

$$(\underset{\underset{H}{|}}{H_3C-CH-CH_2})_3 B \xrightarrow[NaOH]{H_2O_2} (\underset{\underset{H}{|}}{H_3C-CH-CH_2})_3 OB \xrightarrow[H_2O]{NaOH} \underset{\underset{H}{|}\quad\underset{OH}{|}}{H_3C-CH-CH_2}$$

Trialkylborane   Trialkylborate   *anti*-Markovnikov hydration product

Monoborane (BH_3) is a compound that can be easily synthesized under laboratory conditions by the reduction of bortrifluoride (BF_3) with $LiAlH_4$. Because boron hydride is a compound with an electron deficiency in its outer shell, it exists in a dimer form. X-ray diffraction studies show that there are two hydrogen atoms bonded to both boron atoms at the same time. This type of bonding is also called *two-electron, three-center bonding*. However, when BH_3 is synthesized in the presence of compounds such as tetrahydrofuran (THF) or Me_2S, which exhibit Lewis base properties, no dimer structure is formed. Because these compounds are good electron-donating compounds, they form a coordination compound with the boron atom. This complex is in equilibrium with a small amount of free BH_3 and is easier to use. Diborane is a toxic, flammable, and explosive gas.

$$4\ BF_3 + 3\ LiAlH_4 \longrightarrow 2\ B_2H_6 + 3\ LiAlF_4$$

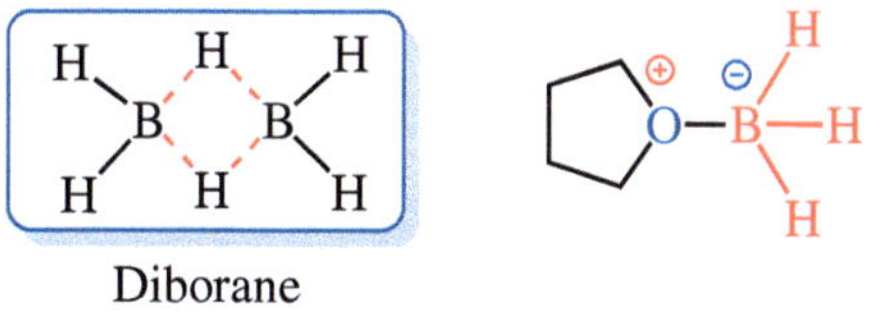

Diborane

Only monomeric boranes react with double bonds. The boron atom is electrophilic because of the empty p orbital and it undergoes an interaction with the π-electrons. First, a complex similar to the bromonium ion is formed between the borane and the double-bond electrons, and the electron density shifts from the alkene to the boron. The carbon atom opposite to the boron becomes slightly electron deficient. This partial charge is more stable on the more substituted carbon atom. Therefore, the boron prefers to bond to the less substituted carbon atom. One of the hydrogen atoms is transferred through a four-center transition state to the electropositive carbon atom. In BH_3, the atom bearing the partial negative charge is hydrogen because hydrogen is more electronegative than boron. The boron atom generally bonds to the less substituted carbon atom, while the proton bonds to the more substituted carbon atom. Furthermore, steric hindrance forces boron to add to the less hindered, less substituted carbon atom. As the hydroboration takes place on the same face of the double bond, this leads to a *cis*-addition (Figure 4.1).

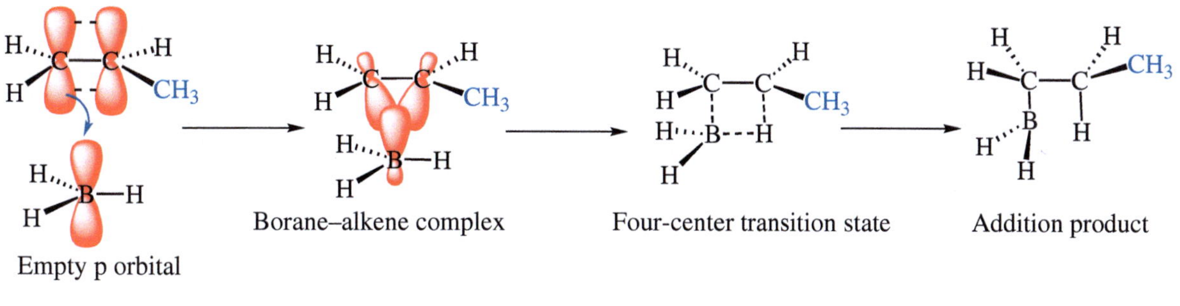

Empty p orbital   Borane–alkene complex   Four-center transition state   Addition product

Figure 4.1 Schematic presentation of addition of BH_3 to a double bond.

The boron atom in alkyl borane formed by the addition of BH_3 to the alkene is again an electron-deficient species. Therefore, it can undergo a further reaction with additional alkenes so that all three hydrogens in BH_3 can be replaced by the alkyl groups to form a trialkyl borane (BR_3). However, depending on the presence of voluminous groups attached to the double bond, all three B—H bonds are not always expected to react. The reaction may stop as a result of one or two additions. The reactivity decreases after the attachment of the first alkyl group. The reactivity is further reduced after the binding of the second alkyl group because of the electron-donating ability of the alkyl groups.

After adding BH_3 to the alkene, the C—B bond formed must be replaced by C—OH. In the second step, a hydroperoxide anion (conjugate base is a better nucleophile) formed by the reaction of hydrogen peroxide with NaOH adds to the electron-deficient alkyl borane to produce the boron hydroperoxide derivative. This intermediate is unstable and rearranges to form a borate ester. An R group bonded to boron migrates to oxygen with the bonding electrons by breaking a C—B and forming a C—O bond, along with cleavage of the O—O bond. Now the charge on boron goes from negative to neutral. This process is performed three times to give a trialkyl borate ($B(OR)_3$) as the product. Finally, the hydroxide anion attacks the formed trialkyl borate and removes alkoxide groups. Protonation of alkoxide ions forms free alcohol. The oxidation does not change the orientation of the product because the *anti*-Markovnikov addition was established in the first step, the addition of BH_3.

Hydroboration reactions are carried out successfully with monoborane (BH_3) as well as with dialkylboranes. One of the most frequently used dialkylboranes is 9-borabicyclo[3.3.1]nonane, which is called 9-BBN in the literature. 9-BBN is a bulky molecule and is prepared by the reaction of 1,5-cyclooctadiene and borane, usually in ethereal solvents [39, 40]. If 9-BBN reacts with asymmetric alkenes, steric factors become more prominent than the electronic ones. Its remarkable selectivity has found wide application in organic synthesis. Hydroboration reactions with 9-BBN are slower than those with borane (B_2H_6). The bulkiness of the alkyl groups affects the reaction rate; however, the selectivity increases in reactions with 9-BBN.

9-BBN = 9-Borabicyclo[3.3.1]nonane

The mechanism of hydroboration is different from all of the addition mechanisms we have discussed so far. The first step in the hydroboration, the addition of the borane to the alkene, is a concerted reaction because bond cleavage and bond formation occur via a four-center transition state at the same time. Therefore, the addition of boron and hydrogen to the double bond occurs on the same side of the molecule and this leads to a syn-addition. The hydroboration reaction of deuterated norbornene provided convincing evidence for *exo*- and syn-addition. The hydroboration of 1,2-dimethyl-cyclohex-1-ene also confirmed the regio- and stereoselectivity of the hydroboration reaction [41].

Asymmetric hydroboration is one of the most effective procedures for the synthesis of chiral products. To achieve asymmetric induction, many alkylborane compounds have been synthesized. In monoalkylborane compounds, if the alkyl group is optically active, asymmetric induction is possible in hydroboration reactions. α-Pinene is an optically active natural compound; its controlled reaction with diborane easily forms the optically active alkylborane, which can be used for asymmetric synthesis. When two α-pinene groups are inserted into the borane, the rate of the reaction of R_2BH with alkenes is significantly affected, as excessive group accumulation occurs around the boron [42, 43]. If the alkene to be reacted is not bulky, the reaction proceeds smoothly. The configuration of the alcohol obtained from the reaction with cis-dimethylbutane is R and its optical purity is 87% [44].

The range of chiral ligands that can be used to affect enantioselective hydroboration has increased considerably. Metal catalysts are successfully applied to asymmetric hydroboration reactions [45].

The application area of the hydroboration reaction is quite broad. The internal alkynes, as well as the terminal alkynes, can also undergo hydroboration reactions such as the alkenes. The boron atom attacks the less substituted carbon, which is also the least hindered one. The oxidation of borane first gives the vinyl alcohol, which immediately tautomerizes to the corresponding aldehyde or ketone [46].

Hydroboration reactions are not only stereospecific (cis-addition); they are also regioselective. An *anti*-Markovnikov product is generally formed in the hydroboration reaction. When hydrogen halides are added to asymmetric double bonds, the proton binds to the carbon atom with the fewer alkyl groups attached. In this way, the most stable carbocation occurs. When the proton binds to the carbon atom with the higher number of the alkyl groups, a less stable carbocation is formed as the intermediate. The product is called the *anti*-Markovnikov addition product. The hydroboration reaction produces an *anti*-Markovnikov product. This does not mean that the product is formed from the less stable intermediate. It should be remembered that the addition of HBr to a double bond in the presence of peroxides also forms the *anti*-Markovnikov product. Even in that case, the *anti*-Markovnikov product is derived from the most stable intermediate. In the case of hydroboration, we are faced with a similar situation. As the boron atom is more electropositive than the hydrogen atom, first the boron atom adds to the double bond to form the most stable intermediate. Finally, we can summarize as follows: by ionic addition, the *anti*-Markovnikov product is derived from the less stable intermediate; however, the *anti*-Markovnikov products formed from the addition of hydrogen halides in the presence of peroxides and hydroboration are always derived from the most stable intermediates.

4.6 Oxidation of Alkenes

One of the most important reactions of alkenes is the oxidation reaction. When we talk about oxidation, we usually mean any reaction that forms a carbon–oxygen bond. Halogenation of an alkene is also an oxidation reaction. In this section, we will talk about the reaction in which we introduce a carbon–oxygen bond into the molecule.

Alkene oxidation takes place very differently. Alkenes can undergo oxidation reactions with various reagents. These reagents can oxidize the alkenes to different degrees. Before we start to discuss the oxidation reaction of an alkene, let us categorize the alkene oxidation reactions. There are four different levels of oxidation. The gentlest and least oxidation is the formation of an epoxide. Moderate oxidation can convert an alkene into a 1,2-diol. However, stronger reagents can cleave the double bond. The strongest reagents can cleave the double bonds to a carboxylic acid. These different oxidation reactions are shown below. Let us examine these oxidation reactions in order.

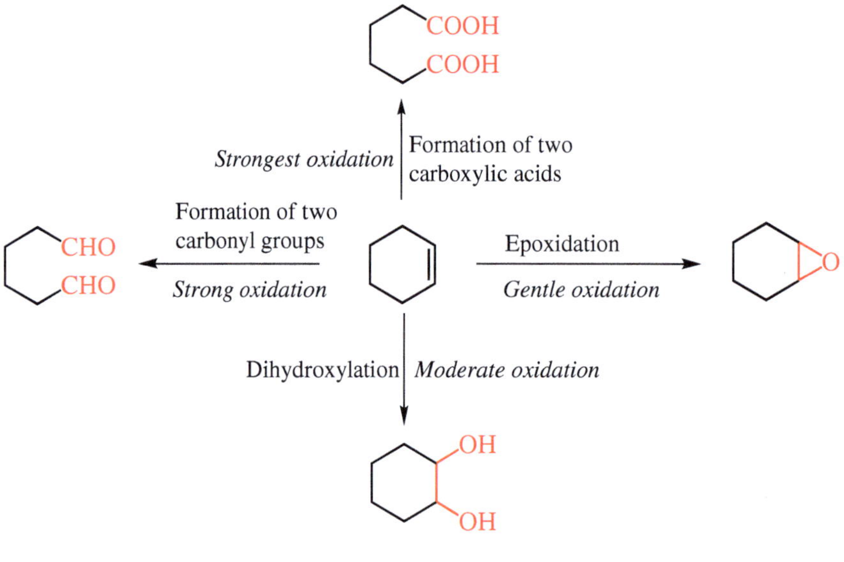

4.6.1 Epoxidation

Epoxidation is a reaction of alkenes with reagents such as hydrogen peroxide and peracids (RCO_3H) to produce cyclic ethers with a three-membered ring containing one oxygen atom. Epoxidation reactions are very frequently used reactions both in industry and in laboratories. The simplest epoxide, ethylene oxide, is produced in millions of tons globally. Ethylene oxide is the basic starting material of compounds such as ethylene glycol, polyethylene glycol, amino alcohol, and acrylonitrile. A general epoxidation reaction is shown below.

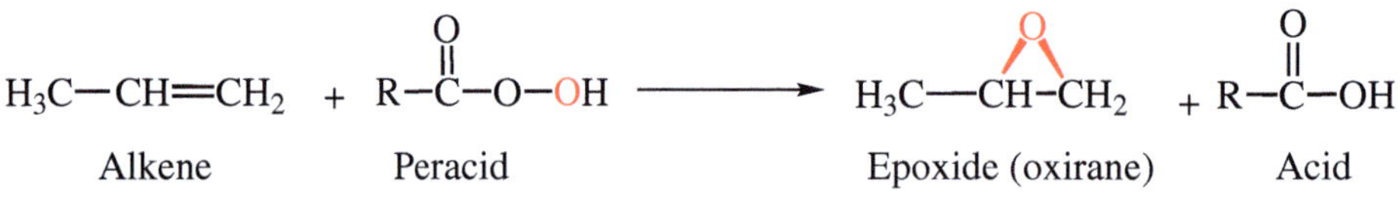

Epoxidation is an electrophilic addition reaction because electron-attracting groups on the peracid and electron-donating groups on the alkene favor the reaction. The epoxidation reaction is an exothermic process. Reaction enthalpy is approximately −38 kcal/mol (159 kJ/mol) [47].

4.6 Oxidation of Alkenes

Epoxidation is a single-step reaction. It has been shown that no ionic intermediates are involved and the reaction rate is not affected by solvent polarity. The oxygen atom of the hydroxyl group of the peracid is an electrophilic species. Bartlett has proposed a mechanism known as the butterfly mechanism as the transition state involving the addition of the hydroxyl oxygen to the double bond and simultaneous shifting of the proton resembles a butterfly [48, 49]. Theoretical calculations also support the existence of such a mechanism [47].

During this reaction, several bonds are broken and several bonds are formed simultaneously. Since this reaction takes place in one step, there is no way for bond rotation during the reaction; therefore, the epoxide ring formed will have the same stereochemistry as the starting alkene. Hence, the epoxidation reactions are stereoselective. The configuration of the epoxide resulting from the oxidation of trans-2-butene with a peracid is trans. Likewise, cis-2-butene is converted into cis-2,3-dimethyloxirane, maintaining its configuration during epoxidation.

Some of the commonly used peracids are given below. m-Chloroperbenzoic acid (m-CPBA) is a strong oxidizing reagent and frequently used in laboratories. However, very pure m-chloroperbenzoic acid is explosive; mixtures containing 15% or more m-chlorobenzoic acid and some water are safer [50]. When excessive amounts are needed or used in the industry, magnesium monoperoxyphthalate (MMPP) is used instead of m-chloroperbenzoic acid. MMPP is more stable and safer to use for large-scale and industrial reactions.

If both sides of alkene are different, peracid approaches the double bond from the less hindered side. For example, the reaction of p-benzoquinone-fused norbornadiene derivative with m-chloroperbenzoic acid (m-CPBA) was performed in methylene chloride at room temperature. Two isomeric epoxides, exo and endo, are formed in almost quantitative yield in a ratio of 82 : 18. Interestingly, epoxidation took place only on the norbornadiene double bond. There are three double bonds, which can be attacked from both sides of the double bonds to give six possible epoxides. The observed regioselectivity was supported by theoretical calculations, which showed that these isomers are thermodynamically the most stable isomers. Furthermore, the activation barriers leading to the formation of these isomers are lower than those of the other unobserved isomers [51].

If the molecule has hydroxyl groups adjacent to the double bond, a neighboring group effect can be observed in epoxidation reactions. The hydroxyl group exerts a directing effect and favors the approach of the peracid from the side at which the hydroxyl group is located. A hydrogen bonding between the hydroxyl group and the peracid is formed in the transition state. Therefore, the peracid attacks the double bond from that side. This directing effect is called the *Henbest rule or syn-effect* [52, 53].

The hydroxyl group, which is in the allyl position in the five-, six-, and seven-membered rings, affects epoxide formation. For example, epoxidation of the following compounds occurs from the same side of the hydroxyl group, despite the steric barrier. While the ratio of syn-product in the five-membered ring is 19 : 1, this ratio goes up to 50 : 1 in the six-membered ring. However, this effect does not operate in medium-sized rings because of their significant flexibility. Therefore, they mainly give *trans*-products.

The reaction of allylic cyclopentenol with *tert*-butylchlorodiphenylsilane gives the corresponding silyl ether. Treatment of this compound with *m*-CPBA produces two epoxides. In contrast to the reaction with the unprotected allyl alcohol, the major product, *anti*-isomer, was formed in 57% yield. The protecting group attached to the oxygen group is bulky; the oxidizing peroxyacid attacks the molecule from the less hindered side [54]. Hydrogen bridging does not operate in this case.

In epoxidation reactions, besides stereoselectivity, regioselectivity is also observed. There are two double bonds in 1,2-dimethylcyclohexa-1,4-diene. Both double bonds can be oxidized. However, when one mole of peracid is used, the double bond to which dimethyl groups are attached is epoxidized. Although these groups create steric hindrance, the electronic effect appears to be dominant here. Epoxidation of both double bonds is possible when peracid is used in excessive amounts, where the syn adduct is the major product because of the directing effect of the first epoxide ring [55–57].

For asymmetric induction, one of the compounds must be optically active or an optically active catalyst must be used. Many new methods have been developed for the synthesis of optically active epoxide. The most frequently applied of these methods is *Sharpless oxidation*. K. Barry Sharpless, an American chemist (born, 1941) received the Nobel Prize in 2001 in part for his role in this discovery. Sharpless oxidation is also known as asymmetric epoxidation of primary and secondary allylic alcohols for the synthesis of enantiomerically pure epoxy alcohols. The epoxidation is carried out with allylic alcohols in the presence of titanium isopropoxide as a catalyst, chiral diethyl tartrate, and *t*-butyl hydroperoxide as the oxidizing reagent [58–60].

Sitagliptin is used as a drug in the treatment of type II diabetes. The precursor for the synthesis of sitagliptin, epoxy alcohol, and ((trifluorophenyl)oxiran-2-yl)methanol, has recently been synthesized by the application of Sharpless asymmetric epoxidation. The allylic alcohol was reacted with *tert*-butylhydroperoxide in the presence of Ti(O*i*Pr)$_4$ and of (−) diisopropyl tartrate. The epoxide was obtained in 80% yield and >90% ee [61].

The reaction mechanism for Sharpless asymmetric oxidation is shown in Scheme 4.1. The reaction starts with substitution of all isopropoxide groups on the titanium by diethyl tartrate, *t*-butylhydroperoxide, and the allylic alcohol. Oxidation takes place on this complex. The chiral diethyl tartrate dictates which face of the double bond will be attacked.

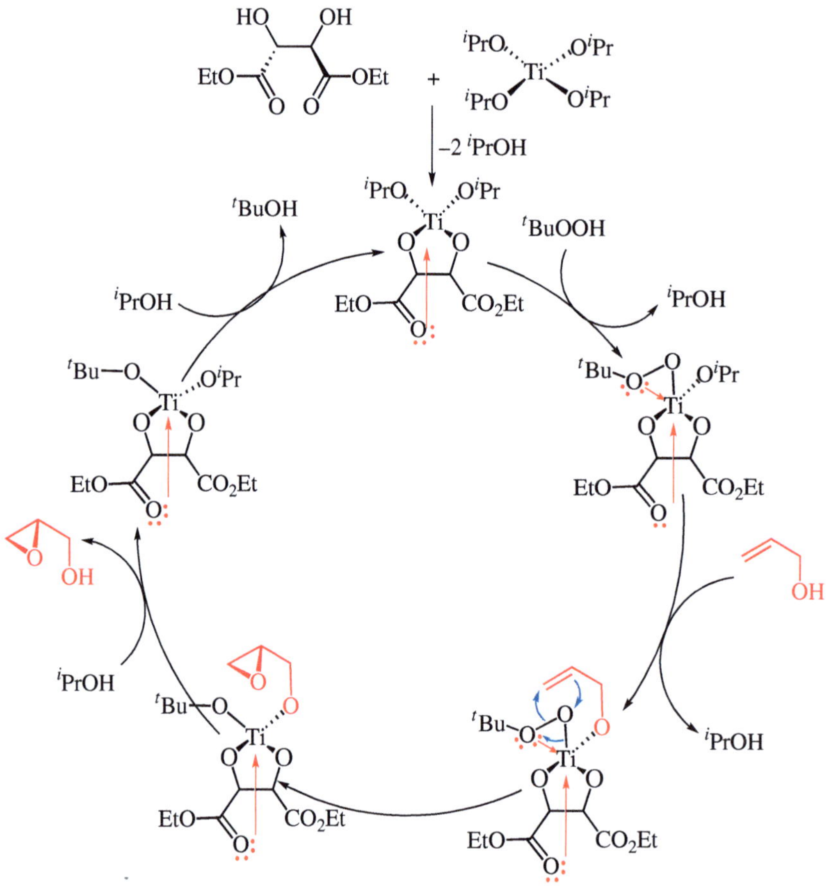

Scheme 4.1 Catalytic cycle for sharpless asymmetric epoxidation of allylic alcohols.

The complex formed between the titanium compound and the allyl alcohol, the hydroperoxide, and the tartrate determines the stereochemistry. The differences in the spatial arrangements of the two enantiomeric tartrates in the complex allow the formation of stereochemically different epoxides, as shown below. The ratio of diastereomers when a catalyst is used varies considerably depending on the configuration of the catalyst. No epoxidation takes place in the absence of the catalyst, Ti(O*i*Pr)$_4$, and no asymmetric induction in the absence of tartrate [62, 63].

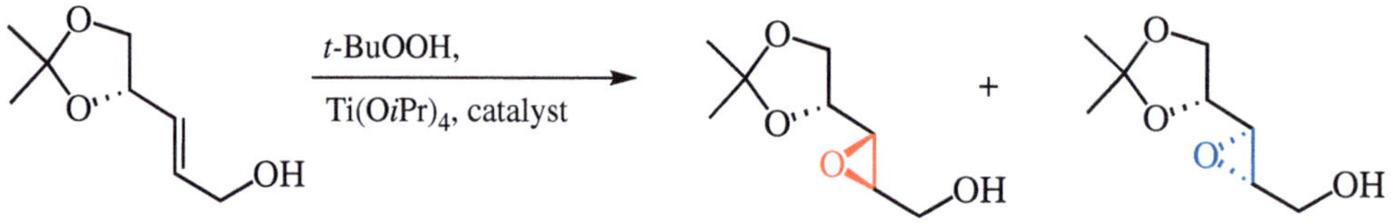

cat.: Diethyltartrate	catalyst: D-(−)-DET	90	:	1
	catalyst: D-(+)-DET	1	:	22
	no catalyst	2.3	:	1

4.6.2 Dioxirane

Another reagent used for the oxidation of double bonds is dimethyldioxirane (DMDO), also called Murray's reagent [64]. Epoxidation reactions are generally carried out in acidic or basic medium. Under neutral conditions, oxygen transfer to a double bond is easily accomplished with dioxirane [65–68]. DMDO is not commercially available because of its instability. The dioxirane reaction is important because of the facile synthesis of epoxides, the comfortable isolation of the products formed, and the lack of by-products in general. As dioxirane is a molecule that reacts entirely in a neutral environment, acid- or base-sensitive functional groups in the molecule are always preserved.

DMDO, generated by the reaction of acetone with potassium hydrogen persulfate, can be distilled directly into the alkene solution to be epoxidized. Because the by-product formed after epoxidation is acetone, the product is easily isolated by removing the solvent from the reaction medium.

When 5-methylenebicyclo[2.2.1]hept-2-ene having two different double bonds is reacted with one equivalent of DMDO, regioselectively the *endo* double bond is epoxidized [69]. Further treatment with excess DMDO forms a mixture of diastereoisomers. Both double bonds are susceptible to ionic reactions and they undergo a particular rearrangement. However, by the reaction with DMDO, the skeleton is preserved.

Epoxidation of hexamethyl Dewar benzene with one equivalent of dimethyldioxirane forms a monoepoxide in 97% yield, while an excess of DMDO produces a bis-epoxide quantitatively [69]. The monoepoxide had been previously postulated as a labile intermediate because of the rapid hydrolysis. As the oxidation with DMDO proceeds under neutral conditions, the epoxides can be isolated. This is a significant advantage of epoxidation with dioxirane.

4.6.3 Epoxide Ring-Opening Reactions

Because of the large ring strain in the three-membered ring (25 kcal/mol, 105 kJ/mol), epoxides are more reactive than ordinary ethers. The ring-opening reaction can proceed by either S_N1 or S_N2 mechanism depending on the reaction conditions as well as the nature of the epoxides. Epoxides can undergo a ring-opening reaction in acidic medium or basic medium. The first step in the acid-catalyzed ring-opening reaction is the protonation of the epoxide oxygen, forming an alkyloxonium ion, which generates a better leaving group. The positive (+) charge on the oxygen pulls the C—O bond electrons toward itself, making the carbon atoms electropositive. The next step is the reaction of the nucleophile with the protonated epoxide. A nucleophile, water, attacks the carbon atom from the back and opens the ring with anti-stereochemistry to form a 1,2-diol.

The anti-stereoselectivity can be followed easily by the ring opening of the cyclic epoxides. For example, cyclohexane epoxide undergoes ring opening in an acidic medium to form exclusively *trans*-1,2-cyclohexanediol. This methodology, epoxidation of a double bond followed by ring opening, is one way to synthesize *trans*-1,2-diols.

158 | *4 Addition Reactions to Alkenes*

When an asymmetrically substituted epoxide undergoes a ring-opening reaction in the presence of a nucleophile other than H₂O, two different products can be formed from a nucleophilic attack. For example, acid-catalyzed methanolysis of 2,2-dimethyloxirane proceeds by exclusive ring opening at the more substituted side. Let us explain why the more substituted carbon is attacked. We should remember that nucleophiles prefer to attack the carbon atoms from the less crowded side. We are faced here with a different case. The protonated oxygen withdraws the bonding electrons, placing a partial positive charge on the carbon atoms. The more positive charge will localize on the tertiary carbon atom because of the electron-donating ability of the alkyl groups. One of the C—O bonds starts to break before the nucleophile attacks. Therefore, the protonated epoxide breaks preferentially in that direction, which puts the positive charge on the more substituted carbon atom. Otherwise, a primary carbocation would be formed.

Ethers do not react with nucleophiles. However, epoxides are so reactive that they undergo ring-opening reactions with nucleophiles such as HO⁻, RO⁻, HS⁻, or NH₃. The S_N2 reaction controls the ring-opening reaction under basic conditions. In contrast to the acid-catalyzed reaction, the nucleophiles attack the less hindered carbon atom. The reaction of 2,3-dimethyloxirane with basic methanol undergoes solvolysis, ring opening occurs by an S_N2 mechanism, and the nucleophile binds to the less substituted carbon atom.

Allylic epoxides are also opened by nucleophiles regio- and stereoselectively. The double-bond carbon (sp²-hybridized carbon atom) attracts bond electrons more strongly than the sp³-hybridized carbon atoms because of the increased s ratio in the sp² orbital. Therefore, the epoxide carbon atom adjacent to the carbon–carbon double bond is more electropositive than the other one, and so the nucleophiles preferentially attack the allylic carbon atom and open the epoxide ring with anti-stereochemistry.

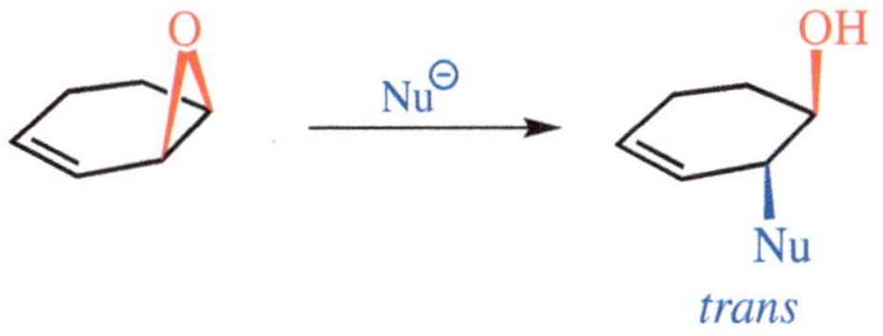

Utilizing regio- and stereoselective opening of allylic epoxides, the natural product conduramine-F₄ (leucanthemitol) was synthesized with high efficiency, as shown below. The epoxyketal was reacted with ammonia in methanol and a single product was formed in quantitative yield. Hydrolysis of the amino alcohol in acidified water gave conduramine-F₄ in high yield [70].

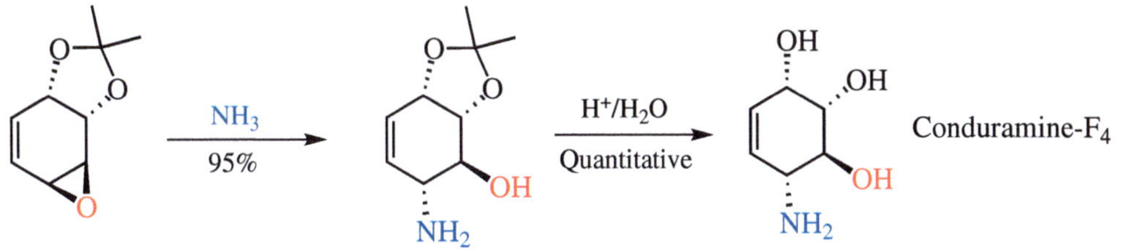

4.6.4 Vicinal cis-Dihydroxylation

Potassium permanganate (KMnO₄) is one of the oldest reagents used to convert alkenes into *cis*-diols. The reaction should be carried out in a cold basic solution to prevent overoxidation. Potassium permanganate can also be used as a test for alkene; the intense purple color disappears upon reaction with alkenes and the reduction product, brown MnO₂, precipitates. KMnO₄ is recommended for use as it is cheaper. KMnO₄ is in a highly positive oxidation state (+7), and therefore, it attracts electrons. KMnO₄ forms a cyclic manganate ester as the intermediate when it reacts with a double bond. Two oxygen atoms can be introduced to the alkene from the same face of the double bond. The configuration of the added oxygen atoms in the manganate ester is *cis*-positioned. The intermediate is usually not isolated and can be converted upon basic hydrolysis into the corresponding free diol with a *cis*-configuration. This reaction is significant in terms of the introduction of *cis*-configured hydroxyl groups into the molecule.

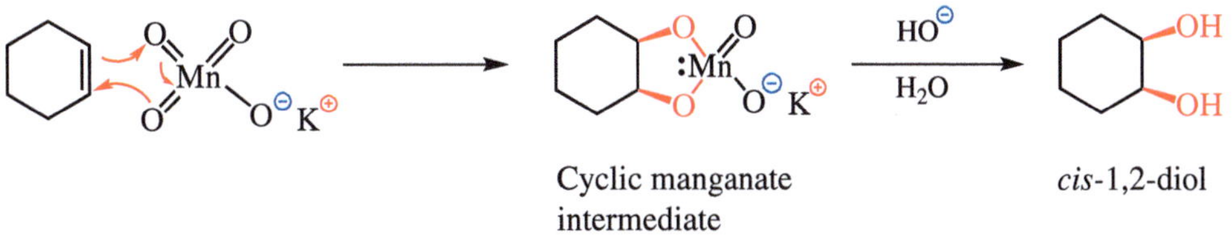

An example given below shows the synthetic application of KMnO₄ oxidation to the synthesis of *talo*-quercitol. Treatment of allylic acetate with KMnO₄ at −5 °C produces surprisingly only one diol in an isolated yield of 56% [71]. KMnO₄ approaches the double bond from the less hindered side of the double bond. To complete the synthesis of *the talo*-quercitol, the ketal functionality is hydrolyzed.

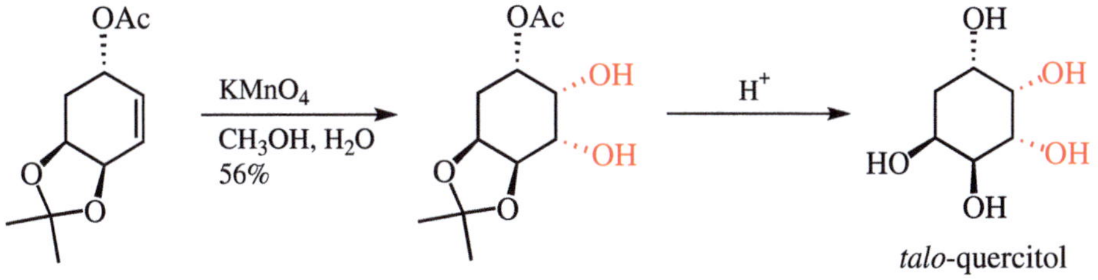

Osmium tetroxide (OsO$_4$) is another reagent that can catalyze the dihydroxylation reaction of alkenes [72]. OsO$_4$ has broad application despite its toxicity and rarity, and it reacts with alkenes to form a cyclic osmate diester, which can then be hydrolyzed to produce vicinal diols with a *cis*-configuration. Only a catalytic amount of OsO$_4$ is used because it is recycled. As the oxidizing agents, hydrogen peroxide or tertiary amine oxides (NMO, *N*-methylmorpholine *N*-oxide) are used. For the formation of cyclic osmate diester, a concerted [3 + 2] pathway addition was suggested in which three electron pairs move simultaneously to form a cyclic ester. The formation of the five-membered osmate ester was experimentally characterized. Sharpless and coworkers [73, 74] postulated a stepwise mechanism. According to this mechanism, first a four-membered ring is formed via [2 + 2] addition of OsO$_4$, which then rearranges to a five-membered ring.

Some products formed by the reaction of osmium tetroxide with various alkenes are given below. Although there are two double bonds in tetrahydroisobenzofuran, only one of them can be converted into a *cis*-diol in a controlled manner [75]. Dihydroxylation proceeds from the opposite side to the furan ring because of steric hindrance. When the diol is reacted with the second mole of OsO$_4$, a mixture of products is formed as both sides of the double bond have steric hindrance. In other examples, hydroxyl groups are always incorporated into the molecule with a *cis*-configuration.

In the past two decades, RuO$_4$ has been increasingly used in organic synthesis as an alternative to OsO$_4$ [76]. RuO$_4$ is often synthesized by using a substoichiometric amount of RuCl$_3$ and a stoichiometric amount of sodium periodate (NaIO$_4$). This system has some advantages over OsO$_4$. While OsO$_4$ generally reacts more easily with double bonds with high electron density, RuO$_4$ reacts with double bonds with low reactivity. Furthermore, RuO$_4$ is less toxic and cheaper than OsO$_4$. The amount used for the reaction is much less than the amount of OsO$_4$. The addition of catalytic amounts of CeCl$_3$ to the reaction mixture decreases the catalyst loading up to 0.25 mol% [77]. Although RuO$_4$ can produce dihydroxylation of alkenes under very controlled conditions, it is a much stronger oxidant than OsO$_4$. Because of the significantly higher redox potential of ruthenium(VIII) compared to osmium(VIII), overoxidation is a common side reaction in ruthenium-catalyzed dihydroxylations. Some examples of RuO$_4$-catalyzed *cis*-dihydroxylation reactions are given below [78–80].

The reaction mechanism is based on a five-membered ring complex, as we discussed in the reaction with OsO_4 and $KMnO_4$ oxidation reactions. It has been experimentally proven that a ruthenate complex is formed as an intermediate product.

4.6.5 Dihydroxylation via PIFA

Because of the high cost and toxicity of OsO_4, there is a need to search for alternative metal catalysts for alkene *cis*-dihydroxylations. Phenyl-iodine(III) bis(trifluoroacetate) (PIFA) [81] reagent has recently received considerable attention because of its low toxicity, easy handling, and reactivity. The oxidation of cyclohexene with PIFA, followed by hydrolysis, affords *cis*-1,2-cyclohexanediol in 95% yield. 1,4-Cyclohexadiene undergoes a dihydroxylation reaction, resulting in the formation of 1,4-adducts [82]. The reaction of benzonorbornadiene with PIFA yields the rearranged compound as the sole product. The formation of 1,4-addition products, as well as the rearranged product derived from the benzonorbornadiene, strongly indicates that an ionic mechanism is operating.

4.6.6 Enzymatic Dihydroxylation

Selective hydroxylation of aromatic compounds is one of the most challenging reactions in synthetic chemistry. The hydroxylated aromatics are essential precursors for various pharmaceuticals. Isolated enzymes can achieve selective hydroxylation. In 1968, Gibson and coworkers [83] studied the microbial oxidation of aromatic hydrocarbons by a soil bacterium (*Pseudomonas putida*) and isolated the first stable *cis*-1,2-cyclohexadienediol.

However, chemists did not pay any attention to this dihydroxylation reaction until Ley and coworkers reported the synthesis of (±)pinotol using the enzymatic oxidation of benzene [84]. Then, chemists generated *meso-cis*-diol derived from benzene on an industrial scale. In the late 1980s, an expansion of the use of these diols started and the *meso-cis*-diol served as the starting material for many syntheses.

In particular, Hudlicky's research group has focused on the application of enzymatically derived arene *cis*-dihydrodiol units as chiral building blocks in synthetic organic chemistry. Starting from cyclohexadiene-1,2-diols as intermediates, various compounds, including inositols, conduritols, and cyclitols, were synthesized [85].

Experiments with labeled oxygen, $^{18}O_2$, showed that both oxygen atoms in the diol were derived from dioxygen. The intermediate may likely be a dioxetane.

4.6.7 Ozonolysis: Oxidative Cleavage of Alkenes

While osmium tetroxide breaks only π-bonds, the reaction of alkenes with ozone can break even σ-bonds to generate carbonyl compounds under very mild conditions. Ozone is a high energy form of oxygen and it can be generated by electrical discharge or UV light by passing through oxygen gas under laboratory conditions. The sun can also convert oxygen into ozone in the upper layers of the atmosphere by UV rays. The ozone layer formed shields the Earth from high-energy UV radiation.

Ozone was applied in organic chemistry between 1903 and 1916 by Carl Dietrich Harries [86]. For many years, ozonolysis was used for structure determination of unknown compounds. By analyzing the structures of the fragments, it was possible to determine the structure as well as the position of the double bonds in the molecule. The alkene, which will react with ozone, will be dissolved in methanol or dichloromethane. Ozone gas (mixed with 3–4% oxygen gas) generated by an ozone generator is passed through solution of the alkene. Ozone has a dipolar structure, as can be seen from the resonance structures shown above. Because of this polarization, ozone undergoes a 1,3-dipolar cycloaddition reaction with the double bonds to form a *primary ozonide* (*molozonide*) in the first step [87, 88].

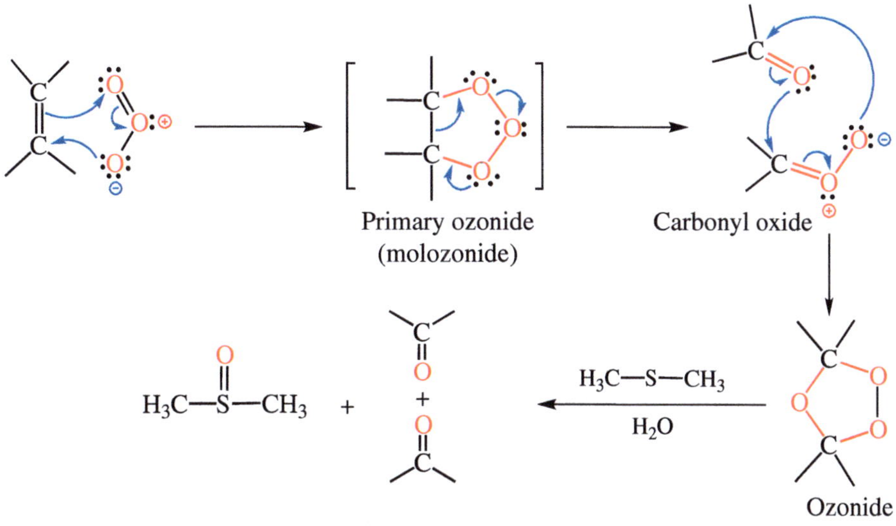

Molozonide is an unstable intermediate as it contains two weak peroxide bonds (oxygen–oxygen bonds), and it immediately breaks into a carbonyl compound and carbonyl oxide. This process is a 1,3-dipolar cycloreversion and takes place immediately, even at very low temperatures. This intermediate product is also called a *Criegee intermediate* since Rudolf Criegee, a Germen organic chemist (1902–1975), proposed its carbonyl oxide structure [88]. The resulting carbonyl oxide and ketone (or aldehyde) react together to form an *ozonide* intermediate, which is more stable than molozonide.

The second step in ozonolysis can proceed in two different ways, the more common one is *reductive workup*. In this step, reducing agents, such as dimethyl sulfide or zinc metal, are added to the reaction mixture. The purpose of the addition of reducing reagents is to cleave the peroxide linkage (O—O bond) in the ozonide and to remove one of the oxygen atoms. The oxygen atom is transferred to dimethyl sulfide. Finally, two carbonyl compounds are formed.

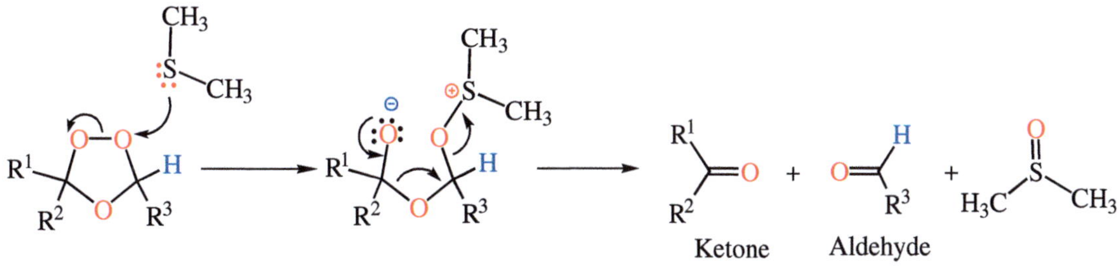

Reductive workup

There is a second type of workup that can be used, namely, *oxidative workup*. The hydrolysis of ozonide will form one mole of hydrogen peroxide and two carbonyl groups. If both carbonyl groups are ketones, there is no problem because ketones do not react with hydrogen peroxide. However, if one or both of the carbonyl groups is an aldehyde, hydrogen peroxide can oxidize one mole of aldehydes to an acid with released hydrogen peroxide. If both carbonyl groups are aldehydes, only one of the two will be oxidized. If it is desired to oxidize both aldehyde groups, hydrogen peroxide must be added to the reaction medium.

The mechanism by which alkenes undergo C=C cleavage into carbonyl groups using ozone is called *the Criegee mechanism* [87]. Numerous conflicting articles have been published on this mechanism. First of all, it was questioned whether ozonide was formed or not. Criegee determined the formation of carbonyl oxide by experiments. Carrying out ozonolysis in methanol allowed the capture of the carbonyl oxide to give a hydroperoxide [88].

Strong evidence for the intermediacy of carbonyl oxide was found by Criegee when the ozonolysis reaction was carried out in the presence of a different aldehyde, benzaldehyde, which is not derived from the starting material, leading to the formation of a new ozonide. A recent experiment by Berger and coworkers [89] showed that crossover experiments with ^{17}O-labeled benzaldehyde lead to the incorporation of ^{17}O label into the molecule. The ^{17}O-NMR spectroscopic measurements indicated the presence of the labeled atom in the ether linkage, not the peroxide linkage. This finding strongly supports the Criegee mechanism.

According to the Criegee mechanism, in ozonolysis reactions with asymmetric alkenes, cross products must be found. For example, the addition of ozone to 3-heptene first gives a five-membered ring adduct, primary ozonide. This ozonide can undergo two different 1,3-dipolar cycloreversions, leading to the formation of pairs **1/2** and **3/4**. In the reaction medium, there are two different aldehydes and two different carbonyl oxides. The zwitterion **1** can react with aldehydes **2** and **3** to give **A** and **C**. On the other hand, the zwitterion **4** can also react with aldehydes **2** and **3** to produce **B** and **A**. Theoretically, three different ozonides can be formed. The expected ozonide has the structure **A**. However, the ozonolysis of 3-heptene was found to give only the ozonide **A** not **B** or **C**. However, this finding does not rule out the Criegee mechanism. The experiments support the Criegee mechanism. Furthermore, theoretical calculations indicate that the carbonyl oxide and aldehyde formed upon dissociation of the primary ozonide are not separated from each other, and they form a strongly bound complex by electrostatic interactions that subsequently transforms into the secondary ozonide [90–92]. However, cross products are observed when an external carbonyl group is added to the reaction medium in high concentration.

Some examples of ozonolysis are given below. The second example shows that one of the aldehydes undergoes an oxidation reaction by hydrogen peroxides released by the hydrolysis of ozonide [93].

4.7 Reduction of Alkenes

It is important to recall that one of the essential reactions of alkenes is the addition reaction, as discussed in the previous sections. As σ-bonds are more stable than π-bonds, the most common reaction of C=C double bonds is the transformation of π-bonds into σ-bonds. When alkene and hydrogen gases are reacted in the presence of a catalyst, such as platinum, palladium, or nickel, two hydrogen atoms add to the C=C double bond to yield alkanes [94]. This reaction is called *hydrogenation*. The reaction is exothermic and the heat released is called *the heat of hydrogenation* and is about −20 to −30 kcal/mol (−80 to 125 kJ/mol), indicating that the product formed is more stable than the alkene. In this section, we will focus only on the reduction of carbon–carbon double bonds as a wide variety of functional groups can be reduced by catalytic hydrogenation.

Depending on the type of the catalyst and reaction conditions, the reduction reactions are categorized into two groups:
1. Heterogeneous catalytic reduction
2. Homogeneous catalytic reduction

4.7.1 Heterogeneous Catalytic Reduction

Heterogeneous catalysts have their advantages and disadvantages. After completion of the hydrogenation reaction, the heterogeneous catalysts are separated by filtration from the reaction medium because they do not dissolve in solvents. Hydrogenation reactions are generally carried out with heterogeneous catalysts. Alkenes do not react directly with hydrogen gas under normal conditions. A temperature of at least 500 °C is required for the reaction to take place. The activation energy required for adding hydrogen is quite high. The hydrogen–hydrogen bond must be weakened for reduction. Otherwise, it is also a challenging process. The activation energy of hydrogenation reactions decreases when a catalyst is used. The task of the catalyst is to lower the activation energy (Figure 4.2).

Generally, the alkene is dissolved in alcohol, hydrocarbon, or acetic acid. A small portion of the catalyst is added to the reaction mixture. The reaction proceeds using hydrogen at atmospheric pressure. The reduction takes place on the catalyst surface.

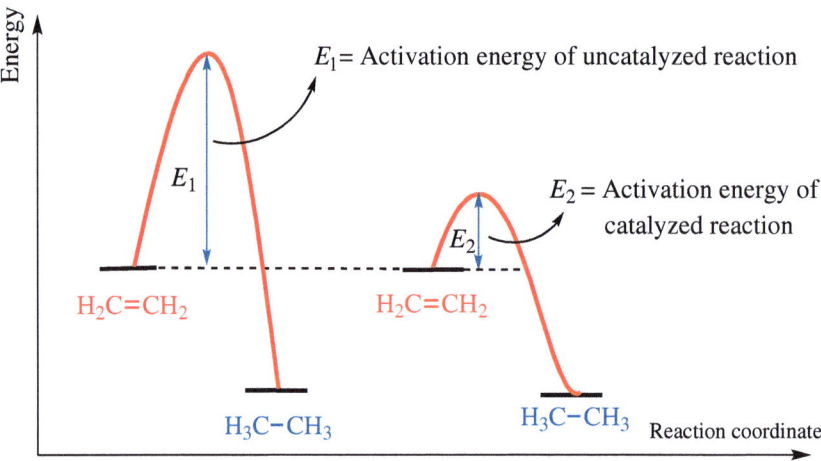

Figure 4.2 Hydrogenation energy diagram for catalyzed and uncatalyzed reactions.

In catalytic hydrogenation, two hydrogen atoms add to the same side of the double bond, displaying a *cis*-addition. The stereochemistry of the addition can be easily determined in cyclic systems. For example, when norbornene is reacted with D_2 over a catalyst, the product is the *cis*-isomer resulting from syn addition. The addition takes place from the *exo* face of the double bond. The reaction of the deuterated norbornene with hydrogen gas results again in the formation of the syn adduct.

The stereoselective addition of hydrogen to the double bond is described as follows. Hydrogen gas is first dissolved in the liquid phase. This process is the most critical part of the hydrogenation reaction. To dissolve the hydrogen gas in the liquid phase, other gases dissolved in the liquid phase must be removed by applying a vacuum. If this process is not carried out, the hydrogenation reaction does not work because of the low solubility of hydrogen. The hydrogen gas and the alkene are adsorbed on the surface of the metal catalyst (Figure 4.3). The hydrogen–hydrogen bond cleaves and each hydrogen forms metal–hydrogen bonds. The alkene forms a complex with the metal by overlapping of the vacant orbitals of the metal with the p orbitals of the alkene. These adsorptions weaken the bonds. In the next step, two hydrogens are transferred from the catalyst surface to the double bond with syn stereochemistry and the resulting saturated hydrocarbon. As the alkane is weakly adsorbed, it diffuses away from the catalyst surface. The exact nature of hydrogenation is not well understood.

It was demonstrated by an experiment that the bond between hydrogen atoms is weakened and the hydrogens are bound to the surface by a covalent bond. When H_2 and D_2 gases are mixed in the presence of a Pt catalyst, these two isotopes are randomly dispersed after a certain time. HD gas is formed, which does not occur when there is no catalyst. This process shows that the gas is adsorbed on the catalyst surface.

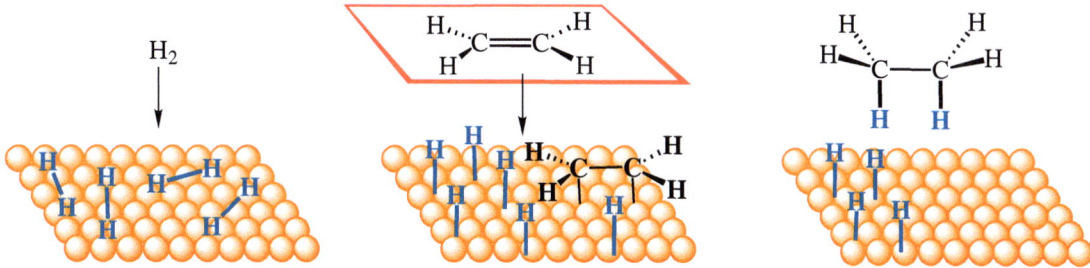

Figure 4.3 Catalytic hydrogenation of an alkene on the metal surface.

Generally, the catalyst used is platinum, palladium, rhodium, or ruthenium. These are highly active catalysts. Nickel (Raney nickel) is also widely used as a catalyst and is cheaper than the others but has low reactivity. Two examples of hydrogenation are given below.

Heterogeneous catalysts occasionally cause double-bond isomerization in the molecule. For example, the expected stereospecificity in the reduction of octahydronaphthalene is not observed. In addition to *cis*-decalin, *trans*-decalin is also formed in the reaction medium. However, this does not mean that reduction does not proceed with *cis*-stereoselectivity. During this reaction, first double-bond isomerization occurs on the catalyst surface. Then, because the two surfaces of the double bond are not equal, the double bond is reduced from both sides, resulting in *cis*- and *trans*-decalin.

The alkynes can also easily undergo hydrogenation reactions. However, using a metal catalyst such as palladium, platinum, or nickel will reduce the alkynes to alkanes. Hydrogenation is a stepwise process; first, an alkene is produced, with further reduction resulting in the final product alkane. It will not be possible to stop the reaction at the alkene stage. To control the reaction, the reactivity of the catalyst must be reduced. However, if the catalyst is modified in some way, the reduction can stop at the alkene stage. A Lindlar catalyst [95, 96] is a "poisoned" metal catalyst that reduces triple bonds to double bonds. This catalyst is palladium precipitated on $CaCO_3$, and it is poisoned with traces of quinoline and lead acetate. The catalytic reduction of alkynes with the Lindlar catalyst always forms *cis*-alkenes, in contrast to Na/NH_3 reduction, which gives *trans*-alkenes. Alkyne is adsorbed on the surface of the metal catalyst, and the hydrogen atoms are transferred to the same side of the alkyne, always forming *cis*-products. Some examples are given below [97].

4.7.2 Homogeneous Catalytic Reduction

So far, we have discussed the reduction of alkenes on a metal surface in the heterogeneous phase. Alkenes can also be reduced in the homogeneous phase with metal catalysts dissolved in organic solvents. One of the essential catalysts is tris(triphenylphosphine)rhodium(I) chloride, known as *Wilkinson's catalyst* [98–100]. Wilkinson's catalyst is named after the British chemist Sir Geoffrey Wilkinson (1921–1996), who received the Nobel Prize in 1973 with E. Fischer for his work. Wilkinson's catalyst can be synthesized by the reaction of rhodium(III) chloride with triphenylphosphine in ethanol. One equivalent of PPh_3 reduces rhodium(III) to rhodium(I), while three equivalents produce the complex. At ambient temperatures, this coordination complex exists as a reddish-brown solid. Wilkinson's catalyst has a square-planar coordination geometry. The catalyst is soluble in solvents and hydrogenation occurs in a homogeneous phase.

$$RhCl_3 \cdot 3H_2O + 4\ P(Ph)_3 \xrightarrow[\text{Heat}]{\text{Ethanol}} Rh[P(Ph)_3]_3Cl + (Ph)_3P=O$$

The reduction mechanism is given in Scheme 4.2. In the first step, one mole of triphenylphosphine dissociates from the complex, forming a three-coordinated rhodium complex. Under a hydrogen atmosphere, the oxidative addition of one mole of hydrogen takes place. Hydrogen atoms are individually bonded to the metal by covalent bonds so that a penta-coordinated dihydrido complex is formed. The oxidation state of Rh is increased from +1 to +3. This 16-electron complex has a still vacant coordination side; therefore, it can bond to the alkene to be reduced, forming a hexa-coordinated complex. The last step is the transfer of the hydrogen atoms to the alkene one by one. While a hydrogen atom transfers to one of the double-bond carbons, the other double-bond carbon atom is covalently bonded to the metal. With the transfer of the second hydrogen atom, the molecule is separated by reductive elimination from the metal, and the oxidation state of Rh is decreased to +1 and the catalyst is regenerated.

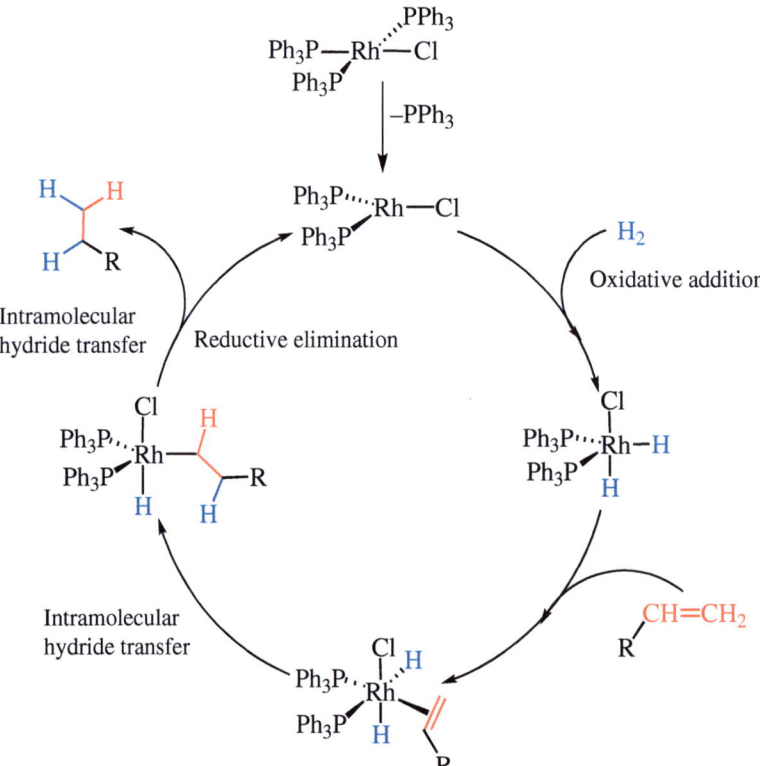

Scheme 4.2 Reaction mechanism for homogeneous catalytic hydrogenation with Wilkinson's catalyst.

4.7 Reduction of Alkenes

Three bulky triphenylphosphine groups are bonded to the metal. This is a large molecule. Therefore, the less substituted and sterically less hindered double bonds are selectively hydrogenated. If there is more than one double bond in the molecule, the least hindered one will be regioselectively reduced. On the other hand, the *cis*-alkenes are reduced faster than the *trans*-alkenes.

Wilkinson's catalyst is not optically active. When triphenylphosphine (PPh$_3$) ligand is replaced by chiral ligands, to produce the chiral ligands that can reduce inactive alkenes into the active compounds, the process is called *asymmetric induction*. One of the essential ligands that is optically active, BINAP (2,2'-bis (diphenylphosphino)-1,1'-binaphthyl) [101], can be attached to the metal and used for a wide variety of purposes. As outlined below, racemic BINAP was first synthesized by Japanese chemist Ryoji Noyori (1938–) and coworkers, and then, the optical resolution was achieved by using the chiral Pd(II) complex through fractional recrystallization. For his outstanding discoveries concerning asymmetric catalysis, Noyori received the Nobel Prize in 2001 shared with William Knowles and Barry Sharpless [102–104].

The asymmetric hydrogenation of geraniol or nerol is the most elegant method for the stereoselective construction of the side chains of vitamins E and K1. Geraniol and nerol are both monoterpenoid alcohols found in many essential oils, and they contain two double bonds with different configurations. Their reaction with Ru-BINAP complex in the presence of hydrogen gas regioselectively reduces only the allylic double bond because of the steric hindrance around the more substituted double bond. The desired compound **A** is formed starting from geraniol or nerol using a different enantiomer of the catalyst. For the reduction of nerol, it is sufficient to take one catalyst in *R* configuration and one in *S* configuration for geraniol [105].

170 | 4 Addition Reactions to Alkenes

As we discussed above, when an alkene is reacted with hydrogen in the presence of a metal catalyst, hydrogen adds to the double bond of the alkene to generate an alkane. During this process, the H—H bond and the C=C bonds are broken, while two new σ-bonds (C—H bonds) are formed. Bond breaking requires energy, while bond formation releases energy.

We can calculate the heat of the hydrogenation of this reaction.
C=C 145.8 kcal/mol
H—H 104.2 kcal/mol
C—C 82.6 kcal/mol
C—H 98.7 kcal mol
$\Delta H° = (145.8 + 104.2) - (82.6 + 2 \times 98.7) = -30$ kcal/mol (126 kJ/mol)

The negative value of $\Delta H°$ shows that the reaction is exothermic. For example, the hydrogenation energies of *cis-* and *trans-*butene are different. While the energy released for the *cis-*isomer is 28.3 kcal/mol, this value for the *trans-*isomer is around 27.4 kcal/mol. These values tell us that the *trans-*isomer is about 0.9 kcal mol more stable than the *cis-*isomer. The lower stability of the *cis-*isomer results from the repulsion between the methyl groups.

$\Delta H° = -27.4$ kcal/mol (−114.6 kJ/mol)

$\Delta H° = -28.3$ kcal/mol (−118.4 kJ/mol)

This heat of hydrogenation can be used to determine the thermodynamic stability of alkenes having different numbers of alkyl substituents attached to the double bond. We discussed in the elimination section that the stability of the double bond increases as the number of alkyl groups bonded to a double bond increases. The hydrogenation energies of some different substituted alkenes are given below.

H_2C=CH_2 H_3C—CH=CH_2 H_2C=$C(CH_3)_2$ $(H_3C)_2C$=$C(CH_3)_2$

$\Delta H° = -32.8$ kcal/mol $\Delta H° = -30.1$ kcal/mol $\Delta H° = -28.5$ kcal/mol $\Delta H° = -26.6$ kcal/mol
$(-137.2$ kJ/mol) $(-125.9$ kJ/mol) $(-119.2$ kJ/mol) $(-111.3$ kJ/mol)

4.8 Addition to Conjugated Dienes

A diene is a hydrocarbon that has two double bonds. If they are separated by a single bond, they are called *conjugated dienes*. For example, butadiene is a conjugated diene. We can imagine that butadiene consists of two ethylene molecules. Because of the conjugation of double bonds, conjugated dienes are thermodynamically more stable than isolated dienes. However, their stability does not mean that their reactivity is low.

Butadiene is thermodynamically more stable and more reactive than ethylene. Let us try to explain why this is the case. The highest occupied molecular orbital (HOMO) and lowest unoccupied molecular orbital (LUMO) of butadiene are formed by the combination of the HOMO and LUMO of ethylene. Figure 4.4 shows only the combination of the HOMOs. A combination of HOMO of ethylene produces two molecular orbitals of butadiene. The in-phase interaction of these two orbitals creates a bonding orbital designated by ψ_1. This bonding molecular orbital ψ_1 is of lower energy than the molecular orbital of ethylene. The HOMOs of ethylene can also interact out-of-phase, which generates an antibonding orbital, ψ_2, which is of higher energy than the ethylene HOMO.

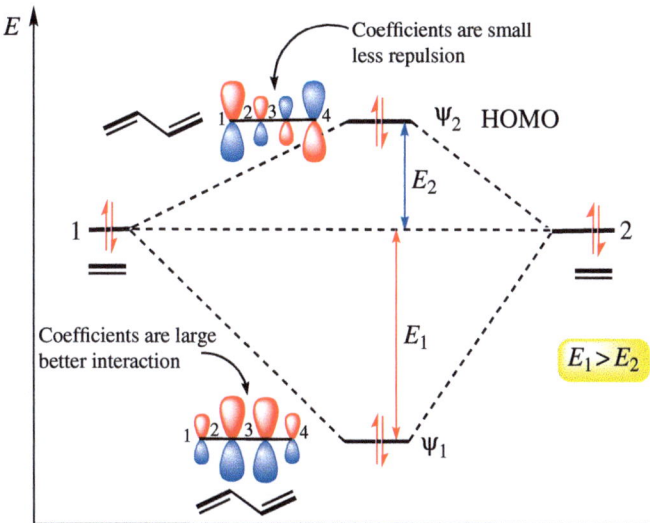

Figure 4.4 HOMO of butadiene created by the combination of HOMO of ethylene. Source: Based on Fleming [106].

The energy E1 released in forming the orbital ψ_1 arises from the overlap between the atomic orbitals of C2 and C3. Such an overlap does not exist in isolated systems. The coefficient of the orbitals on C2 and C3 is large; therefore, it is effective by lowering the energy of the system. By contrast, an increase in energy of ψ_2 caused by repulsion between the orbitals of the different phases is not as significant because the coefficients of these orbitals on C2 and C3 are smaller. Thus, the energy gained from the system in forming ψ_1 is much higher than the energy needed to form ψ_2. The four electrons of butadiene will populate newly formed orbitals. If the decrease in the orbital energy level of ψ_1 were equal to the increase in the energy level of ψ_2, there would be no stabilization. Generally, the decrease in the energy level of ψ_1 is higher than the increase in the energy of ψ_2. In this case, the system is becoming stable. Now, we can understand that the well-known property of conjugated systems is that they are lower in energy compared to unconjugated systems. Therefore, butadiene is thermodynamically more stable than ethylene. However, when the reactivity of butadiene is compared to that of ethylene, butadiene is known to be more reactive. At first glance, it seems that there is a contradictory situation here. However, we should realize that not the total energy of the system determines the reactivity; the energy level of the system's HOMO determines it. As the HOMO of butadiene ψ_2 is higher in energy than the HOMO of ethylene, butadiene is more reactive than ethylene.

Conjugated and isolated dienes undergo electrophilic addition. The electrophilic addition to isolated dienes is similar to what we have seen in the addition reactions of alkenes. However, addition to conjugated dienes is strikingly different and leads to the formation of two products. One of the products is a rearranged product. Let us examine the addition of bromine to 1,3-butadiene. When one mole of bromine is reacted with 1,3-butadiene, two isomeric products are formed: a 1,2-addition product, 3,4-dibromo-1-ene, and a 1,4-addition product, 1,4-dibromobut-2-ene. When the reaction is carried out with HBr, 1,2-addition and 1,4-addition products are similarly formed. In both addition reactions, the normal 1,2-addition products appear as the main products. In rearranged products, the electrophile is bonded to the C1 carbon atom and the nucleophile to the C4 carbon atom. Therefore, we call these products 1,4-addition products. The double bonds in these compounds are shifted.

The formation mechanism of the products is quite simple, similar to the addition reactions we have seen so far. The proton, the electrophile, first adds to one of the double bonds. Both carbon atoms C1 and C2 can be protonated. When we compare the resulting carbocations, the secondary carbocation formed by the protonation of the C1 carbon atom is more favorable. The secondary carbocation formed, which is an allylic cation, is then further stabilized by delocalization of the positive (+) charge over two carbon atoms. Now the nucleophile, a bromide anion, can attack this intermediate at either of the two carbon atoms sharing the positive charge. Attack at the secondary carbon atom forms the 1,2-addition product, a Markovnikov product, while attack at the primary carbocation forms the 1,4-addition rearranged product.

When bromine is added to butadiene, a bromonium ion intermediate will be formed with one of the double bonds. This bromonium ion cannot have a symmetrical structure because of the asymmetric structure of the double bond. The bromonium ion will be so distorted that it can be described as an allylic cation. For that reason, bromide will attack the internal carbon atom C2 as well as the terminal carbon atom C4, providing 1,2- and 1,4-addition products. If the reaction proceeds exclusively over the bromonium ion, the nucleophile would attack C1 and C2 carbon atoms and the rearranged product would not be formed.

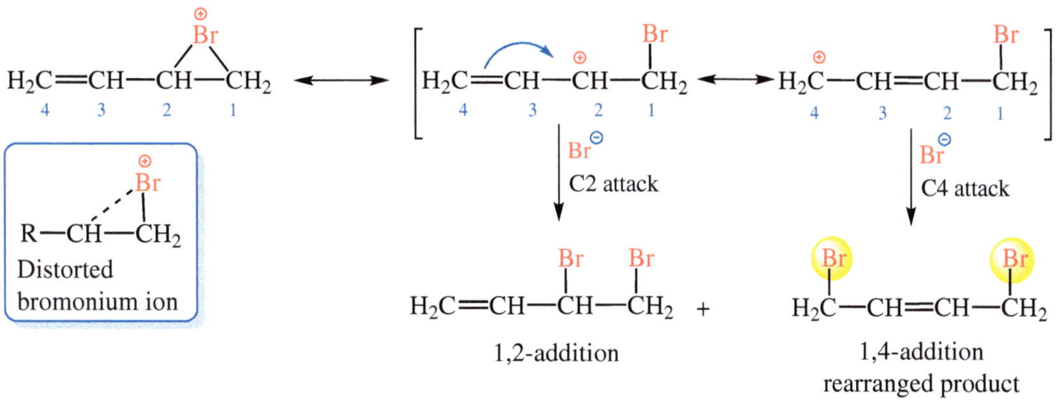

4.8 Addition to Conjugated Dienes

[Diagram showing resonance structures of allylic cation with bromonium ion, C2 attack leading to 1,2-addition product H$_2$C=CH—CHBr—CH$_2$Br, and C4 attack leading to 1,4-addition rearranged product H$_2$CBr—CH=CH—CH$_2$Br. A boxed structure shows the "Distorted bromonium ion" with R—CH—CH$_2$ bridged by Br$^+$.]

According to an alternate mechanism, after the formation of the bromonium ion, the bromide anion can attack the double-bond carbon C4 according to the S$_N$2′ mechanism, shifting the double bond and forming a 1,4-addition product [107]. However, the formation of a 1,4-addition product, where methanol is incorporated into the molecule, excludes this mechanism. Because methanol is a weak nucleophile, it cannot attack a double bond according to the S$_N$2′ mechanism. All of these results prove the formation of the allylic cation as the intermediate.

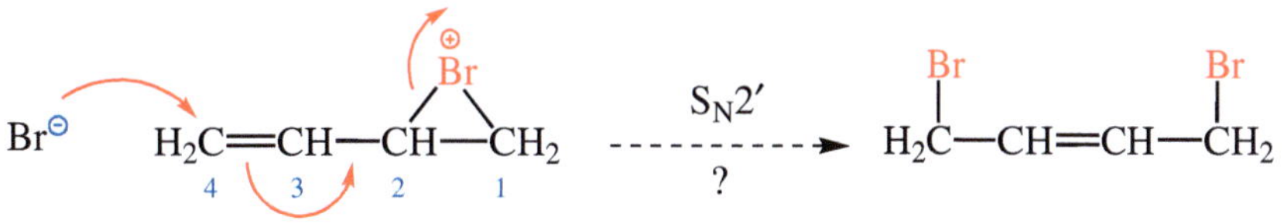

[Mechanism diagram: Br$^-$ attacking H$_2$C=CH—CH—CH$_2$ with Br$^+$, via S$_N$2′, giving H$_2$CBr—CH=CH—CH$_2$Br, labeled with "?"]

So far, we have seen that the conjugated dienes undergo electrophilic addition, forming 1,2- and 1,4-products. Temperature plays a crucial role in determining product distribution. Let us examine the bromination reaction of butadiene at different temperatures. For example, the addition of bromine to 1,3-butadiene at −78 °C produces mainly a 1,2-addition product (80%), while the addition reaction at +40 °C gives mainly a 1,4-addition product (90%). When the mixture obtained at −78 °C is warmed to 40 °C or higher and held there for a particular time, the major product will be a 1,4-addition product. The 1,4-addition product is the thermodynamically more stable product because it has a more substituted double bond. The product formed rapidly at low temperature is called *a kinetically controlled product*. The product, a 1,4-addition product formed at a higher temperature, is the most stable one, and it is called a *thermodynamically controlled product*. We can conclude that kinetically controlled products occur at lower temperatures, while thermodynamically controlled products occur at high temperatures.

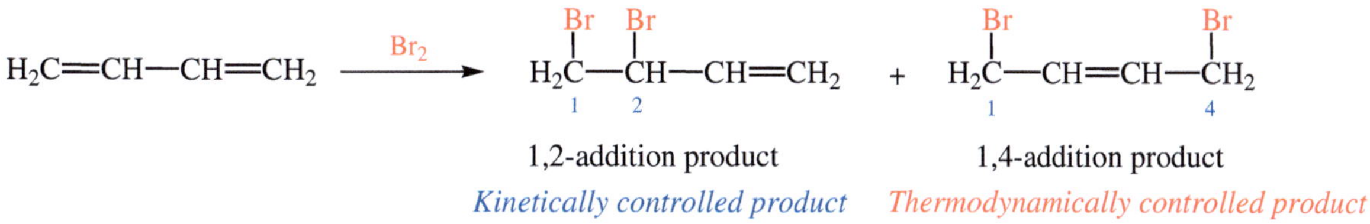

[Reaction: H$_2$C=CH—CH=CH$_2$ + Br$_2$ → H$_2$CBr—CHBr—CH=CH$_2$ (1,2-addition product, Kinetically controlled product) + H$_2$CBr—CH=CH—CH$_2$Br (1,4-addition product, Thermodynamically controlled product)]

	1,2-addition	1,4-addition
−78 °C	80%	20%
+40 °C	10%	90%

At low temperatures, the reaction is irreversible. There is not enough energy for the reverse reaction. The product that formed the fastest will be the major product. The activation energy for the formation of that product, a 1,2-addition product, is the lowest one. However, the reaction at high temperatures is a reverse reaction. The product that formed at low temperature can interconvert to form the thermodynamically more stable product. At high temperatures, the bromine can dissociate and can regenerate the allylic cation. The bromide again can attack both carbon atoms, forming 1,2- and 1,4-addition products. For the formation of a 1,2-addition product, the bromide attacks a secondary carbon atom. To form a 1,4-addition product, an attack on the primary carbocation must take place, which increases the activation energy (Figure 4.5).

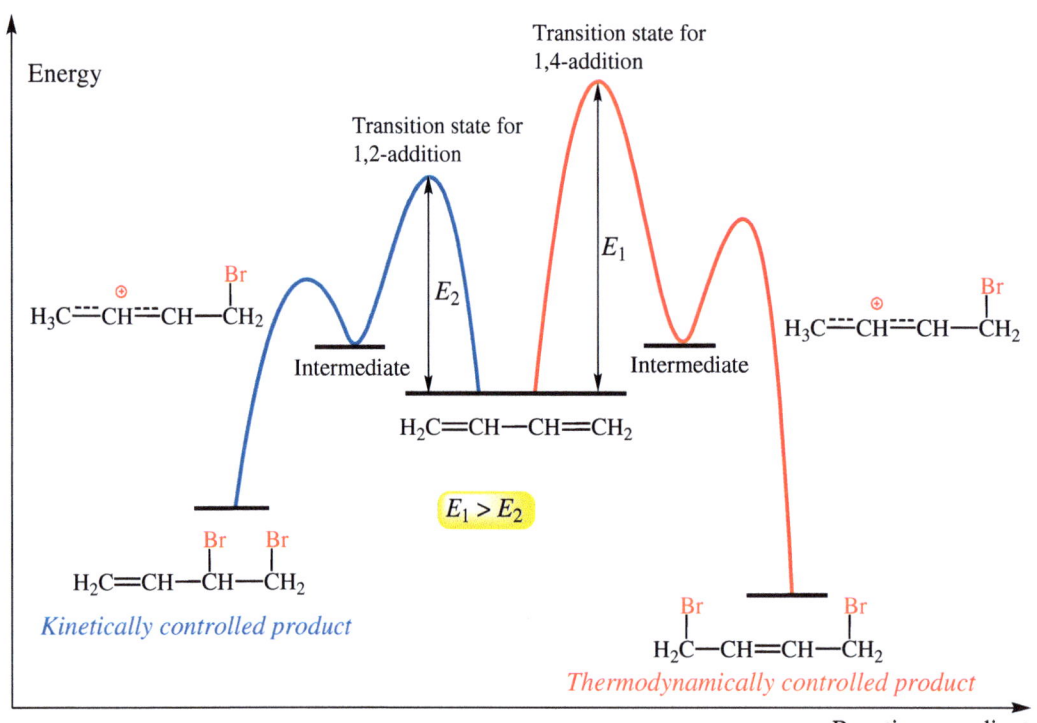

Figure 4.5 Energy diagram for formation of 1,2- and 1,4-adducts.

When the diene is not symmetrical, the structures of the addition products are determined by Markovnikov's rule. For example, let us examine the reaction of 2-methyl-1,3-butadiene with HBr. Both double bonds of the diene system are different. The question is, *Which double bond will be preferably protonated?* Considering that the proton is added to C1 and C4 carbon atoms, let us examine the stability of possible carbocations. Tertiary carbocation occurs when the proton adds to the C1 carbon atom. Additionally, this charge will be further stabilized by the allylic system. By adding the proton to C4, a secondary carbocation will be formed, which will also be stabilized by the allylic system. As the addition to C1 forms a more stable carbocation, the major products of this reaction will derive from this intermediate.

Similar results are also observed in the addition reactions to cyclic diene systems. For example, bromine adds to 1,3-cyclohexadiene at −70 °C to form three isomers of dibromocyclohexene, 1,2-adduct, *trans*-1,4-adduct, and *cis*-1,4-adduct in a ratio of 69 : 8 : 23, respectively. When the mixture is brought to room temperature, the 1,2-adduct forms an equilibrium with the *trans*-1,4-adduct in a ratio of 18 : 64, whereas the amount of *cis*-1,4-adduct remains constant at about 20% [108].

We emphasized that the 1,2-addition product is a kinetic product and is thermodynamically less stable than the 1,4-addition product. In some systems, contrary to this finding, the 1,2-addition product may be more stable than the 1,4-adduct.

Bromination of an exocyclic diene at −10 °C gives a 1,2-addition product in 95% yield. On the other hand, bromination at ambient temperature gives only a symmetrical 1,4-addition product in 94% yield. These results reveal that the activation energy required for the 1,4-addition product is higher than that for the 1,2-addition product [109].

However, when a solution of 1,2-adduct and 1,4-adduct was heated at 77 °C for just 15 minutes, both adducts were converted into the *exo*-1,2-addition product in almost quantitative yield. These experiments implied that the *exo*-1,2-adduct is thermodynamically the most stable compound.

Theoretical calculations also show that the 1,4-adduct is 5.8 kcal/mol higher in energy than the *exo*-1,2-adduct and 4.2 kcal/mol higher than the *endo*-1,2-adduct. These results are in entire agreement with the experimental results. The formation of a 1,4-addition product causes an additional increase in ring strain because the double bond incorporated into the five-membered ring is endocyclic. Therefore, the formation of an exocyclic double bond (1,2-addition) is preferred.

The Diels–Alder reactions are undoubtedly the most critical addition reactions for the dienes. Cyclic compounds are formed as a result of the reaction of dienes with dienophiles. Because this issue is discussed under the Woodward–Hoffmann rules in detail, it will not be covered here again.

Diels-Alder addition

Problems

4.1 Cyclohexene reacts with bromine in a solution containing aqueous NaCl. Write the products that may occur and give configurations of the products

4.2 When the *cis-* and *trans-*2-pentene reacts with bromine, write the structures of the products formed and determine the stereochemistry of the products.

4.3 Write down the structures of the main products formed by HCl addition to the following compounds

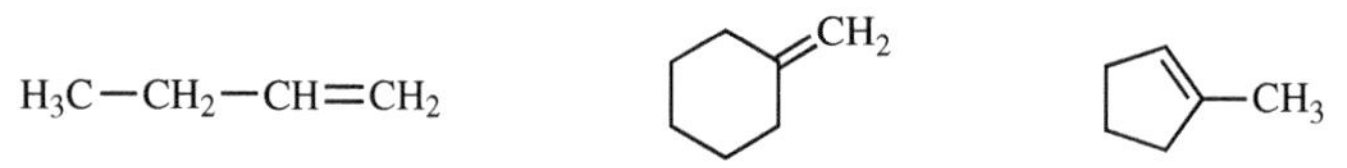

$H_3C-CH_2-CH=CH_2$

(cyclohexane with =CH$_2$ group)

(cyclopentene with -CH$_3$ group)

4.4 Which alkenes are taken as starting compounds for the synthesis of the following halogenated compounds given below?

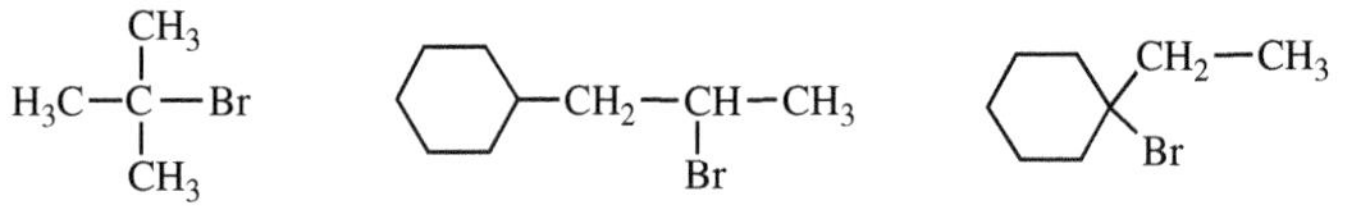

$H_3C-\overset{CH_3}{\underset{CH_3}{C}}-Br$

(cyclohexyl)-CH$_2$-CH(Br)-CH$_3$

(cyclohexyl with CH$_2$-CH$_3$ and Br substituents)

4.5 Find the structures of the products formed in the following reactions.

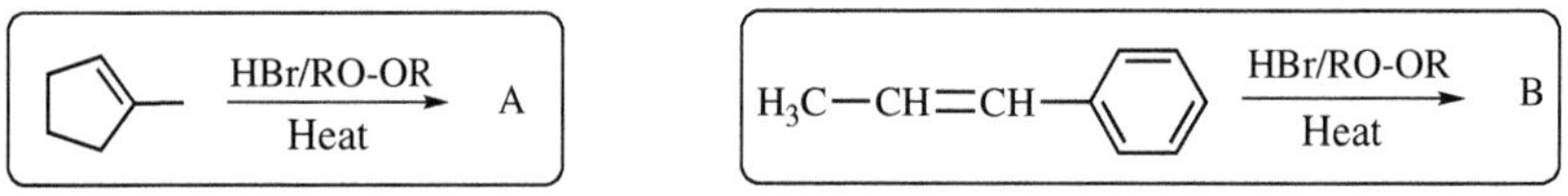

cyclopentene $\xrightarrow[\text{Heat}]{\text{HBr/RO-OR}}$ A

$H_3C-CH=CH-$(phenyl) $\xrightarrow[\text{Heat}]{\text{HBr/RO-OR}}$ B

4.6 Show how you would perform the following transformation?

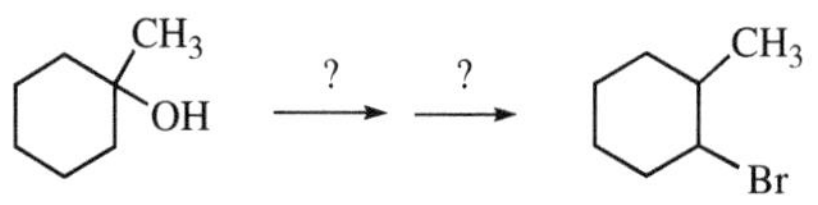

(1-methylcyclohexanol) $\xrightarrow{?}\xrightarrow{?}$ (trans-1-methyl-2-bromocyclohexane)

4.7 3,3-Dimethyl-1-butene reacts with 50% H$_2$SO$_4$. Write down the structure and give the formation mechanism of the main product.

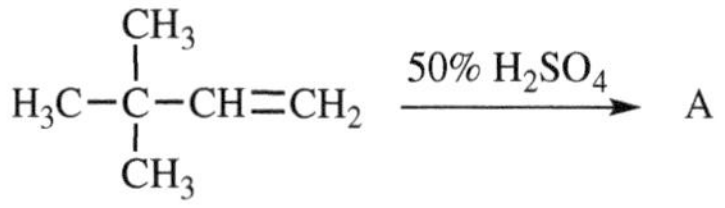

$H_3C-\overset{CH_3}{\underset{CH_3}{C}}-CH=CH_2 \xrightarrow{50\% \text{ H}_2\text{SO}_4}$ A

4.8 Methyl vinyl cyclobutyl forms an alcohol when treated with water in an acidic medium. Discuss the structure and the formation mechanism.

(1-methyl-1-vinylcyclobutane) $\xrightarrow{H^+/H_2O}$ A

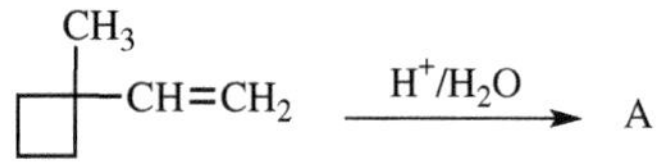

4.9 The alkene used as an intermediate in acetogenins synthesis is reacted with Hg(OAc)$_2$ at room temperature. Write the structure of the product formed.

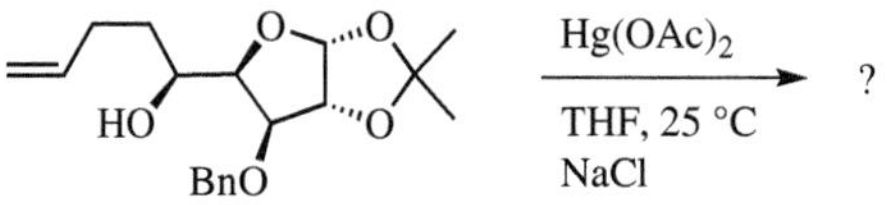

$\xrightarrow[\text{NaCl}]{\text{Hg(OAc)}_2\ \text{THF, 25 °C}}$?

4.10 Write the structure of the main product formed in the hydroboration of 2-methylnorbornene. Discuss regioselectivity and stereoselectivity.

4.11 How can you synthesize the compounds A and B by starting of the alkene given below?

4.12 How can you synthesize 1,2-*trans*-cyclopentanediol starting from the cyclopentene?

4.13 Suggest synthetic steps for the following ketoacid using exocyclic methylene compound?

4.14 The following diene reacts with RCO_3H and then in acid medium forms the cyclic alcohol. Suggest a mechanism for this transformation.

4.15 Cyclohexane-1,4-diene is stereoselectively converted to tetrol. Suggest a method for this transformation and consider the stereochemistry.

4.16 Suggest a method to convert bicyclic alkene into the cyclic diketone.

References

1 Lenoir, D. and Chiappe, C. (2003). *Chem. Eur. J.* 9: 1036.
2 Brown, R.S. (1997). *Acc. Chem. Res.* 30: 131.
3 De La Mare, P.B.D. and Bolton, R. (1982). *Electrophilic Additions to Unsaturated Systems*, 2e, 136. New York, NY: Elsevier.

4 Ruasse, M.-F. (1993). *Adv. Phys. Org. Chem.* 28: 207.
5 Strating, J., Wieringa, J.H., and Wynberg, H. (1969). *Chem. Commun.*: 907.
6 Slebocka-Yilk, H., Ball, R.G., and Brown, R.S. (1985). *J. Am. Chem. Soc.* 107: 4504.
7 Olah, G.A. and Bollinger, J.M. (1967). *J. Am. Chem. Soc.* 89: 4744.
8 Puterbaugh, W.H. and Newman, M.S. (1957). *J. Am. Chem. Soc.* 79: 3469.
9 Shani, A. and Sondheimer, F. (1967). *J. Am. Chem. Soc.* 89: 6310.
10 Chretien, J.R., Coudert, J.D., and Ruasse, M.F. (1993). *J. Org. Chem.* 58: 1917.
11 Markownikoff, V.V. (1870). *Justus Liebigs Ann. Chem.* 153: 256.
12 Tierney, J. (1988). *J. Chem. Educ.* 65: 1053.
13 Beletskaya, I.P. and Nenajdenko, V.G. (2019). *Angew. Chem. Int. Ed.* 58: 4778.
14 Volhardt, K.P.C. and Schore, N.E. (2007). *Organic Chemistry Structure and Function*, 5e, vol. 508. New York: W.H. Freeman & Co.
15 Henne, A.L. and Kaye, S. (1950). *J. Am. Chem. Soc.* 72: 3369.
16 Karasch, M.S. and Mayo, F.R. (1933). *J. Am. Chem. Soc.* 55: 2468.
17 Carroll, F.A. (1998). *Perspectives on Structure and Mechanism in Organic Chemistry*, vol. 594. Pacific Grove, CA.: Brooks/Cole Publishing Company.
18 Isenberg, N. and Grdinic, M. (1969). *J. Chem. Educ.* 46: 601.
19 Cioslowski, J., Hamilton, T., Scuseria, G. et al. (1990). *J. Am. Chem. Soc.* 112: 4183.
20 Cossi, M., Persico, M., and Tomasi, J. (1994). *J. Am. Chem. Soc.* 116: 5373.
21 Galland, B., Evleth, E.M., and Ruasse, M.-F. (1990). *J. Chem. Soc. Chem. Commun.*: 898.
22 Klobukowski, M. and Brown, R.S. (1994). *J. Org. Chem.* 59: 7156.
23 Chatt, J. (1951). *Chem. Rev.* 48: 7.
24 Brown, H.C. and Geoghegan, P. Jr. (1967). *J. Am. Chem. Soc.* 89: 1522.
25 Pasto, D.J. and Gontarz, J.A. (1970). *J. Am. Chem. Soc.* 92: 7480.
26 Olah, G.A. and Clifford, P.R.J. (1971). *Am. Chem. Soc.* 93: 1261–2320.
27 Bach, R.D. and Richter, R.F. (1973). *J. Org. Chem.* 38: 3442.
28 Lewis, A. and Azoro, J. (1981). *J. Org. Chem.* 46: 1764.
29 Bach, R.D. and Henneike, H.F. (1970). *J. Am. Chem. Soc.* 92: 5589.
30 Pasto, D.J. and Gontarz, J.A. (1969). *J. Am. Chem. Soc.* 91: 719.
31 Pecanha, E.P., Verli, H., Rodrigues, C.R. et al. (2002). *Tetrahedron Lett.* 43: 1607.
32 Mohapatra, D.K., Naidu, P.R., Reddy, D.S. et al. (2010). *Eur. J. Org. Chem.*: 6263.
33 Brown, H.C. and Subba Rao, B.C. (1956). *J. Am. Chem. Soc.* 78: 5694.
34 Brown, H.C. and Subba Rao, B.C. (1959). *J. Am. Chem. Soc.* 81: 6423.
35 Vogels, C.M. and Westcott, S.A. (2005). *Curr. Org. Chem.* 99: 687.
36 Wadepohl, H. (1997). *Angew. Chem. Int. Ed.* 36: 2441.
37 Burgess, K. and Ohlmeyer, M.J. (1991). *Chem. Rev.* 91: 179.
38 Matteson, D.S. (1986). *Synthesis*: 973.
39 Soderquist, J.A. and Brown, H.C. (1981). *J. Org. Chem.* 46: 4599.
40 Soderquist, J.A. and Alvin, N. (1998). *Org. Syntheses* 9: 95.
41 Brown, H.C., Bakshi, R.K., and Singaram, B. (1988). *J. Am. Chem. Soc.* 110: 1529.
42 Vishwakarma, L.C. and Fry, A. (1980). *J. Org. Chem.* 45: 5306.
43 Brown, H.C. and Ramachandran, V. (1991). *Pure Appl. Chem.* 63: 307.
44 Streitwieser, A. Jr., Verbit, A., and Bittmen, R. (1967). *J. Org. Chem.* 32: 1530.
45 Crudden, C.M. and Edwards, D. (2003). *Eur. J. Org. Chem.*: 4695.
46 Clay, J.M. and Vedejs, E. (2005). *J. Am. Chem. Soc.* 127: 5767.
47 Plesnicar, B., Tasevski, M., and Azman, A. (1978). *J. Am. Chem. Soc.* 100: 743.
48 Bartlett, P.D. (1950). *Rec. Chem. Prog.* 11: 47.
49 Rebek, J. Jr., Marshall, L., McManis, J., and Wolak, R. (1986). *J. Org. Chem.* 51: 1649.
50 Hussain, H., Al-Harrasi, A., Green, I.R. et al. (2014). *RSC Adv.* 4: 12882.
51 Essiz, S., Dalkilic, E., Sari, Ö. et al. (2017). *Tetrahedron* 73: 1640.
52 Henbest, H.B. and Wilson, R.A.L. (1957). *J. Chem. Soc.*: 1958.

53 Ye, D., Fringuelli, F., Piermatti, O., and Pizzo, F. (1997). *J. Org. Chem.* 62: 3748.
54 Elhalem, E., Comin, M.J., and Rodriguez, J.B. (2006). *Eur. J. Org. Chem.*: 4473.
55 Sureshkumar, D., Maity, S., and Chandrasekaran, S. (2006). *J. Org. Chem.* 71: 1653.
56 Gillard, J.R., Newlands, M.J., Bridson, J.N., and Burnell, D.J. (1991). *Can. J. Chem.* 69: 1337.
57 Mehta, G., Senaiar, R.S., and Bera, M.K. (2003). *Chem. Eur. J.* 9: 2264.
58 Katsuki, T. and Sharpless, K.B. (1980). *J. Am. Chem. Soc.* 102: 5974.
59 Katsuki, T. and Martin, V.S. (1996). *Org. React.* 48: 1–300. (Review).
60 Johnson, R.A. and Sharpless, K.B. (1991). *Comp. Org. Syn.* 7: 389–436. (Review).
61 Anil, D., Altundas, R., and Kara, Y. (2019). *Synth. Commun.* 49: 852.
62 Bruckner, R. (2002). *Advanced Organic Chemistry Reaction Mechanisms*, 115. San Diego/San Fransisco/New York/Boston/London/Sydney/Tokyo: Harcourt, Academic Press.
63 Takano, S., Samizu, K., Sugihara, T., and Ogasawara, K. (1989). *J. Chem. Soc. Chem. Commun.*: 1344.
64 Murray, R.W. (1989). *Chem. Rev.* 89: 1187.
65 D'Accoltia, L., Annese, C., and Fusco, C. (2015). *Curr. Org. Chem.* 19: 45.
66 Curci, R., D'Accolti, L., and Fusco, C. (2006). *Acc. Chem. Res.* 39: 1.
67 Adam, W., Curci, R., and Edwards, J.O. (1989). *Acc. Chem. Res.* 22: 205.
68 Adam, W. and Zhao, C.-G. (2006). *Chemistry of Peroxides*, vol. 2 (ed. Z. Rappoport), 1129. Chichester: Wiley.
69 Asouti, A. and Hadjiarapoglou, L.P. (2000). *Tetrahedron Lett.* 41: 539.
70 Secen, H., Gultekin, S., Sutbeyaz, Y., and Balcı, M. (1994). *Synth. Commun.* 24: 2103.
71 Maraş, A., Seçen, H., Sütbeyaz, Y., and Balcı, M. (1998). *J. Org. Chem.* 63: 2039.
72 Schröder, M. (1980). *Chem. Rev.* 80: 187.
73 Sharpless, K.B. and Teranishi, A.Y. (1977). *J. Am. Chem. Soc.* 99: 3120.
74 Deubel, D.V. and Frenking, G. (2003). *Acc. Chem. Res.* 36: 645.
75 Baran, A. and Balcı, M. (2009). *J. Org. Chem.* 74: 88.
76 Piccialli, V. (2014). *Molecules* 19: 6534.
77 Tiwari, P. and Misra, A.K. (2006). *J. Org. Chem.* 71: 2911.
78 Plietker, B. and Niggemann, M. (2004). *Org. Biomol. Chem.* 2: 2403.
79 Shing, T.K.M., Tai, V.W.-F., and Tam, E.K.W. (1994). *Angew. Chem. Int. Ed.* 33: 2312.
80 Plietker, B. and Niggemann, M. (2003). *Org. Lett.* 5: 3353.
81 Zhdankin, V.V. and Stang, P.J. (2002). *Chem. Rev.* 102: 2523.
82 Celik, M., Alp, C., Gultekin, M.S. et al. (2006). *Tetrahedron Lett.* 47: 3659.
83 Gibson, D.T., Koch, J.R., Schuld, C.L., and Kallio, R.E. (1968). *Biochemistry* 7: 3795.
84 Ley, S.V., Sternfeld, F., and Taylor, S. (1987). *Tetrahedron Lett.* 28: 225.
85 Hudlicky, T., Gonzalez, D., and Gibson, D.T. (1999). *Aldrichimia Acta* 32: 35.
86 Rubin, M.B. (2003). *Helv. Chim. Acta* 86: 930.
87 Criegee, R. (1975). *Angew. Chem. Int. Ed.* 14: 745.
88 Murray, R.W. (1968). *Acc. Chem. Res.* 1: 313.
89 Geletneky, C. and Berger, S. (1998). *Eur. J. Org. Chem.*: 1625.
90 Ponec, R., Yuzhakov, G., Haas, Y., and Samuni, U. (1997). *J. Org. Chem.* 62: 2757.
91 Cremer, D. (1981). *J. Am. Chem. Soc.* 103: 3627.
92 Cremer, D., Kraka, E., McKee, M.L., and Radhakrishnan, T.P. (1991). *Chem. Phys. Lett.* 187: 491.
93 Taber, D.F. and Nakajima, K. (2001). *J. Org. Chem.* 66: 2515.
94 Hudlicky, M. (1996). *Reductions in Organic Chemistry*, 1–429. Washington, DC: American Chemical Society.
95 Lindlar, H. (1952). *Helv. Chim. Acta* 35: 446.
96 Lindlar, H. and Dubuis, R. (1966). *Org. Synth.* 46: 89.
97 Penk, D.N., Robinson, N.A., Hill, H.M., and Turlington, M. (2017). *Tetrahedron Lett.* 58: 470.
98 Young, J.F., Osborn, J.A., Jardine, F.H., and Wilkinson, G. (1965). *J. Chem. Soc. Chem. Commun.* 7: 131.
99 Osborn, J.A., Jardine, F.H., Young, J.F., and Wilkinson, G. (1966). *J. Chem. Soc. A*: 1711.
100 Montelatici, S., van der Ent, A., Osborn, J.A., and Wilkinson, G. (1968). *J. Chem. Soc. A*: 1054.
101 Miyashita, A., Yasuda, A., Takaya, H. et al. (1980). *J. Am. Chem. Soc.* 102: 7932.
102 Knowles, W.S. (2002). *Angew. Chem., Int. Ed.* 41: 1998.

103 Noyori, R. (2002). *Angew. Chem., Int. Ed.* 41: 2008.
104 Knowles, W.S. and Sabacky, M.J. (1968). *J. Chem. Soc. Chem. Commun.*: 1445.
105 Bruckner, R. (2002). *Advanced Organic Chemistry Reaction Mechanisms*, 602. Academic Press.
106 Fleming, I. (1979). *Grenzorbitale und Reaktionen Organischer Verbindungen*, vol. 21. Verlag Chemie.
107 Heasley, V.L. and Chamberlain, P.H. (1970). *J. Org. Chem.* 35: 539.
108 Han, X., Khedekar, R.N., Masnovi, J., and Baker, R.J. (1999). *J. Org. Chem.* 64: 5245.
109 Horasan, N., Kara, Y., Azizoglu, A., and Balcı, M. (2003). *Tetrahedron* 59: 3691.

5
Carbonyl Compounds

The carbonyl group and its chemistry are a significant part of organic chemistry. Most biological and natural compounds contain carbonyl functionality. Particularly with the addition of heteroatoms to carbon–oxygen double bonds (C=O), some new functional groups can be designed. Condensation reactions that provide the formation of new carbon—carbon bonds play an essential role in carbonyl chemistry. The reactions of the carbonyl group can be perceived as very comprehensive and complicated at first. However, when the reactivity of the carbonyl group and the reaction mechanism are studied in depth, it is seen that almost all of these reactions occur via the same mechanism. For example, the commonly known condensation reactions (ester condensation, aldol condensation, Knoevenagel condensation, Perkin condensation, etc.) all proceed by the same mechanism. They are referred to by different names according to the starting compound used. If the characteristic features of these reactions are understood, it will be seen that carbonyl group chemistry is a straightforward and enjoyable subject.

Before starting to examine the reactions of the carbonyl group, it is necessary to understand the reactivity of the carbonyl group. As discussed in previous sections, in order for a chemical reaction to occur, the polarization of bond electrons is required on one side of a molecule with a high electron density and a low electron density on the other. For example, in nucleophilic substitution reactions, we saw that carbon—halogen (C—X) bonds are polarized and nucleophiles attack the carbon atom.

5.1 Reactivity of the Carbonyl Group

The carbonyl group consists of a σ bond and a π bond between the carbon and the oxygen atom. The carbonyl carbon atom and the oxygen atom are sp²-hybridized. Three atoms are connected to the carbonyl carbon atom. All those atoms lie in a plane and the bond angles are approximately 120°. The p orbitals perpendicular to this plane, one on carbon and the other on oxygen, form the π bond. The oxygen atom has two pairs of nonbonding electrons. They occupy the remaining sp² hybrid orbitals of the oxygen atom. As expected, the C=O double bond is shorter than the C—O σ bond and stronger.

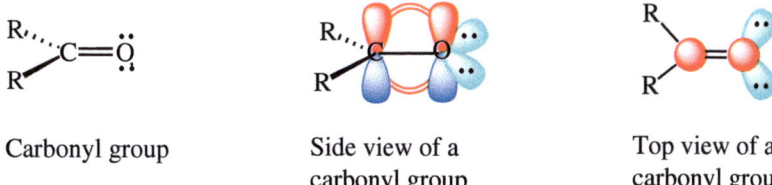

Carbonyl group   Side view of a carbonyl group   Top view of a carbonyl group

Because the electronegativity of the oxygen atom is higher than that of the carbon atom, C—O bond electrons are attracted by the oxygen atom (−I, negative inductive effect). This electron-withdrawing ability of the oxygen atom generates appreciable bond polarization on the carbon—oxygen bond with a partial negative (δ−) charge on oxygen and a partial positive (δ+) charge on carbon. In addition to the inductive effect, the mesomeric effect (−M effect) also causes intense bond polarization.

This polarization of the carbonyl group is responsible for all reactions in which the carbonyl group is involved. As a consequence of this polarization, it is easy to predict that the nucleophiles will attack the carbonyl carbon, which is a Lewis acid, and the electrophiles will attack the carbonyl oxygen atom, which is a Lewis base. Taking into account this polarization, it is possible to understand and predict the carbonyl group reactions.

Reaction Mechanisms in Organic Chemistry, First Edition. Metin Balcı.
© 2022 WILEY-VCH GmbH. Published 2022 by WILEY-VCH GmbH.

Resonance structures of a carbonyl group

Because of the polarization of the C=O double bond, the bond electrons are associated more with the oxygen atom than the carbon atom. This is supported by the measured dipole moments of aldehydes and ketones. The dipole moment is the product of the distance between the positive and negative charge and the magnitude of the charge. The unit of dipole moment is Debye and it is expressed by the letter "μ."

$$\mu = e \times r$$

e, amount of the charge
r, distance between the charges

The distance between the carbonyl carbon and the carbonyl oxygen atom is around 1.2 Å. When we assume that the double bond is fully polarized so that the carbon atom has a positive (+) charge and the oxygen atom has a negative (−) charge, according to the equation given above, the dipole moment should be about 5.9 Debye. However, the dipole moments of aldehyde and ketones measured vary between 2.3 and 2.7 Debye. This value corresponds to the 40%–50% ionic characteristic of the carbonyl group. For this reason, aldehydes and ketones with smaller molecular weight dissolve in water. Hydrogen bonds between the nonbonding electrons on the oxygen atom of the carbonyl group and the H_2O molecules make these compounds soluble in water. As the molecular weight increases, the solubility decreases because the organic part of the compound is becoming more effective. Alkenes do not dissolve in water, regardless of molecular weight, because the polarization of the C=C double bond is very weak.

Hydrogen bonding

Carbon dioxide (CO_2) contains two carbonyl groups. Because the hybridization of the central carbon atom is sp, the molecule has a linear structure, and the total dipole moment μ is zero. The dipole moment is a vector, and there are two equal polarities in the opposite direction in the CO_2 molecule. These vectors in opposite directions cancel each other out. The zero dipole moment does not indicate that the carbonyl group of the CO_2 molecule is not polarized. The carbon atom has a positive charge and the oxygen atoms a negative charge. For that reason, the CO_2 molecule is easily attacked by nucleophiles at the carbon atom.

Dipole moment

The polarization of the carbonyl group is directly responsible for the reactivity of the carbonyl group. When the reactivity of a carbonyl group is compared with that of alkene's C=C double bonds, the former is much more reactive. Alkenes have nucleophilic properties as they have high electron density without significant bond polarization. Therefore, they react primarily with electrophiles according to Markovnikov's rule. However, because of polarization, the carbonyl group can react with both nucleophiles and electrophiles. For example, water can undergo a reversible addition to a carbonyl group quickly, while water cannot react with an alkene double bond in the absence of an acid as a catalyst.

5.1.1 Structure–Reactivity Relationships

An aldehyde carbonyl group is generally more reactive than a ketone carbonyl group. In ketones, two alkyl groups are attached to the carbonyl group. Because these groups are electron donors, they partially reduce the carbonyl group polarization. As a result, the partial positive charge ($\delta+$) on ketone carbonyl carbon is less than that of aldehyde carbonyl carbon. Moreover, steric factors also play an essential role for determining the reactivity. The carbonyl group of an aldehyde is more accessible to a nucleophilic attack than the carbonyl group of a ketone.

When the ketone and aldehyde carbonyl groups are compared with the ester carbonyl group, the reactivity of the ester carbonyl group appears to be less than that of the aldehydes and the ketones. In the case of an ester, one of the alkyl groups in ketone is replaced by an OR group. The oxygen atom will inductively (–I effect) withdraw the σ bond electrons from the C—O bond and increase the polarization. It is expected that the reactivity of the carbonyl group will also increase. However, the non-bonded electron pairs on the oxygen atom will conjugate with the carbonyl group, and oxygen releases its electrons to the carbonyl carbon by resonance. As we mentioned in previous sections, the mesomeric effect is always more dominant than the inductive effect and so the electron density on the ester carbonyl carbon will increase, which will be responsible for the lower reactivity.

One of the most critical indicators showing electron density on the ester carbonyl group is ^{13}C-NMR spectroscopy. Substituents attached to the carbonyl carbon atom influence the carbonyl carbon resonances. Carbonyl carbon resonances appear in two distinct regions. Carboxylic acid and derivatives such as ester, amid, etc., resonate in a range of 160–180 ppm because of the increased electron density at the carbonyl carbon atom. On the other hand, the aldehyde and ketone resonances appear at the lower field of 195–210 ppm.

Double bonds conjugated with the carbonyl group also directly affect the reactivity of the carbonyl group. For example, if a carbonyl group conjugates with a benzene ring, the electron density at the carbonyl carbon atom increases, and the reactivity of the carbonyl group decreases compared to non-conjugated systems. In general, the reactivity of aromatic aldehydes and ketones is lower than that of aliphatics.

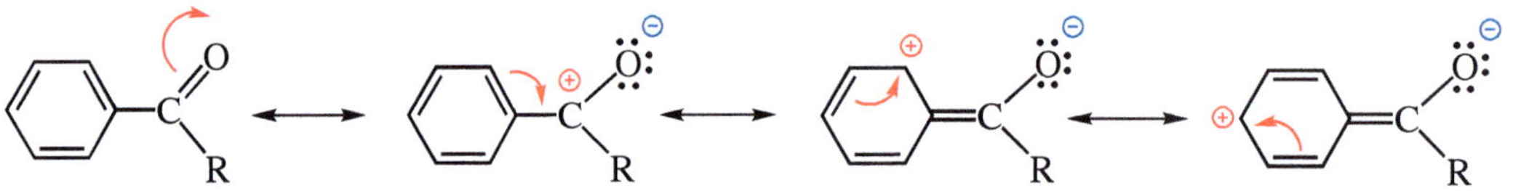

If there are electron-donating groups such as amino and methoxyl groups attached to the aromatic ring, these groups significantly decrease the reactivity of the carbonyl group against nucleophiles by further increasing the electron density on the carbonyl carbon. On the other hand, strong electron-attracting groups, such as the nitro group attached to the aromatic ring, pull electrons from the carbonyl group to the aromatic ring, reducing the electron density on the carbonyl group and significantly increasing the reactivity of the carbonyl group.

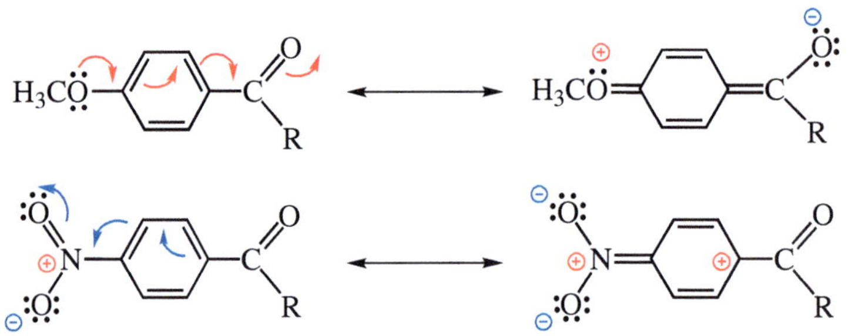

α,β-Unsaturated carbonyl groups are also less reactive than saturated systems. Because of the conjugation of C=C double-bond electrons with the carbonyl group, the reactivity of the carbonyl group decreases, and the nucleophiles prefer to attack more β-carbon atoms. Because the unsaturated carbonyl system contains two electrophilic regions, unsaturated carbonyl compounds are also called ambident electrophiles.

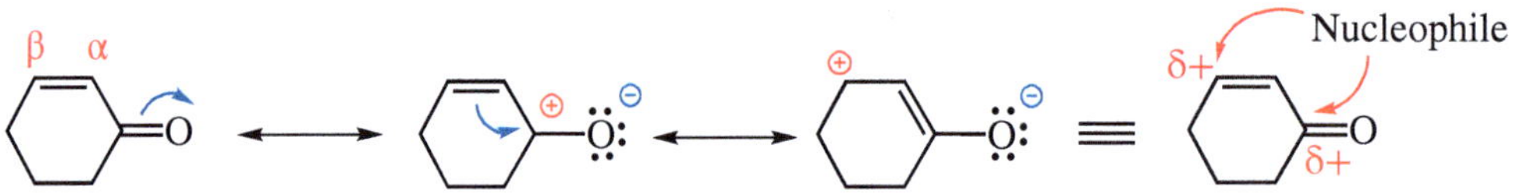

The bulky alkyl groups bonded to the carbonyl carbon atom directly affect the reactivity of the carbonyl group. These groups prevent nucleophiles from attacking the carbonyl group, as shown below, thereby reducing the reactivity of the carbonyl group. The angle between the substituents bonded to the carbonyl group is about 120° because of sp² hybridization. The nucleophilic attack brings these large groups closer together because of the hybridization changes from sp² to sp³ and the bond angle changes from 120° to 109.5°. Thus, large groups reduce the reactivity of carbonyl compounds to nucleophiles.

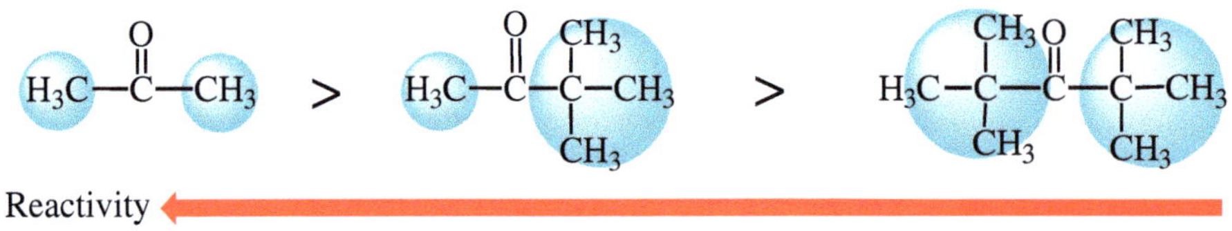

The torsional or angle strain that may occur in the carbonyl group also increases the reactivity of the carbonyl group. In a strain-free carbonyl group, the angle between the carbonyl group and the substituents attached is about 120°. The angle between the alkyl groups bonded to the carbonyl group in cyclobutanone is about 90° and in cyclopropanone is 60°. Consequently, the angle strain in cyclobutanone is about 120°−90° = 30°, and in cyclopropanone, the strain is about 120°−60° = 60°.

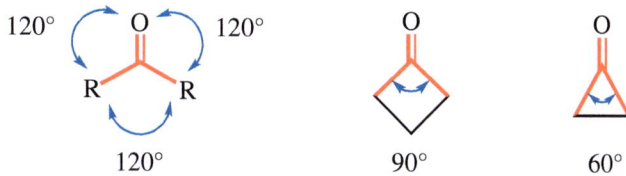

Strain effects contribute to the reactivity of small cyclic carbonyl compounds, and they react faster with nucleophiles to form addition products. The addition of a nucleophile to the carbonyl carbon changes the hybridization from sp² to sp³ and generates a tetrahedral carbon atom with less strain in the ring. For example, as a result of nucleophilic addition to the sp²-hybridized cyclobutanone carbonyl carbon atom, the angle strain will be reduced from 30° to 109.5°−90° = 19.5°. The internal angle in cyclobutanone will not change after the addition. This change in the angle strain increases the reactivity of cyclobutanone compared to a strain-free carbonyl group. Similarly, nucleophilic addition to the cyclopropane carbonyl group decreases the angle strain from 60° to 109.5°−60° = 49.5°. This change in strain makes cyclopropanone more reactive than cyclobutanone.

Any factor that increases the polarization of the carbonyl group also increases the reactivity of the carbonyl group. Electron-attracting groups such as halogens bonded to the α-carbon atom of the carbonyl group increase the reactivity of the carbonyl group toward nucleophiles by the inductive effect.

When one of the methyl protons in acetone is replaced by the chlorine atom, the reactivity of the carbonyl group increases because the chlorine atom is more electronegative than the carbon atom. The carbon–chlorine bond (C—Cl) attracts bond electrons toward the chlorine atom (−I effect). The partially positive (δ+) charged carbon atom attracts the carbon—carbon bond electrons, increasing the polarization on the carbonyl carbon atom. Thus, nucleophiles quickly attack the more positive carbon atom. As the number of electronegative groups attached to the carbon atom increases, because this effect is cumulative, the polarization of the carbonyl group increases; therefore, its reactivity increases even more.

The polarization of the carbonyl group is responsible for all reactions in which carbonyl groups are involved. Because of this polarization, the carbonyl carbon atom is partially positively charged (δ+). Therefore, nucleophiles attack the carbonyl carbon.

Because the carbonyl carbon atom is partially positively (δ+) charged, it withdraws electrons from the groups connected to the carbonyl carbon toward itself to neutralize partially the charge on it. Thus, a partial positive charge (δ+) is then formed on the α-carbon atom. A positive charge on the α-carbon atom attracts electrons from the surrounding bond electrons, and so the bond electrons become polarized. If there is a hydrogen atom attached to the α-carbon atom, the carbon—hydrogen bond electrons will be attracted, reducing the electron density around the hydrogen atom, thereby increasing the acidity of the α-hydrogen atom.

As a result, α-protons can be easily removed by strong bases, such as RO⁻ and HO⁻. This issue is examined in more detail in the condensation section. The bond polarization will continue along the carbon chain; the acidity of the β-protons will also increase. However, this increase is not significant, and it is not possible to remove this hydrogen with a base. The electron-withdrawal effect of the carbonyl group is no longer perceptible on the γ-carbon atom and the bonded protons. Remember that the inductive effect propagates through the σ skeleton and drops exponentially from one carbon to another.

5.2 Reactions of Carbonyl Compounds: Addition Reactions

The basic chemistry of the carbonyl group is based on the addition of a nucleophile to the C=O double bond. We will first discuss the most common reactions of ketones and aldehydes with nucleophiles. When a nucleophile attacks the carbonyl group, the C=O double bond breaks, and a new σ bond is formed between the nucleophile and the carbonyl carbon, forming an alkoxide intermediate. In most of the carbonyl group reactions, there is a hydrogen atom directly attached to the nucleophile, which is then transferred to the alkoxide anion, forming an alcohol. These reactions are significant in organic chemistry because the carbonyl group can be transformed into a variety of functional groups. If there is a hydrogen atom connected to the next adjacent atom, H_2O can easily be eliminated to form a condensation product.

The proton required for elimination may be on the α-carbon atom (H_a) attached to the carbonyl group or the H_b proton directly on the nucleophile. In the first case, a double bond forms between the carbonyl carbon and the α-carbon atom, while in the second case, a double bond is formed between the carbonyl carbon and the nucleophile.

Before proceeding with the reactions of the carbonyl group, it is necessary to examine the energy of the carbonyl group and compare the carbonyl group with the C=C double bonds in terms of energy. We examined the polarization of the carbonyl group. The polarization of the carbonyl group does not indicate that the carbonyl group is weaker than C=C double bonds. The carbonyl group is in fact stronger. The bond decomposition energies of some carbonyl compounds are given in Table 5.1.

Table 5.1 Bonding energies of some carbonyl compounds.

Carbonyl compounds	Bonding energies	
	in kcal/mol	in kJ/mol
:C=Ö ⇌ :C≡O:	257	1077
Ö=C=Ö ⇌ Ö=C−Ö:	192	805
H_2C=C=Ö ketene	185	769
R_2C=Ö ketone	179	749
HRC=Ö aldehyde	176	736
H_2C=Ö formaldehyde	166	694
R_2C=CR_2 olefin	146	610

The data in Table 5.1 show that the C=O double-bond energy is generally greater than the C=C double-bond energy. The carbon–carbon double-bond (C=C) energy is approximately 146 kcal/mol. Because the carbon–carbon (C—C) σ-bond energy is 83 kcal/mol, we can calculate that the π bond energy is about 146−83 = 63 kcal/mol. On the other hand, the C=O double-bond energy is about 178 kcal/mol and the carbon–oxygen σ bond energy is about 86 kcal/mol. We can see that the C=O π bond energy is about 178−86 = 92 kcal/mol. The π bond energies in ketones and aldehydes are found to be 93–96 kcal/mol. These data show that the C=O double bond is stronger than the C=C double bond. According to these values, C=O double bonds cannot easily react with nucleophiles compared with C=C double bonds. However, C=O double bonds react more easily than C=C double bonds. There is a contradictory situation here that needs to be clarified.

$$R_2C=CR_2 + H_2O \underset{\text{Very slow}}{\overset{\text{Slow}}{\rightleftharpoons}} H-CR_2-CR_2-OH \quad \Delta H° = -10\,\text{kcal/mol}$$

$$R_2C=O + H_2O \underset{\text{Very fast}}{\overset{\text{Fast}}{\rightleftharpoons}} R_2C(OH)_2 \quad \Delta H° = +6\,\text{kcal/mol}$$

Geminal diol

The addition of the H$_2$O molecule to the C=C double bond is an exothermic reaction and a stable product, an alcohol, is formed. However, this reaction does not occur under normal conditions. Acid catalysts are needed for the reaction to occur and the activation barrier of this reaction is quite high. However, the addition of H$_2$O to the C=O double bond is an endothermic and rapid reaction. H$_2$O elimination from the resulting geminal diol is an exothermic and fast reaction. In the case of a C=C double bond, both carbon atoms have the same electronegativity. In a C=O double bond, oxygen is more electronegative than carbon. This causes an appreciable polarization of the C=O double bond. This polarization gives the carbon atom some degree of positive charge, which attracts nucleophiles.

We will not examine all kinds of addition reactions to the carbonyl groups in detail. We will focus on mechanisms only by highlighting some types of addition reactions. The structure of the nucleophile plays a significant role in nucleophilic addition reactions. The nucleophile attacking the carbonyl group can be a hydride ion (H$^-$), a heteroatom, or a carbanion. Depending on the nature of the nucleophile, the addition reactions may be reversible or not. For example, when a heteroatom is added to a carbonyl group having the structure HX, the product formed is not very stable and can revert to the starting compounds. In this case, there will be equilibrium between the starting compounds and the addition product. However, if the nucleophile is a carbanion (such as Grignard compounds and organolithium compounds), these reactions will not be reversible as a new C—C bond results from the addition, which cannot be cleaved.

$$R_2C=O + X-H \rightleftharpoons R_2C(OH)(X)$$
X = heteroatom   Reversible reaction

$$R_2C=O + R-H \rightarrow R_2C(OH)(R)$$
R = carbon nucleophile

We will now start to examine the nucleophilic addition reactions of the carbonyl group by classifying them according to the structure of the nucleophile involved (heteroatom, carbon, hydride, etc.). While examining these reactions, we will highlight the common points observed.

5.2.1 Hydration: Addition of Water to Carbonyl Groups

In the previous section, we briefly examined the addition reaction of the H$_2$O molecule to a carbonyl group. The product is a *hydrate*, a geminal diol, which is not a very stable compound. This reaction is called a reversible hydration reaction, and the diol formed can eliminate H$_2$O to generate a ketone or aldehyde. It is straightforward to test whether the H$_2$O molecule adds to the carbonyl group and forms a hydrate that is in equilibrium with the starting carbonyl group and water.

5 Carbonyl Compounds

Using isotopically labeled water enables rapid and reversible addition of water to carbonyl groups. The molecule containing any carbonyl group (with ^{16}O isotope) is mixed with H_2O containing the ^{18}O isotope, and the carbonyl compound is recovered after standing for a while. Spectroscopic examinations confirmed that a significant part or all of the carbonyl group (depending on the duration and amount of the labeled H_2O) was exchanged with the ^{18}O isotope. This change can only be explained by reversible addition of water to the carbonyl group.

$$R_2C=O + H_2O^* \underset{\text{Very fast}}{\overset{\text{Fast}}{\rightleftharpoons}} R_2C(OH)(O^*H) \underset{\text{Very fast}}{\overset{\text{Fast}}{\rightleftharpoons}} R_2C=O^* + H_2O$$

$H_2O^* = H_2{}^{18}O$ — Hydrate

The position of the equilibrium formed between the hydrate and the starting material (carbonyl group and water) depends entirely on the structure of the carbonyl group. The equilibrium generally shifts to the side of the carbonyl compound for steric reasons. However, in some cases, the hydrate is favored. For example, when formaldehyde is mixed with H_2O, nearly 100% of formaldehyde molecules are in hydrate form. This is easy to prove. The carbonyl carbon atom of formaldehyde resonates at 197 ppm. This is the typical region for aldehyde and ketones (190–210 ppm). However, when formaldehyde is dissolved in water, formaldehyde's carbonyl peak appears at 83 ppm, indicating the presence of a tetrahedral carbon atom. Formaldehyde is extremely reactive as there are no alkyl groups attached to the carbonyl carbon atom, which would decrease the polarization of the carbonyl group by electron donation. Moreover, there is no steric effect.

$H_2C=O$ + H_2O ⇌ $H_2C(OH)_2$
0.1%   99.9%

$(H_3C)_2C=O$ + H_2O ⇌ $(H_3C)_2C(OH)_2$
99.9%   0.1%

Acetaldehyde forms around 60% hydrate, while an aqueous solution of acetone consists of only about 0.1% hydrate and 99.9% acetone. With a powerful electron-attracting group, such as a trifluoromethyl group, the equilibrium shifts completely to the side of the hydrate. The strong electron-withdrawing group attached to the carbonyl carbon increases the polarization of the carbonyl group and make it more reactive. Equilibrium constants for hydrate formation of some carbonyl compounds are given in Table 5.2.

Table 5.2 Equilibrium constant K values for some carbonyl compounds.

$$K_a = \frac{[\text{Hydrate}]}{[\text{Carbonyl compound}]}$$

Carbonyl compound	K_a: 25 °C (aqueous)
CH_2O	2280.0
CH_3CHO	1.06
CH_3CH_2CHO	0.85
CF_3CHO	29000.0
CH_3COCH_3	1.4×10^{-3}
$ClCH_2COCH_3$	0.11
CF_3COCH_3	5.0
CF_3COCF_3	1.2×10^6

As discussed earlier, cyclopropanone is a very strained compound because the bond angles are 60° instead of 120° or 109.5°. Such carbonyl compounds easily form hydrates to a significant extent. The driving force for the formation of hydrates is the change in the sp^2 hybridization of the carbonyl carbon to sp^3 hybridization, which releases the strain inherent in the cyclopropanone ring. Therefore, hydration is favored in cyclopropanone and cyclobutanone.

Although the geminal diols formed in some systems are more stable than aldehydes or ketones, it is not possible to isolate hydrates. Hydrates are converted back to original carbonyl compounds by the elimination of water.

Reactive carbonyl compounds readily undergo hydration with H_2O, while unreactive carbonyl compounds require catalysts to form hydrates. Acids or bases catalyze hydrate formation. The acid-catalyzed hydration mechanism is given below.

Acid-catalyzed hydration mechanism

The mechanism given above is a general mechanism that can be applied to all kinds of acid-catalyzed reactions of carbonyl groups. Water is a weak nucleophile. To start the hydration, we have to activate the carbonyl group first. The reaction starts with an electrophilic attack; the protonation of the carbonyl group occurs, which increases the polarization of the carbonyl group. The positive charge on the oxygen atom thereby attracts the bonding electrons and makes the carbonyl carbon more electropositive. The H_2O molecule, a weak neutral nucleophile, quickly attacks the activated carbonyl group, breaks the π bond, and moves the π electrons to the positively charged oxygen atom and forms a protonated intermediate. Hydrate formation is completed by deprotonation of the oxonium ion to form the neutral hydrate.

All chemical reactions can be considered acid–base reactions (Lewis acids and Lewis bases). If there is no reaction between the two compounds, it indicates that the reactivity either of the acid or of the base or both functional groups is low. Two methods can be implemented to ensure that the reaction takes place. The reactivity of a Lewis acid or the compound that acts as a base need to be increased in some way, for which catalysts are needed. In the hydration reaction, we have demonstrated above that acid was used to increase the reactivity of the carbonyl group, which is acting as a Lewis acid. We have here increased the reactivity of the carbonyl group just by protonation. This increase in the Lewis acid property can be seen from the resonance structures of the protonated carbonyl group.

Protonated carbonyl group   Activated carbonyl group

The second method is to increase the nucleophilicity of the nucleophile that attacks the carbonyl group. For this, a base is added to the reaction medium to hydrate the carbonyl group. The hydroxyl ion is used as the catalyst, which is more nucleophilic than the H_2O molecule. The first step starts with the attack of the hydroxyl group (HO^-) on the carbonyl carbon, which pushes the π electrons toward the oxygen atom, forming an anionic intermediate. This intermediate abstracts a proton from water and makes the overall reaction catalytic in the base.

Base-catalyzed hydration mechanism

5.2.2 Hemiacetal Formation: Addition of Alcohols to Carbonyl Groups

Because water adds to at least some carbonyl compounds, it is not surprising that alcohols can also add to the carbonyl groups as they are more nucleophilic than water. As the alkyl groups are electron donors, they increase the electron density on the alcohol oxygen atom, making them more nucleophilic. When aldehydes and ketones are treated with alcohols, as seen in the hydration reaction, alcohols add to aldehyde and ketone carbonyl groups by the same mechanism. The products formed are called *hemiacetals*.

According to the definition used before, hemiketals are formed when alcohols add to ketones, while hemiacetals are formed when alcohols add to aldehydes. Compounds formed by the addition of alcohols to both ketones and aldehydes are today called hemiacetals, without any discrimination. Ketal is an older term for acetal derived from ketone. In hemiacetals, a hydroxyl (OH) group is attached to the carbon atom. Molecules formed by the replacement of this group with one mole of alcohol are called *acetals*. No catalyst is required to add one mole of alcohol to the carbonyl group. Hemiacetal formation is a reversible reaction and the same structural features can stabilize them, as we discussed, by the formation of stable hydrates. However, catalysts are needed for acetal formation. To shift the equilibrium to the side of the acetal, H_2O formed as a result of the reaction must be removed from the reaction medium.

The acid-catalyzed acetal formation mechanism is given above. The mechanism follows precisely the same steps for hydrate formation, but, instead of water, alcohol is used as a nucleophile for the formation of hemiacetals and acetals. According to this mechanism, the proton first binds to one of the free electrons of carbonyl oxygen, forming a positive (+) charge on oxygen. This charge increases the reactivity of the carbonyl group by attracting bond electrons and makes the carbonyl carbon more susceptible to nucleophilic attack. The nucleophile, the alcohol (ROH), adds to the carbonyl carbon and pushes the π electrons toward the oxygen atom. The hydroxyl oxygen atom becomes neutral. Removal of the proton from the tetrahedral intermediate provides the hemiacetal.

Because the reaction is carried out in an acidic medium, it does not stop at this stage. The two oxygen atoms (OH and OR) in the hemiacetal are basic and so either one can be protonated. Protonation of the hydroxyl group in the hemiacetal converts the hydroxyl group into a good leaving group. This facilitates the removal of H_2O from the molecule, forming an oxonium ion as an intermediate. The stabilization of the initially formed carbocation by the formation of the oxonium ion supports the separation of water. The addition of a second molecule of alcohol to the oxonium ion followed by loss of a proton yields the neutral acetal.

All stages shown in acetal formation are reversible. The equilibrium can be shifted toward the acetal side or to the aldehyde or ketone side. As mentioned briefly above, the water eliminated from the hemiacetal must be removed from the reaction mixture in order to shift the equilibrium to the acetal side. The acetals can be hydrolyzed back to the starting ketone or aldehyde upon reaction with an acidic aqueous solution. This reverse reaction of acetal formation is called *acetal hydrolysis*. Acetals are stable compounds in the absence of acids. This stability of acetals makes them extremely important in synthetic chemistry.

We discussed that hemiacetals can be stabilized by the structural features of the ketone or aldehyde. Hemiacetals can also be stabilized by forming a cyclic structure. Five- and six-membered cyclic hemiacetals are stable compounds and they can be isolated. Molecules that contain a carbonyl group and alcohol within the same molecule, such as 4-hydroxybutanal and 5-hydroxypentanal, undergo intramolecular cyclization spontaneously to form the corresponding cyclic hemiacetals.

As five- and six-membered rings are strain-free compounds, five- or six-membered hemiacetals are common in nature, especially in sugar molecules. Sugar molecules have both aldehyde functional groups and hydroxyl groups on the same molecule. Hemiacetals in sugars are formed by cyclization of hydroxy aldehydes. By connecting one of the hydroxyl groups to the aldehyde group intramolecularly, hemiacetals with five- or six-membered rings (pyranose or furanose) are formed. Sugar molecules are generally found in hemiacetal form.

The hemiacetal in glucose, which contains an aldehyde and several hydroxyl groups within the same molecule, is formed by intramolecular cyclization of these groups. When the hydroxyl group attached to the C5 carbon atom in D-glucose attacks the aldehyde, cyclization occurs to yield a six-membered ring. Cyclization generates a new stereogenic center. Because of the sp^2 hybridization of the carbonyl group and its planar structure, the hydroxyl group attacks the aldehyde from either face, forming two different acetals, namely, α-D-glucopyranose and β-D-glucopyranose. These two molecules are diastereomers and are called *anomers*. The carbon atom to which the newly formed hydroxyl group is attached is called *anomeric carbon*. For example, glucose cyclizes in aqueous solution to a mixture of anomers. These anomers are stable compounds in the anhydrous environment. Pure α-D-glucopyranose has a specific rotation of $[\alpha]_D = +112°$ while that of β-D-glucopyranose is $[\alpha]_D = +18.7°$. When any of these pure compounds is dissolved in water, the specific rotation of the solution slowly changes to a fixed value of +52.7°. This observation clearly shows that both isomers generate an equilibrium in the aqueous medium over the open structure according to the equation given above. This phenomenon is called *mutarotation*. After the establishment of this equilibrium, the ratio of these anomers was calculated as 37 : 63 using their specific rotation.

5.2.2.1 Cyclic Acetals and Their Synthetic Application

We emphasized that acetal formation takes place in the presence of acid catalysis. Acetals can be hydrolyzed back to the corresponding ketones or aldehydes by the reaction in an acidic solution. However, the acetals are stable to bases and nucleophiles. When ketones or aldehydes are reacted with 1,2- or 1,3-diols such as ethylene glycol or propane-1,3-diol, in place of two equivalent alcohols, a cyclic acetal is formed. Cyclic acetals are more convenient to synthesize and are highly stable compounds. The acetal formation mechanism is similar to the formation mechanism of a hydrate. In the first step of the reaction, a hemiacetal is formed as an intermediate. The second part of the reaction involves conversion of the hemiacetal into the final product, a cyclic acetal. Cyclic acetals are compounds that play an important role in the protection (masking) of carbonyl groups.

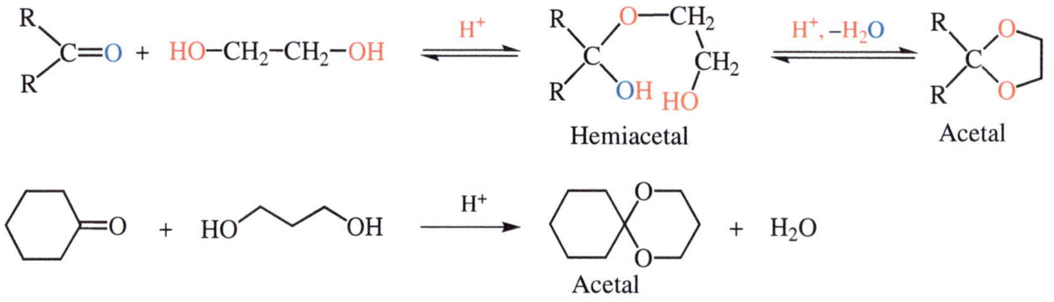

5.2.2.2 What is the Protecting Group?

If a molecule has two different functional groups and both groups react with the reagent used, when one of these groups is desired to react, then the other group must be protected in some way. A group that protects a functional group against a reaction is called a *protecting group*. For example, we have a ketoester and suppose that we want to reduce the ester functionality into an alcohol. However, under the reaction conditions, both functional groups will be reduced. We do not want the ketone to be reduced. Thus, we have to protect the ketone by converting it into a cyclic acetal. Of course, the protecting group must be inert against the reagent used. When the reduction reaction with the ester group is completed, the original carbonyl group can be restored by carrying out the reverse of the reaction, i.e. treatment with an aqueous solution of an acid.

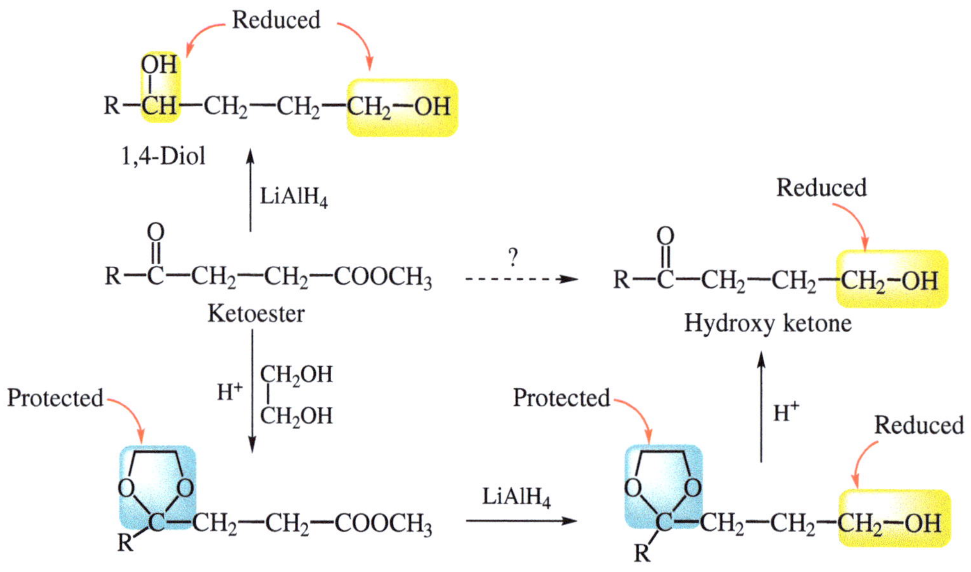

Let us give a concrete example to clarify the issue. The ester group of the ketoester shown above is desired to be reduced to a primary alcohol. For this, the molecule will be reacted with Lithium aluminum hydride (LiAlH$_4$). As we will see in detail later, hydrides reduce esters to primary alcohols. As the ketone carbonyl group is more reactive than the ester carbonyl group, the ketone will also be reduced to an alcohol to form a 1,4-diol. To protect the ketone from being reduced, the carbonyl group is first converted to a cyclic acetal by the reaction with ethylene glycol. Then, the compound will be submitted to the reduction reaction. LiAlH$_4$ cannot react with the acetal group. Then, the acetal group is removed in acidic medium to obtain the targeted keto alcohol.

5.2.2.3 Protection of Diols

Among the most important reactions in organic chemistry are the protection and deprotection of a functional group. Acetal formation is one of the methods applied for the protection of the adjacent hydroxyl groups in carbohydrates, as it is very often applied to the protection of carbonyl groups. Carbohydrates are generally insoluble in organic solvents. In order to increase solubility and perform the desired reactions with other functional groups, diols are reacted with ketones in an acidic environment, and the related acetal is synthesized and alcohol groups are protected. For acetal synthesis, acetone or cyclohexanone is used in the presence of a catalytic amount of acids, such as p-toluenesulfonic acid or camphorsulfonic acid. Yields are higher when 2,3-dimethoxypropane is used instead of acetone. Besides 1,2-diols, 1,3-diols can also be successfully protected as shown below [1].

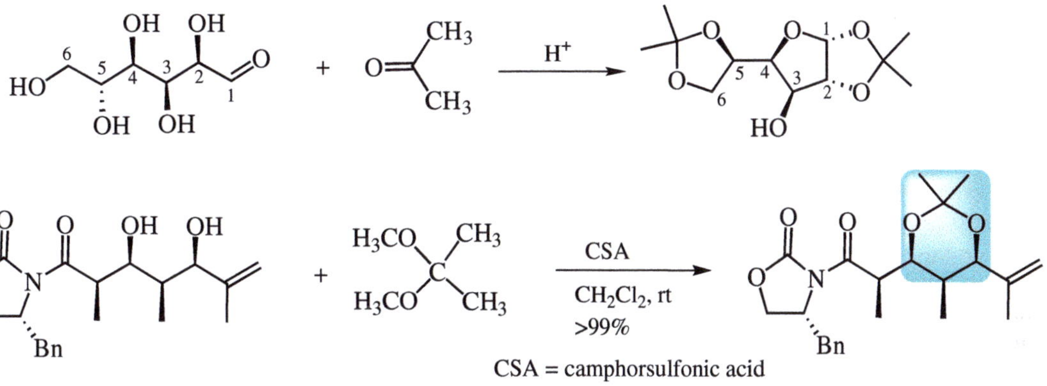

The protection process is often used to prevent the formation of by-products in reactions with other functional groups. For example, cyclohexa-3,5-diene-1,2-diol was protected by acetal formation with 2,2-dimethoxypropane. Photooxygenation of the diene unit affords the cycloaddition product. Selective reduction of the peroxide linkage was performed with thiourea. Removal of the protecting group was carried out in acidified methanol solution to give the natural product conduritol-A in quantitative yield [2].

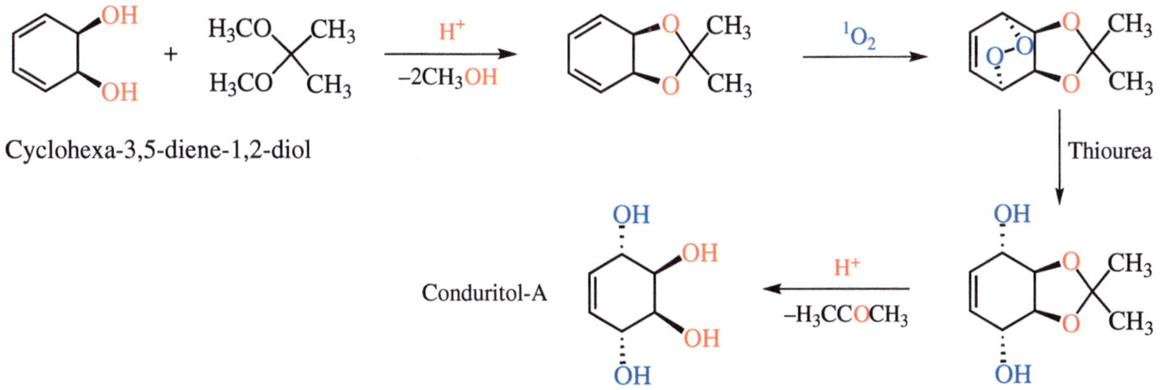

5.2.2.4 Formation of Thioacetals and Their Synthetic Application

Sulfur analogs of acetals are called *thioacetals*. Aldehydes and ketones react with both 1,2-ethanedithiol and 1,3-propanedithiol to form cyclic thioacetals. Monotiols are also widely used. The mechanism for the formation of thioacetals is the same as that for the formation of acetals. The only difference is that the attacking atom is a sulfur atom rather than oxygen. Because the sulfur atom is a stronger nucleophile than oxygen, thioacetals are easier to form. Lewis acids, such as $ZnCl_2$ or BF_3, are used to catalyze the formation of thioacetals. In some cases, protic acids are also used as a catalyst. The equilibrium constant for thioacetal formation is greater than that for acetal formation. Therefore, they can be synthesized easily [3].

Thioacetals are stable in basic and acidic solutions. If a carbonyl group is protected as an acetal, no reaction can be carried out in acidic medium because the acetal group would hydrolyze immediately. However, if a carbonyl group is protected as a thioacetal, reactions can be carried out with other functional groups in acidic and basic conditions. The thioacetal group can be easily removed with mercury salts (Hg^{2+}), which are highly thiophilic and can bind to sulfur to make a good leaving group, and so aldehyde or ketone can be released. The formation of an insoluble mercury(II) sulfide is the driving force behind the reaction. Deprotection of cyclic thioacetals can also be performed with bis(trifluoroacetoxy)iodobenzene [4].

Compared with acetals, thioacetals have different chemical properties, and these properties are of great importance in synthetic chemistry. One of the most important properties is the reduction of thioacetals to hydrocarbons. Thioacetals are easily reduced to hydrocarbons with hydrogen and Raney Ni (nickel–aluminum alloy) as a catalyst under neutral conditions. This reaction is called the *Mozingo reduction*, also known as the *Mozingo reaction* [5]. In contrast, acetals do not react with Raney Ni and hydrogen. This method is successfully applied for the reduction of carbonyl groups to hydrocarbons. If a product containing carbonyl groups has acid- or base-sensitive functional groups, the Clemmensen reduction (under acidic conditions, see Section 6.6) or Wolff–Kishner reduction (under basic conditions) cannot be applied to the reduction of that carbonyl group. Thus, the formation of thioacetals followed by desulfurization is the method of choice.

In the Mozingo reduction it is thought that the reaction takes place on the metal surface. The C—S bond is about 20 kcal/mol weaker than the C—C bond. Absorption of thioacetal on the metal surface weakens the C—S bond. It is assumed that carbon radicals are formed, which react with the hydrogen radicals complexed at the metal surface.

Thioacetals are valuable intermediates for organic synthesis. For example, the ketone functionality of a ketoaldehyde, whose structure is given below, was intended to be converted to an oxime. However, because the aldehyde group is more reactive than ketone, oxime synthesis derived from ketone is not possible. For this reason, the aldehyde needs to be protected first. For the chemoselective protection of aldehyde functionality, the ketoaldehyde was treated with 1,2-ethanedithiol in the presence of SnCl$_2$ to give the cyclic thioacetal. The reaction of the carbonyl group with hydroxylamine provides the ketoxime, which was subjected to hydrolysis using HgCl$_2$–HgO in aqueous methanol to give the desired product [6].

5.2.2.5 Umpolung: Polarity Inversion of the Aldehyde Carbonyl Group

We have discussed the reactivity of carbonyl groups in detail and found that because of bond polarization carbonyl carbon has an electrophilic property and nucleophiles attack the carbonyl carbon atom. If this polarization can be changed in some way, then electrophiles will attack the carbonyl group. A process in which the natural polarity of a functional group is inverted is called *umpolung* [7, 8]. Umpolung is a German word and is the accepted scientific term for reverse polarity.

Of course, electrophiles cannot attack a carbonyl carbon atom. The carbonyl group needs to be masked first. For this, aldehydes are first converted to cyclic thioacetals (dithianes). The acyl proton can be abstracted by organolithium compounds (*n*-BuLi) in tetrahydrofuran (THF) at low temperatures to give the lithio derivative. This carbanion is actually a masked acyl anion. The resulting carbanion can react with various electrophiles such as aldehydes, ketones, epoxides, and acid derivatives. Then, the thioacetal group can be removed by the reaction with Hg^{2+} salts to release the carbonyl group.

The carbanion generated as an intermediate is actually an acyl anion. It is not possible to synthesize acyl anions under normal conditions. They are very unstable intermediates. However, as shown, it is straightforward to synthesize this carbanion starting from 1,3-dithianes. The masked acyl anions readily react with electrophiles, as the acyl carbon atom carries a negative (−) charge. As a result, the electrophiles react with this reverse-polarized carbonyl carbon atom, namely, an acyl anion. Therefore, this reaction is called umpolung.

Here, the following question may immediately spring to mind: *Can this process be performed with normal acetals?* In acetals, oxygen atoms are more electronegative than the sulfur atom in dithianes. Therefore, the acyl proton in acetals is more acidic, and it can be abstracted more easily by the bases than the acyl proton in 1,3-dithianes. However, the stabilization of the carbanion formed after the proton abstraction plays an important role here. Of course, the oxygen atoms can stabilize the carbanion inductively better than the sulfur atoms. However, the sulfur atom has empty d orbitals, which can accept the electrons of the carbanion. Because oxygen does not have empty d orbitals, the carbanion cannot be stabilized as in the case of 1,3-dithianes.

Cyclic 1,3-dithianes are important synthons and are successfully applied in the synthesis of various compounds. For example, it is possible to synthesize α- as well as β-hydroxyketones with this method. First, any aldehyde is converted to 1,3-dithiane and treated with *n*-BuLi to create a carbanion. Because the resulting carbanion is nucleophilic, it can react with an electrophile to give substituted carbonyl compounds.

5.2.3 Reactions of Aldehydes and Ketones with Amines

We examined the reactions of carbonyl compounds with alcohol and thiols, and we saw that acetals and thioacetals are formed. The common feature of thiols and alcohols is that there is only one hydrogen atom attached to the nucleophilic atom (oxygen and sulfur) attacking the carbon atom. If there are two hydrogen atoms (as in amines) on the nucleophile attacking the carbon atom, the reaction usually does not stop at the stage of the addition reaction. The reaction continues with one mole of H_2O elimination, forming a double bond between the carbon atom and the heteroatom. Such reactions are also referred to as *addition–elimination reactions* or *condensation reactions* and the product formed is called an *imine* or *Schiff base*.

The mechanism of imine formation is the same as that of H_2O or ROH addition to carbonyl groups that we have previously shown. The formation of an imine from aldehydes or ketones is reversible and generally takes place under acidic or basic conditions or upon heating. The acid-catalyzed formation of an imine begins with a nucleophilic attack of a primary amine on the carbonyl group, followed by transfer of a proton from the positively charged nitrogen atom to the negatively charged oxygen atom to give the neutral amino alcohol (also called carbinolamine) with a tetrahedral structure. The intermediate, amino alcohol, can be converted into an imine by directly eliminating one mole of H_2O. If the medium contains a catalytic amount of acid, both heteroatoms (oxygen and nitrogen) can be protonated. Because nitrogen is more basic, it is more likely to be protonated. However, the protonation of the nitrogen atom would eliminate the amine group and form the starting material. H_2O elimination becomes easier only when the oxygen atom is protonated, and a double bond is formed between the carbon and nitrogen atom through the electrons on the nitrogen atom. The final product, an imine, occurs when the protonated amine eliminates the proton. Because all of these steps are reversible, the H_2O molecule formed in the reaction must be removed from the reaction medium to shift the equilibrium to the side of the imine. Typically, the dehydration of the amino alcohol is the rate-determining step of Schiff base formation. Acid ion concentration should be controlled when using H^+ as the catalyst in the reaction. If the acid concentration is kept in excessive amounts, the basic nitrogen atom can be protonated and it becomes non-nucleophilic, and no reaction will take place. Therefore, many imine syntheses are best carried out at mildly acidic pH. Schiff bases are stable compounds. However, they can hydrolyze to their ketones and aldehydes by aqueous acid or base.

5.2 Reactions of Carbonyl Compounds: Addition Reactions

[Scheme: cyclopentanone + R–NH₂ → protonated addition intermediate → amino alcohol (carbinolamine) → protonated carbinolamine → (–H₂O) → protonated imine ⇌ imine, Schiff base (C=N–R)]

The mechanism we have described in detail above is the same as the addition mechanism of ROH and H₂O to the carbonyl group. The striking difference is that one mole of H₂O molecule is eliminated after being added. Similar addition and elimination reactions are observed with different amine compounds. We have emphasized that there must be two hydrogen atoms bonded to the nucleophile so that elimination can occur. The second hydrogen atom was necessary for elimination. On the other hand, *how does the reaction proceed when only one hydrogen atom (reactions with secondary amines) is attached to the nitrogen atom?* Let us try to answer this question.

Secondary amines also react similarly with aldehydes and ketones. In the first part of the reaction, the amine reacts with the aldehyde or ketone to give an unstable addition compound with a tetrahedral structure. So far, the reaction mechanism is the same. When we compare the intermediates formed by the addition of primary and secondary amines to carbonyl groups, there is a hydrogen atom required for H₂O elimination on the nitrogen atom on the intermediate product formed by the addition of the primary amine to the carbonyl group.

[Scheme: cyclopentanone + R–NH–R (secondary amine) → protonated intermediate → tetrahedral intermediate with N(R)(R) and OH]

The intermediate formed by the addition of the secondary amine has no hydrogen atom on the nitrogen that can be used for H₂O elimination. Therefore, here an imine cannot be formed. However, if there is any hydrogen atom attached to the carbon atoms adjacent to the carbon bearing the hydroxyl group, H₂O elimination can occur with those α-hydrogen atoms.

[Left structure: Intermediate formed by RNH₂ + carbonyl group — highlighted H on N, labeled "Hydrogen atom for imine formation"]

[Right structure: Intermediate formed by R₂NH + carbonyl group — α-hydrogens labeled, "No hydrogen atom for imine formation"]

Protonation of the hydroxyl group by acid present in the reaction medium converts the hydroxyl group into a better leaving group. The lone pair of electrons on the nitrogen atom eliminate water and form an iminium ion as an intermediate, as in the case of imine formation. However, the nitrogen lacks another hydrogen to eliminate. Therefore, the system goes on an alternative path. With the formation of the iminium ion, the acidity of the protons bonded to the α-carbon atoms is increased because of both inductive and mesomeric effects. Therefore, to generate a neutral compound, one of the protons in the α-position is eliminated to form a C=C double bond and the acid catalyst is regenerated. These compounds are called *enamines*. The suffix "en" indicates the C=C double bond and the "amine" indicates that the molecule is in an amine structure.

Enamines can be converted back to the starting compounds (carbonyl compounds) by hydrolysis in an acidic medium. On the other hand, enamines are stable compounds in a basic environment. This feature of enamines is utilized in the alkylation of carbonyl compounds. This topic is covered in detail in the "alkylation of carbonyl compounds" section.

In the third type of amine, a tertiary amine (NR_3) may also add to carbonyl compounds. Although most tertiary amines are good nucleophiles, they can attack the carbonyl group; however, they cannot form carbinolamines. The addition product formed cannot be stabilized and they can only decompose back to their starting materials. Generally, we can say that tertiary amines do not react with aldehydes and ketones.

Aldehydes and ketones also undergo addition reactions with ammonia, forming relatively unstable imines. They turn into ketones and aldehydes by the reaction with water. Nevertheless, such imines formed by the reaction of ammonia with carbonyl compounds are useful intermediates for the synthesis of primary amines. When the reaction with ammonia is carried out in the presence of hydrogen gas and suitable metal catalysts, the hydrogen molecule adds to the C=N double bond, forming primary amines [9]. The reaction is called *reductive amination*.

Aldehydes and ketones also react with several amine derivatives, such as hydroxylamine, hydrazine, and hydrazine derivatives, such as phenylhydrazine, 2,4-dinitrophenylhydrazine, and semicarbazide, to form imine derivatives. The products formed are called oxime, hydrazone, and semicarbazone. Table 5.3 shows the structures of these products. The formation mechanism of all these compounds is similar to that of imine. The reaction begins as a nucleophilic attack

Table 5.3 Various imine derivatives of aldehydes and ketones.

Carbonyl compound	Amine reagent	Imine derivative	Name of the products
R₂C=O	H₂N—OH Hydroxylamine	R₂C=N—OH	Oxime
R₂C=O	H₂N—NH₂ Hydrazine	R₂C=N—NH₂	Hydrazone
R₂C=O	H₂N—NH—C₆H₅ Phenylhydrazine	R₂C=N—NH—C₆H₅	Phenylhydrazone
R₂C=O	H₂N—NH—C₆H₃(NO₂)₂ 2,4-Dinitrophenylhydrazine	R₂C=N—NH—C₆H₃(NO₂)₂	2,4-Dinitrophenyl-hydrazone
R₂C=O	H₂N—NH—C(O)—NH₂ Semicarbazide	R₂C=N—NH—C(O)—NH₂	Semicarbazone

of the amine group on the carbonyl group, forming carbinolamines, which undergo dehydration to form imine derivatives. In amine derivatives used, the substituents bonded to the nitrogen atom are different; therefore, the products are named differently.

5.2.3.1 Oximes

Oximes are formed as a result of condensation of ketones and aldehydes with hydroxylamine. Oximes that are obtained from aldehydes are called *aldoximes*, while those obtained from ketones are called *ketoximes*. The pH of the medium affects the rate of formation of oximes. The maximum oxime formation rate is observed when the pH is around 4.5. If the pH of the medium is too low (if the medium is acidic), the hydroxylamine will be protonated, and the nucleophilic power of the nitrogen atom decreases and the formation of oxime is prevented. If the substituents bonded to the carbon atom in an oxime are different, then two stereoisomeric isomers, *syn-* and *anti-*oximes, will be formed. Those isomers are stable and can be separated from each other. Oximes are crystalline compounds and are found in the structures of some natural products.

Syn-aldoxime   *Anti*-ketoxime   *Syn*-ketoxime

Oximes are synthetically important intermediates and are used as the starting compound in the synthesis of various compounds. For example, nitriles are readily synthesized by the dehydration of aldoximes. Cyclohexanone oxime can be synthesized from the reaction between cyclohexanone and hydroxylamine. This compound is an essential intermediate in the production of nylon 6, a widely used polymer (see Beckmann and Neber rearrangement). Pralidoxime chloride, which contains an oxime functional group, is an effective oxime derivative used to treat organophosphate poisoning [10].

Cyclohexanone oxime   Pralidoxime chloride

5.2.3.2 Hydrazones

Hydrazone derivatives are compounds formed by the reaction of ketones and aldehydes with hydrazine (Table 5.3). They can be synthesized by the condensation of hydrazine and their substituted derivatives with aldehydes and ketones. They are essential intermediates and they play a key role in heterocyclic chemistry. Hydrazones were used to determine the structures of the unknown aldehydes and ketones before spectroscopic methods were developed. Hydrazone derivatives are generally crystalline compounds. The reaction of the unknown aldehyde or ketone with 2,4-dinitrophenylhydrazine gives crystals of the unknown carbonyl compound. A comparison of the melting point of the hydrazone synthesized with those of the previously characterized compounds can identify the unknown compound. Semicarbazones are also included in the hydrazone group. Although semicarbazide contains three nitrogen atoms, only one of them reacts with carbonyl compounds. Nitrogen atoms directly attached to the carbonyl group do not show nucleophilic properties as the terminal nitrogen atom. The free electrons on these nitrogen atoms attached directly to the carbonyl group are in conjugation with the carbonyl group. Therefore, the nucleophilicity of these nitrogen atoms is reduced.

Semicarbazide

5.3 Reduction of Carbonyl Groups

5.3.1 Wolff–Kishner Reduction (Under Basic Conditions)

One of the most important properties of hydrazones is their reduction to alkanes in a basic medium. Aldehydes and ketones are first converted to the corresponding hydrazones by the reaction with hydrazine. The reaction of hydrazones with a base at high temperatures (200 °C) gives alkanes. These two reactions can be combined into a single reaction called the *Wolff–Kishner reduction* [11, 12]. The reaction is carried out in high boiling point solvents such as ethylene glycol. The water and reduced product formed during the reaction (if the boiling point of the alkane is lower than that of the solvent) are removed from the medium by distillation.

The reaction takes place in two steps. The first step is the reaction of aldehyde or ketone with hydrazine to give hydrazone derivatives. The hydrazine first attacks the carbonyl group as discussed in the case of imine formation, followed by water elimination to give a hydrazone derivative. A base abstracts the acidic N–H proton, forming an anion, which is delocalized over nitrogen and carbon atoms. When the charge is on nitrogen, the molecule abstracts a proton; then, the starting material is formed. However, when the negative charge is on the carbon atom, the protonation of this anion forms the neutral intermediate with a new C—H bond. The double bond is located between two nitrogen atoms. This step is not reversible. In the second step, the base quickly abstracts the acidic N–H proton. The acidity of this proton is increased because of the bonding to an sp^2-hybridized nitrogen atom. The anion formed decomposes irreversibly to give a carbanion and nitrogen gas. Protonation of the carbanion completes the reduction reaction.

The Wolff–Kishner reduction can be applied to those compounds that are not sensitive to the base and contain functional groups resistant to high temperatures. Sterically hindered ketones can also be reduced. Many of the efforts were devoted to improving the Wolff–Kishner reduction.

According to the Huang-Minlon modification [13], the carbonyl compounds are first reacted with an excess of 85% hydrazine hydrate and three equivalents of sodium hydroxide. The excess hydrazine and water are removed by distillation after hydrazone formation and then the temperature is raised to 200 °C. When the reaction time is reduced, the yields are increased.

Cram and coworkers [14] succeeded in performing this reduction reaction at room temperature. They slowly added a solution of hydrazone to a solution of sublimed potassium *tert*-butoxide in anhydrous dimethyl sulfoxide at room temperature. The immediate evolution of nitrogen gas was observed, and the reduced products were formed in 60%–90% yields.

5.3.2 Clemmensen Reduction (Under Acidic Conditions)

One of the methods used to reduce aldehydes and ketones to alkanes is the Clemmensen reduction, which uses hydrochloric acid and zinc amalgam. The compound to be reduced should not have acid-sensitive functional groups. This reaction is particularly useful in aromatic ketones. However, it can also be applied to aliphatic ketones. The Clemmensen reaction may be promoted by ultrasound. Less hindered ketones can be easily reduced. For a detailed mechanism for the Clemmensen reduction, see Chapter 6.

5.3.3 Metal Hydride Reduction of the Carbonyl Groups

One of the most important reactions of ketones and aldehydes is their reduction with metal hydrides [15]. $LiAlH_4$ and sodium borohydride ($NaBH_4$) are generally used as common sources for hydride. The hydride ion is not present during the reaction; it is transferred from the metal to the carbonyl group. Because of the electronegativity difference between aluminum and boron, the Al—H bond in $LiAlH_4$ is more polar than the B—H bond in $NaBH_4$. Therefore, $LiAlH_4$ is a stronger reducing agent than $NaBH_4$. Because hydrides react with water, extreme care must be taken when handling these compounds. One of the essential properties of the two compounds is that they are partially soluble in organic solvents and are not basic like other metal hydrides (LiH, NaH, and KH).

While $LiAlH_4$ reduces aldehydes, ketones, esters, and carboxylic acids into alcohols, $NaBH_4$ is more selective and can only reduce ketones and aldehydes. $NaBH_4$ can also reduce the carbonyl group in α,β-unsaturated carbonyl compounds by the addition of cerium salts (see Luche reduction). $NaBH_4$ is a weak base and reacts slowly with H_2O, and the reduction reactions can also be carried out in protic solvents such as ethanol or an ethanol/water mixture. $LiAlH_4$ can also reduce nitriles and amides to primary amines. Because $LiAlH_4$ reacts violently with H_2O, the reduction reactions are carried out in anhydrous diethyl ether or THF.

Both reagents transfer a hydride anion (H^-) ion to the carbonyl carbon. A second proton is needed to complete the reaction. The proton source may be the solvent used or alcohol or water added to the reaction medium after the reaction. As a result of the reduction, aldehydes produce primary alcohols, while ketones produce secondary alcohols.

The mechanism of reduction with LiAlH$_4$ is given below. The reaction starts with an electrostatic interaction between the Li$^+$ ion and the carbonyl oxygen atom. As a result of this interaction, the polarity of the carbonyl group increases even more. One of the hydrogen atoms attached to the aluminum atom is transferred as a hydride (with the bonding electrons) to the carbonyl carbon, and electrons from the C=O move to the electronegative oxygen, forming a tetrahedral alkoxide ion. Because the released AlH$_3$ is a Lewis acid with an electron deficiency, it immediately collapses to the alkoxide to form the alkyl aluminate. The reaction does not stop at this stage. Because there are three hydrogen atoms on the alkyl aluminate, this intermediate is active and can reduce a second carbonyl compound to form a dialkyl aluminate. The reaction ends with the formation of a tetraalkyl aluminate complex after the third and fourth reductions take place. The last step is the work-up step. A mineral acid (HCl) is added to the reaction medium to convert the tetraalkyl aluminate to an alcohol. The acid has two different functions here. The first is to release the alcohol formed and the second is to dissolve the Al(OH)$_3$ formed during the hydrolysis and to move it into the water phase. Otherwise, Al(OH)$_3$ would collapse and serious problems would occur in the separation process.

Reduction mechanism with LiAlH$_4$

We have demonstrated that one mole of LiAlH$_4$ can reduce four moles of ketone or aldehyde. The reaction rate slows down after the first reduction, while it becomes even slower in the third and fourth stages. Because of the inductive effect of oxygen, the separation of the hydride atom from aluminum decreases in parallel with the increasing number of alkoxide ions bonded to aluminum.

Monoalkyl aluminate > Dialkyl aluminate > Trialkyl aluminate

Rate of the reaction

As the reduction rate decreases according to the number of alkoxides attached to the aluminum atom, the selectivity increases in parallel. Utilizing this feature, a series of lithium alkoxyhydride aluminates with selective properties carrying alkoxide groups were synthesized.

LiAlH$_4$ also efficiently reduces esters to primary alcohols. Let us examine the mechanism of this reaction. As shown in ketone reduction, the reaction begins with the coordination of the Li$^+$ ion to the carbonyl group, followed by transfer of the hydride ion to the carbonyl group and the alkyl aluminate is formed. The alkyl aluminate formed herein is different from the aluminate formed in the reduction of ketones and aldehydes. This intermediate displaces the alcohol portion in the form of RO$^-$, forming an aldehyde. This step is fast. Moreover, because the aldehyde formed is more reactive than the ester, the reducing reagent present in the environment firstly reduces the aldehyde functional group. Therefore, it is not possible to stop the ester reduction reaction at the aldehyde stage with LiAlH$_4$. The aldehyde formed is reduced to the primary alcohol by the same mechanism as we discussed above.

5.3.3.1 Diisobutyl Aluminum Hydride (DIBAL) Reduction

Diisobutyl Aluminum Hydride (DIBAL) is a versatile organometallic hydride and widely accepted reagent and successfully replaced metal hydrides such as LiAlH$_4$ and NaBH$_4$ on a commercial scale [16]. Its selectivity and lower cost make DIBAL an outstanding reducing agent. DIBAL can reduce esters to aldehydes. A similar mechanism is operative in the reduction process. When the esters are treated with DIBAL in apolar solvents such as toluene and hexane at low temperatures (−78 °C), the reaction starts with the transfer of a hydride from aluminum to the carbonyl group, forming a tetrahedral intermediate. This intermediate is different from the one formed in the reduction of ketones and aldehydes with LiAlH$_4$. The intermediate is neutral and quite stable, and it does not decompose immediately to an aldehyde and ROAl(*i*-Bu)$_2$. When water is added to the reaction mixture, the product is hydrolyzed and aldehyde and aluminate are formed. However, in polar solvents such as THF, the tetrahedral intermediate initially formed decomposes faster than it was generated to an aldehyde and ROAl(*i*-Bu)$_2$. At low temperatures, it decomposes quite slowly.

5.3.3.2 Sodium Borohydride (NaBH$_4$) Reduction

As mentioned previously, NaBH$_4$ is one of the most widely used reducing reagents [17]. It is stable to moisture, and it can efficiently reduce aldehydes and ketones to related alcohols in alcoholic and aqueous solutions. Because of its lower reactivity compared to LiAlH$_4$, NaBH$_4$ cannot reduce esters and derivatives such as amides and carboxylic acids. Unlike LiAlH$_4$, NaBH$_4$ shows more selectivity in reduction reactions.

The reaction mechanism is similar to that of the LiAlH$_4$ reduction and proceeds in two steps. First, coordination between a sodium atom and carbonyl oxygen occurs to increase the reactivity of the carbonyl group. Next, hydride transfer take places from the NaBH$_4$ to the carbonyl carbon, forming a new C—H bond and breaking the C=O bond. The alkoxide ion formed binds to the released borohydride to form an alkyl borate. The reaction does not stop at this stage and continues with the transfer of other hydride ions. It is possible to reduce four moles of ketone or aldehyde with one mole of NaBH$_4$. In the last step, alcohol is released by hydrolysis of the complex formed.

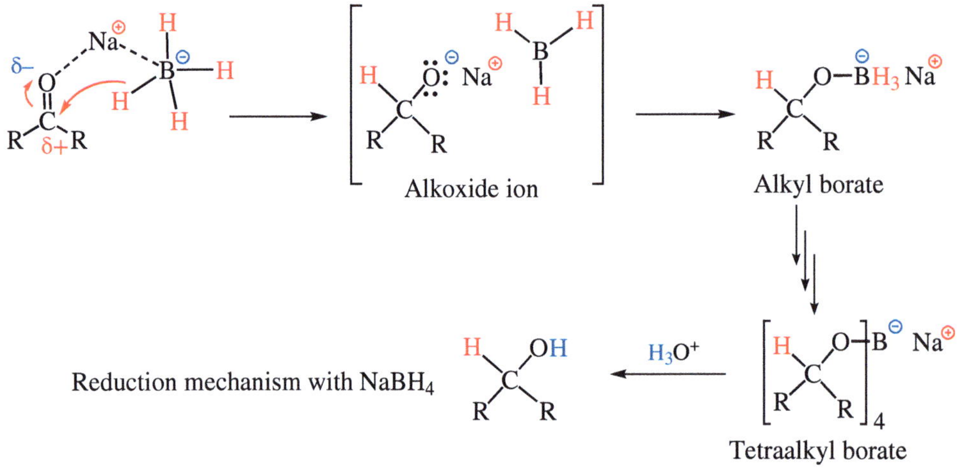

Reduction mechanism with NaBH₄

The reaction rate in LiAlH₄ reduction slows down in the transfer of the second, third, and fourth hydride ions. In NaBH₄ reduction, the situation is quite the opposite, and the first hydride transfer is the step that determines the reaction rate. The rates of other levels are faster. According to the same mechanism for LiAH₄ reduction, it is expected that the reaction rate should also decrease. However, it is thought that the alkoxy borohydride formed in the reduction with NaBH₄ is subjected to disproportionation, as shown in the equation below, and the main reducer is the boron hydride.

$$2\ H_3\overset{\ominus}{B}-OR \rightleftharpoons \overset{\ominus}{BH_4} + H_2\overset{\ominus}{B}(OR)_2$$

An example demonstrating the reactivity and selectivity of NaBH₄ is given below. NaBH₄ has lower reactivity compared to LiAlH₄ and it is also more selective. The ketoester is reduced to the related diol with LiAlH₄, while NaBH₄ reduces only the ketone functionality in that ketoester. The ester group remains intact.

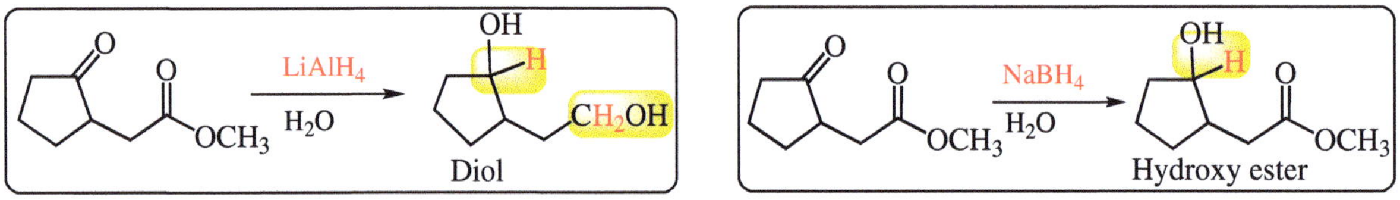

We mentioned that NaBH₄ is usually sluggish toward reducing esters. Only in exceptional cases can electronically modified esters at a high temperature be reduced. Recently, it has been reported that a catalytic amount of NaOMe (5%) in methanol can stabilize NaBH₄. The use of this solution can reduce esters at room temperature to the corresponding alcohols. The intermediate NaBH₃OCH₃ is likely responsible for the reduction. This methodology can be extended for the reduction of aldehydes, ketones, acid chlorides, anhydrides, and imides [18].

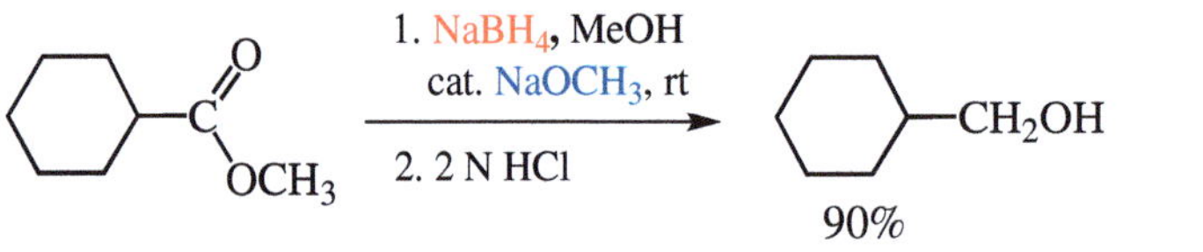

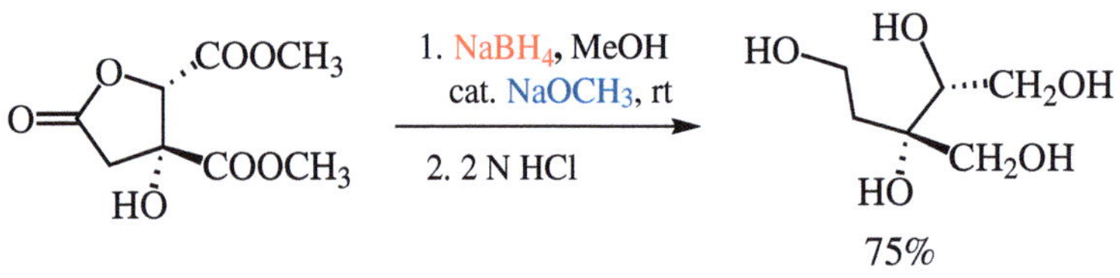

The selectivity of NaBH$_4$ can be further increased. Generally, decreased reactivity increases the selectivity. Hydride transfer must be made difficult in order to reduce reactivity. If an electron-attracting group is attached to the boron atom in the NaBH$_4$ molecule, hydride transfer becomes even more difficult. For example, NaBH$_3$CN is a mild reducing reagent.

Under normal conditions, it is difficult to reduce aldehydes and ketones in water or methanol with NaBH$_3$CN. For the reaction to take place, the pH value of the medium must be around 3–4. The reaction mechanism is the same as that in other mechanisms. If the Na$^+$ cation is replaced with Bu$_4$N$^+$, the Bu$_4$NBH$_3$CN can reduce aldehydes without touching the ketones.

This reduced reactivity allows NaBH$_3$CN to reduce iminium ions to amines at slightly acidic conditions. In order for such a reaction to take place, carbonyl compounds must first be converted into imines. This reduction can also be done in a single step and in the same reaction vessel because the reduction of imines is a faster reaction than the reduction of aldehydes and ketones. For example, the aldehyde given below is first converted into iminium salt in the reaction vessel and rapidly reduced to the relevant amine with the NaBH$_3$CN present in the medium. This reaction is also called *reductive amination* [9]. If a molecule contains both functional groups, carbonyl and amine, intramolecular reductive amination can also be carried out.

5.3.3.3 Reduction of Carboxylic Acids

Reducing the acid carbonyl group is difficult compared to the reduction of ketones and aldehydes. There are two reasons for this sluggish reduction. First, carbonyl carbon polarization is weak compared to that for aldehydes and ketones because of the conjugation of the nonbonding electrons on the hydroxyl oxygen atom with the carbonyl group. This conjugation reduces the reactivity of the carbonyl group. The second reason is that in the first step, a hydride ion reacts with the acid proton, forming a carboxylate anion and hydrogen gas. The negative charge on carboxylate oxygen delocalizes over two oxygen atoms. Therefore, the electron density on the acid carbonyl group increases further, which makes the attack of hydride ions more difficult.

Carboxylic acids can only be reduced with strong reduction reagents because the electron density on the acid carbonyl group is high. As mentioned, the first step is the abstraction of the acidic proton and forming a carboxylate anion. A hydride cannot attack the carboxylate anion because of the negative charge. After the transfer of the first hydride ion from

LiAlH$_4$, AlH$_3$ is formed. Because the released AlH$_3$ is a Lewis acid with an electron deficiency, it immediately binds to the carboxylate anion to generate a new hydride donor (AlH$_3$OCOR). After transfer of a second hydride ion to the carbonyl group, an aldehyde is produced as an intermediate during this reaction. As the aldehyde is more reactive than the carboxylic acid, it cannot be isolated. It is immediately reduced to the corresponding alcohol in the reaction medium according to the same mechanism, as discussed previously. The alcohol formed can be released by adding acid to the reaction medium.

Carboxylic acid derivatives such as esters or acyl chlorides can be more easily converted to primary alcohols using LiAlH$_4$, while NaBH$_4$ is not a strong reducing agent to accomplish this reaction. During the reduction process, a carboxylate anion is not formed as an intermediate. Esters and acyl chlorides have leaving groups such as CH$_3$O$^-$ and Cl$^-$ ions, which can be substituted by hydride ions. The product of the reaction is an aldehyde. A second hydride ion adds to the aldehyde, forming an alkoxide, which can give an alcohol upon protonation.

Intermediates formed in the first step

Aliphatic and aromatic carboxylic acids can be reduced in high yield to the corresponding alcohols using borane (BH$_3$) in THF at room temperature or lower temperatures [19]. Aliphatic carboxylic acids are reduced faster than aromatic carboxylic acids.

In the first step, a reaction takes place between the acid proton and the boron hydride, forming a carboxylate anion, and hydrogen gas is released. A bond is formed between the carboxylate oxygen and boron. It has been established that the remaining two hydrogen atoms bonded to boron undergo a similar reaction and form triacyloxoborane. This product is neutral and the boron atom has electron deficiency. Therefore, the nonbonding electrons on the oxygen atom can form a double bond between the boron and oxygen, making the oxygen positively charged. This charge increases the polarization of the carbonyl group, thereby increasing its reactivity by attracting electrons from the carbonyl group. The reactivity of this intermediate can be compared with that of aldehydes and ketones. Then, analogous to the reduction of an aldehyde, hydride transfer occurs. The reduction reaction continues with borohydride.

5.3 Reduction of Carbonyl Groups

$$3R-\underset{\underset{}{\overset{\overset{O}{\|}}{}}}{C}-OH \xrightarrow[-3H_2]{+BH_3} R-\underset{\underset{}{\overset{\overset{O}{\|}}{}}}{C}-\ddot{O}-B\underset{OCOR}{\overset{OCOR}{\diagup}} \longleftrightarrow R-\underset{\underset{}{\overset{\overset{O}{\|}}{}}}{C}-\overset{\oplus}{O}=B\underset{OCOR}{\overset{\ominus\,OCOR}{\diagup}} \xrightarrow{BH_3} 3R-CH_2-OH$$

Intermediate

Even sterically hindered carboxylic acids, such as adamantane carboxylic acid, can be reduced in high yield. Furthermore, the remarkable selectivity of BH$_3$ was confirmed by the selective reduction of *p*-cyanobenzoic acid to *p*-cyanobenzyl alcohol in 82% yield.

[Adamantane-COOH $\xrightarrow{BH_3, 95\%}$ Adamantane-CH$_2$OH]   [NC-C$_6$H$_4$-COOH $\xrightarrow{BH_3, 82\%}$ NC-C$_6$H$_4$-CH$_2$OH]

Recently, Panda and coworkers [20] reported a simple and straightforward protocol for the reduction of carboxylic acids to the corresponding alcohols under catalyst-free and solvent-free conditions using HBpin (pinacolborane). The reduction is chemoselective and proceeds with high yields. It can be applied to a wide range of aromatic, aliphatic, and heterocyclic carboxylic acids with electron-rich and electron-poor functional groups.

[Kojic acid-COOH $\xrightarrow[\text{Neat, rt, 5-8 h}]{\text{HBpin (3.2 mmol)}}$ -CH$_2$OBpin $\xrightarrow[60\,°C, 3\,h, 72\%]{SiO_2/MeOH}$ -CH$_2$OH ; 99%]   HBpin (pinacolborane structure)

It was mentioned that carboxylic acids could not be reduced with NaBH$_4$ under ambient conditions. When carboxylic acids are treated with NaBH$_4$, first a hydride ion reacts with the acidic proton of the carboxylic acid, forming a carboxylate ion and hydrogen gas. This intermediate is not reactive enough to undergo further hydride transfer. However, when iodine in THF is added to the resulting solution at the same temperature, a new intermediate, RCOOBH$_2$, is formed, which is more reactive (similar to the intermediate formed during the reduction of carboxylic acids with BH$_3$). This intermediate can be reduced to a primary alcohol with NaBH$_4$ [21].

$$R-\underset{\underset{}{\overset{\overset{O}{\|}}{}}}{C}-OH + NaBH_4 \xrightarrow{-H_2} R-\underset{\underset{}{\overset{\overset{O}{\|}}{}}}{C}-\ddot{\overset{\ominus}{O}}-\overset{\oplus}{B}H_3Na \quad \text{This intermediate is not reactive}$$

$$\downarrow \begin{array}{l} 0.5\text{ mol } I_2 \\ -0.5\text{ NaI} \\ -0.5\text{ H}_2 \end{array}$$

$$B(OH)_3 + R-CH_2-OH \longleftarrow\longleftarrow R-\underset{\underset{}{\overset{\overset{O}{\|}}{}}}{C}-\ddot{O}-BH_2 \longleftrightarrow R-\underset{\underset{}{\overset{\overset{O}{\|}}{}}}{C}-\overset{\oplus}{O}=B\underset{H}{\overset{\ominus\,H}{\diagup}}$$

More reactive

One of the essential features of this combination, NaBH$_4$ and I$_2$, is the reduction of a carboxylic acid group, leaving behind an ester group unaffected when both functional groups are present in the compound. This is extremely important in synthetic chemistry.

[o-(COOH)(COOCH$_3$)-C$_6$H$_4$ $\xrightarrow[I_2]{NaBH_4}$ o-(CH$_2$OH)(COOCH$_3$)-C$_6$H$_4$; Selective reduction 82%]

The reduction of α,β-unsaturated carbonyl compounds is often problematic because there are two different centers that can be attacked by nucleophiles, one of which is carbonyl carbon while the other is the positively (+) charged β-carbon atom because of conjugation. If a hydride ion attacks the carbonyl group directly, the carbonyl group will be reduced and the double bond will remain intact. Because of the high electron density of the double bond, the hydride cannot attack the C=C double bond after reducing the carbonyl group. However, if the hydride ion first attacks the positively charged β-carbon atom, an enol will be formed, which will turn into a ketone or aldehyde through the enol tautomer. The saturated carbonyl group can now be attacked by a hydride ion to give the alcohol. In this reaction, both the double bond and the carbonyl group are reduced. As a result, when an α,β-unsaturated carbonyl compound is reduced, saturated and unsaturated alcohols are formed together.

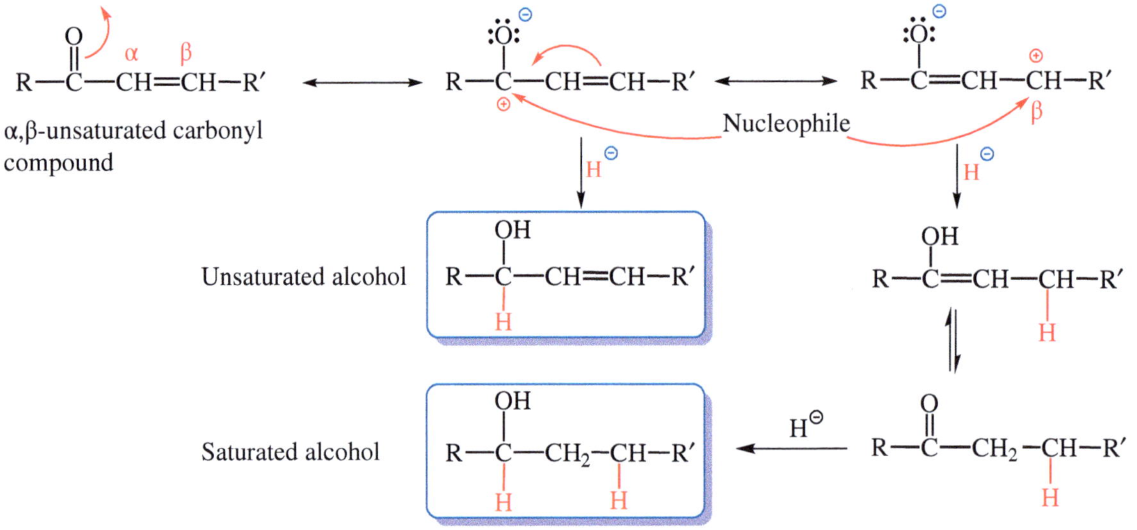

It is possible to reduce selectively the C=O double bond while leaving the C=C double bond intact by choosing the right reducing conditions. As a concrete example of this reaction, we can show the reduction reaction of cyclohexenone with $NaBH_4$. In this reaction, both allylic alcohol and saturated alcohol are formed in 51% and 49% yields, respectively [22]. However, when a mixture of cerium(III) chloride ($CeCl_3$) and $NaBH_4$ is used, the α,β-unsaturated carbonyl compound undergoes 1,2-reduction to the corresponding allylic alcohol. This reaction is known as the *Luche reduction* [23]. When the reduction of cyclohexenone is carried out with $CeCl_3$ and $NaBH_4$, the desired allyl alcohol is produced in 99% yield, while the ratio of completely reduced alcohol remains well below 1%.

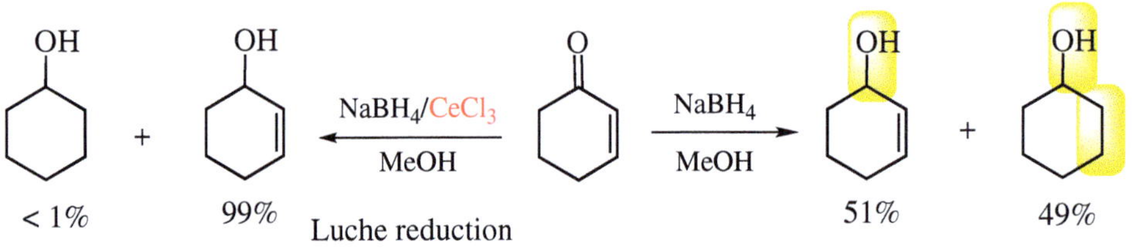

What role does $CeCl_3$ play in this reaction? The thesis that the cerium ion coordinates to the carbonyl oxygen to increase the polarization of the carbonyl group does not receive much support because it is known that lanthanoid ions preferentially bind to alcohol rather than to C=O groups. However, for reduction of the carbonyl group, the polarization of this group needs to be increased so that the nucleophile attacks only the carbonyl group. The cerium coordinates to the alcohol, making its proton more acidic, which allows protonation of the carbonyl oxygen of the ketone to be reduced, as shown below.

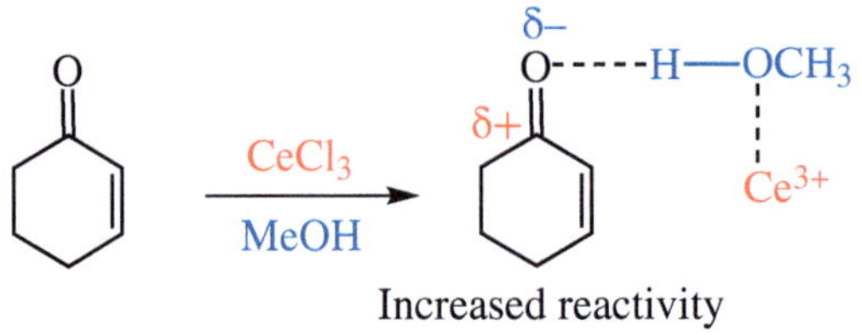

NaBH$_4$ also reacts with the cerium-activated alcohol to form a series of alkoxyborohydrides, which are *hard reagents*. They can attack the carbonyl group and transfer a hydride to the carbonyl group to generate the allylic alcohol. According to the hard and soft acid and base (HSAB) theory, a protonated carbonyl group is a hard acid because of polarization. The nucleophile must also be hard to attack the carbonyl group. The hydride ions in the NaBH$_4$ molecule are not sufficiently hard. To make them hard, OR groups must be bonded to the boron atom. Therefore, the reaction is carried out in methanol.

$$\text{NaBH}_4 + \text{CH}_3\text{OH} \xrightarrow{\text{CeCl}_3} \text{Na}^{\oplus} \text{BH}_3\text{OCH}_3^{\ominus} \xrightarrow[\text{CH}_3\text{OH}]{\text{CeCl}_3} \text{Na}^{\oplus} \text{BH}_2(\text{OCH}_3)_2^{\ominus} \xrightarrow[\text{CH}_3\text{OH}]{\text{CeCl}_3} \text{Na}^{\oplus} \text{BH}(\text{OCH}_3)_3^{\ominus}$$

Hydride transfer

One of the essential properties of the Luche reduction is that it does not reduce the more reactive aldehydes but preferably reduces ketones. We can explain this by the fact that aldehydes are more reactive than ketones and so they quickly form acetals in the presence of methanol, which prevents reduction.

Very recently, Rueping and coworkers [24] reported the chemoselective reduction of α,β-unsaturated ketones by use of the readily available Mg catalyst MgBu$_2$. Excellent yields were obtained for a wide range of ketones under mild reaction conditions, in short times, and with low catalyst loadings (0.2–0.5 mol%).

5.3.3.4 Meerwein–Ponndorf–Verley Reduction and Oppenauer Oxidation

In 1925, Verley [25] and Meerwein [26] discovered that aldehydes could be reduced to alcohols by using a mixture of aluminum ethoxide and an alcohol. In 1926, Ponndorf [27] extended this reaction to the reduction of ketones and used aluminum isopropoxide in isopropanol as the catalyst. The Oppenauer oxidation [28] reaction is the opposite of the Meerwein–Ponndorf–Verley reduction.

A six-membered transition complex is formed during the reaction. In the first stage, because of the interaction between aluminum and carbonyl oxygen, the reactivity of the carbonyl group increases partially. Hydride ion transfer occurs to the carbonyl group to be reduced through this complex. Isopropyl alcohol is used as a source of hydride transfer. Because isopropyl alcohol is oxidized to acetone after hydride transfer, it is easily removed from the reaction medium by distillation and so the equilibrium continually shifts to the side of the products [29, 30].

One of the best examples for the Meerwein–Ponndorf–Verley reduction and Oppenauer oxidation is the reaction of a hydroxy ketone with the catalyst Al(O*i*-Pr)$_3$. Here, while the hydroxyl group is oxidized, the ketone functional group is reduced. This is an intramolecular redox reaction. Generally, the Meerwein–Ponndorf–Verley reduction has been applied to intermolecular systems. This example shows for the first time that such a direct interconversion between hydroxy and ketone groups (intramolecular reduction and oxidation) can also occur within a molecule [31].

The configuration of the hydroxyl group in the resulting product was determined by X-ray analysis. The mechanism shown in the below diagram is proposed for this reaction. First, the hydroxyl group coordinates to the aluminum to form a complex. Next, the carbonyl group coordinates to the aluminum, which activates it for hydride transfer from the alkoxide. The hydride shift from the carbon atom bearing the hydroxyl group proceeds over a six-membered transition state. The desired carbonyl group is formed after hydride transfer.

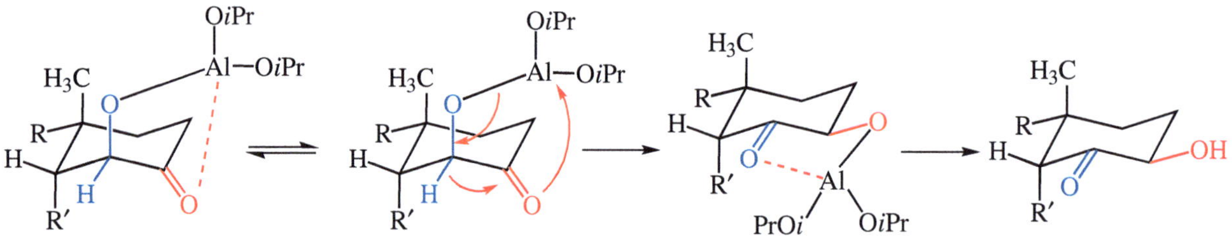

5.4 Reaction of Carbonyl Groups with Organometallic Compounds

The reduction of carbonyl compounds with hydride ions generates alcohols. Alcohols can also be generated by using carbon nucleophiles instead of hydrides, such as alkyl lithium and Grignard reagents, which easily add to the carbonyl groups. Both alkyl lithium and Grignard reagents are strongly polarized compounds. Because the carbonyl group is also polarized, the metal binds to the carbonyl oxygen, while the alkyl group binds to the carbonyl carbon. These reactions are important because of the generation of new C—C bonds. To obtain alcohols, lithium or magnesium alkoxide salts formed are treated with water as depicted below. While aldehydes are reduced to secondary alcohols, ketones are reduced to tertiary alcohols.

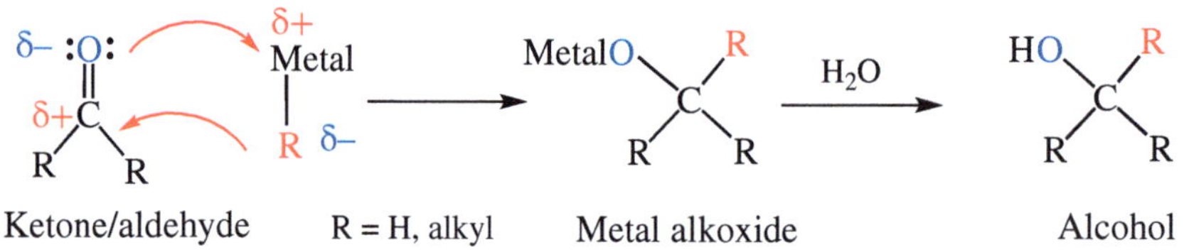

5.4.1 Grignard Reagents

Organomagnesium halides are known as Grignard reagents [32–34]. Grignard reagents were first discovered by French chemist François Auguste Victor Grignard (1871–1935) as a result of the reaction of alkyl halides with magnesium, and he was awarded the 1912 Nobel Prize in Chemistry for his discovery.

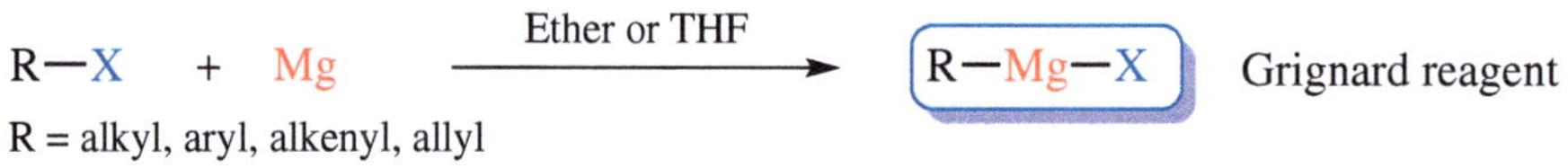

$$R-X + Mg \xrightarrow{\text{Ether or THF}} R-Mg-X \quad \text{Grignard reagent}$$

R = alkyl, aryl, alkenyl, allyl

When alkyl, aryl, or alkenyl halides are reacted with metallic magnesium in ether or THF, magnesium oxidatively enters between the alkyl group (aryl or alkenyl) and halide, and the oxidation number of magnesium increases from 0 to +2. Grignard reagents dissolve well in ethereal solvents (diethyl ether, tetrahydrofuran, dioxane, dimethoxyethane, etc.) because they form a complex with the ether oxygen, which increases the solubility of the Grignard reagents.

For the formation of Grignard compounds, magnesium must first be activated and the oxide layer on the magnesium must be dissolved. For this, magnesium is treated first with a small amount of iodine or diiodo ethane. Removal of MgO allows the Mg and the aryl/alkyl halide to come in contact and react. There must be no water and no functional groups with acidic protons because Grignard reagents are polar compounds and they react instantly with acidic protons.

Because the metal is electropositive in Grignard reagents, bond electrons are polarized toward the carbon atom, and the R group is negatively (−) charged and the metal is positively (+) charged. In the alkyl halide, while the R group is positively (+) polarized, the X group is negatively (−) polarized. However, the polarity of the R group in the Grignard reagent is changed from (+) to (−). This reverse polarization of the alkyl group observed here is called *umpolung*.

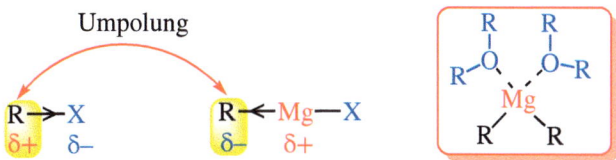

It is well established that Grignard reagents in diethyl ether and tetrahydrofuran solution are best represented by an equilibrium mixture, which is called the *Schlenk equilibrium*. The presence of such an equilibrium is clearly shown by variable nuclear magnetic resonance (NMR) spectroscopic measurements. The ^{1}H-NMR spectrum of CH_3MgX obtained at room temperature shows a single CH_3 signal because of rapid conversion, while measurements at low temperatures (−100 °C) indicate that there are two different methyl signals because of conversion of the dynamic equilibrium into a static one [34].

$$2\ R-Mg-X \rightleftharpoons R-Mg\overset{X}{\underset{X}{\diamond}}Mg-R \rightleftharpoons R-Mg-R + MgX_2$$

Schlenk equilibrium

When dioxane is added to the reaction medium, MgX_2 and $RMgX$, highly associated with diethyl ether, precipitate and leave MgR_2 in the solvent. The reaction of any carbonyl group with $RMgX$ and MgR_2 separately or in a mixture gives the same results. This shows that the reactivities of these compounds with different structures are very similar to each other, albeit with occasional differences in rate [34].

Although the mechanism for the formation of Grignard compounds is controversial, it is accepted that the reaction takes place on the metal surface and starts with transfer of a single electron from magnesium to the carbon—halogen bond. This forms the positively charged Mg^{+1}, which is a radical. The negatively (−) charged RX group dissociates the halogen, which bonds to Mg^{+1}, forming an XMg and R radical. The coupling of these two radicals generates the Grignard reagent, RMgX. The radical that occurs here is controversial. If a free radical was formed, the formation of dimer products such as R–R would be expected. However, dimer products have never been observed in the synthesis of Grignard reagents [35].

$$R-X + Mg \longrightarrow [R-X]^{\bullet -} + Mg^{\bullet +} \longrightarrow [R\bullet + \bullet Mg-X] \longrightarrow R-Mg-X$$

Mechanism for the formation of Grignard reagents

The reaction of 6-bromohex-1-ene with Mg in ether gives (cyclopentylmethyl)magnesium bromide in yields of 3%–10% besides the expected major Grignard reagent hex-5-en-1-ylmagnesium bromide. The formation of the minor product can only be explained by cyclization of the radical formed as an intermediate. Based on this experiment, the following conclusion is reached. Grignard reagents are formed from radicals. However, the lifetime of these radicals formed on the metal surface is rather short, and even if they find enough time for rearrangement, they cannot diffuse out of the solvent cage. Furthermore, the cyclization rate is very slow [36].

To evaluate whether radical intermediates are involved or not, norbornenylethyl bromide was reacted with magnesium in THF at 65 °C [37]. Uncyclized and cyclized products were formed in a ratio of 1 : 2, where the cyclized product predominated. This observation also shows that radicals are involved during the formation of Grignard reagents.

Generally, it is not easy to start Grignard reactions. The solvent used must be anhydrous and prepared just before use, and the alkyl halide must also be dried. Moreover, the glass material to be used in the experiment must be dried beforehand. Today, it is possible to perform Grignard reactions without taking any of these measures. If Grignard reactions are performed in an ultrasonic bath, no particular care is needed to dry either the solvent or the equipment [38]. Doubtless thoroughly drying all compounds and solvents used in a Grignard reaction will give better results. The use of an ultrasonicator will initiate Grignard reactions, the reaction of magnesium with organic bromides, very rapidly with minimal failure.

Alkyl Grignard reagents are obtained through the addition of magnesium metal to alkyl chlorides, bromides, and iodides. Reactivity decreases from iodide to chloride in halides. It is easier to produce Grignard reagents from iodides and bromides. Since the reactivity of alkyl fluorides is low, they are rarely used in Grignard reactions.

$$R-I \; > \; R-Br \; > \; R-Cl \; \gg \; R-F$$

Reactivity

The formation of Grignard reagents is not limited to compounds formed by the reaction between alkyl halides and magnesium. Compounds with the halogen atom bonded to an sp^2 hybridized carbon atom, such as vinyl and aryl halides, also react with magnesium to give organomagnesium halides, which are also called Grignard reagents.

Grignard reagents are polarized compounds. The electronegativity of magnesium is low (1.3). Compared to magnesium, carbon has higher electronegativity (2.5). If it is bonded to magnesium, it behaves almost like a carbanion. Therefore, Grignard reagents can undergo reactions with various functional groups. While magnesium is positively (+) charged, the R groups (alkyl, vinyl, phenyl, etc.) are negatively (−) charged. For a reaction with Grignard reagents, the other compounds must also be polarized. If the polarization of the group to react with Grignard compounds is known (it is straightforward to determine), it is easy to predict the product that will result from the reaction. It is possible to categorize important reactions of Grignard compounds in four different groups.

1. Reactions with compounds containing active hydrogen atoms,
2. Reactions with carbonyl compounds,
3. Reactions with metal halides, and
4. Reactions with alkyl halides.

5.4.2 Reactions with Active Hydrogen-Containing Compounds

Grignard reagents are such strong bases that they will not only react with strong acids (HCl) but they will also immediately react with active hydrogen-containing compounds such as H_2O ($pK_a = 14$), alcohols ($pK_a = 16-18$), phenol ($pK_a = 9.9$), carboxylic acid ($pK_a = 4-5$), mercaptan ($pK_a = 10-11$), primary and secondary amines ($pK_a = 35-38$), amides, acetylene ($pK_a = 25$), and other C–H acidic compounds. When this happens, the Grignard reagent is converted into the corresponding hydrocarbons.

$$\overset{\delta-}{R}-\overset{\delta+}{Mg}-X + \overset{\delta+}{H}-\overset{\delta-}{OH} \longrightarrow R-H + HO-Mg-X$$
Water

$$\overset{\delta-}{R}-\overset{\delta+}{Mg}-X + \overset{\delta+}{H}-\overset{\delta-}{OR} \longrightarrow R-H + RO-Mg-X$$
Alcohol

$$\overset{\delta-}{R}-\overset{\delta+}{Mg}-X + \overset{\delta+}{H}-\overset{\delta-}{O}-\overset{O}{\underset{\|}{C}}-R \longrightarrow R-H + R-\overset{O}{\underset{\|}{C}}-O-Mg-X$$
Acid

Because the hydrogen atom in the compounds containing active hydrogen is positively charged, it will be transferred to the negatively (−) charged R group, forming an alkane or alkene depending on the structure of the starting material. After hydrogen transfer, the remaining group with a negative charge will be bonded to positively charged magnesium. All reactions with different functional groups proceed by the same mechanism. Therefore, it is easy to predict the product that will occur. This reaction is successfully applied to the reduction of alkyl halides or alkenyl halides to alkanes or alkenes, respectively.

5.4.3 Reactions with Carbonyl Compounds

One of the most common reactions of Grignard reagents is with aldehydes and ketones, which lead to the formation of a new C—C bond. As the carbonyl carbon is an electrophile, the nucleophilic part of the Grignard reagent, an R⁻ group, attacks the carbonyl carbon by opening the π bond and forming an alkoxide ion that is bonded to magnesium. The addition of water destroys this complex and forms an alcohol. The type of carbonyl group used determines the type of alcohol formed. A primary alcohol is formed by the reaction of Grignard reagents with formaldehyde, while the reaction with aldehydes and ketones forms a secondary or tertiary alcohol, respectively. For example, ethylmagnesium bromide reacts with benzaldehyde to form a secondary alcohol, while a tertiary alcohol is formed as a result of its reaction with acetone.

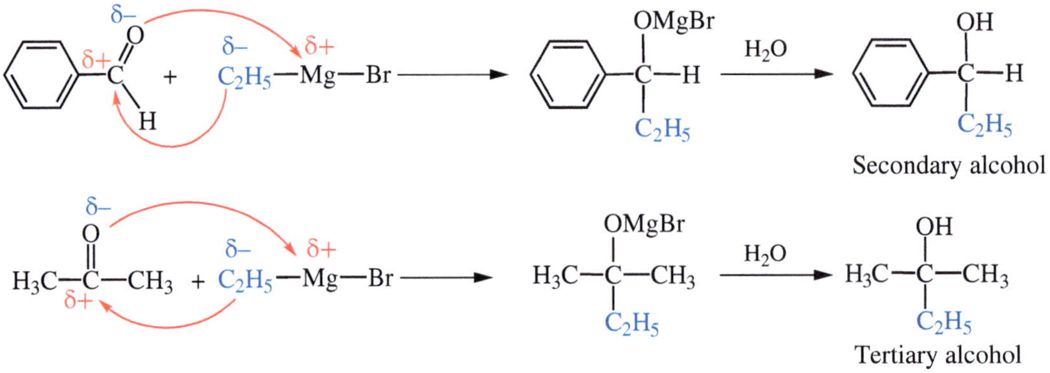

[Reaction scheme: benzaldehyde + C₂H₅–Mg–Br → intermediate with OMgBr → H₂O → secondary alcohol (1-phenyl-1-propanol)]

[Reaction scheme: acetone (H₃C–CO–CH₃) + C₂H₅–Mg–Br → intermediate with OMgBr → H₂O → tertiary alcohol]

The mechanism of addition of Grignard compounds to carbonyl groups has been interpreted in different ways for many years. Once the dimeric structure of the Grignard compounds was determined, a new mechanism for addition was proposed, and it is widely accepted today. According to this mechanism, the Grignard reagent adds to the carbonyl group through a six-membered ring transition state. A six-membered ring is formed between one mole of carbonyl compound and two moles of Grignard compound as shown below. One of the Grignard reagents increases the reactivity of the carbonyl group because of the electrostatic interaction between the carbonyl oxygen and the electropositive magnesium atom, while the second mole of the Grignard reagent transfers the alkyl group to the carbonyl carbon [39, 40].

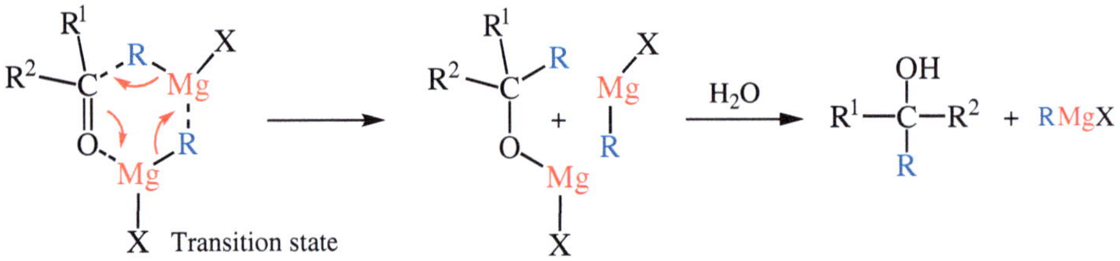

[Transition state scheme showing six-membered ring with R¹, R², Mg, X, R groups → product + RMgX after H₂O]

5.4.4 Stereochemistry

Ketones containing two different substituents are called prochiral compounds. A prochiral carbon atom must be bonded to three different groups. The addition of a fourth group would create a new stereocenter. Because the carbonyl carbon is sp^2-hybridized, the carbonyl carbon and the other substituents attached to it lie on a plane. Therefore, nucleophiles can attack the carbonyl group from either side with the same probability. Because a new asymmetric center is now generated, the two enantiomers are formed in equal amounts, resulting in a racemic mixture.

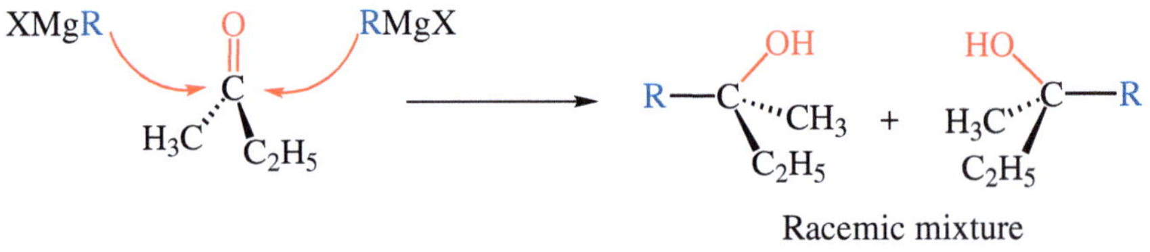

[Scheme showing attack of RMgX on ketone (H₃C, C₂H₅ substituents) from both sides giving racemic mixture of two enantiomers]

The neighboring substituents influence the stereochemistry of nucleophilic addition to ketones or aldehydes. If there is an asymmetric center adjacent to the carbonyl group, two diastereoisomers can be formed, depending on the direction of the attack of the nucleophile. Because the Grignard reagent will attack the carbonyl group from both directions, the configuration of the new stereogenic center will be *R* and *S*. However, the amounts of these diastereoisomers will not be equal. The stereoselectivity of addition can be predicted based on conformational models of the transition state. Cram predicted that the major product is formed from a conformation in which the largest substituent connected to the α-carbon atom must be eclipsed with the other carbonyl substituent. This approach is called *Cram's rule* [41]. When the desired conformation is established, the nucleophiles will approach the carbonyl group from the less hindered side. The major product is called the Cram product. An example is given below.

If one of the enantiomers or diastereomers occurs more than the other as a result of any reaction, this process is called *asymmetric induction*. In the reaction below, one of the possible diastereomers is formed more than the other. Therefore, this reaction is an asymmetric induction reaction.

Cram product   Anti-Cram product

5.4.5 Reaction with Esters

After the addition of a Grignard reagent to the carbonyl groups, if there is a leaving group that is attached to the carbonyl carbon, the reaction will not stop. A second Grignard reagent will also be added and tertiary alcohols will be formed. For example, ester, anhydride, acyl halides, etc., behave similarly. In the first step, Grignard reagents add to the carbonyl group, as we discussed in addition to ketones and aldehydes. The double-bond electrons move to the electronegative oxygen atom, forming a tetrahedral intermediate, an alkoxide ion. In the second step, this tetrahedral intermediate displaces the alkoxide ion (RO⁻), forming a ketone. The reaction does not stop at this stage because the ketone carbonyl group is more reactive than the ester carbonyl group toward the nucleophilic attack. A Grignard reagent will prefer a reaction with the ketone formed. Therefore, a second Grignard reagent adds to the ketone and forms a tertiary alcohol after the protonation of the alkoxide.

5.4.6 Reactions with Different Functional Groups

Grignard reagents undergo addition reactions with different functional groups. We have discussed that the Grignard reagent acts as a nucleophile and reacts with the carbonyl group of aldehydes, ketones, and esters. A similar reaction occurs between the Grignard reagent and CO_2 as it contains a carbonyl group. The carbonyl group in CO_2 is polarized as in ketones and aldehydes. Therefore, Grignard reagents react with CO_2 in two steps. In the first step, addition of the Grignard reagent takes place to the carbonyl group to give the magnesium salt of the carboxylic acid. The addition of a second equivalent Grignard reagent cannot take place because of the decreased reactivity of the carbonyl group. The next step is the protonation of the carboxylate anion with dilute sulfuric acid or hydrochloric acid to form the carboxylic acid. This is one of the methods applied in laboratory conditions for the synthesis of carboxylic acids.

Because Grignard reagents readily react with carbonyl groups, they can also easily add to polarized double bonds. For example, the nitrile group is a highly polarized functional group because of the difference in electronegativity between the carbon and nitrogen atom. Therefore, it is quite easy to predict that Grignard compounds can react with nitriles and even the structure of the product to be formed.

The strongly nucleophilic R group from the Grignard reagent adds to the electrophilic carbon atom in the polar C≡N group similar to that seen by the addition to a carbonyl group and forms a new C—C bond. After addition of one mole of Grignard reagent, one pair of the π electrons move to the electronegative nitrogen, creating an imine salt complex as an intermediate. Therefore, the nitrile carbon atom is no longer positively charged, and a second mole of Grignard cannot be added. The reaction stops at this stage. On the addition of water or aqueous acid, the nitrogen atom protonates, forming an imine, which is not a stable compound. Hydrolysis of imine derivatives with water provides ketones. This methodology is essential because of the conversion of nitriles into ketones. However, the double addition of Grignard reagents to nitriles, leading to tertiary carbinamines, is also possible in some particular cases [42].

The nucleophilic addition of Grignard reagents with sterically crowded substituents, such as tertiary butyl groups, does not proceed efficiently with nitriles. For example, the reaction of *t*-butylmagnesium bromide with benzonitrile produces only 8% of the addition product in 14 hours. However, when the reaction is carried out in the presence of a catalytic amount of CuBr, the reaction proceeds smoothly, giving the addition product in 99% yield after 14 hours [43]. It is well established that copper(I) in catalytic amount activates particular Grignard reactions [44, 45]. Hydrolysis of ketimines gives ketones. The reduction of ketimines with Li in liquid ammonia affords branched primary amines.

During this reaction, first, the oxidative addition of RMgX to CuBr takes place to form a copper complex. The R group bonded to the copper is transferred to the nitrile carbon while the copper is separated from the molecule. The last step is a reductive elimination step, which regenerates the catalyst for the circle [43].

Generally, Grignard reagents add to nitriles more slowly than to carbonyl groups. As described above, catalysts are needed to increase the rate of the reaction. If a molecule contains carbonyl and nitrile groups, it is possible to add the Grignard reagent exclusively to the carbonyl group without protecting the nitrile group.

The hydrogen atom bonded to the terminal alkyne is more acidic than hydrogen atom bonded alkenes and alkanes because of the sp hybridization of the alkyne carbon atom. Remember, the higher the *s* ratio in the hybrid orbitals, the higher the electron-withdrawal ability of the corresponding hybrid orbital. The pK_a value of acetylene is 25, while the pK_a values of Grignard reagents are much higher than 50–60. In terminal alkynes, the C—H bond is polarized. Because Grignard reagents are also strongly polarized compounds, the R group from them abstracts the proton and turns into an alkane, while MgBr binds to the alkyne group to form the alkyne Grignard reagent. Alkynyl Grignard reagents are unique among Grignard reagents for their ability to undergo S_N2 reactions. Of course, they can also add to carbonyl groups. For the protection of terminal alkynes, it is recommended to protect the alkyne hydrogen atom by the trimethylsilyl group.

As we have seen in the nucleophilic substitution part, epoxides possess considerable ring strain. Therefore, they can undergo ring-opening reactions with nucleophiles. As Grignard reagents are also nucleophilic, they tend to react with epoxides similar to the S_N2 reaction mechanism. As discussed, nucleophiles preferentially attack the oxirane ring at the least substituted carbon atom. The new bond is formed with inversion of configuration. An example is given below.

It was mentioned that Grignard reagents react with water to give alkanes. That is the reason why in the synthesis of Grignard reagents everything has to be very dry and the air oxygen must be removed from the reaction medium. It is recommended to run the reactions under nitrogen gas because Grignard reagents react with oxygen in the air and form magnesium alkyl hydroperoxides. When they are hydrolyzed, they turn into alkyl hydroperoxides.

When dry air is bubbled into an ethereal solution of *t*-butylmagnesium bromide at 0 °C, *t*-butyl alcohol is isolated after hydrolysis as the only identifiable product [46]. Two consecutive reactions can explain this reaction. Magnesium hydroperoxides formed in the first step react with a second mole of the Grignard reagent in the reaction medium to give the alcohol. This procedure can be applied to the synthesis of alcohols whose synthesis is difficult by application of classical methods.

When Grignard compounds react with 5% $FeCl_3$ or $MnCl_2$ as a catalyst with air oxygen, chemo- and stereoselectively dimer compounds are formed [47]. Products formed by dimerization of aryl, alkenyl, and alkynyl Grignard reagents are widely used as optical materials, organic conductors, etc.

5.5 Reaction of Carbonyl Groups with Ylides

If an electron-withdrawing group is attached to a methyl group, it is possible to abstract one of the methyl protons using a base. For example, when trialkylmethyl ammonium halides react with strong bases, one of the methyl protons will be abstracted to produce the following ionic compound.

Neutral compounds having a positively charged functional group and a negatively charged functional group within the same molecule are called *betaines*. Betaines are a specific type of zwitterion. If these charges are found on two adjacent atoms, the related compounds are called *ylides*. The positive charge is mainly on the heteroatom such as nitrogen, sulfur, or phosphorus. If the charge is located on the nitrogen atom, the ylides are called ammonium ylides or *N*-ylides. Likewise, it is possible to remove a proton from tribenzylmethyl phosphonium halide, diphenylmethyl sulfonium iodide, and trimethylsulfoxonium iodide using strong bases to form the relevant ylides.

Sulfur and phosphorus ylides are more stable than nitrogen and oxygen ylides. The ability of sulfur and phosphorus to have more than eight valence electrons allows the generation of a double bond between the carbon atom and the heteroatom because sulfur and phosphorus have d orbitals and so they can exceed the octet rule and have five pairs of electrons in their outer orbital and even six pairs in sulfur. These resonance structures make sulfur and phosphorus ylides more stable. Because nitrogen and oxygen cannot increase the number of electrons to more than eight in their outer shell, they cannot generate a double bond with the carbon atom. Therefore, the reactions of sulfur and phosphorus ylides are very different from those of nitrogen and oxygen ylides. In recent years, nitrogen ylides having electron-attracting groups attached to the carbon atom have been synthesized and their reactions have started to be investigated.

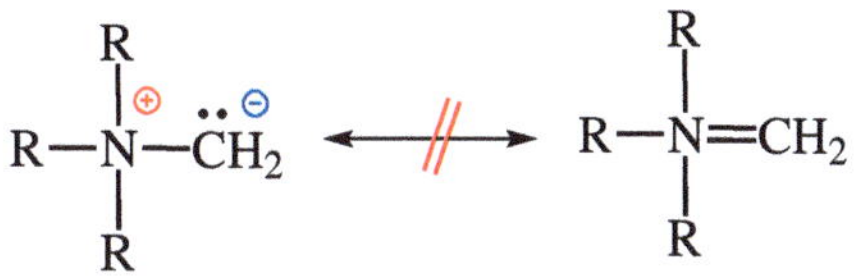

5.5.1 Phosphonium Ylides and Wittig Reactions

The reaction of phosphonium ylides with aldehydes and ketones produces alkenes. This reaction is called *the Wittig reaction* [48, 49], which was first discovered by German chemist Georg Wittig (1897–1987) in 1954, and he was awarded the Nobel Prize in Chemistry in 1979 for his work.

The Wittig reaction is widely used in organic chemistry and it plays a crucial role in the synthesis of prostaglandin, vitamin A, vitamin D_2, etc.

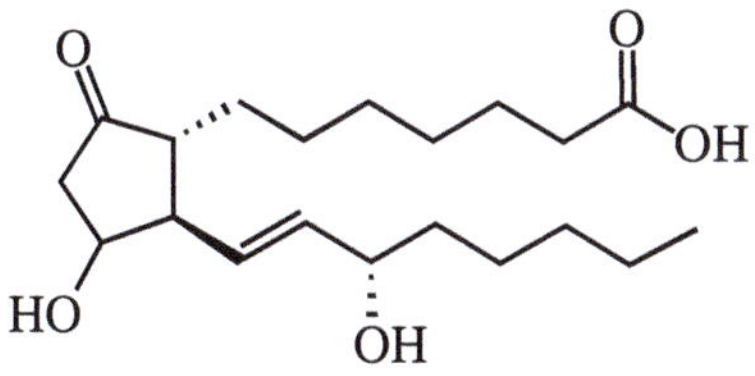

Prostaglandine

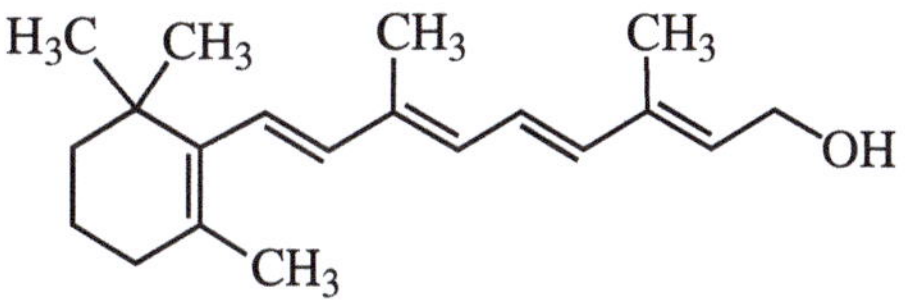

Vitamin A

Phosphonium ylides are stable but quite reactive species. They can be represented by two resonance structures: ylides and ylenes. NMR spectral studies indicate that the dipolar structure is the dominant structure. The π bond between carbon and phosphorus is weak. The ylene structure has a minor contribution. The theoretical calculations also support this finding [50]. A general reaction for the Wittig olefination reaction is given below:

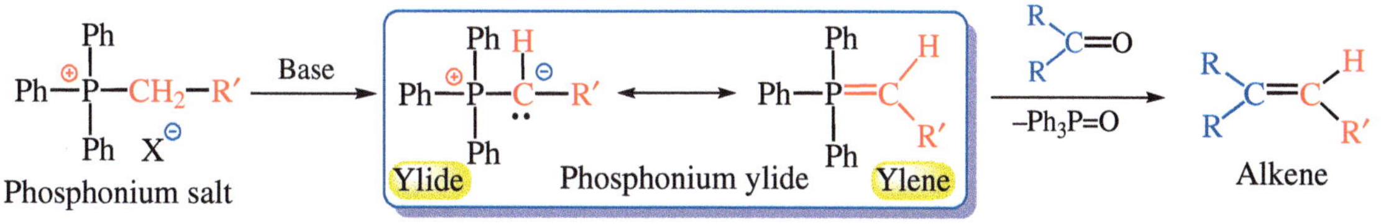

5.5.2 Reaction Mechanism

The Wittig reaction consists of two stages. In the first stage, phosphonium ylide needs to be synthesized. For this purpose, tertiary phosphine compounds (generally triphenylphosphine or trialkylphosphines are preferred) are reacted with alkyl halides. Preferably, the alkyl iodides are heated with triphenylphosphine in THF or chloroform. High temperatures and polar solvents are needed in branched alkyl halides. The nonbonding electrons on the phosphorus atom act as a nucleophile and attack the carbon atom to which the halogen is attached. The substitution of the halogen, according to the S_N2 mechanism, yields the phosphonium salt. The phosphonium salt needs to be converted into phosphonium ylide, which is called a *Wittig reagent*. The alkyl group attached to the phosphorus atom must have a hydrogen atom in the α-position. The positively charged phosphorus atom renders any neighboring proton acidic ($pK_a = 35$).

As we will see later, if the electron-attracting groups are attached to the carbon atom, they further increase the acidity of the protons. For this reason, the base required to form the ylide varies according to the stability of the ylide to be formed. When electron-donating groups (such as alkyls) are attached to the carbon atom bonded to the phosphorus atom, strong bases such as phenyllithium and *n*-butyllithium are used. If an electron-withdrawing group is attached to the carbon atom, bases such as NaOH, NaOMe, NaH, and NEt$_3$ can also be used. It is necessary to examine the ylides in two groups.

5.5.3 Stable Ylides

These are ylides in which the electron-withdrawing groups are attached to the carbon atom carrying the negative (−) charge. The delocalization of the charge stabilizes them and their reactivity is low, and they are commercially available.

5.5.4 Unstable Ylides

These are ylides bearing electron-donating groups attached to the carbon atom having the negative (−) charge. They are highly unstable and react quickly. They must be formed under a nitrogen atmosphere because they react with the oxygen in the air. The stability of the ylides directly affects the configuration of the products formed as a result of the Wittig reaction.

Because of the carbanion character of the carbon atom in the Wittig reagent, the ylide carbon atom is strongly nucleophilic. According to preliminary postulated mechanisms, the nucleophilic ylide carbon atom attacks the electrophilic carbonyl carbon atom of aldehyde or ketone to provide a betaine with a negative charge on oxygen and a positive charge on phosphorus. In the next step, the nucleophilic oxygen atom attacks the positively charged phosphorus atom, forming a four-membered ring that is called an *oxaphosphetane*. The oxaphosphetane quickly collapses to give an alkene and triphenylphosphine oxide. The decomposition proceeds stereoselectively. A *cis*-oxaphosphetane decomposes to give a *cis*-olefin, whereas a *trans*-oxaphosphetane decomposes to a *trans*-olefin. Triphenylphosphine oxide is a very stable compound and its formation is the driving force of the reaction. This whole reaction is called the *Wittig reaction*.

Recent mechanistic studies [51, 52] and theoretical calculations [53] support concerted formation of oxaphosphetane. Phosphonium ylides react with carbonyl compounds via a [2 + 2] cycloaddition rather than via a betaine. It is well established when the reaction is carried out in the presence of lithium salt, and there may be an equilibrium between oxaphosphetane and betaine species [54]. It has been suggested that lithium salts stabilize Wittig intermediates against decomposition to alkene, so that *cis*-oxaphosphetane reverts with much greater facility to the *trans*-oxaphosphetane. The ability of soluble Li salts in the Wittig reaction promote the production of *E* alkene at the expense of *Z* alkene.

Reactions with unstable Wittig reagents mainly produce *cis*-olefins. According to the proposed mechanism, the ylide approaches the carbonyl group crosswise, as depicted below. Thus, the alkyl groups are quite far from each other in the transition complex and do not create any steric barrier. This transition state leads to the formation of *cis*-configured oxaphosphetane. Of course, its decomposition results in the formation of a *cis*-olefin.

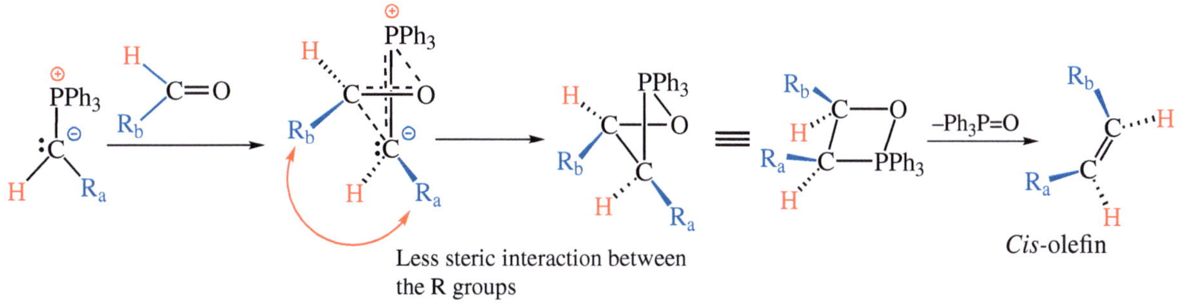

An essential part of the work on Wittig reactions is devoted to the configuration of the olefin formed. If there are electron-attracting groups such as CN, COOR, and Ph bonded to the carbon atom in phosphonium ylides, these ylides are less reactive because they are more stable. Such ylides, because of their low reactivity, do not react with ketones but react with aldehydes to form oxaphosphetanes. The oxaphosphetanes formed from stable ylides are in equilibrium with the starting reagents. Therefore, an equilibration is established between less stable *cis*- and more stable *trans*-oxaphosphetanes. There is enough time for the conversion of *cis*-oxaphosphetanes into *trans*-oxaphosphetanes before the decomposition takes place. Finally, the thermodynamically more stable olefin is formed as the major product by using stable ylides.

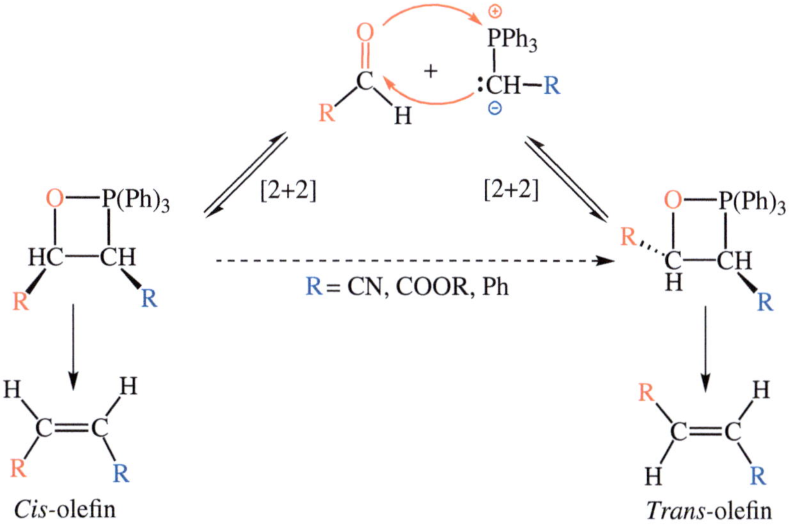

5.5.5 Wittig–Schlosser Reaction

In reactions with unstable Wittig reagents, it was emphasized that *cis*-olefins are formed as the major product. The question of how a *trans*-olefin can be obtained with unstable Wittig reagents may come to mind. It is possible to carry out such a reaction with the Schlosser modification [55–57]. Unstable Wittig reagents were shown to react with aldehydes to form mainly *cis*-oxophosphetanes. When the reaction is carried out at lower temperatures (−78 °C), oxaphosphetanes cannot immediately decompose to the products. They are quite stable at low temperatures.

When lithium halide is added to the reaction mixture, the oxaphosphetanes immediately undergo a ring-opening reaction and form the corresponding *erythro*- and *threo*-betaines. They contain acidic hydrogen in the α-position to the phosphorus atom. The addition of one mole of PhLi to the reaction mixture abstracts the proton in the α-position and forms an *oxido ylide*. Spontaneous pyramidal inversion of α-lithiated betaine generates the thermodynamically more stable *threo*-isomer. The addition of a proton donor occurs with remarkable stereospecificity to give a new intermediate in the *threo*-configuration, which decomposes selectively to furnish pure *trans*-alkene. Another feature of this reaction is that a different electrophile can be attached to the double bond instead of proton.

After the discovery of the Wittig reaction, which made a significant contribution to synthetic organic chemistry, research in this field was intensified and various modifications of the Wittig reaction were developed. One of them is the Schlosser modification, which has been discussed in detail above. Other modifications of this reaction are the Horner–Wittig and Horner–Wadsworth–Emmons reactions. All of these reactions proceed via the same mechanism, but the reagents used are different.

Phosphonium salt
Wittig reagent

Phosphine oxide salt
Wittig–Horner reagent

Phosphonate salt
Horner–Wadsworth–Emmons reagent

M = Metal

5.5.6 Wittig–Horner Reaction

A Wittig–Horner reaction is also a C=C bond formation reaction using alkyldiphenylphosphine oxide and an aldehyde as starting materials [58, 59]. It is mechanistically similar to the Wittig reaction. The Wittig–Horner reagent, phosphine oxides, first must be synthesized. There are several methods for the synthesis of phosphine oxides. Some of them are given below.

The phosphine oxide is deprotonated with *n*-BuLi in an α-position to the phosphorus, generating a phosphorus-stabilized Wittig–Horner reagent. When this phosphine oxide salt reacts with an aldehyde, a betaine is formed in the first step, and the formed betaine turns into alkene over oxophosphetane. When the initially formed betaines are protonated, the alcohols formed can be separated and isolated by chromatography or crystallization. Separated alcohols can be converted into oxophosphetanes at room or higher temperatures and then to the corresponding alkenes when treated separately with the base. The application of this method makes it possible to synthesize the *cis*- and *trans*-alkene in a pure state.

5.5.7 Horner–Wadsworth–Emmons Reaction (HWE Reaction)

This modification of the Wittig reaction utilizing phosphonates transforms aldehydes and ketones into α,β-unsaturated esters [49, 60]. Phosphonate esters are reacted with strong bases to generate stabilized phosphonium anions, which undergo a nucleophilic attack at the carbonyl group of aldehyde or ketone. This step is the rate-determining step. The reaction between a phosphonate ylide and the carbonyl group proceeds via a mechanistic path similar to the Wittig reaction. The elimination to the final products proceeds through oxaphosphetanes. The HWE reaction favors the formation of *trans*-alkenes. One of the advantages of this variation over others is that the phosphonate formed during the formation of an olefin can be easily separated from the product because of its water solubility. The situation is different in phosphine compounds. The resulting phosphine oxide must be separated from the molecule by chromatographic methods.

If there are no electron-withdrawing groups, such as COOR, C≡N, or SO$_2$R, which will stabilize the carbanion in the phosphonate salt, the reaction efficiency is quite low. The stereochemistry of the alkene formed depends primarily on the structure of the phosphonate used. Bulky groups attached to the phosphorus atom and the carbon atom bearing the negative (−) charge generally produce *trans*-alkenes.

We have demonstrated that the configuration of the product (*cis* or *trans*) can be controlled in Wittig reactions. Enantioselective Wittig reactions can also be carried out. The chirality can be transferred to the product by Wittig reactions with suitable chiral Wittig reagents. For this, it is necessary to work with chiral Wittig compounds. Chirality can be directly observed on the phosphorus atom to which different groups are attached or on substituents. Some chiral Wittig reagents are given below. Several chiral reagents have been synthesized, and they have been successfully applied for the desymmetrization of a prochiral ketone or aldehyde.

Chiral Wittig reagents

The reaction of a prochiral ketone with a chiral phosphonate compound carrying an optically active binaphthol (BINOL) group in a basic medium yields an α,β-unsaturated ester with an enantiomeric purity of 91%.

82%, ee 91%

5.5.8 Sulfur Ylides

The most critical use of sulfur ylides in the synthesis of epoxides arises from their reaction with aldehydes and ketones. The sulfur ylides have to be synthesized first. The essential sulfur ylides are dimethylsulfonium methylide and dimethylsulfoxonium methylide, which are *Corey–Chaykovsky reagents*. These ylides are prepared by deprotonation of sulfonium salts, which are readily prepared from the reaction of dimethyl sulfide or dimethyl sulfoxide with methyl iodide. The α-proton can be easily abstracted with NaH because of the increased acidity of the methyl protons.

Trimethylsulfonium salt   Dimethylsulfonium ylide

Trimethylsulfoxonium salt   Dimethyloxosulfonium ylide

The most important use of ylides in synthesis results from the fact that they generate a new carbon—carbon bond upon addition to carbonyl groups. In contrast to the reaction of phosphorus ylides with carbonyl compounds, reactions of sulfur ylides with carbonyl compounds do not usually lead to the formation of four-membered ring species analogous to oxaphosphetanes. In the first step, the ylide adds to the carbonyl group and forms a betaine. The most appropriate reaction is the formation of an epoxide by an intramolecular S_N2 reaction in which the dimethyl sulfide group acts as a leaving group. The reason for this different behavior is easily understood by comparison of the bond energies of the products. In Wittig reactions, as a result of the formation of phosphine oxide ($R_3P=O$), 126.4 kcal/mol energy is released. If the reaction with sulfur ylides formed a four-membered ring followed by the formation of dimethyl sulfoxide, 87.7 kcal/mol of energy would be released. This energy difference prevents the formation of an oxathietane.

5.5 Reaction of Carbonyl Groups with Ylides

The total reaction, the conversion of a carbonyl group into an epoxide, is called *the Corey–Chaykovsky reaction* [61, 62]. When looking at the total reaction, it appears that the epoxide is formed by the addition of a methylene (CH_2) carbene to a carbonyl group. It is not possible to carry out such a reaction because the carbonyl groups, as well as the carbene, are electropositive. However, such a reaction is indirectly accomplished using sulfur ylides.

It is necessary to make a distinction between sulfur ylides. Sulfoxonium ylides are more stable than sulfonium ylides because the negative charge (−) on the carbon atom is stabilized by the resonance with the S=O double bond. Therefore, the carbon atom in sulfoxonium ylides is less nucleophilic and more selective. Consequently, stereo- and regioselectivity are observed in the reactions of these ylides. When a carbonyl group with two different faces, 4-*t*-butylcyclohexanone, is reacted with dimethylsulfonium methylide in THF at 0 °C, two diastereoisomers are formed. One of these is produced in 83% yield and the other in 17%. The different approaches of the ylides can explain the formation of two different products.

When the same reaction is carried out with oxosulfonium ylide in THF at 65 °C, the thermodynamically more stable compound is exclusively formed. Probably, the oxosulfonium ylide adds reversibly to carbonyl groups and forms the most stable product.

The best example of how sulfur ylides behave very differently is their reaction with α,β-unsaturated carbonyl groups. The reactions of sulfonium ylides with α,β-unsaturated carbonyl compounds form epoxides, while those of sulfoxonium ylides form cyclopropanes. The reactions of sulfur ylides with carvone are presented below [61–63].

As fairly established, nucleophilic additions to α,β-unsaturated carbonyl groups can proceed in two different forms: 1,2- and 1,4-addition (Michael-type addition). Because the carbonyl group is more polarized and positively ($\delta+$) charged, it is a hard acid. Sulfonium ylides are unstable and they are hard bases; therefore, they preferentially attack the carbonyl groups. On the other hand, the stable sulfur ylide, oxosulfonium ylide, is a soft base. Therefore, it attacks the β-carbon atom, which is a soft acid.

5.5.9 Julia Olefination

The Julia olefination [64] (also called the *Julia–Lythgoe olefination*) involves the nucleophilic addition of a phenylsulfonyl carbanion to carbonyl compounds. This leads to the formation of an alcohol as an intermediate. Acylation of the alcohol followed by reductive elimination with sodium amalgam or samarium diiodide [65] furnishes *trans*-alkenes. The configuration of the intermediate, β-acyloxysulfone, does not have any influence on the configuration of the alkene. The reaction is named after the French chemist Marc Julia, who discovered this reaction in 1973. The Julia olefination has some advantages over other olefination methods and it provides mild user-friendly reaction conditions [66–68].

The exact mechanism of the reduction step is not known. It has been proposed that some radical intermediates, such as vinyl radicals, are involved. Protonation of the vinylic radical furnishes the alkene. When the reduction is carried out in MeOD as a solvent, deuterium incorporation is observed, which supports the formation of a radical intermediate [65].

There are some modifications of the Julia olefination reaction, such as the *Julia–Kocienski olefination*, which uses heteroaromatic sulfone moieties to complete the transformation. The replacement of the phenylsulfones with certain heteroarylsulfones profoundly changes the reaction mechanism, resulting in a one-pot procedure and evolution of SO_2 gas. The intermediate, β-alkoxysulfone, is unstable and undergoes a facile *Smiles rearrangement* [69] via a spirocyclic intermediate to generate the olefin in a single step. This methodology enables the synthesis of alkenes from benzothiazol-2-yl sulfones and aldehyde in a single step. However, regioselectivity is not observed.

5.5.10 Peterson Olefination

The Peterson olefination (also called the *Peterson reaction*) can be considered a silicon variation of the Wittig-type reaction. The Peterson olefination is a reaction of α-silylcarbanions with aldehydes and ketones to furnish alkenes [70]. In the first step, α-silyl carbanions attack the carbonyl group and form a β-hydroxysilane, which can be isolated. When an α-silyl organomagnesium compound is used as a starting material, the intermediate β-hydroxysilanes can be isolated because of the strong bonding of magnesium to the oxygen atom. Once isolated, the diastereomeric β-hydroxysilanes can be separated.

Treatment of *threo* β-hydroxysilyl intermediate with acid results in protonation of the hydroxyl group, which undergoes an *anti*-elimination to give a *cis*-olefin as the major product in a ratio of $Z/E = 92:8$ [71]. Treatment of the same β-hydroxysilane with a base will furnish the *trans*-isomer in a concerted *syn* elimination in a ratio of $E/Z = 95:5$. It is likely that a four-membered ring, oxasiletanide, is formed as an intermediate. Therefore, the Peterson olefination is an attractive alternative to Wittig reactions [72].

α-Silylcarbanions can be synthesized by either the reaction of the corresponding halotrimethylsilanes with magnesium or deprotonation of $(CH_3)_3SiCH_2Z$ with BuLi, where Z is a carbanion-stabilizing substituent.

$(CH_3)_3SiCH_2Br + Mg \longrightarrow (CH_3)_3SiCH_2MgBr$

$(CH_3)_3SiCH_2Z + n\text{-BuLi} \longrightarrow (CH_3)_3SiCHZ\text{-Li}$

$Z = CN, COOR, SPh, P(OR)_2\text{=O}$

5.6 Reactivity of α-Carbon Atoms

5.6.1 Acidity of α-Hydrogens

Carbonyl compounds display two different types of reactivities. In the previous sections, we discussed the reactivity of the partially positively charged carbonyl group that is attacked by nucleophiles. Carbonyl compounds also show a second type of reactivity. An adjacent carbon atom bonded to the carbonyl group is called an α-carbon atom, and the hydrogen atoms bonded to the α-carbon atom are called α-hydrogen atoms. One of the essential features of the carbonyl group is the acidity of the adjacent α-hydrogens. From the polarization of the carbonyl group (detailed information is given in the introduction of this Chapter), the carbonyl carbon atom is partially positively (+) charged and the carbonyl oxygen atom is partially negatively (−) charged. The carbonyl carbon atom attracts electrons from the neighboring carbons and bond polarization occurs. Therefore, the electron density at the α-carbon atom decreases and around the α-carbon atom, a partial positive ($\delta+$) charge occurs. This charge attracts electrons from adjacent carbon atoms and C—H bonds. Thus, the acidity of the α-hydrogens increases. This effect continues throughout the σ-skeleton.

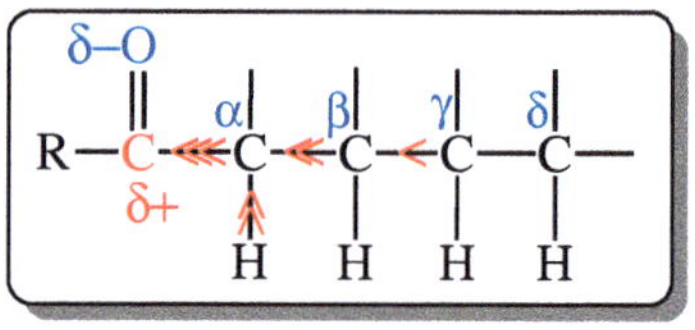

The inductive effect propagates through the σ-skeleton and drops exponentially from one carbon to another. The acidity of the β-protons will also increase. However, this increase is not significant, and it is not possible to remove this hydrogen with a base. The electron-withdrawing effect of the carbonyl group is no longer perceptible on the γ-carbon atom or the bonded protons. It is possible to remove the α-hydrogens using strong bases. Because the pK_a values for C—H bonds in alkanes are high (pK_a = 45–50), the hydrogens of alkanes are in fact not acidic. On the other hand, the pK_a values for these α-protons of carbonyl compounds range from 15 to 20.

The chemistry of the carbonyl group is based on the acidity of the α-protons. Understanding that the α-protons of the carbonyl group are acidic means understanding the chemistry of carbonyl compounds where α-protons are involved. In the previous sections, we discussed the addition of nucleophiles to a carbonyl carbon, followed by the elimination of H_2O. However, when the carbonyl groups are treated with bases, they generally prefer not to attack the carbonyl group, but they abstract the α-proton to form a carbanion.

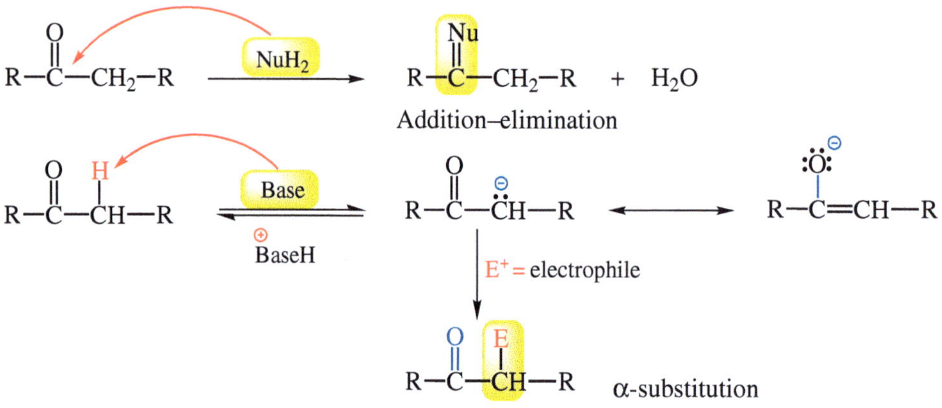

There are two important reasons for the formation of a carbanion. The first is the acidity of the α-proton, as described above, and the second is the resonance stabilization of the product, carbanion, as illustrated in the diagram above. The reaction is terminated by reacting with an electrophile present in the reaction medium. All the carbonyl group reactions we will discuss in this section are based on the acidity of α-protons.

The acidity of α-hydrogens varies depending on the structure of the carbonyl compounds. Evaluating the stability of conjugate bases can give us information about the differences in the relative acidity of α-hydrogens. The acidity values of α-hydrogens with various carbonyl groups are given below. The effect of the carbonyl groups is seen when the pK_a values are compared.

While the pK_a values of saturated hydrocarbons are approximately 45–50, the pK_a value of acetone is 20. α-Protons of an ethyl group attached to the carbonyl group have a pK_a value of about 19–20, while the pK_a values of β-protons are around 44–50. On the other hand, when two carbonyl groups are attached to a methylene carbon atom, the pK_a values of the hydrogens decrease to 9, as can be seen in 2,5-pentadione. As expected, when OR groups replace the methyl groups, the acidity decreases, albeit slightly, i.e. the pK_a values increase.

H_3C-CH_2-H $H_2C=CH-H$ $HC\equiv C-H$ $H_3C-CO-CH_2-H$ $H_3C-CO-CH(\alpha)-CH_2(\beta)-H$

pK_a = 50   pK_a = 44   pK_a = 24   pK_a = 20   pK_a = 19–20   pK_a = 44–50

$H_3C-CO-CH(H)-CO-CH_3$ $H_3C-CO-CH(H)-CO-OCH_3$ $H_3CO-CO-CH(H)-CO-OCH_3$

pK_a = 9   pK_a = 11   pK_a = 13

5.6.2 Keto–Enol Tautomerism

Tautomers are structural isomers (constitutional isomers) with the same molecular formula that interconvert to each other. During this interconversion, a proton attached to any atom shifts to another atom while a double bond also changes its position. One of the most prominent examples of tautomerism in organic chemistry is keto–enol tautomerism.

Ketone ⇌ Enol — Keto–enol tautomerism

Enamine ⇌ Imine — Enamine–imine tautomerism

The acidity of the α-hydrogen atoms of carbonyl compounds is responsible for this process. A ketone exists in equilibrium with its enol isomer. The α-hydrogen atom in the ketone shifts to the oxygen atom when the molecule turns into enol form. During this conversion, the double bond between the carbon and oxygen atoms shifts between two carbon atoms. There is a similar equilibrium between enamine and imine. The hydrogen atom bonded to the nitrogen atom is transferred to the α-carbon atom. The double bond also changes its position. This type of rearrangement is called *tautomerism*, and the two isomers are called *tautomers*. Tautomers should not be confused with resonance forms, where only the electrons are shifting. Tautomers are different compounds.

All ketones and aldehydes are in equilibrium with the related enols. The equilibrium constants for this conversion of carbonyl compounds into enol forms are very small. Only traces of enols are present. Some factors shift the equilibrium to the right or the left. For example, the keto tautomer in acetone comprises more than 99.9% of the mixture, while the enol concentration is less than 0.1%. The reason for the equilibrium lying on the side of the keto form is bond energies. To explain this situation, it is necessary to look at the total energies of the different bonds formed with a tautomer on both sides. The keto form has C=O, C—C, and C—H bonds, whereas the enol has C=C, C—O, and O—H bonds. The sum of the first three bond energies is about 359.6 kcal/mol (1505 kJ/mol), whereas the sum of the three bond energies in the enol is 343.9 kcal/mol (1439 kJ/mol). The keto form is, therefore, more thermodynamically stable by 15.7 kcal/mol (65.6 kJ/mol).

$$H_3C-\underset{\underset{>99.9\%}{}}{\overset{\overset{O}{\|}}{C}}-CH_2-H \quad \rightleftharpoons \quad \underset{<0.1\%}{H_3C-\overset{\overset{O-H}{|}}{C}=CH_2}$$

C=O	178 kcal/mol	Keto-enol tautomerism	C=C	146.7 kcal/mol
C—C	82.9 kcal/mol		C—O	85.6 kcal/mol
C—H	98.7 kcal/mol		O—H	111.6 kcal/mol
	359.6 kcal/mol	$\Delta E = 15.7$ kcal/mol		343.9 kcal/mol

Ketones and aldehydes generally prefer the keto form because they are more stable in the keto form than in the enol form. However, some factors shift this equilibrium to the enol side. One of them is aromaticity. Phenol can theoretically exist in its keto form, cyclohexa-2,4-dien-1-one, but the enol form is more stable because the double bond is a part of the aromatic system. Therefore, the cyclohexa-2,4-dien-1-one cannot be stable; it will immediately convert to the phenol. To stabilize such compounds, the α-protons forming the enol form need to be replaced by alkyl groups. On the other hand, the equilibrium in cyclobutenone is entirely shifted to the side of the keto form because the enol form, cyclobutadienol, is an antiaromatic compound.

The fraction of enol tautomer is higher for 1,3-dicarbonyl compounds. If the –OH group in the enol form can form hydrogen bonds with other groups, the equilibrium will shift to the enol side. The hydrogen bonding between the carbonyl group and the hydroxyl hydrogen stabilizes the enol form. Furthermore, the conjugation of the carbon—carbon bond with the remaining carbonyl group also contributes to the stabilization of the enol form so that the equilibrium shifts to the enol side. Therefore, we can understand why acetylacetone exists as about 85% in enol form. Solvents can also affect the keto–enol equilibrium. For example, in benzene, the enol form of 2,4-pentadione is more stable than the keto form in a ratio of 94 : 6. However, in solvents that can act as hydrogen bond donors, like water, the keto tautomer is more stabilized by hydrogen bonding to the carbonyl groups, making the nonbonding electrons on carbonyl oxygen less available to the hydrogen bond with the enol form. The equilibrium in water is shifted mainly to the side of the 2,4-pentadione. The keto form dominates in a ratio of 84 : 16 [73].

| | Solvent = benzene | 5% | 95% |
| | Solvent = water | 84% | 16% |

When one of the acetyl groups in 2,4-pentadione is replaced by an ester group, the enol ratio in equilibrium decreases. The equilibrium shifts to the side of the ketone and the compound contains about 7% of the enol form under normal conditions. The ester group reduces the electron-withdrawing ability of the carbonyl group. The pK_a value of ethyl acetoacetate is 11 while that of acetylacetone is 9. There is a direct relationship between the pK_a values (acidity of methylene protons) and keto–enol tautomerism; as the pK_a value of the molecule decreases, the probability of forming enols increases.

1,3-Diketone 92.5% ⇌ Enol 7.5%

If the 1,3-dicarbonyl functional group is incorporated in a ring, as in the case of cyclohexane-1,3-dione, the enol ratio will increase compared to the acyclic systems. The carbonyl group and the double bond will form a plane for an ideal conjugation to stabilize the enol form. However, the situation is different in cyclohexane-1,2-dione, which has two carbonyl groups with their dipoles pointing in the same direction. Electrostatic repulsion between these two carbonyl groups will increase the energy of the system. To minimize this energy, the compound will prefer to form an enol. Moreover, the hydrogen bonds formed between the enol hydroxyl group and the other carbonyl group make the system more stable.

Cyclohexane-1,3-dione: 5% ⇌ 95%

Cyclohexane-1,2-dione: δ+ Electrostatic repulsion, δ+ ⇌ 100%

Some molecules cannot enolize. For example, if a 1,3-dicarbonyl compound is incorporated in the norbornane system, theoretically two different enols can be formed. One of them can be formed by the bridgehead hydrogen atom located between the two carbonyl groups. Formation of the enol form is not possible here because the double bond will be located at the bridgehead, which will increase the strain in the molecule (see Bredt's rule). There are methylene protons adjacent to the carbonyl group that can form an enol. The enol formed with these protons is not stable either because the formation of an endocyclic double bond increases the strain present in the ring.

Table 5.4 shows the enol ratios in some keto–enol systems. When cyclohexanone and cyclopentanone are compared, it is seen that the enol ratio is lower in the five-membered ring compared to in the six-membered ring. If a double bond is formed in the five-membered ring, the ring strain increases compared to the six-membered ring. When acetone is compared with

Table 5.4 Enol ratios of some carbonyl compounds.

$H_3C-\overset{O}{\underset{\|}{C}}-CH_3 \rightleftharpoons H_3C-\overset{OH}{\underset{|}{C}}=CH_2 \qquad 6 \times 10^{-5}$

$H-\overset{O}{\underset{\|}{C}}-CH_3 \rightleftharpoons H-\overset{OH}{\underset{|}{C}}=CH_2 \qquad 6 \times 10^{-7}$

cyclopentanone ⇌ cyclopentenol 1×10^{-6}

cyclohexanone ⇌ cyclohexenol 1×10^{-5}

acetaldehyde, it can be seen that the enol ratio increases when the number of alkyl groups bonded to the double bond in the enol form increases. Any factors that stabilize alkenes will stabilize enols as well. Remember from the previous sections that when the number of alkyl groups attached to a double bond increases, the stability of the double bond also increases.

5.6.3 Acid-Catalyzed Enolization

Enol structures play an essential part in carbonyl group reactions. In the examples given in Table 5.4, we see that the equilibrium is mostly on the side of the keto form in cyclic as well as in acyclic systems. It was emphasized that the enol ratio is well below 1% in normal ketones and aldehydes. If we claim that enols are essential intermediates in many carbonyl reactions, then the enol ratio must be increased in some way to carry out chemical reactions. This can be done in two different ways, either in an acidic or in a basic environment.

In the acidic environment, a proton binds to a pair of nonbonding electrons of the carbonyl oxygen, forming a protonated carbonyl group. While the double bond electrons of the carbonyl group open toward oxygen, the positive (+) charge is located on the carbonyl carbon. The positively (+) charged carbon atom inductively attracts the bond electrons and makes α-hydrogens more acidic, and so it is now somewhat easier to abstract the α-hydrogen atom. Water removes a proton from the α-carbon atom, while the bonding electrons generate a double bond between the two carbon atoms to complete enol formation. It should be noted that all reactions shown below are reversible. Keep in mind that a high concentration of enol in acidic conditions will never be formed unless there are some groups that can stabilize the enol form. If there is more than one possible enol product, the formation of the thermodynamically most stable enol will be favored.

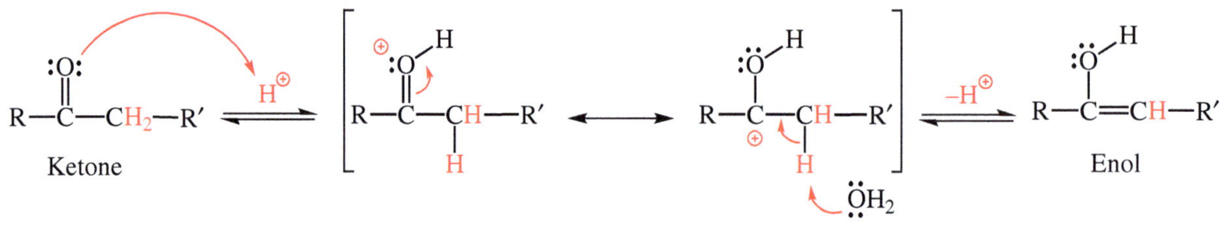

To understand the reactions derived from enols, we should first analyze their structures. Let us take a closer look at the enol form. Unlike a ketone or aldehyde, an enol contains a C=C double bond. This double bond is an electron-rich double bond compared to normal double bonds. The oxygen atom bonded to the double bond inductively attracts double-bond electrons; on the other hand, it increases the electron density on the double bond by resonance, giving electrons to the double bond. Remember that the mesomeric effect is more dominant than the inductive effect. Therefore, the electron density increases on the β-carbon atom. In contrast, the electron density decreases on the α-carbon atom because it is closer to the positively (+) charged oxygen atom, as shown in the resonance structure. Thus, such a double bond acts more nucleophilic and reacts with electrophiles at the β-carbon atom. Because there are two pairs of nonbonding electrons on the oxygen atom, electrophiles can attack either the oxygen atom or the β-carbon atom. Enols are ambident nucleophiles because they are included in the class of compounds that react with two different functional groups.

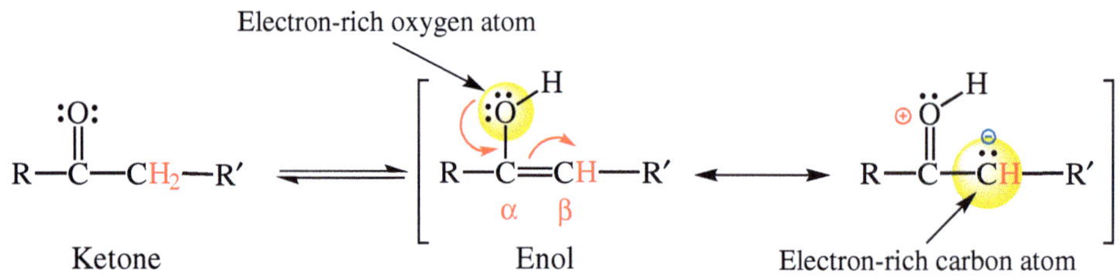

Because the enol double bond is polarized, all electrophilic reactions will proceed through the β-carbon atom. The ^{1}H- and ^{13}C-NMR spectra demonstrate that the enol double bond is significantly polarized. In the ^{1}H-NMR spectrum, the β-protons resonate at about 1.5 ppm higher than the α-protons. The difference in chemical shifts between the α- and β-carbon atoms in the ^{13}C-NMR spectrum is up to 40 ppm. These observed differences reveal that the enol double bond is significantly polarized.

$$\text{R}-\overset{\beta}{\text{CH}}=\overset{\alpha}{\text{CH}}-\text{OR}$$

^{1}H-NMR (ppm)	4.5–5.0	6.0–6.5
^{13}C-NMR (ppm)	100–110	145–150

5.6.4 Base-Catalyzed Enolization

While enolization occurs in basic and acidic conditions, base-catalyzed enolization is more common. In the introduction of this chapter, we discussed the polarization of the carbonyl group and emphasized the acidity of α-protons. The pK_a values of the α-protons in ketones and aldehydes are approximately 18–20. In base-catalyzed enolization, a base abstracts a proton from the α-carbon atom to form an intermediate. Therefore, it is necessary to pay attention to the base to be used to remove the α-proton of the carbonyl groups. The strengths of bases are different. Therefore, the choice of base is important to shift the equilibrium to the enolate side. For example, it is not recommended to generate an enolate ion by using bases such as NaOH or NaOCH$_3$ because the pK_a value of water is ≈15.5 and that of ROH is about >16. With these bases, an enolate can be formed, but the keto–enol equilibrium will be on the side of the keto form and the concentration of the enolate in the solution will be very low.

To perform an irreversible reaction, the base used must have a pK_a value greater than 20. Amine bases are generally used for this purpose. For example, when bases whose pK_a values ranging from 35 to 37, such as lithium diisopropylamide (LDA) or sodium amide, are used, the α-proton can be easily abstracted and the equilibrium shifts entirely to the enolate side. The concentration of the initial ketone or aldehyde will be basically zero.

LDA is easily formed in the reaction medium by the treatment of lithium diisopropylamine with butyllithium (*n*-BuLi). In enolate ions created with strong bases, the equilibrium shifts entirely to the enolate side and the carbonyl group stabilizes the anion formed. In general, the higher the difference between the pK_a value of the base used and the pK_a value for the α-proton of the ketone or aldehyde reacted, the more the equilibrium shifts to the enolate side. Because the difference between the pK_a value (36) of diisopropylamide and the pK_a values of ketones is 18, the equilibrium is completely on the enol side. When bases with a pK_a value of 10–15 are used, the equilibrium is completely on the ketone side. Some of the amine bases that are used frequently in enolate formation are given below.

LDA pK_a = 36
Lithium diisopropylamide

LTMP pK_a = 37
Lithium tetramethylpiperidide

LHMDS pK_a = 26
Lithium bis(trimethylsilyl)amide

We have already mentioned above that the methylene groups to which two electron-withdrawing groups are attached are significantly more acidic than regular ketones and aldehydes. Their pK_a values vary between 9 and 13 depending on the electronic nature of the bonded substituents. Therefore, it is possible to abstract one of the methylene protons with bases such as NaOH or NaOEt and shift the equilibrium to the enol side. The enolate ion formed is stabilized by adjacent carbonyl groups. Because the charge on the enolate ion (−) is distributed on both carbon and oxygen atoms by delocalization, the enolate ion also shows ambident properties. It reacts with electrophiles at two different locations (carbon and oxygen).

5.6.5 Kinetic and Thermodynamic Enolates

If a starting ketone is unsymmetrical and has two different α-protons, as in the case of 2-methylcyclohexanone, and it is treated with a base, two different enolates will be formed. One of them is thermodynamically controlled and the other is kinetically controlled. If the base abstracts the α-proton adjacent to the methyl group, a more thermodynamically stable enolate is formed because more alkyl groups are attached to the double bond. If the other α′-proton is removed, the less stable product, the kinetic-controlled product, is formed.

In this context, we can ask the following question: *Can enolate formation be controlled when the difference between the pK_a values of the α- and α′-protons is minimal (ΔpK_a ≈ 1–2)?* When the reaction with 2-methylcyclohexanone is carried out with LDA, which is bulky and very sensitive toward any steric hindrance, it will remove the α-proton from the less sterically hindered size regardless of the pK_a values of the proton and form the thermodynamically less stable enolate as the major product. It should be emphasized that the enolate ion formed here is no longer in equilibrium with the ketone.

If the reaction is performed with a weaker base, for example, triethylamine (pK_a = 10.8), the α′-proton will be removed from the sterically less crowded side. Of course, the other α-proton will also be removed. However, as the enolate ions formed are in equilibrium with the starting ketone, the kinetically controlled product will return more rapidly to the ketone than the thermodynamically stable enolate. Equilibrium will be established between the enolates that gradually shifts toward the side of the more stable enolate.

The position of the equilibrium is controlled by a variety of factors: time, base, temperature, solvent, and cation.

Factors favoring the formation of kinetically controlled enolates are as follows:

1. Keeping the reaction time short prevents possible reversible conversion.
2. There is no reversible reaction when strong bases are used. With the use of bulky bases, the base will abstract the proton from the sterically less hindered side, forming an enolate with fewer alkyl groups bonded to the double bond.
3. It is more convenient to carry out reactions at low temperatures.
4. Using aprotic solvents such as diethyl ether or THF. There are no acidic protons to encourage the reverse reaction.
5. Using oxophilic cations, e.g. Li$^+$.

Use of these conditions will suppress the equilibrium and ensure that the reaction is irreversible.

Factors favoring the formation of thermodynamically controlled enolates are as follows:

1. Keeping the reaction time long. If there is a reversible transformation, time is needed for the formation of the thermodynamic product.

2. The use of weaker bases creates an equilibrium between ketones and enolates and over time a thermodynamically stable product is formed.
3. Reactions at high temperatures help the equilibrium to shift to the side of the thermodynamic product.
4. Protic solvents with slightly acidic protons favor the reverse reaction, the formation of the most stable product.

Let us give an example of the formation of different enols under different reaction conditions. Enolates formed by the reaction of heptan-2-one with LDA (a strong base) are captured with trimethylchlorosilane and the kinetic controlled enol silyl ether is formed as the major product. On the other hand, the treatment of heptan-2-one with trimethylamine, a weak base, forms the more stable enol silyl ether.

5.6.6 Enol and Enolate Reactions

5.6.6.1 Racemization of Chiral Ketones, α-Epimerization

It was frequently mentioned that a significant proportion of carbonyl group reactions proceed through keto–enol tautomers. Let us try to explain what role this equilibrium plays in the reactions. One of the simplest examples of these reactions is that ketones and aldehydes having a chiral carbon atom at the α-position undergo racemization very easily in an acidic or basic environment. The racemization occurs over enols or enolates. Compounds that do not have any α-hydrogen atoms cannot enolize and do not undergo any of the reactions described below.

Under acidic conditions, first, a protonated carbonyl group is formed. The positively (+) charged carbon atom makes α-hydrogens more acidic and removal of a proton from the α-carbon atom generates the enol. The C=C double bond of an enol has a planar structure, and so any chirality that existed at the α-carbon is lost on enolization. The enol is in equilibrium with the ketone. While the enol turns back into a ketone, the proton bonded to the oxygen atom is then delivered back to the α-carbon atom. Because the double bond has a planar structure, the proton can be bonded to the α-carbon atom from two different directions, from top and bottom, with equal probability. As a result of this reaction, a racemic mixture will be formed and the optical activity will be lost.

Under basic conditions, a base removes a proton from the α-carbon atom, forming a carbanion, which is stabilized by resonance with the carbonyl group. The enolate ion can take a proton from the reaction medium and form the enol. The enols, as well as the enolate ion, are in equilibrium with the ketone. When the enol or enolate reverts to the keto form, a proton can be bonded to the α-carbon atom from both sides, as discussed above, producing equal amounts of the two enantiomers.

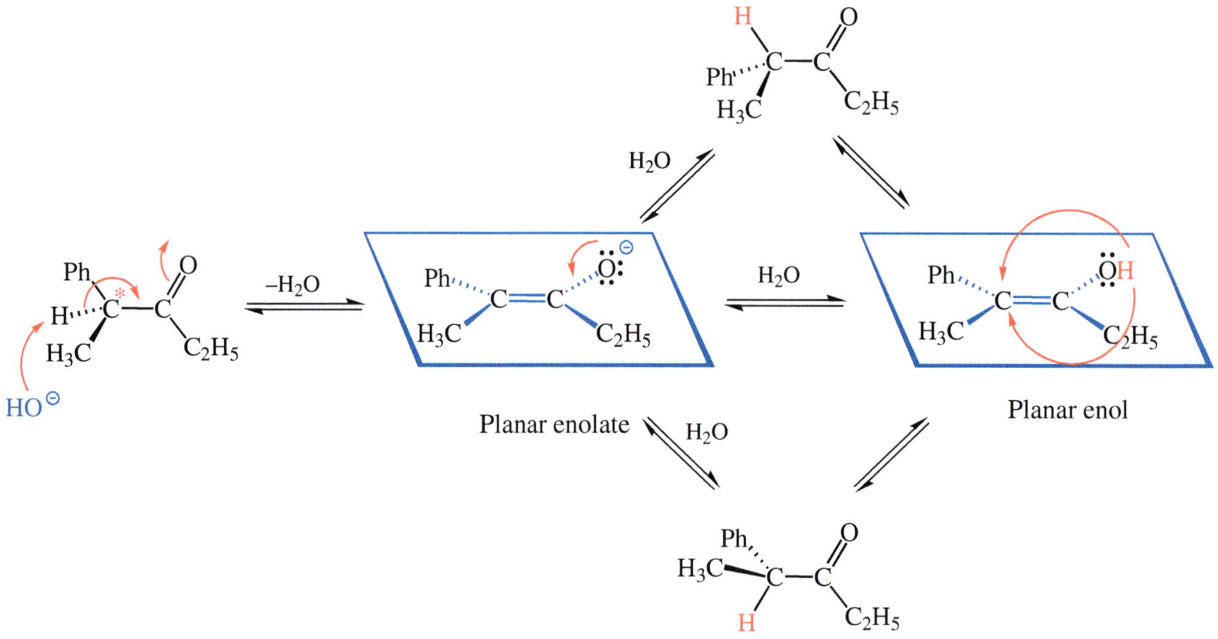

Hydrogen–Deuterium Exchange is a chemical reaction in which covalently bonded hydrogen atoms adjacent to a carbonyl group are exchanged with deuterium atoms. Because of the acidic nature of α-hydrogen atoms, it possible to exchange the α-protons of all carbonyl compounds that can form enols with deuterium. The reaction mechanism is the same as that of the α-racemization reaction discussed above. The synthesis of deuterium-containing compounds in the α-position is absolutely essential for mechanistic studies.

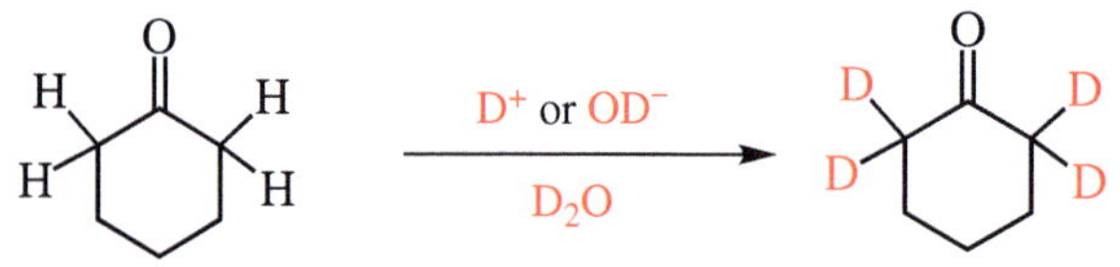

Hydrogen–deuterium exchange takes place via the enol or enolate form. The reaction can be carried out in an acidic or basic environment. Reagents such as the acid or base to be used for this exchange must be deuterated. The process is accelerated by the addition of an acid or base; an excess of D_2O is required. For example, when cyclohexanone is reacted with an excessive amount of D_2O in the presence of a catalytic amount of D^+ or DO^-, the complete exchange of all α-protons is achieved. As shown below, the enol formation is catalyzed by D^+ ions. The hydroxyl group in the enol contains a deuterium atom instead of a hydrogen atom. Because the enol is in equilibrium with a ketone, the deuterium atom bonded to the oxygen is transferred to the α-carbon atom so that the hydrogen–deuterium exchange occurs. The reaction does not stop after the replacement of a single hydrogen; because there is excessive D_2O in the reaction medium, it continues until all hydrogens are replaced by deuterium. Thus, all four α-protons in cyclohexanone are replaced by deuterium. The exchange process can be followed using ^{1}H-NMR spectroscopy because the signals arising from the α-protons slowly disappear.

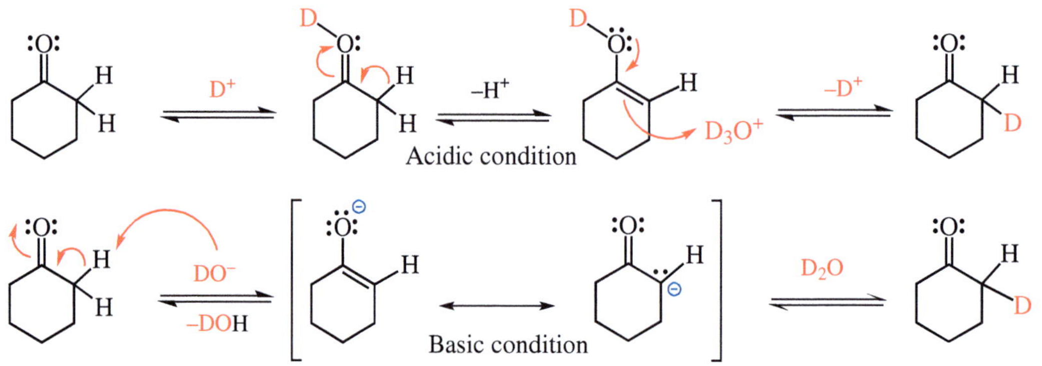

The other hydrogens, such as β- and γ-hydrogens, cannot be exchanged with deuterium. The exchange depends entirely on the pK_a values of the relevant proton. Because the pK_a values of β- and γ-hydrogens in cyclohexanone or other ketones are around 40–45, there is no exchange in acidic or basic medium. If there are two different protons adjacent to a carbonyl group that can form enols, the hydrogens that form the easier enol undergo the exchange first, followed by the other hydrogen atoms.

When 3-phenyl-2-butanone is reacted with D^+ or DO^- and D_2O, the proton adjacent to the phenyl group will first undergo deuterium exchange. When the reaction time is prolonged, the methyl protons will also be replaced with deuterium. We should keep in mind that a deuterium atom can replace only enolizable hydrogen atoms.

Hydrogen–deuterium exchange can also be applied to the structural determination of some compounds. For example, it is not easy to distinguish between the isomers below with a simple ^{1}H-NMR spectrum (of course, with the help of 2D-NMR spectra, one can distinguish between those structures). Deuterium exchange experiments would help us to determine the exact structures. All methylenic protons in compound **A** will undergo deuterium exchange upon treatment with an excess of D_2O in the presence of a catalytic amount of D^+ or OD^-. However, the benzylic protons in compound **B** will not be exchanged because they are not enolizable. This process can be conveniently followed using ^{1}H-NMR spectroscopy. In the case of **A**, all methylenic protons' signals disappear, whereas in the case of **B**, only the signals of the methylenic protons adjacent to the carbonyl groups disappear.

In general, the proton–deuterium exchange decreases in the following order:

1, 3 – Dicarbonyls > Ketones > Esters > Amides > Carboxylate Anion

5.6.6.2 α-Halogenation of Aldehydes and Ketones

One method to make carbonyl group-containing compounds more functional is halogenation at the α-position of the carbonyl group. Aldehydes and ketones react with halogens at the α-carbon atom. For example, when acetone is treated with bromine in acetic acid, monobromoacetone is formed in approximately 45% yield. The reaction stops at the stage of monobromination. In contrast with deuteration, which proceeds until all hydrogen atoms are exchanged, bromination of the second and third hydrogens does not occur when the reaction is carried out under acidic conditions.

We emphasized that in all reactions taking place at the α-position of the carbonyl group, enol or enolate intermediates are involved. The α-position of a carbonyl group can be easily halogenated because of its ability to form an enol in acidic conditions and an enolate in basic conditions. Under acidic conditions, the carbonyl group is first protonated, followed by removal of the α-hydrogen, forming an enol. The distribution of the electron density on the enol double bond is not homogeneous. The negative charge is concentrated more on the β-carbon atom. The enol double bonds are electron-rich and they behave as nucleophiles and react with electrophiles in the same way as alkenes. The double bond attacks the halogen to give an intermediate, an oxygen-stabilized halocarbocation. One bromine atom bonds to the double bond. The removal of the proton bonded to the oxygen atom completes the reaction.

The rate of this reaction, as well as the rate of the base-catalyzed reaction, is independent of the halogen concentration. It was determined that the rate of the reaction was not affected when chlorine, bromine, or iodine was used as a halogen. The reaction rate depends on the ketone concentration; more precisely, the reaction rate depends on the enol or enolate formation rate. As mentioned above, halogenation in acidic conditions stops at the stage of the monohalogenation reaction. The rate of each successive halogenation is slower than that of the first one. At this point, the question may arise, *Why does the reaction stop after the first halogenation?* For a further halogenation reaction, the monohalogenated carbonyl compound must again form an enol. The formation of an enol is likely retarded when a halogen is attached to the α-position. Another question we are faced with is *How can a halogen atom at the α-position hinder the formation of an enol?* The halogen decreases the basicity of the carbonyl oxygen. Halogen inductively attracts bonding electrons and the carbonyl carbon has a more positive characteristic. In this case, the carbonyl carbon pulls the electrons on oxygen toward itself. It reduces the basic feature of the oxygen atom, hindering the protonation of the oxygen atom and consequently the formation of an enol. Finally, we can summarize as follows: under acidic conditions, the rate of the second halogenation is much slower than that of the first one. Therefore, halogenation of carbonyl compounds under acidic conditions gives monohalogenated compounds.

α-Halogenation also occurs when a catalytic amount of acid is not added to the reaction medium. Ketones are in equilibrium with enols. Although the enol concentration in most cases is very low, the halogen reacts with the enol, which is in equilibrium, and an acid is formed as a result of the reaction. The acid concentration gradually increases in the reaction medium, which catalyzes the reaction.

Base-catalyzed halogenation is entirely different. Under basic conditions, enolates are formed as intermediates, and it is not possible to stop the reaction at the monobromination stage. In a basic solution, successive halogenation takes place. In the first step, the base removes one of the α-hydrogens, forming an enolate anion. This anion is much more reactive than the neutral enol. Therefore, it reacts with the electrophilic bromine. After the first bromine atom is attached to the α-carbon atom, the acidity of the adjacent hydrogen increases with respect to the ketone. The electron-withdrawing effect of halogen atom promotes a second α-deprotonation and, thus, the enolization. Finally, all the protons are replaced by a halogen, leading to di- and trihalogenated compounds.

5.6 Reactivity of α-Carbon Atoms | 239

The reaction does not stop at the stage when halogens replace three hydrogens. The bromine atoms attached to the α-carbon atom attract C—C bond electrons and so both the polarization of the carbonyl group (partial (+) charge on the carbonyl carbon atom) increases and the carbon—carbon bond weakens because of polarization. The resulting –CBr$_3$ group bonded to the carbonyl group can be a good leaving group. The hydroxide ion (HO$^-$) present in the reaction medium attacks the electrophilic carbonyl carbon atom and pushes the electrons toward the oxygen atom, making the oxygen atom anionic. This intermediate displaces the tribromomethyl anion (Br$_3$C$^-$), leaving a carboxylic acid. Proton transfer from the carboxylic acid to the tribromomethyl anion forms bromoform (CHBr$_3$). The driving force of this reaction is the formation of a stable anion. Halogens stabilize the negative charge on the carbon atom. The conversion of a methyl ketone to a carboxylic acid using halogens in a basic medium is called the *haloform reaction* [74, 75].

Bromine and chlorine are frequently used in haloform reactions. CHCl$_3$ and CHBr$_3$ are formed as by-products; they are immiscible with water and can be easily separated from the reaction mixture. However, when the reaction is carried out with iodine, a bright yellow solid, CHI$_3$, will precipitate. This reaction is generally used as a chemical test for identifying methyl ketones. The haloform reaction has been used for the classification of acetyl groups (COCH$_3$) and methyl secondary alcohol (–CH(OH)CH$_3$), which are first oxidized to acetyl groups under the reaction conditions. This reaction is called the *haloform test*.

Some compounds that do not contain acetyl groups also give a positive haloform test. For example, when cyclohexane-1,3-dione is treated with chlorine or bromine in a basic medium, glutaric acid is formed. The first two stages of this reaction proceed by the same mechanism as discussed above.

First, the active methylene protons are replaced by bromine under basic conditions to form 2,2-dibromocyclohexane-1,3-dione. In general, when two halogens are attached to the α-carbon atom adjacent to the carbonyl group, it is not possible to cleave the C—C bond in the basic medium because two bromine atoms cannot sufficiently stabilize the carbanion formed by cleavage of the C—C bond. When there are three halogen atoms, the carbanions are stabilized.

In the example above, the situation is somewhat different. When the C—C bond cleaves, the carbon atom bearing the negative (−) charge is stabilized by a carbonyl group as well as two halogen atoms. The carbonyl group here replaces the third halogen atom. Therefore, the C—C bond breaks easily. The dibromide intermediate formed is immediately converted into a tribromide compound, and then a regular haloform reaction continues.

Recently, the haloform reaction has been applied to the direct conversion of acetyl groups to the corresponding amides in the presence of iodine and aqueous ammonia [76].

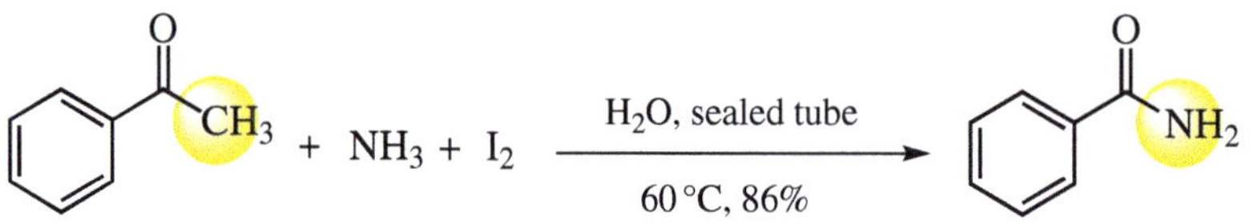

$$\text{PhCOCH}_3 + \text{NH}_3 + \text{I}_2 \xrightarrow[60\,°C,\,86\%]{\text{H}_2\text{O, sealed tube}} \text{PhCONH}_2$$

α-Halogenation of Carboxylic Acids: Hell–Volhard–Zelinsky Reaction: [77] This reaction is used for the halogenation of carboxylic acids at the α-carbon atom and is named after three chemists, Carl Magnus Von Hell, Jacob Volhard, and Nikolay Zelinsky.

The halogenation of the aldehydes and ketones at the α-position was discussed in detail and it was shown that the halogenation reaction proceeds through enols and enolates. Carboxylic acids cannot be halogenated under the same conditions applied to aldehydes and ketones because they cannot form an enol or enolate. Treatment of a carboxylic acid with a base will remove a proton from the hydroxyl group instead of from the α-carbon atom, generating a carboxylate anion, which cannot form an enolate. However, when carboxylic acids are treated with bromine in the presence of a catalytic amount of phosphorus, halogenation at the α-position occurs.

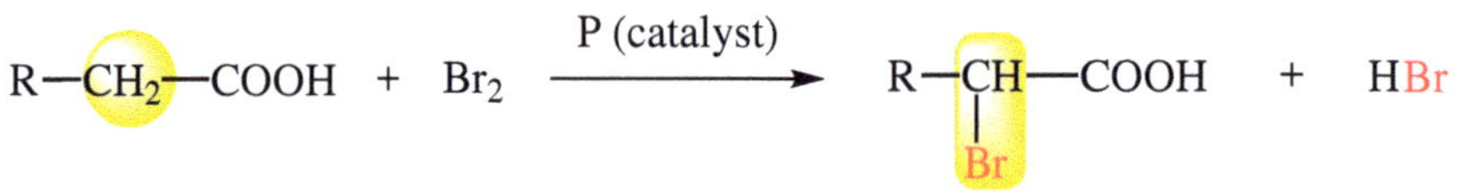

$$\text{R—CH}_2\text{—COOH} + \text{Br}_2 \xrightarrow{\text{P (catalyst)}} \text{R—CHBr—COOH} + \text{HBr}$$

The carbonyl group functionality in acids is not as electropositive as in aldehydes and ketones. Carboxylic acids do not generate stable enols. Therefore, these acids must be first converted into derivatives that can undergo keto–enol tautomerization. For this process, the reactivity of the carbonyl group needs to be increased. If an electron-attracting group, for example, a halogen, replaces the –OH group in carboxylic acids, the reactivity of the carbonyl group increases, thereby forming an enol. To start the reaction, PBr_3 is necessary. PBr_3 is generated in situ because of its highly corrosive properties. Phosphorus reacts with bromine to give phosphorus tribromide, which converts the carboxylic acid into an acyl bromide as shown below.

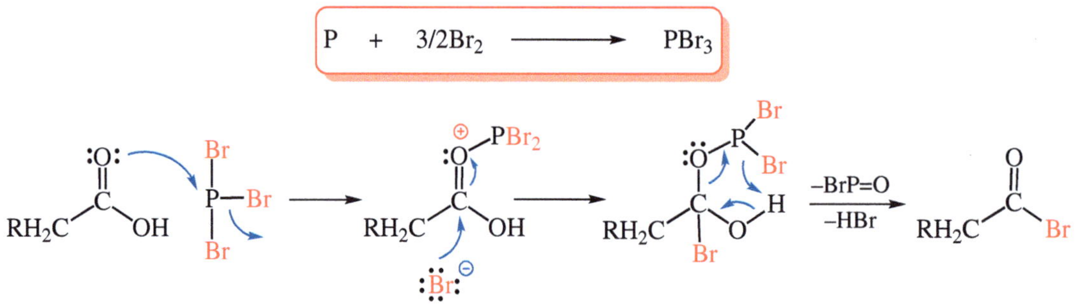

$$P + 3/2\,Br_2 \longrightarrow PBr_3$$

Because bromine is an inductively electron-attracting group, it further increases the reactivity of the carbonyl group and facilitates enol formation. The acyl bromide can then tautomerize to an enol. The enol reacts with bromine at the α-position, forming an α-bromo acyl bromide. The monobrominated compound is much less nucleophilic than the starting acyl bromide. As we discussed in the case of α-halogenation of ketones in acidic conditions, the bromine atom attached to the α-carbon atom retards enol formation. Therefore, the reaction stops at the monobromination stage.

In neutral to slightly acidic aqueous solutions, the α-bromo acyl bromide is hydrolyzed to the α-brominated carboxylic acid (route **a**). In that case, a full molar equivalent of PBr₃ must be used as the catalytic chain is disrupted. If the reaction is carried out in a non-nucleophilic solvent, the α-bromo acyl bromide formed reacts with the hydroxyl group of an unreacted carboxylic acid present in the reaction medium to produce an anhydride. The bromide anion attacks the carbonyl group of anhydride to release the α-bromocarboxylic acid and regenerate the acyl bromide intermediate (route **b**). In this case, a catalytic amount of PBr₃ is used.

5.6.7 α-Alkylation of Carbonyl Compounds

α-Alkylation reactions of carbonyl compounds are significant for the formation of new C—C bonds. To proceed with an α-alkylation reaction, the carbonyl compound must have a hydrogen atom attached to the α-carbon atom. In an overall reaction, hydrogen is replaced with an alkyl group [78].

For an α-alkylation reaction, a hydrogen atom at the α-position of the carbonyl group must be removed using a suitable base. When selecting the base, the pK_a value of the α-hydrogen should be checked. Bases such as NaOH or NaOR are not recommended for alkylation processes. If an alkoxide such as sodium ethoxide is used, deprotonation takes place to the extent of only about 0.1%. The equilibrium largely will stay on the side of the carbonyl group, not on the side of an enolate, and they will produce some side products, such as multiple alkylation products. To avoid side reactions, lithium diisopropyl amide is used for ketones as one of the most suitable bases. After the removal of a proton from the α-position with the base, the appropriate alkyl halide is added to the reaction mixture to complete the alkylation reaction.

Alkylation reactions smoothly proceed when there is only one hydrogen atom adjacent to the carbonyl group. However, if the ketone is unsymmetrical, two different enolate ions will be formed and either α-carbon atom can be alkylated. If one of the formed enolates can be stabilized by conjugation, the product will derive from the most stable enolate. If there is more than one hydrogen atom connected to the same α-carbon atom, this time a dialkylation product can be formed. In the first example given below, an unsymmetrical ketone can be alkylated on both sides. A mixture of products will be formed. In the second example, alkylation can only occur on one side of the carbonyl group. This time mono-, di-, and even trialkylation products can be formed. Let us examine whether these reactions can be controlled or directed.

For example, 2-methylcyclohexanone is an asymmetric ketone and either α-carbon atom can be alkylated [79]. Here, two different products are expected at first glance. When the molecule is reacted with one mole of CH_3I, 2,6-dimethylcyclohexanone and 2,2-dimethylcyclohexanone are formed. The ratio of these products depends on the reaction conditions chosen. When the reaction is carried out at −78 °C with LDA, which is a sterically hindered base, the most accessible and slightly more acidic hydrogen, H-6, will be removed, forming the less stable enolate. The formation of this enolate is not reversible. The less substituted enolate is the kinetic product and it occurs faster because of the lower activation energy.

When the reaction is carried out with a smaller base, such as NaH, the more substituted carbon atom will be deprotonated, forming the thermodynamically most stable enolate because it has a more substituted double bond. Remember, alkyl substituents increase the stability of double bonds.

For the generation of the most stable enolate, equilibrium between both enolates must occur. Then, the equilibrium shifts to the side of the thermodynamically more stable enolate. This is only achieved when the reaction is carried out at high temperatures. Thus, when the alkylation reaction is performed at high temperatures, 2,2-dimethylcyclohexanone is formed as the main product.

5.6.7.1 Stork Enamine Reaction

In previous chapters, we showed the reaction of secondary amines with aldehydes and ketones to form enamines. The reaction of pyrrolidine with aldehydes and ketones in the presence of a catalytic amount of acid produces the corresponding enamine. The electronic structure of enamines is similar to that of enolate anions. As can be seen from the resonance structures of enamines, the negative (−) charge is concentrated on the carbon atom. Enamines can be considered as masked enolate ions in terms of their electronic structure. The β-carbon atoms of enamines have nucleophilic properties. Therefore, enamines easily react with various electrophiles at the β-carbon atom.

One of the essential reactions of enamines is with alkyl halides. For example, when the enamine derivative formed from cyclohexanone and pyrrolidine reacts with ethyl bromide, the lone pair of electrons on the nitrogen atom forms a double bond between the nitrogen and the carbon atom and push electrons from the double bond to attack the carbon atom bonded to the bromine to form an iminium ion. This salt is stable and further alkylation of the iminium salt cannot proceed. The reaction stops at this stage. Hydrolysis of the iminium salt gives a monoalkylated ketone. As the reaction proceeds according to the S_N2 mechanism, only primary alkyl halides or methyl halides should be used. This controlled monoalkylation of aldehydes and ketones is called the *Stork reaction* [80]. Alkylation with enolates forms double and multiple alkylated products. However, alkylation with enamines minimizes the undesired side products; therefore, it is superior to enolate alkylation.

The Stork reaction is not limited to alkylation. For example, the acylation reaction of enamines is successfully applied to the synthesis of 1,3-diketones. The initial acylation forms an acyl iminium salt that hydrolyzes to the corresponding 1,3-diketones. β-Hydroxyketones are synthesized if ketones or aldehydes are used as electrophiles.

Dialkylation of asymmetric carbonyl compounds can also be controlled. For example, alkylation of 2-methylcyclohexanone in basic conditions forms a mixture of 2,6-dimethylcyclohexanone and 2,2-dimethylcyclohexanone. However, when the alkylation is carried out with enamine, the formation of 2,6-dimethylcyclohexanone can be controlled. Although two different enamines can form from 2-methylcyclohexanone and pyrrolidine, the less substituted enamine is formed most rapidly as the major product. There are two reasons for the formation of kinetically controlled product **A**. The formation of the thermodynamically more stable enamine **B** is hindered. The hydrogen atom bonded to the α′-carbon atom is slightly more acidic than the hydrogen atom bonded to the α-carbon atom. The second reason is that because the methyl group in **B** is attached to the sp²-hybridized carbon atom, it is located on the same plane as the double bond carbons and the nitrogen atom. Therefore, steric repulsion occurs between the methyl group and the methylene protons of the pyrrolidine ring. One reason for this greater stability of the less substituted enamine is the decreased repulsion between the methyl group and the methylene protons in **A** by movement of the methyl group out of the plane as the methyl group is connected to an sp³-hybridized carbon atom. Alkylation of enamine **A** will, of course, produce 2,6-dimethylcyclohexanone [80].

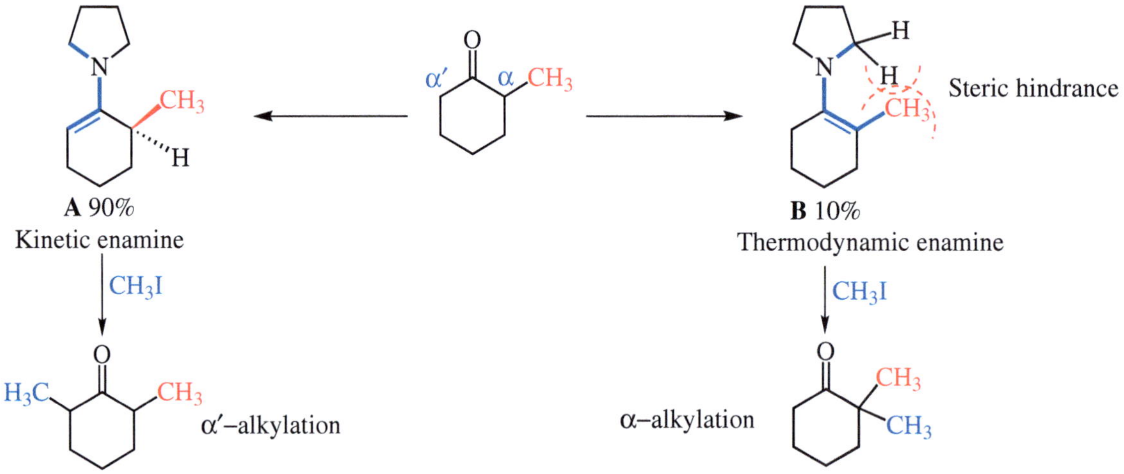

Enantioselective alkylation of ketones is possible using optically active pyrrolidine derivatives, (S)-1-amino-2-methoxymethylpyrrolidine (SAMP) and (R)-1-amino-2-methoxymethylpyrrolidine (RAMP); it was first developed by Corey and Enders [81] in 1976 and was further developed by Enders [82]. This method is usually a three-step sequence. The first step is the formation of SAMP/RAMP hydrazones. The hydrazones formed are reacted with LDA to form azaenolates. Alkylation proceeds with excellent yields. The chiral auxiliary controls the stereochemistry. After the alkylation reaction is completed, the imine derivative is converted into a carbonyl compound by hydrolysis or ozonolysis [82, 83].

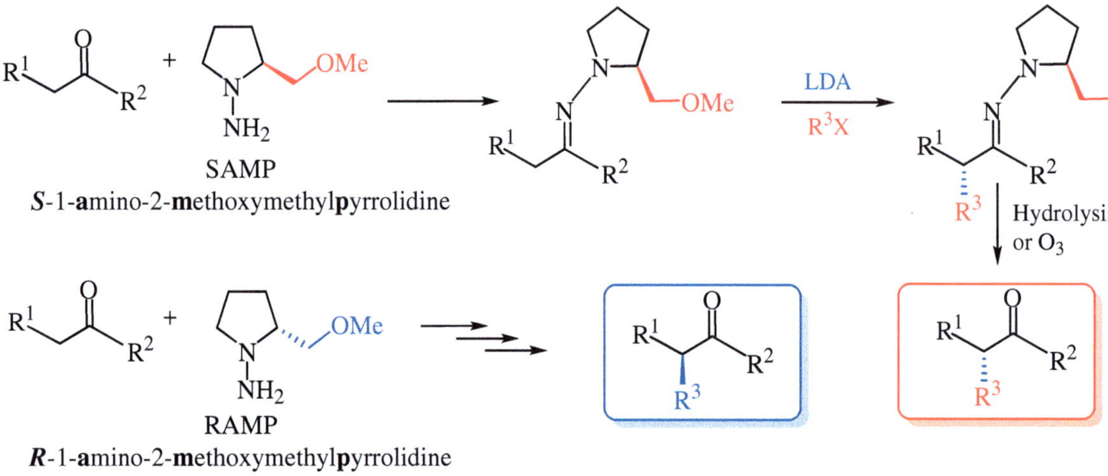

The configuration of the pyrrolidine derivative determines the configuration of the alkylation product. After removal of an α-proton with LDA, the imine turns into an azaenolate with a lithium cation chelating both the nitrogen and oxygen, forming a rigid structure and making the two faces of the complex highly inequivalent as shown below. The electrophiles attack from the bottom to produce the desired product with enantiomeric purity.

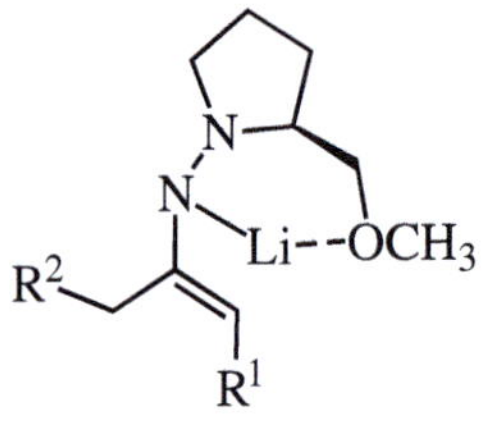

Direct alkylation of aldehydes in a basic medium is a rare reaction. Because the aldehyde carbonyl group is more reactive than ketone and ester carbonyl groups, the enolate anion formed instantly reacts with a second aldehyde carbonyl group to give condensation products. Therefore, the reactivity of the aldehyde carbonyl group must first be reduced in order to perform alkylation reactions of aldehydes. Alkylation of aldehydes can be smoothly carried out with SAMP or RAMP-hydrazones.

Imines are useful intermediates for the alkylation of aldehydes. Tertiary butylamine is usually used for the formation of imines. Because the tertiary butyl group forms a steric barrier after the formation of the imine, it prevents the nucleophiles from attacking the imine carbon atom and the formation of by-products. Stronger bases are needed to remove the α-proton as the formation of imines reduces the acidity of the α-proton. Therefore, Grignard or alkyl lithium compounds are used as the base. Alkylation is carried out with alkyl halides followed by hydrolysis of imines.

5.6.7.2 Enolates Derived from 1,3-Dicarbonyl Compounds

The 1,3-dicarbonyl moiety is an essential building block for the synthesis of important molecules [84]. The inductive effect of the carbonyl groups causes the α-protons to be more acidic. The acidity of the CH_2 or CH groups between the two carbonyl groups is much higher than that of the α-protons of ketones and aldehydes. Because these protons are acidic, they can easily be removed with NaOEt because ethanol's pK_a value is around 16.

Acetylacetone has two types of α-protons: methyl protons and methylene protons. When acetylacetone is reacted with NaOEt, one of the methylene protons will be removed as they are more acidic than the methyl protons forming the enolate ion. The stability of the enolate is due to the distribution of the negative (−) charge on both carbonyl groups, and most of the charge resides on the electronegative oxygen atoms. When any electrophile is introduced into the reaction medium, the enolate ion reacts rapidly with it to give an α-substituted acetylacetone. If the electrophile is ethyl bromide, the ethyl group is attached to the carbon atom.

Resonance-stabilized carbon nucleophiles such as enolates derived from 1,3-dicarbonyl compounds can react with α,β-unsaturated carbonyl compounds to form addition products. The enolate ion reacts with the electropositive β-carbon atom, leading to the formation of a new C—C bond. Thus, the enolate ion undergoes a 1,4 addition on the α,β-unsaturated carbonyl compound, forming 1,5-dicarbonyl compounds, which are important synthons for further functionalization. This reaction is known as the *Michael addition* [85].

If one of the acetyl groups in acetylacetone is replaced with an ester group, alkylation reactions can be carried out as discussed above. However, the synthetic potential of the resulting product differs from that formed by the alkylation of acetylacetone. Thus, ethyl acetoacetate is easily converted into its enolate by reaction with NaOEt in ethanol. The resulting carbanion is stabilized by resonance. The enolate ion reacts rapidly with benzyl bromide to form an α-substituted dicarbonyl compound.

When alkylated ethyl acetoacetate is hydrolyzed in an acidic medium, a β-ketoacid is formed. One of the most important properties of β-ketoacids is that they undergo decarboxylation very easily when they are heated and they turn into related ketones. Decarboxylation is not a general reaction of carboxylic acids. However, if the carboxylic acids have a carbonyl group in the β-position, decarboxylation occurs by the formation of a six-membered transition complex. The resulting product is an alkylated acetone. It is also possible to perceive the ethyl acetoacetate molecule as a masked acetone.

5.6 Reactivity of α-Carbon Atoms | 247

After alkylation of one of the methylene protons in ethyl acetoacetate, the compound has one acidic α-hydrogen remaining, and so the alkylation process can continue to produce a dialkylated product. The second alkyl group will be bonded to the same carbon atom bearing the first alkyl group. On heating the compound with aqueous hydrochloric acid, the dialkylated compound undergoes hydrolysis, followed by decarboxylation to give a disubstituted acetone derivative.

One of the oldest and best-known alkylation reactions is the *malonic ester synthesis*. A malonic ester is used in a reaction that converts alkyl halides to carboxylic acids. Although this reaction was discovered many years ago, it is successfully applied in the synthesis of many systems even today. This reaction is essentially identical to the process applied to the synthesis of alkylated acetones as discussed above. The only difference between the two processes is in the choice of the starting material. After alkylation of the methylene protons of diethyl malonate, ester groups are hydrolyzed to form a substituted malonic acid. 1,3-Dicarboxylic acids, when heated like ketoacids, are readily decarboxylated to produce the corresponding monoacid. Here too, decarboxylation takes place via a six-membered transition complex. This methodology allows the synthesis of substituted carboxylic acids at the α-position.

As shown above, malonic ester and acetoacetic ester synthesis can produce dialkylated products. This particular property of these reactions can be applied to the synthesis of cyclic carboxylic acids using alkyl dihalides. For this reaction, 1,3-, 1,4-, or 1,5-dihaloalkanes or ditosylalkanes can be used, depending on the size of the ring to be synthesized. The method applied here is the same as that applied in the dialkylation reaction. The only difference is that both halogens or tosylate groups are on the same molecule.

In the presence of a base, first, a monosubstituted malonic ester will be formed. Because the base is still present in the reaction medium, it will remove the remaining acidic proton, followed by intramolecular substitution to form a cyclic system. Hydrolysis of the product, followed by acidification and slight heating furnishes a cyclic carboxylic acid.

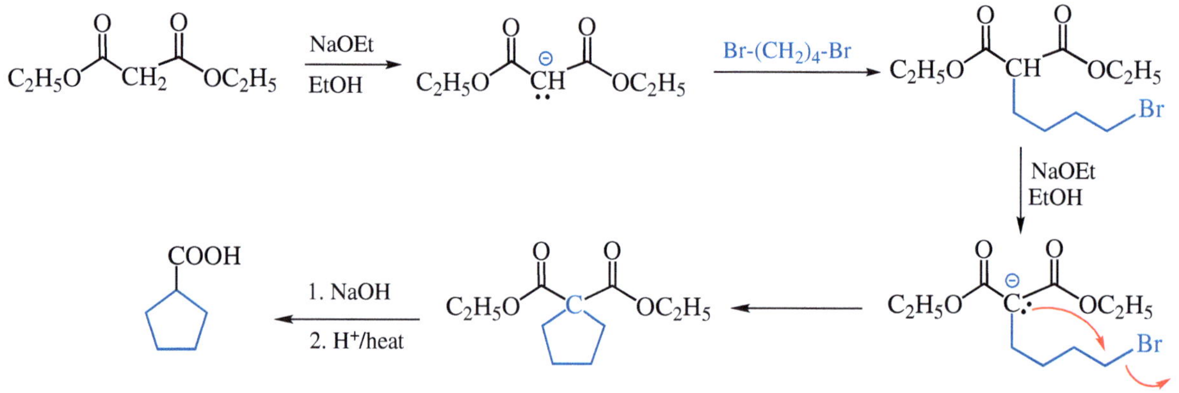

In the examples we have given so far, we have focused on the alkylation of a methylene group located between two carbonyl groups. There is also a methyl group attached to the carbonyl group in the 1,3-dicarbonyl compounds we examined, but we may wonder whether it is possible to alkylate this methyl group.

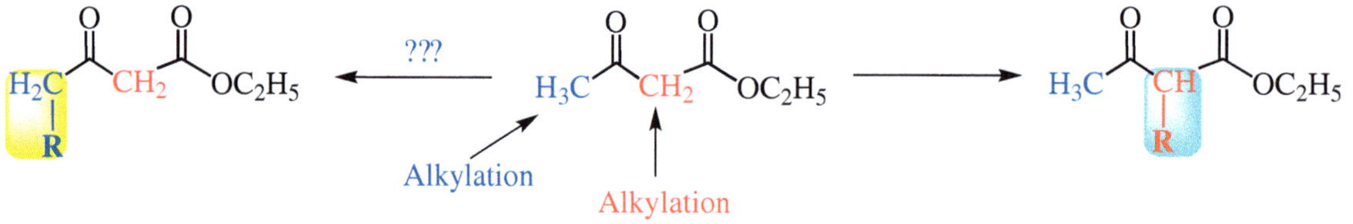

In order to perform such a reaction, a dianion intermediate must first be formed by adding two moles of a base to one mole of a 1,3-dicarbonyl compound. If bases such as alkoxylates are used, only one of the methylene protons will be removed. However, when strong bases such as alkyl lithium, potassium amide, or LDA are used, a 1,3-dicarbonyl compound can be converted into a dianion by sequential abstraction of two protons. When 3-oxobutanal is reacted with potassium amide, the base first abstracts one of the more acidic methylene protons between the two carbonyl groups [86]. The second equivalent of base cannot abstract the remaining methylene proton because there will be two negative (−) charges on the same carbon atom. For that reason, the base abstracts one of the outer methyl protons, forming a dianion [87].

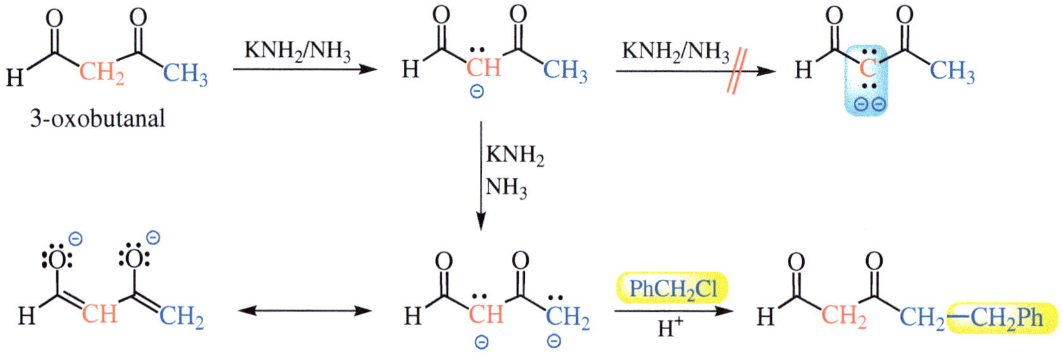

Because methyl hydrogens are less acidic than methylene hydrogens, the carbanion formed by the removal of a hydrogen from the methyl group reacts faster with electrophiles than the other because it is more unstable and more basic. This methodology allows us to perform regiospecific alkylation of 1,3-dicarbonyl compounds at the less acidic position.

This reaction is also successfully applied to β-ketoaldehyde and β-esters. For example, the intermediate product used to complete synthesis of a ten-membered natural product, lactone Diplodialide A, was synthesized by alkylation of the methyl group after generation of a dianion [88].

5.6.7.3 Enolates, Ambident Nucleophiles: C vs. O Alkylation

After the enolate is formed, the negative (−) charge is delocalized over the oxygen atom and the α-carbon atom. Enolates are ambident nucleophiles and they may attack the alkylating agent via either the oxygen or α-carbon atoms. If the alkylation takes place on the carbon atom, substituted carbonyl compounds (ketones, aldehydes, and esters) are formed. In contrast, enol ethers are formed as a result of alkylation via an oxygen atom. When these two reactions are examined in terms of bond formation energies, it becomes clear that C-alkylation is preferred because of preserving the carbonyl group (C=O) and a carbon–carbon (C—C) bond. Additionally, a new C—C bond is formed. The table below shows that the total formation energy of these three bonds is around 345 kcal/mol. On the other hand, while maintaining the C—O bond in O-alkylation, a new carbon–carbon double bond (C=C) and a new carbon–oxygen (C—O) bond are formed. The total formation energy of these bonds is 317 kcal/mol. This shows that the C-alkylation product is 28 kcal/mol more stable.

C=O	179.0 kcal/mol
C—C	83.0 kcal/mol
C—C	83.0 kcal/mol
	345.0 kcal/mol

C—O	85.5 kcal/mol
C—O	85.5 kcal/mol
C=C	146.0 kcal/mol
	317.0 kcal/mol

The ratio of the products depends on several factors. Whether C-alkylation or O-alkylation occurs depends on the reagents and reaction conditions, in particular, the solvent, the associated counter cation, and the electrophile. In the enolate ion, the negative (−) charge is delocalized over the carbon and the oxygen atom. Because the oxygen stabilizes the negative (−) charge better than the carbon atom (because of the higher electronegativity), the charge is mostly on the oxygen atom. If we start from that point, we can think at first glance that the alkylation reactions should proceed mostly over the oxygen atom. However, this is not the case at all; the alkylation reactions take place mostly on the carbon atom. Let us try to explain this outcome and the controllability of these two different alkylations.

Various factors, such as the density of negative charge, solvation, cation coordination, and product stability in the enolate ion, directly control these two different alkylation reactions. The reactivity of the oxygen atom will be decreased when working in protic solvents as hydrogen bonding between the oxygen atom and solvent molecules will occur. In this case, C-alkylation will be favored. However, polar aprotic solvents cannot cause hydrogen bonding. Then, O-alkylation will be preferred [89–91]. When the reaction is carried out in polar aprotic solvents such as dimethylsulfoxide (DMSO) or dimethylformamide (DMF), the O-alkylation product occurs in 70% and 79% yields, respectively. When the reaction is carried out in *t*-butyl alcohol or ethanol, the C-alkylation products are formed in high yields.

Solvent	Yield	
t-BuOH	0%	100%
EtOH	8%	92%
DMSO	70%	30%
DMF	79%	21%

Because small cations are tightly bonded to the oxygen atom by coordinating with the oxygen atom, they will promote C-alkylation. Larger cations, such as K, favor O-alkylation. Lithium enolate forms a cluster in the tetrahydrofuran. The structures of these compounds were determined by X-ray analysis [92]. Because the solvation properties of solvents such as tetrahydrofuran and dimethoxyethane are not good, more stable cluster structures are preferred. This cluster structure hinders the approach of the electrophile to the oxygen atom. On the other hand, the nucleophilic carbon is more exposed to the electrophiles. O-Alkylation is also increased when crown ethers or cation-complexing agents are added to the reaction mixture.

For 1,3-dicarbonyl substrates, the two oxygen atoms may chelate to the metal cation, resulting in exclusive C-alkylation. The leaving group in the alkylating compound also has an essential effect on the ratio of C- or O-alkylation. In general, C-alkylation products are obtained in alkylation reactions with alkyl iodides and alkyl bromides. Changing the leaving groups to sulfonates or sulfates favors O-alkylation. With the help of the HSBA theory, it is easy to understand these different alkylations. The carbon atom is softer than the oxygen atom. The carbon atoms in alkyl halides (R–I and R–Br) are also soft. Therefore, they prefer to react with the carbon atom because of carbon's lower hardness. Because the sulfonate or sulfates attract the bond electrons, the carbon atom becomes highly polarized and turns into a hard acid. Therefore, the reaction with hard oxygen is preferred. In summary, the best conditions for C-alkylation will be the use of a lithium enolate in a protic solvent with an alkyl iodide as the alkylating agent. The reaction of enolates with silyl electrophiles generally results in O-silylation because of the formation of a strong Si—O bond.

5.7 Condensation Reactions of Carbonyl Compounds

X	Yield		Dialkyl product
X = OTs	88%	11%	1%
X = Cl	60%	32%	8%
X = Br	39%	38%	23%
X = I	13%	71%	16%

Ellison and coworkers [93] reacted the cyclohexanone enolate anion and methyl bromide in the gas phase, and they found the formation of O-methylated products exclusively. Houk and Paddan-Row [94] calculated the energy barriers for the gas-phase reaction between acetaldehyde enolate and methyl fluoride, and they found that the energy barrier for O-methylation is lower than that for C-methylation, even though the C-methylation product is thermodynamically favored. These results were recently supported by further calculations in which solvent and counterion effects were also considered [95].

5.7 Condensation Reactions of Carbonyl Compounds

In a condensation reaction, two carbonyl compounds react with each other in the presence of an acid or a base to form a new compound (formation of a new C—C bond) and release a small molecule, such as water or alcohol. So far, we have learned about the addition of various nucleophiles to carbonyl groups and we have shown enolate formation. Enolates can react with various electrophiles. In this section, we will again study enolate formation and the reaction of enolates as nucleophiles with a carbonyl group, which acts as an electrophile. We will discuss similar reactions with different electrophiles.

There are condensation reactions with different names, including the following:
Aldol condensation
Dieckmann condensation
Claisen ester condensation
Knoevenagel condensation
Perkin condensation
Stobbe condensation, etc.

As we will see, all of these reactions proceed through a similar mechanism. The common point of all these reactions is the abstraction of hydrogen from the α-carbon atom to generate a nucleophile that adds to the electrophilic carbonyl group of a second molecule. These reactions have a special place in synthetic chemistry and are extremely important. An significant proportion of these reactions in living cells are condensation reactions. These reactions have become particularly important as a new carbon–carbon (C—C) bond is formed during the reaction.

5.7.1 Aldol Condensation

An aldol condensation is a reaction between two aldehydes or ketones according to the reaction mechanism shown above. The reaction occurs under both acidic and basic conditions. One carbonyl group is converted into a nucleophilic enolate, which adds to the carbonyl group of the second carbonyl compound. If the starting compound is an aldehyde, as a result of the reaction, a β-hydroxy aldehyde is formed, and, in the case of a ketone, a β-hydroxy ketone is formed.

Under basic conditions, the base first abstracts one of its α-protons and forms an enolate. In the second step, the nucleophilic enolate attacks the carbonyl group of a second aldehyde with the enolate carbon atom forming a new C—C bond. The double bond electrons move toward the oxygen atom, forming an alkoxide ion. The alkoxide is transformed into a β-hydroxy aldehyde by taking a proton from H_2O in the reaction medium. Because the newly formed molecule contains an aldehyde and an alcohol functional group, this reaction is called the *aldol addition*.

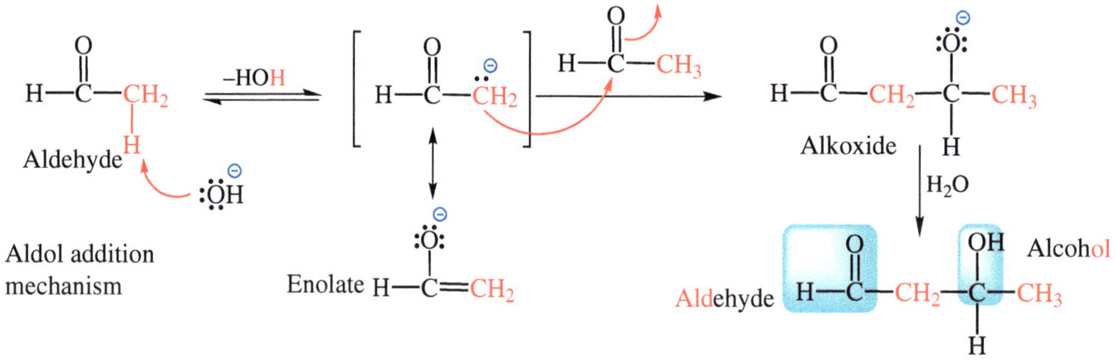

The aldol products, β-hydroxy aldehydes or β-hydroxy ketones, can be easily dehydrated into α,β-unsaturated carbonyl compounds when heated under the same reaction conditions. The hydrogens in the α-position to the carbonyl group in the aldol product are acidic. Under basic conditions, the α-hydrogen can be easily removed. The carbanion formed is stabilized by conjugation with the carbonyl group. The enolate ion can expel the HO⁻ group from the molecule according to the E1cB mechanism. The hydroxide ion is a poor leaving group. The driving force of this reaction is the formation of a double bond that is conjugated with the carbonyl group. The overall reaction, the addition of enolate to the carbonyl group followed by water elimination, is called the *aldol condensation*.

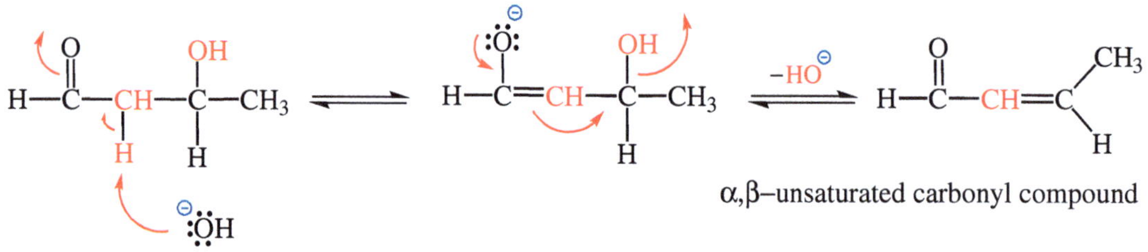

Aldol condensation reactions take place under either basic conditions or acidic conditions. The mechanisms are significantly different. While bases activate the nucleophile by forming an enolate, acids activate the electrophile by increasing the polarization of the carbonyl group. In the first step, the acid protonates the carbonyl oxygen and forms an enol. In the next step, the enol attacks the protonated carbonyl group of another molecule with the β-carbon atom and an aldol addition product is furnished. In the acidic medium, the hydroxyl group is protonated and eliminated to give an α,β-unsaturated aldehyde or ketone. The product is called an enone, "en" for the double bond and "one" for the carbonyl group.

5.7 Condensation Reactions of Carbonyl Compounds

[Scheme: Aldol condensation under acidic conditions, showing acetaldehyde protonation, enol formation, nucleophilic attack, and dehydration to form the enone H-C(=O)-CH=C(CH$_3$)(H).]

Aldol condensation under acidic conditions

Aldol addition and aldol condensation reactions are generally reversible. Consequently, if an aldol product or aldol condensation product is treated with a catalytic amount of base or acid, the reverse reaction can occur, which is called the *retro-aldol reaction*. When a β-hydroxy aldehyde or ketone is treated with a base, the base first abstracts the most acidic proton bonded to the oxygen atom, forming an alkoxide ion; then, the reverse reaction occurs in which the aldol product decomposes into two carbonyl compounds. The mechanism of the retro-aldol reaction is the reverse of that of the corresponding aldol reaction.

[Scheme: Retro-aldol mechanism with $^-$OH, showing β-hydroxy aldehyde decomposing via $-H_2O$ to enolate + acetaldehyde, then with H_2O giving 2 H-C(=O)-CH$_3$.]

As mentioned in the introduction, the aldol condensation is not limited to the reaction between an aldehyde and a base. Condensation reactions occurring between ketones are also called aldol condensations. In reactions with aldehydes, the equilibrium usually shifts to the side of the condensation products because of the increased reactivity of the aldehyde carbonyl group. However, in condensation reactions with ketones, the equilibrium moves more to the side of the starting reactants.

The reaction can be controlled by choosing appropriate reaction conditions. The product formed must be removed from the reaction medium to shift the equilibrium to the side of the product. For example, a Soxhlet extractor containing Ba(OH)$_2$ can be placed on the reaction balloon and the acetone in the balloon heated under reflux. When acetone comes in contact with Ba(OH)$_2$, the reaction takes place, and the aldol adduct formed returns to the balloon. Because the boiling point of the condensation product will be much higher than that of acetone, it cannot evaporate, the contact between the product and the base is cut, and the equilibrium completely shifts to the side of the product.

[Scheme: Acetone with $^-$OH → enolate → attack on second acetone → alkoxide intermediate → protonation to β-hydroxy ketone → dehydration to H$_3$C-C(=O)-CH=C(CH$_3$)$_2$ (mesityl oxide).]

When the aldol condensation is performed with an unsymmetrical ketone, butenal, that has different protons in the α- and α′-positions, two different enolates are conceivable. The enolate with the higher substituted double bond is more stable than the other one. However, when a bulky base, such as LDA, is used, the base will abstract a proton from the sterically less hindered side, from the methyl group, forming a kinetic enolate. As a result, the product will be formed from the less stable enolate. However, when the reaction is carried out in the presence of a weaker base, such as a hydroxide ion, then two products will be formed.

Based on the examples given above, we can generalize the following points regarding aldol reactions:

- Aldol reactions occur between aldehydes and ketones, forming the addition products β-hydroxy aldehyde and β-hydroxy ketone, respectively.
- The reaction is reversible.
- Acids or bases catalyze addition and condensation.
- Aldol products can be turned into α,β-unsaturated carbonyl compounds by removal of water.

5.7.2 Crossed Aldol Condensation

An aldol reaction between two different aldehydes or ketones that have a proton in the α-position (which can form enol) can proceed. This type of reaction is called the *crossed aldol reaction* and *crossed aldol condensation*. Let us assume a reaction between two different aldehydes: acetaldehyde and propionaldehyde. The reaction of the aldehydes with a base generates enolates, which act as nucleophiles. The aldehydes will act as electrophiles and carbanion acceptors. In the reaction medium, four compounds will be present: acetaldehyde, propionaldehyde, enolate derived from acetaldehyde, and enolate derived from propionaldehyde. In such a case, theoretically four products arising from self-condensations and crossed condensations can be formed.

These four products are formed by the addition of each enolate with two carbonyl compounds whose structures are given below. It should be noted that the products **3** and **4** will also have stereoisomers. Therefore, the aldol reaction of two different carbonyl groups is quite complicated.

5.7 Condensation Reactions of Carbonyl Compounds

[Structure 1: H−C(=O)−CH₂−C(OH)(H)−CH₃] — **Self-condensation product**

[Structure 2: H−C(=O)−CH₂−C(OH)(H)−CH₂CH₃] — **Crossed condensation products**

[Structure 3: H₃C−C(OH)(CH₃)−CH(H)−C(=O)−H]

[Structure 4: H₃CH₂C−C(OH)(CH₃)−CH(H)−C(=O)−H] — **Self-condensation product**

These products are not statistically distributed; their distribution entirely depends on the reactivity of the carbonyl compounds used. For example, compounds **1** and **2** are more likely to form because of steric reasons and the initial enolate being more reactive. When these reactions are performed in acidic and basic environments, the products formed are dehydrated and each product forms two geometric isomers (*cis* and *trans*). Therefore, eight different products can form as a result of dehydration. Because of the variety of products, such reactions are not synthetically useful.

The question of whether it is possible to synthesize a single aldol product by using two different carbonyl compounds may arise at this point. If one of the reacting carbonyl compounds contains no α-hydrogen and it cannot form an enolate ion, it cannot self-condense. The number of products theoretically drops from 4 to 2. For example, because benzaldehyde has no α-proton, two products can be formed from the condensation reactions of any aldehyde or ketone with benzaldehyde. When benzaldehyde is reacted with acetaldehyde, the formation of the self-condensation product can be avoided by adding acetaldehyde slowly to a solution of excess benzaldehyde and the base. Under these conditions, the concentration of the enolate derived from acetaldehyde will be always low. As a consequence, the crossed aldol product will be formed as the sole product.

The condensation reaction between an aldehyde and ketone will also theoretically furnish four products. The reaction between a ketone having an α-hydrogen with an aromatic carbonyl compound having no hydrogen at the α-position can form two products. Ketones do not tend to undergo self-condensation because of the steric hindrance in the aldol addition product. Those reactions of aromatic aldehydes with ketones or aldehydes are called the *Claisen–Schmidt condensation*, which is named after two investigators, Rainer Ludwig Claisen and J. G. Schmidt, who independently published on this topic in 1880 and 1881. They reacted acetophenone with benzaldehyde in the presence of a base and obtained chalcone in high yield. Dehydration occurs readily because of the double bond formed in conjugation with the carbonyl group as well as with the benzene ring.

Claisen–Schmidt condensation

Chalcone

It is possible to synthesize interesting cyclization products by aldol condensation when different carbonyl groups are on the same molecule. For example, when the 5-oxohexanal derivative reacts with KOH, the cyclohexenone derivatives are formed. The ketoaldehyde can form three different enolates. However, the enolate formed from the ketone side (by the abstraction of a methyl proton) adds to the aldehyde carbonyl group. As the reactivity of the aldehyde is greater than that of the ketone, addition takes place between the enolate formed from the ketone and aldehyde group to form the product.

5.7.3 Robinson Annulation

This reaction, named after Robert Robinson who in 1935 discovered it first, is one of the most critical reactions applied to steroid synthesis [96]. Robinson received the Nobel Prize in 1947. The reaction proceeds in two main steps: the first one is the *Michael addition*, the addition of an enolate to the α,β-unsaturated carbonyl groups, followed by aldol condensation. As a result, a new six-membered ring is formed. The whole reaction is called the *Robinson annulation* [97]. The reaction of 2-methylcyclohexane-1,3-dione with methyl vinyl ketone (but-3-en-2-one) in the presence of a base furnishes the bicyclic enone that is one of the essential building blocks of steroids [98].

The reaction starts with the removal of the most acidic proton between the two carbonyl groups. The enolate can be stabilized by resonance by two carbonyl groups. The next step is a conjugate addition of the enolate to α,β-unsaturated carbonyl compounds, yielding a 1,5-diketone (Michael addition). Under the same reaction conditions, 1,5-diketones can undergo an intramolecular aldol condensation reaction to yield the final product.

As mentioned above, the Robinson annulation has been successfully applied to the synthesis of a steroid, estrone. 2-Methylcyclopenta-1,3-dione is used to generate an enolate, which adds to an aryl-substituted α,β-unsaturated carbonyl to furnish the Michael addition product. The intramolecular aldol condensation of the products results in the formation of the Robinson annulation product. The product is transformed into estrone, a female sex hormone, after a few steps.

5.7.4 Claisen Ester Condensation

So far, we have seen that the aldol reaction is a reaction of aldehydes and ketones in the presence of a base. In the first step, a base removes the acidic α-hydrogen, forming an enolate, which adds to the carbonyl group of a second aldehyde or ketone. All condensation reactions proceed according to this mechanism. In the Claisen ester condensation, an ester is used instead of an aldehyde or a ketone, unlike in the aldol condensation. The selection of a base is critical. For example, in the aldol reaction, we used NaOH as a base. In the case of the Claisen condensation, we cannot use NaOH as a base because the ester functionality would undergo hydrolysis to form a carboxylate anion. The detailed mechanism of this reaction is given in the ester hydrolysis section. The formation of a carboxylate will decrease the carbonyl group's reactivity as well as the acidity of the α-hydrogens.

It is recommended to choose the same alkoxide as a base as is present in the ester group. For example, sodium methoxide should be used for methyl benzoate and sodium ethoxide for ethyl acetate.

Of course, the alkoxide ions can also attack the carbonyl groups. Let us first try to answer the question, *What kind of reaction occurs between ethyl acetate and sodium ethoxide?"* There are two types of reactions. In the first one, the ethoxide anion can attack the ester carbonyl group, forming a tetrahedral alkoxide intermediate. While reforming the carbonyl double bond, one of the ethoxide groups will be expelled from the molecule, forming the same molecule. Because the attacking group and the group previously attached to the carbonyl group are the same, there will be no change in the molecule. However, if the alkoxide group used is different from the group in the ester group, a similar reaction will still occur, but this time, transesterification may occur and a mixture of esters can be formed (methyl and ethyl ester mixture). In order to prevent the formation of a mixture, the same alkoxide is used.

In the second type of reaction, the ethoxide ion abstracts one of the α-hydrogens from the methyl group, forming an enolate ion. The enolate is more basic than the ethoxide ion. Therefore, the concentration of enolate is very low. This enolate acts as a nucleophile and attacks the carbonyl group of a second ester molecule, forming a tetrahedral intermediate. This intermediate is different from the tetrahedral intermediate formed during the aldol reaction. The intermediate formed in the aldol reaction takes a proton and furnishes the aldol product. However, the fate of the intermediate formed in the Claisen condensation is different. The intermediate here has a leaving group attached to the carbonyl carbon. Therefore, during the formation of the C=O double bond, the electrons expel the ethoxide group, form a β-ketoester, and regenerate the alcohol's conjugate base.

Claisen ester condensation

The following question may arise regarding the Claisen condensation: *Can the resulting product, a β-ketoester, undergo further reaction with the base still present in the reaction mixture?* As the methylene protons ($pK_a = 11$) in the β-ketoester are more highly acidic than the methyl protons in the starting compound, ethyl acetate ($pK_a = 25$), as soon as the β-ketoester is formed in the reaction medium, one of the α-methylene protons will be abstracted by the base, forming a stable carbanion. This process requires a full equivalent of the base rather than the catalytic amount. The formation of this anion will decrease the reactivity of the carbonyl group bearing the ethoxy group. Therefore, a further condensation with the ester group cannot take place.

The deprotonation reaction is irreversible and shifts the equilibrium to the side of the condensation product. This carbanion is stabilized by both the carbonyl group and the sodium ion. The product, a β-ketoester, is recovered by adding H_2O to the reaction mixture. Therefore, the Claisen condensation should only be carried out with those esters having two or three hydrogen atoms at the α-position. Esters with one α-hydrogen atom do not undergo a normal Claisen condensation reaction with weak bases. However, with strong bases, such as LDA, they can be converted into β-ketoesters.

Although we have always emphasized that the aldol condensation reaction is closely related to the Claisen condensation reaction, there are some differences between them. Let us examine them briefly.

- In the aldol reaction, an enolate adds to the carbonyl group of an aldehyde or ketone, whereas in the Claisen condensation, an enolate formed from an ester adds to the carbonyl group of an ester.
- While aldol addition is catalyzed by an acid or a base, the Claisen reaction generally proceeds with bases. When an acid is used, the ester group is hydrolyzed.
- While the aldol addition proceeds with a catalytic amount of base, an equivalent amount of base is required for the Claisen condensation because the β-ketoester formed is more acidic, and it reacts with the base and forms an enolate.
- In aldol reactions, NaOH can be used as a base. However, if NaOH is used in the Claisen condensation, ester hydrolysis occurs and the reaction does not take place. Alkoxides are generally used.
- In order to obtain the final product in the Claisen condensation, it is necessary to add acidic H_2O to the medium at the end of the reaction. For the aldol reaction, it is not required.

5.7.5 Crossed Claisen Ester Condensation

As can be seen with the aldol condensation, a condensation between two different esters is called a *crossed Claisen ester condensation*. The problem faced here too is the theoretical formation of four different products. There are two types of crossed Claisen ester condensations. One is the condensation reaction between two different esters, both having α-hydrogen atoms, and the other is when only one of the esters has α-hydrogens. In the first case, the reaction is not synthetically useful because four condensation products will be formed. However, when the esters have different reactivities, the reaction can be useful. On the other hand, when one of the two esters has no α-hydrogen (an ester that cannot form an enolate), the condensation reactions are successful. Among the esters that cannot form enols (no protons in the α-position), formic acid esters, dialkyl carbonates, oxalic acid esters, and benzoic acid esters are the most commonly used.

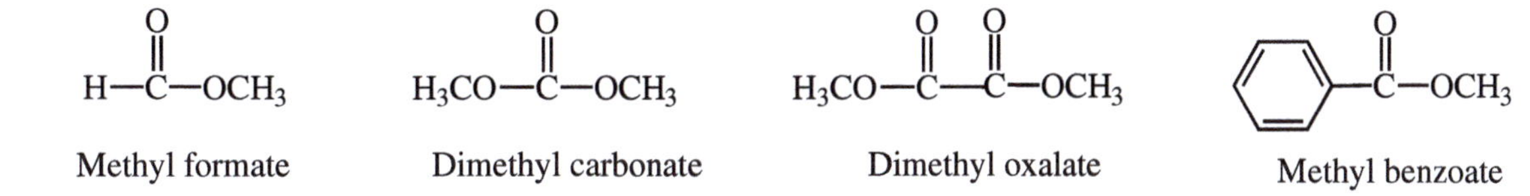

Methyl formate   Dimethyl carbonate   Dimethyl oxalate   Methyl benzoate

Some examples of crossed Claisen ester condensation reactions are given below. For example, the reaction of ethyl benzoate with ethyl acetate furnishes ethyl 3-oxo-3-phenylpropanoate as the major product. However, the formation of the self-condensation product of ethyl acetate cannot be eliminated. It is possible to minimize the amount of the self-condensation product by keeping the concentration of ethyl acetate, an enolizable compound, as low as possible. This can be done by the slow addition of ethyl acetate to a solution of ethyl benzoate and the base.

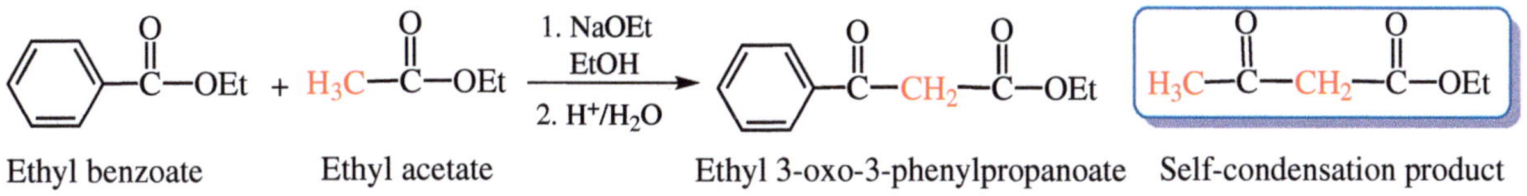

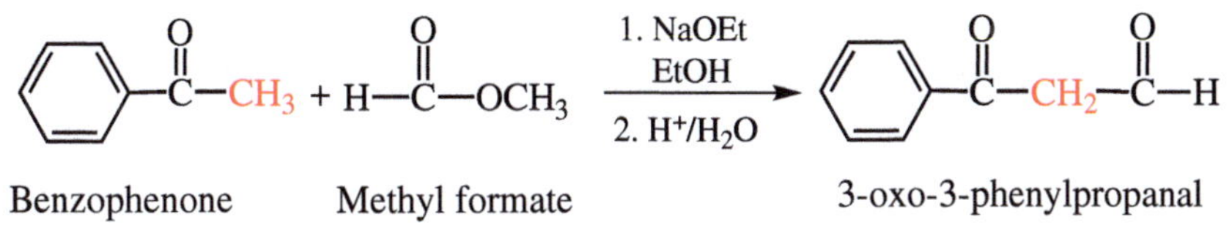

Ethyl benzoate   Ethyl acetate   Ethyl 3-oxo-3-phenylpropanoate   Self-condensation product

Benzophenone   Methyl formate   3-oxo-3-phenylpropanal

The crossed Claisen ester condensation can also be carried out with strong and sterically hindered bases such as LDA. One of the esters can be selectively reacted with LDA, which will abstract the α-proton irreversibly to form an enolate. Then, this enolate can be reacted with another ester as shown below.

Another variation of the Claisen condensation is the reaction of esters with ketones. The reaction proceeds well when the ester does not have any hydrogen atoms at the α-position. For example, a reaction of 2,2,-dimethylcyclohexanone with ethyl formate gives a β-ketoaldehyde in high yield.

5.7.6 Dieckmann Condensation

The Dieckmann condensation is a variation of the Claisen condensation. The Claisen ester condensation takes place intermolecularly between two ester groups. If the ester groups are on the same molecule, an intramolecular reaction occurs between the ester groups in the presence of a base to furnish a cyclic β-ketoester. This reaction is called the *Dieckmann reaction* and is named after the German chemist Walter Dieckmann (1869–1925). The reaction has been successfully applied to the synthesis of five- and six-membered compounds. For this, 1,6- and 1,7-diesters are ideal starting compounds.

The mechanism of the Dieckmann condensation is the same as that of the Claisen condensation. First, the base removes one proton from the α-carbon atom next to the ester group, forming an enolate, which adds to the carbonyl group of the other ester group, forming a cyclic tetrahedral intermediate. During the carbonyl group's reformation, the electrons on the oxygen atom displace the ethoxy group to produce a β-ketoester. Since the protons attached to C2 and C6 carbon atoms in this system are identical, the result does not change which proton is removed.

Starting with a 1,6-diester, five-membered ring, β-ketoesters can be easily synthesized [99]. The rate of formation of five-membered rings is higher than that of six-membered rings. Although six-membered rings are more stable, their slow formation rate depends on entropy. As the ring grows, their formation becomes more and more difficult. It is challenging to synthesize 8-, 9-, and 10-membered rings by Dieckmann condensation.

Construction of small rings by Dieckmann condensation is also difficult. For example, the expected cyclopropane derivative is not formed when diethyl succinate is reacted under Dieckmann condensation conditions. During this reaction, first, a Claisen ester condensation takes place. The resulting intermediate undergoes a Dieckmann condensation to give a six-membered ring, diethyl 2,5-dioxocyclohexane-1,4-dicarboxylate.

Recently, the Dieckmann condensation has been successfully applied to the synthesis of spirocyclic pyrrolidine derivatives starting from commercial cyclic α-amino acids. Acylation of an amino ester with methyl malonyl chloride gives an amide in high yield. The Dieckmann condensation of diester with NaOH in methanol furnishes the cyclization product, which undergoes hydrolysis, followed by decarboxylation to give the desired product [100].

5.7.7 Knoevenagel Condensation

The Knoevenagel condensation is a modified aldol condensation in which compounds having active methylene groups add to an aldehyde or ketone in the presence of a weak amine base, followed by the removal of water, resulting in C=C double bond formation [101]. This reaction was explored and established by the German chemist Emil Albert Knoevenagel (1865–1921) in 1890 and was named after him as the Knoevenagel condensation reaction. Two carbonyl groups or two nitrile groups or even nitro groups can be attached to the active methylene group.

The condensation reaction of diethyl malonate and benzaldehyde in the presence of piperidine is given below. First, an enolate must be formed to start the reaction, as we discussed in the case of other condensation reactions. The base, piperidine, abstracts a proton from the active methylene group to generate a carbanion, an enolate. The reaction proceeds similarly to the other condensation reactions. The enolate acts as a nucleophile and attacks the carbonyl group of the benzaldehyde. The condensation reaction is usually followed by dehydration of one mole of H_2O from the intermediate to produce α,β-unsaturated carbonyl compounds.

A second mechanism is suggested for the Knoevenagel condensation reaction. Let us examine the reaction between malononitrile and isobutyraldehyde in the presence of piperidine. According to this mechanism, the secondary amine, piperidine, first reacts with the carbonyl group acting as an electrophile to form an iminium salt. The protonated imine is actually more reactive than the corresponding carbonyl compound, and with the formation of this salt, the reactivity of the carbonyl group increases.

On the other hand, some part of the base removes one of the acidic protons from the methylene group, forming a stabilized carbanion by resonating with the adjacent nitrile groups. The carbanion attacks the carbon atom of the iminium ion to form a tetrahedral intermediate. Then, a proton is removed from the intermediate product by the base, and the elimination of piperidine forms the condensation product. A straightforward variation of the mechanism indicates that piperidine acts as an organocatalyst, forming an iminium intermediate as the acceptor. As can be seen from the below examples, electron-withdrawing groups, such as carbonyl groups and nitriles attached to the methylene group, increase the methylene hydrogens' acidity.

A nitro group also activates the α-hydrogens enough to allow deprotonation under the weakly basic conditions because they are strongly electron attracting. These types of compounds also undergo Knoevenagel-type condensation reactions. For example, a nitronate is formed when nitromethane reacts with basic aluminum oxide (Al_2O_3). The nitronate attacks the carbonyl group of the benzaldehyde, forming a hydroxy nitro compound. The condensation reaction completes when a hydroxyl group is removed from the molecule.

Having an acidic proton is essential for two reasons. First, a high concentration of carbanions can be formed even with weak bases. The second reason is that it facilitates water elimination to complete the condensation.

The Knoevenagel condensation forms the product when compounds containing active methylene groups are reacted with ketones or aldehydes in the presence of amines such as piperidine or pyridine. However, when malonic acid is used in place of an active methylene compound in the presence of pyridine or piperidine, the reaction does not stop at the stage of the condensation product. The compound undergoes decarboxylation in a subsequent step, forming α,β-unsaturated carboxylic acid. This reaction is called the *Doebner modification* [102].

The mechanism is similar to that of the Knoevenagel condensation. First, the base abstracts one of the acidic methylene protons, forming an enolate. The anion is stabilized because of the extensive delocalization of the negative (−) charges and the occurrence of intramolecular hydrogen bonding. The dianion attacks as a nucleophile the carbonyl group, forming the addition product. The final step is decarboxylation and dehydration via a cyclic transition state. The product formed has a *trans* configuration.

Very recently, Schijndel and coworkers [103] presented a green Knoevenagel procedure for the synthesis of various cinnamic acid derivatives under solvent-free conditions. The most important part of the procedure is the use of environmentally benign amines or ammonium salts as catalysts instead of pyridine and piperidine, which are used in the traditional Knoevenagel condensation. The decarboxylation step was carried out, giving the desired products in high yield and purity.

5.7.8 Perkin Condensation

The Perkin condensation involves the reaction between aromatic aldehydes and carboxylic acid anhydrides in the presence of a base, the sodium or potassium salt of the acid or triethylamine, to give α,β-unsaturated carboxylic acid [104]. The reaction was first developed by the English chemist William Henry Perkin (1838–1907) in 1868 and is used to make cinnamic acids. The reaction, the *Perkin condensation*, is named after its discoverer. The first example of this reaction is the synthesis of Coumarin by heating the sodium salt of salicylaldehyde with acetic anhydride in the presence of a sodium or potassium salt.

The base to be used in this reaction is very important. Generally, the salt of the acid forming the anhydride is chosen as the base. For example, if the reaction is carried out with the anhydride of acetic acid, sodium acetate, the salt of acetic acid should be used as the base. Bases such as pyridine and triethylamine can also be used in this reaction. Generally, the reaction takes place at very high temperatures (150–200 °C).

5.7 Condensation Reactions of Carbonyl Compounds

The reaction mechanism is similar to that of the aldol condensation. In the first step, the base abstracts one of the acetic anhydride methyl protons, yielding an enolate ion. The enolate ion attacks the carbonyl carbon of the benzaldehyde, forming a tetrahedral alkoxide intermediate. As in the aldol condensation, one mole of H_2O is eliminated from the molecule (due to the high temperature); thus, the condensation reaction is terminated. When the product is treated with acid, the anhydride is hydrolyzed and cinnamic acid is formed.

5.7.9 Stobbe Condensation

In 1893, the German chemist Hans Stobbe (1860–1938) reacted acetone with diethyl succinate in the presence of sodium ethoxide, but the expected result, a Claisen type of condensation, was not observed; instead, tetraconic acid was surprisingly formed as an aldol type of condensation product. The reaction was named the *Stobbe condensation* after its discoverer. Extensive studies carried out by Stobbe and coworkers revealed that both ketones and aldehydes generally undergo similar condensation reactions with succinic esters. Aromatic, aliphatic ketones are generally used. The diester is mainly limited to succinic esters. The primary product is the salt of the half ester.

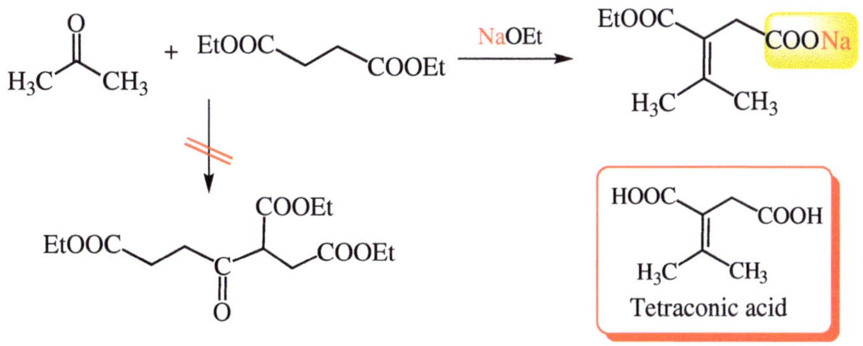

Johnson and coworkers suggested the following reaction mechanism for this conversion [105]. The reaction starts as usual with the proton abstraction of the α-carbon atom next to the ester group. The enolate formed adds to the carbonyl group of acetone, forming an alkoxide intermediate. This alkoxide undergoes an intramolecular cyclization reaction by attacking the ester carbonyl group and forming a γ-lactone after removal of the ethoxy group. The lactone intermediate has been isolated and supports the suggested mechanism. Furthermore, the formation of a γ-lactone is likely the driving force of this reaction. The lactone undergoes a ring-opening reaction by further proton abstraction, followed by ring-opening results in the final product. If the esters have only one α-hydrogen atom, the lactone intermediate can be isolated. This reaction has wide applications in organic synthesis, especially for the preparation of estrone derivatives.

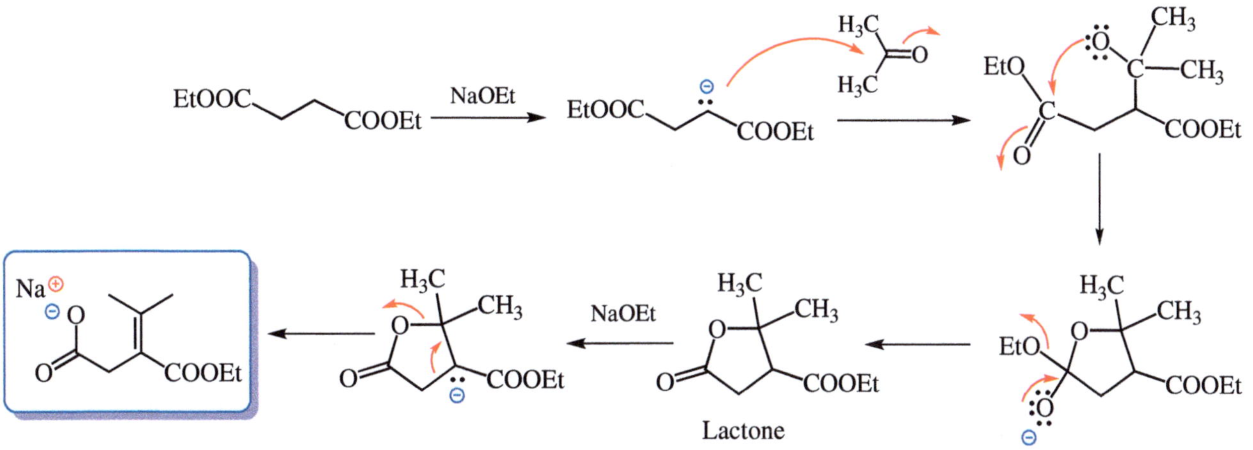

5.7.10 Role of Condensation Reactions in Synthetic Chemistry

After examining the carbonyl group's various condensation reactions, let us explore the synthetic role and importance of these reactions via a specific example. One of the most important fields of modern synthetic organic chemistry is the synthesis of natural compounds in nature and living cells under laboratory conditions. Most of these compounds have complex structures. There are several stereogenic centers within the molecule. One has to pay considerable attention to these centers during the synthetic steps. Therefore, natural product synthesis is one of the most challenging areas of organic chemistry. Total synthesis requires excellent organic knowledge, a strong background, skill, and patience. The synthetic strategy must be designed with great care and all alternatives must be considered. There are total syntheses containing up to 50 steps in the literature today. Possible blockage at any stage prevents the continuation of the total synthesis. Care should be taken to ensure that the designed synthesis steps are as few as possible and highly efficient. Synthesis with many steps and low yield will be very time-consuming and uneconomical.

The natural product testosterone belongs to the steroid group. It is present in all mammals, birds, reptiles, etc., and is the male sex hormone. Testosterone was first isolated in 1935 by Lacquer and coworkers [106], who obtained only about 10 mg from 100 kg from bull testes. This amount was sufficient to establish its chemical structure. Several formal, totally synthetic pathways were published. We will discuss the total synthesis published by Johnson and coworkers in 1956 in which mostly condensation reactions of the carbonyl groups were used [107, 108]. It is extraordinary that Johnson and coworkers and other groups achieved the total synthesis of testosterone before NMR spectroscopy had become common.

1,6-Dimethoxynaphthalene (**1**) is the starting compound for the synthesis. This naphthalene derivative was first reduced to diene with Na in ethanol (Birch reduction) to give enol ether **2**. The reaction of enol ether **2** with a mineral acid produced the corresponding ketone **3**, generating two different enolates by treatment with a base. Because one of the enolates is conjugated with the benzene ring, it is thermodynamically more stable. Thus, when ketone **3** is treated with sodium methoxide, enolate **4** is formed as the preferred one.

5.7 Condensation Reactions of Carbonyl Compounds

Treatment of enolate **4** with ethyl vinyl ketone results in the addition product **5** (Michael addition), a 1,5-diketone, an ideal compound for an intramolecular aldol condensation reaction (see Robinson annulation). When the diketone is reacted with the base, first, the aldol adduct **6** is formed, followed by dehydration to give the condensation product **7**. Although there are four different α-protons in diketone **5**, the product is formed from that carbanion generated by the α-hydrogen's abstraction from the ethyl group. Of course, the base can also abstract the other α-protons, but these do not turn into products because of the strain and entropy factors of the product to be formed.

The next step is the annulation of the fourth ring required for the testosterone skeleton. When ketone **7** is treated with a base, the conjugated enolate **8** is preferably formed. The reaction of the enolate with methyl vinyl ketone (Michael addition) followed by condensation (Robinson annulation) results in the formation of the desired condensation product **10**. Testosterone has three six-membered rings and one five-membered ring in its structure. The synthesized compound, enone **10**, has four six-membered rings. One of these rings, the benzene ring, has to be converted into a five-membered ring. During these processes, other functional groups, such as enones, must be protected so that they do not react. For this, the carbonyl group is converted into acetal **11**. After protecting the carbonyl group, the double bond conjugated with the aromatic ring is reduced with K/NH$_3$ to give the product **12**. The Birch reduction of **12** converts the benzene ring into a 1,4-cyclohexadiene derivative, which undergoes hydrolysis upon treatment with acid to give α,β-unsaturated carbonyl compound **13**. Catalytic hydrogenation of **13** was carried out selectively without reduction of the other double bond to give **14**. The aldol condensation of **14** with benzaldehyde formed the product **15**. The α-hydrogen atom next to the carbonyl group was replaced by a methyl group by the reaction of **15** with potassium *t*-butoxide followed by the addition of CH$_3$I. The ring opening of **16** was carried out with ozonolysis to give a dicarboxylic acid **17**, which was converted into the dimethyl ester **18**.

The diester **18** is a very suitable compound for the Dieckmann condensation, and it is also a suitable compound to annulate a five-membered ring required for the structure of testosterone. As a result of the diester's reaction with a base, the desired Dieckmann condensation product **19** was formed, and the ketone **20** was synthesized by the hydrolysis of the ester group followed by decarboxylation. Testosterone synthesis was completed by reducing the ketone to an alcohol and removing the protecting acetal group and subsequent double bond shift.

With this example, we tried to demonstrate how essential condensation reactions are in the total synthesis of natural products with complex structures. The full details of the synthesis were not discussed. Only condensation reactions were highlighted.

5.8 Ester Hydrolysis Reactions

The products formed by the reaction of a carboxylic acid with alcohols in the presence of a catalytic amount of a mineral acid or Lewis acid and removal of one mole of H_2O from the molecule are called *esters* and the related reaction is called *esterification*. The reaction of esters with H_2O to form one mole of acid and one mole of alcohol is called *ester hydrolysis*. These two reactions, esterification and hydrolysis, are equilibrium processes.

$$R-\underset{\text{Acid}}{COOH} + \underset{\text{Alcohol}}{R'OH} \underset{\text{Hydrolysis}}{\overset{\text{Esterification}}{\rightleftharpoons}} R-\underset{\text{Ester}}{COOR'} + \underset{\text{Water}}{H_2O}$$

Hydrolysis is a concept that students often confuse with solvolysis. These concepts will be clarified here. Hydrolysis is the reaction of a molecule with H_2O to break some chemical bonds that exist in the molecule. The prefix "*hydro*" means water, while the suffix "*lysis*" means break down and so hydrolysis means to break down the molecule with water. During hydrolysis, the H_2O molecule splits into H^+ and HO^- ions. These ions bind to the appropriate regions of the molecule they react with. When we look at the total reaction in ester hydrolysis, where an acid and an alcohol are formed, the H^+ ions of the H_2O molecule is bonded to the RO– group, while the HO^- group is bonded to the carbonyl group of the acid.

Hydrolysis is not an oxidation or reduction reaction. There is no change in the atoms' oxidation numbers to which H^+ and HO^- ions are attached. Hydrolysis is not limited to the reaction of esters with H_2O alone. Some examples for hydrolysis are given below. Acids or bases often catalyze hydrolysis reactions.

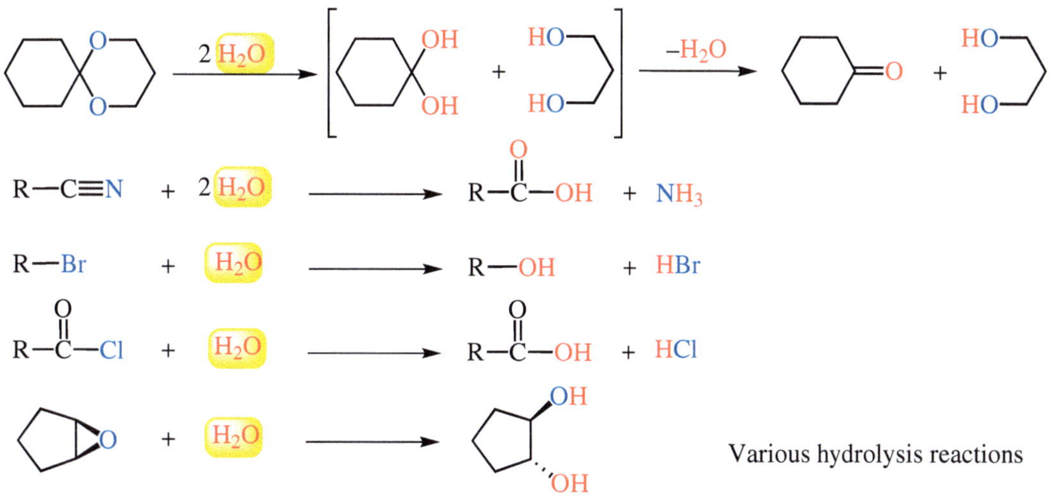

5.8 Ester Hydrolysis Reactions

[Various hydrolysis reactions shown:]

- Spiro acetal + 2 H$_2$O → [cyclohexane-1,1-diol + HOCH$_2$CH$_2$OH] → cyclohexanone + HOCH$_2$CH$_2$OH + H$_2$O (removed)
- R—C≡N + 2 H$_2$O → R—C(=O)—OH + NH$_3$
- R—Br + H$_2$O → R—OH + HBr
- R—C(=O)—Cl + H$_2$O → R—C(=O)—OH + HCl
- Epoxide (cyclopentene oxide) + H$_2$O → trans-cyclopentane-1,2-diol

Various hydrolysis reactions

The conversion of an acetal to a ketone or an aldehyde and alcohol with H$_2$O is a hydrolysis reaction. Like the conversion of acyl chloride to a carboxylic acid and HCl, the conversion of a nitrile to a carboxylic acid and NH$_3$ is also a hydrolysis reaction. The conversion of alkyl halides to alcohols and ring-opening of epoxides to diols are also hydrolysis reactions. As can be seen from these examples given above, the H$_2$O molecule dissociates into H$^+$ and HO$^-$ ions, which bind to the appropriate atoms in the molecule. Of course, acid or base catalysts are often required for the above-mentioned reactions to take place.

Solvolysis is the reaction of a compound with solvent molecules, such as H$_2$O, NH$_3$, ROH, and RCOOH, in which the solvent acts as a nucleophile. According to this definition, hydrolysis is a solvolysis reaction with water. Solvolysis reactions usually take place with excessive amounts of solvent, and they are not an oxidation or reduction reaction. The oxidation number of elements is not changed during solvolysis reactions. Some examples of solvolysis reactions are given below.

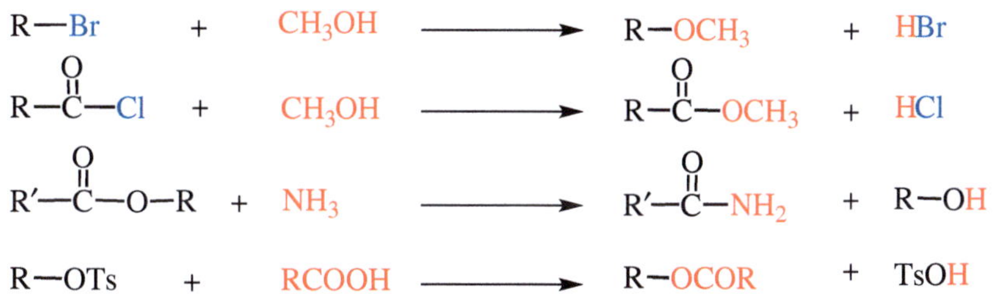

- R—Br + CH$_3$OH → R—OCH$_3$ + HBr
- R—C(=O)—Cl + CH$_3$OH → R—C(=O)—OCH$_3$ + HCl
- R'—C(=O)—O—R + NH$_3$ → R'—C(=O)—NH$_2$ + R—OH
- R—OTs + RCOOH → R—OCOR + TsOH

Solvolysis reactions

5.8.1 Ester Hydrolysis

Ester hydrolysis reactions should be examined in three groups:

1. According to the broken bond
   a) Cleavage of acyl—oxygen bond
   b) Cleavage of alkyl—oxygen bond

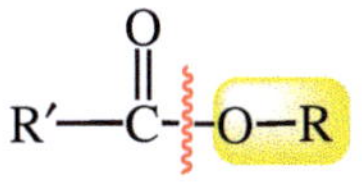

R'—C(=O)⋮O—R   R'—C(=O)—O⋮R
Acyl-oxygen cleavage 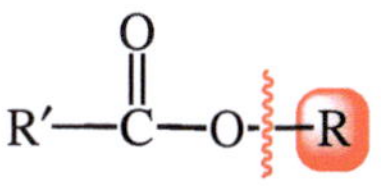   Alkyl-oxygen cleavage

2. According to the pH of the reaction medium
   a) Hydrolysis under acidic conditions
   b) Hydrolysis under basic conditions

3. According to the reaction molecularity
   a) Monomolecular hydrolysis
   b) Bimolecular hydrolysis

When these alternatives are combined, esters could, theoretically, be hydrolyzed by eight possible mechanisms. However, only six have been found so far.

5.8.2 Ester Hydrolysis Under Basic Conditions

Although there are two different cleavages (acyl–oxygen cleavage and alkyl–oxygen cleavage), the most common decomposition is acyl–oxygen cleavage. Ester hydrolysis in a basic medium is an irreversible reaction.

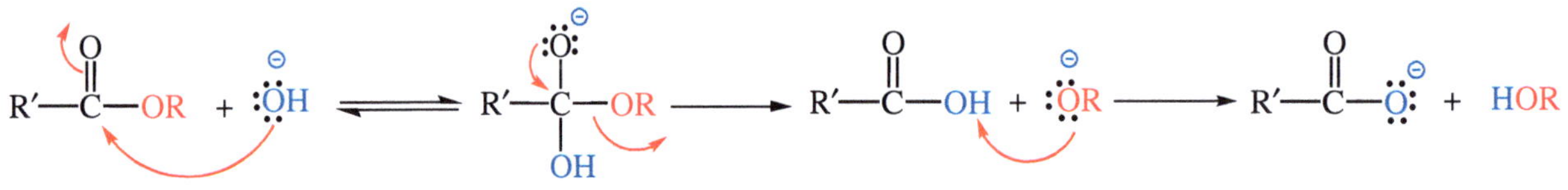

This reaction is also briefly referred to as $B_{AC}2$. The letter "B" indicates that the reaction is base-catalyzed, the symbol "AC" indicates that the acyl—oxygen bond breaks, and the number 2 indicates that the reaction is bimolecular. Under basic conditions, the hydroxide ion first attacks the ester carbonyl carbon atom, forming a tetrahedral alkoxide intermediate. While the carbonyl group is reforming, the electrons on the oxygen atom expel the alkoxide group (RO$^-$) from the molecule, generating a carboxylic acid. Because the alkoxide ion is more basic than the carboxylate anion, it abstracts the acidic proton from the carboxylic acid, forming the carboxylate anion and alcohol. During this reaction, an acyl–oxygen cleavage occurs. It is relatively easy to determine which bond (acyl–oxygen or alkyl–oxygen) is cleaved in such ester hydrolysis. Isotope labeling experiments can support the mechanism. When the reaction is carried out using a hydroxide ion labeled with the ^{18}O-isotope, the labeled oxygen will be incorporated in the carboxylic acid if an acyl–oxygen cleavage takes place; otherwise, the label will appear in the alkoxide ion as shown below.

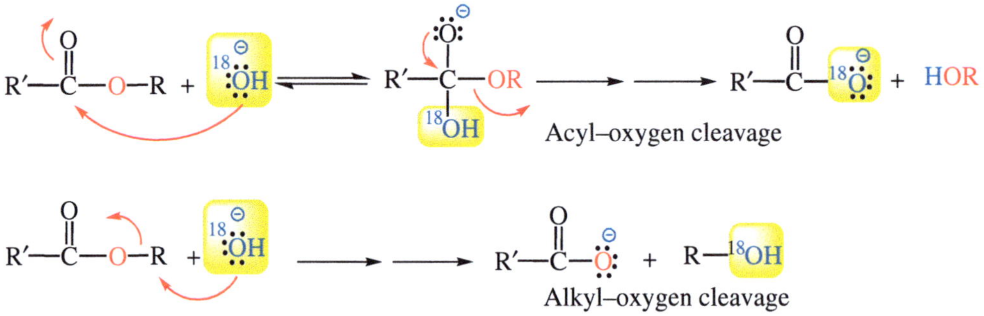

Another method used to determine the reaction mechanism under basic conditions is to use optically active compounds. For example, suppose that the carbon atom directly connected to the oxygen atom is optically active. In that case, it is also easy to determine the mechanism by which the hydrolysis reaction proceeds from analysis of the products formed after hydrolysis.

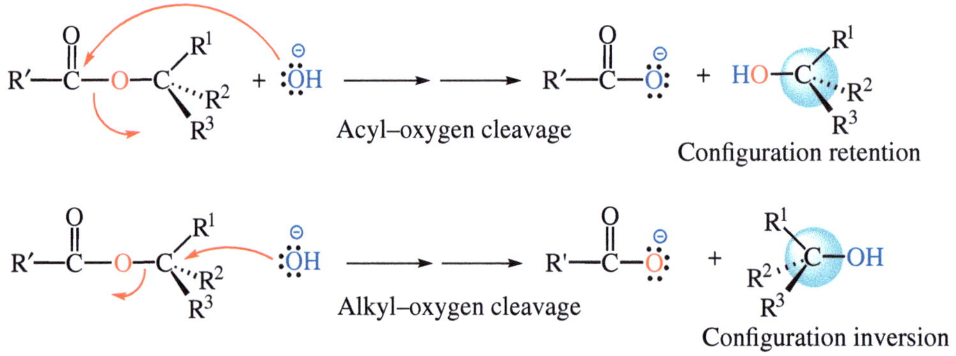

If the alcohol configuration is not changed, it is understood that the alkyl–oxygen bond is not involved in the hydrolysis reaction. This observation indicates that the base attacks the carbonyl carbon atom and an acyl–oxygen cleavage occurs. If the base attacks the alkoxy group's carbon atom according to an S_N2 reaction, an inversion of the configuration will take place that would support alkyl–oxygen cleavage.

We emphasized that the esters underwent an acyl–oxygen cleavage in a basic medium and suggested that they form a tetrahedral intermediate. However, this mechanism, acyl-oxygen cleavage, was questionable. There are two mechanisms here:

1. Substitution
2. Addition–elimination

According to the substitution mechanism, the nucleophile, the hydroxide ion, attacks the carbonyl carbon atom and substitutes the alkoxy group similar to the S_N2 reaction mechanism without forming a tetrahedral intermediate. To address this question, Bender [109] labeled the carbonyl group oxygen atom with O^{18}. If the reaction proceeds according to the substitution mechanism, all label O^{18} should appear in the carboxylic acid.

S_N2 substitution mechanism

The second mechanism is the addition–elimination mechanism. According to this mechanism, the base first attacks the carbonyl group, the carbon–oxygen double bond opens, and a tetrahedral intermediate is formed. During the formation of the double bond, the alkoxide ion leaves the molecule. Let us examine the structure of the intermediate product. Two oxygen atoms, beside the alkoxy group, are attached to the carbon atom. While one of them is bonded to a hydrogen atom, the other has a negative charge on it. The proton can be attached to either one of the oxygen atoms (blue box). As both intermediates are in equilibrium with the ester, the labeled oxygen atom can be eliminated from this intermediate because of a reversible reaction. When the reaction is interrupted before completion and the unreacted ester is isolated, the analysis revealed that some part of the ester was no longer labeled. This finding supports that the reaction proceeds by addition–elimination mechanism.

5.8.3 $B_{AL}2$ Mechanism

In this notation, again the letter B indicates that the reaction is base-catalyzed and the symbol "AL" indicates that the alkyl–oxygen bond cleaves. The $B_{AL}2$ mechanism is an extraordinarily rare process. Barclay and coworkers [110] reported in one case that the $B_{AL}2$ mechanism predominates. They reacted a highly hindered ester, methyl 2,4,6-tri-t-butylbenzoate, with H_2O^{18} and no incorporation of ^{18}O into the acid was detected. Bulky *ortho* substituents impede the attack on the carbonyl group to start the reaction by the $B_{AC}2$ mechanism. The H_2O^{18} attacks the relatively unhindered methyl group instead. They proposed a $B_{AL}2$ mechanism for this hydrolysis process.

Douglass and coworkers [111] investigated the hydrolysis of two sterically hindered esters, methyl trimethylacetate, and methyl triphenylacetate that they labeled at the alkoxy oxygen atom with ^{18}O. Hydrolysis by the $B_{AL}2$ mechanism will lead to remain of the ^{18}O isotope in the carboxylic acid, whereas the $B_{AC}2$ mechanism will result in loss of the ^{18}O isotope in the carboxylic acid. They found that the hydrolysis of methyl trimethylacetate proceeds entirely through the $B_{AC}2$ mechanism. The three methyl groups are probably insufficient to hinder the base's attack on the carbonyl group. In the case of methyl triphenylacetate, they found that 95% of the compound undergoes hydrolysis through the $B_{AC}2$ mechanism, whereas only 5% proceeds through the $B_{AL}2$ mechanism.

a $R' = -C(CH_3)_3$, $B_{AC}2$ 100% $B_{AL}2$ 0%
b $R' = -CPh_3$ $B_{AC}2$ 95% $B_{AL}2$ 5%

5.8.4 Ester Hydrolysis Under Acidic Conditions

Acid-catalyzed ester hydrolysis can also proceed by more than one mechanism: acyl–oxygen cleavage ($A_{AC}2$) and alkyl–oxygen ($A_{AL}2$) cleavage. Dilute acids catalyze the reaction. The first step is the protonation of the carbonyl oxygen, which activates the carbonyl group. The resulting positive (+) charge is distributed on the oxygen and carbon atoms. The water attacks the electrophilic carbonyl carbon atom with the nonbonding electrons generating a protonated tetrahedral intermediate. The proton on the $-OH_2^+$ group is transferred to the more basic alcohol oxygen, which makes the –OR group an excellent leaving group. The removal of alcohol furnishes a carboxylic acid and generates the acid catalyst.

Acid-catalyzed ester hydrolysis

Ester hydrolysis reactions usually proceed according to the $B_{AC}2$ and $A_{AC}2$ mechanisms. However, esters of tertiary alcohols and groups that can form stable carbocations (allylic, benzylic systems, etc.), analogs to the ionization step in an S_N1 or E1 reaction, are hydrolyzed in an acidic medium according to the alkyl–oxygen cleavage mechanism. For example, let us examine the hydrolysis of an optically active tertiary alcohol ester in acidic medium by the $A_{AL}1$ mechanism. First, the carbonyl group is protonated and the reactivity of the carbonyl carbon increases. The positive charge (+) on the carbonyl carbon atom withdraws the bonding electrons from the alkoxy group. During the reformation of the carbonyl group, the stable tertiary carbocation is separated from the molecule according to the S_N1 mechanism. The carbocation reacts with H_2O to form the corresponding tertiary alcohol. If the alkoxy carbon atom is optically active, it undergoes racemization. The racemization of the alcohol reveals that the alkyl–oxygen bond is broken and the reaction proceeds according to the S_N1 mechanism [112].

$A_{AL}1$ Alkyl–oxygen cleavage

5.8.5 Asymmetric Ester Hydrolysis

The synthesis of enantiomerically pure compounds is a major topic in contemporary organic synthesis. The chemoenzymatic approach for asymmetric synthesis has found widespread application in preparative organic chemistry. Lipases can catalyze asymmetric hydrolysis of esters. Hydrolysis of esters with pig liver esterase enantioselectively converts an ester into a carboxylic acid. One of the enantiomers is hydrolyzed faster than the other, leading to an excess of hydrolyzed product from one enantiomer (kinetic resolution). In the case of having two symmetrical ester groups within a molecule, the enzyme can discriminate between the enantiotopic groups and hydrolyze one of the ester groups (desymmetrization).

5.8.6 Transesterification

Transesterification is an organic reaction in which a carboxylic acid ester is converted into a different carboxylic acid ester by reacting with different alcohols in the presence of an acid or a base [113]. Transesterification is frequently used in the laboratory and in industry.

$$R'-\overset{O}{\underset{\|}{C}}-OR + R'OH \rightleftharpoons R'-\overset{O}{\underset{\|}{C}}-OR' + ROH \quad \text{Transesterification}$$

For example, some acids are poorly soluble in organic solvents. Thus, ester synthesis from such acids can be difficult. In this case, it is recommended to use methyl or ethyl esters of the corresponding acids because the esters' solubility is much better than that of the acids. Methyl and ethyl esters are generally starting compounds often used in transesterification reactions because they are easily synthesized and available. We discussed that the solvolysis or hydrolysis of esters is very slow because of the low nucleophilicity of water and alcohol. Therefore, transesterification reactions are performed under acidic or basic conditions.

274 | 5 Carbonyl Compounds

[Scheme: Transesterification under basic conditions]

The applicability of transesterification is not restricted to the laboratory scale. Several industrial processes use this reaction to produce different types of compounds. Polyethylene terephthalate is synthesized by the reaction of dimethyl terephthalate with ethylene glycol in the presence of zinc acetate as a catalyst [114]. Some additional examples of transesterification are given below.

One of the best examples of transesterification is biodiesel production, which is widely performed. As is known, biodiesel is a type of fuel that can be an alternative to fossil fuels. Biodiesel can be produced from vegetable oil, animal oil/fats, or waste cooking oil by transesterification. Because the density of vegetable oils is high, it causes engine fouling in a short time. However, transesterification generates a mixture of methyl esters of long-chain fatty acids with lower molecular weights, which can burn more easily.

Transesterification can also be carried out under acidic conditions. We examined in detail the hydrolysis reactions of esters in an acidic medium. Hydrolysis takes place with H_2O. When an excessive amount of alcohol (usually used as solvent) is used instead of H_2O in hydrolysis reactions, transesterification reactions occur in an acidic medium by a mechanism similar to that of hydrolysis.

Acid-catalyzed transesterification

Problems

5.1 The following reactions are carried out in the presence of an acid and at 50–60 °C, the H_2O formed is removed from the reaction medium. Write the structures of the products

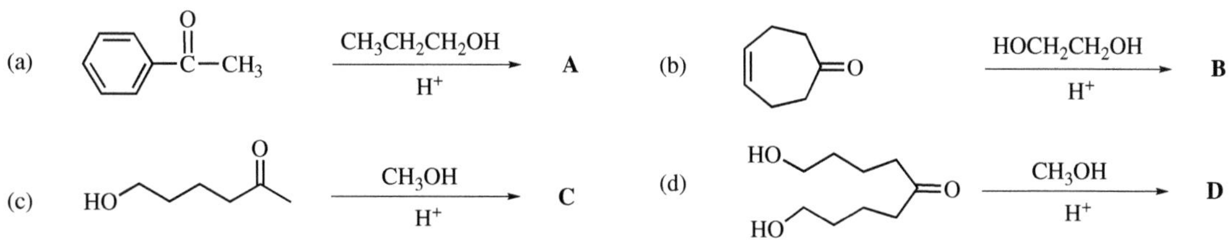

(a) PhC(=O)CH₃ + CH₃CH₂CH₂OH $\xrightarrow{H^+}$ **A**

(b) cyclohept-4-enone + HOCH₂CH₂OH $\xrightarrow{H^+}$ **B**

(c) HO-(CH₂)₄-C(=O)-CH₃ + CH₃OH $\xrightarrow{H^+}$ **C**

(d) HO-CH₂-CH₂-CH₂-C(=O)-CH₂-CH₂-CH₂-OH + CH₃OH $\xrightarrow{H^+}$ **D**

5.2 Propose efficient synthesis of the following compound beginning with the indicated starting compound.

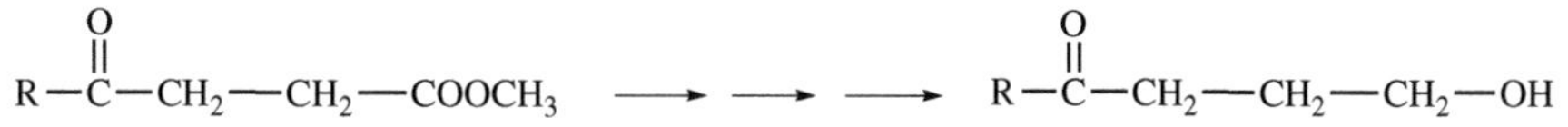

$$R-\overset{O}{\underset{\|}{C}}-CH_2-CH_2-COOCH_3 \longrightarrow \longrightarrow \longrightarrow R-\overset{O}{\underset{\|}{C}}-CH_2-CH_2-CH_2-OH$$

5.3 How can you synthesize the β-hydroxy ketone starting from propanal and styrene oxide?

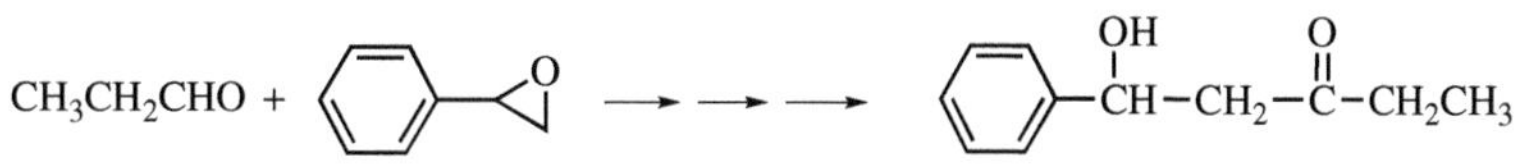

$$CH_3CH_2CHO + \text{PhCH(O)CH}_2 \longrightarrow \longrightarrow \longrightarrow \text{Ph-CH(OH)-CH}_2-\overset{O}{\underset{\|}{C}}-CH_2CH_3$$

5.4 The synthetic pathways applied to the synthesis of Conduritol-E is given below. Write down the structures of the intermediates.

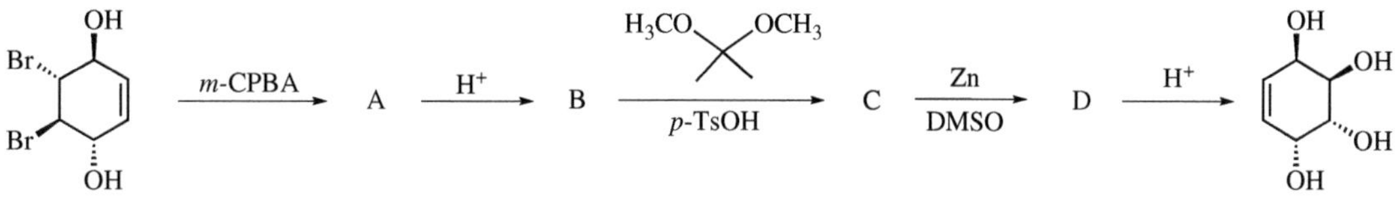

dibromo-cyclohexenol $\xrightarrow{m\text{-CPBA}}$ **A** $\xrightarrow{H^+}$ **B** $\xrightarrow[p\text{-TsOH}]{H_3CO\text{-C(OCH}_3)\text{-}}$ **C** $\xrightarrow[DMSO]{Zn}$ **D** $\xrightarrow{H^+}$ Conduritol-E

5.5 Develop a synthetic strategy leading to the following compound. Hint: use 1,3-dithiane.

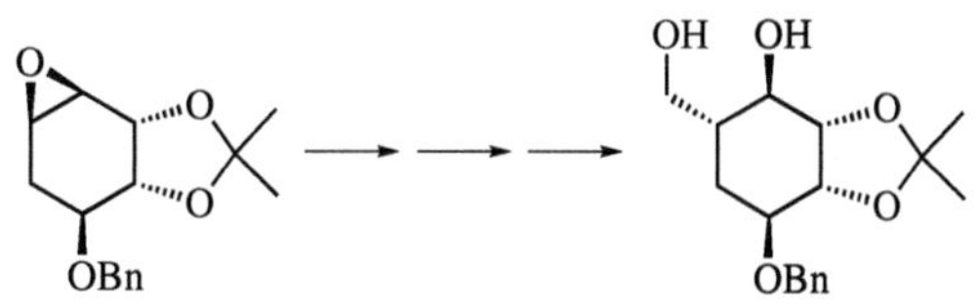

5.6 Propose a synthesis for selective reduction of the ketone functionality of the ketoaldehyde by a multistep process.

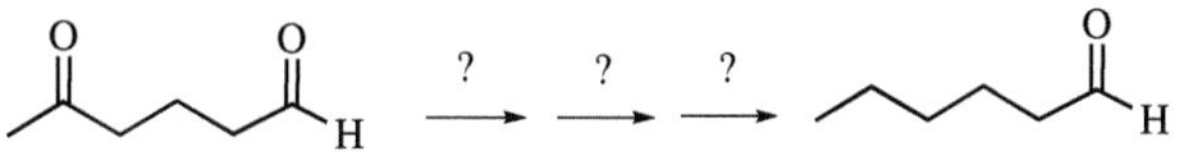

ketoaldehyde $\xrightarrow{?}\xrightarrow{?}\xrightarrow{?}$ aldehyde

5.7 Show how the following conversions might be accomplished.

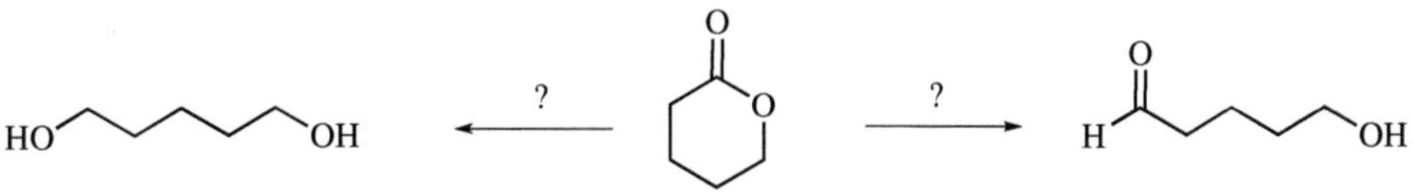

HO-(CH₂)₅-OH $\xleftarrow{?}$ δ-valerolactone $\xrightarrow{?}$ OHC-(CH₂)₃-CH₂-OH

5 Carbonyl Compounds

5.8 The following conversion is carried out in two steps. Write down the structure of the intermediate product and give the formation mechanism.

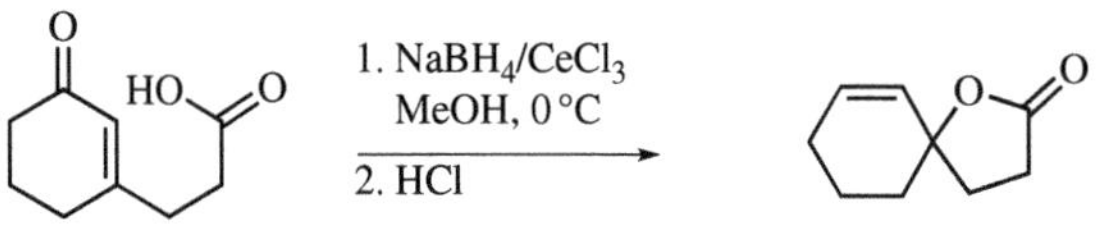

[Structure: cyclohexenone with -CH₂CH₂COOH substituent] → (1. NaBH₄/CeCl₃, MeOH, 0 °C; 2. HCl) → [spirolactone structure]

5.9 *R*-2-Phenylpropanal reacts with ethyl magnesium bromide. Write the products and indicate which product is the main product.

5.10 Give structures of A, B, and C compounds that are necessary for the synthesis of the following alcohol.

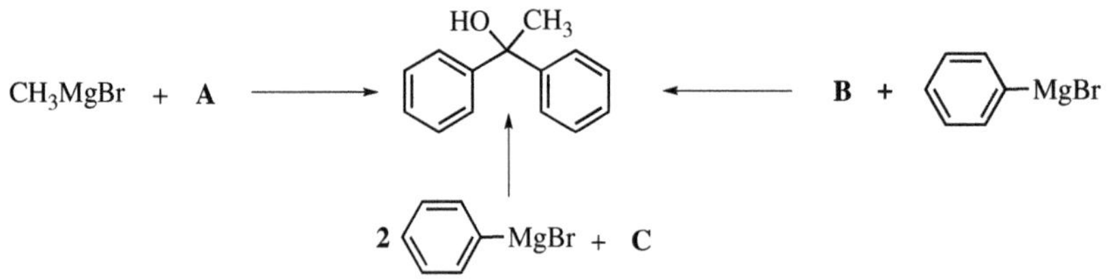

CH₃MgBr + A → [Ph₂C(OH)CH₃] ← B + Ph–MgBr
                     ↑
              2 Ph–MgBr + C

5.11 Write the product of the reaction below.

HC≡C–CH₂CH₂Br + Mg ⟶ A

5.12 Propose a suitable method for the synthesis of cyclohexyl ethylbenzene using Wittig's reagents.

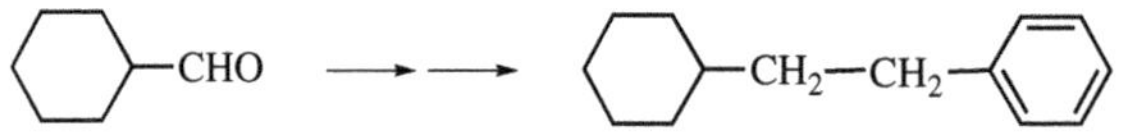

Cyclohexyl–CHO → → Cyclohexyl–CH₂–CH₂–Ph

5.13 Write the structures of the product that may occur in the following reaction. Explain which product preferentially will be formed and why.

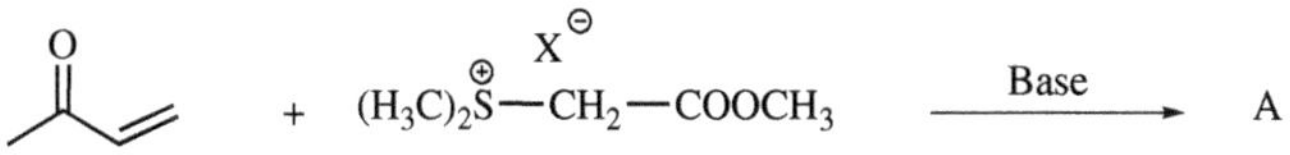

[methyl vinyl ketone] + (H₃C)₂S⁺–CH₂–COOCH₃ X⁻ —Base→ A

5.14 Draw the structures of the possible enol tautomers of the following compounds and show which one will be more stable.

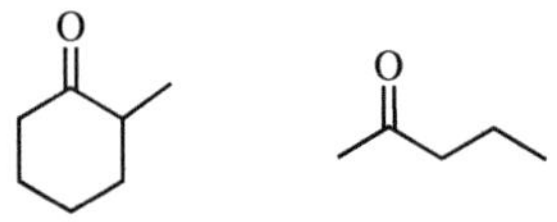

[2-methylcyclohexanone] [2-pentanone]

5.15 Write the keto tautomers of the following compounds.

[2-(hydroxymethylene)cyclohexanone] [1,2-dihydroxycyclohexene] [phenol-like cyclohexadienol] [hydroquinone]

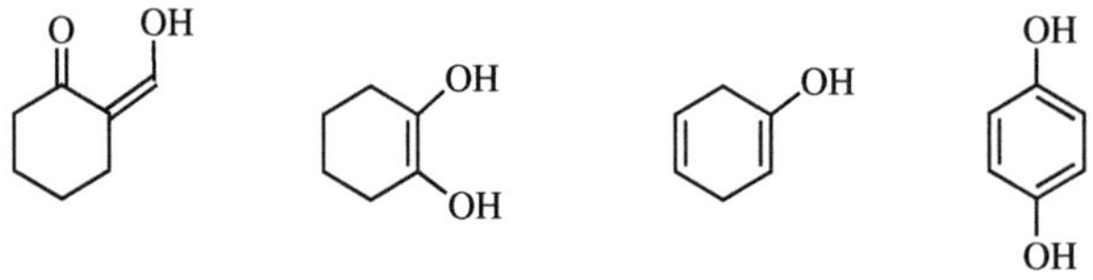

Problems

5.16 2-Ethyl-3-methylcyclopentanone is an optically active compound. Discuss the configurations of the products formed after treatment with a catalytic amount of acid or base and left for a while.

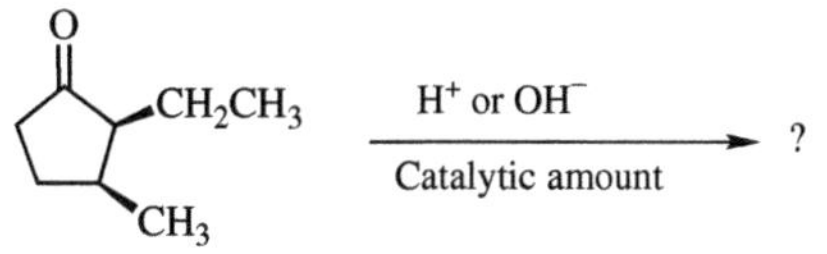

$$\text{2-ethyl-3-methylcyclopentanone} \xrightarrow[\text{Catalytic amount}]{H^+ \text{ or } OH^-} ?$$

5.17 The following compounds are mixed with D_2O and catalytic amount of D^+ and left at room temperature for a while. Write down the structures of the products that are formed.

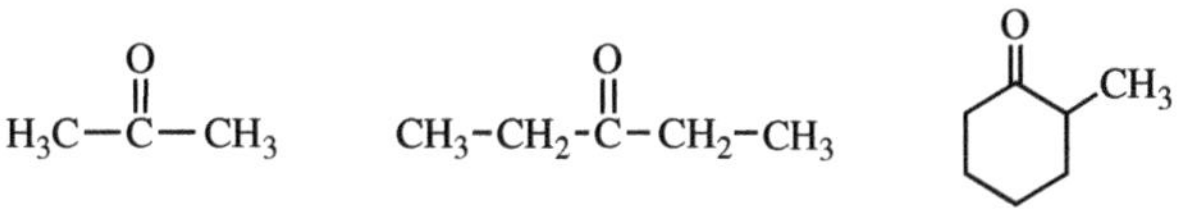

$$H_3C-\overset{O}{\underset{\|}{C}}-CH_3 \quad\quad CH_3-CH_2-\overset{O}{\underset{\|}{C}}-CH_2-CH_3 \quad\quad \text{2-methylcyclohexanone}$$

5.18 How can you differentiate between 1,3-diketone and 1,4-diketone by taking advantage of proton/deuterium exchange?

5.19 1-(4-Chlorophenyl) ethanone reacts with chlorine in acidic and basic media. Write down the products formed.

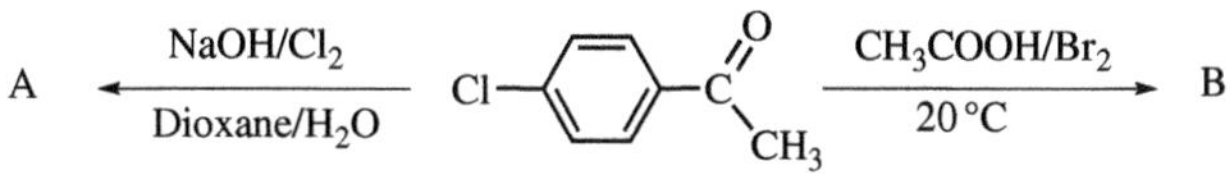

$$A \xleftarrow[\text{Dioxane/H}_2\text{O}]{\text{NaOH/Cl}_2} \text{Cl-C}_6\text{H}_4\text{-COCH}_3 \xrightarrow[20\,°C]{\text{CH}_3\text{COOH/Br}_2} B$$

5.20 Isomeric cyclohexane-1,3-dione derivatives are reacted with chlorine gas in a basic environment. Write the products.

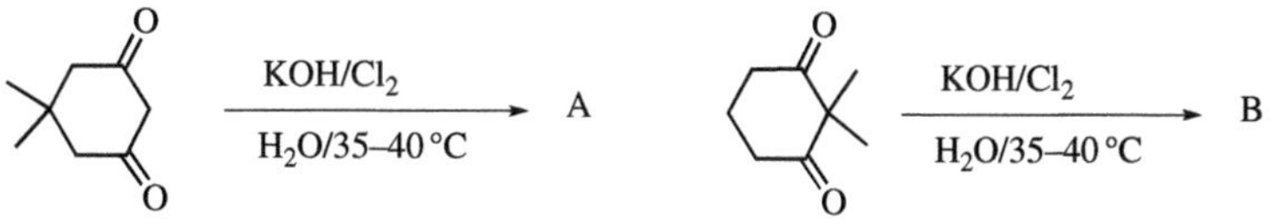

$$\text{5,5-dimethylcyclohexane-1,3-dione} \xrightarrow[\text{H}_2\text{O/35-40\,°C}]{\text{KOH/Cl}_2} A \quad\quad \text{4,4-dimethylcyclohexane-1,3-dione} \xrightarrow[\text{H}_2\text{O/35-40\,°C}]{\text{KOH/Cl}_2} B$$

5.21 Write the products of each of the following reactions.

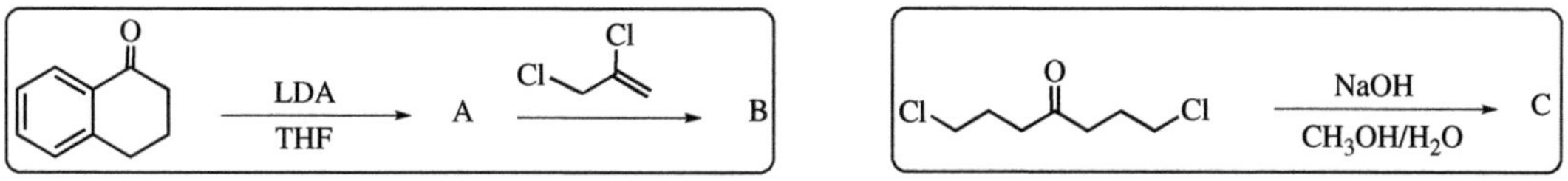

$$\text{α-tetralone} \xrightarrow[\text{THF}]{\text{LDA}} A \xrightarrow{\text{Cl-CH}_2\text{-C(Cl)=CH}_2} B \quad\quad \text{Cl-(CH}_2)_3\text{-CO-(CH}_2)_3\text{-Cl} \xrightarrow[\text{CH}_3\text{OH/H}_2\text{O}]{\text{NaOH}} C$$

5.22 How can you synthesize 7-methyl-oct-7-en-2-one starting from ethyl acetoacetate?

$$H_3C\overset{O}{\underset{\|}{\text{-}}}CH_2\overset{O}{\underset{\|}{\text{-}}}OC_2H_5 \longrightarrow \text{7-methyl-oct-7-en-2-one}$$

5.23 Write the structures of the products and final products in the equations given below.

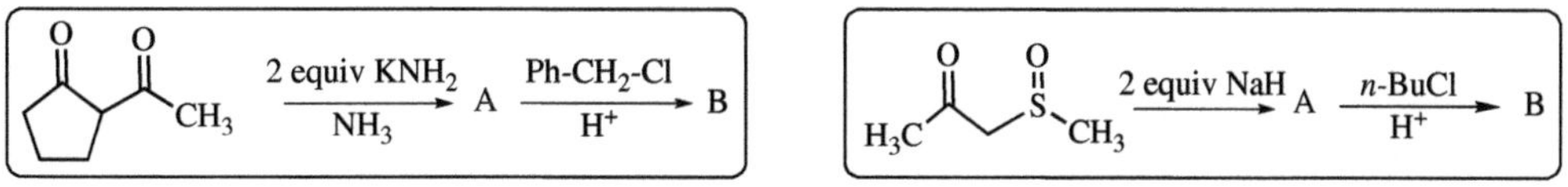

$$\text{2-acetylcyclopentanone} \xrightarrow[\text{NH}_3]{\text{2 equiv KNH}_2} A \xrightarrow[\text{H}^+]{\text{Ph-CH}_2\text{-Cl}} B \quad\quad H_3C\text{-CO-CH}_2\text{-SO-CH}_3 \xrightarrow{\text{2 equiv NaH}} A \xrightarrow[\text{H}^+]{n\text{-BuCl}} B$$

5.24 Write down the structures of possible cyclic condensation products in the following reaction that could theoretically occur.

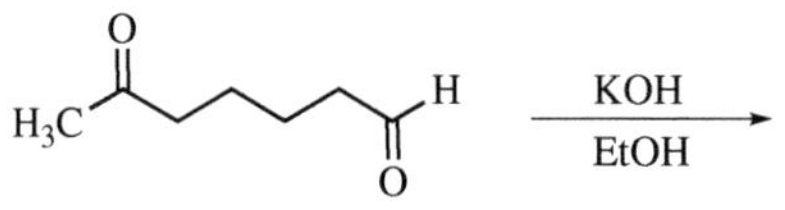

$$H_3C-CO-CH_2CH_2CH_2-CHO \xrightarrow{KOH, EtOH}$$

5.25 Write the structures of the products formed in the reactions given below.

2-methylcyclohexanone + methyl vinyl ketone $\xrightarrow{KOH, EtOH}$ A

4-(2-methyl-2-acetyl)cyclohexanone $\xrightarrow{K_2CO_3}$ B

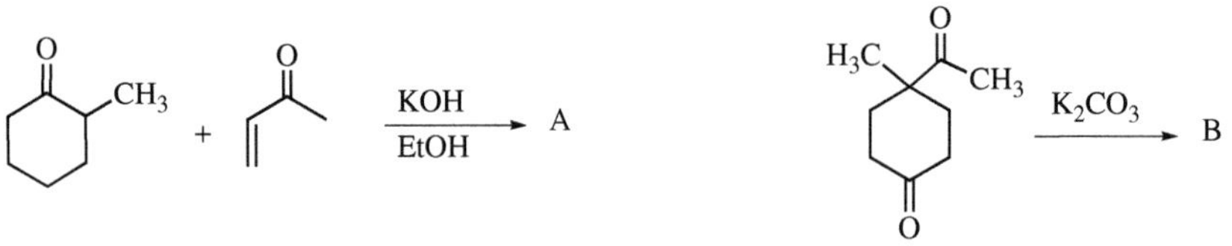

5.26 Write the structures of the products formed in the reactions given below.

$$CH_3-\overset{O}{\underset{\|}{C}}-OCH_3 \xrightarrow{NaOCH_3/CH_3OH} A$$

$$H_3C-CH_2-\overset{O}{\underset{\|}{C}}-OCH_3 \xrightarrow{NaOCH_3/CH_3OH} B$$

$$Ph-CH_2-\overset{O}{\underset{\|}{C}}-OCH_3 \xrightarrow{NaOCH_3/CH_3OH} C$$

$$Ph-\overset{O}{\underset{\|}{C}}-OCH_3 \xrightarrow{NaOCH_3/CH_3OH} D$$

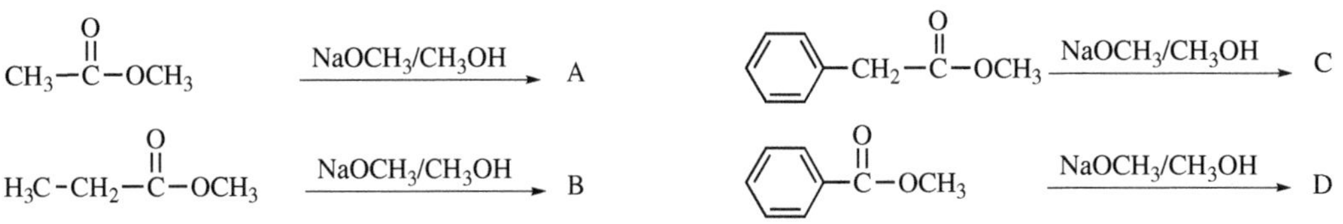

5.27 Suggest a reasonable mechanism for the formation of the following lactone.

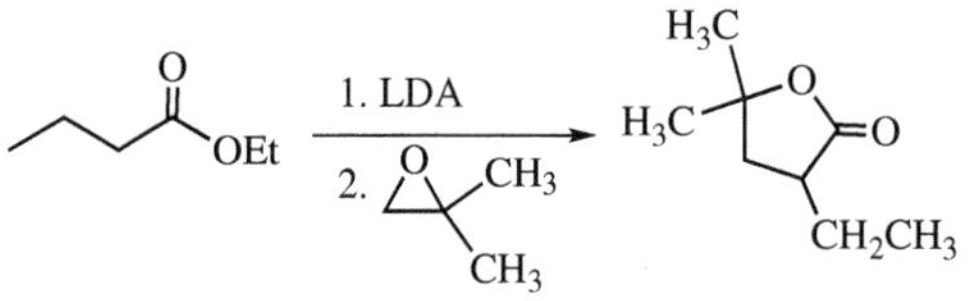

$$CH_3CH_2CH_2COOEt \xrightarrow[\text{2. } \underset{CH_3}{\overset{O}{\triangle}}\!\!\!\!\!\!\!\!\!\!\!\!\!\!-CH_3]{\text{1. LDA}} \text{γ,γ-dimethyl-α-ethyl-γ-butyrolactone}$$

5.28 Write the structures of the products formed as a result of the following condensation reaction.

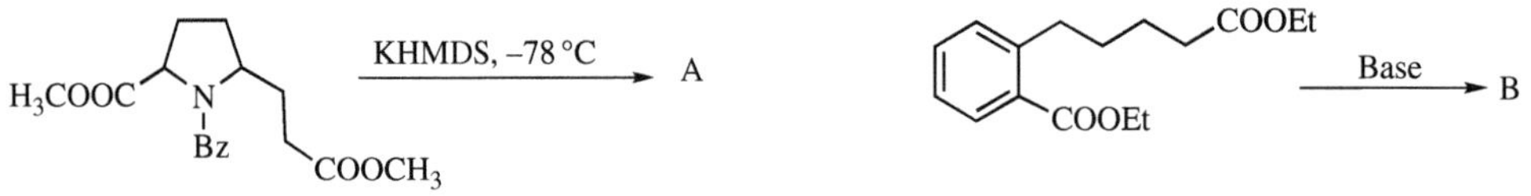

N-Bz-pyrrolidine with H₃COOC and CH₂CH₂COOCH₃ substituents $\xrightarrow{KHMDS, -78\,°C}$ A

ortho-disubstituted benzene with –CH₂CH₂CH₂CH₂COOEt and –COOEt $\xrightarrow{\text{Base}}$ B

References

1. Evans, D.A., Kim, A.S., Metternich, R., and Novack, V.J. (1998). *J. Am. Chem. Soc.* 120: 5921.
2. Sütbeyaz, Y., Seçen, H., and Balcı, M. (1988). *J. Chem. Soc. Chem. Commun.*: 1331.
3. Corey, E.J. and Seebach, D. (1988). *Org. Synth. Coll. Vol.* 6: 556.
4. Fleming, F.F., Funk, L., Altundas, R., and Yu, T. (2001). *J. Org. Chem.* 66: 6502.
5. Mozingo, R., Wolf, D.E., Harris, S.A., and Folkers, K.J. (1946). *J. Am. Chem. Soc.* 65: 1013.
6. Guven, S., Ozer, M.S., Kaya, S. et al. (2015). *Org. Lett.* 17: 2660.
7. Seebach, D. and Corey, E.J. (1975). *J. Org. Chem.* 40: 231.
8. Seebach, D. (1979). *Angew. Chem. Int. Ed.* 18: 239.
9. Dangerfield, E.M., Plunkett, C.H., WinMason, A.L. et al. (2010). *J. Org. Chem.* 75: 5470.
10. Jokanovic, M. and Prostran, M. (2009). *Curr. Med. Chem.* 16: 2177.
11. Kishner, N. (1911). *J. Russ. Chem. Soc.* 43: 582.
12. Wolff, L. (1912). *Justus Liebigs Ann. Chem.* 394: 86.
13. Minlon, H. (1949). *J. Am. Chem. Soc.* 71: 3301.

14 Cram, D.J., Sahyun, M.R.V., and Knox, G.R. (1962). *J. Am. Chem. Soc.* 84: 1734.
15 Ashby, E.C. and Boone, J.R. (1976). *J. Am. Chem. Soc.* 98: 524.
16 Webb, D. and Jamison, T.F. (2012). *Org. Lett.* 14: 568.
17 Lane, C.F. (1975). *Synthesis*: 135.
18 Prasanth, C.P., Ebbin, J., Abhijith, A. et al. (2018). *J. Org. Chem.* 83: 1431.
19 Yoon, N.M., Pak, C.S., Brown, H.C. et al. (1973). *J. Org. Chem.* 38: 2786.
20 Harinath, A., Bhattacharjee, J., and Panda, T.K. (2019). *Chem. Commun.* 55: 1386.
21 Kanth, J.V.B. and Periasamy, M. (1991). *J. Org. Chem.* 56: 5964.
22 Johnson, M.R. and Rickborn, B. (1970). *J. Org. Chem.* 35: 1041.
23 Gemal, A.L. and Luche, J.-L. (1981). *J. Am. Chem. Soc.* 103: 5454.
24 Jang, Y.K., Magre, M., and Rueping, M. (2019). *Org. Lett.* 21: 8349.
25 Verley, A. (1925). *Bull. Soc. Chim. Fr.* 37: 537.
26 Meerwein, H. and Schmidt, R. (1925). *Justus Liebigs Ann. Chem.* 444: 221.
27 Ponndrof, W.Z. (1926). *Angew. Chem.* 39: 138.
28 Oppenauer, R.V. (1937). *Recl. Trav. Chim. Pays-Bas* 56: 137.
29 Cha, J.S. (2006). *Org. Process Res. Dev.* 10: 1032.
30 De Graauw, C.F., Peters, J.A., van Bekkum, H., and Huskens, J. (1994). *Synthesis*: 1007.
31 Shing, T.K.M., Lee, C.M., and Lo, H.Y. (2004). *Tetrahedron* 60: 9179.
32 Westerhausen, M., Koch, A., Gorls, H., and Krieck, S. (2017). *Chem. Eur. J.* 23: 1456.
33 Knappke, C.E.I. and von Wangelin, A. (2011). *J. Chem. Soc. Rev.* 40: 4948.
34 Orchin, M. (1989). *J. Chem. Educ.* 66: 586.
35 Garst, J.F. and Soriaga, M.P. (2004). *Coord. Chem. Rev.* 248: 623.
36 Garst, J.F., Deutsch, J.E., and Whitesides, G.M. (1986). *J. Am. Chem. Soc.* 108: 2490.
37 Ashby, E.C. and Pham, T.N. (1984). *Tetrahedron Lett.* 25: 4333.
38 Smith, D.H. (1999). *J. Chem. Educ.* 76: 1427.
39 Ashby, E.C.Q. (1967). *Rev. Chem. Soc.* 21: 259.
40 Yamazaki, S. and Yamabe, S. (2002). *J. Org. Chem.* 67: 9346.
41 Cram, D.J. and Elhafez, A. (1952). *J. Am. Chem. Soc.* 74: 5828.
42 Pearson-Long, M.S.M., Boeda, F., and Bertus, P. (2017). *Adv. Synth. Catal.* 359: 179.
43 Weiberth, F.J. and Hall, S.S. (1987). *J. Org. Chem.* 52: 3901.
44 Kharasch, M.S. and Tawney, P.O. (1941). *J. Am. Chem. Soc.* 63: 2308.
45 Erdik, E. (1984). *Tetrahedron* 40: 641.
46 Walling, C. and Buckler, S.A. (1955). *J. Am. Chem. Soc.* 77: 6032.
47 Cahiez, G., Moyeux, A., Buendia, J., and Duplais, C. (2007). *J. Am. Chem. Soc.* 129: 13788.
48 Maercker, E. (1965). *Org. React.* 14: 270.
49 Maryanoff, B.E. and Reitz, A.B. (1989). *Chem. Rev.* 89: 863.
50 Carey, F.A. and Sundberg, R.J. (2007). *Advanced Organic Chemistry, Part B: Reactions and Synthesis*. Springer.
51 Vedejs, E. and Marth, C.F. (1990). *J. Am. Chem. Soc.* 112: 3905.
52 Vedejs, E. and Peterson, M.J. (1994). *Top. Stereochem.* 21: 1–157.
53 Robiette, R., Richardson, J., Aggarwal, V.K., and Harvey, J.N. (2006). *J. Am. Chem. Soc.* 128: 2394.
54 Maryanoff, B.E., Reitz, A.B., Mutter, M.S. et al. (1986). *J. Am. Chem. Soc.* 108: 7664.
55 Schlosser, M. and Christmann, K.F. (1966). *Angew. Chem., Int. Ed.* 5: 126.
56 Wang, Q., Deredas, D., Huynh, C., and Schlosser, M. (2003). *Chem. Eur. J.* 9: 570.
57 Schlosser, M. (1970). *Top. Stereochem.* 5: 1–30.
58 Horner, L., Hoffmann, H.M.R., and Wippel, H.G. (1958). *Chem. Ber.* 91: 61.
59 Horner, L., Hoffmann, H.M.R., Wippel, H.G., and Klahre, G. (1959). *Chem. Ber.* 92: 2499.
60 Wadsworth, W.S. Jr., and Emmons, W.D. (1961). *J. Am. Chem. Soc.* 83: 1733.
61 Corey, E.J. and Chaykovsky, M. (1965). *J. Am. Chem. Soc.* 87: 1353.
62 Aggarwal, V.K. and Richardson, J. (2003). *Chem. Commun.*: 2644.
63 Ciaccio, J.A. and Aman, C.E. (2006). *Synt. Commun.* 36: 1333.
64 Julia, M. and Paris, J.-M. (1973). *Tetrahedron Lett.* 14: 4833.
65 Keck, G.E., Savin, K.A., and Weglarz, M.A. (1995). *J. Org. Chem.* 60: 3194.
66 Blakemore, P.R. (2002). *J. Chem. Soc. Perkin Trans. 1*: 2563.

67 Aissa, C. (2009). *Eur. J. Org. Chem.*: 1831.
68 Dussart, N., Trinh, H.V., and Gueyrard, D. (2016). *Org. Lett.* 18: 4790.
69 Alonso, D.A., Fuensanta, M., Najera, C., and Varea, M. (2005). *J. Org. Chem.* 70: 6404.
70 Peterson, D.J. (1968). *J. Org. Chem.* 33: 780.
71 Hudrlik, P.F. and Peterson, D. (1975). *J. Am. Chem. Soc.* 97: 1464.
72 van Staden, L.F., Gravestock, D., and Ager, D. (2002). *J. Chem. Soc. Rev.* 31: 195.
73 Spencer, J.N., Holmboe, E.S., Kirshenbaum, M.R. et al. (1982). *Can. J. Chem.* 60: 1180.
74 Fuson, R.C. and Bull, B.A. (1934). *Chem. Rev.* 15: 275.
75 Kmentová, I., Almássy, A., Feriancová, L., and Putala, M. (2020). *J. Chem. Educ.* 97: 1139.
76 Cao, L., Ding, J., Gao, M. et al. (2009). *Org. Lett.* 11: 3810.
77 a Gleason, J.G. and Harpp, D.N. (1970). *Tetrahedron Lett.* 11: 3431. b Zhang, L.H., Duan, J., Xu, Y., and Dolbier, W.R. Jr., (1998). *Tetrahedron Lett.* 39: 9621.
78 Cano, R., Zakarian, A., and McGlacken, G.P. (2017). *Angew. Chem. Int. Ed.* 56: 9278.
79 Caine, D. (1868). *J. Org. Chem.* 1964: 29.
80 Stork, G., Brizzolara, A., Landesman, H. et al. (1963). *J. Am. Chem. Soc.* 85: 207.
81 Corey, E. and Enders, D. (1976). *Tetrahedron Lett.* 17: 3.
82 Job, A., Janeck, C.F., Bettray, W. et al. (2002). *Tetrahedron* 58: 2253.
83 Lazny, R. and Nodzewska, A. (2010). *Chem. Rev.* 110: 1386.
84 Kudyakova, Y.S., Bazhin, D.N., Goryaeva, M.V. et al. (2014). *Russ. Chem. Rec.* 83: 120.
85 Krause, N. and Hoffmann-Roder, A. (2001). *Synthesis*: 171.
86 Harris, T.M., Boatman, S., and Hauser, J.R. (1963). *J. Am. Chem. Soc.* 85: 3273.
87 Thompson, C.M. and Green, L.C. (1991). *Tetrahedron* 47: 4223.
88 Ishida, T. and Wada, K. (1979). *J. Chem. Soc. Perkin Trans. 1*: 323.
89 Kurts, A.L., Macias, A., Genkina, N.K. et al. (1969). *Dokl. Akad. Nauk. SSSR* (Engl. Transl.) 187: 595.
90 Kurts, A.L., Genkina, N.K., Macias, A. et al. (1971). *Tetrahedron* 27: 4777.
91 Carey, F.A. and Sundberg, R.J. (2007). *Advanced Organic Chemistry, Part A: Structure and Mechanism*, vol. 615. Springer.
92 Williard, P.G. and Carpenter, P.G. (1986). *J. Am. Chem. Soc.* 108: 462.
93 Jones, M.E., Kass, S.R., Filley, J. et al. (1985). *J. Am. Chem. Soc.* 107: 109.
94 Houk, K.N. and Paddon-Row, M.N. (1986). *J. Am. Chem. Soc.* 108: 2659.
95 Seitz, C.G., Zhang, H., Mo, Y., and Karty, J.M. (2016). *J. Org. Chem.* 81: 3711.
96 Rapson, W.S. and Robinson, R. (1935). *J. Chem. Soc.*: 307.
97 Jung, M.E. (1976). *Tetrahedron* 32: 3.
98 Zoretic, A., Bendiksen, B., and Branchaud, B. (1976). *J. Org. Chem.* 41: 3767.
99 Coates, R.M. and Chung, S.K. (1973). *J. Org. Chem.* 38: 3677.
100 Fominova, K., Diachuk, T., Sadkova, I.V., and Mykhailiuk, P.K. (2019). *Eur. J. Org. Chem.*: 3553.
101 Heravi, M.M., Janati, F., and Zadsirjan, V. (2020). *Monatsh. Chem.* 151: 439.
102 Zacuto, M.J. (2019). *J. Org. Chem.* 84: 6465.
103 van Schijndel, J., Canallea, L.A., Molendijka, D., and Meuldijk, J. (2017). *Green Chem. Lett. Rev. B* 10: 404.
104 Rabbani, G. (2018). *Org. Chem. Curr. Res.* 7: 1.
105 Johnson, W.S., McCloskey, A.L., and Dunnigan, D.A. (1950). *J. Am. Chem. Soc.* 72: 514.
106 David, K., Dingemanse, E., Freud, J., and Laqueur, E.Z. (1935). *Physiol. Chem.* 223: 281.
107 Johnson, W.S., Bannister, B., Pappo, R., and Pike, J.E. (1956). *J. Am. Chem. Soc.* 78: 6354.
108 Johnson, W.S., Korst, J.J., Clement, R.A., and Dutta, J. (1960). *J. Am. Chem. Soc.* 82: 614.
109 Bender, M.L. (1951). *J. Am. Chem. Soc.* 73: 1626.
110 Barclay, L.R.C., Hall, N.D., and Cooke, G.A. (1962). *Can. J. Chem.* 40: 1981.
111 Douglas, J.E., Campbell, G., and Wigfield, D.C. (1993). *Can. J. Chem.* 71: 1841.
112 Carroll, F.A. (1998). *Perspective on Structure and Mechanism in Organic Chemistry*, 434. Brooks/Cole Publishing Company.
113 Otera, J. (1993). *Chem. Rev.* 93: 1449.
114 Weissermel, K. and Arpe, H.-J. (1993). *Industrial Organic Chemistry*, 2nde, 396. Weinhein: VCH Verlagsgesellschaft. Translated by Charlet R. Lind.

6
Aromaticity

6.1 Aromatic Compounds

6.1.1 Discovery and Structure of Benzene

Benzene was first isolated by Faraday in 1825 from the flammable gases used in street lamps at that time. In 1834, Mitscherlich determined that the liquid obtained by heating benzoic acid in the presence of limestone was the same as the liquid isolated by Faraday. It was determined that the carbon content of this isolated compound was much higher than that of the other organic compounds. This ratio was initially determined to be 1 : 1 (CH), and then, the structure of the benzene molecule was determined to be $(CH)_6$. In 1845, Hofmann and Mansfield isolated benzene from coal tar. Although Mitscherlich called this compound "benzin," other chemists did not accept it in the first place because of the similarity with the names of other compounds, and the German chemist Liebig called it "benzol," which is still used in German today. This name was not adopted by other countries such as France and Britain because it could be confused with alcohols. Therefore, they changed the name "benzin" to "benzene," During this confusion, Laurent proposed the word "pheno," which derives from Greek and means "giving light." This word was also not accepted, but today, it is used as phenyl (phenol for alcohol) for benzene derivatives.

Parallel to the name confusion, very different structures were proposed for benzene, in accordance with its formula, $(CH)_6$. In 1861, an Austrian chemistry teacher, Loschmidt, suggested that benzene has a cyclic structure. Later, in 1865, Kekulé proposed a cyclic arrangement of six carbons with alternating single and double bonds and suggested the following structure.

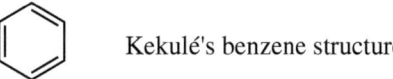

Kekulé's benzene structure

In 1867, Dewar proposed what we now know as "Dewar benzene" for benzene structure. Ladenburg proposed the prismane structure in 1874 after experimentally demonstrating that carbon and hydrogen atoms in benzene are identical. The common point of these proposed compounds is that all of their closed formula is $(CH)_6$.

Dewar benzene   Prismane

Studies have shown that none of these proposed structures are correct after a short time. These compounds were then synthesized by various methods. Although Kekulé's proposed structure was the perfect one at that time, it brought with it some problems. *What were these?* Because there are three double bonds in this structure, such a compound must react like alkenes. For example, alkenes react with bromine to form addition products. However, benzene does not undergo such addition reactions; it forms electrophilic substitution products in which one of the ring protons is replaced by an electrophile.

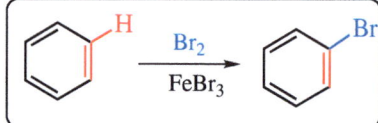

Electrophilic addition · Electrophilic addition · Electrophilic substitution

According to the proposed structure, the benzene molecule cannot have a hexagonal structure with equal C—C distances because there are three double bonds and three single bonds in the structure. The length of the carbon–carbon double bond is 1.34 Å, while the length of the carbon–carbon single bond is about 1.54 Å. This shows that the bond lengths are different and the benzene ring cannot have a hexagonal structure with equal C—C distances. As a result of studies carried out between 1869 and 1874, it was determined that the hydrogen and carbon atoms in the benzene molecule were equal.

If the benzene structure proposed by Kekulé with alternating single and double bonds had been correct, there would have been two different isomers of 1,2-disubstituted benzene derivatives. For example, the isomers of 1,2-dibromobenzene should be different according to the carbon–carbon bonds between the bromine atoms. However, even spectroscopic investigations at very low temperatures have shown that there are not two different structures.

1,2-Disubstituted benzene isomers

Indeed, in 1872, Kekulé proposed that the forms of benzene are in a state of equilibrium, and this equilibrium is so rapid that it prevents isolation of these 1,2-dibromo isomers. Thus, the two 1,2-dibromobenzenes would also be rapidly in equilibrium.

Kekulé structures of benzene

However, even this suggestion was not sufficient to explain why benzene did not undergo the electrophilic addition reaction characteristic for alkenes. Later, the stability of benzene as well as the lack of reactivity toward electrophilic addition was explained by the lack of localization of the double-bond electrons in the benzene ring.

Controversy over the structure of benzene continued until 1930. Experimental studies, especially those employing X-ray diffraction, showed that benzene has a planar structure with equal C—C distances of 1.39 Å, which is shorter than that of a carbon–carbon single bond (1.54 Å) but longer than that of a C=C double bond (1.34 Å). In other words, benzene does not have alternating single and double bonds. These intermediate distances are explained by electron delocalization. The carbon atoms in benzene are sp^2 hybridized. The p orbitals from each carbon atom overlap to form a delocalized molecular orbital (MO) that extends around the ring, giving stability and decreased reactivity to the benzene ring. To reflect the delocalized nature of the electrons, benzene is often depicted by a circle inside a hexagonal arrangement of carbon atoms as shown below.

After the structure of the benzene ring was determined, the resistance of the benzene ring to electrophilic addition reactions was investigated over a long period. In the first step, the heat of hydrogenation of the benzene ring (the amount of energy released by hydrogenation of the ring) was determined (Figure 6.1).

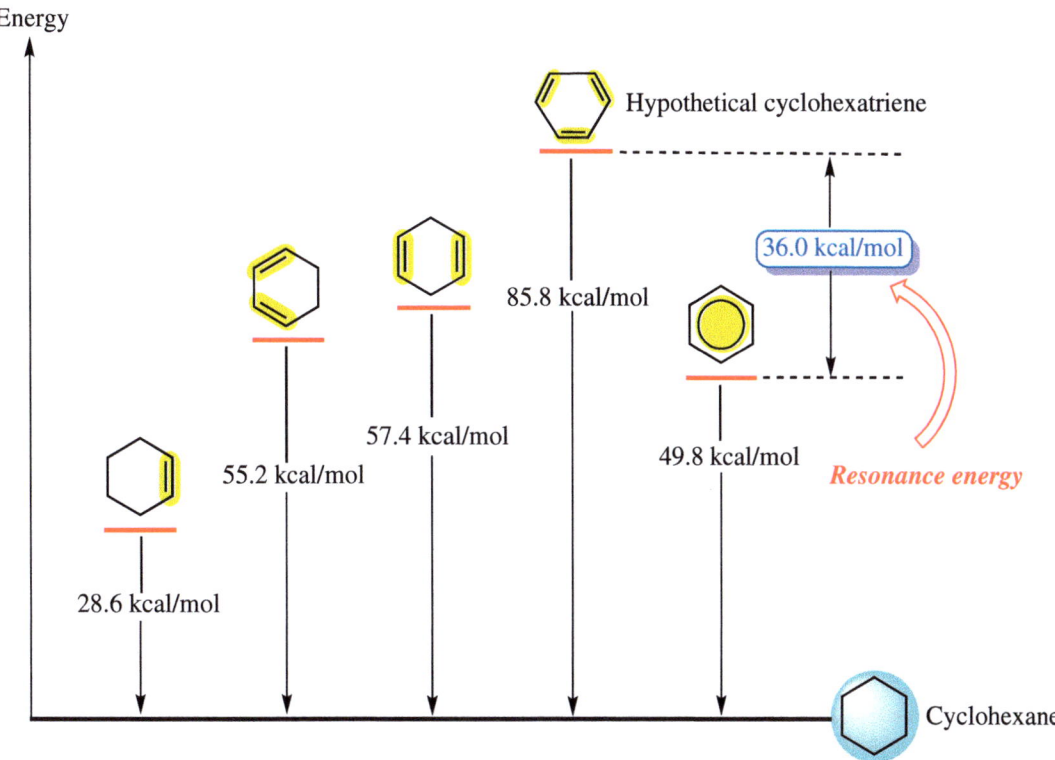

Figure 6.1 Heat of hydrogenation of benzene, cyclohexadienes, and cyclohexene.

The heat of hydrogenation of the benzene ring is exothermic by $\Delta H° = 49.8$ kcal/mol. In order to make sense of this value, it is necessary to compare it with the heat of hydrogenation of cyclohexene and cyclohexadienes. The heat of hydrogenation of cyclohexene is $\Delta H° = -28.6$ kcal/mol, a value expected for the hydrogenation of a *cis* double bond. The heat of hydrogenation of cyclohexa-1,4-diene is -57.4 kcal/mol. This value is about twice ($-28.6 \times 2 = -57.2$ kcal/mol) the heat of hydrogenation of cyclohexene. Cyclohexa-1,4-diene has two nonconjugated independent double bonds, and the resonance energy of the isolated double bonds is about zero. The heat of hydrogenation of cyclohexa-1,3-diene is -55.2 kcal/mol, which is ($-57.4 - 55.2 = 2.2$ kcal/mol) 2.2 kcal/mol less than the heat of hydrogenation of cyclohexa-1,4-diene. This means that cyclohexa-1,3-diene is about 2.2 kcal/mol more stable than cyclohexa-1,4-diene. The stability of cyclohexa-1,3-diene results from the conjugation of the double-bond electrons. We can think of the benzene molecule as hypothetical cyclohexa-1,3,5-triene. Because there are three double bonds in the molecule, the heat of hydrogenation of hypothetical cyclohepta-1,3,5-triene (3×-28.6) is expected to be -85.8 kcal/mol. Considering the energy that may arise from the conjugation of double bonds, this energy should be a maximum of (3×2.2) 6.6 kcal/mol. Then, the hydrogenation energy of benzene should be around ($-85.4 + 6.6$) -78.8 kcal/mol. However, the measured value for benzene is -49.8 kcal/mol, much less than the expected value of -78.8 kcal/mol. When we compare this value with the calculated value for hypothetical cyclohexa-1,3,5-triene (-85.8 kcal/mol), we find that there is a difference of 36.0 kcal/mol. This means when benzene is hydrogenated, 36.0 kcal/mol less energy is released than expected. The difference is the *resonance energy of benzene*, about 36.0 kcal/mol. This finding means that benzene is 36 kcal/mol lower in energy than hypothetical cyclohexa-1,3,5-triene, i.e. benzene is 36.0 kcal/mol more stable. By delocalizing of the double-bond electrons in the benzene ring, the benzene ring is stabilized and all carbon–carbon bonds are equal because the electrons are distributed evenly on the molecule. Other terms are also used to describe this quantity (36.0 kcal/mol) such as *delocalization energy*, *aromatic stability*, or simply *aromaticity of benzene*.

Prior to the discovery of the structure of benzene, a number of compounds containing the benzene ring were already known. These include benzoic acid, benzaldehyde, isolated from almond tree, *p*-cymene isolated from cumin, methyl salicylate, anethole, and vanillin. All of these compounds are fragrant compounds and all have benzene rings. For this reason, these compounds were called "aromatic," inspired by the Greek word "aroma," used for compounds with nice smell.

284 | *6 Aromaticity*

Benzaldehyde Benzoic acid Methyl salicylate Anethole *p*-Cymene Vanillin Menthol

Over time, it was found that the expression "aromatic" was not very accurate; there were many compounds containing benzene rings that did not smell pleasant, as well as many aliphatic compounds with a nice smell. For example, menthol isolated from the mint plant was included in the class of aromatic compounds as it was fragrant, while a number of nonfragrant compounds with benzene rings were not included in this group.

Then, it was found that the number of hydrogens in the compounds included in the aromatic class was less than the number of carbon. For example, it was determined that hydrogen content was higher in oils that did not belong to the aromatic group. Compounds with a low hydrogen content for some time were called aromatic compounds.

6.1.2 Aromatic, Antiaromatic, and Nonaromatic Compounds

In 1925, Robinson proposed the *sextet electron rule*. According to this concept, the association of six free electrons on the benzene ring was responsible for the stability of the benzene ring and not undergoing an addition reaction as double bonds. In 1931, the German physicist Hückel tried to explain for the first time the stability of the benzene ring by quantum mechanics by separating the bond electrons in the benzene ring into σ and π electrons.

6.1.2.1 Aromatic Compounds

According to Hückel's rule, a compound can be aromatic if it is

(1) cyclic,
(2) conjugated,
(3) monocyclic coplanar, and
(4) with $(4n + 2)$ π electrons ($n = 0, 1, 2, 3, 4,\ldots$).

Compounds that meet these conditions are more stable and they are called *aromatic compounds*. The π electron count is defined by the series of numbers generated from $4n + 2$ where n = zero or any positive integer.

For a molecule to be aromatic, it should have 2, 6, 10, 14, 18, 22, etc., electrons. According to this rule, a conjugated, monocyclic coplanar organic compound with 10 π electrons is also aromatic and has a stable structure. Hückel's rule has been known since 1931, but it did not receive the necessary attention at that time. According to Hückel's rule, for example, compounds with 10, 14, 18, and 22 π electrons are also aromatic compounds. However, at the time of the publication of Hückel's rule, there was no known compound with more than six π electrons to use to test its validity. Therefore, it could not be established to what extent this rule was valid. However, 25 years later, some compounds having more than six π electrons have been synthesized to determine the validity of the rule. A few examples are given below. In the following, these compounds will be discussed in detail.

10π   14π   18π

Before examining this rule in detail, let us focus on electron numbers. In order for a compound to be aromatic, it is not necessary for all electrons involved in delocalization to be π bond electrons. For example, while all electrons in the benzene ring are π electrons, in thiophene, only four of the electrons are π electrons. Because the hybridization of the sulfur atom is sp², only two of the electrons in sulfur are in the p orbital involved in delocalization. The remaining two electrons are in the sp² orbital perpendicular to the p orbitals and do not participate in the π orbital system.

Benzene   Thiophene

6.1.2.2 Antiaromatic Compounds

According to Hückel's rule, a compound can be antiaromatic if it is

(1) cyclic,
(2) conjugated,
(3) monocyclic coplanar, and
(4) with $(4n)$ π electrons ($n = 1, 2, 3, 4,...$).

Compounds that meet these conditions are very unstable and more reactive than others and are called antiaromatic compounds. The number of electrons varies according to n as n is an integer. For a molecule to be antiaromatic, it should have 4, 8, 12, 16, etc., electrons. The energy levels of these compounds are much higher than those of acyclic unsaturated compounds and they have negative resonance energy.

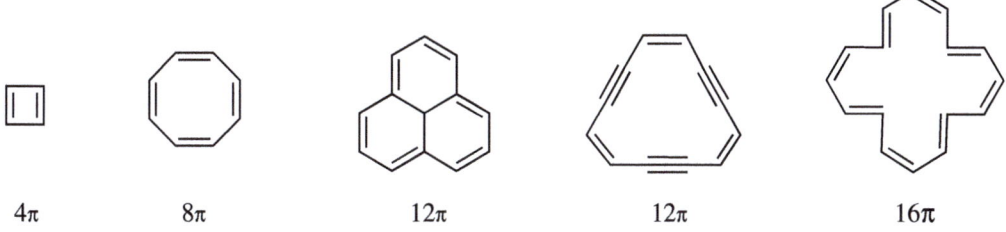

4π   8π   12π   12π   16π

The compounds given above are antiaromatic in terms of electron number. Cyclobutadiene, with four π electrons, is an antiaromatic compound, while cyclooctatetraene (COT), with eight π electrons, is not antiaromatic. Therefore, if COT were planar, it would be antiaromatic, a destabilizing situation. Molecules tend to try to adopt a more stable tube-shaped conformation. Thus, COT exhibits polyolefin properties.

Cyclooctatetraene

Two different 12 π systems are given above. As both of these have planar structures, they are classified as antiaromatic compounds. If one of the conjugated bonds is a triple bond, only two of the bond electrons participate in delocalization because the other two electrons, as shown below, cannot participate because of the perpendicular orientation of the π orbital system to the p orbitals. Thus, triple-bond electrons are considered as double-bond electrons.

They are not involved in the delocalization

6.1.2.3 Nonaromatic Compounds

All compounds with a nonplanar structure and with an electron number of $(4n+2)\,\pi$ or $4n\,\pi$ are called nonaromatic compounds. The electron number of nonaromatic compounds may be 2, 4, 6, 8, 10, 12, 14, 16, 18, etc. For example, COT and cyclodeca-1,3,5,7,9-pentaene are nonaromatic compounds because they do not have planar structures. *Why is COT not antiaromatic?* COT must have a planar structure to be antiaromatic. The reason for this lack of planarity is that the internal angles in a regular planar octagonal structure are 134°, while sp² hybrid angles of 120° are most stable. To avoid angle strain, the molecule adopts a nonplanar tub-shaped conformation. Furthermore, if COT has a planar structure, the π electrons will delocalize because of the parallel orientation of the orbitals. Consequently, the compound will be antiaromatic. The energy levels of antiaromatic compounds (negative resonance energy) are considerably higher than those of the nonconjugated ones. Therefore, the energy level of the molecule will also increase because of the negative resonance energy. In summary, antiaromaticity and angle strain will make the molecule very unstable. That is the reason why COT does not have a planar structure and tends to have a tub-shaped conformation.

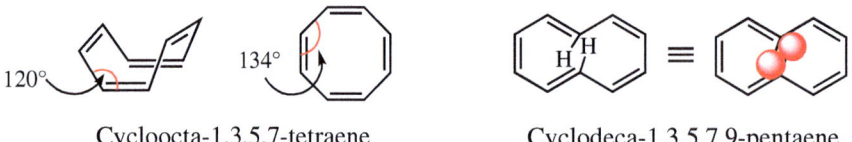

Cycloocta-1,3,5,7-tetraene   Cyclodeca-1,3,5,7,9-pentaene

Cyclodeca-1,3,5,7,9-pentaene, with 10 π electrons, is a cyclic and conjugated compound. If the structure were planar, according to Hückel's rule, the molecule would be aromatic. However, while this compound contains 10 π electrons, it is nonaromatic. Let us try to explain why this compound does not have a planar structure. Because of its molecular structure, eight of the hydrogen atoms are directed out of the ring, while the other two hydrogens are directed into the ring because of sp² hybridization. Because there is not enough space for these two hydrogens, they will push each other and one will be directed downward while the other will be directed upward of the ring. The molecular structure will deviate from the planar structure, and parallel orientation of the p orbitals will be prevented. This compound is therefore a nonaromatic compound.

We have tried to describe aromatic, antiaromatic, and nonaromatic compounds with the information we have given so far. We can also define these terms roughly as shown in the diagram in Figure 6.2. Imagine an open carbon chain with

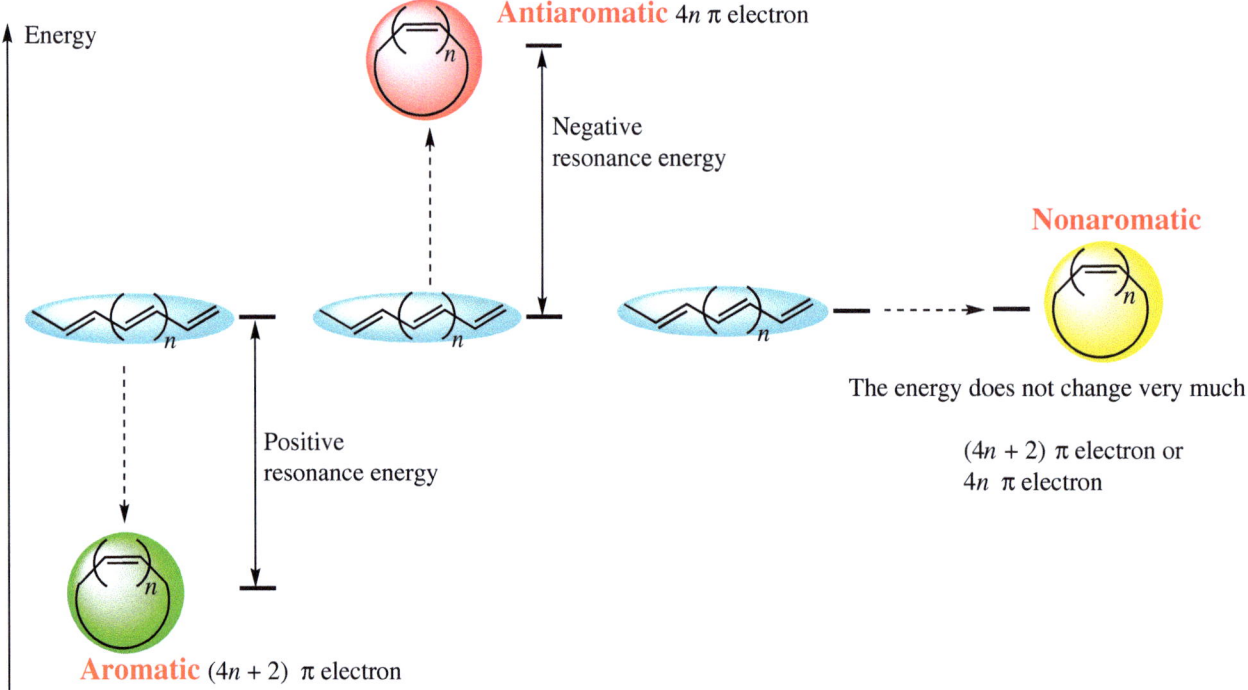

Figure 6.2 Relative energies of aromatic, nonaromatic, and antiaromatic compounds.

conjugated double bonds. Suppose that the molecule is conjugated by connecting it at both ends and turning it into a ring. If the number of electrons in the ring is $(4n + 2)$ ($n = 0, 1, 2, 3, 4$, etc.), then the energy of the system will decrease relative to the open-chain system; these compounds are called *aromatic*. If the number of electrons is $4n$ ($n = 1, 2, 3, 4$, etc.), then the energy of the system will increase; these compounds are called *antiaromatic*. If the number of electrons is $4n + 2$ and/or $4n$ and the energy of the system does not change much compared to the open-chain system when the ring is formed, these compounds are called *nonaromatic*. Some examples of aromatic, antiaromatic, and nonaromatic compounds are given in Section 6.3.

6.1.3 Determination of the Molecular Orbitals of Aromatic Compounds

Before examining the aromatic compounds, let us qualitatively examine the formation of a π bond as a result of the interaction of two p orbitals by the MO theory. When examining the formation of π bond in the ethylene molecule, we will not consider the formation of σ bonds. According to MO theory, atomic orbitals interfere in two different ways to form new MOs. The number of the new MOs is based on the number of atomic orbitals involved. Because two p orbitals are required for the formation of the ethylene double bond, two new MOs are formed. As the energy level of one of these orbitals is lower than that of the atomic orbitals, this orbital is called the bonding orbital (ψ_1 orbital), and the other orbital is called as the antibonding orbital ((ψ_2 orbital) because the energy level of this orbital is higher than that of the atomic orbitals. The resulting MOs are shown in Figure 6.3. Two π electrons occupy the bonding orbital, which is lower in energy compared to the unhybridized p orbitals (atomic orbitals).

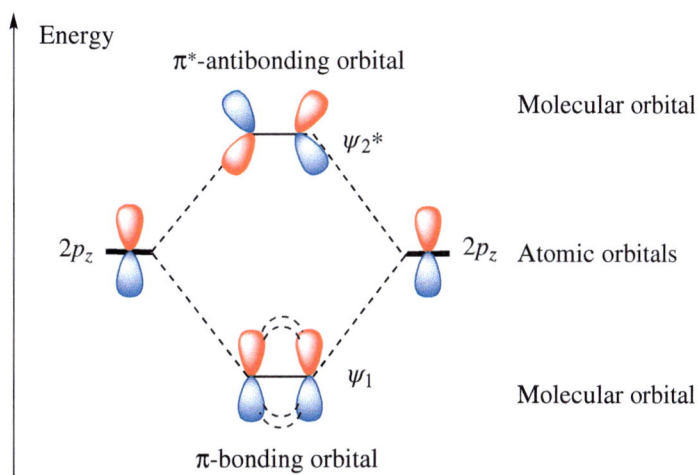

Figure 6.3 Relative energies of atomic orbitals and molecular orbitals of ethylene.

Hückel's MO calculations can be applied to cyclic, conjugated, and planar systems to determine the energy levels of all orbitals. However, the relative energies of the MOs of benzene can also be determined by a simplified approach developed by Frost. The Frost circle is a simple method to estimate the relative π orbital energies of both planar and cyclic compounds with an uninterrupted π electron system: It involves the following steps: first we draw a circle and place the ring representing the compound of interest (in the case of benzene, a hexagon) in the circle so that a vertex is pointing down. The other vertices of the hexagon (polygon) must touch the circle as shown in Figure 6.4. Each point where the polygon touches the circle represents an energy level. Now, the energy levels of the six MOs are placed at each vertex. We draw a line in the middle of a hexagon. From the geometry of a regular hexagon, there should be equal separation of the three energy levels. Orbitals ψ_1, ψ_2, and ψ_3 below the dotted reference line in Figure 6.4 are bonding orbitals, while orbitals ψ_4, ψ_5, and ψ_6 above the reference line are antibonding orbitals. Now, let us populate these MOs with six electrons of benzene starting from the lowest energy orbital ψ_1 first in accordance with Hund's rule. After adding six electrons to this array, the final electronic configuration shows that the electrons in the bonding orbitals are paired. This is a stable electronic configuration and this configuration is known as a *closed-shell electronic configuration*. Generally, electrons populating the bonding orbitals stabilize the molecule. On the other hand, electrons in the antibonding orbitals destabilize the molecule. Aromatic compounds will have all occupied MOs completely filled whereas antiaromatic compounds will have incompletely filled orbitals.

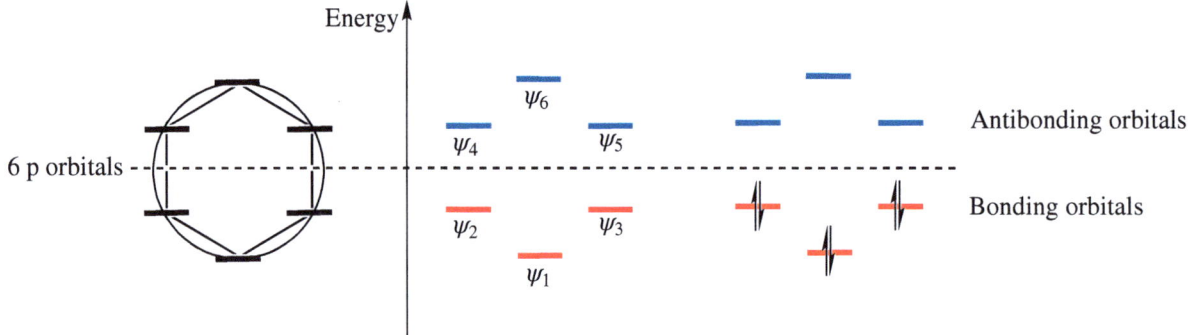

Figure 6.4 Molecular orbitals (MOs) of benzene.

6.1.4 What Are the Criteria for Aromaticity? How Does One Quantify Aromaticity?

There are several criteria that are used to determine whether a given compound is aromatic, nonaromatic, or antiaromatic. The oldest criterion to quantify the aromatic character of a molecule is thermodynamic. Aromatic compounds are unusually stable and less reactive. The second criterion is structural; the electrons are delocalized in a cyclic system so that there is no alternation of the σ bonds observed in open-chain conjugated systems. The third criterion is the magnetic properties of aromatic compounds.

6.1.4.1 Thermodynamic and Aromatic Resonance Stabilization Energy

In the previous section, we already discussed the enhanced thermodynamic stability of benzene, which was obtained from measurements of the heat released when double bonds in a six-carbon ring are hydrogenated to give cyclohexane. We showed that benzene is 36.0 kcal/mol more stable than expected.

Isodesmic Reaction Pople and coworkers [1] introduced the term *isodesmic*. An isodesmic reaction is a hypothetical chemical process in which the numbers and types of bonds in the reactants and products are the same. For example, formation of penta-1,4-diene and prop-1-yne from pent-1-en-4-yne and prop-1-ene is an *isodesmic* reaction as shown below. It is a hypothetical reaction and so may never occur [2–4].

Here, the products and reactants must have equal numbers of each type of carbon–carbon bond: they must contain equal numbers of atoms in the same state of hybridization groups containing the same number of H atoms. Using isodesmic reactions, one can calculate the strain energies in cyclic compounds, the resonance energies of aromatic compounds, the stability of carbocations and radicals, the heat of formation of compounds, etc.

One of the most popular types of isodesmotic reaction is hydrogen transfer from one reactant to another. For example, one such reaction involves conversion of cyclohexa-1,3-diene and cyclohexane to benzene and cyclohexene. The reactants and products must contain equal numbers of carbons in the same state of hybridization and hydrogens bound to identical atom types. For example, the reactants contain 12 sp²- and 12 sp³-hybridized carbon atoms. The products must also contain the same number of hybridized carbon atoms. Using this reaction would allow us to calculate the resonance stabilization energy of benzene. To calculate the resonance energy, we have to know the heat of formation energies of the reactants as well as the products. These values may be experimental or calculated ones.

$\Delta H_{f_o} = 106.2$ kJ/mol $\Delta H_{f_o} = -123.4$ kJ/mol $\Delta H_{f_o} = 82.6$ kJ/mol $\Delta H_{f_o} = -5.0$ kJ/mol

The next step will be equalization of the values on the left and the right side of the equation. The values of the heats of formation ΔH_{fo} for the gas phase are taken from the literature [5].

$$(3 \times 106.2) + (-123.4) = 82.6 + (3 \times -5.0) + Q$$
$$318.60 - 123.4 = 82.6 - 15.0 + Q$$
$$195.20 = 67.6 + Q$$
$$Q = 127.6 \text{ kJ/mol} \, (Q = 30.5 \text{ kcal/mol})$$

To equalize the values on the left and right, we add 30.5 kcal/mol to the right and this value, designated by Q, represents the heat of the isodesmic reaction. Its positive sign indicates stabilization of the molecule on the right side as compared with the compounds on the left. This energy is exactly the resonance stabilization energy of benzene and is in line with the values obtained by other methods.

6.1.4.2 Structural Evidence for Aromaticity

Benzene has a perfectly hexagonal structure with equal C—C bond lengths of 1.39 Å and all C—H bond lengths are equal to 1.08 Å, with each bond angle being 120°. In butadiene, the single C—C bond length is 1.48 Å and the C=C double-bond length is 1.32 Å. These values can be compared to 1.532 Å for ethane and 1.32 Å for ethylene. Because of the aromaticity of benzene, full delocalization of the π electrons is responsible for equal C—C bonds. It is important to note that there are no distinct single or double bonds within benzene. There is no bond alternation. Experiments demonstrate that the actual bond length is somewhere between that of a single bond and that of a double bond.

This loss of C—C bond alternation is generally regarded as an important diagnostic feature for aromatic compounds. In naphthalene and other polycyclic aromatic compounds, C—C bond distances are not equal, but they deviate slightly in length (1.37 and 1.42 Å) from those in benzene (1.39 Å). There is a certain degree of bond localization. However, this variation does not affect the aromaticity or the resonance stability of naphthalene.

6.1.4.3 Magnetic Evidence for Aromaticity

Magnetic Susceptibility When an aromatic compound is placed in a strong magnetic field, the compound will generate a magnetic field under the influence of this external magnetic field [6–8]. Magnetic susceptibility is the degree of magnetization generated in the presence of an external magnetic field. The magnetic susceptibility of a material, commonly symbolized by χ^m, is equal to the ratio of the magnetization M to the strength of the external magnetic field H.

$$\chi^m = \frac{M}{H}$$

Diamagnetic susceptibility exaltation (Λ) is defined as the difference between the measured magnetic susceptibility of a compound and a calculated value. Large negative values are aromatic, for example, benzene ($\Lambda = -13.4$). Values close to zero are nonaromatic, for example, cyclohexane ($\Lambda = 1.1$), while large positive values are antiaromatic, for example, cyclobutadiene ($\Lambda = +18$).

$$\Lambda^m = \chi^m - \chi_{cal}$$

The magnetic exaltation in aromatic compounds arises from the ring current of the π electrons. When an aromatic compound is placed in a strong external magnetic field, the π electrons of the ring circulate around the ring and generate a ring current. This ring current in turn induces a local secondary magnetic field. This leads to current loops: one above and one below the plane of σ bonds. The induced magnetic field opposes the external magnetic field in the center (above and below) of the aromatic ring but reinforces the external magnetic field at the periphery (Figure 6.5).

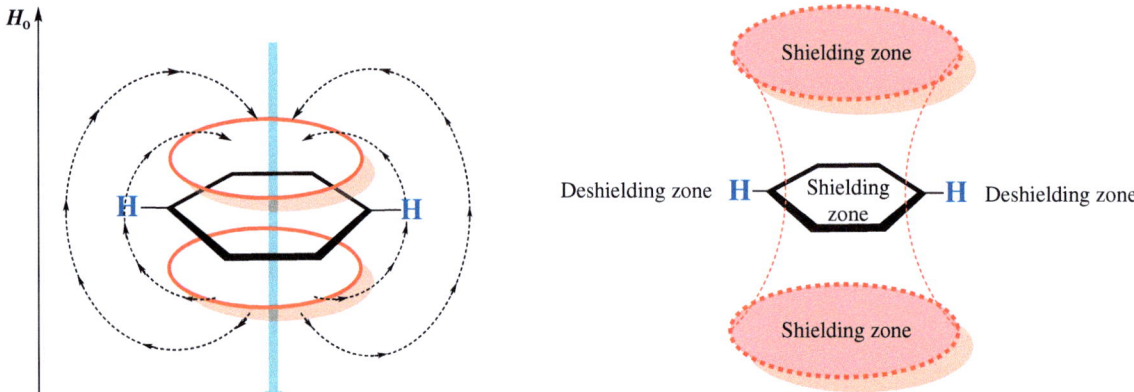

Figure 6.5 The schematic presentation of an aromatic ring current, shielding zones, and deshielding zones.

The ring current generated by delocalization of π electrons is called the *diamagnetic ring current*. As a result, protons in the molecular plane and outside the ring attached to the aromatic ring are deshielded because of the influence of an increased magnetic field. The resonances of these protons are shifted to a lower field. Conversely, protons in the center of the ring and above and below the ring are under the influence of a decreased magnetic field and are shielded. Therefore, the resonances of these protons appear at a high field. John Anthony Pople (Nobel Prize in Chemistry, 1998), a British theoretical chemist (1925–2004), suggested that the ca. 2 ppm greater deshielding of the benzene protons resonating at 7.27 ppm relative to those of ethylene resonating at 5.28 ppm may be an indication of the ring current induced by an external magnetic field. The effects inside the rings are much larger than those on the outside. For example, in [14]annulene (for definition, see Section 6.3), it is noteworthy that the signal of the methyl protons (located in the middle of the ring) is found at −4.25 ppm. In the absence of the specific shielding effects of the ring, one would expect a signal for these protons at about 1.0 ppm. The outer protons (12 protons) of aromatic [18]annulene resonate at 9.28 ppm, whereas the inner protons (6 protons) resonate at −2.99 ppm because of the strong shielding.

On the other hand, the situation is reversed in the case of antiaromatic compounds, which generate a *paramagnetic ring current* upon placement in an external magnetic field that reinforces the external magnetic field at the center of the ring.

Li-NMR as a Criterion for Aromaticity Numerous attempts have been made to quantify aromaticity. The chemical shift of lithium ions (Li^+) in complexes of lithium with aromatic compounds can also be used as a criterion for aromaticity. Lithium ions tend to bond as a π coordinate complex to the face of aromatic compounds. The chemical shifts of ^7Li salts are generally near zero and show little variation among different compounds. However, the chemical shift of a Li^+ ion in a π coordinated complex will be affected by electron delocalization in aromatic and antiaromatic compounds. For example, the Li^+ resonance of lithium cyclopentadienide (LiCp) appears unusually at a high field, −8.60 ppm because of the shielding effect [9].

Hückel's rule predicts that the cyclobutadiene dianion will be aromatic. The high field resonance of the lithium ion (−5.07 ppm) in tetrakis(trimethylsilyl)cyclobutadiene dianion clearly supports the presence of a diamagnetic ring current resulting from the six π electron system. The four-membered ring is planar and almost square, as determined by X-ray analysis.

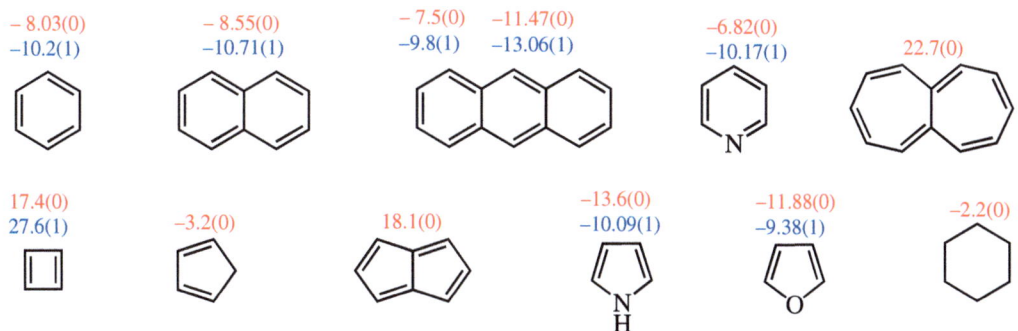

Reduction of 1,2,4,5-tetrakis(trimethylsilyl)benzene with lithium metal results in the formation of a dark brown solution of the benzene dianion. In the ^{1}H NMR (nuclear magnetic resonance) spectrum, benzene protons resonate at 5.11 ppm (2H) [10]. Of particular interest is ^{7}Li resonance at 10.7 ppm, being observed at the lowest magnetic field among the organolithium compounds reported so far. The unusual low-field shift is due to the strong deshielding of lithium atoms by the paratropic ring current arising from the eight π antiaromatic system. The X-ray analysis shows that the benzene dianion has a nearly planar structure.

Nucleus-independent Chemical Shift (NICS) The nucleus-independent chemical shift (NICS) method allows the evaluation of the aromaticity, antiaromaticity, and nonaromaticity of monocyclic as well as polycyclic systems. This concept was introduced by Paul von Ragué Schleyer, an American physical organic chemist (1930–2014) and coworkers in 1996 [11]. NICS is a computational method that calculates the absolute magnetic shielding at the center of a cyclic system using quantum mechanics programs. NICS is typically computed at ring centers. Negative NICS values in the center of the ring indicate the presence of diamagnetic ring current (aromaticity), whereas positive values indicate a paramagnetic ring current (antiaromaticity). At the beginning, the NICS(0) values were calculated at the centers of the rings, but later, it was realized that the σ frame affects these values. Then, it was suggested to calculate NICS values 1.0 Å above the plane of the molecule, which are denoted as NICS(1). The NICS values of selected compounds are given below [12].

6.1.5 Homoaromaticity

In 1959, Winstein [13] introduced the term "*homoaromatic*" to describe some specific compounds that display aromaticity in which conjugation of π electrons is interrupted by one or more sp^3-hybridized carbon atoms. Although this sp^3-hybridized carbon atom disrupts the continuous overlap, it is still possible for a significant interaction between the p orbitals to occur, provided that the orientation of the p orbitals allows a significant overlap. The homoaromatic compounds are known to exist as cationic and anionic species. Some studies support the existence of neutral homoaromatic compounds, although these are less common. Homoaromaticity [14, 15] requires electron delocalization in a cyclic system and magnetic properties (i.e. magnetic field-induced ring current) of aromatic compounds. Positive (+) and negative (−) charges should delocalize throughout the cyclic system. Moreover, there should be a large degree of bond equalization and a stabilizing resonance energy. Of course, the resulting stabilization will be reduced because of the poorer overlap of the orbitals.

The tropylium cation is an aromatic compound with six π electrons and a planar structure. The homotropylium ion can be derived by the insertion of a CH$_2$ group into the seven-membered ring of the tropylium ion. However, the homotropylium ion cannot have a planar structure. It adopts a boat-shaped conformation with the "seven-membered" ring (C1–C7) not greatly deviating from planarity.

Tropylium ion   Homotropylium ion

The homotropylium cation $(C_8H_9)^+$ (for ionic aromatic compounds, see Section 6.2) is one of the best studied example for homoaromatic compounds. The parent compound was first synthesized by Petit and coworkers in 1962 by reacting COT with strong acids [15, 16]. Evidence for homoaromaticity came from observations of an unusual chemical shift of the methylene protons in the ^{1}H NMR spectrum. There is a large chemical shift difference between the methylene proton resonances ($\Delta\delta = 5.86$ ppm). The proton located over the ring resonates at -0.73 ppm, clearly indicating the existence of a strong diamagnetic ring current [17]. The ring protons resonate in a range of 6.4–8.5 ppm. An X-ray crystal structure analysis of a derivative yielded a homoaromatic bond length (C1–C7) of 1.775 Å. The NICS value -11.1 ppm calculated for the homotropylium cation supports the aromaticity.

The ^{1}H NMR spectra of metal carbonyl complexes from the homotropylium ion were particularly important, showing the existence of homoaromaticity [18]. The molybdenium complex with the six π electrons was determined to be homoaromatic because of the large chemical shift difference between the resonances of methylene protons. However, a localized structure was proposed for the iron complex demanding only four electrons.

$Mo(CO)_3$ $Fe(CO)_3$

A wide range of homoaromatic ions are shown below. The first three carbocations are all based on aromatic two-electron cyclopropenium ions. In the first case, one methylene group is inserted in the cyclopropenium ion. In the case of the next two ions, two and three methylene groups are inserted.

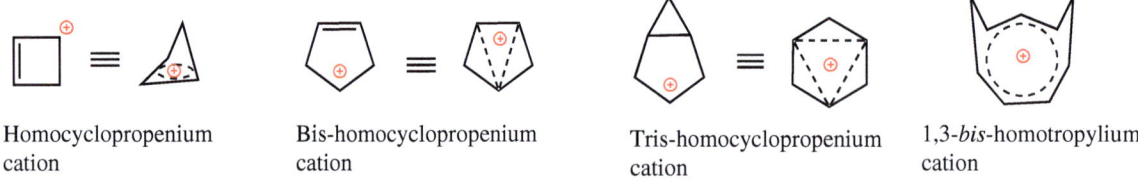

Homocyclopropenium cation   Bis-homocyclopropenium cation   Tris-homocyclopropenium cation   1,3-*bis*-homotropylium cation

In contrast to the homoaromatic cations, efforts to show the presence of an homoaromatic anion failed. The bicyclic anion shown below has been the best studied of the potential homoaromatic anions, and interest in and controversy over this system continue today. On the basis of the high field shift of the methylene proton resonances, it was suggested that there is a diamagnetic ring current in the anion, which would support the homoaromaticity [19]. Jiao and coworker [20] calculated the magnetic properties of this anion and found a large diamagnetic susceptibility exaltation and stability. They concluded that this compound is bishomoaromatic. Contrary to this result, Werstiuk and coworker [21] found that there is no evidence for a ring current. The NMR chemical resonances calculated by Werstiuk and Ma were in good agreement with the experimental values. They argued against homoaromaticity in this anion.

6.1.6 Möbius Aromaticity

Hückel's rule is applicable to planar or nearly planar cyclic systems without any apparent twist of the p frameworks. However, in 1964, Heilbronner conceived *Möbius aromaticity* in which 4n electron fully conjugated cyclic molecules with an odd number of orbital phase inversions will be stable if they possess a closed-shell structure [22–24]. To introduce a node in the orbitals, a series of p orbitals can be twisted by contorting the nuclear framework (Figure 6.6). Furthermore, Heilbronner also predicted that larger annulenes should tolerate a twist without any apparent strain. In normal Hückel annulenes, all π electrons populate the bonding orbitals. In the Möbius system, the electrons also populate the bonding orbitals. However, there is at least one sign inversion.

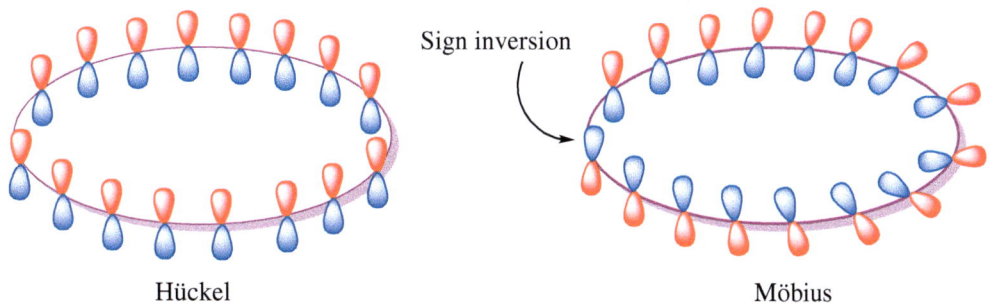

Figure 6.6 The orbitals of a Hückel and Möbius annulene.

Let us try to examine Möbius aromaticity with COT. First, we assume that COT has a planar structure (hypothetical COT). We can roughly estimate the MO energy levels by placing one point of the ring at the bottom and cutting the ring exactly in half as shown in Figure 6.7. Now, the orbitals at the bottom are the bonding orbitals, those in the middle of the ring are nonbonding orbitals, and those at the top are antibonding orbitals. Six π electrons occupy the bonding orbitals, which are lower in energy compared to the p atomic orbitals. However, the electrons in the nonbonding orbitals are not paired. There

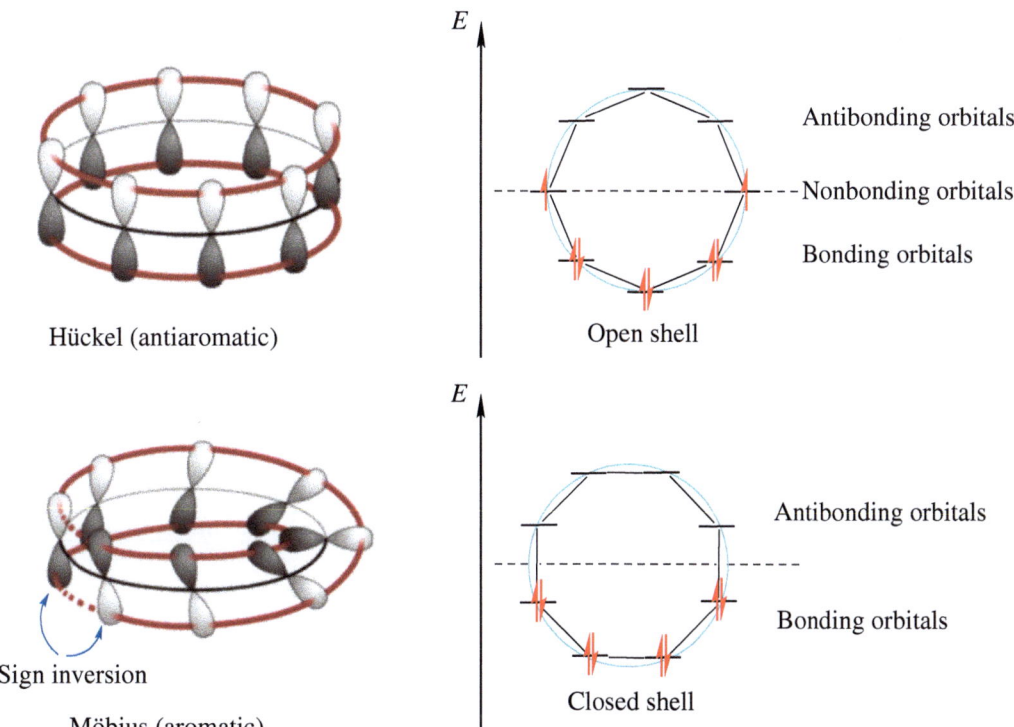

Figure 6.7 Population of orbitals in the case of hypothetical cyclooctatetraene and Möbius cyclooctatetraene. Representing the sign inversion. Source: Herges [24]. Reproduced with permission of American Chemical Society.

is no closed shell and two unpaired electrons in each of two nonbonding orbitals make the system unstable and reactive. Therefore, COT does not adopt the planar structure (see Section 6.1.2.3).

To estimate the Möbius orbitals, we draw COT so that one bond touches the bottom and cuts the ring exactly in half as shown in Figure 6.7. Now, the orbitals at the bottom are the bonding orbitals and those at the top are the antibonding orbitals. The eight electrons of COT populate the bonding orbitals and generate a closed shell. In conclusion, cyclic conjugated systems with $4n$ π electrons are open shell (diradicals) and antiaromatic. On the other hand, in a Möbius annulene with $4n$ π electrons, the electrons also fill the bonding orbitals and they are aromatic. However, Möbius annulenes with $(4n + 2)$ π electrons are diradicals and antiaromatic and they exhibit the reverse characteristic.

In 1971, von Ragué Schleyer and coworkers reported that 9-chlorobicyclo[6.1.0]nona-2,4,6-triene labeled with deuterium at the 9-position undergoes solvolysis to form dihydroindenol with deuterium completely scrambled to all positions [25, 26]. They postulated the formation of the cyclononatetraenyl cation as the intermediate. The ^{1}H NMR spectrum of this cation showed that the ring protons resonate at 8.5 ppm. Calculations showed that this cation with 8 π-electrons is a Möbius aromatic rather than nonaromatic [27]. A large negative NICS value of −13.4 of $(CH)_9{}^+$ also supports the aromaticity. This value is more negative than that of benzene (−10.2).

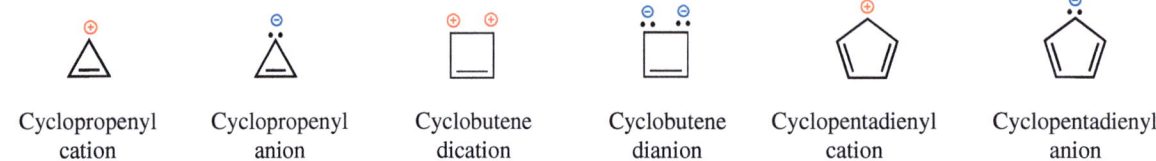

The cyclononatetraenyl cation is easily formed, and it does not have a planar structure (Figure 6.8). In the case of having a planar structure, it should be antiaromatic and unstable. The compound has a Möbius structure and is aromatic. According to calculations (B3LYP 6-311+G*), the energy difference between the Möbius structure and planar Hückel conformation is 26.3 kcal/mol. The aromatic structure is responsible for deuterium scrambling (Figure 6.8).

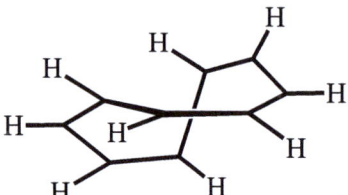

Figure 6.8 Side view of the cyclononatetraenyl cation.

6.2 Aromatic Ions

Is it possible for an ion to be aromatic? Yes, Hückel's rule also applies to charged molecules, if they meet certain criteria. They must have a planar structure and a cyclic delocalized π electron system containing $(4n + 2)$ electrons, where n is any whole number. According to Hückel's rule, compounds having 2 ($n = 0$), 6 ($n = 1$), 10 ($n = 2$), 14 ($n = 3$), 18 ($n = 4$), and so on, π electrons should be aromatic. A number of cations and anions with completely conjugated cyclic planar structures are shown below.

Cyclopropenyl cation · Cyclopropenyl anion · Cyclobutene dication · Cyclobutene dianion · Cyclopentadienyl cation · Cyclopentadienyl anion

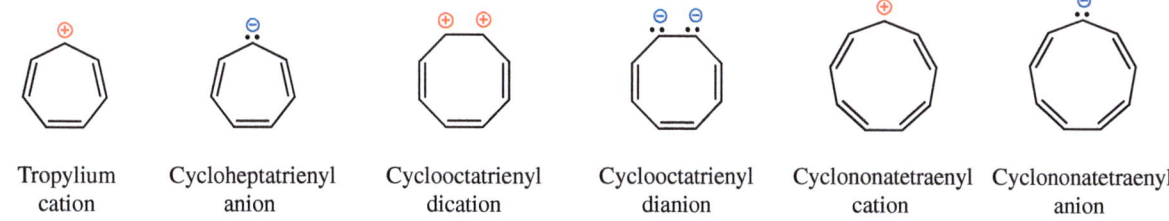

| Tropylium cation | Cycloheptatrienyl anion | Cyclooctatrienyl dication | Cyclooctatrienyl dianion | Cyclononatetraenyl cation | Cyclononatetraenyl anion |

Cyclopropene has a planar structure as three points define a plane. However, cyclopropene with two π electrons is not aromatic. It does not have an uninterrupted ring and so the electrons can delocalize. The hybridization of one of the ring atoms is sp^3. For delocalization of the electrons, the next carbon must have an empty p orbital. Only sp^2- and sp-hybridized carbons have p orbitals. Therefore, cyclopropene does not fulfill the Hückel criterion for aromaticity.

Breslow and coworker reacted 3-chlorocyclopropene by mixing it with antimony pentachloride, aluminum trichloride, or silver fluoroborate and obtained the cyclopropenyl cation [28]. The ^{1}H NMR spectrum of the cyclopropenyl cation shows a sharp singlet at 11.1 ppm, clearly indicating the presence of a strong diamagnetic ring current and the aromaticity of this compound.

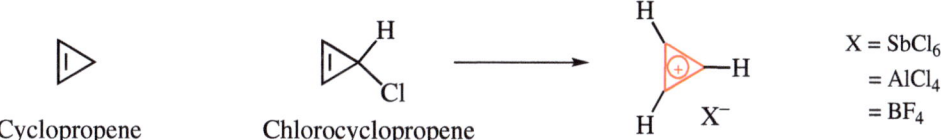

Cyclopropene　　　Chlorocyclopropene

X = SbCl$_6$
= AlCl$_4$
= BF$_4$

The electron configuration of the cyclopropenyl cation is shown in Figure 6.9. As one can see, the cyclopropenyl cation has two π electrons in the bonding orbital. In a fully delocalized form, this cation is aromatic.

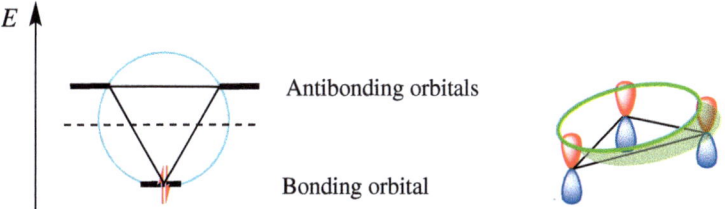

Figure 6.9 Orbitals of the cyclopropenyl cation.

The triphenyl derivative of the cyclopropenyl cation has been crystallized and its structure determined by X-ray crystallography showing that all C—C bonds are equal, 1.40 Å, similar to benzene (1.39 Å) [29].

The smallest neutral molecule considered potentially aromatic is cyclopropenone [30, 31]. If the cyclopropenone carbonyl group is polarized where both π electrons of the carbonyl group are shifted toward the oxygen atom, then the three-membered ring has two π electrons. The chemical properties of cyclopropenone are dominated by strong polarization of the carbonyl group. The carbonyl oxygen is strongly basic. Cyclopropenone can be approximately considered moderately aromatic. The high dipole moment of cyclopropenone ($D = 4.7$) is a strong evidence in favor of polarization of the carbonyl group. The chemical shifts of the ring protons in cyclopropenone appear at 9.08 ppm.

A general idea in organic chemistry over the past 50 years was that the cyclopropenyl anion is antiaromatic. However, a correlation between cycloalkene acidities and allylic bond angles reveals that energetically this is not the case, cyclopropenyl anion is nonaromatic [32]. Furthermore, the antiaromatic destabilization energy of the cyclopropenyl anion estimated by the following isodesmic reaction was found to be only 4.1 kcal/mol. This value is substantially lower than the corresponding value for cyclobutadiene (31.0 kcal/mol) [33]. This is also indicative of the nonaromaticity of the cyclopropenyl anion.

Cyclobutene is not aromatic because of the presence of two sp³-hybridized carbon atoms. For delocalization of the electrons, the next carbon atoms must have two p orbitals. In the case of having two sp²-hybridized carbon atoms with empty p orbitals, delocalization of two electrons is possible.

According to MO theory and Hückel's rule, the cyclobutenyl dication with two π electrons ($(4n+2)$, $n = 0$) should be aromatic because it is cyclic and has a planar structure. On the other hand, considerable charge–charge repulsion arising from two units of positive charge over only four carbon centers may destabilize the system. The tetraphenyl derivative of this cation was synthesized by vigorous stirring of 3,4-dibromotetraphenylcyclobutene in SO_2 at −60 °C in the presence of SbF_5 [34]. When trans-3,4-dibromotetramethylcyclobutene was treated with SbF_5, the corresponding dication was formed. The low field chemical shift (3.68 ppm) of the methyl protons and special stability are attributable to its aromaticity.

The cyclobutenyl dianion has a closed shell with six electrons. Actually, it should be stabilized by forming an aromatic system. However, the strong electron repulsion in a planar system may be the dominant destabilizing effect. The tetraphenylcyclobutenyl dianion was synthesized by the reaction of 1,2,3,4-tetraphenylcyclobutene with the strongest base $(CH_3)_3SiCH_2K$ in tetrahydrofuran (THF) [35]. The NMR spectral studies showed that the phenyl groups take over a considerable part of the negative charges. This result was in agreement with the theoretical calculations. The NMR data indicated that this anion should not be characterized as aromatic.

Sekiguchi and coworkers [36] synthesized the 1,2-diphenyl-3,4-bis(trimethylsilyl)-cyclobutenyl dianion and 1,3-diphenyl-2,4-bis(trimethylsilyl)cyclobutenyl dianion by the reaction of the corresponding cyclobutadiene cobalt complexes with lithium metal in THF. According to the X-ray structural analysis, the two Li^+ ions are located above and below the plane of the four-membered ring and are bonded to the four quaternary carbon atoms. The ^7Li NMR spectra indicate that the lithium atoms are shielded by a diamagnetic ring current, arising from the six π electron system.

Cyclopentadiene is not aromatic because the electron delocalization is interrupted by the presence of the sp³-hybridized ring carbon atom. By removal of a hydride anion (H⁻) from the sp³-hybridized ring carbon, the cyclopentadienyl cation will be formed and the hybridization will change to sp². Consequently, the cyclopentadienyl cation has a planar structure and possesses cyclic uninterrupted π electrons. Very little experimental information is available on the cyclopentadienyl cation. It has long been questioned whether the cyclopentadienyl cation is antiaromatic and recent discoveries have suggested that it may not be.

Cyclopentadiene	Cyclopentadienyl cation	Cyclopentadienone	Cyclopentadienyl anion
Nonaromatic	Antiaromatic?		Aromatic

At the five-position halosubstituted cyclopentadiene derivatives are attractive precursors for the synthesis of the cyclopentadienyl cation [37]. However, reaction of 5-iodocyclopentadiene with silver perchlorate failed. On the other hand, Saunders and coworkers [38] reacted 5-bromocyclopentadiene with SbF₅, and they claim to have generated unsubstituted cyclopentadienyl cation with triplet electronic configuration [39] determined by electron magnetic spin resonance spectroscopy (Figure 6.10).

Lambert and coworkers [40] synthesized the pentamethylcyclopentadienyl cation by hydride abstraction from commercially available pentamethylcyclopentadiene with the trityl cation (Ph₃C⁺). X-ray crystallographic analysis showed serious bond alternation in the molecule.

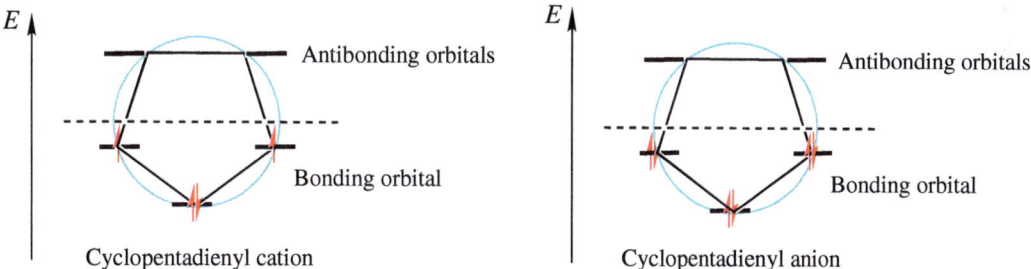

Figure 6.10 Orbitals of the cyclopentadienyl cation and cyclopentadienyl anion.

Cyclopentadienone [41] is an extremely unstable compound and spontaneously undergoes a [4 + 2] Diels–Alder reaction even at very low temperatures. The reactivity of cyclopentadienone has been attributed to its antiaromatic character. However, some substituted cyclopentanone derivatives are quite stable. NICS calculations show the antiaromatic character of the parent as well as the substituted compounds [42].

Recently, the heat of hydrogenation of cyclopentanone was determined to be 14.0 kcal/mol. On the basis of this value, it was suggested that cyclopentanone is a nonaromatic compound in contrast to previous views [43]. Furthermore, the authors claim that the NICS calculations and NMR data lead to different conclusions and these methods do not correlate with the energetic criterion for antiaromaticity.

Tetraphenylyclopentadienone is kinetically stable even at modestly high temperatures. The aryl groups provide enough steric hindrance to Diels–Alder cycloaddition and make these derivatives stable. However, at high temperatures in the presence of appropriate dienophiles, it does undergo cycloaddition reactions.

Cyclopentadiene has three carbon–carbon single bonds and two double bonds. Double bonds are shorter and more reactive than single bonds. The cyclopentadiene anion generated by a proton abstraction from cyclopentadiene contains six π electrons delocalized over a planar five-membered ring. Six π electrons populate the bonding orbitals as shown in Figure 6.10, and this electronic configuration forms a closed shell. In the anion, all C—C bonds become equal in strength. The stability of the anion is evidenced by the unusually low pK_a = 15 value of cyclopentadiene compared to the pK_a value of trimethylcyclopropene (pK_a = 60) [44]. The resonance energy has been estimated to be 24–27 kcal/mol [45].

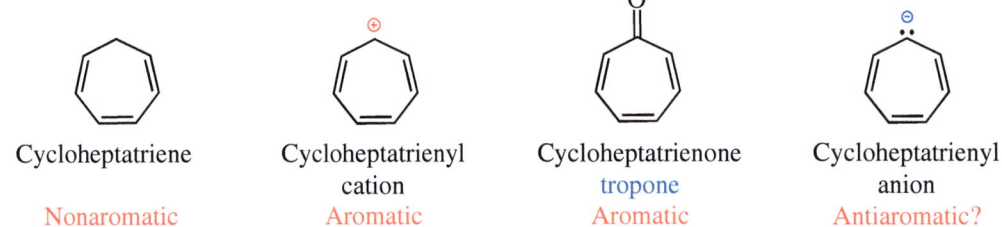

pK_a = 16 pK_a = 19

The reaction of cyclopentadiene with sodium hydride and bases such as amines, hydroxides, alkyllithium compounds, or metals (Na, K) furnishes the reactive cyclopentadienyl anion by loss of a proton. The chemical shift of the protons of the cyclopentadienyl anion appears at a high field, δ = 5.6 ppm, compared to benzene (δ = 7.27 ppm) because of the increased shielding caused by the negative charge [46].

1,3,5-Cycloheptatriene is not an aromatic compound because it does not meet the criterion of aromaticity. Cycloheptatriene contains one sp³-hybridized carbon atom that does not have a p orbital, so that it is not completely conjugated.

Cycloheptatriene	Cycloheptatrienyl cation	Cycloheptatrienone tropone	Cycloheptatrienyl anion
Nonaromatic	Aromatic	Aromatic	Antiaromatic?

Furthermore, an aromatic compound must have a planar structure. Cycloheptatriene exists in a boat conformation with the CH₂ group and C3–C4 carbon–carbon double bond bent out of the plane formed by the carbon atoms C1, C2, C5, and C6. Cycloheptatriene undergoes two dynamic processes, ring inversion and valence tautomerization, as shown below [47, 48].

The cycloheptatrienyl cation, also known as the tropylium ion, contains six π electrons delocalized over the planar seven-membered ring. The tropylium ion has been generated for the first time by Merling [49] in 1891 by the reaction of cycloheptatriene with bromine. The compound formed was insoluble in organic solvents, but it was soluble in water, in MeOH, or in EtOH. However, Merling could not realize the generation of an aromatic ion soluble in water. More than 60 years later, von Eggers Doering and coworker discovered that the compound formed was the tropylium ion formed by thermal elimination of hydrogen bromide from dibromotropilidene obtained by the addition of bromine to cycloheptatriene [50]. The tropylium ion can also be generated by hydride exchange reactions with the trityl cation rapidly and quantitatively only in solvents affect the dissociation of this salt.

All molecules are aromatic if all bonding MOs are fully occupied, while all nonbonding and antibonding orbitals are unoccupied. In the tropylium ion, all the three bonding orbitals are occupied by six π electrons, while none of the antibonding orbitals is occupied by an electron (Figure 6.11). Furthermore, an equalization (or close to it) of the bond length in the tropylium cation is an additional criterion for aromaticity. This is revealed by X-ray crystallographic analysis of the tropylium ion, showing that the bond lengths are 1.47 Å. Benzene protons resonate at 7.27 ppm, whereas the tropylium cation resonates at 9.17 ppm because of the decreased electron density. Remarkably, the NICS value of the tropylium ion (−7.6) does not deviate very much from that of benzene (−9.6).

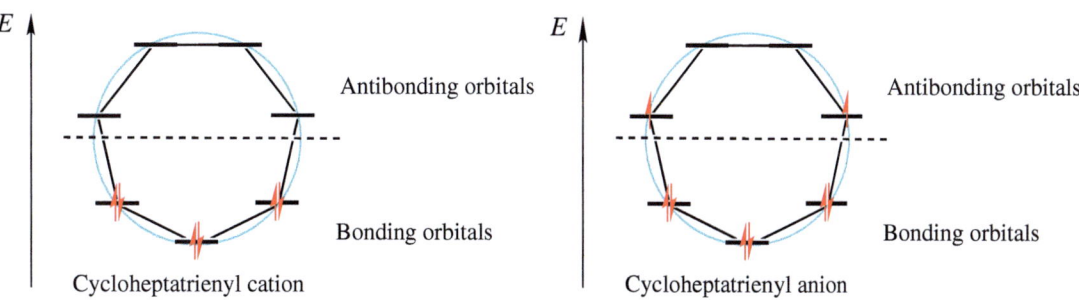

Figure 6.11 Orbitals of the cycloheptatrienyl cation and cycloheptatrienyl anion.

Tropone can be easily synthesized by selenium dioxide oxidation of cycloheptatriene. Dewar in 1945 proposed that tropone could have aromatic properties. The carbonyl group is polarized with a partial positive charge on the carbon atom and a partial negative charge on oxygen. Tropone possesses some aromaticity owing to the contribution of its electronically polarized resonance form. The dipole moment of tropone is 4.17 D, while that of cycloheptanone is 3.04 D, which may be regarded as evidence for aromaticity.

The tropone has three different protons and they resonate at 6.40, 6.55, and 6.93 ppm. The ^{1}H NMR average shifts indicate that the diamagnetic ring current is enhanced in tropone [51]. The X-ray analysis of tropone carried out at −60 °C shows pronounced bond alternation, but it is nevertheless approximately planar [52]. The C=C double bonds appear marginally longer and the single bonds slightly shorter than the usually accepted values. The NICS(1) value of −2.9 indicates little resonance stabilization in this system [53].

In contrast to the tropylium ion, the cycloheptatrienyl anion has eight π electrons [53]. Two of these electrons are unpaired and occupy a pair of double degenerate antibonding MOs (Figure 6.12). The cycloheptatrienyl anion was synthesized by the treatment of 7-methoxycycloheptatriene with K–Na alloy in THF at −20 °C. The NMR spectral studies of some monosubstituted derivatives showed that they have nonplanar structures, in contrast to the cation.

The pK_a values of cyclic system with interrupted conjugation are very important. For example, for cyclopentadiene itself, the pK_a is 16, while for cyclopropene and cycloheptatriene, they are 61 and 36, respectively [54]. These values clearly indicate that cyclopentadiene forms an aromatic ion by removal of a proton, whereas cyclopropene and cycloheptatriene do not form a stable system.

After the discovery of the benzene structure, chemists became interested in the synthesis of the lower vinylogue cyclobutadiene, as well as the higher vinylogue COT with the expectation that they would be as stable as benzene. COT was first synthesized by Willstätter in 1911 starting from the alkaloid, *pseudopelletierine*, in a multistep process [55]. The steps of the synthesis are given in detail in Section 3.4.2. Later, Reppe and coworkers described a one-step method for the synthesis of COT by tetramerization of acetylene in the presence of nickel cyanide and calcium carbide at 60° and under pressure [56].

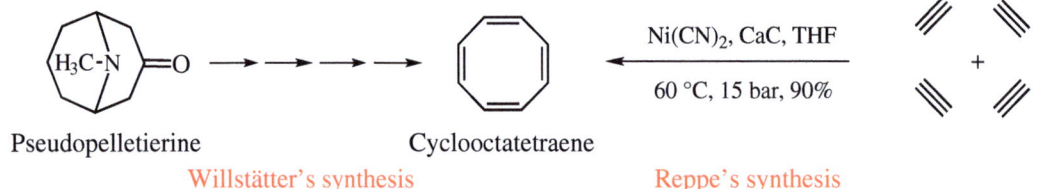

COT is an eight π electron system (4n, n = 2). In the case of a planar structure, this molecule will become antiaromatic. Eight π electrons do not form a stable electronic configuration (Figure 6.12). With an unstable (open-shell) electronic configuration, there is no energy advantage to a planar molecule. Furthermore, we know that the sp^2-hybridized carbon atoms have 120° bond angles.

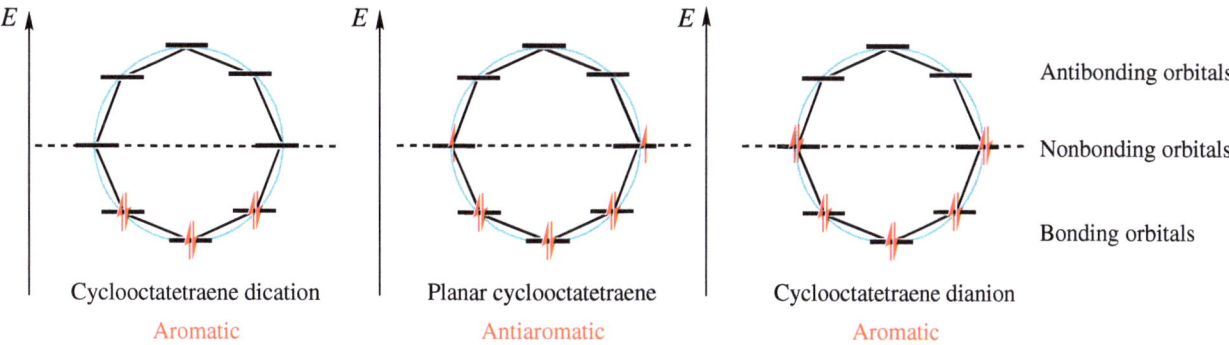

Figure 6.12 Orbitals of the cyclooctatrienyl dication, cyclooctatetraene, and cyclooctatetraene dianion.

In the case of a planar structure, the CCC bond angles will be 134°, which will introduce considerable ring strain into the molecule. To avoid this strain as well as antiaromaticity, the molecule escapes the planar structure and assumes a nonplanar shape, in which orbital overlap is greatly diminished (Figure 6.13).

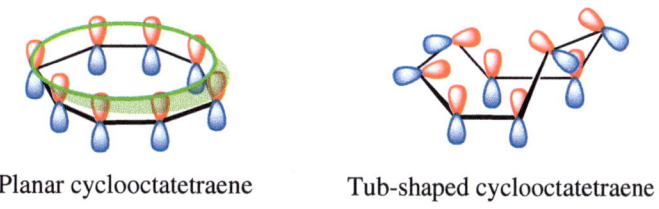

Figure 6.13 Planar and tub-shaped cyclooctatetraene.

Moreover, unlike benzene, COT was found to be highly reactive to electrophiles just like other alkenes. As determined by X-ray structural analysis [57], COT is nonplanar and adopts a tub-shaped conformation to which Hückel's rule cannot be applied. The C—C bond lengths alternate as expected. The C—C bonds are 1.46 Å, whereas the C=C bonds are 1.32 Å in length. As the dihedral angles between the vicinal bonds are 56° (determined in the crystal structure), the π bonds cannot conjugate and the p orbitals cannot overlap properly [58]. COT resonates at 5.78 ppm, indicating that it is in the typical alkene region. The NICS(0) value of tub-shaped COT was found to be +1.2 ppm, while the NICS(0) value of planar COT with a triplet configuration was reported to be 30.1 ppm. All these data clearly indicate that COT is nonaromatic.

COT undergoes three fundamental structural changes [59]. The first of these processes is termed ring inversion and the second one is the double-bond shift. Both of these processes take place presumably via planar transition state. The ring inversion barrier has been measured by various methods and was found to be 10–13 kcal/mol while the barrier for double-bond shift is 2–4 kcal/mol higher than the inversion barrier. The third process is valence isomerization to [4.2.0]bicyclooctatriene, which does not require a planar COT for transition state.

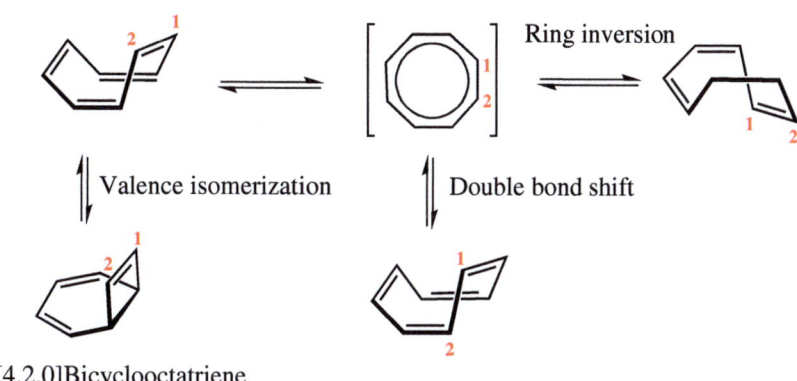

According to Hückel's MO theory, compounds with $(4n + 2)$ π electrons and a planar structure should possess relative electronic stability as a consequence of their having closed-shell MO configurations with substantial electron delocalization. As one can see from Figure 6.12, the COT dication has six π electrons with a closed-shell configuration. Therefore, it is expected that this dication will be aromatic.

Olah and coworkers [60, 61] reported the first successful generation of a COT dication by the reaction of 1,4-dimethylcyclooctatetraene with SbF_5 in SO_2ClF (an effective two-electron oxidizing agent) at $-78\,°C$ to result in the formation of the corresponding 1,4-dimethylcyclooctatetraene dication.

Cyclooctatetraene dication

The ^1H NMR spectrum of the dication consists of three resonances appearing at 9.96, 9.90, and 3.98 of relative areas 2 : 1 : 3, respectively. The low field shift of the protons is due to the presence of two (+) charges on the molecule as well as the diamagnetic ring current. We should be able to distinguish between these two effects in some way. It has been well established that the chemical shifts in a series of aromatic molecules are a linear function of the π electron densities of the carbon atoms [62]. The effect of a (+) charge localized on a single carbon atom on the chemical shift is approximately 10 ppm. Since there are two positive (+) charges loaded on the molecule, the total effect on the chemical shift will be around 20 ppm. These charges are distributed over eight carbon atoms. Thus, we can easily estimate that the effect of these charges on a single carbon atom will be (20/8) 2.5 ppm. The observed NMR chemical shift difference is approximately 4.2 ppm. This means that the proton resonances are shifted about 1.7 ppm to a lower field. The observed chemical shift difference (4.2 ppm) clearly occurs because of the location of the ring protons in the deshielding region of the diamagnetic ring current generated from the aromatic dication.

According to Hückel's theory, the COT dianion, which has 10 π electrons, should be aromatic if it has a planar structure. However, we know that COT has tub-shaped geometry. Katz [63] reacted COT with potassium metal (a good electron donor, $K \rightarrow K^+ + e^-$) in ether or liquid ammonia and obtained an aromatic dianion.

Cyclooctatetraene dianion

Using the Frost–Musulin diagrams based on Hückel's MO theory, it is easy to demonstrate that all π electrons of the COT dianion are located in doubly occupied bonding and nonbonding MOs, which makes these systems particularly stable. Therefore, the COT dianion has a singlet ground state (Figure 6.12).

The structure of the potassium salt of a 1,3,5,7-tetramethylcyclooctatetraene dianion has been determined by X-ray analysis [64]. The eight-membered ring is planar, with C—C bond lengths of 1.407 Å without significant bond alternation. Spectroscopic and structural studies show that the COT dianion is stabilized by delocalization of the π electrons and it is aromatic.

Recent calculations show that the COT dianion has a substantial aromatic stabilization energy of 25 kcal/mol approaching that of benzene (36 kcal/mol). Despite being aromatic, COT highly unstable thermodynamically because of the strong Coulomb repulsion between the two excess electrons [65]. The protons of the COT dianion resonate at 6.4 ppm. This value is higher than the resonance frequency of the other aromatic protons. The existence of two negative (−) charges contributes to shielding of the ring protons. However, the formation of an aromatic system shifts the proton resonances to lower field. Here, two effects influence the chemical shift of the protons operating at the opposite direction. The calculated NICS(0) = −13.9 value also denotes aromaticity, (NICS(0) value for benzene is −9.7) [11]. As the COT dianion has a planar structure, the energy of the system is enhanced because of the introduced angle strain. However, the putative aromatic stabilization is large enough to overcome the angle strain in such system.

6.3 Annulenes

Monocyclic, completely conjugated hydrocarbons with $(4n + 2)$ and/or $4n$ π electrons are generally named as *annulenes*. Annulenes can be aromatic (benzene), antiaromatic (cyclobutadiene), and nonaromatic (such as COT). The geometric structure of annulenes may be planar or not planar. When the annulenes have planar structures, they can be aromatic or antiaromatic according to the number of electrons. All of the nonplanar annulenes are nonaromatic compounds, regardless of the number of electrons. The discovery of the structure of benzene and Hückel's MO theory prompted chemists to prepare molecules designed to provide experimental evidence for their predictions.

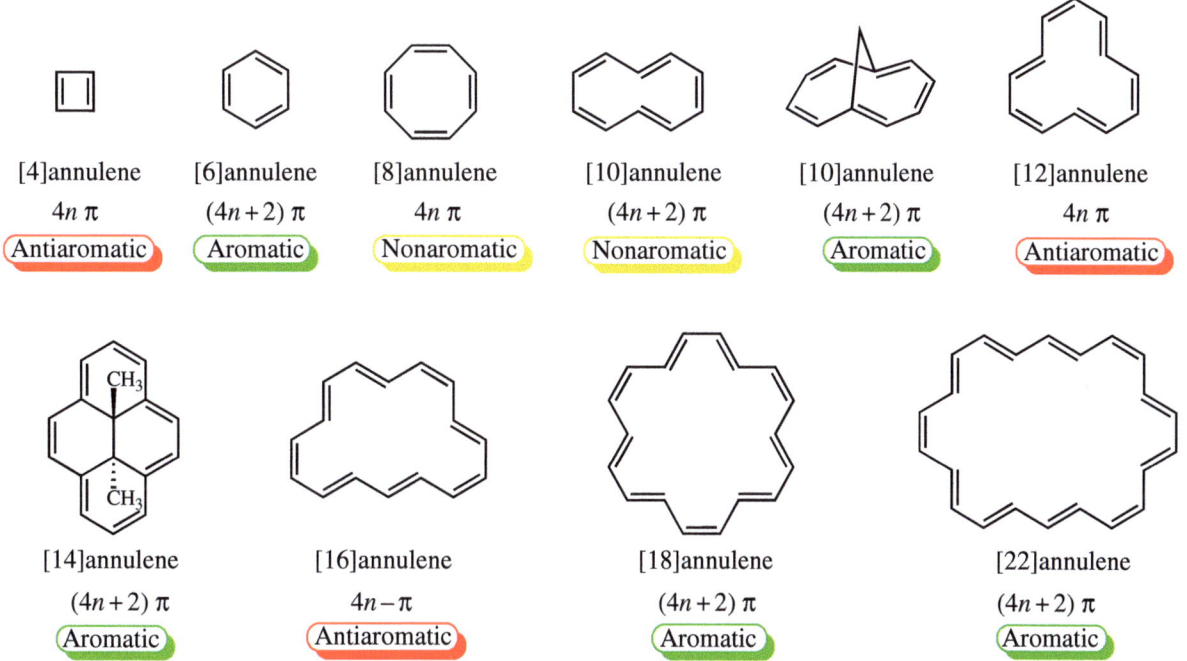

The criteria for aromaticity that we discussed can be applied to higher annulenes as well. However, achieving planarity is a challenge when it comes to many larger rings because of the potential steric clashes or angle strains. Only cyclobutadiene and benzene are fully planar. In this section, we will discuss cyclobutadiene as the smallest neutral annulene as well as some higher annulenes.

6.3.1 Cyclobutadiene

More than 100 years ago, chemists strived to synthesize cyclobutadiene. They expected that cyclobutadiene would exhibit pronounced stability. Unfortunately, all attempts were unsuccessful. The first successful synthesis of cyclobutadiene was reported by Pettit and coworkers in 1965 [66]. They reacted *cis*-3,4-dichlorocyclobutene with excess $Fe_2(CO)_9$ and obtained cyclobutadiene iron tricarbonyl complex in 40% yield. Cyclobutadiene was released from the stable iron complex by reaction with ceric ammonium nitrate. In the absence of trapping agents, cyclobutadiene acts as diene and dienophile and undergoes a [4 + 2] cycloaddition reaction to form the dimer. This dimerization is an extremely fast reaction. When the reaction was conducted in the presence of methyl acetylene carboxylate, a derivative of Dewar benzene was obtained [67]. This methodology has general utility for the synthesis of Dewar benzene derivatives. Cyclobutadiene can also be generated by photolysis of different precursors at low temperature in a matrix [68].

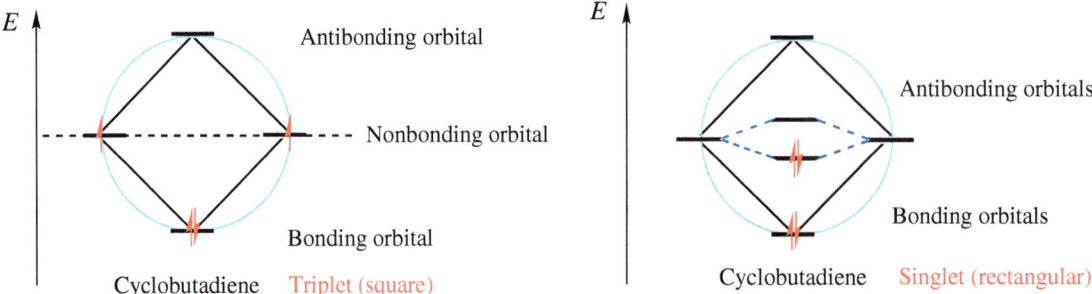

By using a Frost circle for cyclobutadiene, we can easily determine the MOs. There are one bonding orbital, two degenerate nonbonding orbitals, and one antibonding orbital (Figure 6.14). Two electrons populate the bonding orbital, while the remaining electrons are in the nonbonding orbitals and they are unpaired. This is a triplet electronic configuration and will not contribute to the stabilization of cyclobutadiene. It is expected that cyclobutadiene will be antiaromatic with this configuration.

Figure 6.14 Orbitals of cyclobutadiene (square and rectangular).

According to the Jahn–Teller theorem, the degeneracy of these orbitals can be destroyed by a slight perturbation of the molecular symmetry so that the two nonbonding group orbitals split into one with more bonding character and one with more antibonding character. Now, the electrons will move to the newly formed bonding orbital [69]. Therefore, cyclobutadiene is probably not square but rectangular. The analysis of the infrared (IR) spectrum of cyclobutadiene and deuterated analogs in a matrix confirmed the theoretical calculations that cyclobutadiene is a rectangular molecule. The NICS(1) value for cyclobutadiene is 17.4 and this value indicates a strong antiaromatic character for cyclobutadiene units [70].

A number of stable and highly substituted cyclobutadiene derivatives have been synthesized. For example, X-ray analyses have been reported for the stable cyclobutadiene derivatives (shown below) that showed pronounced bond alternations [71, 72]. The double bonds, 1.344 and 1.339 Å, are normal in length, but the 1.600 and 1.597 Å single bond lengths are much longer than usual. These data show that the ground states are singlet and the cyclobutadiene rings are not square but rectangular [73].

Tri-*t*-butylcyclobutadiene is of particular significance, representing the first demonstration of the chemical shift of a proton directly attached to the cyclobutadiene ring. It should provide important information about the ring current in a four-membered ring. The precursor, the diazo compound, was irradiated at −70 °C to give the desired compound. The olefinic proton resonates at 5.38 ppm. Using the C2 proton resonance (6.42 ppm) of cyclopentadiene as a reference, the ring proton is shifted 6.42 − 5.38 = 1.04 ppm upfield. This indicates that the compound is antiaromatic [74]. Recent theoretical calculations also confirm the antiaromaticity of cyclobutadiene [75].

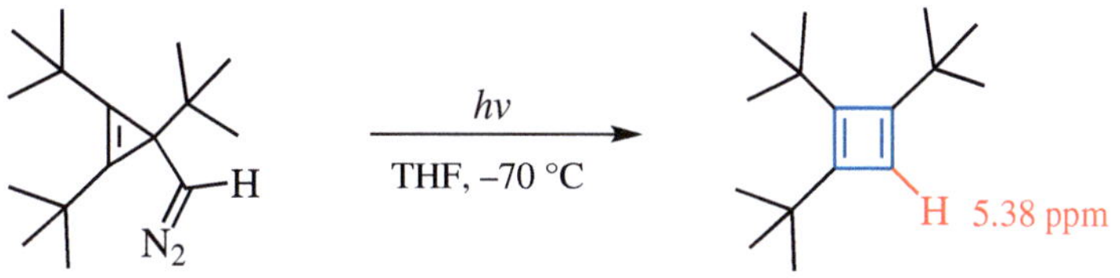

Higher homologs of cyclobutadiene, benzene and COT, were discussed in detail. Therefore, next, we will discuss [10]annulene and other higher annulenes.

6.3.2 [10]Annulene

According to Hückel's rule, [10]annulene will be aromatic if it has a conjugated and planar structure. There are three geometrical isomers of [10]annulene, cyclodecapentaene: all-*cis*, mono-*trans*, and *trans–cis–cis–trans–cis*, as shown below.

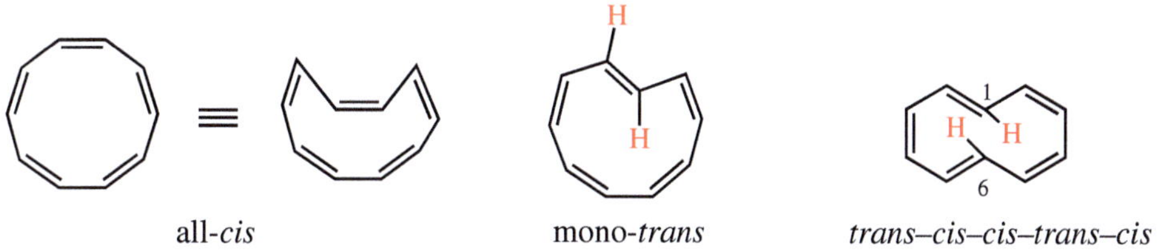

all-*cis*   mono-*trans*   *trans–cis–cis–trans–cis*

All-*cis* and mono-*trans* isomers were synthesized by irradiation of *cis*-decaline at −70 °C as crystalline solids. The ^{1}H NMR spectra show that all the proton resonances lie in the alkene region. The all-*cis* isomer resonates as a singlet at 5.66 ppm. It was concluded that those compounds are not aromatic, they are nonaromatic [76]. Both compounds are thermally unstable and they isomerize to stereoisomeric dihydronaphthalenes [77]. In all-*cis* planar structures of [10]annulene (regular decagon), the angles would have to be 144° between carbon atoms, which is too large to accommodate in a sp^2 framework. Therefore, the system tends toward a boat-like conformation to relieve the angle strain. In fact, the angle strain should not be a problem in forming an aromatic compound. In the case of the COT dianion, we have seen that the resonance energy can overcome the strain energy. In the case of the all-*cis* isomer, probably the resonance energy will not compensate for the strain energy. Therefore, this molecule tends to assume a nonplanar conformation and is not aromatic [78, 79].

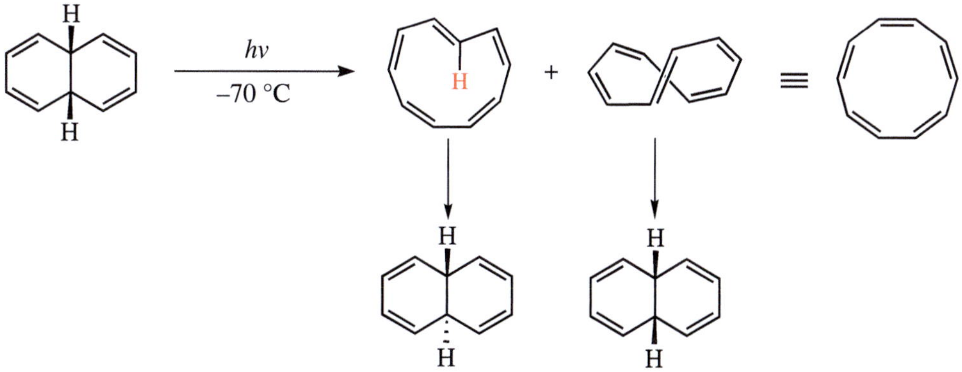

Unfortunately, the *trans–cis–cis–trans–cis* (naphthalene-like isomer) isomer could not be synthesized despite many attempts [80]. This isomer has a minimal bond angle strain, but it suffers a severe nonbonded transannular interaction (repulsion) between the two internal hydrogens at positions C1 and C6. These two internal hydrogen atoms were replaced by a transannular methylene bridge by a German chemist Emanuel Vogel (1927–2011) and coworker in 1964 in an elegant way. This should result in the formation of an almost if not completely planar cyclodecapentaene skeleton. Thus, 1,6-methano[10]annulene has been prepared starting from naphthalene [81, 82]. The Birch reduction of naphthalene to isotetralin, followed by dichlorocarbene addition results in the formation of a dichloroadduct. The chlorine atoms were removed by Na in liquid ammonia. Bromination of the tricyclohydrocarbon followed by hydrogen bromide elimination gave the [10]annulene in good yield. Oxidation of the tricyclohydrocarbon by 2,3-dichloro-5,6-dicyano-1,4-benzoquinone (DDQ) also gave the title compound. This compound was the first [10]annulene to test the validity of Hückel's rules.

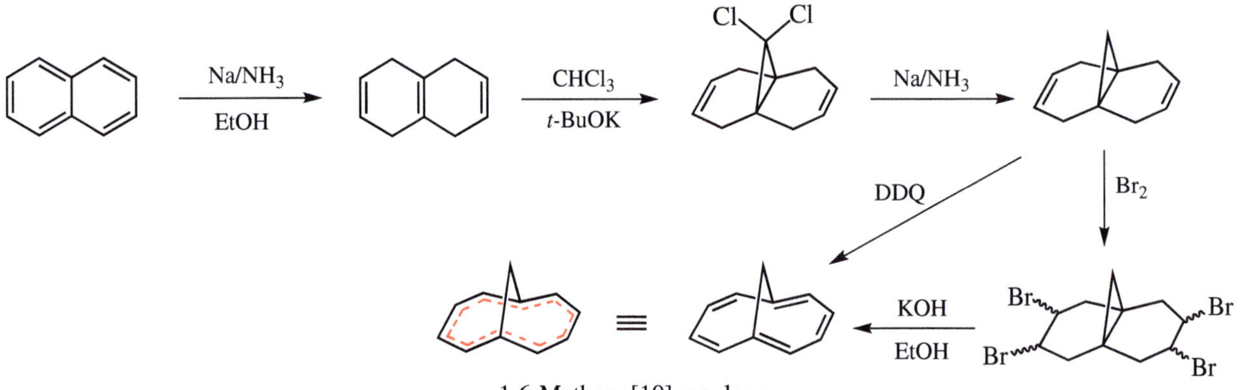

1,6-Methano[10]annulene

The perimeter protons of this bridged [10]annulene are deshielded and resonate at a field (7.27–6.95 ppm) lower than that of normal olefinic protons. The methylene protons are shielded, and they resonate at a high field, −0.5 ppm than normal. The methylenic protons are bonded to two vinyl groups. According to the substituents' constants, groups such as –CH_2– should resonate at approximately 3 ppm. However, the resonance signal of methylene protons is shifted around 3.5 ppm to a high field. This outcome can only be explained by the presence of a strong diamagnetic ring current. The 1,6-methano[10]annulene is an aromatic compound. The X-ray crystal structure shows that the perimeter is not completely planar and bond distances are in the range of 1.37–1.42 Å [83]. A small distortion from planarity does not prevent aromaticity [84].

It was examined whether [10]annulene is in equilibrium with its valence isomer or not. The measured distance between the carbon atoms C-1 and C-6 was reported to be 2.24 Å, which implies that there is effectively no bonding between these two atoms. The calculated NICS(1) value for [10]annulene is −17.7, which also supports the presence of strong aromaticity [85].

Valence isomerization

The results obtained by the reduction of COT with K suggested that the reduction of 1,6-methano[10]annulene might form an antiaromatic dianion from [10]annulene. Indeed, the reaction of [10]annulene with Li sand at −80 °C gave a dark brown solution of a species that was characterized as a dilithium salt. The chemical structural proof is based on the oxidation by air, which affords the hydrocarbon in 88% yield. The dianion exhibits remarkable thermal stability. The methylene proton resonances were shifted about 12.0 ppm downfield. On the other hand, the perimeter protons were shifted about 4–6 ppm to a higher field [86]. All these clearly indicates the existence of a strong paramagnetic ring current.

[10]Annulene, aromatic

[12]Annulene, antiaromatic

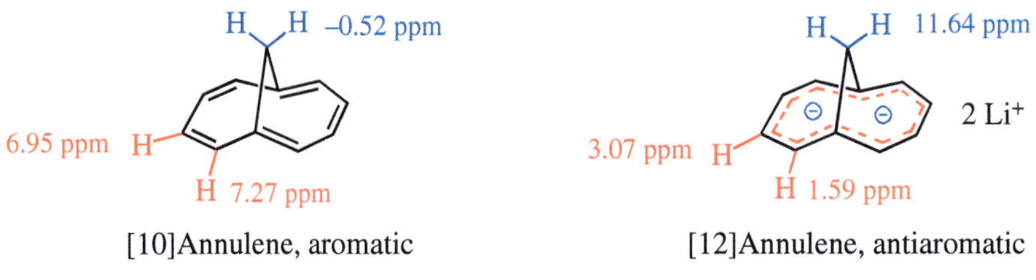

Homoazulene or 1,5-methano[10]annulene, an isomer of 1,6-methano[10]annulene, was synthesized by Scott and coworkers starting from dihydrocinnamic acid via a multistep process [87]. The ^{1}H NMR spectral studies reveal the presence of an induced diamagnetic ring current in this [10]annulene comparable to that of Vogel's 1,6-methano[10]annulene.

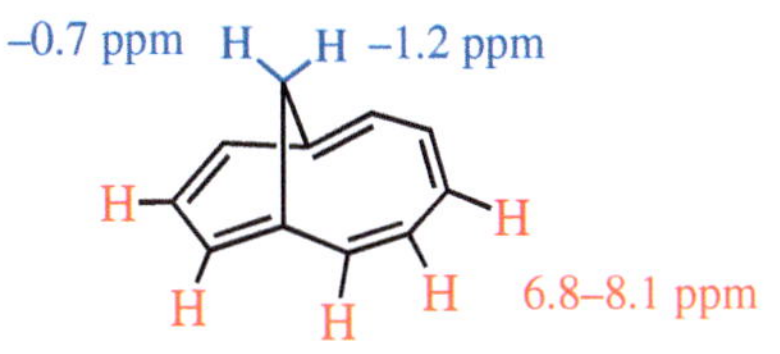
–0.7 ppm
H
H –1.2 ppm
H
H
H
H
H
6.8–8.1 ppm

1,5-Methano[10]annulene

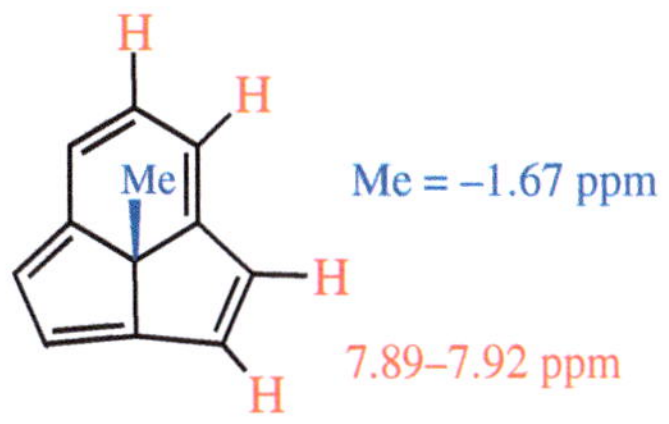
H
H
Me
Me = –1.67 ppm
H
H
7.89–7.92 ppm

11-Methyl-1,4,7-methano[10]annulene

11-Methyl-1,4,7-methano[10]annulene is another type of bridged [10]annulene that is a stable compound [88]. The ^{1}H NMR spectrum is consistent with a symmetrical structure and the existence of a diamagnetic ring current. The resonance frequency of the central methyl group is shifted to a high field and it appears at −1.67 ppm. The perimeter protons resonate at a lower field. Moreover, the compound is inert toward the powerful dienophiles at room temperature.

6.3.3 [12]Annulenes

1,7-Methano[12]annulene, a 12 π analog of 1,6-methano[10]annulene, has been synthesized starting from tricyclic diene. Treatment of the dibromocarbene adduct obtained by addition of dibromocarbene to the diene with silver acetate results in the formation of the ring-opening product. Removal of the acetate groups with $LiAlH_4$ followed by reduction of the bromine atoms with Na in liquid ammonia gives the diol. Elimination of the hydroxyl groups and consequent DDQ oxidation furnishes the [12]annulene [89].

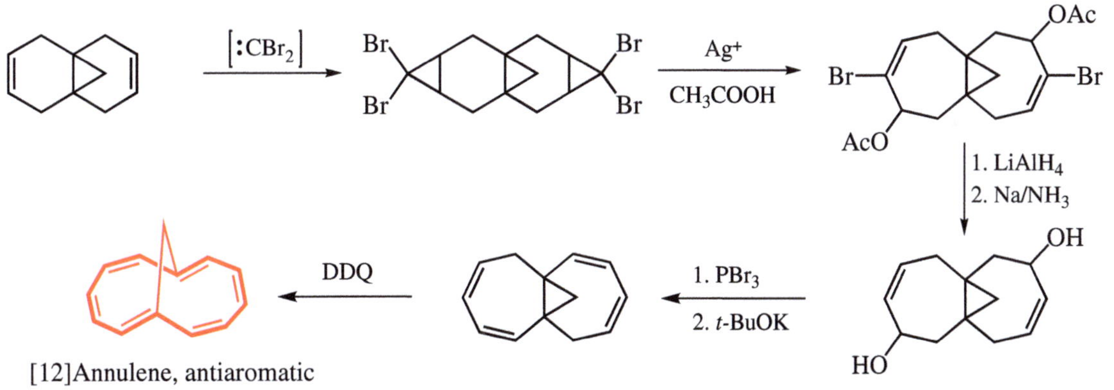
[:CBr2]
Br
Br
Br
Br
Ag+
CH3COOH
OAc
Br
Br
AcO
1. LiAlH4
2. Na/NH3
OH
HO
1. PBr3
2. t-BuOK
DDQ
[12]Annulene, antiaromatic

The perimeter protons resonate in a range of 5.2–5.5 ppm, and they are shifted to a high field compared to the olefinic proton resonances. Remarkably, the signals of the bridge protons are shifted to 6.06 ppm. Therefore, [12]annulene sustains a relatively strong paramagnetic ring current.

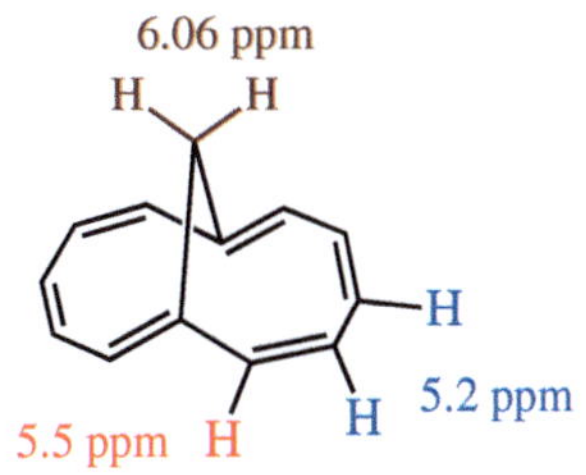
6.06 ppm
H
H
H
5.2 ppm
5.5 ppm H
H

1,7-Methano[12]annulene

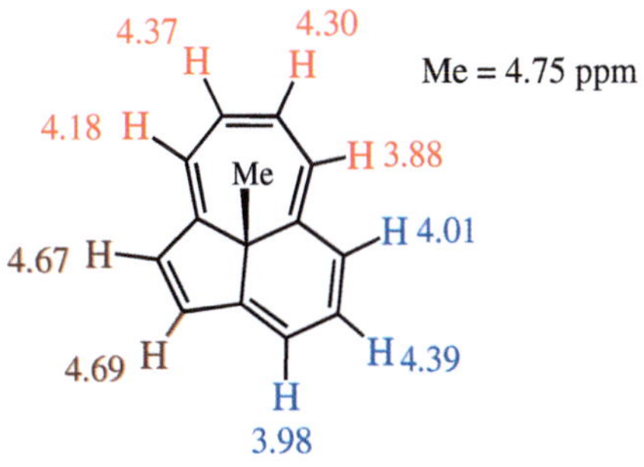
4.37
H
4.30
H
Me = 4.75 ppm
4.18 H
H 3.88
Me
H 4.01
4.67 H
4.69 H
H 4.39
H
3.98

Tricyclic[12]annulene

In 1986, the first tricyclic [12]annulene with a rigid geometry was synthesized by Hafner and coworker as a blue hydrocarbon [90]. The ring protons exhibit a strong upfield shift of about 2 ppm in the range of 3.88–4.69, whereas the methyl group experiences a substantial downfield shift of about 4 ppm and appears at 4.75 ppm as a singlet. These chemical shifts are consistent with a pronounced paramagnetic ring current in the molecule.

6.3.4 [14] and Higher Annulenes

According to Hückel's rule, [14]annulenes ($4n + 2$; $n = 3$) with planar structures should be aromatic. The first [14]annulene, cyclotetradecaheptaene, was synthesized by Sondheimer and coworker in 1960 [91]. The NMR spectrum of the [14]annulene shows the presence of a significant diamagnetic ring current. The outer protons resonate at 7.6 ppm, whereas the internal proton resonances are shifted to a very high field, −0.61 ppm. It has been shown that [14]annulene in solution exists as two species, **A** and **B**, in fast equilibrium at room temperature. Oth demonstrated that these two isomers are configurational isomers [92]. The activation barrier for this process is about 10 kcal/mol. The X-ray crystal structure shows that there are no alternating single or double bonds. The bond lengths around the ring range from 1.35 to 1.41 Å [93]. There is some distortion from planarity distributed throughout the molecule. The significant nonplanarity arises from the steric repulsion of the four internal hydrogen atoms located in the center of the molecule.

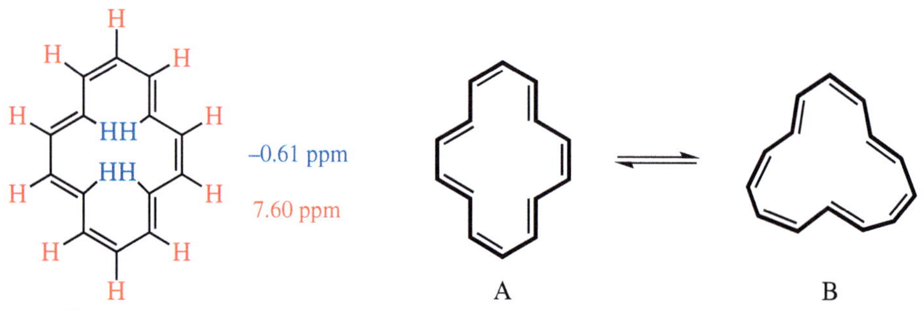

The steric repulsion associated with the internal hydrogen in [14]annulene **A** has been removed by replacement of the internal hydrogens with an ethylene bridge bearing two methyl groups. Boekelheide and coworker [94, 95] synthesized the *trans*-15,16-dimethyl-dihydropyrene and several derivatives. The dihyropyrene nucleus has an almost planar and completely delocalized 14 π electron perimeter. In addition, the internal methyl groups are ideally situated for NMR studies. The X-ray structural analysis shows that all of the aromatic C—C bonds are essentially of equal length (1.388–1.398 Å) [96]. The methyl groups, located above and below the plane of the ring, resonate at $\delta = -4.25$ ppm. In the absence of specific shielding of the ring current, the methyl protons would resonate at about $\delta = 1.0$ ppm. This dramatic shift of the methyl resonances of around 5.5 ppm supports strong aromaticity and the location of the methyl group in the most effective region of the shielding cone. The perimeter protons resonate at 7.98–8.67 ppm.

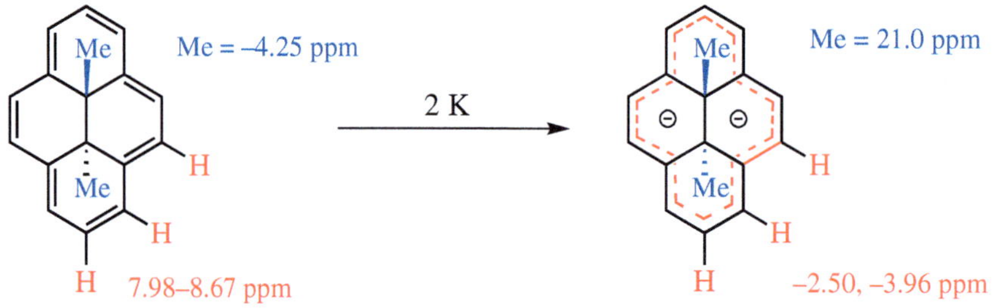

trans-15,16-Dimethyldihydropyrene
[14]annulene, aromatic, diamagnetic

[16]Annulene, antiaromatic, paramagnetic

As discussed before, a neutral aromatic hydrocarbon can be converted to the corresponding dianion. This corresponds to a change from a $(4n+2)$ system to a $4n$ system or, alternatively, a $4n$ system to $(4n+2)$ system. Thus, it is predicted that there may be a sharp reversal in the resonance of the ring protons as well as the resonances of the internal protons. The reduction of *trans*-15,16-dimethyldihydropyrene with potassium gives the corresponding dianion [97]. The signal for the methyl protons appears at $\delta = 21.0$ ppm compared to the resonance frequency of the neutral hydrocarbon (-4.25 ppm). A remarkable downfield shift of 25 ppm shows that there is a very strong paramagnetic ring current effect. On the other hand, again, consistent with the prediction, the ring protons are observed at -2.50 and -3.96, a high field shift of about 12 ppm.

After successful synthesis of 1,6-methano[10]annulene with a naphthalene perimeter, Vogel and coworkers proceeded to a new family of [14]annulenes with an anthracene perimeter that exist as syn- and anti-isomers. The ^{1}H NMR spectrum of syn-isomer is in almost complete agreement with an aromatic structure [98]. The ring protons resonate in a range of 7.4–7.9 ppm. The outer bridge protons resonate at -1.2 ppm, indicating that these protons are located in the cavity of the generated diamagnetic ring current. On the other hand, the inner bridge proton resonance appears at 0.9 ppm. The occurrence of this resonance at relatively low field is very striking. These protons are in close proximity because of the molecular geometry. A steric interaction between those protons will arise from the overlapping of van der Waals radii. The steric perturbation of the C—H bond involved causes the charge to drift to the carbon atom. As a result, the chemical shift of the protons moves to a lower field. Furthermore, the syn-isomer achieves a conjugated system with angles of up to 35° between the adjacent p orbitals [99]. The aromaticity is maintained despite this angle.

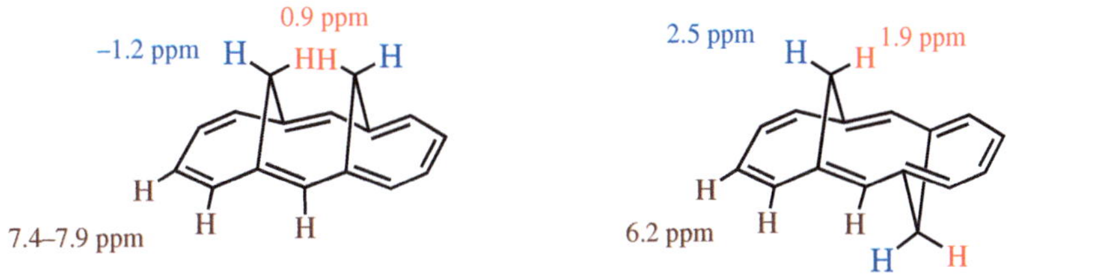

syn-[14]Annulene, aromatic

anti-[14]Annulene, nonaromatic

The properties of the syn-isomer differ fundamentally from those of the anti-isomer [100, 101]. As confirmed by X-ray structure analysis [102], the torsional angles between the p orbitals of neighboring carbon atoms reach values up to 70°. Therefore, the anti-isomer is no longer an aromatic compound. It behaves like a polyolefin.

Dioxo-*syn*-[14]annulene

Dioxo-*anti*-[14]annulene

In order to remove the steric interaction between the inner bridge protons in *syn*-[14]annulenes, the methylene bridges were replaced by carbonyl groups. The dioxo-*syn*-[14]annulene was synthesized in a multistep process as shown below [103]. The syn-isomer is very stable and resists flash vacuum pyrolysis at 500 °C. The X-ray structure shows that the perimeter is virtually planar, despite repulsion between the two adjacent carbonyl groups. The proton resonances appear between 8.53 and 7.81 ppm, clearly indicating the existence of strong aromaticity.

On the other hand, the dioxo-*anti*[14]annulene [104] with a strongly puckered ring is sensitive to light and it polymerizes readily at room temperature. The spectra show that the compound is a polyalkene and there is rapid valence tautomerism between the two Kekulé forms. The great difference between these syn- and anti-isomers provides strong evidence of the effects of molecular geometry on the electronic structure of cyclic conjugated molecules.

[18]Annulene belongs to a subgroup of [4n + 2]annulenes, where n = 4. [18]Annulene has 12 outside protons and 6 inside protons. Although all protons are bound to sp^2-hybridized carbon atoms, the inner protons are strongly shielded, and their resonances appear at −2.99 ppm, whereas the outer protons resonate at 9.30 ppm [105]. The unusual chemical shift difference of about 12.3 ppm between the inner and outer proton resonances clearly shows the presence of a strong diamagnetic ring current. In particular, the high-field resonance of the inner proton supports the presence of a strong shielding effect. On the other hand, a single signal at 5.45 ppm is observed at 120 °C. This is consistent with rapid exchange of the outer and inner protons at that temperature. Furthermore, the X-ray analysis [106] shows that the molecule is close to planarity. The bond lengths are in the range of 1.385–1.405 Å. These values are close to the bond length of benzene.

Annulenes up to and including [30]annulene have been synthesized. However, increasing the ring size decreases the π electron delocalization and consequently the ring current due to the conformational flexibility. The aromatic stabilization energy also decreases as the ring size increases. The energy gap between the highest occupied molecular orbital (HOMO) and lowest unoccupied molecular orbital (LUMO) is reduced.

6.4 Aromaticity in Fused Systems

Many conjugated systems can be built up from benzene. However, in this part, we will mainly discuss compounds built from nonbenzenoid compounds, such as pentalene, azulene, heptalene, and octalene.

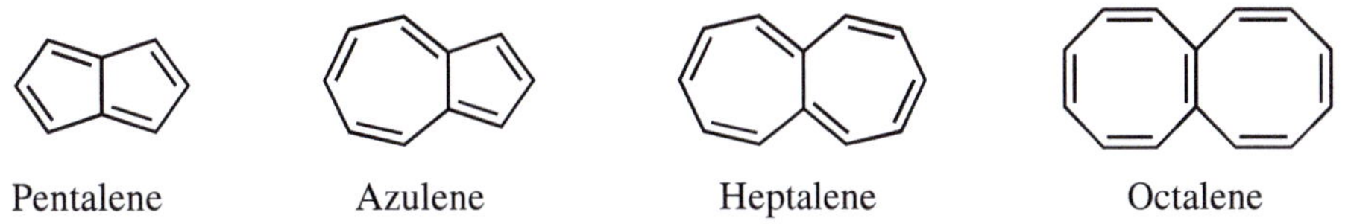

Pentalene   Azulene   Heptalene   Octalene

Pentalene is a polycyclic hydrocarbon composed of two fused cyclopentadiene rings. The question of the nature of bonding in pentalene has motivated numerous experimental and theoretical studies. *N*-Bromosuccunimid (NBS)-bromination of 1,5-dihydropentalene followed by base-induced HBr-elimination of the monobromide (or elimination of NR_3 from the corresponding amine salt) gives the parent compound, which rapidly undergoes a dimerization reaction because of the antiaromatic character of pentalene [107, 108]. Pentalene is generated for the first time in an argon matrix by the photocleavage of dimers [109]. The results show the lack of stabilization.

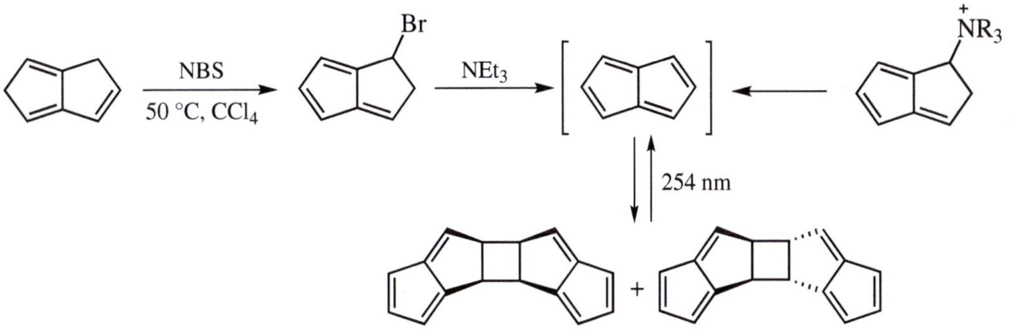

The first stable pentalene, 1,3,5-tri-*tert*-butylpentalene, was synthesized as a deep blue, thermally stable solid [110]. The ring proton resonances are shifted upfield, 4.72 and 5.07 ppm, exhibiting the presence of a paramagnetic ring current.

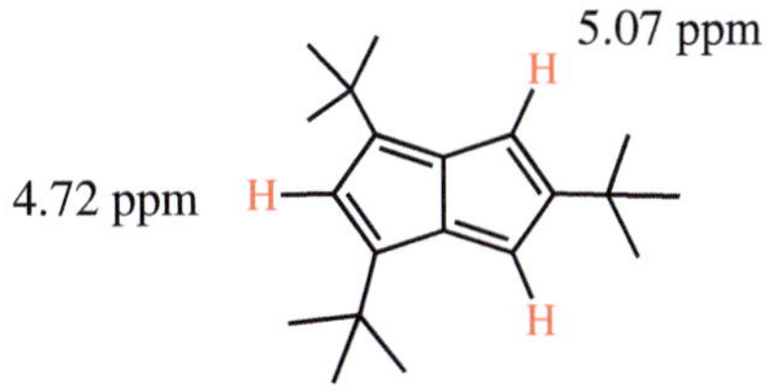

Azulene, one of the few nonbenzenoid aromatic compounds, is isomeric with naphthalene and appears to have significant aromatic stabilization. However, the stabilization energy is only about half that of naphthalene (33 vs. 61 kcal/mol) [111]. It acts like a combination of the cyclopentadienyl anion and cycloheptatrienyl cation. It has a large dipole moment (1.08 D) for a hydrocarbon. Although the structural isomer naphthalene is a colorless compound, azulene is a deep blue compound. The reason for the blue color as well as for the large dipole moment is explained by the intramolecular charge transfer derived from the polarized resonance structure as shown below.

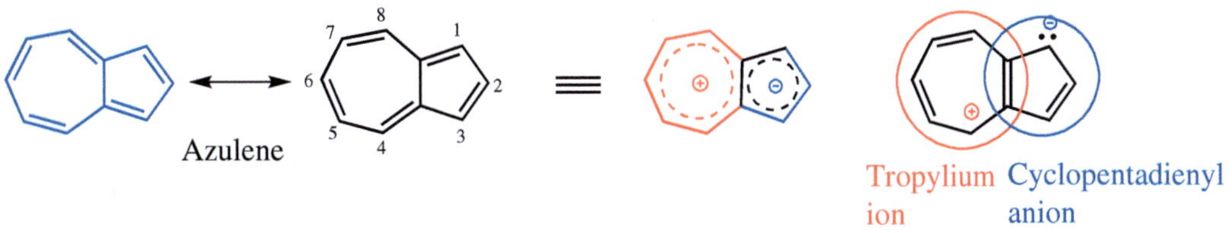

Azulene can be synthesized on a large scale using the Ziegler–Hafner method with pyridine and 2,4-dinitro-chlorobenzene as starting materials in which pyridinium salt is used as a starting material in a reaction with the cyclopentadienide ion

[112–114]. Recently, azulene derivatives were synthesized by the platinum(II)-catalyzed intramolecular cycloisomerization of 1-en-3-yne with *ortho*-disubstituted benzene derivatives in high yield [115]. Electrophilic substitution reactions take place at the 1- and 3-positions, and nucleophilic addition occurs at the 4-, 6-, and 8-positions.

Heptalene is a polycyclic hydrocarbon composed of two fused cycloheptatriene rings [116]. The 12 π analog heptalene synthesis starts from 1,6-methano[10]annulene. Reaction with diazomethane affords the ring-enlarged product. Pyrolysis of the ring-enlarged product at 400 °C (in vacuo) forms the dihydroheptalene mixture. Hydride abstraction with triphenylmethyl tetrafluoroborate followed by proton elimination gives heptalene which is an unstable, nonplanar, and nonaromatic compound [117]. The compound undergoes polymerization.

It is interesting to note that the conjugate acid of heptalene is very stable, arising from the stability of the resulting tropylium cation. Moreover, some derivatives of heptalene such as 3,8-dibromoheptalene [118] and dimethyl 3,8-heptalenedicarboxylate [119] are stable compounds at room temperature.

Heptalene can be reduced by lithium at −80 °C into the corresponding dianion [120]. The ^{1}H and ^{13}C NMR studies indicate that this dianion, contrary to the neutral molecule, is aromatic. The dianion of heptalene is most likely planar, while the resonance energy is probably large enough to counterbalance the strain energy and maintain the perimeter planar.

Annulenoannulenes, bicyclic ring systems, formally consist of two annulenes fused together. For example, naphthalene is a [6]annuleno[6]annulene made from two benzene rings. **Octalene** is a polycyclic hydrocarbon composed of two fused COT rings. Octalene, a [8]annuleno[8]annulene is a 14 π system. Octalene has been synthesized to address the question of whether octalene has an aromatic or an olefinic structure (for the synthesis of octalene, see Section 8.2.2.3) [121]. It was predicted that the resonance stabilization of octalene (50 kcal/mol) is not sufficient to overcome the strain energy on planarization of the two 8-membered rings [122]. Consequently, a nonplanar olefinic structure was proposed.

The ^{1}H NMR spectrum of octalene shows only signals in the range of olefinic protons at $\delta = 5.65$ and 6.30 ppm. However, the ^{13}C NMR spectrum exhibits seven signals, in agreement with structure **B**. In the case of structure **A**, one would expect only four signals.

Octalene can be readily reduced by lithium in THF to a dianion as well as to a tetraanion. The reduction can be interrupted at a stage of the dianion [123]. Two negative charges can be delocalized over the entire molecule, forming a 16 π electron system that would make the system antiaromatic. On the other hand, two negative charges can be localized in one ring, converting that ring into a 10 π electron aromatic system similar to the dianion of COT. However, the ^{13}C NMR spectrum of the dianion shows only four signals. In such a case, the four signals observed in the ^{13}C NMR spectrum can only be rationalized by assuming a fast charge transfer from one ring to another.

Further reduction converts the dianion into the tetraanion, which exhibits four signals in the ^{13}C NMR spectrum. Ongoing from neutral hydrocarbon to the dianion and the tetraanion, the carbon resonances undergo an upfield shift as expected. Of course, the ^{1}H NMR chemical shifts are also affected by the reduction. However, the surprisingly small upfield shift in the ^{1}H NMR resonances seen on going from the neutral hydrocarbon to the dianion $\Delta\delta = 0.41$ ppm and from the dianion to the tetraanion $\Delta\delta = 0.07$ ppm indeed shows that the charge-induced upfield shift is largely compensated for by the formation of a strong diamagnetic ring current with a delocalized 18 π-system.

Fulvenes are cyclic cross-conjugated molecules with an odd number of carbon atoms in the ring [124–127]. According to the number of carbon atoms in the ring, they are named triafulvene, pentafulvene, heptafulvene, etc.

The ^{1}H NMR spectrum of triafulvene displays a signal in the region of aromatic proton resonances, indicating a significant contribution of the resonance form. The molecule has a significant dipole moment (1.90 D). However, the ^{1}H NMR spectra of pentafulvene and heptafulvene show resonances in the range of olefinic protons. Furthermore, microwave spectroscopy shows that the dipole moment of pentafulvene is 0.44 D and of heptafulvene is 0.48 D.

Fulvalenes are cyclic cross-conjugated molecules with two conjugated rings connected by an *exo*-cyclic central double bond. They represent another interesting class of compounds to investigate potential aromaticity. Among the possible symmetrical structures, pentafulvalene [128, 129] and heptafulvalene [130] have been prepared, but they were found to exhibit a polyene character. Triafulvalene however has not been isolated.

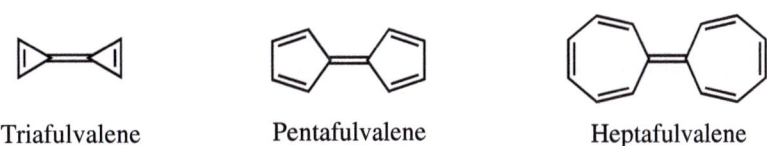

Fulvalenes with different rings are much more interesting in view of their aromatic properties. Calicene formed by combination of cyclopentadiene and cyclopropene is a very interesting compound because of the generation of aromatic units by polarization of the electrons. Unfortunately, no synthesis of this parent compound has been reported. Calculation shows that the compound has a large dipole moment of 5.6 D [131]. However, several substituted derivatives have been prepared. Hexaphenyltriapentafulvalene [132] has been synthesized. The measured dipole moment of 6.3 D demonstrates the strong contribution of a dipolar structure. This is the highest dipole moment ever reported for a hydrocarbon. This large dipole moment can be explained by charge transfer from the three-membered ring to the five-membered ring, affording a partial aromatic character to both rings. The resonance stabilization of two aromatic rings compensates for the charge separation energy.

Sesquifulvalene, or pentaheptafulvalene, was synthesized starting from acetoxytropylium fluoroborate. Reaction of sodium cyclopentadienide with the tropylium ion gives a mixture of coupling products whose thermolysis at 360 °C provides the title compound [133].

NMR spectral analysis shows that sesquifulvalene has mainly olefinic protons and it is not aromatic with little charge separation [134]. Actually, pentaheptafulvalene was expected to show considerable stabilization from charge separation that forms a six π tropylium ion and a six π cyclopentadienyl anion as shown above. Calculation shows that the energy needed for the seven-membered ring to attain the planarity of a cation exceeds that gained by charge separation and formation of independent aromatic rings.

6.5 Aromaticity in Heterocyclic Compounds

6.5.1 Heteroaromatic Compounds with Three-Membered Ring

Borirene is one of the smallest heterocycles. The methylene group of cyclopropene is replaced by a BH group and the compound is isoelectronic with the cyclopropenium cation. X-ray analysis of some derivatives as well as theoretical calculations show that this compound is aromatic [135, 136]. Comparison of the C=C double-bond length in cyclopropene (1.304 Å) with the C=C double bond in borirene shows lengthening of 0.08 Å. On the other hand, the ring C—B bonds are shortened by 0.10 Å compared with the reported C—B bond length in trivinylborane (1.558 Å). Therefore, the borirene ring can truly be considered an aromatic 2 π system [137].

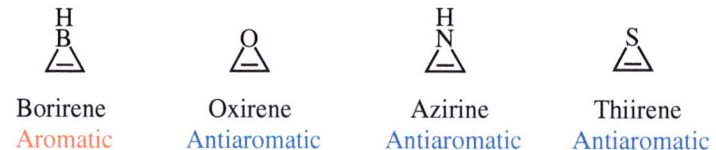

Borirene	Oxirene	Azirine	Thiirene
Aromatic	Antiaromatic	Antiaromatic	Antiaromatic

Oxirene [138] contains an unsaturated three-membered ring with two carbon and one oxygen atoms. It has never been observed and so the substance is mainly studied by quantum chemical computation. As the structure is extremely strained and proposed to be an antiaromatic four π electron system, it is expected to be very high energy and unstable according to the calculations. Experiments have indicated that substituted oxirenes (as intermediates or transition states) may be involved in carbonylcarbene rearrangements observed in the Wolff rearrangement [139]. Actually, the labeled carbon isotope that is originally in the carbonyl position will be in the –COOH group of acetic acid. For diazoacetaldehyde, photolysis causes only 8% of migration of labeled carbon atoms, which would correspond to the formation of 16% of the product through the oxirene [140]. This outcome clearly indicates that only the intermediate oxirene can be responsible for the distribution of the label atom.

Replacement of the methylene group of cyclopropene by an NH group gives 1*H*-**azirine**, while replacement by a sulfur atom results in **thiirene**. 1*H*-Azirine is inherently antiaromatic because of its four π electrons [141]. Because the 1*H*-azirine molecule is rather highly reactive and unstable, no experimental results concerning it are available. Delocalization of the lone pair electrons on the nitrogen atom is thought to destabilize the ring to an extent that precludes isolation. Calculations show that 1*H*-azirine is 33.5 kcal/mol higher in energy than the isomeric 2*H*-azirine [142].

Thiirene is a 4 π-electron system usually considered to be an antiaromatic species. Ultraviolet (UV) irradiation of 1,2,3-thiadiazole-4,5-dicarboxylate leads to the formation of bis(carboethoxy)thiirene as the only photoproduct, which is successfully trapped with furan [143].

6.5.2 Heteroaromatic Compounds with a Five-Membered Ring

Conjugated heterocyclic compounds with a five-membered ring are also subjected to Hückel analysis for aromaticity. The lone pairs on heteroatoms provide a potential source for the missing π electrons. However, only one lone pair from a heteroatom can contribute to the aromatic π electron system. Furan, pyrrole, and thiophene are isoelectronic with the cyclopentadienyl anion.

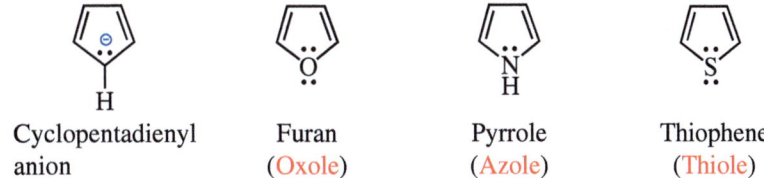

Cyclopentadienyl anion   Furan (Oxole)   Pyrrole (Azole)   Thiophene (Thiole)

The hybridization of heteroatoms bearing lone pairs of electrons is not sp³. For maximum overlapping, the heteroatoms are sp² hybridized. In the case of pyrrole, the lone pair of electrons populates the p orbital. The proton is bonded to the sp² hybrid orbital, and it is located in the plane of the molecule (Figure 6.15). In the case of furan and thiophene, the second lone electron pair is located in the remaining sp² hybrid orbital, again in the plane of the molecule. Because of geometrical reasons, there is no opportunity for overlapping. The lone pair of nitrogen electrons in pyrrole cannot be protonated. On the other hand, in furan, one pair localized on the oxygen atom can be protonated by strong acids.

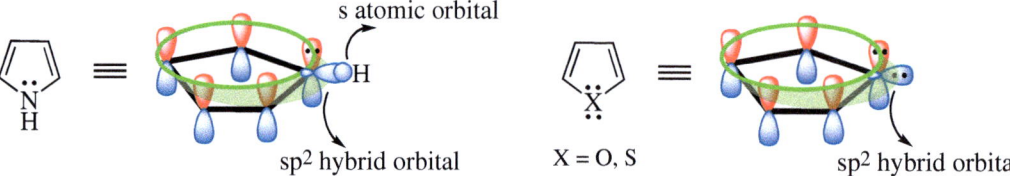

Figure 6.15 The p orbitals of furan, pyrrole, and thiophene.

Furan, thiophene, and pyrrole obey Hückel's $(4n + 2)$ rule and therefore they are aromatic because of the planarity and the uninterrupted cycle of p orbitals containing six electrons: four from the two double bonds and two from a lone pair of the heteroatom.

MOs are formed by combining the atomic orbitals on the atoms in the molecule. The lower the energy difference between the overlapping orbitals, the lower the energy level of the formed MO. In the case of benzene, all overlapping p orbitals are on carbon atoms. Therefore, all of them have the same energy and high stabilization energy is released by the formation of MOs (Figure 6.16). However, in the heteroaromatic system, one of the p orbitals is located on the heteroatom. As the heteroatoms are more electronegative than carbon, their p orbitals are lower in energy. Therefore, by forming the MOs, lower stabilization energy will be formed. Consequently, the resonance energies (thermodynamic stabilities) of these heteroaromatic systems will be less than that of benzene.

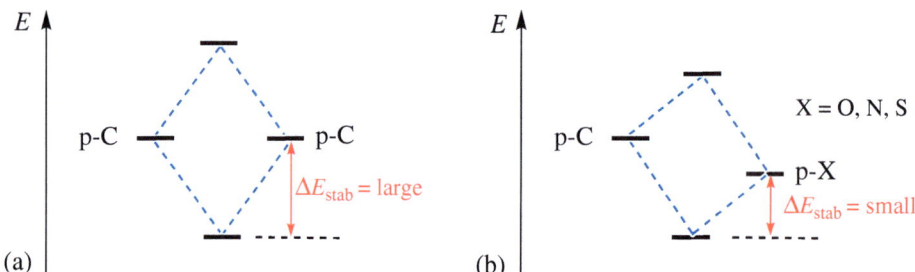

Figure 6.16 The schematic presentation of atomic and molecular orbitals: (a) p orbitals with equal energy and (b) p orbitals with different energies.

The resonance energies reflect the difference in the electronegativities of S (2.6), N (3.0), and O (3.4) relative to C (2.5). Thiophene, with the lower electronegative heteroatom, has the greatest resonance energy of these five-membered ring heterocycles. In the case of furan, the electrons are tightly bonded to oxygen (the most electronegative atom), which has the lowest resonance energy. Oxygen cannot release the electrons as well as sulfur for delocalization.

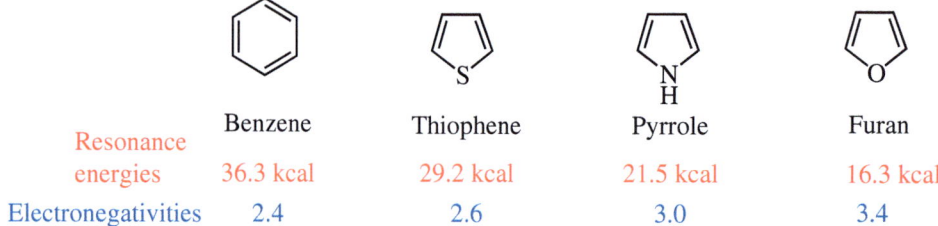

	Benzene	Thiophene	Pyrrole	Furan
Resonance energies	36.3 kcal	29.2 kcal	21.5 kcal	16.3 kcal
Electronegativities	2.4	2.6	3.0	3.4

Pyrrole is an extremely weak base because the lone pair of electrons is used for delocalization to generate aromaticity. When pyrrole is protonated, its aromaticity will be destroyed. Therefore, the conjugated acid of pyrrole is a very strong acid. On the other hand, pyrrolidine, a saturated derivative of pyrrole, is a base and can bond a proton. The dipole moment of pyrrole is 1.8 D, which is slightly greater than that of pyrrolidine (1.57 D). The dipole moments of pyrrole and pyrrolidine are in opposite directions. Pyrrole's nitrogen donates electrons to the ring for aromaticity (see the resonance structures of pyrrole). In pyrrolidine, the nitrogen atom withdraws electrons because of the electronegativity.

6.5 Aromaticity in Heterocyclic Compounds

Furan, pyrrole, and thiophene have six π electrons distributed over five atoms and so the carbon frameworks are electron-rich relative to benzene, in which six π electrons are distributed over six carbon atoms. Consequently, all react faster than benzene with electrophiles.

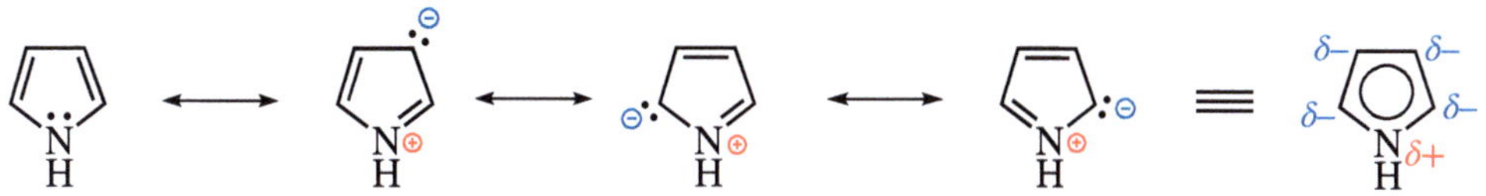

Resonance structures of pyrrole

Pyrrole can easily undergo an electrophilic substitution reaction and electrophiles attack preferentially at the C2 carbon atom because the intermediate obtained is more stable because of distribution of the positive (+) charge over three atoms. However, the intermediate formed by C3 carbon attack can distribute the positive (+) charge only over two atoms.

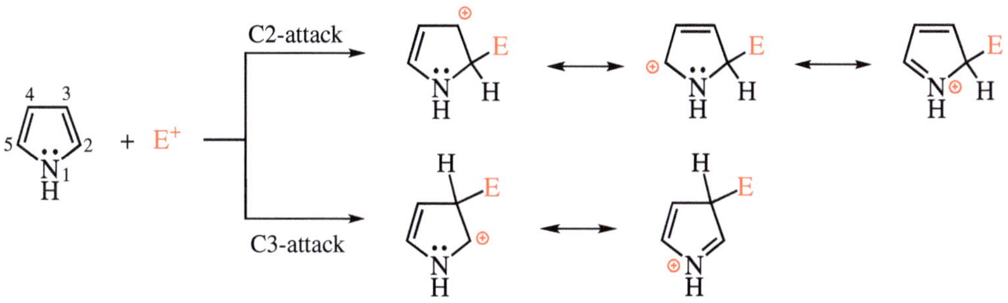

Some physical properties of pyrrole such as bond lengths, ^{1}H and ^{13}C NMR chemical shifts, and coupling constants are given below.

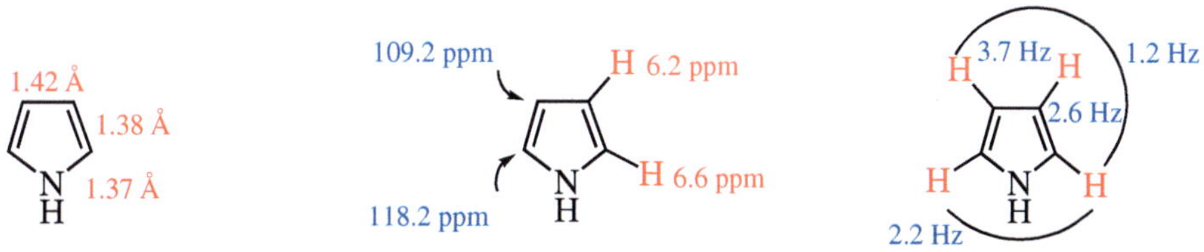

Bond length

^{1}H- and ^{13}C-NMR shifts

Coupling constants

As expected for aromatic compounds, pyrrole undergoes an electrophilic substitution reaction. The reaction of pyrrole with acetyl nitrate (formed by mixing nitric acid with acetyl anhydride) occurs smoothly at −10 °C and gives the substitution products in a ratio of 4 : 1, where the α-nitration product is the major product [144]. Pyrrole reacts with halogens so rigorously that only tetrahalo-pyrroles can be isolated. Unstable 2-chloropyrrole can be synthesized by halogenation with SOCl$_2$ [145].

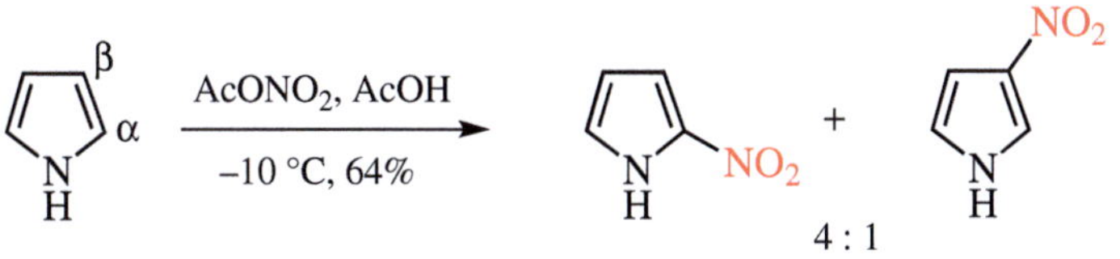

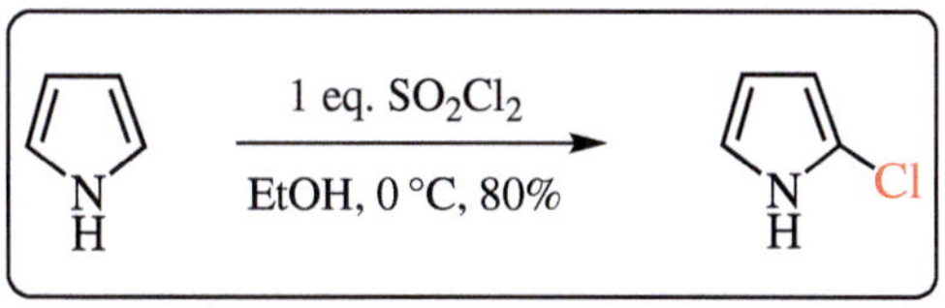

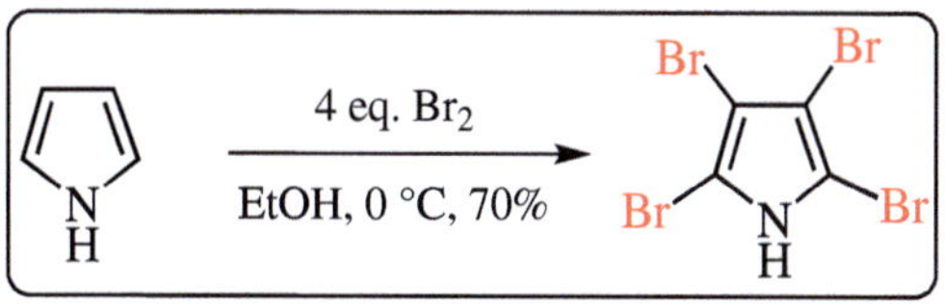

318 | 6 Aromaticity

Formylation of pyrrole with dimethylformamide (DMF) and phosphoryl chloride (Vilsmeier–Haack reaction) is a generally applicable process [146, 147]. The Vilsmeier–Haack reaction is used to convert electron-rich aromatic compounds to the corresponding aromatic aldehydes using DMF, phosphorus oxychloride, and aqueous work-up. The reaction begins with the reaction of DMF with the phosphorus oxychloride to form an iminium salt known as the *Vilsmeier reagent*. The electron-rich aromatic compound then attacks the iminium ion with loss of aromaticity. A deprotonation step restores aromaticity, followed by the release of a chloride ion to form an iminium intermediate. Aqueous work-up then leads to the pyrrole aldehyde as the final product in 83% yield.

Substitution takes place at the C2 carbon atom. The α-reactivity can be completely blocked by a large substituent such as the triisopropyl silyl group attached to the nitrogen atom. Reaction of *N*-triisopropyl pyrrole with 1 equiv NBS exclusively gives the product substituted at the C-3 carbon atom [148]. The silyl group can easily be removed by tetra-*n*-butylammonium fluoride.

Pyrrole is a poor diene for the [4 + 2] Diels–Alder cycloaddition reaction. For example, pyrrole reacts with acetylenic dicarboxylic acid derivatives and gives Michael addition products [149]. In the case of a [4 + 2] cycloaddition reaction, the aromaticity will be lost. When an electron-withdrawing group is placed on the nitrogen atom of pyrrole, the aromaticity of the ring will be reduced. The compound will then gain more of a diene character than an aromatic one. The reaction of pyrroles substituted with electron-withdrawing groups with dienophiles gives cycloaddition products in the range of 30–54% [150]. However, Michael addition products are still found to be the major products [151].

6.5 Aromaticity in Heterocyclic Compounds

Furan has a low boiling point (31 °C; boiling point of pyrrole is 139 °C) and is a toxic compound. As discussed above, with six π electrons, it is an aromatic compound. It has a planar structure. Furan also undergoes an electrophilic substitution reaction as well as a Diels–Alder-type cycloaddition reaction. Chemical shifts of furan are consistent with aromatic compounds, but resonances are shifted to higher fields (compared to benzene) as expected from the increased electron density on the carbon atoms.

Bond length — 1.44 Å, 1.35 Å, 1.37 Å

^{1}H and ^{13}C NMR shifts — 110.0 ppm, H 6.2 ppm, H 7.3 ppm, 142.0 ppm

Coupling constants — 3.5 Hz, 0.4–1.0 Hz, 1.8, 1.0–2.0 Hz

Resonance structures of furan show that the oxygen atom donates the electrons depicted as a lone pair to the five-membered ring. Furan reacts slowly with hydrogen chloride. Hot, aqueous mineral acids cause hydrolytic ring opening.

Resonance structures of furan

Because of the relatively low aromatic stabilization energy (16.3 kcal), the chemistry of furan differs from that of pyrrole. Furan also undergoes enol ether chemistry and a cycloaddition reaction beside a normal electrophilic substitution reaction. Reaction of furan with bromine or chlorine at room temperature results in polyhalogenated compounds. Under milder conditions, reaction of bromine in DMF gives 2-bromo- or 2,5-dibromofurans [152, 153].

Vilsmeier formylation of furan is a good synthetic route to α-formylfuran, furfural [154]. Furfural is the most commonly produced industrial chemical because its production is very flexible. It can be produced on a large industrial scale by the acid-catalyzed dehydration of five-carbon sugars (pentoses), particularly xylose.

Furfural — Vilsmeier reaction

Pyridine–sulfur trioxide complex can be used for sulfonation of the furan ring. An electrophilic substitution reaction takes place at the C2 carbon atom as well as at the C5 carbon atom [155].

Furan undergoes a ring-opening reaction with water in the presence of strong acids. Protonation at the C2 carbon atom is energetically (7 kcal/mol more stable) preferred to that at the C3 carbon atom. C2 protonation leads to the formation of 2,5-dihydro-2-furanol, which is subject to protonation on the ring oxygen atom. The ring opening of this intermediate produces 4-hydroxy-2-butenal. Of course, 2,3-dihydro-3-furanol can also be formed. However, recent calculations show that the ring-opening of 2,3-dihydro-3-furanol is energetically not favored because of the high energy barrier [156].

320 | *6 Aromaticity*

Furan also undergoes [4 + 2] cycloaddition reactions with various dienophiles. The reaction of furan with maleic anhydride gives the *exo*-adduct [157]. However, ^{1}H NMR spectral studies indicate the initial formation of the *endo*-adduct. The initial rate of formation of the *endo*-adduct was found to be greater than that for the formation of the *exo*-adduct because of the secondary orbital interactions (see Section 10.4.1). After 24 min, the concentrations of *endo*- and *exo*-adducts were the same. After 48 h, the *endo*-adduct disappeared. Furthermore, it was found that the rate constant for formation of the *endo*-adduct was actually 500 times larger than the *exo*-adduct formation rate constant. This rate constant difference arises from the different activation energies (3.8 kcal favoring the *endo*-adduct). Moreover, it was determined that the *exo*-adduct is 1.9 kcal/mol more stable than the *endo*-adduct [158]. As the formation of these adducts is reversible, the *exo*-adduct is the final product.

At higher temperatures, even electron-rich alkenes such as vinylene carbonate are added to furan to give a mixture of *endo*- and *exo*-cycloadducts in which the *endo*-product was formed as the major product [159]. These adducts are potential intermediates for the synthesis of polyhydroxylated cyclohexane derivatives. The *endo*-adduct was converted into diacetate derivatives by hydrolysis. Boron tribromide-assisted ring-opening of the *endo*-diacetate followed by hydrolysis of the acetate group results in the formation of a halo-conduritol derivative with conduritol-A configuration [160].

Furan also undergoes a cycloaddition reaction with singlet oxygen. Tetraphenylporphyrin-sensitized photooxygenation of furan in an aprotic solvent yields the corresponding unsaturated bicyclic endoperoxide characterized by ^{1}H and ^{13}C NMR spectra. In a nonpolar aprotic solvent, the ozonide undergoes thermal rearrangement to the corresponding *cis*-diepoxide and epoxylactone [161]. This ozonide has an interesting structure. It looks like an ozonide derived from the ozonolysis of cyclobutadiene.

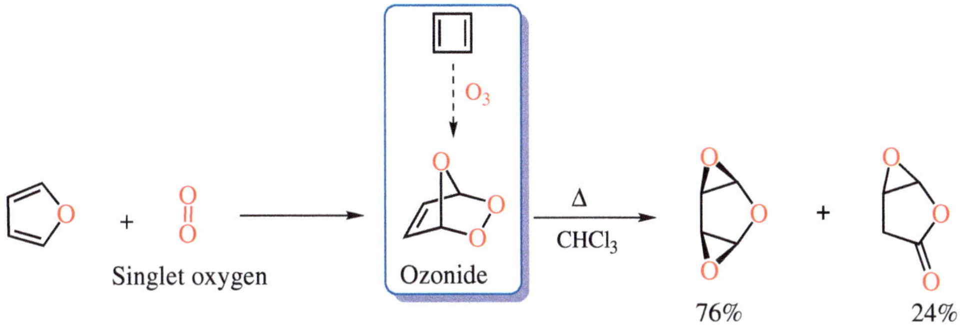

Directed lithiation of furan with alkyllithium in anhydrous diethyl ether at room temperature or at the reflux temperature gives the mono-lithiated furan derivative [162]. Applying forcing conditions can generate 2,5-dilithiation of furan. The preference for α-lithiation is clearly demonstrated by the rearrangement of 3-lithiofuran, synthesized from 3-bromofuran by metal–halogen exchange at −78 °C, to the more stable 2-lithiofuran if the temperature rises over −40 °C. Reaction of lithiofuran compounds with various electrophiles produces substituted furan derivatives [163].

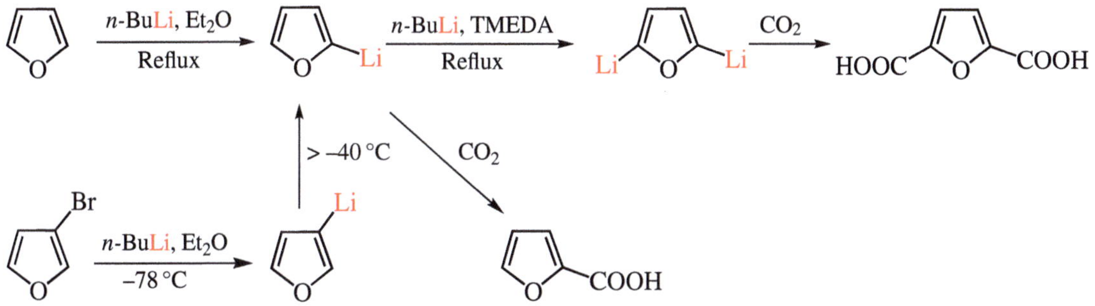

Thiophene is a colorless liquid with a benzene-like odor. It is an aromatic compound with six π electrons. As the resonance energy of thiophene (29.2 kcal/mol) is closer to that of benzene (36.3 kcal/mol), thiophene resembles benzene in most of its reactions. One of the electron pairs on sulfur is significantly delocalized in the ring. Like furan and pyrrole, in thiophene, six π electrons are distributed over five atoms and so the carbon frameworks are electron-rich. Therefore, thiophene also undergoes electrophilic substitution reactions faster than benzene does.

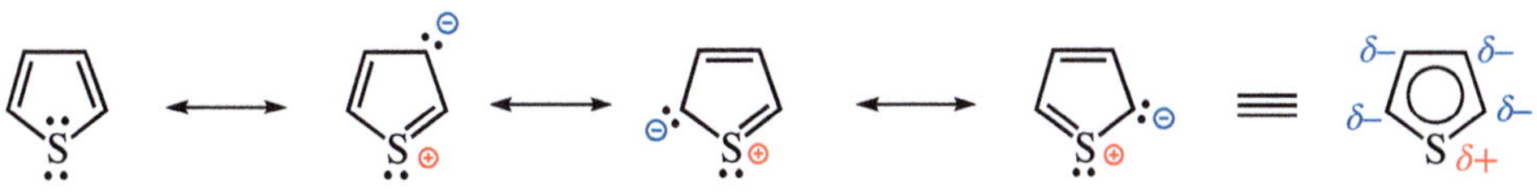

Resonance structures of thiophene

As a consequence of its aromaticity, thiophene does not exhibit the properties seen in sulfides. For example, the sulfur atom resists alkylation reactions. Some physical properties of thiophene are given below.

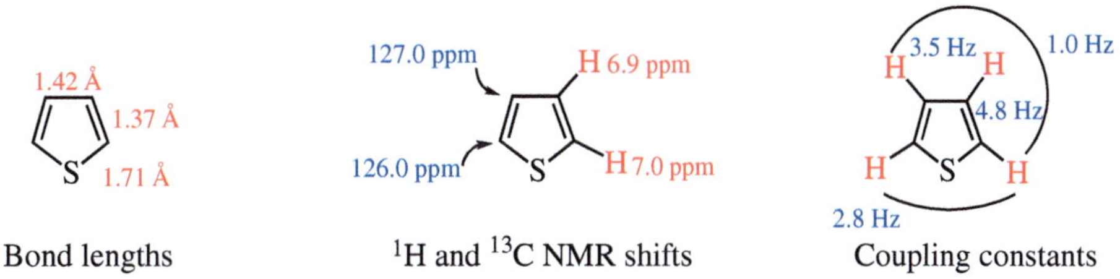

Bond lengths

^{1}H and ^{13}C NMR shifts

Coupling constants

Thiophene is fairly reactive toward electrophilic substitution and unreactive toward nucleophilic substitution. For example, thiophene undergoes electrophilic substitution: halogenation, sulfonation, alkylation, Friedel–Crafts acylation, nitration, etc. The Friedel–Crafts acylation of thiophenes is a generally used reaction, and it gives good yields under controlled conditions. Using aluminum trichloride as a catalyst generates tars. To avoid tar formation, tin tetrachloride is used as a catalyst [164]. Halogenation of thiophene occurs at room temperature as well as at lower temperatures (−30 °C) in the dark. Because of the increased reactivity, the rate of halogenation at room temperature is 10^8 faster than that of benzene [165]. Iodination with I_2 in aq. HNO_3 produces 2-iodothiophene in 70% yield [166].

For a long time, it has been known that unactivated thiophene is highly aromatic and hence would not have any tendency to undergo the Diels–Alder reaction, even with strong dienophiles. However, reaction of thiophene with maleic anhydride under high pressure (0.8 GpA) and at 100 °C under the solvent-free conditions gave the *exo*-addition product in 93% yield [167]. On the other hand, electron-rich thiophenes such as 3,4-methoxythiophenes react with acetylene dicarboxylate to produce phthalates arising from sulfur extrusion from the initially formed cycloaddition product in 35% yield [168].

While organic sulfides can be selectively oxidized to the corresponding sulfoxides, this method completely fails at low temperatures for oxidation of thiophene. The reaction of thiophene with peracids leads to the formation of thiophene-*S*-oxide dimers. The initial oxidation at the sulfur atom first yields thiophene-S-oxide, which will then gain a more diene character because of the reduced aromaticity. Thiophene-*S*-oxide behaves as a diene and dienophile. Cycloaddition results in the formation of the dimers [169].

Thienothiophenes are bicyclic aromatic compounds with two thiophene rings fused to each other. They are mainly used as electron-rich building blocks for semiconducting polymers. They are compared with naphthalene. Therefore, the synthesis of thienothiophenes gains importance. Thieno[3,4-*b*]thiophene has been synthesized starting from 3,4-dibromothiophene. Reaction of 3,4-dibromothiophene with trimethylsilylacetylene in the presence of catalytic amounts of Pd/CuCl catalyst gives the cross-coupling product. Bromine–lithium exchange followed by reaction with elemental sulfur and ring-closure in aqueous medium produces the thienothiophene [170].

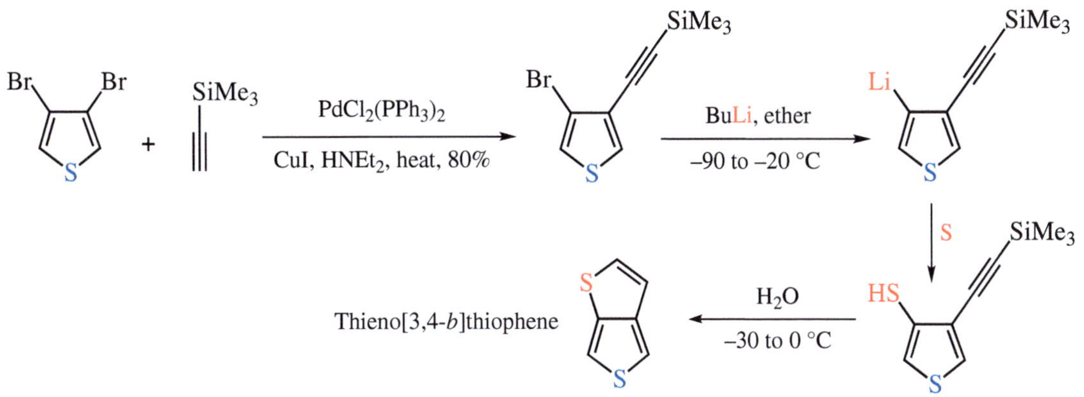

6.5.3 Heteroaromatic Compounds with Six-Membered Ring

Pyridine is a colorless liquid. Its boiling point is 115 °C. Pyridine, with six π electron, is electronically related to benzene and thus meets Hückel's criteria for aromatic systems. The six π electrons are delocalized over the ring. The nitrogen atom is sp^2 hybridized. Two of the three sp^2 hybrid orbitals are overlapping with the sp^2 orbitals of the neighboring carbon atoms forming the σ bonds (Figure 6.17). The lone pair of electrons populates the remaining sp^2 hybrid orbital. The lone electron pairs are directed away from the ring and in the same plane of the ring. The lone pair of electrons is not involved in the aromatic system because of geometry. The p orbitals are perpendicular to the plane, whereas the sp^2 hybrid orbitals are located in the plane of the ring. Unhybridized p orbitals with a single electron participate in the delocalization.

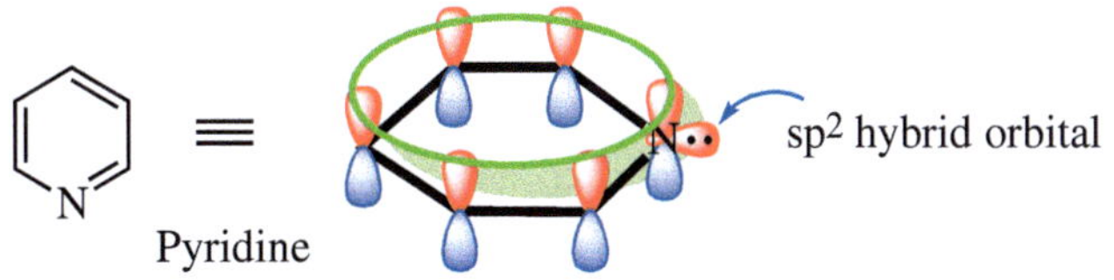

Figure 6.17 The p orbitals of pyridine.

In benzene, the electron density is equally distributed over the ring. In contrast to benzene, the electron density in pyridine is not equally distributed over the ring, arising from the negative inductive and mesomeric effects (−I and −M effects) of the nitrogen atom. For this reason, pyridine has a dipole moment and resonance energy of pyridine (28 kcal/mol) is lower than that of benzene (36.0 kcal/mol). Some spectral data for pyridine are given below.

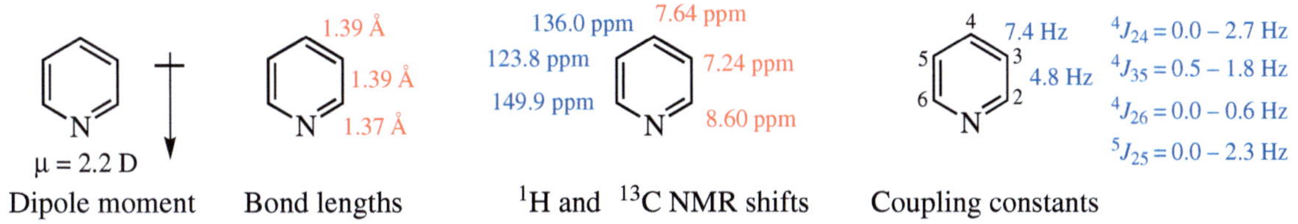

Six-membered heterocycles with an electronegative heteroatom are generally electron-deficient compounds. Such compounds are classified as p-deficient. The electron-deficient nature of pyridine, as compared to benzene, can be explained by the electron-withdrawing effects (inductive and mesomeric) of the nitrogen atom. In particular, the mesomeric effect is more pronounced at the C-2 and C-4 carbon atoms as compared to that at the C-3 carbon atom. The resonance structures of pyridine are given below.

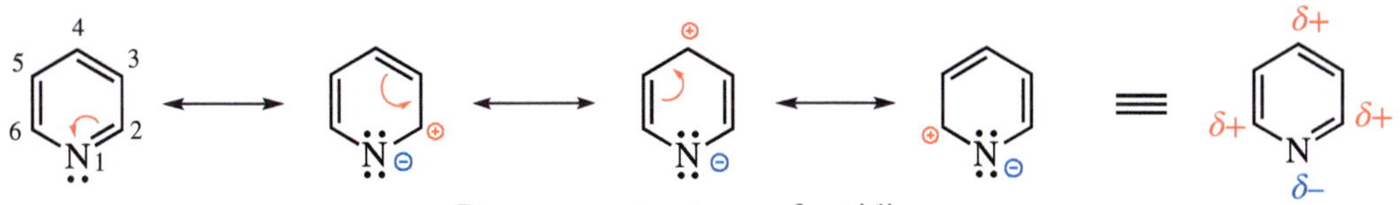

Resonance structures of pyridine

6.5.3.1 Electrophilic Aromatic Substitution

Even with this electron-deficient nature, pyridine undergoes electrophilic substitution reactions under vigorous conditions and yields are often quite low. If the nitrogen atom of pyridine is protonated under the reaction conditions, the reactivity against electrophiles decreases further because a positive charge on the nitrogen atom would make the carbocation intermediates formed during electrophilic substitution reaction even less stable. The reactivity of pyridine has been well studied by theoretical calculations and it was found that the electrophilic reactivity order of pyridine is C3 > C2 > C4. The electropiles mainly attacks the C3 carbon atom. Actually, this carbon atom is also deactivated. Compared to benzene, the C3 position of pyridine is considerably less reactive toward electrophilic attack because of deactivation by the ring nitrogen. However, it is less deactivated compared to the C2 and C4 carbon atoms. The most stable intermediates are formed by attack of an electrophile on the C3 carbon atom as shown below. When the electrophiles are attacking the C2 or C4 positions, one of the resonance contributors is unstable because the nitrogen atom has an unfavorable sextet and a positive charge.

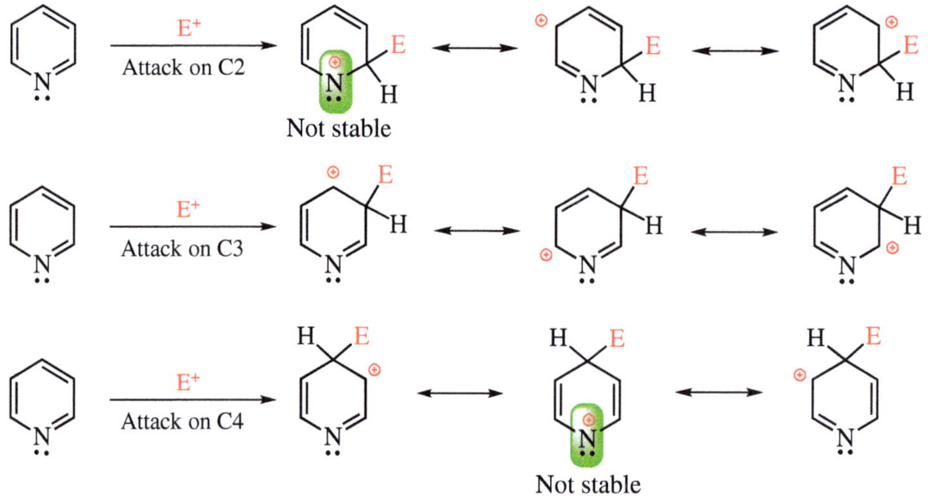

Electrophilic substitution of aromatic compounds proceeds via two steps: addition of an electrophile to one of the double bonds and then elimination of the proton (H$^+$). The first step is slower and the rate-determining step. Relative rates of substitution at different positions can be predicted by analyzing the stability of the σ-complex formed in the first step.

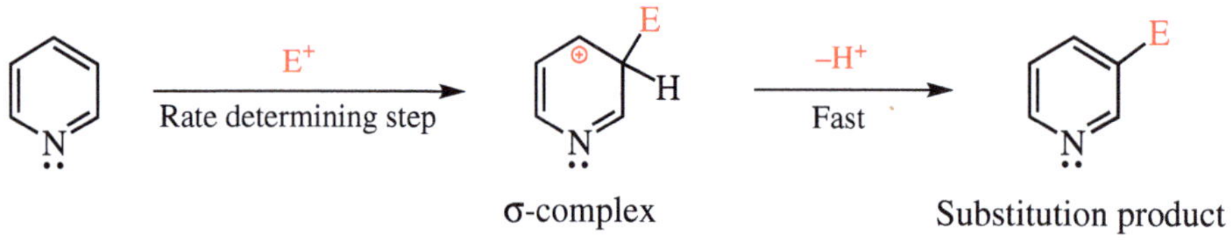

As pyridine is an electron-deficient aromatic compound, it does not undergo Friedel–Crafts alkylation or acylation reactions. Friedel–Crafts alkylation and acylation usually fail for pyridine because they lead only to addition at the nitrogen atom. Pyridine also reacts with acids and forming a pyridinium ion. Pyridinium salts are still aromatic and the aromaticity is not destroyed. Pyridines form stable salts with strong acids. Pyridine itself is often used as a base to neutralize acid formed in a reaction. The basicity of pyridine (pK_a = 5.2) is less than that of aliphatic amines. This reduced basicity can be explained by the hybridization of the nitrogen atom. In saturated amines (such as NH$_3$ or NR$_3$), the lone electron pair is in an sp^3-hybridized orbital. However, in pyridine, the lone pair of electrons are in an sp^2 orbital. The higher the s character of an orbital, the more the lone pair of electrons are tightly bonded to the nitrogen atom. This of course decreases the basicity. For example, nitriles in which the lone pair is in an sp orbital are of low basicity.

6.5 Aromaticity in Heterocyclic Compounds

Forcing reaction conditions are required for direct halogenation of pyridine. 3-Bromopyridine can be synthesized by the reaction of bromine in oleum with pyridine in 86% yield [171]. On the other hand, 3-chloropyridine can be synthesized in the presence of chlorine and aluminum chloride at 100 °C [172].

2-Bromo and 2-chloro-pyridines have been efficiently synthesized by treatment of pyridine–palladium complex with bromine or chlorine at lower temperatures and shorter reaction times in high yields. The ratio of the products strongly depends on the reaction temperature. For example, bromination at 25 °C gave 2-bromopyridine in 82% yield, whereas 3-bromopyridine was formed in only 12% yield. On the other hand, when the reaction temperature was increased to 50 °C, these isomers were formed in a ratio of 1 : 1 [173].

Typically, nitration of pyridine was carried out at 350 °C. 3-Nitropyridine was obtained under very harsh conditions only in 12% yield [174]. Pyridine was reacted with N_2O_5 dissolved in SO_2 at 0 °C to give 3-nitro pyridine under very mild condition in 69% yield [175].

The reaction mechanism is not electrophilic aromatic substitution. Investigation of the mechanism by ^1H NMR spectroscopy showed that N-nitropyridinium nitrate was the initial product formed by the reaction of pyridine with N_2O_5. After addition of an SO_2 group to the α-position, the nitro group undergoes a [1,5] sigmatropic shift from the nitrogen atom to the C-3 carbon atom. Elimination of HSO_3^- results in the formation of 3-nitropyridine [176].

N-nitropyridinium nitrate

Pyridine cannot be sulfonated at temperatures below 200 °C because of the formation of a sulfur trioxide complex. However, at higher temperatures, the complex probably undergoes an intermolecular rearrangement to produce the 3-sulfonic acid. An improvement on this procedure is to use excess sulfur trioxide and $HgSO_4$ as a catalyst to lower the reaction temperature and increase the yield. It is likely that N-mercuration takes place in the first step [177].

6.5.3.2 Nucleophilic Aromatic Substitution

We discussed that pyridine is an electron-deficient aromatic compound. Therefore, it undergoes an electrophilic aromatic substitution reaction only with great difficulty as pyridine resembles a strongly deactivated benzene. Aromatic electrophilic substitution on pyridine is not a useful reaction. By contrast, pyridine is highly activated toward the attack by electron-rich nucleophiles. The nitrogen atom makes pyridines more reactive toward nucleophilic substitution, particularly at the 2- and 4-positions, because attack at these positions leads to the most stable intermediates since the negative charge is located on the nitrogen atom, the most electronegative of the ring atoms.

L = leaving group

Nucleophilic aromatic substitutions occur more easily when the aromatic compound contains good leaving groups such as chlorine, fluorine, or sulfonic acid fragments. In 1914, Aleksei Chichibabin, a Russian organic chemist (1871–1941) published a seminal paper describing the first general example of the nucleophilic aromatic substitution of hydride [178, 179]. Generally speaking, the hydride ion is a poor leaving group. *The Chichibabin reaction* may be defined as the nucleophilic substitution by an amino group of a hydride ion attached to an aromatic nitrogen heterocycle. The Chichibabin reaction is carried out either at boiling temperatures in an aromatic hydrocarbon or in dialkyl aniline. In the case of having $NaNH_2$ in excess, 2,4-diaminopyridines are formed. If these positions are already substituted, the amide attacks the C5 carbon atom. Beside the formation of 2-aminopyridine, hydrogen gas is evolved.

Chichibabin reaction

The mechanism of the Chichibabin reaction is quite complex. First, complexation of pyridine's nitrogen atom with the sodium ion increases the positive charge at the C2 carbon atom. Therefore, the amide ion attacks preferentially the C2 carbon atom, forming a σ-complex-1. NaNH$_2$ can now abstract a proton from the σ-complex, forming a new σ-complex-2. The hydride transfer and formation of hydrogen gas probably involve interaction of σ-complex-1 acting as an acid with the anionic intermediate as shown below [180].

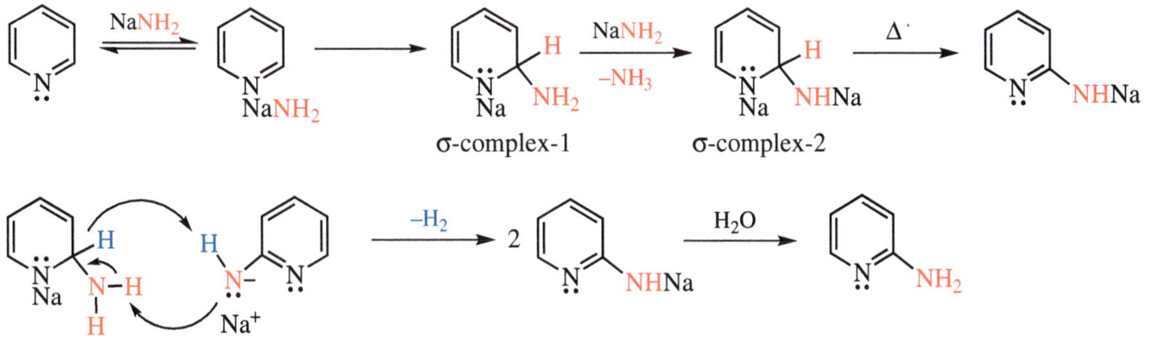

Substituents such as halogen, nitro, and alkoxy sulfonyl groups at the α- and β-positions can be relatively easily displaced by a number of nucleophiles via an addition–elimination mechanism. Cherng reacted various halopyridine derivatives substituted at different positions with sulfur, oxygen, and carbon nucleophiles under microwave irradiation and found that the reactions were complete within several minutes in high yield [181]. Furthermore, it has been detected that the relative reactivity of halopyridines in nucleophilic substitution reactions follows the order 2-halopyridine > 4-halopyridine > 3-halopyridine.

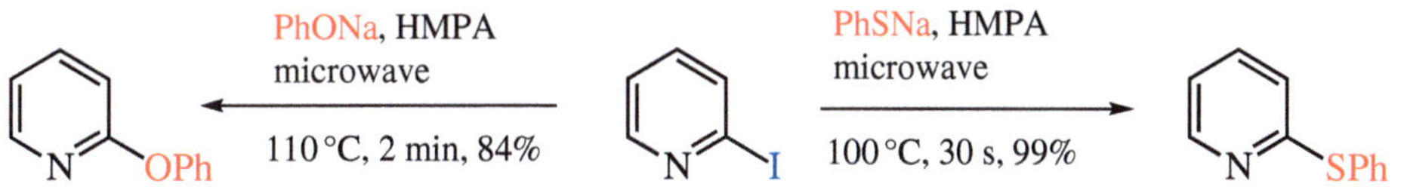

The C2- and C4-substituted halopyridine derivatives undergo a nucleophilic substitution reaction with ammonia at 200 °C to give the corresponding aminopyridine derivatives, while 3-bromopyridine does not form a substitution product under the same conditions. However, when the reaction was carried out under high pressure with CuSO$_4$ as a catalyst, the desired product was formed in 88% yield [182].

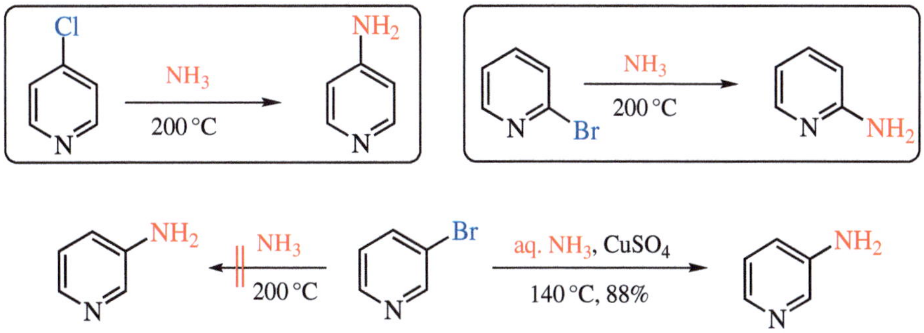

6.5.3.3 Pyridine N-Oxides

Pyridine N-oxides are very useful synthetic intermediates. They are widely applied in the activation and functionalization of pyridine. Their reactions differ significantly from those of neutral pyridines. Various resonance structures can be drawn for pyridine N-oxides as shown below. These resonance contributors show that N-oxide moiety here acts as a σ electron-withdrawing group as well as a π back electron-donating group. In summary, the N-oxide function facilitates the addition of electrophiles and nucleophiles to C2 and C4 carbon atoms. Under extremely acidic conditions, electrophilic substitution can occur at the 3-position.

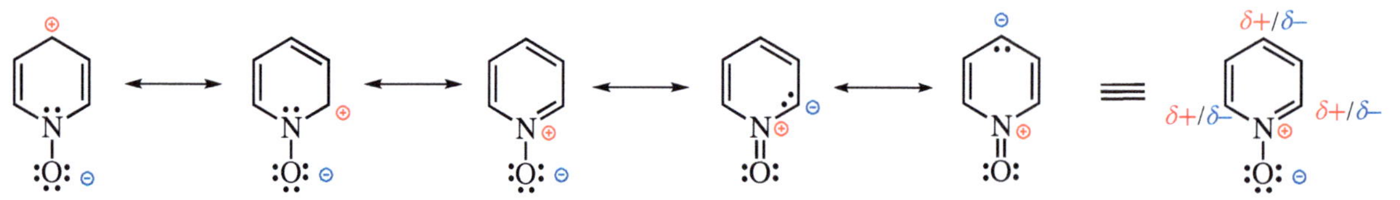

Several methods have been described for the synthesis of pyridine N-oxides. The oxidation of pyridine can be achieved with hydrogen peroxides, trifluoroperacetic acid, m-chloroperbenzoic acid, MeReO$_3$/H$_2$O$_2$, etc. However, peracids are perhaps the most common reagents employed. The deoxygenation (or deprotection) of the pyridine N-oxides has been widely studied [183].

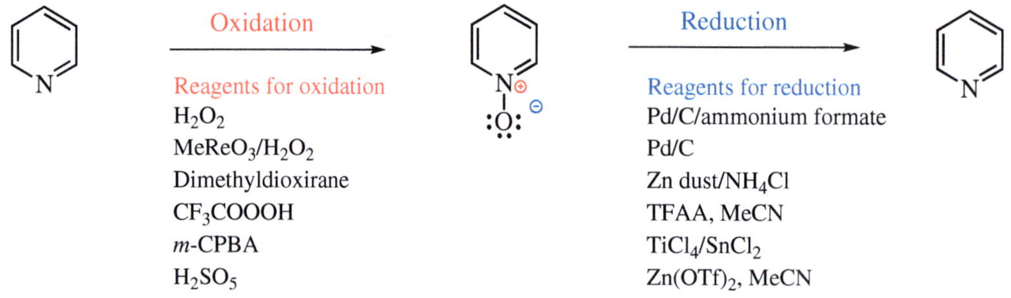

Reagents for oxidation
H$_2$O$_2$
MeReO$_3$/H$_2$O$_2$
Dimethyldioxirane
CF$_3$COOOH
m-CPBA
H$_2$SO$_5$

Reagents for reduction
Pd/C/ammonium formate
Pd/C
Zn dust/NH$_4$Cl
TFAA, MeCN
TiCl$_4$/SnCl$_2$
Zn(OTf)$_2$, MeCN

6.5.3.4 Electrophilic Substitution

3-Bromopyridine N-oxide was successfully synthesized by treatment of pyridine N-oxide with bromine in oleum at 70 °C as the major product [184]. 2,5- and 3,4-Dibromo-pyridine-N-oxides were also formed as side products. HgSO$_4$-catalyzed sulfonation of pyridine N-oxide also produces at the C3 carbon atom substituted 3-sulfonic acid [185].

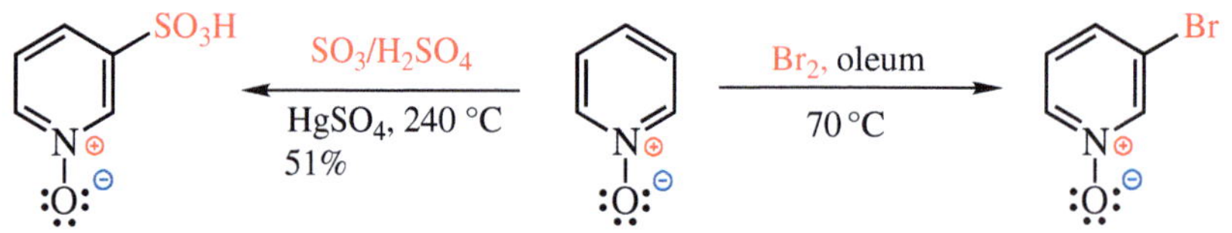

6.5.3.5 Nucleophilic Substitution

Generally, for direct chlorination of pyridine, acetyl hypochlorite is commonly used giving a low yield. However, chlorination of pyridine-N-oxide with phosphorus oxychloride in the presence of a stoichiometric amount of triethylamine gives 2-chloropyridine in 90% yield with 99% selectivity [186].

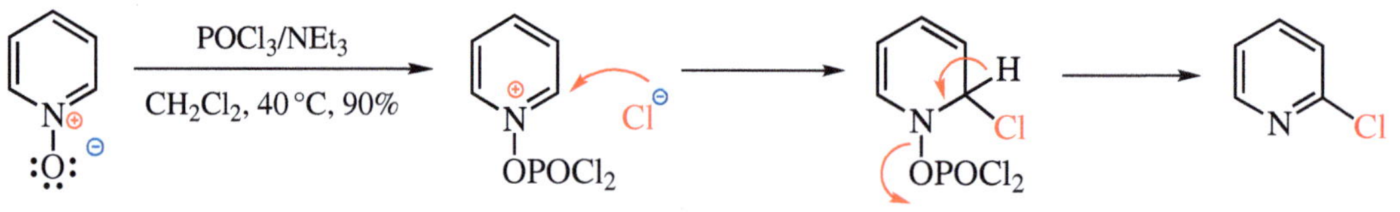

Reaction of pyridine N-oxide with PhMgBr in THF followed by quenching with water gives phenyl-N-hydroxydihydropyridine isolated in 60–80% yields. Treatment of phenyl-N-hydroxydihydropyridine with Ac$_2$O (acetic anhydride) at 100 °C results in 2-phenylpyridine in 43% yield [187].

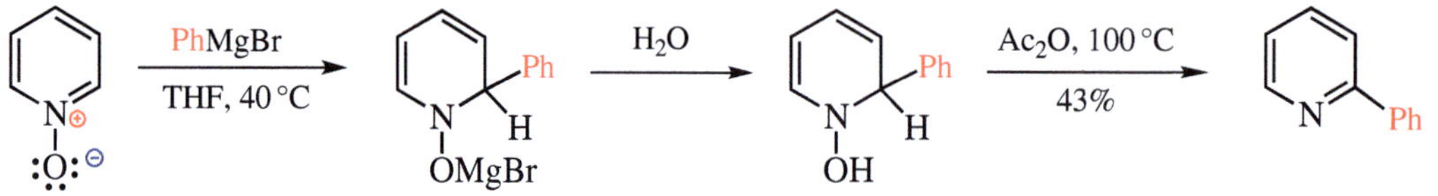

6.5.3.6 Six-Membered Ring Heteroaromatic Compounds with Two Nitrogen Atoms

Pyridazine, pyrimidine, and pyrazine (diazines) contain an additional nitrogen atom compared to pyridine. Those two heteroatoms withdraw electrons from the ring even more than in pyridine. These effects also decrease the basicity. Therefore, diazines are even more resistant to electrophilic substitution reactions. These compounds are also electron-deficient aromatic compounds.

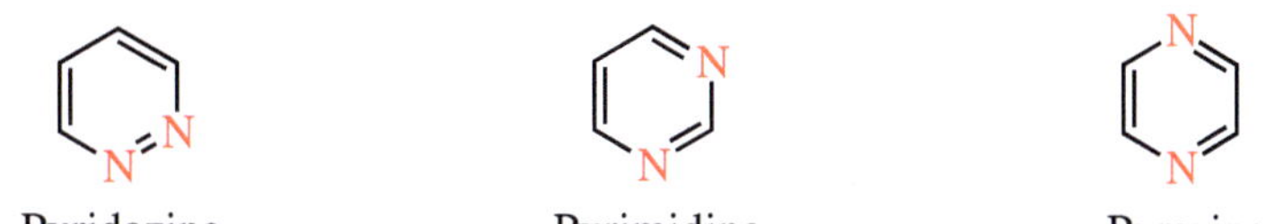

Pyridazine   Pyrimidine   Pyrazine

As a consequence of the electron-withdrawing effect of nitrogen atoms, diazines are more easily attacked by nucleophiles than pyridine is. Pyridazine is a weak base and reacts with mineral acids to form a quaternary salt. The protonation of the second nitrogen atom is hindered because of the high energy required to generate an additional positive charge on the neighboring nitrogen atom. Some spectral data for azines are given below. The chemical shifts clearly show that all three compounds are aromatic.

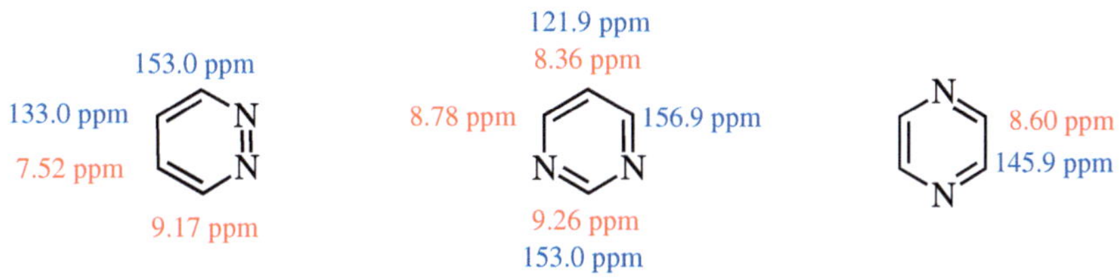

Pyridazine: 153.0 ppm, 133.0 ppm, 7.52 ppm, 9.17 ppm

Pyrimidine: 121.9 ppm, 8.36 ppm, 8.78 ppm, 156.9 ppm, 9.26 ppm, 153.0 ppm

Pyrazine: 8.60 ppm, 145.9 ppm

Upon reaction with alkyl halides, they form *N*-alkyl quaternary salts. This alkylation reaction takes place at one nitrogen atom not at the second nitrogen atom [188]. These failures arise from the expected reduction in nucleophilicity of the second nitrogen atom upon quaternization of the first. However, diquaternary salts can be synthesized in all diazines with the more reactive trialkyloxonium tetrafluoroborate as shown below [189]. Actually, the presence of two positive charges in the ring should enhance reactivity against nucleophilic attacks.

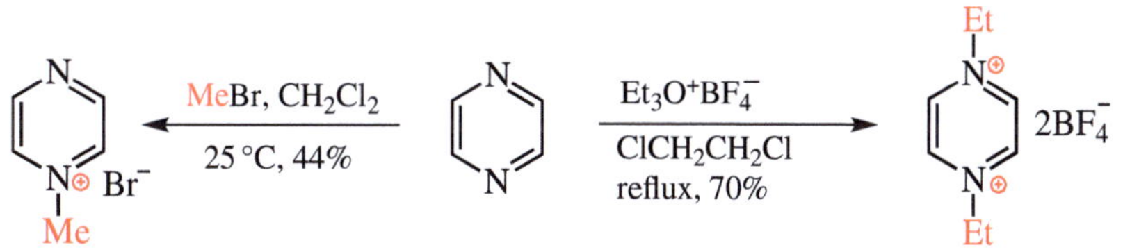

Pyridazine can readily be alkylated and arylated with organolithium compounds as well as with Grignard reagents. For example, reaction of pyridazine with phenylmagnesium bromide undergoes substitution at the C4 carbon atom, whereas with *t*-butyllithium, it gives the C3-substituted product [190].

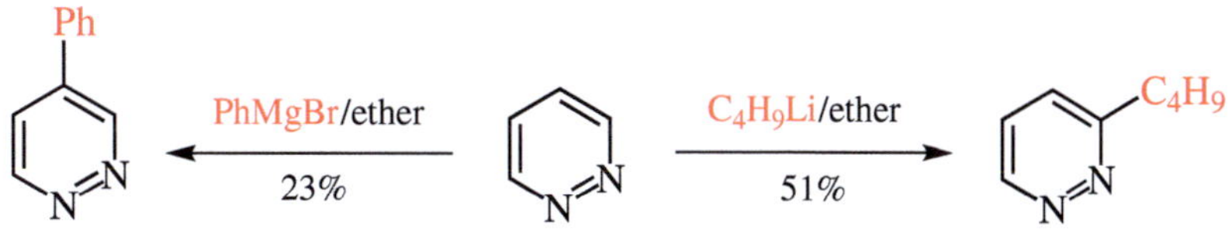

Pyrimidine is less aromatic compared to pyridine and benzene. The resonance energy of pyrimidine is 26 kcal/mol. Diprotonation takes place with the strong acids as the nitrogen atoms are far from each other. Pyrimidine cannot easily undergo electrophilic substitution reactions.

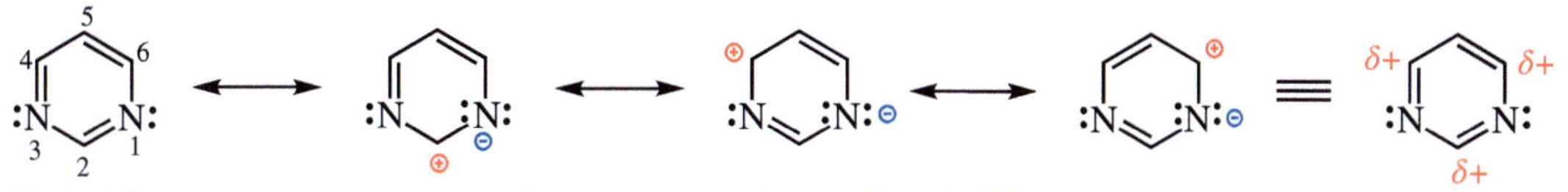

Pyrimidine   Resonance structures of pyrimidine

Electrophilic substitution with pyrimidine is very difficult. Substitution occurs at the C5 position, which is the least electron-deficient carbon atom. Nucleophilic substitution takes place at the C2, C4, and C6 carbon atoms. Nucleophiles can selectively add to pyrimidine. For example, 2-chloro-6-lithiopyridine, synthesized by the reaction of 2-chloropyridine with BuLi, adds to the C4 carbon atom of pyrimidine. Actually, this is an addition–elimination reaction. The hydride anion is substituted [191]. Pyrazine also undergoes a similar nucleophilic substitution reaction.

Pyridazine, pyrimidine, and pyrazine undergo an hydrogen/deuterium (H/D) exchange reaction at all positions with MeONa/MeOD at high temperatures [192]. This exchange reaction is faster than that for pyridine. This indicates that the ring protons of azines are more acidic than the protons in pyridine. Probably, it arises from the presence of the second nitrogen atom because of the electron-withdrawing effect of nitrogen atoms.

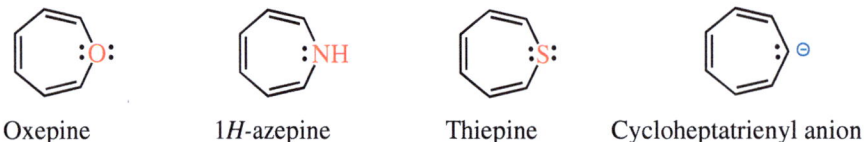

To functionalize azines, nonnucleophilic lithium tetramethylpiperidide (LTMP) is used. The diazine lithiums formed is very unstable. Reaction of such an intermediate with iodine forms the corresponding iodo-diazine compound [193]. Iodo-pyrazine has also been synthesized by nucleophilic substitution of chloropyrazine with sodium iodide in a mixture of acetic acid, sulfuric acid, and acetonitrile in 80% yield [194].

6.5.4 Heteroaromatic Compounds with a Seven-Membered Ring

Heterocyclic compounds with a seven-membered ring, oxepine, azepine, and thiepine are homologues of furan, pyrrole, and thiophene, and they are isoelectronic with the unstable cycloheptatrienyl anion. Consequently, heterotropilidenes may provide valuable information for the theory of modern aromaticity.

Oxepine 1H-azepine Thiepine Cycloheptatrienyl anion

6.5.4.1 Oxepine

Dehydrohalogenation of 1,2-epoxy-4,5-dibromocyclohexane with sodium methoxide in ether gives benzeneoxide, which is in equilibrium with the corresponding valence isomer oxepine [195, 196]. Oxepine is stable at room temperature and readily isomerizes to phenol upon treatment with acids. The big difference in the ^{1}H NMR spectra of benzene oxide and oxepine should be in the resonance frequency of α-protons. Actually, the α-protons of oxepine should resonate at approximately 6.00 ppm. On the other hand, the resonance frequency of α-protons in benzene-epoxide should appear in the region of 3.0–3.5 ppm. However, the resonance signals of α-protons appear between the expected values for benzene-oxide and oxepine, clearly indicating fast valence isomerization between the two isomers.

Oxepine is certainly not antiaromatic and it has a boat-like conformation. The chemical and spectral properties indicate that oxepine may be regarded as a typical polyolefin. In the case of a planar structure, one of the lone pairs would be a part of an eight π electron antiaromatic system. To avoid antiaromaticity, oxepine exists in a boat-like conformation, with incomplete conjugation of the π system.

Furthermore, Vogel and Günther determined by using ^{1}H NMR spectroscopy that bicyclic benzene oxide is 1.7 kcal/mol more stable than oxepine in apolar solvents with an activation barrier for the conversion from benzene-oxide to oxepine of 9.1 and 7.2 kcal/mol for the reverse reaction.

Reaction of oxepine with maleic anhydride gives in a few minutes a cycloaddition product. The adduct is formed by the addition of maleic anhydride to benzene-oxide, which is in equilibrium with oxepine.

2,7-Dimethyloxepine [197], synthesized by dehydrobromination of 4,5-dibromo-1,2-epoxy-1,2-dimethylcyclohexane with sodium methoxide, is not in equilibrium with the corresponding benzene-oxide. The ^{1}H NMR spectral studies show that the compound is in a boat conformation similar to the structure of cycloheptatriene. Methyl substitution at the 2- and 7-positions changes the stability order, rendering oxepine the energetically favored isomer because of the repulsion of the two methyl groups in benzene oxide.

6.5.4.2 1H-Azepine

1H-Azepine was first synthesized by Hafner by hydrolysis of ethyl 1H-azepine-N-carboxylate with potassium hydroxide followed by acidification. However, the compound could not be characterized because of the instability of the N-carboxylic acid. The acid undergoes decarboxylation to produce 1H-azepine, which rapidly rearranges into the more stable 3H-azepine [198].

Theoretical calculations (at B3LYP/6-31G(d) level) show that 1H-azepine is 7.9 kcal/mol more stable than the valence isomer, benzene-imine. Vogel and coworkers succeeded in isolating and characterizing 1H-azepine at low temperatures [199]. A solution of trimethylsilyl carboxylate, synthesized by the reaction of azepine N-methyl ester with iodotrimethylsilane, reacts with methanol at −78 °C to give N-carboxylic acid. The decarboxylation of carboxylic acid in CDCl$_3$ at room temperature forms 1H-azepine, which is stable at −60 °C for a couple of hours. NMR spectral studies show that the spectrum is analogous to that of oxepine.

6.5.4.3 Thiepine

Despite the successful synthesis of oxepine and 1*H*-azepine, the parent thiepine has eluded synthesis. This is probably due to the thermal instability of thiepine. It is 7.0 kcal/mol less stable than benzene-sulfide [200]. The energy difference is much larger than for the oxepine-benzene-oxide system because sulfur is better accommodated in three-membered rings. Benzene-sulfide is unstable because of the low activation barrier for the extrusion of sulfur. Loss of sulfur from thiepine occurs by valence isomerization to the corresponding thianorcaradiene followed by irreversible cheletropic loss of sulfur.

Thiepine can be stabilized by either $Fe(CO)_3$-complexation or introduction of substituents into the seven-membered ring. Reaction of stable thiepine 1,1-dioxide with $Fe_2(CO)_9$ gives the iron tricarbonyl complex in high yield. Reduction of the sulfone group to sulfide furnishes the iron–thiepine complex [201].

The first isolated metal-free thiepine was synthesized in 1974 using electron-withdrawing substituents to stabilize the thiepine ring [202]. Reaction of 3-pyrrolidinethiophene with dimethyl acetylenedicarboxylate gives the cycloaddition product, which undergoes an electrocyclic ring opening process, giving the thiepine derivative. Even this stabilized thiepine derivative eliminates sulfur at room temperature to give the corresponding benzene derivative. The authors predicted an antiaromatic character for thiepine.

The instability of thiepine and its derivative is attributed to ready sulfur extrusion. It is well accepted that sulfur extrusion proceeds through the valence isomer benzene-sulfide intermediates. Introduction of bulky substituents, such as *t*-butyl groups at the C2 and C7 positions of thiepine, will hinder the formation of benzene-sulfide because of the repulsion between these bulky groups so that the thiepine structure will be favored. Therefore, synthesis of the thermally stable 2,7-di-*t*-butylthiepine was accomplished. The 2,6-di-*t*-butylthiopyrylium ion was first converted to the hydroxymethyl derivative. The carbinol was converted to the corresponding mesylate. Solvolysis of the mesylate was accomplished with acetic anhydride and sodium acetate in glacial acetic acid at 90 °C to give the stable thiepine derivative [203].

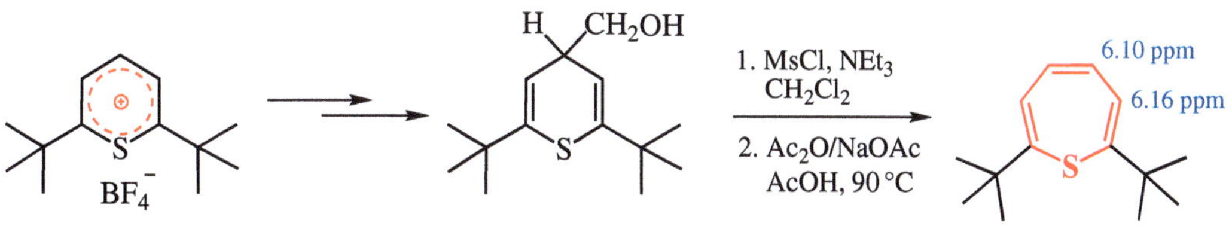

The ¹H NMR spectrum of the thiepine derivative shows that the ring protons resonate in the range of olefinic protons. The X-ray crystal structure analysis shows that the thiepine ring has a boat conformation, and there is considerable bond alternation [203, 204]. It is interesting to note that the thiepine derivative was stable in decalin at 130 °C for 20 h. However, when the thermolysis of thiepine derivative was carried out in benzene in the presence of an equimolar amount of triphenylphosphine at 110 °C for 42 h, *o*-di-*t*-butylbenzene was quantitatively formed.

6.6 Electrophilic Aromatic Substitution: Chemistry of Benzene

We already have discussed the electronic properties of benzene, its aromaticity, and its remarkable stability. The chemical reactivity of benzene contrasts with the reactivity of the alkenes in that substitution reactions occur in preference to addition reactions. Although aromatic compounds have multiple double bonds, these compounds do not undergo cycloaddition reactions. The lack of reactivity toward addition reactions results from the resonance stability of benzene. In this section, we will discuss the following:

1. Electrophilic substitution reactions of benzene, nucleophilic substitution reactions, and addition reactions.
2. How the substituents attached to the benzene ring can change the reactivity for additional substitution reactions and the regioselectivity observed in the products?
3. Conversion of the substituents into new functional groups.

Benzene undergoes *electrophilic aromatic substitution reactions* in which an electrophile substitutes one of the hydrogen atoms attached to the benzene ring and the aromaticity of the ring system is preserved as shown below.

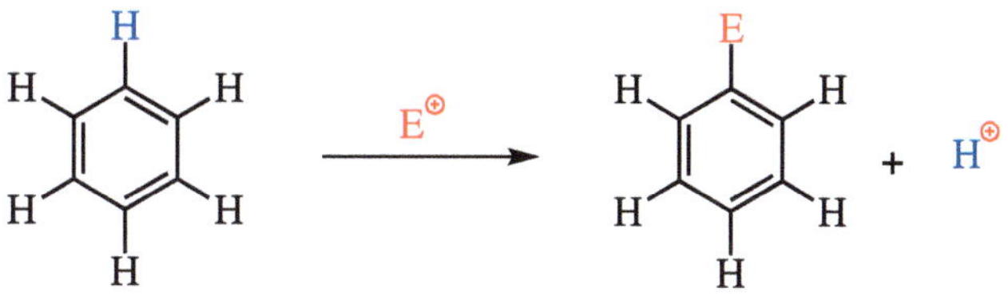

Electrophilic aromatic substitution

A number of substituents can be introduced into the benzene ring through electrophilic substitution reactions. The benzene ring can be substituted by halogens, a nitro group, sulfonic acid, an alkyl group, an acyl group, etc.

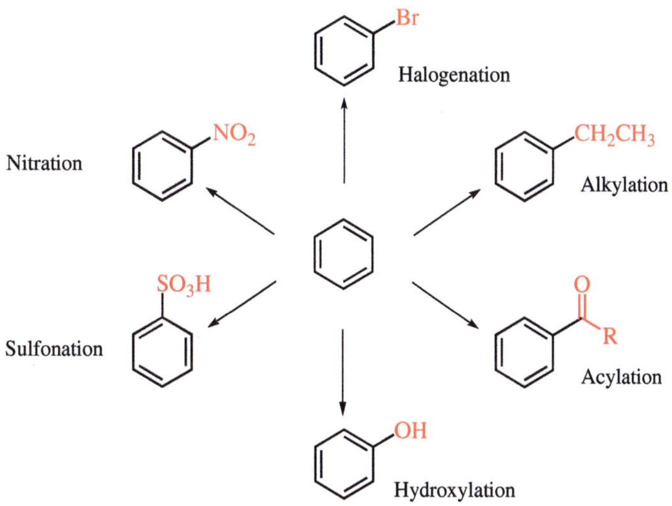

Because of the π electron clouds above and below the benzene ring, benzene is a nucleophile. Therefore, it will react with electrophiles. The electrophilic substitution reaction proceeds in two steps. In the first step, an electrophile approaches the π electrons of benzene and forms a bond to one of the carbon atoms generating a positive charge on the other carbon atom. The formed cation, σ-complex, is a nonaromatic cyclohexadienyl carbocation also called an arenium ion. Although this cation is stabilized by delocalization on the ring, this step is not favored thermodynamically as the resonance energy of the benzene ring is lost. Of course, delocalization over the entire ring is interrupted because of the formation of a sp^3-hybridized carbon atom bearing the electrophile. Finally, the arenium ion loses its proton from the sp^3-hybridized carbon atom regenerating the aromaticity. The overall substitution reaction is exothermic because the new bond formed between the electrophile and the carbon atom is stronger than the broken bond (C—H bond). Furthermore, the formation of a C=C double bond will restore the aromaticity. The intermediate may be trapped by a nucleophile. Such trapping would form a 1,2-addition product and lose the aromaticity. Therefore, the electrophilic substitution reaction is much more favored.

Electrophilic aromatic substitution involves two steps, each of which has its own transition states. This means that there is a "double-humped" reaction energy diagram (Figure 6.18). It is experimentally well established that the arenium ion is a true intermediate in electrophilic substitution reactions. It is not a transition state. The arenium ion lies in an energy valley between two transition states. The energy for the first activation E_1 is higher than that for the energy for the second activation E_2. Therefore, the first step is the rate-determining step (slower step) while the second step deprotonation step is the fast step.

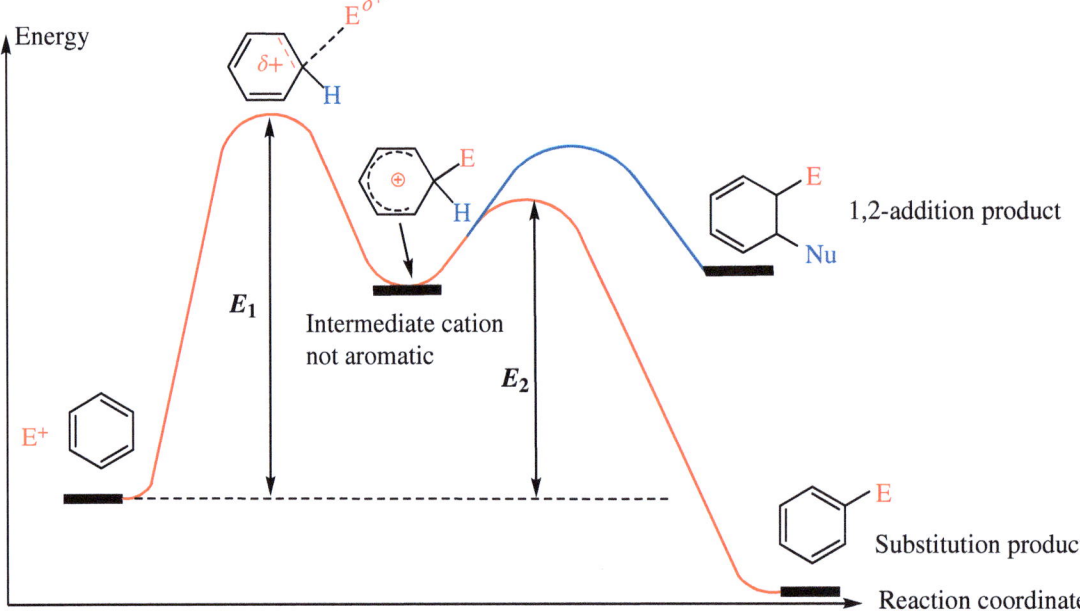

Figure 6.18 Potential energy diagram describing the course of electrophilic substitution and addition.

6.6.1 Halogenation of Benzene

Let us recall first the addition of bromine to cyclohexene. Alkenes react rapidly with bromine at room temperature or lower temperatures to give a 1,2-addition product. The reaction of an alkene with bromine or chlorine does not require a catalyst. The double bond breaks and a bromine atom attaches to each carbon. The reaction is exothermic by about 29 kcal/mol. Bromine is a very polarizable molecule, and approaching any double bond induces a dipole in it because of the repulsion between the π electrons and the electrons bonding two bromine atoms. In the first stage of the reaction, one of the bromine atoms becomes attached to both carbon atoms, with the positive charge being found on the bromine atom. A bromonium ion is formed. The bromonium ion is then attacked from the back by a bromide anion, forming the 1,2-*trans*-addition product.

Benzene is aromatic and much more stable and therefore less reactive than an alkene. Therefore, for bromination or chlorination of benzene, iron bromide ($FeBr_3$) or iron chloride ($FeCl_3$) is required as a catalyst.

As a bromine or chlorine molecule approaches the benzene ring, the delocalized π electrons of the benzene ring repel electrons in the bromine–bromine (chlorine–chlorine) bond as discussed above. The positive end of bromine or chlorine acts as an electrophile. However, this polarization is not enough to attack the benzene ring and disrupt the aromaticity. Therefore, the bromine or chlorine must be further activated by Lewis acid catalysts. Lewis acids such as $FeBr_3$ or $FeCl_3$ have the ability to accept electron pairs because of the electron deficiency. Bromine first reacts with $FeBr_3$ and forms a complex so that the bromine becomes more electrophilic. Then, the π electrons of the benzene ring acts as a nucleophile and attacks the electrophilic bromine and substituting iron tetrabromide. Now, the aromaticity is destroyed. The $FeBr_4^-$ formed in this step acts as a base and abstracts the proton attached to the sp^3-hybridized carbon atom, restoring the aromaticity. This reaction not only furnishes bromobenzene and hydrogen bromide but also regenerates $FeBr_3$ as the active catalyst. Chlorination also requires the presence of an active catalyst, such as $AlCl_3$ or $FeCl_3$. The mechanism of chlorination is identical with that of bromination.

Direct fluorination of benzene with fluorine is not feasible at room temperature. Fluorine reacts extremely rapid with benzene, making it very difficult to control the reaction. The high energy released in the process can break several bonds of the products and hence the reaction creates an uncontrolled product mixture. The classical method for introducing fluorine into an aromatic compound is via the *Balz–Schiemann reaction*, heating controlled thermal decomposition of aryldiazonium tetrafluoroborate. This reaction will be discussed later in Section 6.8.2. However, there are several compounds in which the fluorine atom is bonded to a positively charged nitrogen atom. These reagents are generally less reactive than other fluorinating agents and are quite stable and selective in electrophilic fluorinations. With these compounds, fluorination of benzene can be carried out [205]. The reaction of benzene with 1-(chloromethyl)-4-fluoro-1,4-diazabicyclo[2.2.2]octane bis(tetrafluoroborate) (a commercial compound), which is an effective fluorinating agent, in trifluoromethanesulfonic acid furnishes fluorobenzene in 83% [206]. This reaction can also be applied to other aromatic compounds as well as to substituted aromatic compounds.

Iodine is the least reactive of the halogens in electrophilic substitution reactions. Iodination cannot be achieved using the conventional method like using Lewis acid catalyst. Strong oxidizing agents must be used to oxidize iodine (I_2) to the iodonium ion, I^+. HNO_3, SO_3, hypervalent iodine compounds such as $PhI(OTf)_2$, ceric ammonium nitrate ($Ce(NH_4)_2(NO_3)_6$), etc., can be used as the oxidizing agents. Nitric acid is used to generate the iodonium ion, which acts as an electrophile. The iodination reaction is an endothermic reaction (2.9 kcal/mol).

The reaction of iodine with benzene is a reversible reaction generating HI. This equilibrium lies very much to the left side and no detectable amounts of iodo compound are formed. It is suggested that in the presence of nitric acid, hydrogen iodide is removed by oxidation so that the equilibrium shifts to the side of iodobenzene [207].

Another way to shift the reaction equilibrium to the side of the product is removal of HI for forming a salt. The best reagent for iodination of benzene is iodine chloride by using a catalytic amount of $Cp_2FeB[3,5-(CF_3)_2C_6H_3]_4$ in the coexistence of DDQ or ZnO. As the chlorine is more electronegative than the iodine, it withdraws the bonding electrons toward itself, making iodine a positively charged species [208].

6.6.2 Nitration

Actually, benzene reacts with nitric acid to furnish nitrobenzene. However, the reaction is sluggish and not convenient. Furthermore, the nitrogen atom in HNO_3 is not electrophilic enough to attack the benzene ring; it must somehow be activated. Therefore, a mixture of nitric acid and sulfuric acid is used for nitration reactions. First, sulfuric acid protonates the hydroxyl group of nitric acid. Now, the protonated nitric acid can remove water to form a nitronium ion, which is a stronger electrophile. The nitronium ion has a linear structure.

6.6 Electrophilic Aromatic Substitution: Chemistry of Benzene

Formation of nitronium ion

The electrophilic nitronium ion reacts with the nucleophilic π electrons of benzene to form a resonance-stabilized arenium ion, a σ-complex. This is the rate-determining slower step and it destroys the aromaticity of benzene. The final step is the deprotonation of the σ-complex, which is fast and has no effect on kinetics. Aromaticity is restored in the second step by elimination of H⁺.

σ-complex (arenium ion)

6.6.3 Sulfonation

Sulfonation of aromatic compounds is a very important chemical transformation in the synthesis of pharmaceuticals, surfactants, detergents, dyes, and pesticides. The most commonly used sulfonating agents are sulfur trioxide (SO_3), oleum, sulfuric acid, and chlorosulfuric acid. Benzene does not undergo a sulfonation reaction with concentrated sulfuric acid at room temperature. However, at higher temperatures SO_3, which can be formed by the loss of water from the sulfuric acid, can sulfonate the benzene ring. Removal of water will shift this equilibrium to the side of the sulfonation product.

Mainly, fuming acid (oleum), which is made by dissolving of 8% of sulfur trioxide in concentrated sulfuric acid, is used as a sulfonation reagent. The sulfur atom in sulfur trioxide is electrophilic because three oxygens withdraw electrons, making the sulfur electrophilic and the oxygens nucleophilic. The sulfur in SO_3 is electrophilic enough to attack the benzene ring directly. Under these conditions, the equilibrium will be shifted to the side of the products and benzenesulfonic acid will be formed in high yield.

The mechanism for sulfonation is similar to the mechanism for nitration. Again, the π electrons of the benzene ring act as a nucleophile and they attack the sulfur atom, which is the electrophilic part of sulfur trioxide, forming a cyclohexadienyl cation as the intermediate. Removal of proton from the sp³-hybridized carbon atom and formation of the C=C double bond restore the aromaticity. Protonation of sulfonate anion produces sulfonic acid. Recent DFT calculations support this mechanism [209].

Unlike the other electrophilic substitution reactions, sulfonation of aromatic compounds is a reversible reaction. Sulfonic acid functionality can be removed from the benzene ring by heating in dilute sulfuric acid at 100 °C. The mechanism for desulfonation is identical to the sulfonation mechanism, except that it is in the reverse order. This reaction can be used to control aromatic substitution reaction further. The ring carbon atom bearing the sulfonic acid group can be blocked from any attack, and electrophiles are directed to attack the other positions.

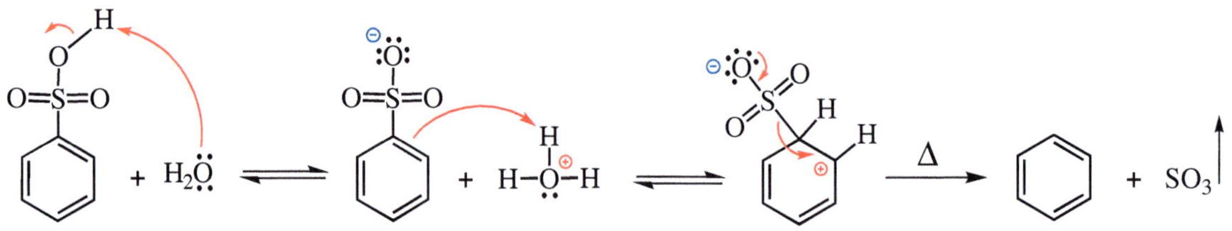

6.6.4 Friedel–Crafts Acylation

Introduction of an RCO- group into a molecule is called an acylation or alkanoylation reaction. It is named after the French chemist and mineralogist Charles Friedel (1832–1899) and the American chemist James Mason Crafts (1839–1917) who discovered the reaction in 1877. Acylation is based on the reaction between an acyl group and an aromatic ring.

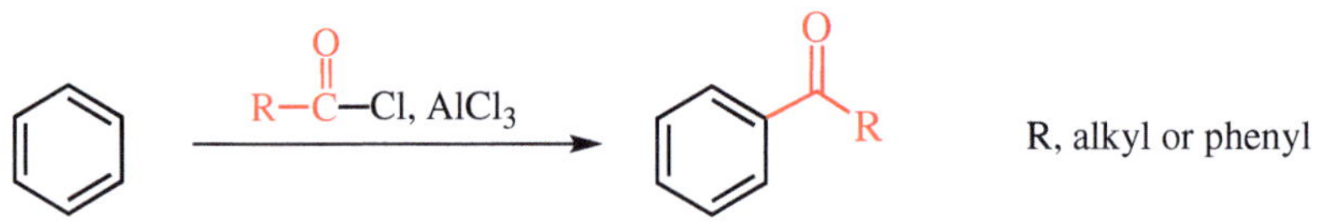

Friedel–Crafts acylation reaction

Anhydrides or a carboxylic acid together with a mineral acid may also be used as acylation reagents. The usual method for generating an acyl cation involves the reaction between an acyl halide and a Lewis acid. Acyl chlorides can be easily synthesized by the reaction of carboxylic acids with thionyl chloride. The common acyl groups are the benzoyl group and acetyl group.

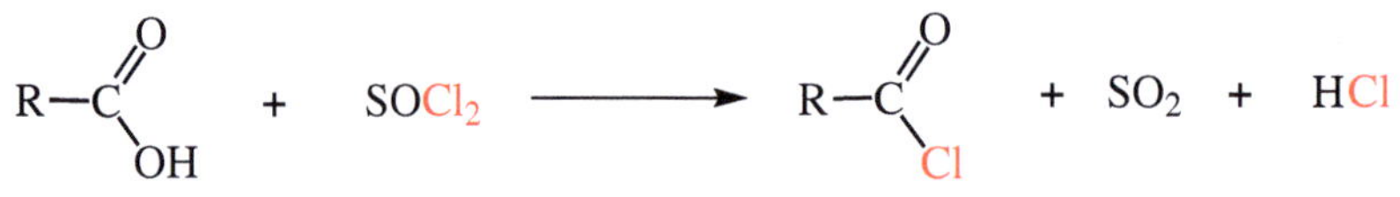

Acyl chloride

The key intermediate in the Friedel–Crafts acylation reaction is the acylium cation, which is formed by the reaction of acyl chlorides and aluminum chloride ($AlCl_3$). Aluminum chloride initially coordinates with the nonbonding electrons of carbonyl oxygen. This complex is in equilibrium with an isomer in which aluminum chloride is bound to the halogen. Dissociation of this complex generates the acylium cation. The electrophile formed is resonance-stabilized because the neighboring oxygen atom donates the nonbonding electrons to the vacant p orbital on carbon as shown below.

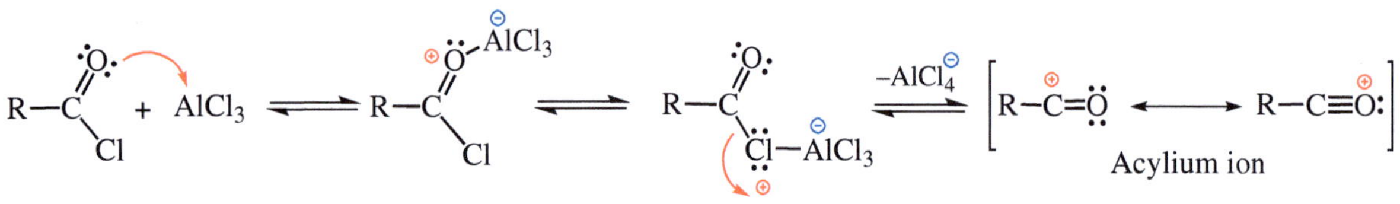

Acylium ion

In the next step, the π electrons of the aromatic ring act as a nucleophile and attack the electrophilic acylium ion to form the cyclohexadienyl cation. This cation is a resonance-stabilized high-energy intermediate as the aromaticity is destroyed. In the next step, the aromaticity is regenerated by removing the proton attached to the sp^3-hybridized carbon atom.

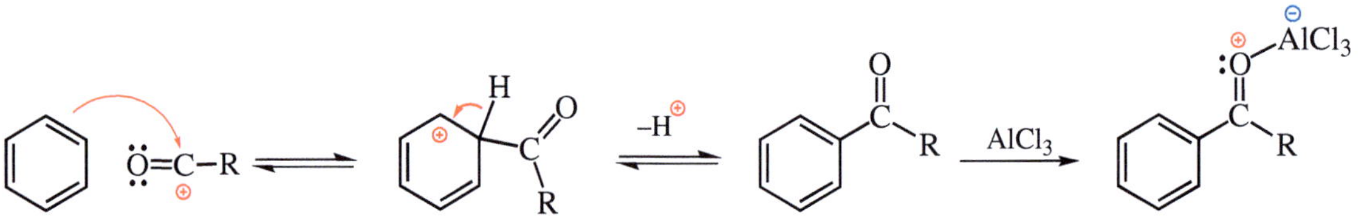

The reaction requires the use of at least 1 equiv of Lewis acid such as AlCl$_3$ because the resulting ketone binds 1 equiv of AlCl$_3$ in a Lewis acid complex. The complex is destroyed upon aqueous work-up to give the desired ketone. Lewis acids such as SbF$_5$, TiCl$_4$, SnCl$_4$, and BF$_3$ can also promote the reaction.

Unfortunately, Friedel–Crafts formylation has not yet been successful because halides and anhydrides of formic acid are considerably less stable. The only known stable halide of formic acid is formyl fluoride. Formylation of an aromatic compound was accomplished with the stable formyl fluoride, which was synthesized by the reaction of acetic formic anhydride with anhydrous hydrogen fluoride. A mixture of formyl fluoride, acetyl fluoride, and the corresponding acids is formed. Formyl fluoride can easily be separated from the reaction mixture.

Formyl fluoride forms a complex with boron trifluoride at low temperatures that is used as the formylating reagent in electrophilic aromatic substitution reactions. Various aromatic aldehydes were synthesized in yields from 56% to 78%. Use of aluminum halide as a catalyst results in the decomposition of formyl fluoride [210]. Vilsmeier–Haack formylation cannot be applied to benzene as the Vilsmeier reagent is a weak electrophile. Vilsmeier–Haack formylation is possible only on nucleophilic aromatic compounds such as aniline, phenol, and their derivatives.

6.6.5 Friedel–Crafts Alkylation

One of the most useful electrophilic aromatic substitution reactions is Friedel–Crafts alkylation, the introduction of an alkyl group into the aromatic ring. Aromatic compounds can be alkylated by carbocations, forming a new carbon–carbon bond. Suitable reagents are alkyl halides and alkyl sulfonates, which generate carbocations as electrophiles in the presence of AlCl$_3$. Friedel–Crafts alkylation is not restricted to the use of alkyl halides. Carbocations can also be generated from alcohols and olefins in the presence of catalytic amounts of acids.

Friedel–Crafts alkylation reaction

Although the Friedel–Crafts alkylation reaction works generally well, there are some limitations that restrict its use. This reaction works only with benzene, alkylbenzenes, and activated benzene derivatives (benzene derivatives having strong electron-donating groups). On the other hand, benzene derivatives having strong electron-withdrawing groups (–NO$_2$, –SO$_3$H, and –C=O) are less susceptible to alkylation reactions. The alkyl groups increase the reactivity of the benzene ring, making it more reactive than the starting material. For example, as some i-propylbenzene is formed, the ring is activated, reacting even faster than benzene itself. As a consequence, di- or trisubstituted benzene derivatives will be formed as side products. Overalkylation can be hindered by using a large excess of benzene.

Recall that a primary carbocation undergoes rearrangement to a more stable carbocation. Only certain alkylbenzenes, t-butylbenzene, i-propylbenzene, and ethylbenzene can be synthesized because the corresponding cations are not prone to rearrangement. However, when *n*-butylchloride is used as an alkylation reagent, the expected product, *n*-butylbenzene, is formed as the minor product in 35% yield. The major product is the rearranged product, 2-phenylbutane (65%).

The alkyl halides can react via an S_N1 reaction as well as an S_N2 reaction. The reaction of *S*-2-chlorobutane with benzene in the presence of $AlCl_3$ gives the alkylated product with complete racemization. This means that alkylation proceeds through an S_N1 mechanism with a solvent separated ion pair. On the other hand, the reaction of mesylate of *S*-methyl lactate and $AlCl_3$ with benzene forms the alkylation product with complete inversion of configuration. This means that the mesylate anion is substituted by benzene in an S_N2 reaction. In this case, a carbocation cannot be generated. The cation formed would be strongly destabilized by the strong electron-withdrawing COOMe group.

Friedel–Crafts alkylations can also be carried out with compounds generating carbocations as the intermediates. Protonation of alkenes with HF forms carbocations. HF is used as the protonating acid because the fluoride ion is a weak base and does not attack the carbocations. The alkenes are first protonated according to Markovnikov's rule, forming the most stable carbocations. The electrophilic attack on the benzene forms the σ-complex. The fluoride ion picks up the hydrogen from the σ-complex to restore the aromaticity.

Besides $AlCl_3$, many other Lewis acids including $BeCl_2$, BF_3, $TiCl_4$, $SbCl_5$, or $SnCl_4$ are also used as catalysts for Friedel–Crafts alkylation. The major drawbacks are always the use of stoichiometric amounts of Lewis acids and toxic alkyl halides. Using alcohols as the alkylating reagents is the major advantage of alkylation reactions as water is formed as the only side product [211]. Employing small amounts of a Mo(IV)-complex, allyl alcohols can be introduced into electron-rich arenes such as phenol and anisole. Reaction of prenyl alcohol with *p*-cresol in the presence of Mo-complex forms the chromane derivative in 28% yield. By using the tertiary alcohol, the chromane derivative is formed in 46% yield of the "open" *o*-allyl intermediate (∼5%) [212].

In the beginning, we discussed how the alkylation reaction with primary alkyl halides gives rearranged products as the major compounds. Unlike Friedel–Crafts alkylation, no rearrangement occurs with the Friedel–Crafts acylation. The question arises *How can we control the rearrangement during alkylation?* In other words, *how can we synthesize the nonrearranged products starting from the primary alkyl halides?* For example, let us look at how we can synthesize n-propylbenzene, which could not be produced by the Friedel–Crafts alkylation reaction of benzene with 1-propylchloride. In the first step, we perform a Friedel–Crafts acylation reaction with propionyl chloride in the presence of AlCl$_3$. Rearrangement is out of the question. The next step will be the reduction of the carbonyl group to the corresponding hydrocarbon.

There are several methods available for reduction of a ketone to a hydrocarbon. The Wolff–Kishner reduction of ketones utilizes hydrazine (NH$_2$NH$_2$) as the reducing agent in the presence of a strong base such as KOH. The mechanism of the reduction is shown in Section 5.1.3. Clemmensen reduction is a second way to reduce the carbonyl groups of an aromatic ketone to hydrocarbons.

6.6.6 Clemmensen Reduction

Clemmensen reduction reduces acylbenzenes to alkylbenzenes by using amalgamated zinc (zinc dissolved in mercury) in the presence of hydrochloric acid. Aromatic alkyl ketones are stable to heating with zinc and HCl, but Clemmensen discovered that the use of a zinc–mercury amalgam as a catalyst instead of zinc gave the corresponding hydrocarbons. The mechanism for Clemmensen reduction has not been completely clarified [213], but it is well known that reduction takes place at the surface of the zinc catalyst via a radical intermediate. It is well established that the alcohol is not an intermediate. Treatment of the corresponding alcohols under the same reaction conditions does not give alkanes [214]. After protonation of the carbonyl group by acid, electron transfer from zinc to the carbonyl group takes place. The carbon radical formed binds to the Zn through the oxygen atom of the carbonyl group. Proton abstraction from water forms the intermediate, which undergoes homolytic cleavage of the carbon–oxygen bond, forming a new carbon radical. Further proton abstraction from water results in the formation of the corresponding methylene group.

The Wolff–Kishner reduction is carried out under strongly basic conditions using high temperatures and protic solvents, whereas Clemmensen reduction is performed in strongly acidic conditions. In the case of having some protective groups in the molecule, they can be removed. Then, alternative methods should be considered. The catalytic hydrogenation using a metal catalyst such as Pd/C or Pt/C with hydrogen can be applied to reduce the carbonyl group. Usually, reduction of ketones stops at the stage of alcohol. However, if the alcohol is at a benzylic position, reduction can continue further to the alkane. The second method is first conversion of the ketone to a thioacetal using ethane-1,2-dithiol and BF$_3$. Treatment of the thioacetal with the reducing agent Raney/Ni in the presence of hydrogen gives the hydrocarbon (see Section 5.2.2).

6.6.7 Reactivity of Monosubstituted Benzene Derivatives

When a monosubstituted benzene undergoes an electrophilic aromatic substitution reaction, the new substituent can be directed primarily to the *ortho*, *meta*, or *para* position, and three possible disubstitution products might be obtained. The rate of the reaction compared to the rate of the reaction with benzene may be slower or faster. The group attached to the benzene ring will determine the position of the incoming substituent as well as the rate of the reaction. Substituents that increase the rate of the electrophilic aromatic substitution reaction are called *activating*, and those that slow the rate of the reactions are called *deactivating*. Substituents that donate electrons to the benzene ring by inductive or mesomeric effect are called *activating substituents*. In contrast, substituents that withdraw electrons from the benzene ring by inductive or mesomeric effect are called *deactivating substituents*.

If a substituent is more electron withdrawing than a hydrogen atom, this substituent will withdraw the σ electrons (−I effect) away from the benzene ring and decrease the electron density at the benzene ring. For example, a protonated amine group attached to the benzene ring is an inductively electron-withdrawing group. On the other hand, alkyl groups attached to the benzene ring donate σ electrons (+I effect) to the benzene ring better than a hydrogen atom. A carbon atom is slightly more electronegative than H. Therefore, hydrogen atoms of the methyl group push the electrons toward the carbon atom. The *ipso* carbon atom (the carbon atom to which the substituent is bonded) of the benzene ring withdraws the σ electrons from the methyl group because of the sp² hybridization.

If a substituent directly attached to the benzene ring has a lone pair of electrons on the next atom, they will delocalize over the ring; these substituents will activate the benzene ring by increasing the electron density at the benzene ring by mesomeric effect (+M effect). Of course, the rate of substitution will also be increased. For example, the lone pair of electrons on oxygen in phenol delocalize into the benzene ring and increase the electron density.

If a substituent is attached to the benzene ring by an atom that is doubly or triply bonded to a more electronegative atom, this group will withdraw the π electrons of the ring by inductive and mesomeric effect and delocalize the ring electrons onto the substituent. As a consequence, the electron density at the benzene ring will be decreased and the ring will be deactivated. The rate of the second substitution will also be lowered. For example, the π electrons of the benzene ring of benzaldehyde will delocalize onto the aldehyde group.

Y is more electronegative than X

Strongly activating substituents donate electrons to the aromatic ring by resonance and withdraw electrons from the benzene ring inductively. However, the mesomeric effect always dominates the inductive effect. The lone pair of electrons from the substituents flow from the substituent to the ring, making the ring negatively charged. The substituents with mesomeric electron-donating ability have the general structure −X:, where the X atom has a lone pair of electrons for donation.

Strongly activating substituents

Moderately activating substituents also donate electrons to the aromatic ring by resonance and withdraw electrons from the ring inductively. For example, N-acetylated aniline and O-acetylated phenol derivatives belong to this group. These substituents moderately activate the aromatic ring. The substituents can donate the lone pairs on the nitrogen or oxygen atoms in two different directions: into the ring and into the acetyl group. As these substituents increase the reactivity of the benzene ring, this means that the decreased donation of the electrons to the ring overwhelms the inductive effect.

Weakly activating substituents also donate electrons to the ring by resonance. Alkyl, aryl, and C=C double bonds belong to this group. We have already discussed that an alkyl group donates electrons by induction (+I effect). The resonance structures of 1,1'-biphenyl and styrene are given below.

Weakly activating substituents

Strongly deactivating substituents withdraw electrons from the benzene ring by resonance and inductively. Besides the $-NO_2$, $-C\equiv N$, and $-SO_3H$ substituents, ammonium ions such as $-NH_3^+$, $-NH_2R^+$, $-NHR_2^+$, and $-NR_3^+$ substituents strongly deactivate the benzene ring. Ammonium salts have no resonance effect, but they withdraw electrons through the σ bonds.

Strongly deactivating substituents

Moderately deactivating substituents are generally carbonyl groups directly attached to the benzene ring. Aldehydes, ketones, esters, and carboxylic acids belong to this group. They withdraw electrons by resonance and inductively. They decrease the reactivity as well as the rate of the electrophilic substitution.

Moderately deactivating substituents

Weakly deactivating substituents are the halogens. The halogens also contain nonbonding electrons like nitrogen and oxygen. Like activating groups, the halogens also donate electrons to the ring by resonance. On the other hand, they withdraw electrons from the ring inductively. For example, oxygen (3.4) and chlorine (3.2) have similar electronegativities. Therefore, they withdraw electrons from the ring inductively.

Weakly deactivating substituents

However, the donation ability of oxygen by resonance is different from that of chlorine. The resonance interaction of chlorine electron pairs with the ring is much less effective than the interaction of oxygen electron pairs. In the case of nitrogen and oxygen substituents, the 2p orbitals are overlapping. However, chlorine uses the 3p orbital to overlap with the 2p orbital of carbon. As the sizes of these orbitals are different, they do not overlap so effectively. Therefore, the resonance effect of chlorine is weak (Figure 6.19).

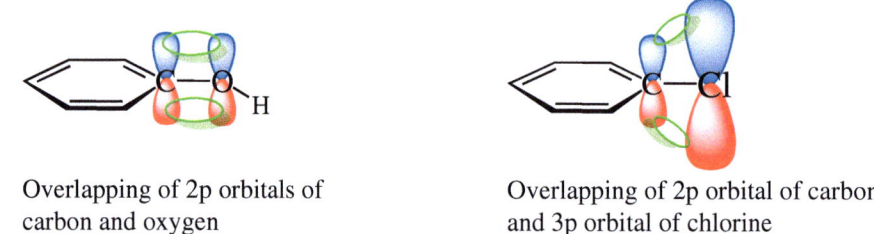

Overlapping of 2p orbitals of carbon and oxygen

Overlapping of 2p orbital of carbon and 3p orbital of chlorine

Figure 6.19 The schematic presentation of overlapping of 2p orbital of carbon with 2p and 3p orbitals of oxygen and chlorine atoms.

Bromine and iodine exert weaker polar effects than chlorine, but their resonance effects are also weaker because they use 4p and 5p orbitals. These halogens also deactivate the benzene ring. What about fluorine? Fluorine has 2p orbitals and donates electrons by resonance better than the other halogens do. However, as fluorine is the most electronegative element, it withdraws electrons from the ring. All halogens withdraw electrons inductively more strongly than they donate electrons by resonance to the benzene ring. Fluorine is also a deactivating group.

6.6.8 Directing Effects of Substituents: Activating Groups

When a monosubstituted benzene undergoes an electrophilic aromatic substitution reaction, the question is *Which position of the benzene ring will be attacked by the new substituent?* For example, the nitration of anisole could in principle give three substitution products: *ortho-*, *meta-*, and *para-*nitroanisole. If the substitution were totally random, the *ortho-*, *meta-*, and *para-*products ratio would be 2:2:1. However, experiments show that this substitution is not random but regioselective. From the product distribution, one can easily recognize that the methoxyl group of anisole preferentially activates the *ortho-* and *para-*positions.

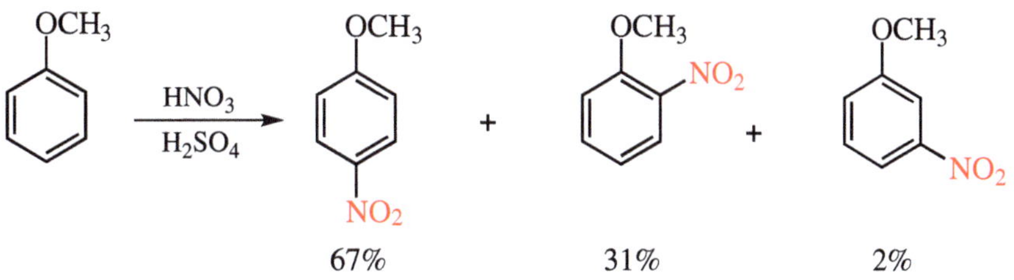

6.6 Electrophilic Aromatic Substitution: Chemistry of Benzene

We already have discussed that the methoxyl group is an electron-donating substituent by resonance and increases the electron density at the benzene ring. The resonance structures of anisole show that the electron density at the *ortho*- and *para*-positions is more increased. This does not mean that the electron density at the *meta*-position is decreased. The electron density is increased around the entire benzene ring. However, the increase in the electron density at the *meta*-position is not as much as that at the other positions.

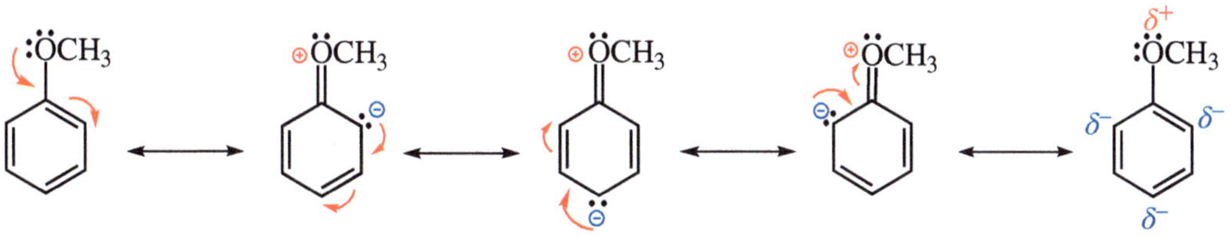

The ^{1}H NMR spectrum of anisole and aniline clearly shows the electron density distribution on the benzene ring. Comparison of the chemical shifts of aromatic protons of anisole and aniline with those of benzene clearly indicates that the resonances of all protons are shifted to the higher field because of the increased electron density at the benzene ring [215]. The *meta*-position is almost unaffected by both kinds of substitution. The ring protons of aniline are more shifted to the higher field than the protons of anisole. This can be explained by the different electronegativities of oxygen and nitrogen atoms. Because nitrogen is less electronegative than oxygen, it can release the nonbonding electrons more easily and increase the electron density at the ring much better than the oxygen atom can. Now, we can understand why the substituents preferentially attack the *ortho*- and *para*-positions.

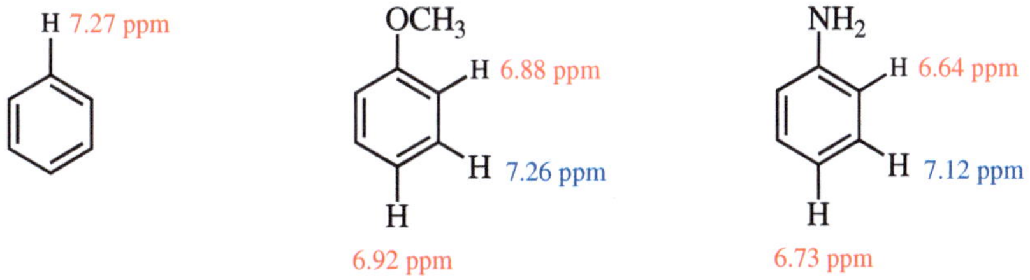

Furthermore, to understand the *ortho* and *para* directing effect of the methoxyl group, we have to analyze the relative stabilities of the three intermediates that would arise from *ortho*, *meta*, and *para* attacks on anisole. The resonance effect of the substituent must also be considered to understand activation. The resonance effect of a substituent is its ability to stabilize the carbocation intermediate by delocalization of electrons from the substituent into the ring. The three possible ions formed by the attack of the NO_2^+ cation at the *ortho*-, *meta*-, and *para*-positions are shown below. In all cases, the ring is positively charged. Four reasonable resonance structures can be written for the arenium ions resulting from *ortho* and *para* attacks. On the other hand, only three resonance structures can be written for the arenium ion resulting from the *meta* attack. This means that the *ortho* and *para* intermediates are more stable than the *meta* intermediate because there is one particularly favorable intermediate that allows the positive charge to be stabilized by donation of nonbonding electrons from the substituent oxygen atom. The intermediates formed by *meta* attack do not have such stabilization.

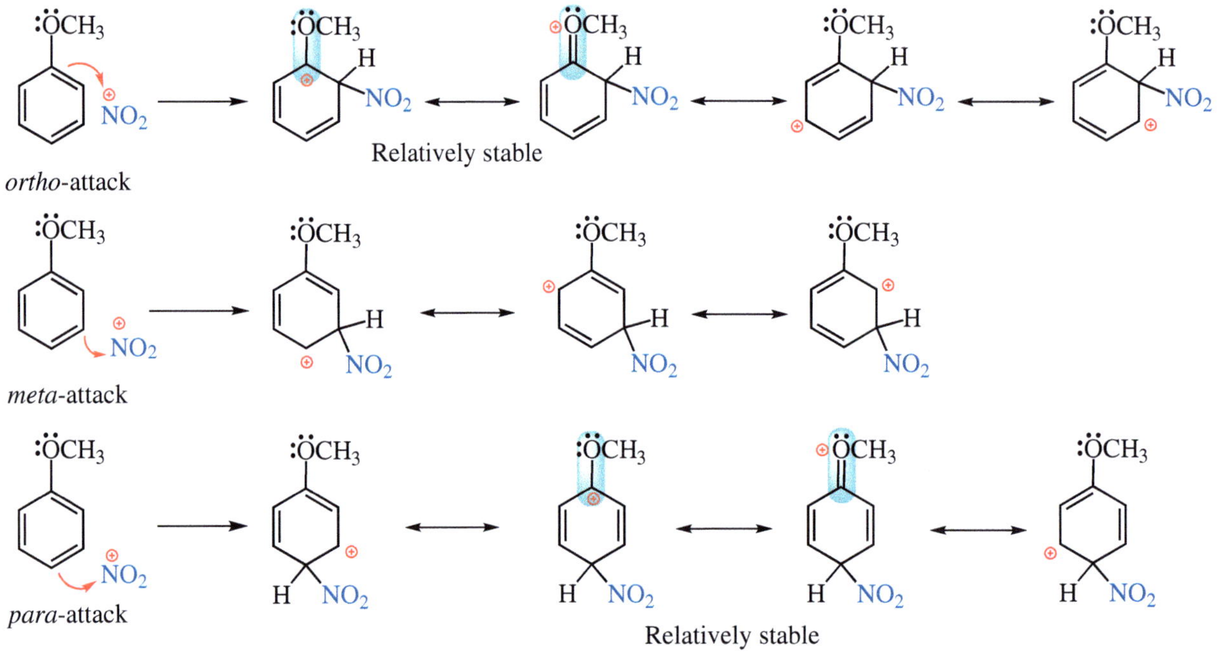

ortho-attack

meta-attack

para-attack

We can conclude *that all substituents that donate electrons by resonance (+M effect) are ortho and para directors*. However, it will be difficult to predict how much of the product will be the *ortho* isomer and how much the *para* isomer. Because there are two *ortho* positions and one *para* position, it is expected on the basis of probability that there should be 67% *ortho* isomer and 33% the *para* isomer. The *ortho*-position is sterically hindered, whereas the *para*-position is not. If either the substituent attached to the ring or the electrophile is large, steric hindrance will inhibit the formation of the *ortho* product and increase the ratio of the *para* product. For example, the nitration of toluene, ethylbenzene, and *t*-butylbenzene under the same reaction conditions demonstrates a decrease in the *ortho–para* ratio with an increase in the size of the substituent.

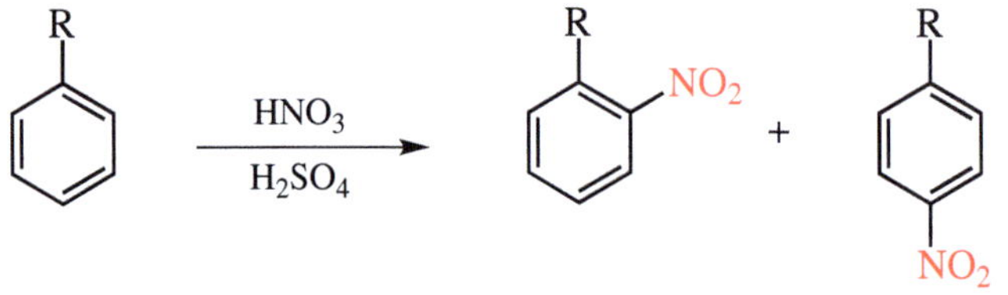

R = CH_3	61 :	39
R = C_2H_5	50 :	50
R = $C(CH_3)_3$	18 :	82

Halogenation of phenol and aniline is a very interesting reaction. The halogenation takes place in the absence of catalysts, and it is difficult to stop the reaction at the stage of single substitution. In aqueous solution, the bromination of phenol and aniline proceeds rapidly and in the presence of excess bromine, it leads quickly to the formation of 2,4,6-tribromophenol and 2,4,6-tribromoaniline, respectively.

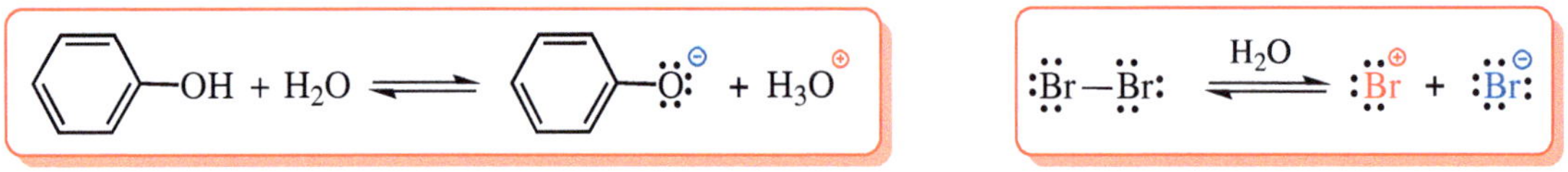

The solvent has a great influence on the reaction. In a water solution, phenol undergoes dissociation and affords an appreciably sufficient concentration of phenoxide anion that is more activated to electrophilic attack [216]. Bromine also gets ionized to a larger extent and forms bromonium ions, which are highly stabilized in water. Ionization of bromine enhances the formation of tribromophenol.

6.6 Electrophilic Aromatic Substitution: Chemistry of Benzene

Mechanism for bromination

The monobromophenol is more active than its precursor, phenol, and the monobromophenoxide ion rapidly undergoes a second bromination, followed by an even more rapid third bromination. We can explain this situation as follows. We have shown that the halogens deactivate the benzene ring. In such a case, incorporation of a bromine atom into the ring should decrease the reactivity of the ring. However, we know that the reactivity of the benzene increases with the number of bromine atoms. The pK_a values of the brominated phenols also vary with the number of bromine atoms attached to the benzene ring.

pK_a values of brominated phenol derivatives.

Phenol	10.0
4-Bromophenol	9.17
2,4-Dibromophenol	7.79
2,4,6-Tribromophepol	6.08

The pK_a value of the phenol is 10. Introduction of a bromine atom into the *para*-position decreases the pK_a value to 9.17. This increases the acidity of the molecule with respect to the number of bromine atoms. When the acidity is increased, the concentration of the phenoxide ion is also increased. In other words, increasing the concentration of the phenoxide ion means increasing the electron density in the ring, which in turn increases the reactivity.

In order to control the bromination of the phenol, dissociation of the hydroxyl group should be hindered. For this purpose, nonpolar solvents are needed. When phenol is reacted with bromine in apolar solvents such as $CHCl_3$ or CS_2 at low temperatures, monosubstituted phenol is formed. At this point, one might raise the question: *If the brominated version of phenol is more active than phenol, why do we not get tribrominate in apolar solvents?* The answer is that tribromide will also be formed in apolar solvents, but it will take longer time because of the lower concentration of phenoxide ion in nonpolar solvents.

We discussed that the strongest activating and *ortho* and *para* directing substituents are the hydroxyl and amino groups. Therefore, it is difficult to control substitution reactions. Furthermore, the reaction of aniline with electrophiles is initiated by attack of the electrophiles by the nitrogen atom instead of the ring. In order to carry out electrophilic substitution reactions with aromatic amines, the electron density at the nitrogen atom should be decreased. For this reason, aniline is first converted into acetanilide by the reaction with acetic anhydride in the presence of pyridine. Phenol can also be converted into phenyl acetate by the same reaction. Now, the lone pairs on nitrogen and oxygen atoms are less available for delocalization with the ring because the lone pairs on nitrogen and oxygen will also delocalize within the carbonyl group. The acetyl groups act to decrease the overall electron-donating character of oxygen and nitrogen. Now, the nonbonding lone electron pairs responsible for the high reactivity are diverted to the acetyl carbonyl groups. However, the overall influence of the modified substituents is still activating and *ortho/para* directing.

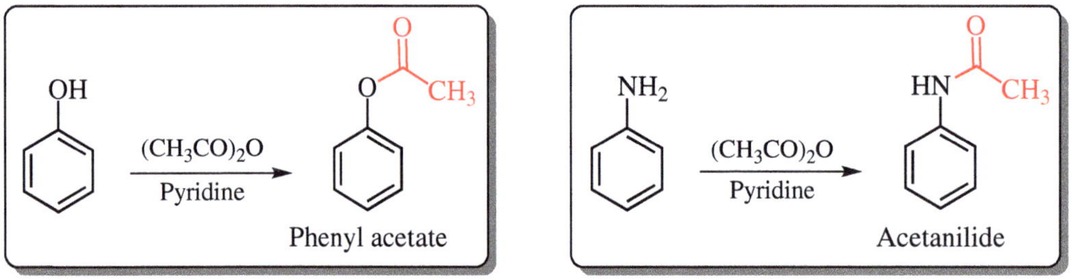

For the synthesis of substituted aniline derivatives, the amine group first must be protected with an acetyl group. Thus, the reactivity of the ring is reduced and the attack of the electrophiles on the nitrogen atom is prevented. For example, for the synthesis of *p*-nitro aniline, first aniline will be converted to acetanilide. Nitration of acetanilide mainly gives the *p*-nitroaniline derivative because of the *N*-acetyl group, which sterically hinders (blocks) the *ortho*-position, making *para* substitution preferable.

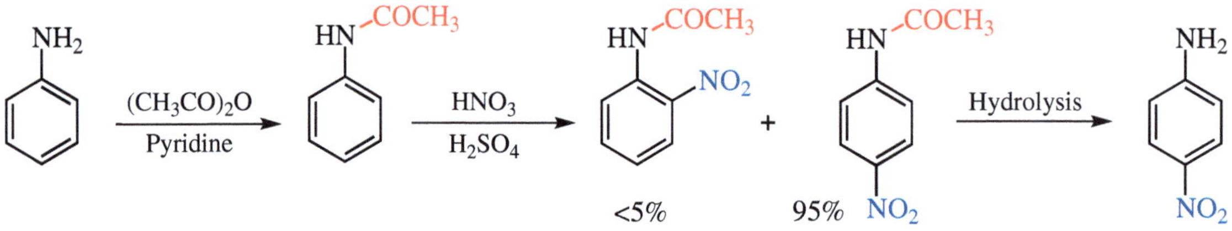

By drawing the resonance structures of aniline or phenol, we were able to determine which carbon atoms would be preferentially attacked by the incoming substituent. However, when an alkyl group is attached to the benzene ring, it is not possible to draw resonance structures. Now, we may raise the question *Which position in toluene will be attacked by the substituents?* For example, bromination of toluene is regioselective and gives *ortho* and *para* substitution products.

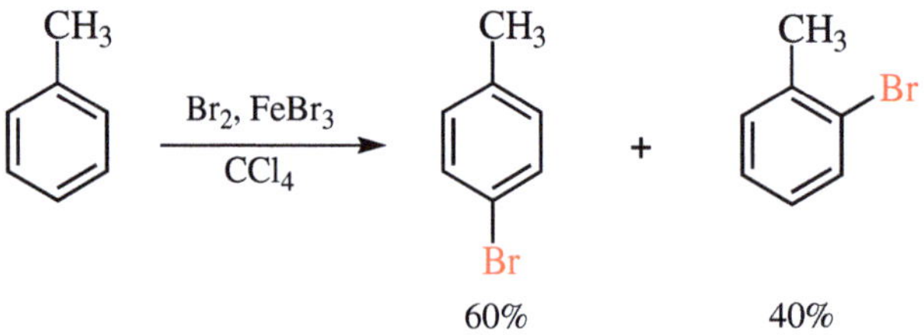

The bromination takes place in the presence of iron(III) bromide ($FeBr_3$). The methyl group acts as an activating group and increases the electron density on the aromatic ring. The rate of the reaction is 25 times faster than that of benzene. To understand the *ortho* and *para* directing effect of the methyl group, we need to examine the stability of the resonance structures formed by the attack of bromine at the different positions of toluene. These resonance structures are shown below.

In all cases, the positive charge is spread over the ring. However, in the case of *ortho* and *para* attacks, the positive charge is spread over two secondary carbons and one tertiary carbon atom. On the other hand, by *meta* attack, all resonance forms are secondary carbocations. The σ-complexes with tertiary carbons are of course more stable than the secondary carbons are. Thus, toluene reacts faster than benzene does at the *ortho* and *para* positions.

6.6.9 Directing Effects of Substituents: Deactivating Groups

The influence of *meta* directing substituents can be explained by using the same arguments that were used for *ortho* and *para* directing substituents. The resonance structures of nitrobenzene show that the electron density at the *ortho* and *para* positions is decreased. This does not mean that the electron density at *meta* position is increased. The electron density is decreased around the entire benzene ring. However, the decrease in electron density at the *meta* position is not as great as that at the other positions.

The ^{1}H NMR spectra of nitrobenzene and benzaldehyde also demonstrate the electron density distribution on the benzene ring. Comparison of the chemical shifts of aromatic protons of nitrobenzene and benzaldehyde with those of benzene clearly indicates that the resonances of all protons are shifted to the lower field because of the decreased electron density at the benzene ring [215]. The *meta* positions are less affected by both kinds of substitution. Now, we can understand why the substituents preferentially attack the *meta* positions.

Nitrobenzene reacts in electrophilic aromatic substitution to give mostly a *meta* substitution product. The bromination of nitrobenzene furnishes exclusively *meta*-bromonitrobenzene. Other electrophilic substitution reactions of nitrobenzene also give mostly *meta* isomers.

This regioselectivity can also be explained by inspection of all resonance structures of the intermediates generated after the electrophile, the bromonium ion, attacks the benzene ring. Although all the three intermediates are resonance-stabilized, the *meta* intermediate is more stabilized than the *ortho* and *para* intermediates. For both *ortho* and *para* reactions, a resonance structure places the positive charge directly on the nitro-substituted carbon atom, where it is disfavored by the repulsive interaction with the positive charge on the nitrogen atom. Therefore, the *meta* intermediate is more favored and it is formed faster than the other intermediates are. Thus, while the electron-withdrawing substituents by resonance deactivate *all* positions, it occurs to a greater extent at the *ortho* and *para* positions than at the *meta* positions.

All *meta* directing groups have either a partial positive charge or a full positive charge on the atom directly attached to the ring. They deactivate the benzene ring. For example, strongly electronegative fluorine atoms in trifluoromethyl benzene strongly withdraw electrons and make this group inductively electron-withdrawing. The benzene ring is strongly deactivated. Therefore, the electrophilic substitution reactions of trifluoromethyl benzene are very sluggish. Protonated aniline derivatives behave similarly. As there is a positive (+) charge on the nitrogen atom, which is directly attached to the aromatic ring, it strongly withdraws electrons from the ring as in the trifluoromethyl group and reduces the reactivity of the ring and directs the substituents to the *meta* position.

The *meta*-directing effect of these groups can be explained using the same arguments used for the *meta*-directing nitro group. In the case of *ortho* and *para* intermediates, a single resonance structure is not favorable because the positive charge is directly on the carbon atom bearing the strong electron-withdrawing group, the trifluoromethyl group, because of the repulsive interaction between the carbocation and positively polarized carbon atom. There is no such situation in the case of *meta* substitution. Thus, the most stable intermediate is formed when the electrophile attacks the *meta* position. Therefore, these and similar substituents are *meta* directors.

The nitration reaction of aniline gives very interesting results in terms of product distribution. In an electrophilic substitution reaction, the *meta* and *para* substitution products are formed in approximately the same ratio.

The formation of *m*-nitroaniline is understandable because aniline undergoes protonation in the strongly acidic solution forming an ammonium ion. As we discussed above, a protonated aniline acts as a *meta* director. On the other hand, the formation of *p*-nitroaniline arises by the rapid nitration of the small amount of highly reactive unprotonated aniline in the reaction mixture.

We have discussed that the deactivating groups by resonance (NO_2, CHO, COR, SO_3H, etc.) and inductively (CF_3 and NH_2R^+) are all *meta*-directing groups. Halogens are also slightly deactivating substituents. They inductively withdraw electrons from the benzene ring more strongly than they donate electrons by resonance. However, contrary to the other deactivating substituents, halogens are *ortho* and *para* directing. Electronegativity does not mean that they will keep their own lone pairs bound to themselves; they can also delocalize over the benzene ring. Thus, a halogen atom attached directly to the ring can stabilize a positive charge on the carbon atom bearing the halogen atom in the same way that hydroxyl or amino groups can. These intermediates are generated only by *ortho* and *para* attacks. In general, halogens are deactivators, but, contrary to the other deactivators, they are *ortho* and *para* directors.

Inductive effect destabilizes the intermediate carbocation

Mesomeric effect stabilizes the intermediate carbocation

6.6.10 Electrophilic Aromatic Substitution on Disubstituted Benzenes

If a disubstituted benzene is subjected to an electrophilic substitution reaction again, we can ask the following question: *To which position will the new substituent be attached?* Of course, the activating and directing effects of substituents discussed for the substitution of monosubstituted benzenes also hold for disubstituted benzenes, except that the directing influences now come from two groups. Each of the substituents exerts an influence on subsequent substitution reactions. The activation or deactivation of the ring can be predicted more or less by the sum of the individual effects of these substituents.

Let us discuss, as an example, the orientation of the electrophilic substitution reaction of *p*-methylacetanilide. Both of the substituents are activating groups. However, the two substituents direct the new substituent to different positions. As the amide functionality is much more activating than the methyl group, the amide group has dominant influence and determines the position of the substituent. Substitution occurs mainly at the *ortho* position to the amide group.

p-methylacetanilide → Br$_2$, FeBr$_3$ → Major product + Minor product

Chlorination of *p*-chlorotoluene occurs preferentially at the two equivalent positions that are *ortho* to the methyl group rather than the two equivalent positions *ortho* to the chlorine atom. These groups are *ortho* and *para* directing. The *para* positions for each group are blocked by the other substituent. As the methyl group is a more activating group, the major product arises from the attack at the *ortho* position next to the methyl group.

p-chlorotoluene → Cl$_2$, FeCl$_3$ → 71% + 29%

In contrast to the *para* isomer, the directing effects of the substituents in the *meta* isomer cooperate with each other. Chlorination mainly takes place at the *ortho* positions of the substituents but mainly at the site of the methyl group. However, only a small amount of reaction occurs at the sterically hindered *ortho* position between the two groups. No product is formed at the *meta* position of the substituents.

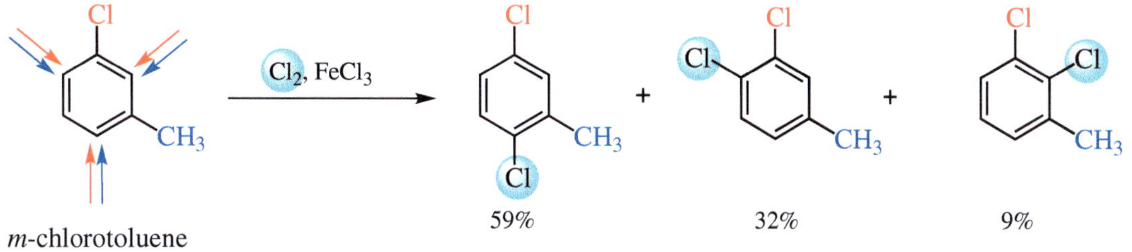

m-chlorotoluene → (Cl$_2$, FeCl$_3$) 59% + 32% + 9%

1,3-Dimethylbenzene is activated toward the electrophilic substitution reaction because the two methyl groups are both activating. Both of the methyl groups activate the same positions. Substitution does not occur to an appreciable extent between *meta* substituents if another position is open. Therefore, bromination of 1,3-dimethylbenzene exclusively gives 1-bromo-2,4-dimethylbenzene as the sole product.

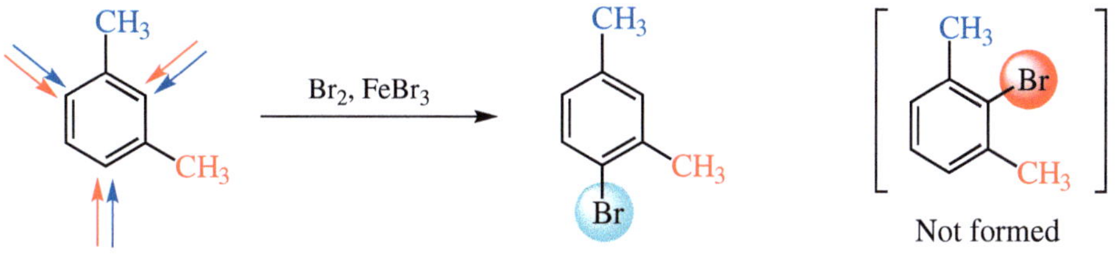

1,3-dimethylbenzene → (Br$_2$, FeBr$_3$) product + [Not formed]

The product distribution is also easy to predict for 1,3-disubstituted benzene derivatives where both substituents are *meta* directors. In contrast to the results for 1,3-dimethylbenzene, electrophiles do not react at the C4 position of 1,3-dinitrobenzene. There is only one unsubstituted position that is *meta* to each, which can be attacked by electrophiles.

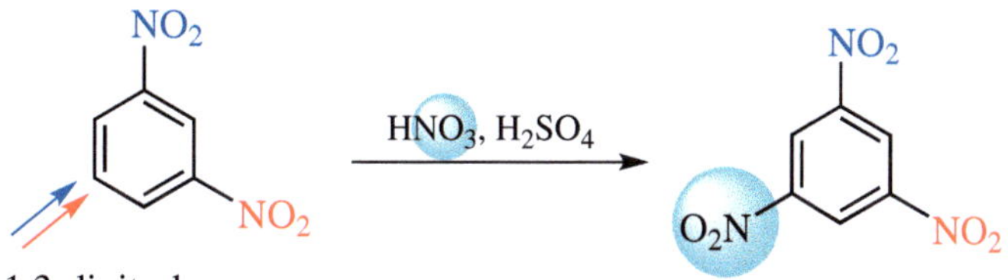

1,3-dinitrobenzene → (HNO$_3$, H$_2$SO$_4$) 1,3,5-trinitrobenzene

If the substituents direct the electrophiles to different positions, the substituent with the stronger activity will dominate. In the following reaction, chlorination of 1-*t*-butyl-3-nitrobenzene, the *t*-butyl group is the activating group and is dominating. The substitution occurs at the *para* position to the *t*-butyl group. The steric effect hinders substitution at the crowded position *ortho* to both substituents.

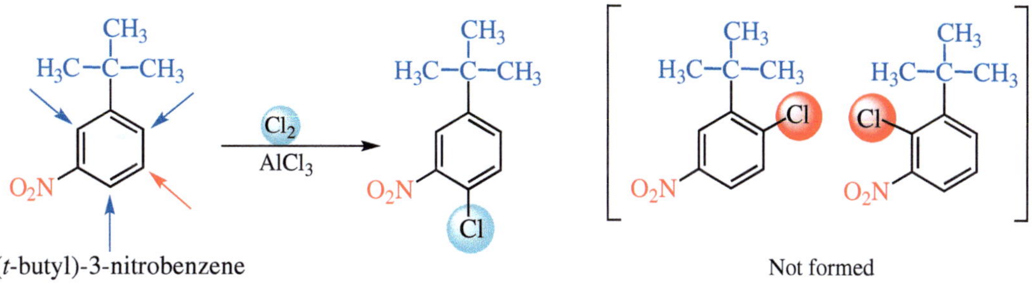

1-(*t*-butyl)-3-nitrobenzene → (Cl$_2$, AlCl$_3$) product + [Not formed]

When two substituents with activating and deactivating ability direct the new substituent to the same position, the position of the new substituent can be easily predicted. For example, nitration of *p*-nitrotoluene exclusively forms a single product, 2,4-dinitrotoluene.

p-nitrotoluene → 2,4-dinitrotoluene (HNO₃, H₂SO₄)

In *o*-nitrotoluene, the substituents are activating and deactivating as discussed above. They direct the incoming electrophile to the same positions. The product arises from the sum of individual group effects. The major product of electrophilic substitution is *para* to the activating methyl group.

o-nitrotoluene

6.7 Functionalization of the Side-Chain Substituents of Benzene

6.7.1 Oxidation of the Side Chain of Alkylbenzenes

Oxidation of the benzylic C—H bonds of alkyl benzene is a very important reaction because the oxidative products serve as vital building blocks for the synthesis of special chemicals. Alkyl groups are usually fairly resistant to oxidation. However, when they are attached to a benzene ring, they are easily oxidized by a basic solution of $KMnO_4$ or other oxidation reagents such as CrO_3. For example, oxidation of toluene gives benzoic acid in almost quantitative yield. Two factors contribute to this high yield procedure, despite the use of a strong oxidant. First, the aromatic ring is resistant to attack by $KMnO_4$, which oxidizes the side chain. Second, the benzylic position is susceptible to hydrogen abstraction by the oxidant.

Toluene → Benzoic acid (1. $KMnO_4$, HO^-, heat; 2. H_3O^+)

Alkylbenzenes with a longer chain than methyl are ultimately degraded to benzoic acid. For example, when a long carbon chain such as an *n*-butyl group is attached to the benzene ring, oxidation continues and benzoic acid is formed. An interesting variation is on a cyclic molecule like tetralin; the oxidation of tetralin gives dicarboxylic acid.

n-butylbenzene → benzoic acid (1. $KMnO_4$, HO^-, heat; 2. H_3O^+)

Tetralin → phthalic acid (1. $KMnO_4$, HO^-, heat; 2. H_3O^+)

The mechanism of side-chain oxidation of aromatic rings is complex. The first step is almost certainly homolytic cleavage of the benzylic C—H bond by permanganate to form a benzylic radical, which can be stabilized by resonance. Permanganate oxidizes the benzylic group first to an alcohol, followed by oxidation to a carbonyl group. Then, the remaining carbon atom next to the carbonyl group is cleaved. This mechanism is strongly supported by the fact that the *t*-butylbenzene is resistant to oxidation as it does not contain any benzylic C—H bond.

tert-butylbenzene —1. KMnO₄, HO⁻, heat; 2. H₃O⁺→ No reaction

6.7.2 Halogenation of Side Chains of Alkylbenzenes

Benzylic halogenation is an important reaction when a leaving group is needed for substitution or elimination reactions. We have shown that benzene does not undergo any reaction with bromine or chlorine in the absence of Lewis catalysts. In contrast, alkylbenzene bearing at least one hydrogen atom in the benzylic position can be chlorinated or brominated by heating or irradiating. The mechanism of benzylic halogenation also proceeds through radical intermediates. Heat or light induces homolytic cleavage of the halogen–halogen bond into radicals. Then, in the first step, a halogen radical (bromine radical) abstracts a hydrogen atom from the benzylic position forming a carbon-centered benzylic radical and HBr. In the next step, the benzylic radical abstracts a bromine radical from the unreacted bromine molecule producing benzylbromide as well as a new bromine radical.

This process is a chain reaction because the newly formed bromine radical can enter into a new cycle of propagation. The benzyl radical can be stabilized by delocalization of the radical around the entire benzene ring as the p orbital is parallelly aligned with the p orbitals of the benzene ring (Figure 6.20).

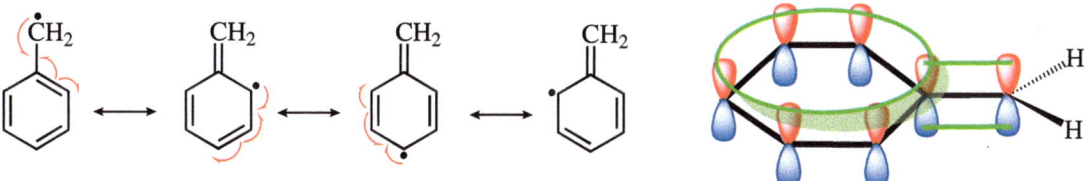

Figure 6.20 Resonance structures of benzyl radical and its stabilization.

The bond dissociation energy of the benzylic C—H bond is about 85 kcal/mol, which is considerably weaker than that of tertiary C—H (bond strength 93 kcal/mol), secondary C—H (95 kcal/mol), and primary C—H (104 kcal/mol) bonds. Phenyl C—H bonds are about 110 kcal/mol, reflecting the instability of phenyl radicals.

Bonds	Hydrocarbon radicals	Bond dissociation energies in kcal/mol
CH₃—H	Methyl	104
C₂H₅—H	Ethyl	98
(H₃C)₂HC—H	iso-Propyl	95
(H₃C)₃C—H	tert-Butyl	93
H₂C=CH—H	Vinyl	112
HC≡C—H	Ethynyl	133
C₆H₅—H	Phenyl	110
C₆H₅H₂C—H	Benzyl	85

In addition to bromine, *N*-bromosuccinimide is also used as an important reagent for benzylic bromination reactions. Chlorination of benzylic positions proceeds by the same mechanism. The chlorination reaction cannot be controlled as the bromination reactions. The chlorine radical is too reactive to give entirely benzylic substitution. The selectivity is decreased. Abstraction of primary, secondary, and tertiary hydrogens by the chlorine atom is *exothermic*. Therefore, the stability of the formed radicals has less influence on the activation energy. For example, chlorination of ethylbenzene forms β-chloroethylbenzene beside the major product, α-chloroethylbenzene. The bromine radical is not as reactive as the chlorine radical. Therefore, bromination reaction occurs exclusively at the benzylic position.

6.7.3 Arenediazonium Ion as an Electrophile

Arenediazonium ions can be used as electrophiles in electrophilic substitution reactions to synthesize substituted benzene derivatives. Primary arylamines react with nitrous acid to produce arenediazonium salts. First, the unstable nitrous acid must be prepared by the treatment of sodium nitrite with aqueous HCl. Nitrous acid can be formed by protonation of a nitrite ion. Protonation of nitrous acid generates an oxonium ion, which has a good leaving group (water). Removal of the water forms a nitrosonium ion. Now, the amine group of aniline can attack the nitrosonium ion to generate *N*-nitrosoammonium salt. The deprotonation of this intermediate generates a stable *N*-nitrosoamine, which tautomerizes to a diazohydroxide in a reaction that is similar to keto-enol tautomerization. Protonation of the hydroxyl group generates a good leaving group, water. This intermediate loses water to form the diazonium ion.

Only primary alkyl and aryl amines can form diazonium salts. As secondary amines have just one hydrogen on the nitrogen atom, they yield the relatively stable *N*-nitrosoamines. The reaction stops at this stage. Tertiary amines do not have hydrogens bonded to nitrogen. Therefore, when a tertiary amine is reacted with nitrous acid, equilibrium is established between the tertiary amine and an *N*-nitrosoammonium compound.

N-nitroso-N-methylaniline

N-nitroso-ammonium salt

Formation of diazonium salts is very important because the diazonium group can be readily replaced with many other groups that are otherwise difficult to install. Diazonium ions are weak electrophiles, and, therefore, they can undergo aromatic electrophilic substitution reactions with activated aromatic compounds such as aryl amines and phenols. For example, N,N-dimethylaniline reacts with the diazonium ion almost at the *para* position. However, careful control of the pH of the reaction medium is necessary for the success of the process. The product is an azo compound. The mechanism for the formation of this product is the same as the one we discussed for other electrophiles.

p-dimethylaminoazobenzene

Aromatic azo-compounds are colored. Several of those compounds synthesized by diazo-coupling are widely used as dyes for textiles because of their extended conjugated π electron system causes them to absorb in the visible region of the electromagnetic spectrum. Azo-compounds that contain both an acidic and a basic group can be utilized as indicators because the colors of the conjugate acid and the conjugate base are different. *Para red* is the first azo dye, discovered in 1880. *Para red* was synthesized by diazotization of *p*-nitroaniline, followed by coupling with β-naphthol.

β-naphthol

β-naphthol orange

6.8 Nucleophilic Aromatic Substitution Reactions

Nucleophilic aromatic substitution (S_NAr) is one of the most widely applied reactions providing a broadly useful platform for the modification of aromatic ring scaffolds. In Chapter 2, we discussed nucleophilic substitution reactions and the major mechanisms S_N1 and S_N2. *Can we apply these two mechanisms to aromatic compounds?* In fact, neither of them can be applied to nucleophilic aromatic substitution reactions. First of all, the S_N1-type mechanism cannot be involved because a cation directly on the benzene ring is very unstable. A phenyl cation is less stable than even a primary carbocation because of the hybridization of the aromatic carbon atom. Furthermore, as the positive charge of a phenyl cation is localized in a sp^2 hybrid orbital that is orthogonal to the π system of the aromatic ring, there is no stabilization by the ring electrons. Aryl halides also cannot undergo an S_N2-type reaction because, as the nucleophile approaches the back of the sp^2 carbon atom bearing the leaving group, it will be repelled by the π electron cloud of the aromatic ring. Furthermore, the back lobe of the sp^2 carbon is directed toward the center of the ring. An inversion mechanism is hindered by the geometry of the ring.

In order for nucleophiles to attack the aromatic ring, the electron density in the ring must first be decreased. This is only possible by attaching powerful electron-withdrawing substituents to the ring. In other words, reducing the electron density makes the ring more reactive toward nucleophilic attack. However, the same substituents deactivate the ring toward electrophilic aromatic substitution reactions. There are four principal mechanisms by which nucleophilic aromatic substitution can occur.

1. Addition–elimination mechanism
2. Reaction of arenediazonium salts with nucleophiles
3. Elimination–addition mechanism: benzyne mechanism
4. Free radical mechanism

6.8.1 Addition–Elimination Mechanism ($S_N Ar$ Mechanism)

According to this mechanism, the following criteria must be fulfilled for a reaction to occur.

(a) The aromatic ring must have at least one strong electron-withdrawing group.
(b) The aromatic ring must have a leaving group.
(c) The leaving group must be positioned *ortho* or *para* to the electron-withdrawing group.

The greater the number of electron-withdrawing substituents, the more easily the nucleophilic aromatic substitution will occur. The presence of electron-withdrawing groups on the aromatic ring makes the system more susceptible to nucleophilic attack. The mechanism is similar to that of electrophilic substitution except that an anion rather than a cation intermediate is involved.

Let us consider the reaction of *p*-fluoronitrobenzene with NaOH. First, hydroxide (the nucleophile) attacks the carbon atom bearing the fluorine atom forming a negatively charged intermediate, called a *Meisenheimer complex*. Because this destroys the aromaticity of the ring, it is also the rate-limiting step. The negative charge is delocalized over the ring and further delocalized into the electron-withdrawing group (nitro group). The intermediate is electron-rich and is stabilized by electron-withdrawing substituents, such as a nitro group (NO_2). The negative charge can be stabilized by the electron-withdrawing group only by positioning of the nitro group in an *ortho* or *para* position. Removal of fluoride from the σ-complex gives the product *p*-nitrophenol.

In the case of aliphatic substitution reactions, the bond strength of C–X (X = F, Cl, Br, and I) is the dominating factor so that the order of reactivity increases from fluorine to iodine. Fluorine is a very poor leaving group in $S_N 1$ as well as in $S_N 2$ reactions. In nucleophilic aromatic substitution reactions, the rate-determining step attacks the aromatic ring, not breaking the C–X bond. The more electronegative halogen, fluorine, pulls the electron density out of the ring and activates it toward the attack. Therefore, the reactivity order is as follows: F > Cl > Br > I [217].

The halo-2,4-dinitrobenzene derivatives are converted into the corresponding substitution products at different rates in nucleophilic aromatic substitution reactions. As one can see, fluorine reacts most rapidly, although it is a very poor leaving group. The fluorine derivative was observed to be 3100 times faster than iodine [217].

F-2,4-(NO₂)₂-C₆H₃	Cl-2,4-(NO₂)₂-C₆H₃	Br-2,4-(NO₂)₂-C₆H₃	I-2,4-(NO₂)₂-C₆H₃
3100	5.2	3.4	1.0

Relative rates for aromatic nucleophilic substitution

The important differences between electrophilic aromatic substitution and nucleophilic aromatic substitution reactions are given below in Table 6.1.

Table 6.1 The differences between the electrophilic and nucleophilic aromatic substitution reactions.

Electrophilic aromatic substitution	Nucleophilic aromatic substitution [218]
The aromatic ring is electron-rich: nucleophilic	The aromatic ring is electron-poor: electrophilic
The electron-donating groups increase the rate of the reaction	The electron-withdrawing groups increase the rate of the reaction
The ring electrons attack the electrophiles	Nucleophiles attack the ring
The leaving group is H^+	The leaving group is Cl^-
The substituents determine the position of the attack (electronic and steric effects)	The leaving groups determine the position of the attack

Source: Based on Kwan and coworkers [218].

As we discussed above, in the case of nucleophilic aromatic substitution reactions, the electron-withdrawing group must be in the *ortho* or *para* positions relative to the leaving group so that the negative charge on the ring can be distributed over the electron-withdrawing group. However, when the electron-withdrawing group is at the *meta* position, the negative charge cannot be delocalized over the electron-withdrawing group. Therefore, in that case, either the reaction does not work or it will work very slowly.

Substituents other than halogens can also serve as leaving groups. Although alkoxy groups are very poor leaving groups in S_N2 reactions, they can undergo substitution reactions as leaving groups. They can be easily converted to phenols. Sulfonyl groups can also be converted, through their salts, to phenols. Nitro groups can be displaced if other nitro groups are present to activate the aromatic nucleus.

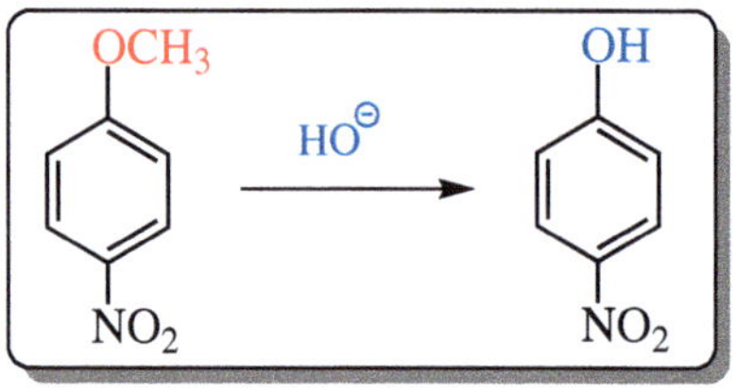

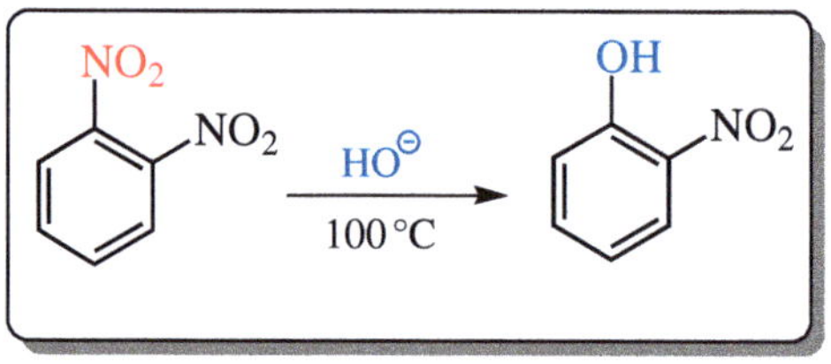

6.8.2 Reaction of Arenediazonium Salts with Nucleophiles

In the previous chapter, we discussed the generation of aryl diazonium salts by the reaction of an aniline derivative with nitrous acid. The aliphatic diazonium ions rapidly undergo decomposition to molecular nitrogen and carbocations. However, the aromatic diazonium salts are stable enough to exist in solution. They are stabilized by resonance so that one electron pair from the ring can delocalize into the functional group as shown below.

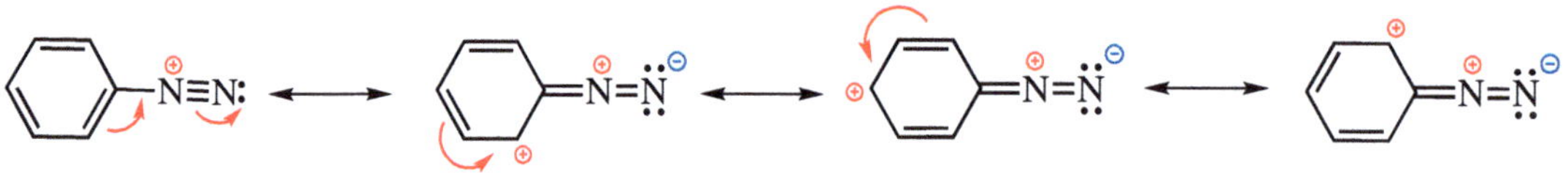

Resonance structures of benzenediazonium ion

The arenediazonium can also be isolated as salts with nonnucleophilic anions such as tetrafluoroborate (BF_4^-) or hexafluorophosphate (PF_6^-). The arenediazonium salts play a very important role in the synthesis of various substituted aromatic compounds because they have an excellence leaving group, N_2. At higher temperatures (>50 °C), nitrogen extrusion takes place and a very reactive phenyl cation is formed. The phenyl cation is unstable and highly unselective. Therefore, it can undergo various nucleophile substitution reactions.

6.8.2.1 Reductive Dediazonization

The diazonium group can be replaced with a hydrogen atom. In fact, this is a reductive deamination reaction and the suitable reagent for this reaction is hypophosphorous acid (H_3PO_2) [219]. The reduction can also be performed with $NaBH_4$ [220].

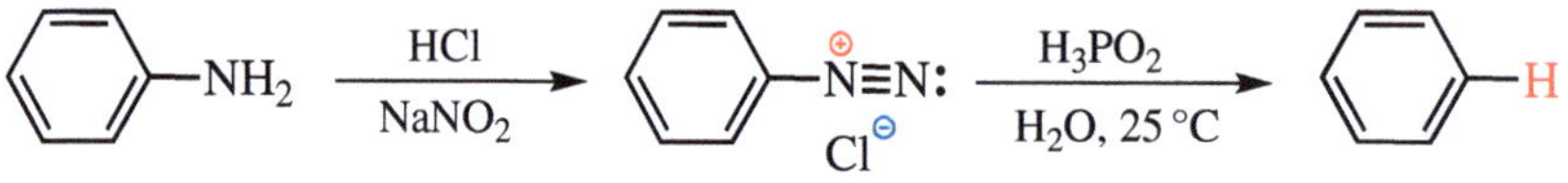

The reduction by H_3PO_2 proceeds by one-electron reduction, followed by extrusion of nitrogen and formation of a phenyl radical. The reduction is improved by using cuprous oxide as a catalyst [221]. Then, the hypophosphorous acid transfers a proton to the phenyl radical.

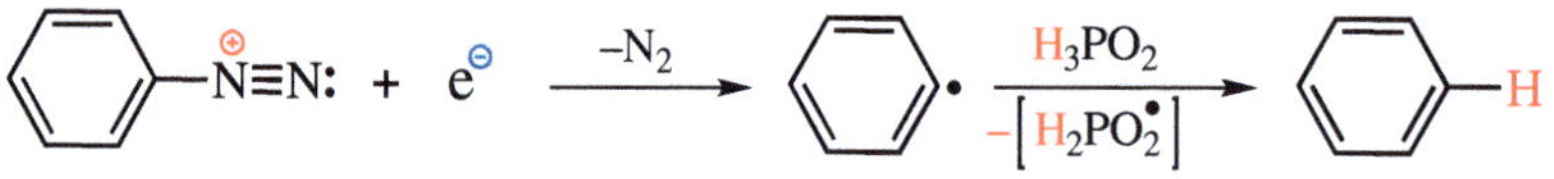

Initiation

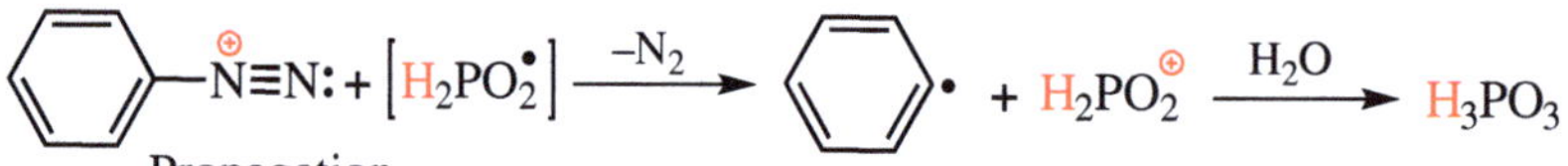

Propagation

Reductive deamination can be successfully applied to the synthesis of some substituted aromatic compounds that are difficult to synthesize. The directing effects of an amino or a nitro group can be used and then the amino group subsequently can be removed. For example, the synthesis of 1,3,5-tribromobenzene is difficult. The product cannot be made by direct bromination of benzene. Starting with aniline, however, tribromination occurs *ortho* and *para* to the strongly directing amino substituent. Diazotization followed by treatment with H_3PO_2 to remove the diazo group yields the product with the desired constitution.

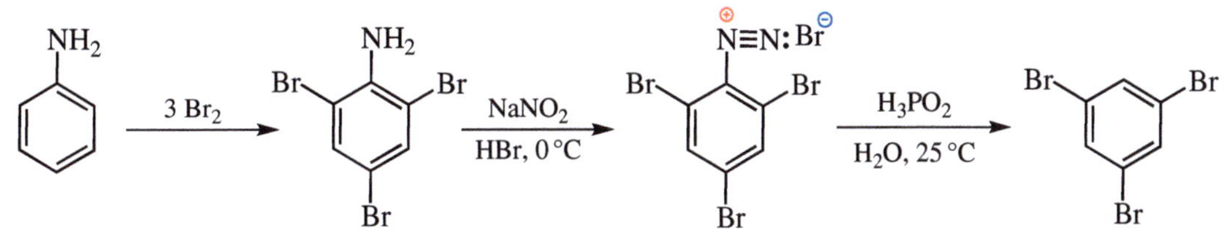

The diazonium group can also be replaced by a hydroxyl group to give a phenol. When an acidified solution (H_2SO_4) of an arenediazonium salt is heated, the hydroxyl group of water will capture the initially formed phenyl cation to form a phenol. Phenols can also be synthesized under milder conditions in the presence of Cu_2O in an aqueous solution of $Cu(NO_3)_2$ [222].

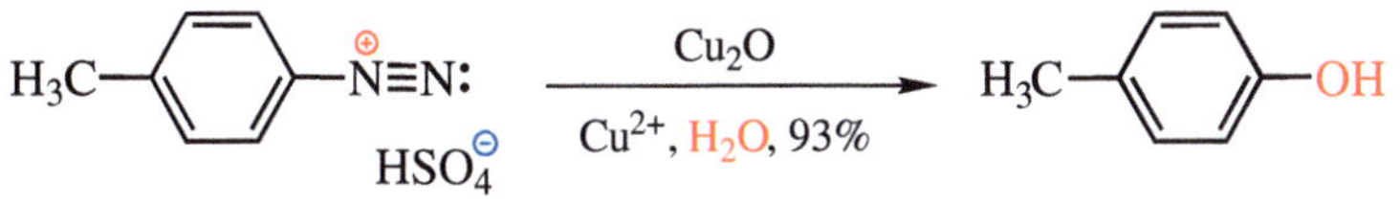

The replacement of the diazonium group of arenediazonium salts by a halogen substituent is a well-known alternative to direct halogenation of aromatic compounds. It is difficult to synthesize aryl fluorides by the direct fluorination of aromatic compounds because of the violent nature of the reaction and the difficulty in controlling the reaction. Aryl fluorides can be easily prepared via diazonium ions by the *Balz–Schiemann reaction* [223]. First, aryl diazonium tetrafluoroborate salt (it can be isolated) is prepared by the reaction of amine derivatives with $NaNO_2$ in the presence of HBF_4. This aryl diazonium salt is subjected to heat to cause thermal decomposition and to give the aryl fluorides. It is likely that first the aryl cation is formed, which abstracts the fluoride ion from the tetrafluoroborate anion as shown below.

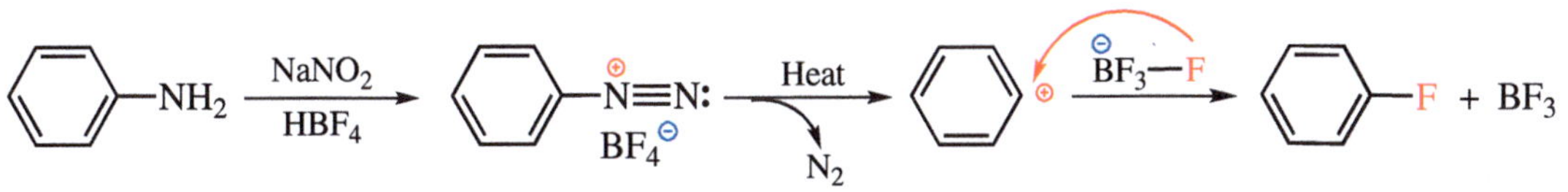

Iodobenzene derivatives are valuable intermediates in the synthesis of a wide variety of organic compounds via reactions involving C—C bond formation. Aryl diazonium ions are converted to the corresponding iodides by reaction with iodide salts. Treatment of the following diazonium hexafluorophosphate salt with sodium iodide in acetone gives a mixture of uncyclized and cyclized iododediazoniation products [224]. The formation of the cyclized product is evidence of the formation of radical intermediates. It is suggested that the key step is the reduction of the diazonium ion by a single electron transfer from iodide to form the aryl radical. The aryl radical then abstracts iodine from either I_2 or I_3^-.

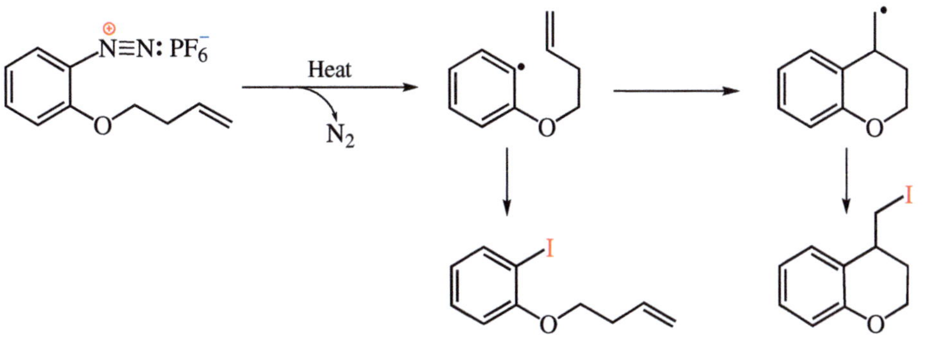

We discussed the synthesis of aryl fluorides and aryl iodides by using arenediazonium ions. Aryl bromide or aryl chlorides cannot be synthesized by heating diazonium salts in the presence of KBr or KCl. Nucleophiles, such as Cl⁻, Br⁻, or CN⁻, can substitute the diazonium group if the corresponding copper(I) salts are added to a solution containing the arenediazonium salt. This reaction is known as the *Sandmeyer reaction* [225].

The mechanism starts with a single electron transfer from the copper(I) to the diazonium group to form a neutral diazo radical and copper(II) halide. Then, the C—N bond of the diazo radical undergoes homolytic cleavage to form an aryl radical and release nitrogen gas. The aryl radical abstracts a halide from the copper(II) halide to regenerate the copper(I) halide catalyst and furnish the final aryl halide product.

As discussed above, these aryl diazonium substitution reactions significantly expand to the synthesis of polysubstituted benzene derivatives. The usual precursor to an aryl amine is the corresponding nitro compound. A nitro group deactivates the aromatic ring and directs substituents to the *meta* position. A nitro group can be easily reduced to an amine, which activates the aromatic ring and directs the substituents to *ortho* and *para* positions. Conversion of an aryl amine to a diazonium ion intermediate allows us to introduce various substituents into the aromatic ring. The intermediates in some of these reactions are the highly reactive aryl cations and some aryl radicals. The transformation of arene diazonium salts allows us to achieve the regioselective construction of substituted benzene derivatives. A summary of these reactions is given below.

6.8.3 Elimination–Addition Mechanism

An early commercial method for the phenol (the Dow process) involved the reaction of chlorobenzene with concentrated sodium hydroxide and a small amount of water in a pressurized reactor at temperatures above 350 °C. Similarly, chlorobenzene reacts with sodium amide (NaNH$_2$) in liquid ammonia even at very low temperature, −33 °C, to give aniline. These nucleophilic substitution reactions were surprising, as we know that aryl halides are generally incapable of reacting by an S$_N$1 or S$_N$2 mechanism.

In the previous chapter, we discussed that the addition–elimination mechanism for nucleophilic aromatic substitution requires strong electron-withdrawing groups on the aromatic ring. As the benzene ring in both cases is not activated, direct substitution does not seem to be the mechanism of these reactions. Both reactions take place through an *elimination–addition reaction* that involves the formation of a highly unstable intermediate known as *benzyne* or *dehydrobenzene*.

The mechanism of this reaction has been studied extensively and much evidence has accumulated in support of a stepwise process. Sodium hydroxide or sodium amide initiates elimination by abstracting one of the *ortho*-protons because they are the most acidic because of the electron-withdrawing ability of the chlorine atom. The negative charge generated on the *ortho*-carbon atom is partly stabilized by the inductive effect of the chlorine atom. Then, the anion eliminates a chloride anion to produce the highly unstable and reactive *benzyne*. Now, the incoming nucleophile can attack the triple bond of benzyne. Protonation of the resulting anion generates the substitution products.

The first evidence for an elimination–addition mechanism involving benzyne generation was published by Roberts in 1953 [226, 227]. Roberts synthesized a chlorobenzene with an isotopically labeled ^{14}C atom to which the chlorine was attached. The reaction of this ^{14}C-labeled chlorobenzene with KNH$_2$ gave aniline. However, analysis of the product indicated that half of the product had the amino group attached to the ^{14}C-labeled carbon atom as expected, but the other half, the amino group, was bonded to the neighboring carbon atom.

* = ^{14}C-labeled carbon atom

Normal substitution product

Cine substitution product

When o-bromotoluene was treated with potassium amide (KNH$_2$) in liquid ammonia, two products were formed. The occurrence of a normal substitution product (o-toluidine) along with a cine-substitution product (m-toluidine) is a strong indication that a benzyne mechanism is operating. Substitution at the adjacent carbon atom is called *cine-substitution*.

Direct substitution product

Cine substitution product

The rate of the elimination–addition reactions depends on the type of halogens. There are two different reactions that influence the rate of the reaction: proton removal and the leaving group's ability to depart. When the leaving group is Br or I, removal of the proton is the rate-determining step. When Cl or F is the leaving group, cleavage of the C—Cl or C—F bond is the rate-determining step. The following order has been determined from the reaction of aryl halides with KNH$_2$ in liquid ammonia: Br > I > Cl > F.

Benzyne is an extremely reactive intermediate. The triple bond in benzyne cannot be a conventional triple bond. In Section 1.2, we saw that a triple bond has a linear geometry in R—C≡C—R. Therefore, four linear atoms cannot be incorporated into a six-membered ring. Rather, it is in fact a distorted triple bond among more sp^2-hybridized carbons. The π bonds of the benzene ring are not changed. The two remaining electrons of the "triple bond" occupy the two sp^2 hybrid orbitals at the "triple bond" carbons that are not involved in any σ bond. These sp^2 hybrid orbitals form a new bond. For ideal overlapping, the sp^2 orbitals should be parallel to one another. However, because of the geometry, these two sp^2 orbitals are directed 60° away from each other and so their overlapping is not very effective. This makes the triple bond in benzyne highly reactive (Figure 6.21). On the other hand, these orbitals are perpendicular to the p orbitals of the ring and therefore they cannot participate in the resonance with the ring.

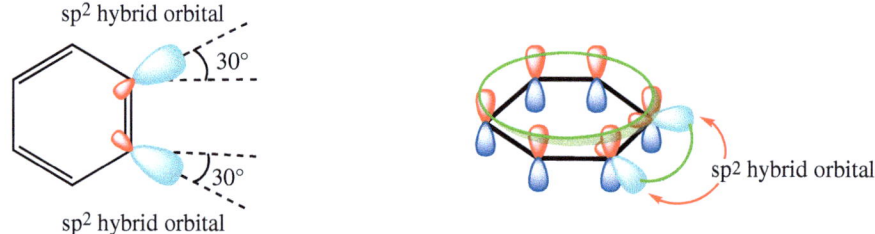

Figure 6.21 The structure of benzyne.

Benzyne was trapped at 6 K in frozen argon and studied through various spectroscopic methods. The triple bond is about 1.26 Å longer (and weaker) than that in ethyne (1.20 Å) and the adjacent C—C bond is about 1.39 Å, slightly longer than that in benzene. The other C—C bond lengths are within 0.01 Å of those of benzene [228–230].

Besides dehydrohalogenation from aromatic halides with strong bases, benzyne can be prepared by a wide variety of methods, some of which are given below. The most convenient and practical method is diazotization of o-aminobenzoic acid. The zwitterion generated by diazotization of anthranilic acid with isoamyl nitrite can cause loss of nitrogen and carbon dioxide to give benzyne [231]. Benzyne is also generated by flash pyrolysis from benzocyclobutenedione. Oxidation of 1-aminobenzotriazole with lead tetraacetate also generates benzyne under very mild conditions. Another method is based on *o*-trimethylsilylaryl triflates. Removal of the trimethylsilyl group induces the elimination of triflate and release of benzyne.

Benzyne is too unstable to be isolated, but it can be trapped with various dienes [232, 233]. Benzyne can undergo [4 + 2] cyclization reactions. Generation of benzyne in the presence of dienes leads to the formation of Diels–Alder products. When benzyne is generated in the presence of anthracene, triptycene is formed by a cycloaddition reaction. When furan is added to a reaction that forms benzyne, furan traps the benzyne intermediate to give benzooxanorbornadiene. In the absence of any dienophile or nucleophile, benzyne undergoes rapid dimerization to form biphenylene. Biphenylene is quite stable both chemically and thermally and behaves in many ways like a traditional polycyclic aromatic hydrocarbon (PAH).

6.9 Polycyclic Aromatic Compounds

6.9.1 Naphthalene

Naphthalene is the simplest fused aromatic compound consisting of two benzene rings. Hückel's rule can also be extended to polycyclic aromatic compounds. The naphthalene protons resonate as two sets of multiplets (AA′BB′-system) at 7.81 and 7.46 ppm, even lower than the benzene protons, strongly indicating the presence of a diamagnetic ring current. The α-protons absorb at lower field than the β-protons because the α-protons are closer to both rings. The X-ray crystallographic measurements also confirm the aromaticity of the naphthalene ring. Naphthalene has a planar structure, like benzene. Unlike benzene, the carbon–carbon bonds in naphthalene are not of the same length. The C—C bonds deviate only slightly from those of benzene (1.39 Å), and they are different from the pure single C—C bond (1.54 Å) and C=C double bonds

(1.33 Å). The resonance energy of naphthalene, 60 kcal/mol (30 kcal/mol per aromatic ring), is less than twice that of benzene (the resonance energy of benzene is 36 kcal/mol).

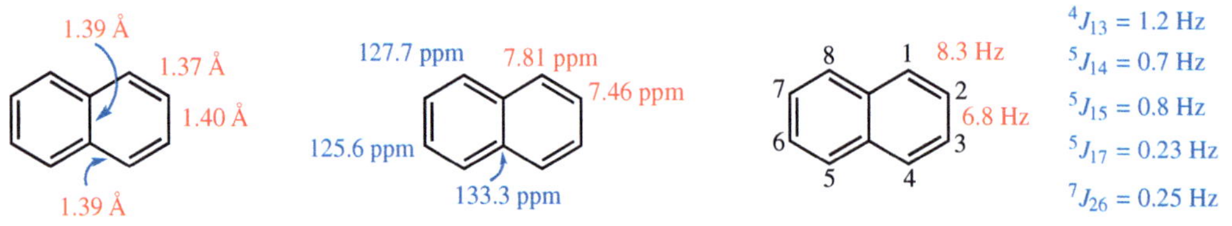

Bond lengths: 1.39 Å, 1.37 Å, 1.40 Å, 1.39 Å

^{1}H and ^{13}C NMR shifts: 127.7 ppm, 7.81 ppm, 7.46 ppm, 125.6 ppm, 133.3 ppm

Coupling constants: $^3J_{12}$ = 8.3 Hz, $^3J_{23}$ = 6.8 Hz, $^4J_{13}$ = 1.2 Hz, $^5J_{14}$ = 0.7 Hz, $^5J_{15}$ = 0.8 Hz, $^5J_{17}$ = 0.23 Hz, $^7J_{26}$ = 0.25 Hz

All polycyclic aromatic compounds can be represented by a number of different resonance structures. Naphthalene, for instance, has three Kekulé resonance structures. In naphthalene, the vicinal coupling, $^3J_{12}$ = 8.3 Hz, is greater than the vicinal coupling, $^3J_{23}$ = 6.8 Hz. Keep in mind that the bond lengths correlate with the vicinal coupling constants. The difference between these vicinal coupling constants arises from the different bond lengths, which can be understood in terms of partial bond localization. The X-ray measurement also shows that the C1—C2 bond is shorter than the C2—C3 bond. Therefore, the resonance structure **B** is the dominant one. The two structures **A** and **C** have one discrete benzene ring each, but they may also be viewed as 10 π electron annulene having a bridging single bond. The structure **B** has two benzene rings that share a common double bond.

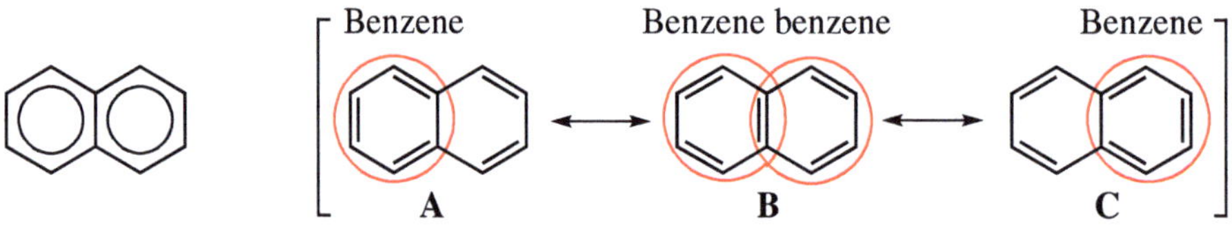

Resonance structures of naphthalene

Naphthalene is more reactive than benzene, both in substitution and in addition reactions. The electrophilic substitution reaction in naphthalene is quite similar to the benzene reaction. In naphthalene, electrophilic substitution reactions can occur at two different positions: at the α-carbon atom (C1) or at the β-carbon atom (C2). The carbocation intermediate formed by the attack of electrophile at the α-carbon atom of naphthalene is much more stabilized by resonance. Because the resulting carbocation is delocalized over the entire ring and stabilized, the naphthalene is more easily attacked by electrophiles than benzene is. In order to understand why electrophiles prefer to attack the C1 carbon atoms, we should analyze the stability of the carbocation formed as the intermediates. The number of resonance structures for an ion increases the stability, which actually decreases the internal energy of the ion and makes the ion more stable. As shown below, attacks of an electrophile at C1 carbon as well as at C2 carbon have equal numbers of resonance structures. Closer inspection of the resonance structures reveals the difference. An attack at C1 carbon atoms allows two resonance structures to keep an intact benzene ring. However, an attack at the C2 carbon atom allows only one intact benzene structure. Therefore, the electrophiles prefer to attack at the C-1 carbon atom.

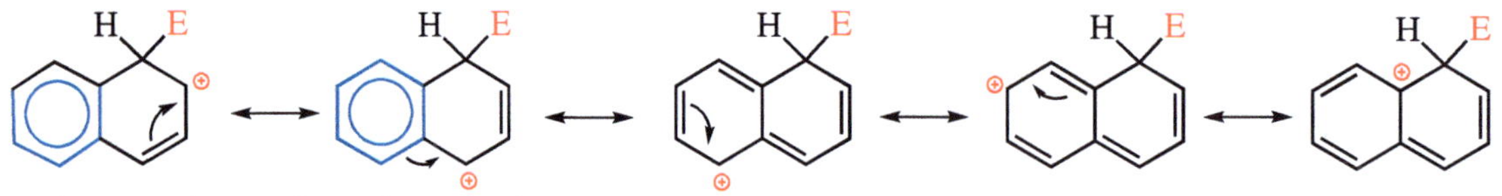

Intact benzene ring
C1 attack   Intact benzene ring

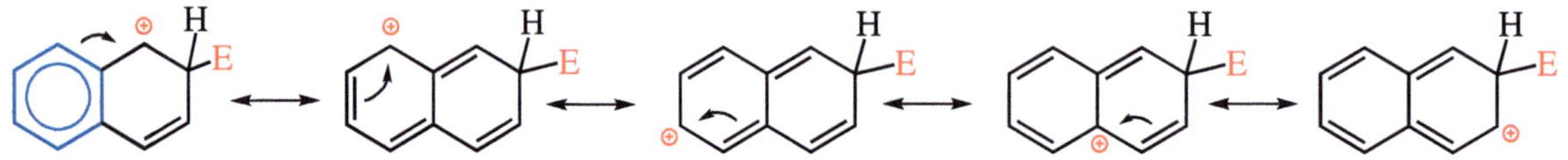

Intact benzene ring
C2 attack

Bromination and chlorination of benzene proceed only in the presence of a catalyst. However, bromination and chlorination of naphthalene proceed without a catalyst as naphthalene is more reactive than benzene. For example, reaction of naphthalene with bromine results in smooth conversion into 1-bromonaphthalene. 2-Bromonaphthalene is formed in a yield of 1%. Selectivity was also observed in the nitration reaction. The major product, 1-nitronaphthalene, is formed in 84% yield, whereas 2-nitronaphthalene is formed as the side product in 8% yield. Naphthalene can be alkylated using Friedel–Crafts reactions.

Sulfonation of naphthalene is a very important reaction because the naphthalene and sulfonic acid are used as dyestuff intermediates. The usual sulfonating agents are oleum and sulfuric acid. The substitution normally takes place at the C1 (α-position) carbon atom. As we discussed above, the intermediate formed by the attack on the α-position is more stable than the intermediate formed by the attack on the β-position. Therefore, the transition state leading to the α-isomer will be lower in energy and the α-isomer will be formed fastest as *the kinetically favored product*. However, the β-isomer is a few kcal/mol more stable than its α-isomer. This is due to the steric repulsion between the hydrogen atom in the adjacent aromatic ring and the sulfonic group. When the reaction is carried out at higher temperatures, the β-isomer, a *thermodynamically controlled product*, is formed.

Because the sulfonation reaction is reversible, at higher temperatures, the α-isomer rearranges to the thermodynamically more stable β-isomer. At low temperatures, there is no sufficient energy to overcome the activation barrier for the reverse process, desulfonation of the α-isomer. At higher temperatures, there is sufficient energy to overcome this energy barrier, and the product can revert to the reactant and form the more stable β-isomer (Figure 6.22).

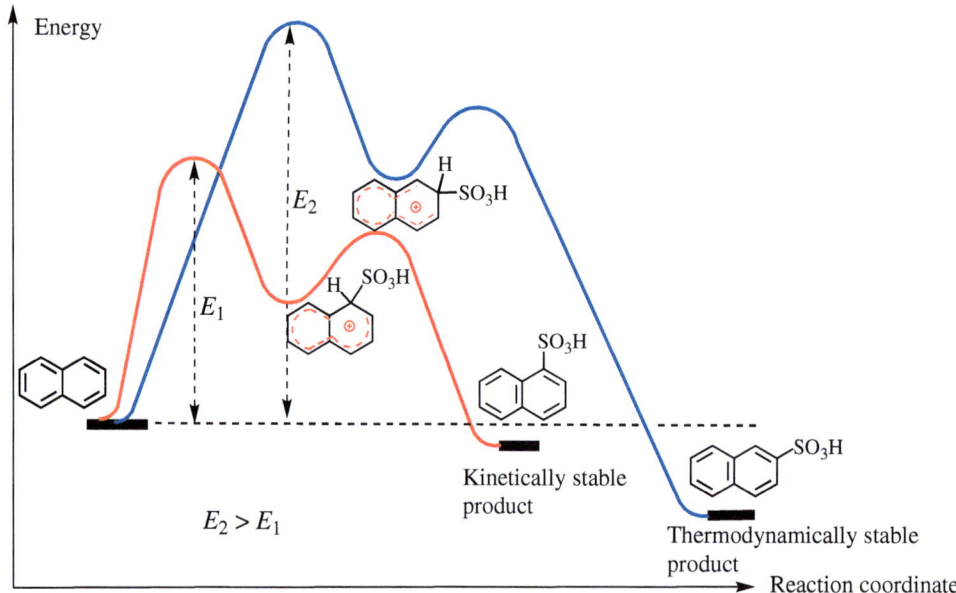

Figure 6.22 The schematic presentation of sulfonation of naphthalene.

6.9.2 Benzenoid Aromatic Compounds

PAHs are compounds having two or more condensed benzene rings in their structure. Some examples of PAHs are anthracene, phenanthrene, pyrene, benzo[e]pyrene, and coronene. Anthracene and phenanthrene are isomeric compounds. In anthracene, the three rings are fused in a linear way, while in phenanthrene, they are fused so that they produce an angular molecule. As the number of fused aromatic compounds increases, the resonance energy per benzene ring decreases and the compounds become more reactive. The resonance energy of anthracene is 84 or 28 kcal/mol per benzene ring. The resonance energy of phenanthrene is 91 or 30 kcal/mol per benzene ring.

As these compounds are not as strongly stabilized as benzene, they can undergo addition reactions. For example, anthracene undergoes 1,4-addition at the C9 and C10 positions to give an addition product with two separate benzene rings. The resonance energy of these two separate benzene rings is $36 \times 2 = 72$ kcal/mol. If this value is compared with the resonance energy of anthracene (84 kcal/mol), one can see that the loss of the resonance energy is only $84 - 72 = 12$ kcal/mol. That is the reason why anthracene undergoes a cycloaddition reaction at the C9 and C10 positions. Anthracene can also theoretically undergo a cycloaddition reaction at the C1 and C4 positions. When we examine the product formed in such a case, a naphthalene ring is left behind. The resonance energy of the naphthalene ring is 60 kcal/mol. Subtracting this value from the resonance energy value of anthracene gives $84 - 60 = 24$ kcal/mol. This means that if anthracene undergoes cycloaddition at these positions, the energy loss of the system will be 24 kcal/mol. When this value is compared to with 12 kcal/mol, it is better understood why anthracene undergoes cycloaddition at the C9 and C10 carbon atoms.

6.9 Polycyclic Aromatic Compounds

Resonance energy 72 kcal/mol
Energy loss : 12 kcal/mol

Resonance energy 60 kcal/mol
Energy loss : 24 kcal/mol

Anthracene and phenanthrene can undergo electrophilic substitution reactions. However, the C9–C10 bond in phenanthrene is as reactive as an alkene double bond. Halogenation occurs readily at this double bond to give 9,10 addition as well as 9-substituted products. Anthracene behaves similarly and it is even more reactive than phenanthrene and has a greater tendency to add bromine at the 9,10 positions than to undergo substitution.

Pyrene is a polycyclic aromatic compound consisting of four fused benzene rings, forming a planar aromatic system. Pyrene is formed during the combustion of organic compounds. The pyrene molecule is highly symmetrical, and despite having 16 π electrons and not following Hückel's $4n + 2$ rule, it is aromatic because the π bond in the middle of the ring does not participate in the resonance. Hückel's $4n + 2$ rule works best with monocyclic ring systems. The monocyclic periphery with 14 π electrons forms an aromatic system.

Two isomeric species of benzopyrene are benzo[a]pyrene and the less common benzo[e]pyrene. They belong to the chemical class of PAHs. Benzo[a]pyrene is a widespread environmental contaminant formed during incomplete combustion or pyrolysis of an organic material. It is present in cigarette smoke, automotive exhaust, and roasting meat.

Pyrene   Benzo[a]pyrene   Benzo[e]pyrene   Coronene

Benzo[*a*]pyrene is the most potent carcinogenic substance among the PAHs. Tobacco smoke in particular contains high concentrations of PAHs. *What is the mechanism of the carcinogenic action of this hydrocarbon?* Its diol epoxide metabolites formed by the oxidation of benzo[*a*]pyrene are responsible for the carcinogenic effect. The oxidation of benzo[*a*]pyrene with enzyme of the liver first forms the epoxide. Another enzyme catalyzes the ring-opening reaction of the epoxide, followed by further oxidation of the remaining double bond to give the diol epoxide, which reacts with deoxyribonucleic acid (DNA), resulting in mutations and eventually cancer. It is assumed that the amine nitrogen in guanine attacks the cyclopropane ring as a nucleophile, opening the ring. The attached guanine disrupts the DNA double helix. This observation shows that any alkylation reagent can also be carcinogenic to DNA. Indeed, that is found to be the case.

Coronene is also a PAH composed of one central and six adjacent benzene rings. It is fully aromatic. The aromatic protons resonate at 8.72 ppm as singlet, indicating the presence of a strong diamagnetic ring current. Coronene maps onto the basal plane of graphite and graphene and therefore can be used as a finite model for carbon allotropes with 2D extended structures.

Graphite has a giant covalent structure consisting of planar layers of carbon atoms forming a hexagonal mesh pattern. A graphite crystal consists of these layers of carbon atoms with an interplanar distance of 3.4 Å, while the bond length between the carbon atoms is 1.42 Å. This bond distance is very similar to that found in the benzene ring (1.39 Å) (Figure 6.23). Each carbon atom uses three of its electrons to form simple bonds to its three close carbon atoms. The fourth electron is delocalized over the whole of the sheet of atoms in one layer. The important thing is that the delocalized electrons can move anywhere within the sheet. These electrons are not fixed to a particular carbon atom. However, there is no direct contact between the delocalized electrons in one sheet and those in the neighboring sheets. These free electrons allow graphite to conduct electricity and heat. *So what holds the sheets together?* There are no covalent bonds between the layers. The van der Waals dispersion forces hold the layers together. As the delocalized electrons move around in the sheet, temporary dipoles are generated, which will induce opposite dipoles in the sheets above and below.

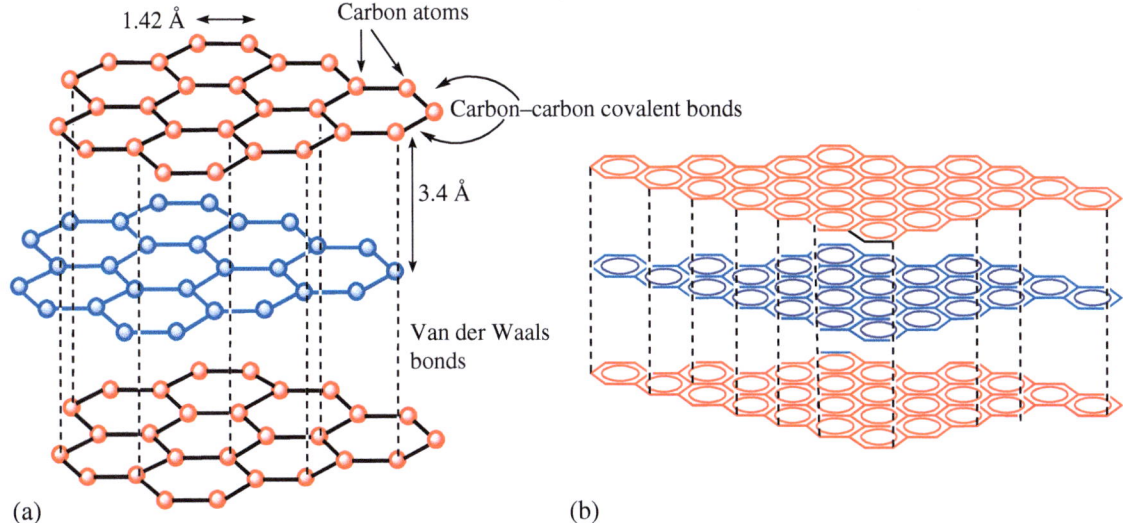

Figure 6.23 (a) The structure of graphite. (b) The fused-ring planar-layer structure of graphite.

Graphene is a single layer of carbon atoms, tightly bound in a hexagonal honeycomb lattice as discussed above in the case of graphite layers. It is an allotrope of carbon in the form of a plane of sp^2 bonded atoms with a molecular bond length of 1.42 Å (Figure 6.24). It can also be considered as an aromatic molecule, the family of flat PAHs. Graphene is the building block of graphite. Layers of graphene stacked on top of each other form graphite, with an interplanar spacing of 3.4 Å. There are about three million layers of graphene in 1 mm of graphite. Graphene was originally observed in electron microscopes in 1962. However, Andre Geim and Konstantin Novoselov at the University of Manchester rediscovered, isolated, and characterized graphene in 2004 [234]. These scientists were awarded the 2010 Nobel Prize in Physics "for groundbreaking experiments regarding the two-dimensional material graphene."

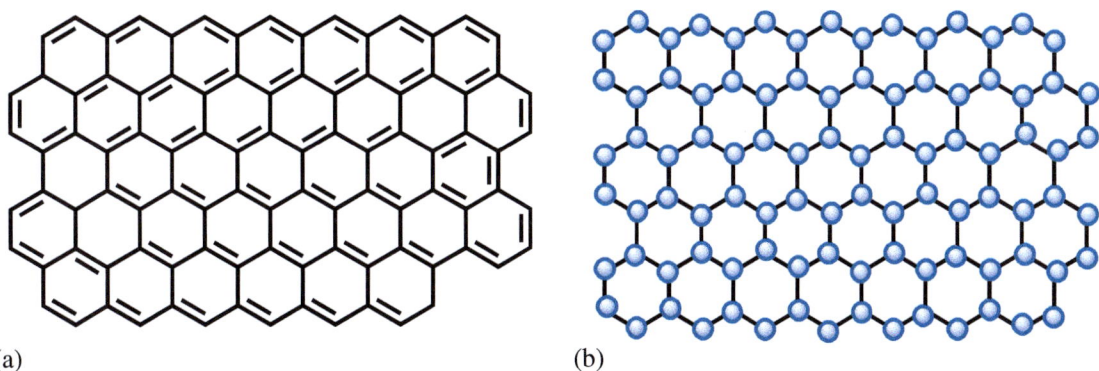

Figure 6.24 (a) The structure of graphene. (b) The fused-ring planar-layer structure of graphene.

Carbon nanotubes are cylindrical molecules that consist of graphene sheets. Carbon nanotubes are one of the allotropes of carbon, an intermediate between fullerenes (closed shells, see below) and graphene (flat sheets). Carbon nanotubes can be single walled with a diameter of less than 10 Å, double-walled, or multiwalled carbon nanotubes consisting of several concentrically interlinked nanotubes, with diameters reaching more than 1000 Å (Figure 6.25) [236]. The structure of the simplest carbon nanotube can best be visualized as the wrapping of a graphene sheet into a hollow cylinder. Their length can reach several micrometers or even millimeters. Like in graphene, the carbon atoms are sp^2 hybridized. The nanotubes naturally align themselves into "ropes" held together by relatively weak van der Waals forces. In 1991, Sumio Iijima, a senior researcher from Japan, found an extremely thin needle-like material when examining carbon materials under an electron microscope [237, 238]. The material was proved to have basically a graphite structure. These materials were named "carbon nanotubes" as they had a tubular structure of carbon atom sheets. Carbon nanotubes are highly chemically stable unless they are simultaneously exposed to high temperatures and oxygen. This property makes them extremely resistant to corrosion.

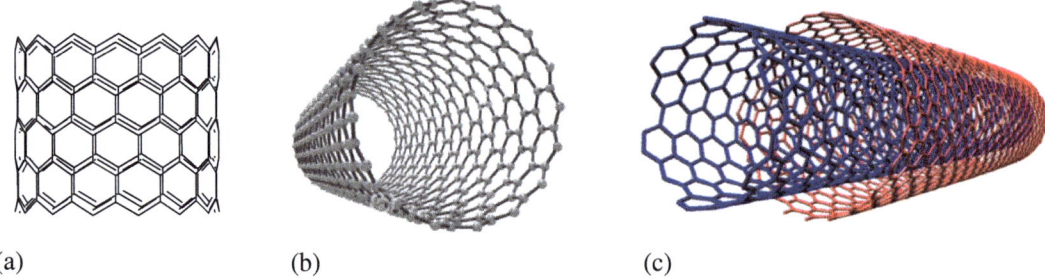

Figure 6.25 (a) The structure of a carbon nanotube, (b) a single-walled carbon nanotube, and (c) a double-walled carbon nanotube. (c) Source: Wang [235].

Carbon nanotubes possess highly conductive electrical and thermal properties. This makes them highly attractive for numerous applications in many areas of technology, such as electronics, optics, composite materials, and nanotechnology. Furthermore, they are very light and their mechanical tensile strength can be 400 times that of steel. Their hollow interior can be filled with various nanomaterials, separating and shielding them from the environment. This property is extremely important for drug delivery.

Fullerene is another allotrope of carbon and its molecule consists of 60 carbon atoms connected by single and double bonds so as to form a closed or partially closed mesh having a sphere with 12 pentagonal and 20 hexagonal faces, resembling a soccer ball (Figure 6.26). Fullerene was discovered in 1985 by Harold W. Kroto, Richard E. Smally, and Robert F. Cuerl, Jr. by an experiment using a laser to vaporize graphite rods in an atmosphere of helium gas [240]. In 1996, they were awarded the Nobel Prize for their pioneering discovery. Another fairly common fullerene has the empirical formula C_{70}, but fullerenes with 72, 76, 84, and even up to 100 carbon atoms are also known. The C_{60} molecule was named *buckminsterfullerene* after the American architect R. Buckminster Fuller, whose geodesic dome is constructed on the same structural principles. Studies indicated that C_{60} and also C_{70} were indeed exceptionally stable and provided convincing evidence for the cage structure proposal.

Each carbon atom is sp^2 hybridized and bonded to three carbon atoms as in graphene. There are two types of bond lengths in fullerene: the C—C bonds in the pentagons are 1.45 Å, whereas the C—C bonds in the hexagons are shorter, 1.40 Å. At first glance, C_{60} fullerene would appear to be aromatic because of its benzene-like-structure. However, the bending of the ball prevents aromaticity because of the nonplanar structure. C_{60} fullerene behaves like an electron-deficient alkene and reacts readily with electron-rich species. Fullerene C_{60} has the interesting electrical property of being a very good electron acceptor, which means it accepts loose electrons from other materials. Fullerene conducts electricity. It has stronger conductivity than copper and is harder than diamond.

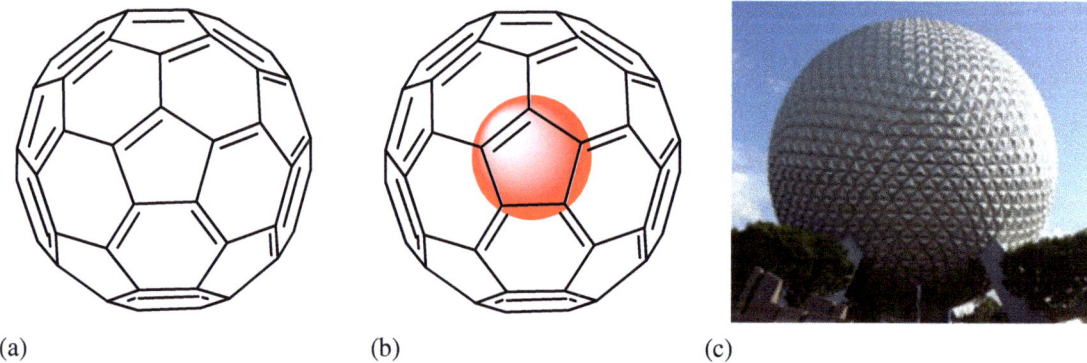

Figure 6.26 (a) The structure of fullerene C_{60}, (b) a doped fullerene, Source: Based on Elsasser [239]. and (c) a geodesic dome. The figure is printed with the permission of from Ref. [239]. Source: Pixabay.

The C_{60} molecule undergoes a wide range of chemical reactions. It readily accepts and donates electrons, a behavior that suggests possible applications in batteries and advanced electronic devices. The double bonds in fullerene can easily undergo hydrogenation and halogenation. The halogen atoms can be replaced by other functional groups, thus opening useful routes to a wide range of novel fullerene derivatives.

Problems

6.1 Show the direction of the dipole moments in the following compounds.

6.2 Which of the following compounds are aromatic, nonaromatic, and antiaromatic?

6.3 Cyclopentadienone is an unstable compound and undergoes a Diels–Alder reaction with itself. On the other hand, cycloheptatrienone is a stable compound. Write the structure of the adduct and explain the different stabilities of these two compounds.

6.4 Which of the following compounds undergoes S_N1 solvolysis in the presence of silver ion faster?

6.5 Pentalene is an elusive compound; however, the pentalene dianion is a stable compound. Explain.

6.6 4-Pyrone generates a stable carbocation by protonation. Show which oxygen atom is preferentially protonated and why?

6.7 The reaction of *p*-bromotoluene with KNH_2 in ammonia yields three products. Write the structures of these compounds.

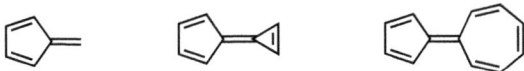

6.8 Propose a synthesis for 1-isobutyl-4-methoxybenzene starting from anisole.

6.9 At what position and on what ring is chlorination of the benzoate expected to occur?

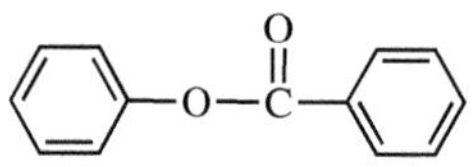

6.10 1-Fluoro-4-nitrobenzene and 1-iodo-4-nitrobenzene are reacted with KNH_2. Give the structures of the product. Which product undergoes a faster reaction?

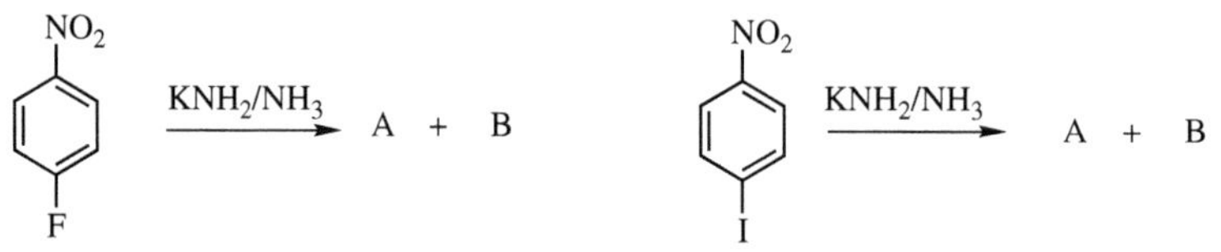

6.11 How can you synthesize the 3-bromoaniline starting from benzene?

6.12 How can you synthesize the methyl 2,4-dichlorobenzoate starting from ethylbenzene?

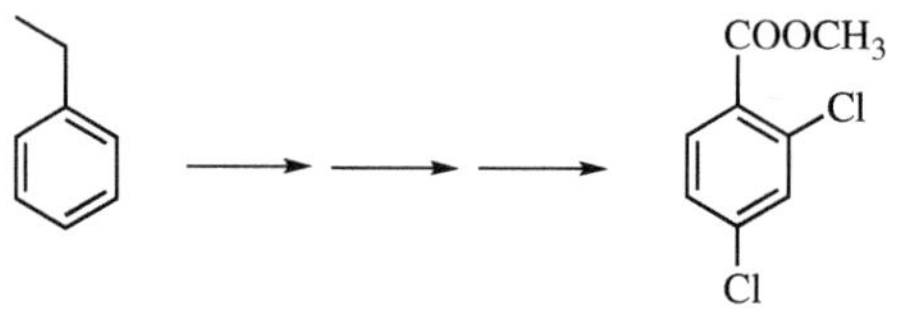

6.13 Predict the product of the following reaction and propose a mechanism for its formation.

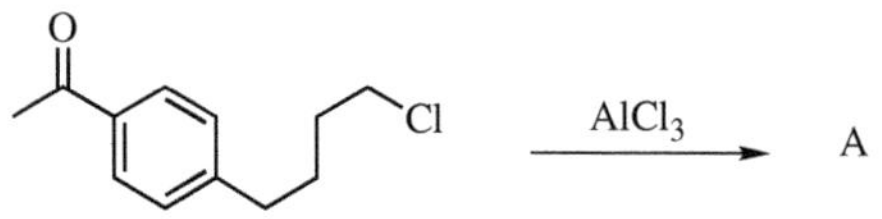

6.14 How can you synthesize 1,3-dibromo-2-methoxybenzene starting from anisole?

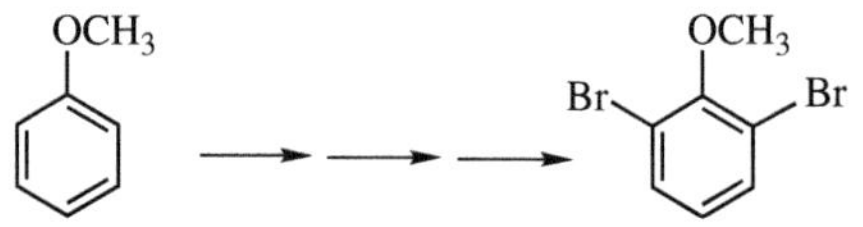

6.15 How can you synthesize 1-bromo-4-fluorobenzene starting from aniline?

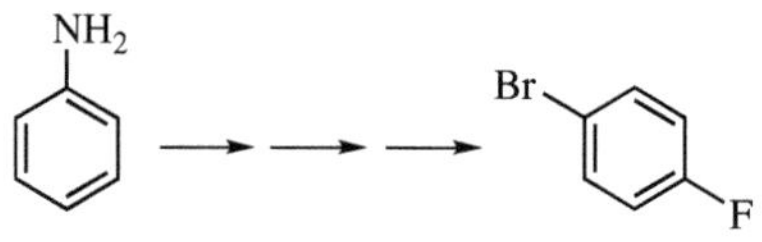

6.16 Propose a mechanism to account for the following reaction and give the structure of the reactant to use.

6.17 Show how you would synthesize 3-bromobenzoic acid starting from benzene?

6.18 Starting with benzene as the only aromatic compound and using the other reagents, give a rational synthesis of the following compound.

6.19 Propose structures for all compounds.

6.20 Propose structures for all compounds.

References

1 Hehre, W.J., Ditchfield, R., Radom, L., and Pople, J.A. (1970). *J. Am. Chem. Soc.* 92: 4796.
2 Ponomarev, D.A. and Takhistov, V.V. (1977). *J. Chem. Educ.* 74: 201.
3 Wheeler, S.E., Houk, K.N., von Ragué Schleyer, P., and Allen, W.D. (2009). *J. Am. Chem. Soc.* 131: 2547.
4 von Ragué Schleyer, P. and Pühlhofer, F. (2002). *Org. Lett.* 4: 2873.
5 Pedley, J.B., Naylor, R.D., and Kirby, S.P. (1986). *Thermochemical Data of Organic Compounds*. London: Chapman and Hall.
6 von Ragué Schleyer, P. and Jiao, H. (1996). *Pure Appl. Chem.* 68: 209.
7 Pople, J.A. (1956). *J. Chem. Phys.* 24: 1111.
8 Gershoni-Poranne, R. and Stanger, A. (2015). *Chem. Soc. Rev.* 44: 6597.
9 Sekiguchi, A., Matsuo, T., and Watanabe, H. (2000). *J. Am. Chem. Soc.* 122: 5652.
10 Sekiguchi, A., Ebata, K., Kabuto, C., and Sakurai, H. (1991). *J. Am. Chem. Soc.* 113: 7081.
11 von Ragué Schleyer, P., Maerker, C., Dransfeld, A. et al. (1996). *J. Am. Chem. Soc.* 118: 6317.
12 Chen, Z., Wannere, C.S., Corminboeuf, C. et al. (2005). *Chem. Rev.* 105: 3842.
13 Winstein, S. (1959). *J. Am. Chem. Soc.* 81: 6524.
14 Williams, R.V. (2001). *Chem. Rev.* 101: 1185.
15 Childs, R.F. (1984). *Acc. Chem. Res.* 17: 347.
16 Rosenburg, J.L. Jr., Mahler, J.E., and Pettit, R. (1962). *J. Am. Chem. Soc.* 84: 2842.
17 Warner, P., Harris, D.L., Bradley, C.H., and Winstein, S. (1970). *Tetrahedron Lett.*: 4013.
18 Kaesz, H.D., Winstein, S., and Kreiter, C.G. (1966). *J. Am. Chem. Soc.* 88: 1319.
19 Winstein, S., Ogliaruso, M., Sakai, M., and Nicholson, J.M. (1967). *J. Am. Chem. Soc.* 89: 3656.

20 Jiao, H. and von Ragué Schleyer, P. (1995). *AIP Conference Proceedings 330: E.C.C.C. 1, Computational Chemistry* (eds. F. Bernardi and J.-L. Rivail), 107. Woodbury, NY: American Institute of Physics.
21 Werstiuk, N.H. and Ma, N.H. (1999). *Can. J. Chem.* 77: 752.
22 Heilbronner, E. (1964). *Tetrahedron Lett.*: 1923.
23 Kawase, T. and Oda, M. (2004). *Angew. Chem. Int. Ed.* 43: 4396.
24 Herges, R. (2006). *Chem. Rev.* 106: 4820.
25 von Ragué Schleyer, P., Baborak, J.C., Su, T.M. et al. (1971). *J. Am. Chem. Soc.* 93: 279.
26 Anastassiou, A.G. and Yakali, E. (1972). *J. Chem. Soc., Chem. Commun.*: 92.
27 Mauksch, M., Gogonea, V., Jiao, H., and von Ragué Schleyer, P. (1998). *Angew. Chem. Int. Ed.* 37: 2395.
28 Breslow, R. and Groves, J.T. (1970). *J. Am. Chem. Soc.* 92: 984.
29 Sundaralingam, M. and Jensen, L.H. (1966). *J. Am. Chem. Soc.* 88: 198.
30 Potts, K.T. and Baum, J.S. (1974). *Chem. Rev.* 74: 189.
31 Komatsu, K. and Kitagawa, T. (2003). *Chem. Rev.* 103: 1371.
32 Kass, S.R. (2013). *J. Org. Chem.* 78: 7370.
33 Glukhovtsev, M.N., Laiter, S., and Pross, A. (1996). *J. Phys. Chem.* 100: 17801.
34 Olah, G.A. and Staral, J.S. (1976). *J. Am. Chem. Soc.* 98: 6290.
35 Boche, G., Etzrodt, H., Marsh, M., and Thiel, W. (1982). *Angew. Chem. Int. Ed.* 21: 133.
36 Sekiguchi, A., Matsuo, T., and Tanaka, M. (2002). *Organometallics* 21: 1072.
37 Breslow, R. and Hoffman, J.M. (1972). *J. Am. Chem. Soc.* 94: 2110.
38 Saunders, M., Berger, R., Jaffe, A. et al. (1973). *J. Am. Chem. Soc.* 95: 3017.
39 Wörner, H.J. and Merkt, F. (2007). *J. Chem. Phys.* 127: 034303.
40 Lambert, J.B., Lin, L., and Rassolov, V. (2002). *Angew. Chem. Int. Ed.* 41: 1429.
41 Ogliaruso, M.A., Romanelli, M.G., and Becker, E.I. (1965). *Chem. Rev.* 65: 261.
42 Pal, R., Mukherjee, S., Chandrasekhar, S., and Row, T.N.G. (2014). *J. Phys. Chem. A* 118: 3479.
43 Fattahi, A., Liebman, J.F., Miranda, M.S. et al. (2014). *Int. J. Mass Spectrom.* 369: 87.
44 Streitwieser, A. Jr. and Nebenzahl, L.L. (1976). *J. Am. Chem. Soc.* 98: 2188.
45 Bordwell, F.G., Drucker, G.E., and Fried, H.E. (1981). *J. Org. Chem.* 46: 632.
46 Panda, T.K., Gamer, M.T., and Roesky, P.W. (2003). *Organometallics* 22: 877.
47 Balcı, M. (1992). *Turk. J. Chem.* 16: 42.
48 McNamara, O.A. and Maguire, A.R. (2011). *Tetrahedron* 67: 9.
49 Merling, G. (1891). *Chem. Ber.* 24: 3108.
50 von Eggers Doering, W. and Knox, L.H. (1954). *J. Am. Chem. Soc.* 76: 3203.
51 Ohkita, M., Sano, K., Suzuki, T., and Tsuji, T. (2001). *Tetrahedron Lett.* 42: 7295.
52 Barrow, M.J. and Mills, O.S. (1973). *J. Chem. Soc., Chem. Commun.*: 66.
53 Wiberg, K.B. (2001). *Chem. Rev.* 101: 1317.
54 Breslow, R. (2014). *Chem. Rev.* 14: 1174.
55 Willstätter, R. and Waser, E. (1911). *Chem. Ber.* 44: 3423.
56 Reppe, W., Schlichting, O., Klager, K., and Toepel, T. (1948). *Justus Liebigs Ann. Chem.* 560: 1.
57 Claus, K.H. and Krüger, C. (1988). *Acta Crystallogr. C* C44: 1632.
58 Nishinaga, T., Ohmae, T., and Iyoda, M. (2010). *Symmetry* 2: 76.
59 Paquette, L.A. (1982). *Pure Appl. Chem.* 54: 987.
60 Olah, G.H., Staral, J.S., and Paquette, L.A. (1976). *J. Am. Chem. Soc.* 98: 1267.
61 Olah, G.H., Staral, J.S., Liang, G. et al. (1977). *J. Am. Chem. Soc.* 99: 3349.
62 Fraenkel, G., Carter, R.E., McLachlan, A., and Richards, J.H. (1960). *J. Am. Chem. Soc.* 82: 5846.
63 Katz, T. (1960). *J. Am. Chem. Soc.* 82: 3784.
64 Goldberg, S.Z., Raymond, K.N., Harmon, C.A., and Templeton, D.H. (1974). *J. Am. Chem. Soc.* 96: 1348.
65 Sokolov, A.Y., Magers, D.B., Wu, J.I. et al. (2013). *J. Chem. Theory Comput.* 9: 4436.
66 Watts, L., Fitzpatrick, J.D., and Pettit, R. (1965). *J. Am. Chem. Soc.* 87: 131.
67 Watts, L., Fitzpatrick, J.D., and Pettit, R. (1965). *J. Am. Chem. Soc.* 87: 3253.
68 Maier, G. (1974). *Angew. Chem. Int. Ed.* 13: 425.
69 Masamune, S. (1980). *Tetrahedron* 36: 343.
70 McKee, M.L., Balcı, M., Kílíç, H., and Yurtsever, E. (1998). *J. Phys. Chem. A* 102: 2351.
71 Irngartinger, H. and Rodewald, H. (1974). *Angew. Chem. Int. Ed.* 13: 740.

72 Delbaere, L.T.J., James, M.N.G., Nakamura, N., and Masamune, S. (1975). *J. Am. Chem. Soc.* 97: 1973.
73 Balcı, M., McKee, M.L., and von Ragué Schleyer, P. (2000). *J. Phys. Chem. A* 104: 1246.
74 Masamune, S., Nakamura, N., Suda, M., and Ona, H. (1973). *J. Am. Chem. Soc.* 95: 8481.
75 Karadakov, P.B., Hearnshaw, P., and Horner, K.E. (2016). *J. Org. Chem.* 81: 11346.
76 Masamune, S., Hojo, K., Bigam, G., and Rabenstein, D.L. (1971). *J. Am. Chem. Soc.* 93: 4966.
77 van Tamelen, E.E., Burkoth, T.L., and Greeley, R.H. (1971). *J. Am. Chem. Soc.* 93: 6120.
78 Masamune, S. and Darby, N. (1972). *Acc. Chem. Res.* 5: 272.
79 Castro, C., Karney, W.L., McShane, C.M., and Pemberton, R.P. (2006). *J. Org. Chem.* 71: 3001.
80 Smith, M.B. (2013). *March's Advanced Organic Chemistry: Reactions, Mechanisms, and Structures*, 72. Wiley.
81 Vogel, E. and Roth, H.D. (1964). *Angew. Chem. Int. Ed.* 3: 228.
82 Vogel, E., Klug, V., and Breuer, A. (1974). *Org. Synth.* 54: 11.
83 Bianchi, R., Pilati, T., and Simonetta, A. (1980). *Acta Crystallogr. B* 36: 3146.
84 Haddon, R.C. (1988). *Acc. Chem. Res.* 21: 243.
85 Nendel, M., Houk, K.N., Tolbert, L.M. et al. (1998). *J. Phys. Chem. A* 102: 7191.
86 Schmalz, D. and Günther, H. (1988). *Angew. Chem. Int. Ed.* 27: 1692.
87 Scott, L.T., Brunsvold, W.R., Kirms, M.A., and Erden, H. (1981). *J. Am. Chem. Soc.* 103: 5216.
88 Gilchrist, T.L., Tuddenham, D., McCague, R. et al. (1981). *J. Chem. Soc., Chem. Commun.*: 657.
89 Vogel, E., Königshofen, H., Müllen, K., and Oth, J.F.M. (1974). *Angew. Chem. Int. Ed.* 13: 281.
90 Hafner, K. and Kühn, V. (1986). *Angew. Chem. Int. Ed. Engl.* 25: 632.
91 Sondheimer, G. and Gaoni, Y. (1960). *J. Am. Chem. Soc.* 82: 5765.
92 Oth, J.F.M. (1971). *Pure Appl. Chem.* 25: 573.
93 Chiang, C.C. and Paul, I.C. (1972). *J. Am. Chem. Soc.* 94: 4741.
94 Boekelheide, V. and Phillips, J.B. (1967). *J. Am. Chem. Soc.* 89: 1695.
95 Mitchell, R.H. and Boekelheide, V. (1974). *J. Am. Chem. Soc.* 96: 1547.
96 Williams, R.V., Edwards, W.D., Vij, A. et al. (1998). *J. Org. Chem.* 63: 3125.
97 Mitchell, R.H., Klopfenstein, C.E., and Boekelheide, V. (1969). *J. Am. Chem. Soc.* 91: 4931.
98 Vogel, E., Sombroek, J., and Wagemann, W. (1975). *Angew. Chem. Int. Ed.* 14: 564.
99 Destro, R., Pilati, T., and Simonetta, M. (1977). *Acta Crystallogr. B* 33: 940.
100 Vogel, E., Haberland, U., and Günther, H. (1970). *Angew. Chem. Int. Ed.* 9: 513.
101 Gramaccioli, C.M., Mimun, A.S., Mugnoli, A., and Simonetta, M. (1973). *J. Am. Chem. Soc.* 95: 3149.
102 Vogel, E. (1982). *Pure Appl. Chem.* 54: 1015.
103 Balcı, M., Schalenbach, R., and Vogel, E. (1981). *Angew. Chem. Int. Ed.* 20: 809.
104 Vogel, E., Nitsche, R., and Krieg, H.-U. (1981). *Angew. Chem. Int. Ed.* 20: 811.
105 Sondheimer, F., Wolovsky, R., and Amiel, Y. (1962). *J. Am. Chem. Soc.* 84: 274.
106 Bregman, J., Hirshfeld, F.L., Rabinovich, D., and Schmidt, G.M.J. (1965). *Acta Crystallogr.* 19: 227.
107 You, S. and Neuenschwander, M. (1996). *Chimia* 50: 24.
108 Hafner, K., Dönges, R., Goedecke, E., and Kaiser, R. (1973). *Angew. Chem.* 85: 362.
109 Bally, T., Chai, S., Neuenschwander, M., and Zhu, Z. (1997). *J. Am. Chem. Soc.* 119: 1869.
110 Hafner, K. and Süss, H.U. (1973). *Angew. Chem. Int. Ed.* 12: 575.
111 Lemal, D. and Goldman, G.D. (1988). *J. Chem. Educ.* 65: 923.
112 Langhals, H. and Eberspächer, M. (2018). *Synthesis* 50: 1862.
113 Ziegler, K. and Hafner, K. (1955). *Angew. Chem.* 67: 301.
114 Hafner, K. and Meinhardt, K.-P. (1984). *Org. Synth.* 62: 134.
115 Usui, K., Tanoue, K., Yamamoto, K. et al. (2014). *Org. Lett.* 16: 4662.
116 Hafner, K. (1971). *Pure Appl. Chem.* 28: 153.
117 Vogel, E., Wassen, J., Konigshofen, H. et al. (1974). *Angew. Chem. Int. Ed.* 13: 733.
118 Vogel, E. and Ippen, J. (1974). *Angew. Chem. Int. Ed.* 13: 734.
119 Vogel, E. and Hogrefe, F. (1974). *Angew. Chem. Int. Ed.* 13: 735.
120 Oth, J.F.M., Müllen, K., Königshofen, H. et al. (1974). *Helv. Chim. Acta* 57: 2387.
121 Vogel, E., Runzheimer, H.V., Hogrefe, F. et al. (1977). *Angew. Chem. Int. Ed.* 16: 871.
122 Allinger, N.L. and Gilardeau, C. (1967). *Tetrahedron* 23: 1569.
123 Müllen, K., Oth, J.F.M., Engels, H.-W., and Vogel, E. (1979). *Angew. Chem. Int. Ed.* 18: 229.
124 Neuenschwander, M. (1986). *Pure Appl. Chem.* 58: 55.

125 Halton, B. (2005). *Eur. J. Org. Chem.* 16: 3391.
126 Neuenschwander, M. (2015). *Helv. Chim. Acta* 98: 763.
127 Kleinpeter, E. and Fettke, A. (2008). *Tetrahedron* 49: 2776.
128 Coşkun, N. and Erden, I. (2011). *Tetrahedron* 67: 8607.
129 Bergmann, E.D. (1968). *Chem. Rev.* 68: 41.
130 Asao, T., Morita, N., and Kitahara, Y. (1972). *Synth. Commun.* 2: 353.
131 Stanger, A. (2013). *J. Org. Chem.* 78: 12374.
132 Ghigo, G., Shahi, A.R.M., Gagliardi, L. et al. (2007). *J. Org. Chem.* 72: 2823.
133 Schenk, W.K., Kyburz, R., and Neuenschwander, M. (1975). *Helv. Chim. Acta* 58: 1099.
134 Hollenstein, R., Mooser, A., Neuenschwander, M., and Von Philipsborn, W. (1974). *Angew. Chem.* 86: 595.
135 Eisch, J.J., Shafii, B., Odom, J.D., and Rheingold, A.L. (1990). *J. Am. Chem. Soc.* 112: 1853.
136 Eisch, J.J., Shafii, B., and Rheingold, A.L. (1987). *J. Am. Chem. Soc.* 109: 2526.
137 Braunschweig, H., Damme, A., Dewhurst, R.D. et al. (2013). *J. Am. Chem. Soc.* 135: 1903.
138 Leward, E.G. (1983). *Chem. Rev.* 83: 519.
139 Kirmse, W. (2002). *Eur. J. Org. Chem.* 2002: 2193.
140 Zeller, K.-P. (1977). *Tetrahedron Lett.*: 707.
141 Dickerson, C.E., Bera, P.P., and Lee, T.J. (2018). *J. Phys. Chem. A* 122: 8898.
142 Csaszar, A.G., Demaison, J., and Rudolph, H.D. (2015). *J. Phys. Chem. A* 119: 1731.
143 Burdzinski, G., Luk, H.L., Reid, C.S. et al. (2013). *J. Phys. Chem. A* 117: 4551.
144 Cooksey, A.R., Morgan, K.J., and Morrey, D.P. (1970). *Tetrahedron* 26: 5101.
145 Gilow, H.M. and Burton, D.E. (1981). *J. Org. Chem.* 46: 2221.
146 Jones, G. and Stanforth, S.P. (1997). *Org. React.* 49: 1.
147 Ilyin, P.V., Pankova, A., and Kuznetsov, M.A. (2012). *Synthesis* 44: 1353.
148 Bary, B.L., Mathies, P.H., Naef, R. et al. (1990). *J. Org. Chem.* 55: 6317.
149 Noland, W.E. and Lee, C.K. (1980). *J. Org. Chem.* 45: 4573.
150 Kitzing, R., Fuchs, R., Joyeux, M., and Prinzbach, H. (1968). *Helv. Chim. Acta* 51: 888.
151 Chen, Z. and Trudell, M.L. (1996). *Chem. Rev.* 96: 1179.
152 Baciocchi, E., Clementi, S., and Sebastiani, G.V. (1975). *J. Chem. Soc., Chem. Commun.*: 875.
153 Keegstra, M.A., Klomp, A.J.A., and Brandma, L. (1990). *Synth. Commun.* 20: 3371.
154 Downie, I.M., Earle, M.J., Heaney, H., and Shuhaibar, K.F. (1993). *Tetrahedron* 49: 4015.
155 Skully, J.F. and Brown, E.V. (1954). *J. Org. Chem.* 19: 894.
156 Liang, X., Haynes, B.S., and Montoya, A. (2018). *Energy Fuel* 32: 4139.
157 Newman, M.S. and Addor, R.W. (1955). *J. Am. Chem. Soc.* 77: 3789.
158 Lee, M.W. and Herndon, W.C. (1978). *J. Org. Chem.* 43: 518.
159 Yur'ev, Y.K. and Zefirov, N.S. (1961). *Zhur. Obshchei Khim.* 31: 685.
160 Baran, A., Kazaz, C., Seçen, H., and Sütbeyaz, Y. (2003). *Tetrahedron* 59: 3643.
161 Gollnick, K. and Griesbeck, A. (1985). *Tetrahedron* 41: 2057.
162 Ramanathan, V. and Levine, R. (1962). *J. Org. Chem.* 27: 1216.
163 Chadwick, D.J. and Willbe, C. (1977). *J. Chem. Soc., Perkin Trans.* 1: 887.
164 Johnson, J.R. and May, G.E. (1943). *Org. Synth. Coll.* II: 8.
165 Marino, G. (1965). *Tetrahedron* 21: 843.
166 Minnis, W. (1943). *Org. Synth. Coll.* II: 357.
167 Kumamoto, K., Fukada, I., and Kotsuki, H. (2004). *Angew. Chem. Int. Ed.* 43: 2015.
168 Corral, C., Lissavetzky, J., and Manzanares, I. (1997). *Synthesis* 29: 29–31.
169 Treiber, A. (2002). *J. Org. Chem.* 67: 7261.
170 Brandsma, L. and Vwekruijsse, H.D. (1990). *Synth. Commun.* 20: 2275.
171 den Hertog, H.J., den Does, L.V., and Laandheer, C.A. (1962). *Recl. Trav. Chim. Pays-Bas* 91: 864.
172 Pearson, D.E., Hargreave, W.W., Chow, J.K.T., and Suthers, B.R. (1961). *J. Org. Chem.* 26: 789.
173 Paraskewas, S. (1980). *Synthesis*: 378.
174 den Hertog, H.J. and Overhoff, J. (1930). *Recl. Trav. Chim. Pays-Bas* 49: 552.
175 Bakke, J.M. (2003). *Pure Appl. Chem.* 75: 1403.
176 Bakke, J.M. (2005). *Heterocycl. Chem.* 42: 463.
177 van Gastel, M.A.J.P. and Wibaut, J.P. (1934). *Recl. Trav. Chim. Pays-Bas Belg.* 53: 1031.

178 Chichibabin, A.E. and Zeide, O.A. (1914). *Zhur. Russ. Fiz. Khim. Obshch (J. Russ. Phys. Chem. Soc.)* 46: 1216.
179 Lewis, D.A. (2017). *Angew. Chem. Int. Ed.* 56: 9660.
180 McGill, C.K. and Rappa, A. (1988). *Adv. Heterocycl. Chem.* 44: 1.
181 Cherng, Y.-J. (2002). *Tetrahedron* 58: 4931.
182 Hashimoto, S., Otani, S., Okamoto, T., and Matsumoto, K. (1988). *Heterocycles* 27: 319.
183 Bull, J.A., Mousseau, J.J., Pelletier, G., and Charette, A.B. (2012). *Chem. Rev.* 112: 2642.
184 van Ammers, M., den Hertog, H.J., and Haase, B. (1962). *Tetrahedron* 18: 227.
185 Moscher, H.S. and Welch, F.J. (1955). *J. Am. Chem. Soc.* 77: 2902.
186 Jung, J.-C., Jung, Y.-J., and Park, O.-S. (2001). *Synth. Commun.* 31: 2507.
187 Kato, T. and Yamanaka, H. (1965). *J. Org. Chem.* 30: 910.
188 Darby, W.L. and Vallarino, L.M. (1983). *Inorg. Chim. Acta* 75: 65.
189 Curphey, T.J. and Prasad, K.S. (1972). *J. Org. Chem.* 37: 2259.
190 Letsinger, R.L. and Lasco, R. (1956). *J. Org. Chem.* 21: 812.
191 Choppin, S., Gros, P., and Fort, Y. (2000). *Org. Lett.* 2: 803.
192 Zoltewicz, J.A., Grahe, G., and Smith, C.L. (1969). *J. Am. Chem. Soc.* 91: 5501.
193 Blair, V.L., Blakemore, D.C., Hay, D. et al. (2011). *Tetrahedron Lett.* 52: 4590.
194 Ple, N., Turck, A., Heynderickx, A., and Queguiner, G. (1998). *Tetrahedron* 54: 9701.
195 Vogel, E., Böll, W.A., and Günther, H. (1965). *Tetrahedron Lett.* 6: 609.
196 Vogel, E. and Günther, H. (1967). *Angew. Chem. Int. Ed.* 6: 385.
197 Vogel, E., Schubart, R., and Böll, W.A. (1964). *Angew. Chem. Int. Ed.* 3: 510.
198 Hafner, K. (1964). *Angew. Chem. Int. Ed. Engl.* 3: 165.
199 Vogel, E., Altenbach, H.-J., Drossard, J.-M. et al. (1980). *Angew. Chem. Int. Ed.* 19: 1016.
200 Jansen, H., Slootweg, J.C., and Lammertsma, K. (2011). *Beilstein J. Org. Chem.* 7: 1713.
201 Nishino, K., Takagi, M., Kawata, T., and Murata, I. (1991). *J. Am. Chem. Soc.* 113: 5059.
202 Reinhoudt, D.N. and Kouwenhoven, C.G. (1974). *Tetrahedron* 30: 2093.
203 Yamamoto, K., Yamazaki, S., Kohashi, Y. et al. (1982). *Tetrahedron Lett.* 23: 3195.
204 Gleiter, R., Krennrich, G., Cremer, D. et al. (1985). *J. Am. Chem. Soc.* 107: 6874.
205 Lal, G.S., Pez, G.P., and Syvret, R.G. (1996). *Chem. Rev.* 96: 1737.
206 Shamma, T., Buchholz, H., Prakash, G.K.S., and Olah, G.A. (1999). *Isr. J. Chem.* 32: 207.
207 Butler, A.R. (1971). *Chem. Educ.* 48: 508.
208 Mukaiyama, T., Kitagawa, H., and Matsuo, J.-I. (2000). *Tetrahedron Lett.* 41: 9383.
209 Shi, H. (2017). *Comput. Theor. Chem.* 1112: 111.
210 Olah, G.A. and Kuhn, S.J. (1960). *J. Am. Chem. Soc.* 82: 2380.
211 Rueping, M. and Nachtsheim, B.J. (2010). *Beilstein J. Org. Chem.* 6: 6.
212 Malkov, A.V., Spoor, P., Vinader, V., and Pavel Kočovský, P. (1999). *J. Org. Chem.* 64: 5308.
213 Di Vona, M.L. and Rosnati, V.J. (1991). *Org. Chem.* 56: 4269.
214 Vedejs, E. (1975). *Org. React.* 22: 401.
215 Balcı, M. (2005). *Basic ^{1}H- and ^{13}C-NMR Spectroscopy*. Elsevier.
216 Tee, O.S., Paventi, M., and Bennett, J.M. (1989). *J. Am. Chem. Soc.* 111: 2233.
217 Bartoli, G. and Todesco, P.E. (1977). *Acc. Chem. Res.* 10: 125.
218 Kwan, E.E., Zeng, Y., Besser, H.A., and Jacobsen, E.N. (2018). *Nat. Chem.* 10: 917.
219 Kornblum, N. (1944). *Org. React.* 2: 262.
220 Hendrickson, J.B. (1961). *J. Am. Chem. Soc.* 83: 1251.
221 Korzeniowski, S., Blum, L., and Gokel, G.W. (1977). *J. Org. Chem.* 42: 1469.
222 Cohen, T., Dietz, A.G. Jr., and Miser, J.R. (1977). *J. Org. Chem.* 42: 2053.
223 Cresswell, A.J., Davies, S.G., Roberts, P.M., and Thomson, J.E. (2015). *Chem. Rev.* 115: 566.
224 Abeywickrema, A.N. and Beckwith, A.L.J. (1987). *J. Org. Chem.* 52: 2568.
225 Galli, C. (1988). *Chem. Rev.* 88: 765.
226 Roberts, J.D., Semennow, D.A., Simmons, H.E. Jr., and Carlsmith, L.A. (1956). *J. Am. Chem. Soc.* 78: 601.
227 Roberts, J.D., Simmons, H.E. Jr., and Carlsmith, L.A. (1953). *J. Am. Chem. Soc.* 75: 3290.
228 Kukolich, S.G., McCarthy, M.C., and Thaddeus, P. (2003). *J. Phys. Chem. A* 119: 4353.
229 Winkler, M. and Sander, W. (2010). *Aust. J. Chem.* 63: 1013.
230 Wentrup, C. (2010). *Aust. J. Chem.* 63: 979.

231 Friedman, L. and Longullo, F.M. (1963). *J. Am. Chem. Soc.* 85: 1549.
232 Wenk, H.H., Winkler, M., and Sander, W. (2003). *Angew. Chem. Int. Ed.* 42: 502.
233 Tadross, P.M. and Stoltz, B.M. (2012). *Chem. Rev.* 112: 3550.
234 Novoselov, K.S., Geim, A.K., Morozov, S.V. et al. (2004). *Science* 306: 666.
235 YuHang Wang http://www2.chem.umd.edu/groups/wang/
236 Sinnott, S.B. and Andreys, R. (2001). *Crit. Rev. Solid State* 26: 145.
237 Iijima, S. (1991). *Nature* 354: 56.
238 Iijima, S. and Ichihashi, T. (1993). *Nature* 363: 603.
239 Katherina Wong (2011). Disney Land Epcot Ball Epcot Spaceship Earth. Pixabay. http://michelsasser.blogspot.com/2011/03/buckminsterfullerene.html (accessed December 2020)
240 Kroto, H.W., Heath, J.R., O'Brien, S.C. et al. (1985). *Nature* 318: 162.

7

Reactive Intermediates: Carbocations

New products are formed as a result of chemical reactions. *How does the conversion of starting compounds (reactants) to products occur?* It is not always easy to answer this question. There are some reactions in which reactants are directly converted into products. On the other hand, most chemical reactions do not proceed in a single step. In a multistep reaction, the reactants are first converted into short-lived and highly unstable *intermediates* that quickly convert into products or new intermediates.

$$\text{Reactants} \underset{k_{-1}}{\overset{k_1}{\rightleftharpoons}} \text{Intermediates} \underset{k_{-2}}{\overset{k_2}{\rightleftharpoons}} \text{Products}$$

As shown in the above equation, there is an equilibrium between the intermediates formed and the reactants. The intermediates can be converted back into the reactants as well as into products. However, if the intermediates are converted into products, they do not turn back into the intermediates. In general, the conversion of intermediates to products is faster ($k_2 > k_1$) than their formation rate, and so it is often not possible to isolate intermediates and examine their structures.

Figure 7.1 shows the energy diagrams of two different reactions. The first one shows the energy diagram of a reaction in which no intermediate is involved. The second one shows the formation of an intermediate. First, let us look at the energy distribution of a single-step system. As an example, we can take an S_N2 reaction proceeding in a single step.

$$\text{Nu} + \text{R-X} \longrightarrow \text{Nu---R---X} \longrightarrow \text{Nu-R}$$
$$\text{Reactant} \qquad \text{Transition state} \qquad \text{Product}$$

In an S_N2 reaction, the forming of a bond between the carbon atom and the nucleophile and the breaking of the bond between the carbon atom and the leaving group occur simultaneously. The system's free energy increases, forming the *transition state*, which is the highest point on the reaction coordinate diagram (Figure 7.1). As can be seen from the energy diagram, the transition complex cannot be isolated; it turns directly into a product in a single-step reaction.

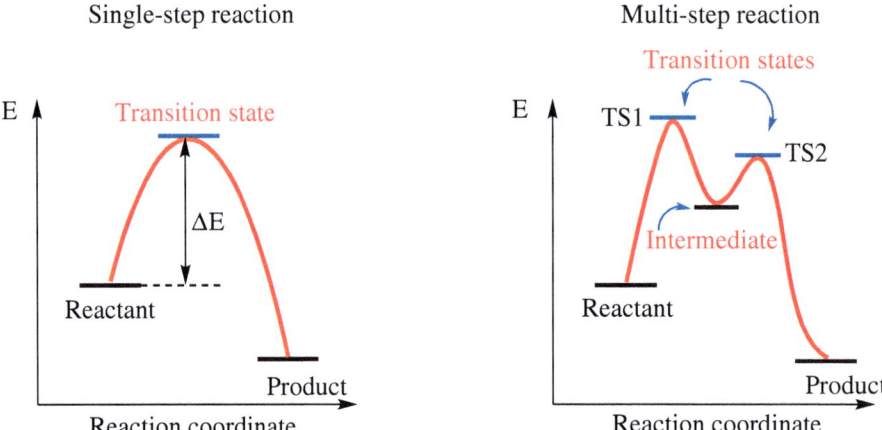

Figure 7.1 The energy diagrams of a single-step and multistep reactions.

Let us now analyze a reaction proceeding according to the S_N1 mechanism. The reaction occurs in two steps. In the first step, the bond between the carbon atom and the leaving group breaks. Meanwhile, the energy of the system reaches a maximum value (first transition state, TS1). This structure cannot be isolated and continues to change until the carbocation is formed. With the formation of the carbocation, the energy of the system partially decreases. The carbocation formed here is called an *intermediate product*. In the next step, the nucleophile attacks the carbocation to form the product. For this process, the carbocation needs to be reactivated to form a product. Therefore, a second transition state (TS2) is formed. An energy valley is built between TS1 and TS2, where the intermediate is located.

$$R-X \longrightarrow \underset{\substack{\text{Transition}\\\text{state TS1}}}{R\text{-}\text{-}\text{-}X} \xrightarrow{-X^{\ominus}} \underset{\text{Intermediate}}{\left[R^{\oplus} \right]} \xrightarrow{Nu^{\ominus}} \underset{\substack{\text{Transition}\\\text{state TS2}}}{R\text{-}\text{-}\text{-}Nu} \longrightarrow R-Nu$$

Reactant · Product

As the intermediates are highly reactive, generally, they cannot be isolated. Only in exceptional cases can the reactive intermediates be isolated at low temperatures or by matrix isolation. In some cases, the existence of the intermediate can be observed by spectroscopic methods. We have shown that at least two transition complexes must form in reactions to obtain an intermediate. The energy levels of these transition complexes determine whether the intermediate product can be isolated or not. For example, as can be seen from the energy diagrams given in Figure 7.2, if the activation energy of the second transition complex is lower than that of the first ($\Delta E_2 < \Delta E_1$), this intermediate product cannot be isolated. However, if the second's activation energy is greater than the first's ($\Delta E_2 > \Delta E_1$), it may be possible to isolate the intermediate product. When reactive intermediates are not observable, their existence must be inferred through experiments by changing the reaction conditions such as concentration and temperature and studying the chemical kinetics. Once information is obtained about the intermediate, the reaction mechanism is much better understood.

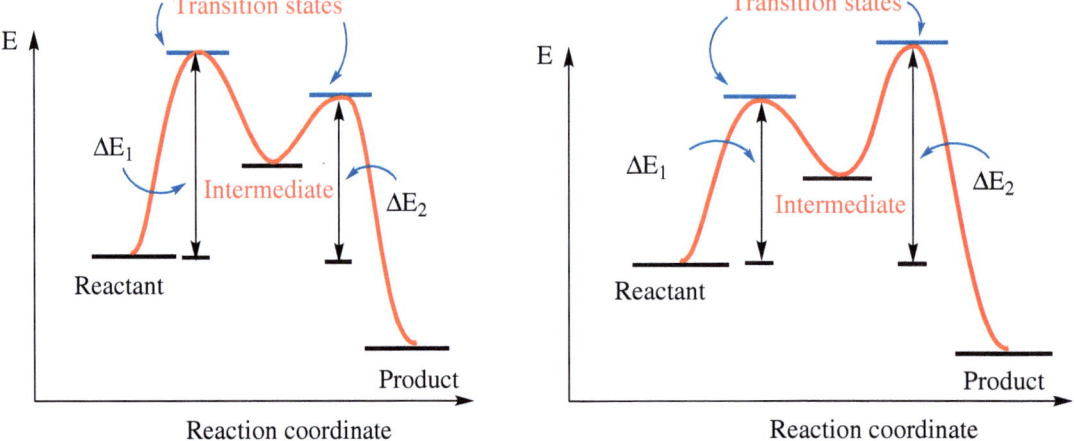

Figure 7.2 The energy diagrams of a two-step reaction.

Intermediates are generally electron-deficient compounds. When atoms have fewer than eight electrons in the valence shell, they tend to react with nucleophiles to increase the number of electrons in the valence shell to eight and form more stable compounds. Electron-deficient carbon intermediates can be neutral, such as carbene and radicals, or positively (+) charged, such as carbocations. On the other hand, intermediates can also be negatively charged (−) species such as carbanions that carry a negative (−) charge on an atom, following the octet rule. However, those compounds also belong to the class of reactive intermediates because of the negative charge. In addition to these, there are some reactive intermediates that are neutral and satisfy the octet rule. These compounds are generally highly strained molecules. For example, dihydroaromates, such as benzyne, fall into this group and are unstable compounds. Electronically excited molecules are also classified as reactive intermediates. For example, the excited state of the oxygen molecule, singlet oxygen, is also a reactive intermediate product.

Reactive intermediates are classified according to the central atom as follows:

According to the central carbon atom:

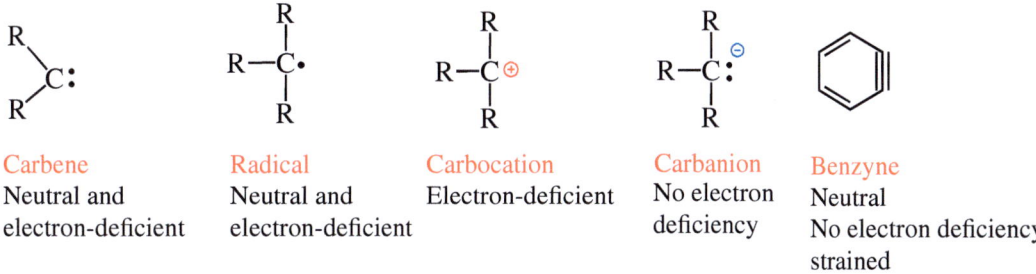

Carbene	Radical	Carbocation	Carbanion	Benzyne
Neutral and electron-deficient	Neutral and electron-deficient	Electron-deficient	No electron deficiency	Neutral No electron deficiency strained

According to the central nitrogen atom:

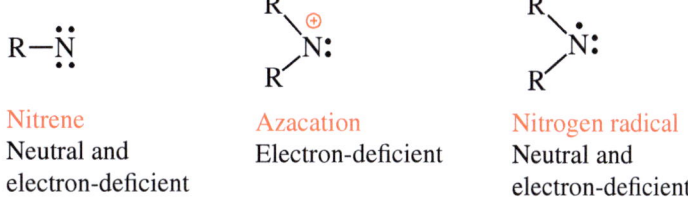

Nitrene	Azacation	Nitrogen radical
Neutral and electron-deficient	Electron-deficient	Neutral and electron-deficient

According to the central oxygen atom:

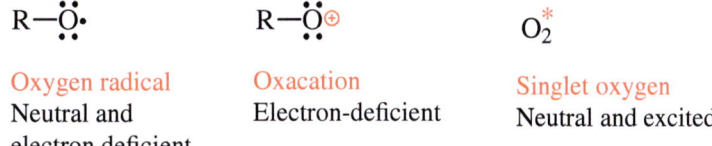

Oxygen radical	Oxacation	Singlet oxygen
Neutral and electron deficient	Electron-deficient	Neutral and excited

The intermediates listed above will be examined in three separate groups. Ionic intermediates will be discussed first, followed by neutral intermediates (such as carbene and nitrene), and in the last section, radicals and excited intermediates.

During a chemical reaction, some bonds are broken and new bonds are formed. Breaking of a bond can occur in different ways. For example, let us consider an A–B molecule that reacts with any reagent. The A—B bond can be broken differently depending on the reaction conditions. There are two types of bond cleavages: homolytic and heterolytic. When the bond cleavage between A and B is homolytic, two electrons in the bond are divided equally between A and B and two free radicals are formed as the reactive intermediates. The chemical reactions of the radicals and their properties will be examined in the radicals section. The heterolytic bond cleavage creates ion pairs with positive (+) and negative (−) charges, whereas the bonding electrons remain on one of the separated atoms. The electronegativity of elements and the electronic structure of groups A and B determine on which atom the electrons will remain. Radicals and ions formed due to both homolytic and heterolytic bond cleavages are classified as reactive intermediates because they are electron-deficient species or have a charge.

384 | 7 Reactive Intermediates: Carbocations

$$A-B \longrightarrow A\bullet + B\bullet \quad \text{Homolytic cleavage}$$

$$A-B \longrightarrow A:^{\ominus} + B^{\oplus} \quad \text{Heterolytic cleavage}$$

$$A-B \longrightarrow A^{\oplus} + B:^{\ominus} \quad \text{Heterolytic cleavage}$$

It is necessary to know the electronic structures and chemistry of intermediates well in order to understand chemical reactions and predict which products may form as a result of a reaction. Someone who knows and understands the chemistry of intermediates knows chemical reactions and reaction mechanisms.

If the positive (+) charge is on a carbon atom in a heterolytic bond cleavage of a molecule A–B, the formal charge of that carbon atom is +1, and such compounds are called carbocations. For many years, these compounds were called methyl, ethyl carbonium ions [1].

$CH_3^{\oplus}$	$H_3C-CH_2^{\oplus}$	$H_3C-\overset{CH_3}{\underset{}{C}}H^{\oplus}$	$H_3C-\overset{CH_3}{\underset{CH_3}{C^{\oplus}}}$
Methyl carbocation	Ethyl carbocation	i-Propyl carbocation	t-Butyl carbocation

The carbocations can be classified into two groups: classical carbocations and nonclassical carbocations. Classical carbocations contain a carbon atom having a sextet of electrons with three σ bonds. Inductive or mesomeric effects stabilize the positive charge. However, nonclassical carbocations have a three-center, two-electron structure and they are penta-coordinated. An example of a nonclassical carbocation is the 2-norbornyl carbocation. The CH_2 group is bonded to three carbon atoms (C1, C2, and C3) and two hydrogen atoms. Furthermore, the CH_2 group is bonded to the carbon atoms C1 and C2 through two electrons. Therefore, it is called a three-center, two-electron structure.

$$H_3C-CH_2^{\oplus} \qquad \text{Ph}-CH_2^{\oplus} \qquad \text{Nonclassical norbornyl cation}$$

Classical carbocations　　　　Nonclassical norbornyl cation

7.1 Structure and Stability of Carbocations

A carbocation is a molecule in which a carbon atom has a positive charge and is bonded to three substituents. Carbocations have a trigonal planar structure because of their sp^2 hybrid orbitals. The vacant p orbital is perpendicular to the plane formed by the substituents and indicates its electron-deficient nature [2].

Carbocation

Both NMR spectroscopic [3] and crystallographic studies [4] performed on the t-butyl carbocation show that it has a planar structure and the angle between the methyl groups (C–C$^+$–C) is 120°. The racemization of optically active alkyl halides during the solvolysis reaction indirectly also indicates that carbocations have a planar structure.

$$H_3C\cdots\overset{\oplus}{C}-CH_3 \quad Sb_2F_{11}^{\ominus}$$
$$H_3C$$

t-Butyl carbocation

The stability of carbocations varies according to the nature of substituents attached to the carbon atom carrying the positive (+) charge. If these groups are alkyl groups, both the number of alkyl groups and the carbocation's stability will be higher. Among the alkyl-substituted carbocations, tertiary carbocations are the most stable.

Tertiary > Secondary > Primary > Methyl   R = alkyl group

⟵ Increasing stability

One method used to determine carbocations' stability is to measure the energy required to form the carbocation from the corresponding alkyl halide. However, the carbocation's relative stability should also be determined in the gas phase rather than in solution because the solvation influences carbocations' stabilities in solution.

The hydride ion affinity (HIA) shows the relative stability of carbocations. The HIA is defined as the negative value of the reaction enthalpy (ΔH) of the reaction between a carbocation and a hydride ion in the gas phase.

For example, the gas-phase reaction of methyl carbocation and a hydride ion has the formation enthalpy of $\Delta H = -313.4.0$ kcal/mol. The HIA values of some selected carbocations are given in Table 7.1.

Table 7.1 The hydride ion affinities for selected carbocations in the gas phase. Source: Based on Kim [5].

Carbocation	HIA (R^+) in the gas phase (kcal/mol)
Methyl	313.4
Ethyl	270.7
i-Propyl	249.7
t-Butyl	236.9
Allyl	255.6
Vinyl	287.0
Benzyl	234.0

Recent theoretical calculations are also in agreement with the values given in Table 7.1. Based on this table, we can provide the following order for the stability of carbocations [5]:

$$(CH_3)_3C^+ \approx C_6H_5CH_2^+ > CH_2=CH-CH_2^+ \approx (CH_3)_2CH^+ > CH_3CH_2^+ > CH_3^+$$

⟵ Carbocation stability

In basic organic chemistry books, in general, it is mentioned that alkyl groups stabilize carbocations by electron donation. However, the carbon atoms always have the same electronegativity. Therefore, the question arises: *How can a carbon atom be electron-donating?* The increased stability of carbocations with the increased number of alkyl groups is due to the stronger attraction of the bonding electrons of the carbon atom carrying the positive (+) charge rather than the electron-providing properties of the alkyl groups.

As shown in the figure below, the C—C bond between the methyl carbon atom and the carbon atom bearing the positive (+) charge is formed by the interference of sp^2 hybrid orbitals and sp^3 hybrid orbitals. The electronegativity of an sp^2 hybrid orbital is about 2.75 while that of an sp^3 hybrid orbital is around 2.5. The higher the s ratio in hybrid orbitals, the higher the electronegativity. Therefore, the sp^2 hybrid orbital attracts bond electrons stronger than the sp^3 hybrid orbital because of the increased s ratio. Thus, the bonding electrons are attracted more strongly by the carbon atom carrying the (+) charge and the stability of the carbocation increases. Such a situation makes the methyl group an electron donor [6].

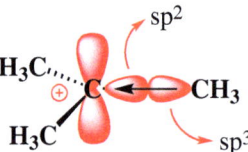

Another factor that stabilizes carbocations is *hyperconjugation*, which is the principal mechanism by which alkyl substituents stabilize carbocations [7]. According to the classical resonance theory, electrons can delocalize by the parallel overlap of p orbitals. However, electron delocalization can also occur by parallel alignment of an unoccupied p orbital with an adjacent hybridized orbital in a σ bond. The unoccupied p orbital of the *t*-butyl cation with the positive (+) charge overlaps with the C—H σ bond orbitals. An increased number of C—H bonds adjacent to the carbocation center increases the number of stabilizing interactions by transferring σ electrons into the empty *p* orbital (Figure 7.3). Both NMR spectroscopic and crystallographic measurements reveal that the C—C bond is shortened in the *t*-butyl carbocation, which can be explained by hyperconjugation [3, 8].

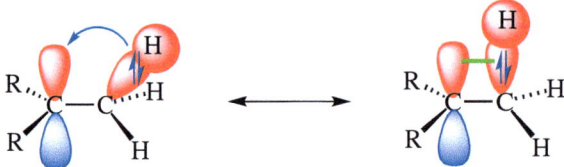

Figure 7.3 Schematic presentation of hyperconjugation.

The interaction of the empty p orbital with the filled C–H σ orbital results in the formation of two new molecular orbitals. The energy level of the filled σ bond is lower than that of the empty p orbital. The σ electrons settle in the newly formed molecular orbital and reduce the energy of the system; in other words, they stabilize the carbocation (Figure 7.4). The more the C—H or C—C σ bonds are adjacent to the carbocation center, the more stable the carbocation is. According to this theory, tertiary carbocations are more stable than secondary and primary cations.

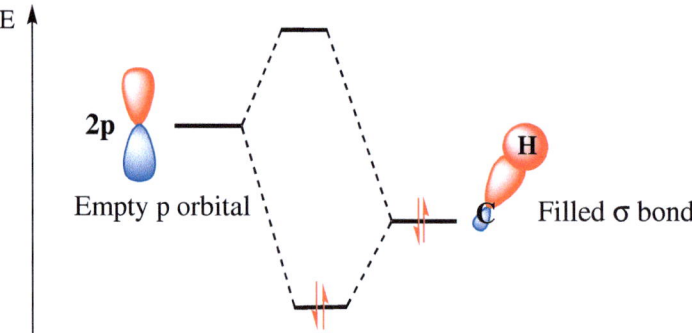

Figure 7.4 Presentation of hyperconjugation by orbital interaction.

The stability of carbocations depends on many factors. We discussed several factors influencing carbocations' stability in the introduction part of this book in detail. Carbocations adjacent to a carbon–carbon double bond are stabilized because of the overlap between the empty p orbital and the π bond electrons. Stabilization of carbocations does not mean that they are isolable at room temperature. With increased stability, the lifetime of carbocations is partially extended.

If a heteroatom (such as oxygen or nitrogen) is directly attached to the carbon atom carrying a positive (+) charge, the carbocation becomes stable. One would expect that electron-withdrawing atoms, such as oxygen or nitrogen, would destabilize the carbocation. Heteroatoms, such as oxygen and nitrogen, are electron-withdrawing groups by inductive effect (−I); on the other hand, they are electron-donating groups (+M) by resonance. Remember, if two effects are operating simultaneously in opposite directions, the resonance effect is always the dominant one. Therefore, the heteroatoms adjacent to the carbocation center stabilize the carbocation.

If the carbocation center is more than one bond away from the heteroatom, there is no resonance effect. The heteroatom cannot stabilize the carbocation by electron donation. However, the heteroatom can destabilize the carbocation center by the electron-withdrawing effect.

7.2 Generation of Carbocations

Carbocations generally occur by either of the following fundamental steps:

1. Cleavage of a bond between the carbon atom and a leaving group (ionization mechanism).
2. Electrophilic addition to the π bonds.

7.2.1 Ionization Mechanism

As a result of the heterolytic cleavage of the bond between the carbon atom and the leaving group, carbocations are formed. The leaving group takes away the bonding electrons and it is separated from the molecule as an anion. The remaining carbon atom with a positive (+) charge is an electron-deficient species. The more easily the group leaves the molecule, the greater the likelihood of carbocation formation is. For example, the –OH group is not a good leaving group, but in acidic conditions, it is easily separated from the molecule as H_2O and a carbocation is formed.

In S_N1 and E1 reactions, a carbocation is formed as an intermediate after the separation of a halide ion or a suitable leaving group from the molecule. Since carbocations are generally short-lived intermediates, they react in various ways. George Andrew Olah, a Hungarian and American chemist (1927–2017), concentrated a significant part of his research on the formation of long-lived alkyl cations in solvents. As a result of these studies, long-lived carbocations were synthesized for the first time by treatment of alkyl fluorides with the Lewis acid SbF_5 [9] in polar solvents such as SO_2ClF, SO_2F_2, or SO_2 [10]. For his outstanding contribution to carbocation chemistry, Olah was awarded the Nobel Prize in Chemistry in 1994 [11].

With this method, various carbocations were generated at room and lower temperatures, and the structures were directly examined by NMR spectroscopy. While the lifetime of the cations formed under normal conditions varies between 10^{-10} and 10^{-6} seconds, the carbocations formed by this method remain stable at room temperature for a long time. In addition to SbF_5, mixtures such as $HSO_3F\text{-}SbF_5$ and $HF\text{-}SbF_5$ are also used to form carbocations from alcohols and halides. These acids are called *magic acids* or *super acids* [12].

7.2.2 Electrophilic Addition to π Bonds

Electrophiles' attack on the π electrons of a double or triple bond results in cleavage of the π bond, which bonds to one of the carbon atoms with a σ bond, while a positive (+) charge is formed on the other carbon atom. The stronger the electrophile, the lower the activation barrier required for the formation of the carbocation is and the faster the addition takes place. The carbocation formed reacts in various ways. These reactions will be discussed in detail in the following sections.

E = electrophile   Carbocation

7.3 Detection of Carbocations

When alkyl halides such as propyl, butyl, and pentyl fluorides are treated with SbF_5/FSO_3H in SO_2ClF between −40 and −150 °C, remarkably stable carbocations are formed. The stability of these carbocations allows their structures to be investigated by nuclear magnetic resonance (NMR) spectroscopy as well as by the other spectroscopic techniques.

When isopropyl fluoride is treated with a superacid, serious changes are observed in both its ^1H- [13] and ^{13}C-NMR [14] spectra. The methyl (CH_3) and methine (CH) protons in isopropyl fluoride resonate at 1.23 and 4.64 ppm, respectively. After formation of the carbocation, abnormal shifts in the resonances of these protons are observed. The resonance of the methine proton carrying the positive (+) charge shifts to 13.5 ppm ($\Delta\delta \approx 9$ ppm), while the resonance of methyl protons shifts to 5.06 ppm ($\Delta\delta \approx 3.8$ ppm). These values clearly indicate the formation of a carbocation. Similar shifts are also observed in the carbon resonances. In particular, the chemical shift of the carbon atom bearing the positive (+) charge is drastically affected and it appears at 319.6 ppm ($\Delta\delta \approx 266$ ppm), proving the existence of a carbocation. The shifts observed in both proton and carbon values show a parallelism and are due to the completely changing electron density.

δ_H 1.23 ppm
δ_C 27.0 ppm

δ_H 5.06 ppm
δ_C 61.8 ppm

δ_H 4.64 ppm
δ_C 53.9 ppm

δ_H 13.5 ppm
δ_C 319.6 ppm

The coupling constant, $^1J_{CH}$, observed between the carbon atom and the proton varies according to the hybridization of the C—H bond. There is an empirical correlation between the carbon–proton coupling constant J_{CH} and the *s* character in the hybrid orbitals of the carbon–hydrogen bond [15].

$$J_{CH} = 500 \ (s)$$

Because the *s* ratio in an sp^3-hybridized bond is 0.25 (25%), the coupling constant J_{CH} should be about 125 Hz according to this equation. For alkenes, the coupling constants for sp^2-hybridized carbon atoms are expected to be around 160–170 Hz.

The coupling constants measured in the isopropyl carbocation (J_{CH} = 171.3 Hz) indicate that hybridization of the carbocation bearing the positive charge is sp^2 and so the fluoride ion is completely separated from the molecule and a planar structure is formed. The measured coupling constant J_{CH} in the methyl group (J_{CH} = 131.7 Hz) reveals that there is no change in hybridization in the methyl group.

7.4 Reactions of Carbocations: Rearrangements

As carbocations contain six electrons in their valence shell, they do not obey the octet rule, and they are unstable intermediates and undergo various reactions to increase the number of electrons in the valence shell from six to eight. Carbocations generally undergo three different reactions in order to stabilize.

7.4.1 Reactions with Nucleophiles

Because carbocations are electron-deficient compounds, the positive charge is stabilized by the addition of nucleophiles to form a σ bond as discussed for S_N1 reactions.

7.4.2 Double-bond Formation by Proton Elimination

The removal of the nearby hydrogen, followed by the formation of a C=C double bond will complete its octet and make the compound neutral.

7.4.3 Rearrangement

A group attached to the adjacent carbon atom can be shifted with the bonding electrons to the adjacent atom and so a more stable carbocation is formed. This shift is called a *1,2-shift* as the migrating group is shifting from the carbon atom C1 to the adjacent carbon atom C2.

The migrating group in this example can be an alkyl group or a hydride ion (H$^-$). As given in the example above, a more stable tertiary carbocation occurs as a result of the shift of a hydride from the primary carbocation. The carbocation formed can also undergo further rearrangement.

7.4.4 Carbocation Rearrangement

The driving force for the rearrangement is the formation of a more stable carbocation, such as a tertiary carbocation from a primary one as discussed above. If a carbocation with the same structure is formed because of the rearrangement, these systems are called *degenerate systems* and the process is called *degenerate rearrangement*.

7.4.5 Ethyl Carbocation

The question as to whether an ethyl carbocation can rearrange may arise. If the ethyl carbocation undergoes rearrangement by hydride shift, the same cation will be formed. The carbocation formed as a result of the rearrangement does not become more stable.

$$H_2C-CH_2^+ \rightleftharpoons {}^+H_2C-CH_2$$
$$\text{Degenerate rearrangement}$$

However, the labeling experiments conducted show that such a rearrangement exists. In experiments performed with a trideutero ethyl carbocation, it was determined by spectroscopic methods that deuterium was evenly scrambled over the molecule after a certain time [16].

$$D_3C-CH_2^+ \underset{\sim D^-}{\rightleftharpoons} D_2C^+-CH_2D \underset{\sim H^-}{\rightleftharpoons} D_2HC-C^+HD \underset{\sim D^-}{\rightleftharpoons} DHC^+-CHD_2$$

What is the mechanism of this rearrangement? It is possible that the deuterium (D^+) is first eliminated to form ethylene, followed by the addition of the deuterium (D^+) to the ethylene double bond to generate the ethyl carbocation again. To confirm this mechanism, the ethyl cation was generated in the presence of DF/SbF_5. According to this proposed mechanism, deuterium should be incorporated into the ethyl cation. However, it was determined that there was no deuterium atom in the molecule, indicating that the elimination–addition mechanism is not operating.

$$H_3C-CH_2F \xrightarrow[DO_3SF]{DF/SbF_5} H_3C-CH_2^+ \ SbF_6^-$$
$$\text{No deuterium incorporation}$$

These experiments show that the hydrogen is not completely separated from the molecule during the rearrangement. In the light of these results, it was determined that the ethyl cation has a nonclassical carbocation structure with a three-center, two-electron bond in agreement with theoretical calculations [17].

7.4.6 Isopropyl Carbocation

If the isopropyl carbocation undergoes rearrangement, a secondary carbocation will be converted into a primary carbocation by a hydride shift. This process is endothermic. Saunders and Hagen showed the scrambling of the hydrogen atom within the isopropyl cation and determined the activation energy, E_a, to be 16.4 kcal/mol by ^{1}H-NMR spectroscopy [18]. More interesting than the proton scrambling was the rearrangement of the carbon skeleton. Olah generated the isopropyl cation from 2-chloropropane labeled at the C-2 carbon atom with ^{12}C-isotope and determined by spectroscopic methods that the labeled carbon atom was distributed over the entire skeleton [19]. Saunders and Hagen suggested that a protonated cyclopropane was responsible for proton scrambling. The protonated cyclopropane might conceivably be formed directly from the isopropyl cation by simultaneous ring closure and hydride shift.

7.4.7 Cyclopentyl Carbocation

The cyclopentyl cation was prepared starting from chlorocyclopentane in SbF_5/SO_2ClF solution at $-70\,°C$. The cyclopentyl carbocation should have three different proton signals in the ^{1}H-NMR spectrum and three different carbon signals in the ^{13}C-NMR spectrum. The ^{1}H-NMR and ^{13}C-NMR spectra recorded at $-70\,°C$ consist of a single resonance at 4.48 and 95.4 ppm, respectively [20]. As shown for the ethyl carbocation, a rearrangement, a 1,2-hydride shift, takes place so fast on the NMR time scale that all hydrogens and all carbon atoms become equivalent. This situation can be explained in two different ways.

1. There is hydrogen scrambling by rapid 1,2-hydrogen shifts. The detection time of NMR spectroscopy is so slow that all protons and carbon atoms are perceived equally; thus, the relevant atoms resonate as singlets.

    Very fast 1,2-hydride shift

2. Instead of a very fast rearrangement, there may be an intermediate product that makes all protons and carbon atoms identical. Remember the equivalency of the protons and the carbon atoms in benzene.

In that case, electron spectroscopy for chemical analysis (ESCA) also called X-ray photoelectron spectroscopy (XPS) can distinguish between these two processes: a fast hydrogen shift or a highly symmetrical intermediate. The most crucial difference between NMR and ESCA is that NMR can only observe processes occurring for longer than 10^{-7} seconds. On the other hand, ESCA is much more sensitive than NMR and determines processes that take place over as little as 10^{-16} seconds.

ESCA gives us information about the binding energies of the atoms. *How does it work?* A molecule is irradiated with a specific electromagnetic wave ($h\nu$). This wave removes the electrons from the various orbitals of the molecule and the electrons separated from the molecule gain specific kinetic energy. This kinetic energy depends on the binding energy (E_B) of the electrons separated from the molecule. If the energy sent to the molecule is $E = h\nu$, the ejected electrons' kinetic energies are T. By measuring the kinetic energies of electrons in the inner shell of atoms, the number of different orbitals is determined.

$$T = h\nu - E_B$$

Analysis of the cyclopentyl carbocation with ESCA reveals that it contains four uncharged carbon atoms and a positively (+) charged carbon atom with relative separations of electron binding energy of $4.3 \pm 0.5\,eV$. This shows that the cyclopentyl carbocation undergoes rearrangement by a very rapid hydride shift. This process cannot be detected by NMR spectroscopy [21].

Subsequently, the solid-state ^{13}C-NMR spectrum of the cyclopentyl cation at $-203\,°C$ showed three peaks at 320.0, 71.0, and 28.0 ppm [22]. The peak at 320.0 ppm was assigned to the carbon atom bearing the positive (+) charge. The degenerate hydride shift was frozen at that temperature.

7.4.8 The Wagner–Meerwein Rearrangement

The *Wagner–Meerwein rearrangement* is an extremely common reaction in organic chemistry. A carbocation can undergo rearrangement in which a hydrogen, alkyl, or phenyl group can migrate with bonding electrons to the neighboring carbon atom. This reaction results in the formation of a new carbocation, which is generally more stable, and a skeletal change

7 Reactive Intermediates: Carbocations

in the molecule occurs. This rearrangement was first discovered by the Russian chemist Georg Wagner [23] (also known as Igor Igorevich Wagner, 1849–1903) and German chemist Hans Meerwein [24] (1879–1965) by the conversion of isoborneol to camphene by heating in the presence of an acid. Therefore, such reactions are named after the two scientists. The mechanism of similar reactions will be discussed in more detail in the nonclassical carbocations section.

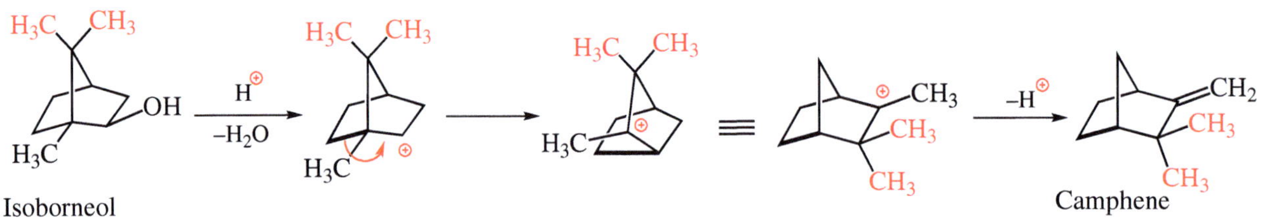

As discussed in the previous section, the thermodynamic driving force for rearrangements is the formation of a more stable carbocation. In order to allow an intermolecular rearrangement to occur, the bonding orbital of the migrating group must be coplanar with the empty p orbital of the carbocation (Figure 7.5). During the rearrangement, a transition complex is formed in which the migrating group (a hydride ion or an alkyl group) is partially bonded to two carbon atoms. This type of rearrangement is called the *nucleophilic rearrangement* as the migrating group shifts to the carbocation center with the bonding electrons (as a nucleophilic group). Wagner–Meerwein rearrangement reactions are among the most typical examples of nucleophilic rearrangements.

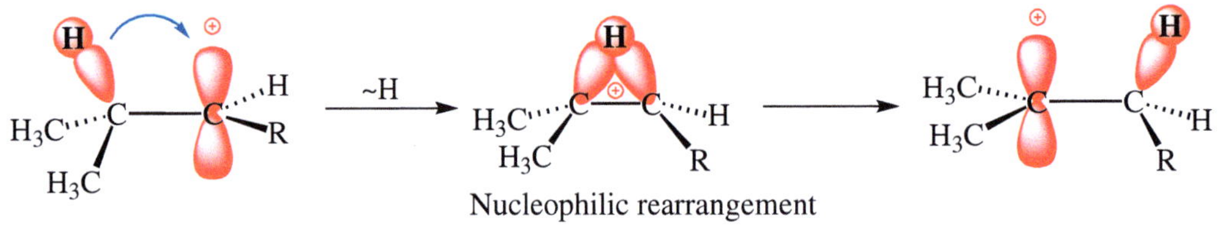

Figure 7.5 Schematic presentation of migration of a hydride.

When 2,2-dimethyl-1-bromopropane (also called neopentyl bromide) is treated with aqueous formic acid, 2-methylbutan-2-ol is formed as a single product. According to the reaction mechanism, the carbocation (neopentyl cation) formed in the first step is a primary carbocation and is unstable and prone to rearrangement. To form a more stable carbocation, one of the methyl groups attached to the neighboring carbon atom migrates to the carbon atom bearing the positive (+) charge, forming a tertiary carbocation that is much more stable. The carbocation reacts with water to form a stable product, 2-methylbutan-2-ol. Neopentyl alcohol is never obtained in this reaction.

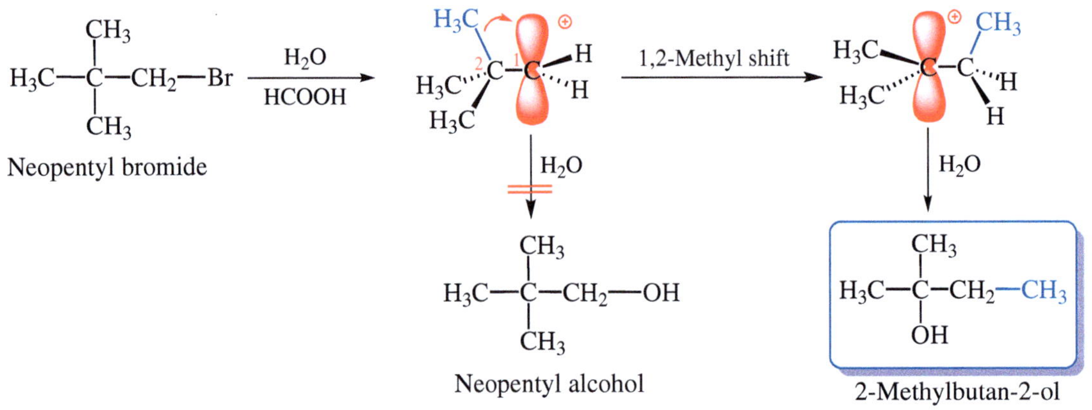

Because the alkyl group shifts from the carbon atom C2 to the carbon atom C1, this is generally referred to as a 1,2-shift. At this point, the following question may arise: *Is a shift such as a 1,3- or 1,4-shift larger than a 1,2-shift possible?* In general, no other types of shifts are encountered in the shift of alkyl groups. For example, only the 1,2-hydride shift is observed in the 3,3-dimethyl-1-butyl carbocation and there is no 1,3-alkyl shift product. However, long-distance hydride shifts are encountered in some molecules with suitable geometries.

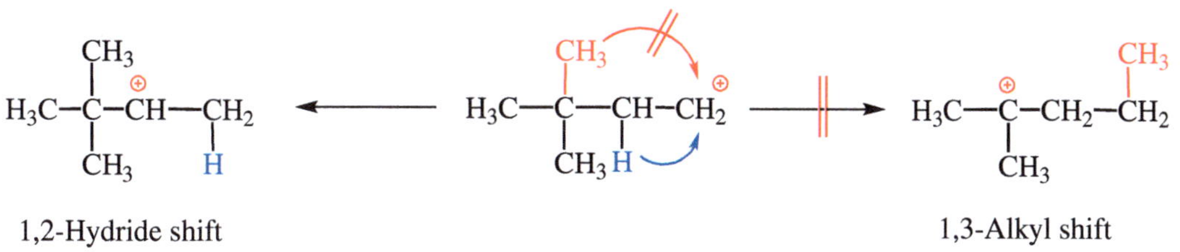

1,2-Hydride shift   1,3-Alkyl shift

It was emphasized that one of the most effective factors in the rearrangement of carbocations is the formation of a more stable carbocation. With the formation of one, the energy level of the system decreases. The increasing stability of carbocations is not the only factor that leads to skeletal rearrangement. If angle strain, steric crowding, or torsional strain in the molecule decreases by a hydride or alkyl shift, a rearrangement from a stable carbocation to a less stable carbocation can occur. For example, the cyclobutane derivative shown below is converted into the cyclopentane derivative when treated with HBr. The most interesting feature of this reaction is the rearrangement of an initial tertiary carbocation formed to a less stable secondary carbocation. This rearrangement occurs because the energy gained by decreasing the strain in the molecule is greater than that required to transform the tertiary carbocation into the secondary carbocation.

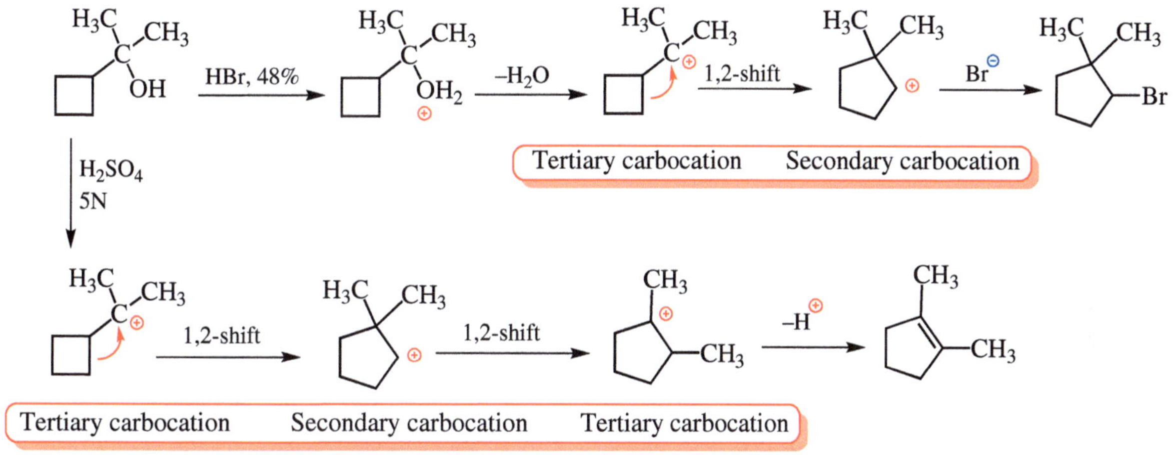

In the reaction carried out in H_2SO_4, the secondary carbocation is further rearranged to the tertiary carbocation, followed by proton elimination to give a 1,2-dimethylcyclopent-1-ene.

7.4.9 Pinacol Rearrangement

Pinacol is a compound with two hydroxyl groups, each attached to the adjacent carbon atom [25]. Pinacol was first synthesized in 1859 by the German chemist Wilhelm Rudolph Fittig (1835–1910) by the combination of ketyl radicals formed in the reaction of acetone with magnesium. Fittig reacted 2,3-dimethylbutane-2,3,-diol with acid and obtained 3,3-dimethyl-2-butanone, whose structure was identified by Butleron in 1874. This conversion of a diol into a ketone is called the *pinacol rearrangement*. As the compound contains a carbonyl group, the product is called *pinacolone*.

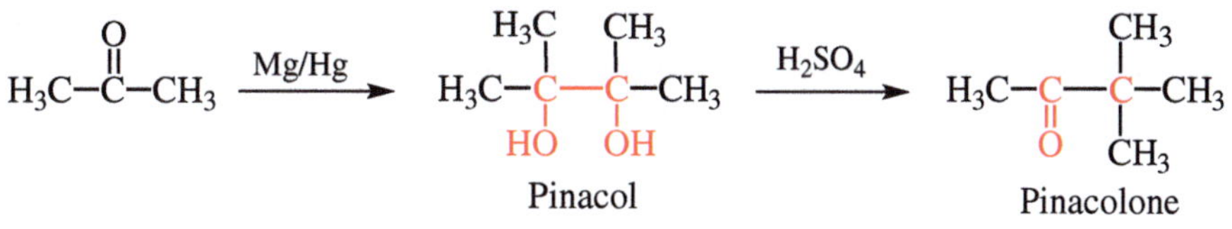

The acid-catalyzed pinacol rearrangement starts with the protonation of a hydroxyl group and subsequent loss of water results in a tertiary carbocation. The methyl group's migration is assisted by the hydroxyl group to generate a more stable carbocation in which the oxygen atom stabilizes the charge. This cation is simply the conjugate acid of the ketone, pinacolone.

Because pinacol is a symmetrical molecule, the reaction mechanism is quite simple. If the diol used is asymmetrical and the groups attached to the carbon atoms are different, then we are faced with some questions.

Which hydroxyl group will be eliminated or which carbocation will be formed?
What is the migration tendency of the different substituents?
Is there any effect arising from strain or steric hindrance?

Let us first start with symmetrical diols with different substituents to check their migration ability. Butane-2,3-diol has a symmetrical structure. When either of the two hydroxyl groups is eliminated, the same carbocation will be formed. The question is which substituent (hydride or methyl) will migrate preferentially? Experiments show that only the hydride group migrates; the methyl group does not. Here, we come to the following conclusion: hydride groups migrate more easily than alkyl groups.

Let us now examine the competition between the phenyl group and an alkyl group in a symmetrical diol, 2,3-diphenylbutane-2,3-diol. Because the molecule is symmetrical, a single carbocation will form. The phenyl group generally shifts faster than the alkyl group to form the rearranged product, 3,3-diphenylbutan-2-one, a kinetic product. This experiment shows that the phenyl group migrates more easily than the alkyl group. The compound formed from the methyl shift may be thermodynamically more stable because of the conjugation between the carbonyl group and the phenyl ring. Keep in mind that migration ability does not consider the stability of the final product. After these two reactions with symmetrical diols, we can conclude that the phenyl group migrates more easily than the alkyl group and the hydride group migrates more easily than the alkyl group.

In the case of an asymmetric diol with different substituents, the reaction becomes more complicated; which hydroxyl group will be eliminated and which group will migrate? For example, when 2-methyl-1,1-diphenyl-propane-1,2-diol is treated with an acid, two tertiary carbocations can be formed. One of them is resonance stabilized by the phenyl groups. As this carbocation will have lower activation energy, it will be formed faster. After rearrangement, the kinetic product will be formed as the major product. In this system, a migrating group is a methyl group. However, this does not mean that the methyl group is a better migrating group than the phenyl group.

7.4 Reactions of Carbocations: Rearrangements

In the example given below, 1,1-diphenylethane-1,2-diol rearranges rapidly to an aldehyde that is a kinetically controlled product. However, the thermodynamically controlled product is formed slowly. First, the hydroxyl group attached to the carbon atom bearing the phenyl groups will be eliminated, forming the more stable carbocation. A hydride shift will produce the aldehyde. When the aldehyde carbonyl group is protonated, it will behave like a pinacol intermediate; the successive phenyl shift and hydride shift result in the thermodynamic product.

Bachmann and Ferguson [26] synthesized symmetrical aromatic pinacols having the structure $R^1R^2C(OH)–C(OH)R^1R^2$ and subjected them to acid-catalyzed rearrangement to determine their migration aptitudes. They found that the aromatic ring with electron-donating groups migrates faster than the aromatic groups with electron-withdrawing groups. The relative migration aptitudes of some substituted aromatic rings are given in Table 7.2 [27].

Table 7.2 Relative migration aptitudes of some substituted aromatic rings. Source: Based on Zaczek [27].

	R = p-OEt	p-OMe	p-Me	p-Ph	m-OMe	H	p-Cl	m-Cl
	500	500	15.7	11.5	1.6	1.0	0.66	0

The pinacol rearrangement is a type of reaction with a wide range of synthetic applications and is successfully applied to the synthesis of spiro compounds. The spiro compound given below is easily synthesized from two different diols.

7.4.10 Tiffeneau–Demjanov Rearrangement

The Tiffeneau–Demjanov reaction [28–31] is mechanistically very similar to the pinacol rearrangement. The same intermediate is formed in both rearrangement reactions. The only difference between these two reactions is that the starting

7 Reactive Intermediates: Carbocations

compounds used are different. While 1,2-amino alcohols are used as the starting compounds in the Tiffeneau–Demjanov reaction, 1,2-diols are used in the pinacol rearrangement. Ambiguity in determining which hydroxyl group will be removed in the pinacol rearrangement is eliminated by the Tiffeneau–Demjanov reaction as the amino group is only removed by nitrous acid deamination reaction.

To convert an amino group into a carbocation, the diazonium salt must first be formed. For this, an amino alcohol is treated with sodium nitrite in an acidic medium, and the resulting diazonium salt is then converted into the carbocation by deazotation (for a detailed mechanism for the formation of the diazo compound, see Section 6.7.3). Because the resulting intermediate has the same structure as the intermediate formed in the pinacol arrangement, this reaction is also called the *semipinacol rearrangement*.

The Tiffeneau–Demjanov reaction has high synthetic potential and is mostly applied to transfer cyclic ketones into higher homologs. Because the synthesis of cyclic aminoalcohols is easy, the ring expansion reaction is successfully applied to many systems. 1,2-Aminoalcohols are readily obtained by catalytic reduction of cyanohydrins formed by adding hydrogen cyanide to the corresponding ketones.

In the case of an asymmetrical ketone, two different groups can migrate. The electronic structures of the groups determine which group will migrate. As secondary carbon atoms can migrate better than primary carbon atoms (path a), 3-methyl-cyclohexanone is formed as the major product.

Another method used for the synthesis of 1,2-aminoalcohols is the reaction of a cyclic ketone with nitromethane in a basic medium, followed by the reduction of a hydroxy nitro compound with Raney Ni. The application of this method to cyclohexanone furnishes a ring-enlarged product, cycloheptanone, in 41% overall yield.

7.4 Reactions of Carbocations: Rearrangements

[Reaction scheme: cyclohexanone → (CH₃NO₂/NaOEt) → 1-(nitromethyl)cyclohexanol → (H₂/Raney Ni) → 1-(aminomethyl)cyclohexanol (1,2-Amino alcohol) → (HONO, 0 °C) → cycloheptanone]

(3R)-1-Amino-2,3-dimethylpentan-2-ol rearranges under the Tiffeneau–Demjanov reaction conditions to form two products. These products are the result of the 1,2-shift of the methyl and secondary butyl groups. An exciting feature of this reaction is that the secondary butyl group migrates with retention of the configuration [32]. This indicates that the migrating group is not completely separated from the molecule during migration. Cross experiments, reacting two different configurated systems in the same reaction medium, showed no group transfer from one molecule to another. If the migrating group was wholly detached from the molecule, complete or partial racemization would be observed in the migrating group's configuration. Therefore, it is likely that the migrating group creeps from one carbon atom (C1) to the other (C2).

[Reaction scheme: (3R)-1-amino-2,3-dimethylpentan-2-ol → (HNO₂, –N₂) → Major product and Minor product]

It was shown that the intermediate product formed in the pinacol rearrangement has the same structure as that formed in the Tiffeneau–Demjanov reaction. The intermediate product that starts the rearrangement in both reactions is the hydroxyl group carrying a positive (+) charge on the adjacent carbon atom. This intermediate can also be generated, starting from the other functional groups. For example, the acid-catalyzed ring-opening reaction of epoxides will also form the same intermediate. Thus, epoxides may also be precursors of potential pinacol-type rearrangements. For this reason, in order to comprehend organic chemistry well, it is necessary to find the common points in mechanisms.

[Scheme showing: Epoxide ring-opening (H⁺) from cyclohexene oxide → cation intermediate; Pinacol rearrangement from diol (H⁺) → same intermediate; Tiffeneau–Demjanov rearrangement from amino alcohol (NaNO₂/H⁺) → same intermediate]

Ketones containing tertiary alkyl groups can undergo rearrangement in acidic medium by a similar mechanism, as discussed above. When the carbonyl group is protonated, an alkyl group shifts from the neighboring carbon atom, generating an intermediate similar to that seen in the pinacol rearrangement. This intermediate will further rearrange to form the thermodynamically most stable ketone.

[Mechanism scheme: R^2–C(R^1)(R^3)–C(=O)–R^4 → (H⁺) → protonated ketone → 1,2-shift → rearranged carbocation → (–H⁺) → R^2–C(R^3)–C(=O)–R^1 with R^4]

7.4.11 Dienone–Phenol Rearrangement

The dienone–phenol rearrangement converts 4,4-disubstituted cyclohexadienone into a stable 3,4-disubstituted phenol upon treatment with an acid [33, 34]. The first step is the protonation of the oxygen atom of the carbonyl group. The carbocation formed undergoes an alkyl shift to form a resonance-stabilized carbocation. Removal of a proton from the intermediate forms the phenol derivative. The driving force of this reaction is the aromatization of the system.

Spirodienone systems are also easily rearranged according to a similar mechanism to form tricyclic compounds. For this purpose, the acid-catalyzed rearrangement of the spirodienone forms 1,2,3,4-tetrahydro-9-phenanthren-9-ol [35].

7.4.12 Neighboring Group Participation in Molecular Rearrangement (Anchimeric Assistance)

In Section 2.2.5, we discussed the participation of a neighboring group in a nucleophilic substitution reaction. When a carbocation is formed in a molecule, a group attached to the neighboring carbon atom can interact with the reaction center with an electron pair, σ bond electrons, or π bond electrons on it. Neighboring groups can increase the reaction rate, affect the stereochemistry, or lead to rearrangement. The involvement of a neighboring group in a nucleophilic substitution reaction is called *neighboring group participation (anchimeric assistance)*. In this section, we will discuss how the neighboring groups affect the molecular rearrangement.

Neighboring group participation in molecular rearrangement will be examined in three separate groups.

1. Involvement of σ bond electrons: The participation of σ-bonds is mostly observed in strained cyclic systems. This subject will be discussed in detail in the nonclassical carbocations section.
2. Involvement of nonbonding electrons.
3. Involvement of π bond electrons.

It is possible to perceive the rearrangement reactions as nucleophilic substitution reactions. As discussed in the nucleophilic substitution section, there are generally two different nucleophilic substitution mechanisms. According to the first mechanism (S_N1), the leaving group first separates from the molecule to form a carbocation, which reacts with a nucleophile to give the product. In an S_N1-type rearrangement, the migrating group (nucleophile) leaves the neighboring carbon atom and shifts to the carbocation center.

The second mechanism is an S_N2-type mechanism. The nucleophile attacks the carbon atom from the back and displacement occurs with inversion of configuration. In an S_N2-type rearrangement, the migrating group attacks the neighboring carbon atom bearing the leaving group from the back, making it easier for the leaving group to separate from the molecule, and causes configuration inversion.

Neighboring group participation with nonbonding electrons plays an essential role in rearrangement [36, 37]. Let us examine the acetolysis reaction of two isomeric acetoxy tosylates. The *cis-isomer* undergoes a normal S_N2 reaction with potassium acetate in acetic acid and forms the *trans*-diacetate. A configuration inversion takes place during the substitution reaction. However, the acetolysis of the *trans*-isomer gives a mixture of two products. The nonbonding electrons on the carbonyl oxygen atom attack the neighboring carbon atom from the back and substitute the tosylate group, forming an acyloxonium ion [38], which is stabilized by delocalization of the positive (+) charge. Now the acetate anion can attack at the C1 carbon atom, forming a substitution product with configuration retention. On the other hand, an attack on the C2 carbon atom forms a rearranged product. The neighboring group here controls the reaction.

Winstein observed neighboring group participation in the reaction of 3-bromobutan-2-ol with HBr [39]. The reaction of *erythro*-3-bromo-2-butanol with HBr gave pure *meso*-2,3-dibromobutane, while the *threo*-isomer yielded DL-2,3-dibromobutane under the same reaction conditions. After protonation of the hydroxyl group, the bromine atom attacks with the nonbonding electrons the neighboring carbon atom from the back, forming a cyclic intermediate, a bromonium ion. A negatively charged ion, such as a bromide ion, can attack from the opposite side of either carbon atom, giving rise to a single isomer, the *meso*-isomer. The reaction with the *threo*-isomer proceeds similarly, forming the DL-pair.

A phenyl group can also involve in carbocation rearrangement reactions and it can affect the configuration or cause skeletal rearrangement of the molecule. For example, the solvolysis reaction of *p*-bromobenzenesulfonate in acetic acid in the presence of sodium acetate produces two rearranged products [40].

For this rearrangement, the phenyl ring is responsible. It attacks the carbon atom bearing the leaving group with the π electrons from the back, forming a phenonium ion intermediate. The acetate attacks the carbon atom bearing the methyl groups, forming the rearranged product. The rate is increased by electron-donating substituents in the p-position of the phenyl ring and decreased by electron-withdrawing substituents.

7.4.13 Nonclassical Carbocations

In the section entitled "neighboring group participation", we discussed how nonbonding electrons and π bond electrons affect the reaction mechanism, change the products' configuration, and form rearranged products. In this section, we will discuss the involvement of σ bond electrons in neighboring group participation.

What is the difference between classical and nonclassical carbocations? A classical carbocation has a positively charged carbon atom with six electrons that bond three substituents. A nonclassical carbocation involves three carbons with two electrons. This means that three atoms share two electrons.

Nonclassical carbocations have been extensively studied with 2-norbornyl systems. In the very first studies, Saul Winstein, a Canadian chemist (1912–1969), revealed that the solvolysis of optically active *exo*-2-brosylate (*p*-bromobenzenesulfonates) with acetic acid gave the *exo*-norbornyl acetate exclusively with complete loss of optical activity [41, 42]. On the other hand, the *endo*-isomer was solvolized under the same conditions with complete inversion of the configuration. It forms the *exo*-acetate. One of the enantiomers was formed in 3–13% excess, depending on the solvent used. The optical activity was also significantly lost. Another essential point observed in the reaction is that the solvolysis rate of the *exo*-isomer was 350 times as rapid as that of the *endo*-isomer. Furthermore, the solvolysis rate of the *endo*-isomer was similar to that of the solvolysis of a cyclohexyl derivative.

Winstein has proposed the following mechanism for these transformations. The ionization of the *endo*-brosylate is considered normal. However, the ionization of the *exo*-brosylate is assisted by C1—C6 σ bond electrons, leading directly to a *nonclassical carbocation*. The σ bond electrons between the C1–C6 carbons act like a nucleophile and attack the C2 carbon atom to which the brosylate group is attached from the back (as in the S_N2 reaction mechanism) and this allows this group to be easily removed.

If the C1—C6 σ bond electrons separated from the C1 carbon atom and connected to the C2 carbon atom, the resulting carbocation would be classical. In the case of nonclassical carbocation, C1—C6 σ bond electrons at the same time bind the C1 and C2 carbon atoms. These two electrons bind three centers at the same time. The C6 carbon atom is bonded to five atoms: three carbons (C1, C2, and C5) and two hydrogens. This nonclassical carbocation is achiral and has a plane symmetry passing through the C4, C5, and C6 plain. In this symmetrical structure, C1/C2 and C3/C7 carbon atoms are identical.

In the second step of the reaction, the nucleophile, an acetate anion, attacks the nonclassical carbocation to terminate the reaction. Attack by the acetate at C1 and C2 is equally likely because of the symmetrical structure of the carbocation. Attack at the C2 carbon atom yields the original configuration (*exo*-configuration); attack at the C1 carbon atom yields the enantiomer, of course, with the *exo*-configuration. The loss of optical activity observed in the starting compound is explained by the formation of a racemic mixture.

In the case of the *endo*-norbornyl brosylate, no group can provide anchimeric assistance as in the *exo*-isomer. The C1—C7 σ bond electrons can assist in removing the brosylate group in the *endo*-isomer by a similar mechanism and forming a nonclassical carbocation. However, with this rearrangement, the molecule's intermolecular strain will increase further as the five-membered ring will turn into a four-membered ring. For this reason, such anchimeric support is not possible in this system. The *endo*-isomer ionizes slowly to a classical secondary carbocation, which rapidly converts into the more stable nonclassical carbocation, which reacts as described above to form a mixture of enantiomers.

We discussed above that during solvolysis of the *endo*-isomer, one of the enantiomers formed was 3–13% in excess depending on the solvent. When one of the enantiomers is in excess, the optical activity will not be lost entirely. Because the leaving group departs slowly from the molecule, it will remain closer to the C2 carbon atom, creating an asymmetric environment. Nucleophiles cannot attack C1 and C2 carbon atoms equally. Attack on the C1 carbon atom will be slightly greater.

Winstein suggested that the involvement of the σ bond plays a significant role in norbornyl systems and explains the different solvolysis rates due to anchimeric assistance. These findings were entirely rejected by H. C. Brown, 1979 Nobel Prize winner, who discovered the hydroboration reaction [43]. Brown claimed that the solvolysis of the *exo*-isomer is not rapid and the solvolysis of the *endo*-isomer is retarded. According to Brown's theory, the separation of the brosylate group from the *endo*-isomer is prevented by the *exo*-hydrogen atom attached to the C6 carbon atom decreasing the reaction rate.

Steric hindrance
endo-isomer

Brown's theory

No steric hindrance
exo-isomer

H. C. Brown has also stated his objection to the formation of a nonclassical carbocation developed by Winstein. According to Brown's theory, the intermediate formed is a rapidly equilibrating classical carbocation. The bonding electrons between C1 and C6 are continually shifting between C1–C6 and C2–C6. The 1,2 alkyl shift interconverts a classical carbocation so fast that NMR cannot detect it. Of course, this fast equilibrating secondary carbocation will form the enantiomers. Brown compares this fast 1,2-alkyl shift to an automobile windshield wiper.

Brown's theory: classical carbocation rearrangement

During these claims, various compounds were synthesized by other groups, and their solvolysis rates were examined in detail [44–47]. Replacement of the hydrogen atom at C6 by a methyl group increases the solvolysis ratio k_{exo}/k_{endo} from 100 to 180 (Table 7.3). On the other hand, the attachment of electron-withdrawing groups to the carbon atom C6 decreases the solvolysis rate of the *exo*-isomer. All these data demonstrate that the C1—C6 bond provides anchimeric assistance. However, even these findings did not deter Brown from his claims.

Table 7.3 Relative solvolysis rates of substituted *exo*- and *endo*-norbornyl tosylates.

R_1	R_2	k_{exo}/k_{endo}
H	H	100
H	CH_3	180
H	CO_2CH_3	3.7
H	CN	0.4
CN	H	1.1

While the classical and nonclassical carbocation debate between Brown and Winstein was growing in ferocity with each passing year, in the early 1960s, Olah first participated in this discussion with the synthesis of long-lived carbocations via magic acids. The norbornyl cation was formed by dissolving 2-fluoronorbornane in SbF_5/SO_2 or SO_2ClF [48, 49]. The ^{13}C-NMR spectrum of the norbornyl cation at $-159\,°C$ shows five peaks in accordance with the proposed structure. However, observing five separate carbon signals here does not indicate that the ion has a nonclassical carbocation structure. According to Brown's claim, the bond between C1 and C6 shifts rapidly between C2 and C6. In that case, five signals will also be observed as the carbon atoms C1/C2 and C3/C7 are equal. However, rapid equilibration of a classical carbocation is not possible at this temperature. NMR spectroscopic studies of the 2-norbornyl cation provide further conclusive evidence for the formation of a nonclassical carbocation [50].

C1=C2	C1=C2=C6		
C3=C7	C3=C5=C7		
C4	C4		
C5 Five different	Three different		
C6 Carbon atoms	Carbon atoms		

The spectrum's exciting feature is that when the temperature is raised to $-70\,°C$, the number of carbon signals decreases from 5 to 3 (91.7, 37.7, and 30.8 ppm). This situation can only be explained by a rapid hydride shift between the C1, C2, and C6 carbon atoms. As shown in the equation above, one of the hydrogens attached to the C6 carbon atom shifts as hydride to the C1 carbon atom and then to the C2 carbon atom. In such a case, the C1, C2, and C6 carbon atoms become equal to each other. Simultaneously, the C3, C5, and C7 carbon atoms also become equal to each other. Therefore, the ^{13}C-NMR spectrum of the norbornyl cation consists of three signals. When the molecule is brought to room temperature, a new hydride shift between C2 and C3 is disclosed, which makes all carbon atoms equal. Therefore, the ^1H-NMR spectrum of the norbornyl cation in the temperature range -50 to $20\,°C$ gives a single peak at 3.10 ppm for protons and a single peak for the carbon atom.

As mentioned above, these observations do not prove that the norbornyl cation is a nonclassical carbocation. Later, the solid-state ^{13}C-NMR spectrum of the norbornyl cation was recorded in the temperature range -160 to $-268\,°C$ (5 K) [51]. The spectrum remained unchanged down to 5 K and showed five peaks. According to Brown's claim, if the norbornyl cation was a classical carbocation, the activation energy required for equilibrium would be less than 0.2 kcal/mol at 5 K. This activation barrier is well below the energy barrier required for a molecular arrangement. In classical carbocations, seven different signals should be observed in the ^{13}C-NMR spectrum of a conventional norbornyl cation. Therefore, the measurement around 5 K demonstrates that the structure is a nonclassical carbocation. Subsequent ESCA measurements also revealed that the norbornyl cation is a nonclassical carbocation rather than a classical one [21]. Finally, X-ray crystallographic measurements carried out at 40 K in 2013 confirmed the nonclassical structure [52].

7.4 Reactions of Carbocations: Rearrangements

The norbornyl cation is a nonclassical carbocation in which σ bond electrons are involved in its formation. The delocalization of π bond electrons can also generate nonclassical carbocations. 7-Norbornenyl systems exhibit interesting behavior according to the position of the substituent attached to the bridge atom. For example, *anti*-norbornenyl tosylate solvolysis is 10^7 times faster than that of its *syn*-isomer [53–56].

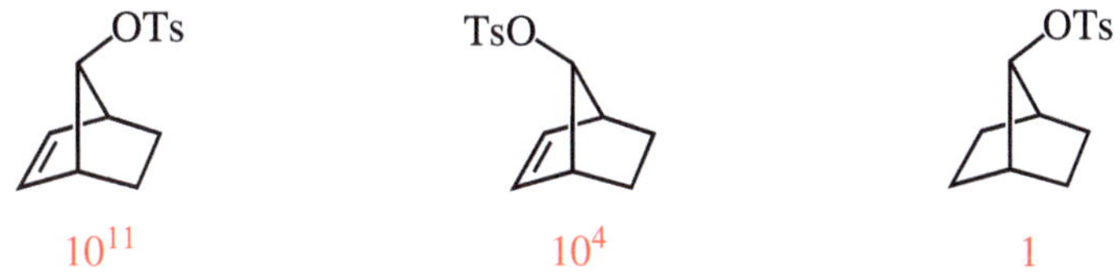

10^{11} 10^4 1

Relative rates for solvolysis

This remarkable difference observed in the solvolysis rate is explained only by the support of neighboring groups. The double-bond electrons attack the tosylate group from the back, facilitating removal of the tosylate group (increasing the reaction rate) and forming the pentacoordinate nonclassical carbocation. Because the nucleophile attacks the molecule at its *anti*-position, no change is observed in the acetate configuration. Solvolysis of the *syn*-tosylate led to the formation of bicyclo[3.2.0]hept-3-en-2-yl acetate. When the same solvolysis reaction was performed with a saturated system to reveal the effect of the double bond, it was determined that 7-norbornenyl system solvolysis was 10^{11} times faster than that of the 7-norbornanyl system.

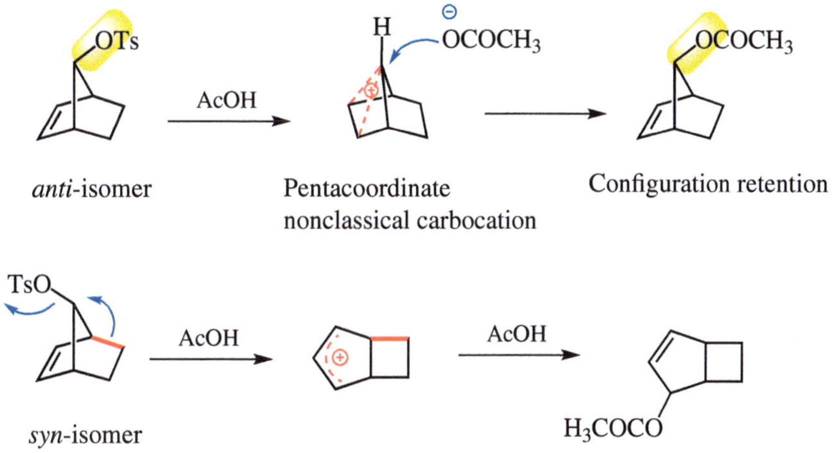

anti-isomer → Pentacoordinate nonclassical carbocation → Configuration retention

syn-isomer

Wagner–Meerwein rearrangements are more prominent than normal addition reactions in bicyclic systems and rearranged products are formed as the main products. For example, the electrophilic addition of bromine to benzonorbornadiene furnishes the rearranged product in quantitative yield. The normal 1,2-addition product was not observed [57, 58].

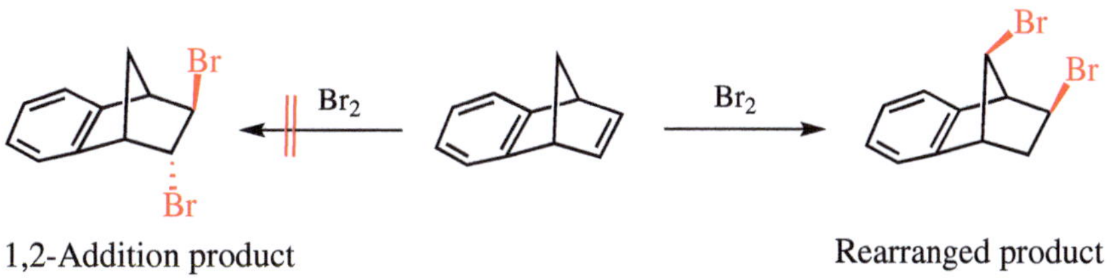

1,2-Addition product   Rearranged product

When looking at the product formed as a result of the reaction, one of the bromine atoms is perceived as attached to the bridge carbon atom. However, this is not the case. The bridge carbon has changed its position. The formation of this product can be explained only by the formation of a nonclassical carbocation as the intermediate.

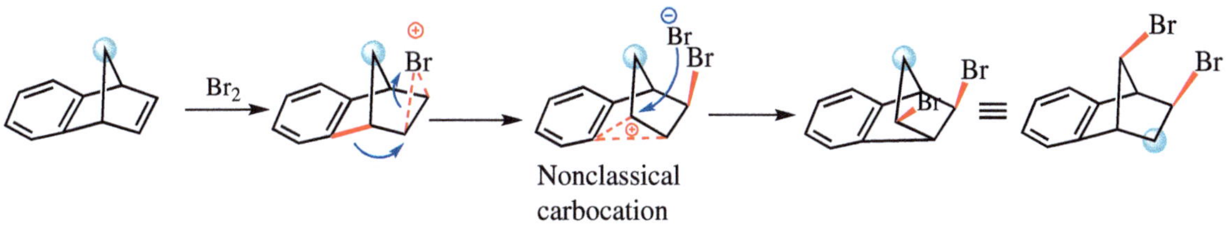

Nonclassical carbocation

7.4.14 Nametkin Rearrangement

The Nametkin reaction is a variation of the Wagner–Meerwein reaction and is generally observed in monoterpenes with acid treatment. Terpene chemists call the migration of a methyl group from the C3 carbon atom to the C2 carbon atom the Nametkin rearrangement [59, 60].

For example, the first step reaction in the conversion of 4-methylcamphene to 4-methylisoborneol is the Nametkin reaction. Protonation of the exocyclic double bond leads to the tertiary carbocation. In general, the reaction follows the mechanistic pathway of a 1,2-methyl shift. Then, the carbocation formed produces a nonclassical carbocation and the Wagner–Meerwein rearrangement follows. The reaction with a nucleophile furnishes 4-methyl-isoborneol.

The Nametkin rearrangement is not always followed by the Wagner–Meerwein rearrangement. Sometimes, a double bond can be formed by proton elimination after the Nametkin rearrangement.

7.4.15 Hydride Shift

A hydride shift is as crucial as an alkyl shift in carbocation chemistry. Hydride shifts can occur if there is a carbocation intermediate. In general, a hydride shifts from one carbon to the adjacent carbon atom, forming a more stable carbocation. Hydride shifts over larger distances are also possible. We briefly discussed 1,2-hydride shifts in the norbornenyl system.

For example, when benzene is reacted with *n*-propyl bromide in the Friedel–Crafts reaction, isopropyl benzene is formed instead of the expected product, *n*-propylbenzene, which results from a 1,2-hydride shift of the initially formed propyl carbocation to the more stable isopropyl carbocation.

In rearrangement reactions, we have seen so far that the migrating group shifts from a C1 carbon atom to a C2 carbon atom, called a 1,2-shift. If there are one or more atoms between the carbon atom to which the migrating group is first attached and the carbon atom to which it is last bonded, such shifts are called long-distance shifts (*transannular shifts*). For such shifts to occur, the carbon atom to which the migrating group is attached and the atom to which it will migrate must be close to each other. Therefore, such shifts are more common in medium-sized ring systems (8- to 11-membered).

Solvolysis of *cis*-cyclooctene epoxide in aqueous formic acid gave four different products [61]. The *trans*-1,2-diol is the expected ring-opening product. However, the other products are derived from an intermediate carbocation formed by a 1,5-hydrogen shift. The carbocation formed as a result of the hydride shift either reacts with H_2O to give a 1,4-diol or eliminates to form isomeric cyclooctenol derivatives.

Various experiments demonstrated that the methyl group does not migrate longer distances. For example, the reaction of 1-methylcyclodecane-1,6-diol in an acidic medium undergoes a transannular 1,6-hydride shift to give 6-methylcyclodecan-1-one. Labeling experiments clearly showed that the deuterium atom shifts from the C6 carbon atom to the C1 carbon atom. On the other hand, similar experiments carried out with 1,6-dimethylcyclodecane-1,6-diol did not show any shift of the methyl group [62].

7.4.16 Base-induced Nucleophilic Rearrangements

In all the rearrangement reactions we have examined so far, a carbocation center is formed first. Then, a group attached to the neighboring carbon atom migrates with bonding electrons to the carbocation center. This mechanism is similar to the S_N1 mechanism.

In base-induced nucleophilic rearrangements, the nucleophilic group migrates to the neighboring carbon atom bearing the leaving group and substitutes this group. This mechanism is much more similar to the S_N2 mechanism. Because the group migrates with bonding electrons in both rearrangement reactions, these reactions are called nucleophilic rearrangements. One of the most common nucleophilic rearrangements with significant synthetic potential is the Favorskii rearrangement.

7.4.17 Favorskii Rearrangement

The Favorskii rearrangement is a base-catalyzed conversion of cyclic or acyclic α-haloketones to the corresponding carboxylic acids. Russian chemist Alexei Yevgrafovich Favorskii (1860–1945) reacted 2-chlorocyclohexanone with KOH in methanol and obtained the ring-contracted product cyclopentanecarboxylic acid [63].

Ring-contracted product

Later, the applicability of this reaction to α-halogen ketones, in general, was determined. It is widely used to synthesize cage compounds and the highly branched carboxylic acid. Various studies have been conducted on the reaction mechanism, and it has been determined that the mechanism varies according to the system [64–66]. Two major mechanisms are considered for the Favorskii rearrangement.

1. Cyclopropanone mechanism: Compounds having α-protons rearrange by the cyclopropanone mechanism.
2. Semibenzilic rearrangement: Compounds having no α-proton rearrange by the semibenzilic mechanism.

According to the cyclopropanone mechanism, the base first abstracts a proton from the α-carbon atom, forming an enolate stabilized by the carbonyl group. The resulting carbanion acts as a nucleophile and intramolecularly attacks the α′-carbon atom from the back and substitutes the halogen to form a cyclopropanone intermediate. The ethoxide anion attacks the carbonyl carbon of the cyclopropanone and then, with the reformation of the carbonyl group, one of the bonds of the cyclopropane ring opens and forms a carbanion. The carbanion captures a proton from the reaction medium and becomes neutral. Opening the cyclopropane ring reduces the strain in the molecule. If the reaction is carried out in an aqueous medium, the corresponding carboxylic acids will be formed instead of the ester.

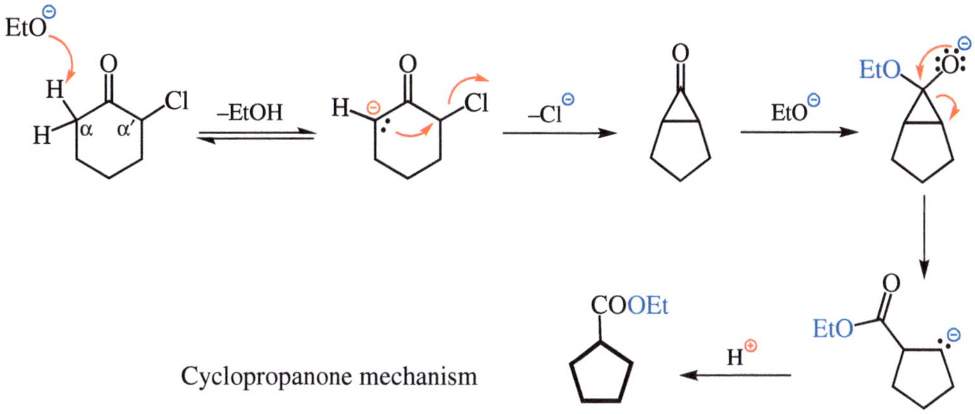

Cyclopropanone mechanism

As an alternative to the above mechanism, the semibenzilic rearrangement mechanism is suggested. As this mechanism resembles the "benzil–benzilic acid rearrangement," it is called the semibenzilic rearrangement. According to this mechanism, the base directly attacks the carbonyl carbon and opens the C=O double bond. Then, while the carbonyl group is reformed, the carbon atom directly connected to the carbonyl group shifts with the bonding electrons to the neighboring carbon atom, removing the halogen and forming the cyclopentane derivative. Various mechanistic studies have been conducted to distinguish between these two mechanisms.

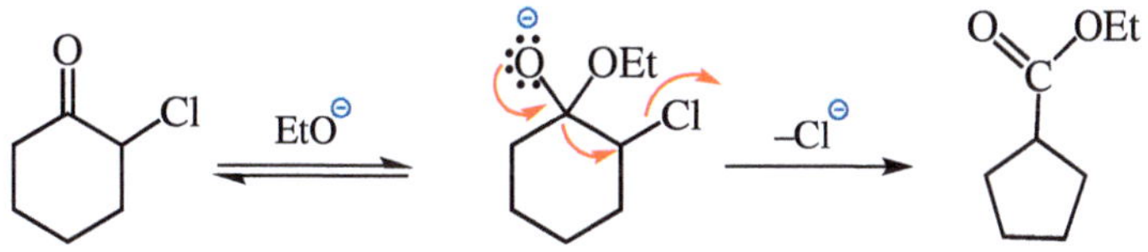

Semibenzilic rearrangement mechanism

Bordwell and coworkers [64] reacted isomeric 1-chloro-3-phenylpropan-2-one and 1-chloro-1-phenylpropan-2-one with sodium methoxide in methanol to give the same product, 3-phenylpropanoic acid, as a result of the Favorskii rearrangement. The resulting formation of the same product from different starting compounds can only be explained by the formation of a common intermediate, namely, the cyclopropanone intermediate.

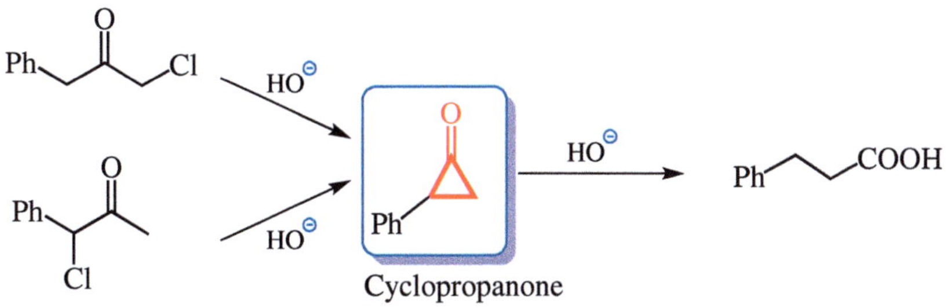

Cyclopropanone

More detailed studies were carried out with labeling experiments in chlorocyclohexanone. The experiment performed with 2-chlorocyclohexanone labeled with ^{14}C-carbon in the C-2 position produced cyclopentanecarboxylic acid with the label equally distributed in the cyclopentane derivative at the C1 and C2 carbons [67]. This is evidence for the formation of the cyclopropanone intermediate with a symmetrical structure. The cyclopropanone can undergo the ring-opening reaction with equal probability on either side of the carbonyl group, distributing the label equally between the C1 and C2 carbon atoms. The cyclopropanone intermediates have been isolated in the event of having bulky substituents attached to the α-carbon atoms. This reveals that the mechanism runs entirely on cyclopropanone.

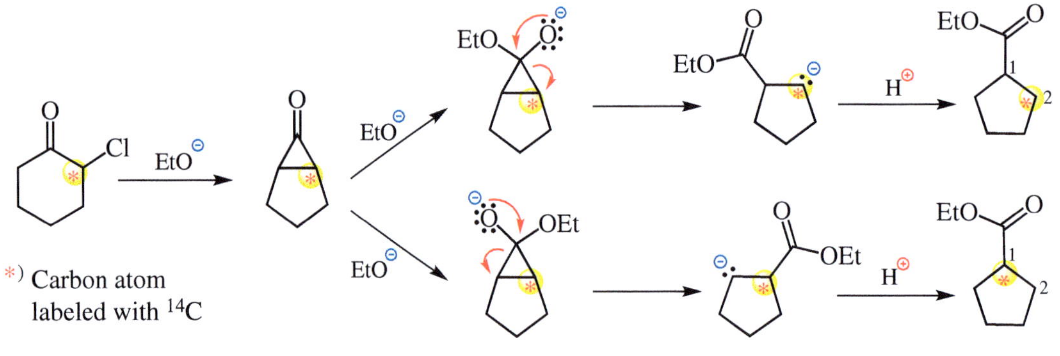

*) Carbon atom labeled with ^{14}C

Cyclopropanone mechanism

If this reaction had run according to the semibenzilic rearrangement mechanism, the ^{14}C-label at the C2 position would not be dispersed in the cyclopentane derivative formed. All of the label would be at the C1 carbon atom.

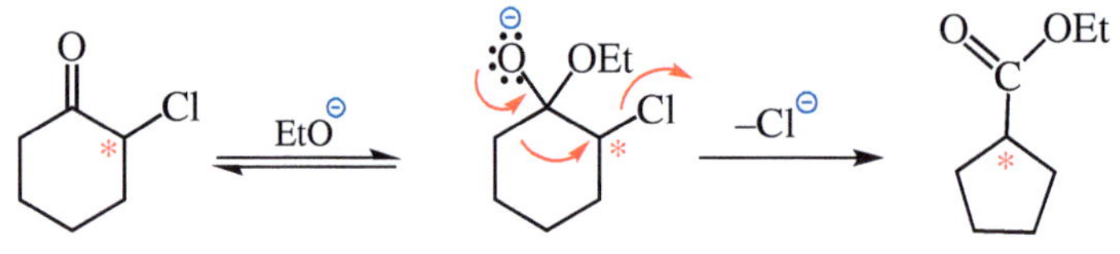

Semibenzilic rearrangement mechanism

As discussed, α-haloketones, not having an α-proton, can also undergo the Favorskii rearrangement. The product cannot be formed according to the cyclopropanone mechanism. The semibenzil rearrangement plays an essential role in the formation of the product, as shown below [68].

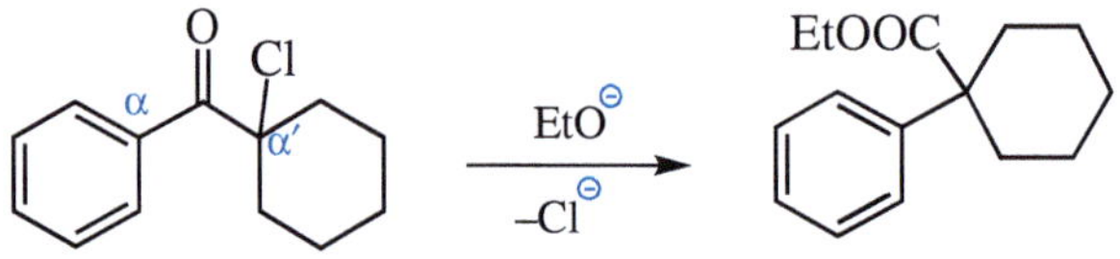

The Favorskii rearrangement has been successfully applied to the synthesis of cubane, an interesting compound that is difficult to synthesize by other methods. Decomposition of cyclobutadiene–iron complex in the presence of 2,5-dibromobenzoquinone yields the Diels–Alder adduct. Irradiation of the adduct followed by treatment with aqueous KOH affords cubane 1,3-dicarboxylic acid. Decarboxylation of the diacid gives cubane [69]. It is likely that the ring-contraction process proceeds through the semibenzilic mechanism.

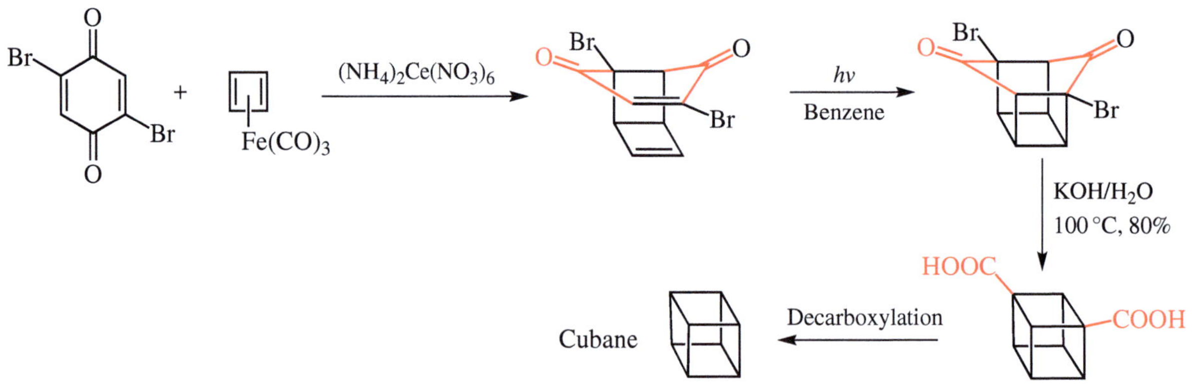

7.4.18 Ramberg–Bäcklund Rearrangement

The Ramberg–Bäcklund reaction is a base-induced conversion of α-halosulfones into *cis* and *trans* alkenes [70–73]. The reaction mechanism is similar to that of the Favorskii rearrangement (cyclopropanone mechanism). The anionic mechanism is generally accepted for the Ramberg–Bäcklund reaction. The first step is the deprotonation of α-hydrogen with a weak base followed by cyclization to form a thiirane dioxide (also called episulfone) ring. The resulting thiirane dioxide intermediate turns into an olefin by removing sulfur dioxide under reaction conditions. This reaction mechanism is very similar to the Favorskii reaction. It is especially crucial for the synthesis of some olefins that are difficult to construct under normal conditions.

If the reaction is carried out at lower temperatures, it is possible to isolate the derivatives of thiirane dioxide. When α-iodosulfone was reacted with potassium *t*-butoxide at −78 °C, besides the expected olefin, thiirane dioxide was isolated in 69% yield as the main product [74, 75]. When episulfone was heated or treated with a base, it was converted into the cyclopentene derivative. The X-ray analysis of the thiirane dioxide revealed the proposed structure.

The Ramberg–Bäcklund rearrangement has been successfully applied especially to the synthesis of small rings. Na$_2$S was used to generate cyclic sulfides. The chlorination of the sulfide with *N*-chlorosuccinimide followed by oxidation with *m*-CPBA gave the precursor for the Ramberg–Bäcklund rearrangement. The reaction of the chlorosulfone with potassium *t*-butoxide in dry tetrahydrofuran afforded [4.4.2]propella-3,8,11-triene in good yield [76, 77].

7.4.19 Benzil–Benzilic Acid Rearrangement

The benzil–benzilic acid rearrangement was discovered by the German chemist Justus Freiherr von Liebig (1803–1873) in 1838 by the reaction of benzil with potassium hydroxide to produce benzilic acid. This reaction is a classic reaction in organic chemistry. Generally, treatment of 1,2-diketones with strong bases yields α-hydroxycarboxylic acids [78–80].

If the aryl groups are different, the aryl groups with electron-donating substituents might be expected to migrate faster than the phenyl group. The experimental results demonstrate that precisely the reverse occurs. Phenyl groups with electron-withdrawing substituents migrate faster than the unsubstituted phenyl group. The base prefers to attack the carbonyl group to which a phenyl group with electron-attracting groups is bonded. After the hydroxide ion is added to the carbonyl group, there is no competition between the phenyl groups for migration. Of course, in this case, the phenyl group with the electron-withdrawing group will migrate. The intermediate formed first determines which group will migrate.

When we compare this reaction with the Favorskii rearrangement, we see that both proceed by a similar mechanism. In the Favorskii reaction, there is a leaving group, halogen. In the benzil–benzilic acid rearrangement, the carbonyl group behaves like a leaving group. When the aryl group shifts, the double-bond electrons move to the oxygen atom, forming an alkoxide.

The benzil–benzilic acid rearrangement is also applied to the 1,2-dicarbonyl compound being part of a ring. The reaction of 9,10-phenanthraquinone with a base converts it into 9-hydroxyfluorene-9-carboxylic acid. Ring contraction occurs as we saw in the Favorskii rearrangement. The ring contraction reaction has also been observed in smaller rings. For example, the reaction of cyclobutan-1,2-dione with a base produces 1-hydroxycyclopropane-1-carboxylic acids. The benzil–benzilic acid rearrangement is observed in aliphatic systems as well as in aromatic systems.

Comisar and Savage reacted benzil at high temperature (300–380 °C) with water. No catalyst (base) was added to this more environmentally benign medium. They obtained benzilic acid in high yield [81].

7.5 Rearrangement to Electron-deficient Nitrogen

In the examples we have seen so far, an alkyl group, an aryl group, or a hydride shifts with the bonding electrons to the adjacent carbon atom carrying a positive charge, forming a new carbocation. The molecules undergo rearrangement. In this section, we will discuss the migration of these groups to an electron-deficient nitrogen atom with a positive charge. Nitrogen compounds can easily undergo protonation to form ammonium salts. Although ammonium and iminium cations have a positive charge on the nitrogen atom, a shift from the neighboring carbon atom to the nitrogen atom cannot occur because of the electron octet. However, two azacations shown below with unfilled valence shells are prone to shifting from the neighboring carbon atoms. The driving force for these rearrangements is, of course, the electron deficiency of the nitrogen atoms. The shifts to electron-deficient neutral intermediates such as carbene and nitrene are described in the carbene section in detail.

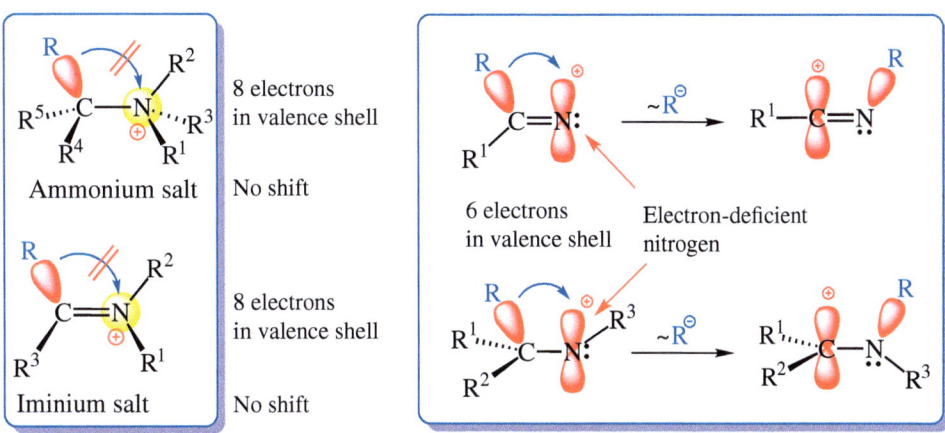

7.5.1 Beckmann Rearrangement

The Beckmann rearrangement is an essential reaction in organic chemistry to convert an oxime into an amide under acidic conditions. The reaction was first discovered by the German chemist Ernst Otto Beckmann (1853–1923) in 1886.

Oximes are easily obtained by condensation of ketones or aldehydes with hydroxylamine. The Beckmann rearrangement generally requires a strong acid. The reaction begins by protonation of the alcohol group, forming a better leaving group. The hydroxyl group can also be removed by Lewis acids, such as PCl_5 or $SOCl_2$. The group *anti* to the hydroxyl group migrates to the nitrogen atom, resulting in a carbocation and release of a water molecule. The nitrilium ion formed is trapped by water to give the amide [82–84]. The migration process is likely a synchronous process and stereospecific. The migration of the R group and the removal of water take place at the same time.

Kenyon and Young examined the stereochemical course of the Beckmann rearrangement of an optically active ketoxime, 3-ethylheptan-2-one oxime, into the substituted acetamide. They determined that the amide formed in the example below preserves 99% of its optical activity. This observation shows that the migrating group is not entirely detached from the molecule. Otherwise, racemization would be observed in the amide formed [85].

The Beckmann rearrangement has a wide range of applications in industry. For example, the Beckmann rearrangement of cyclohexanone oxime obtained by the reaction of cyclohexanone and hydroxylamine gives ε-caprolactam, which is the feedstock in the production of nylon-6. When the resulting caprolactam is heated at about 260 °C in an inert atmosphere, the ring breaks and it undergoes polymerization to form nylon-6. Nylon-6 is widely used in fibers and plastics.

The *Beckmann fragmentation* is a reaction that competes with the Beckmann rearrangement. Certain oximes having a quaternary carbon atom *anti* to the hydroxyl group produce nitriles instead of the expected normal Beckmann rearrangement product [86–88]. Those products are called *abnormal* Beckmann rearrangement products. The alkyl group that needs to shift does not migrate here. Instead, the bonding electrons shift to the nitrogen atom to form a nitrile. A stable carbocation is formed that can undergo an elimination or substitution reaction.

Tertiary carbocations are not always required for the formation of abnormal Beckmann products. The oximes having substituents at the α-position, such as hydroxy, alkoxy, or amino groups, also undergo fragmentation under Beckmann rearrangement conditions [82].

7.5.2 Neber Rearrangement

In 1926, Neber and Friedolsheim, while examining the Beckmann rearrangement of O-sulfonyl ketoxime in the presence of a base, obtained an unexpected product, an aminoketone [89]. The Neber rearrangement is the conversion of oxime tosylates in the presence of a base to α-aminoketones. The starting compounds required for the Neber rearrangement are easily obtained from reacting ketones with hydroxylamine under acidic conditions. O-sulfonation or O-acylation is conducted using tosyl chloride or acyl halides, respectively. The reaction is generally carried out in ethanolic sodium ethoxide. The base abstracts the α-proton and the carbanion formed attacks the nitrogen atom, removing the tosylate group and forming an azirine ring. Azirines are susceptible to nucleophilic addition. Therefore, the ethoxide anion in the reaction medium adds to the azirine ring, converting it into an aziridine ring. Hydrolysis of aziridines produces α-aminoketones. The overall reaction is the conversion of ketones into the corresponding α-aminoketones. Therefore, the Neber reaction is a reaction with high synthetic potential for the synthesis of α-aminoketones [90].

In the Beckmann rearrangement, the migrating group must be in the *anti*-position to the hydroxyl group. Such an orientation of the migrating group and the leaving group is not required in the Neber reaction. The major product from the Neber rearrangement is determined by the relative acidities of the α-protons and not by the stereochemistry of the oxime. The position of the tosylate group does not affect the reaction. Therefore, an alternate mechanism was suggested. Studies show that an aziridine ring can also be formed through a nitrene intermediate. The abstraction of an α-proton by a base followed by removal of the tosylate moiety furnishes an unsaturated nitrene, which is converted into an azirine [91, 92].

The Neber reaction has been successfully applied in the synthesis of some very important intermediates. For example, in the pyridine derivative given below, the amino group is readily introduced into the α-position via the Neber rearrangement. This intermediate is used as an important starting compound in the synthesis of some natural compounds [93].

7.5.3 Stieglitz Rearrangement

The Stieglitz rearrangement was first discovered by the American chemist Julius Stieglitz (1867–1937) by the reaction of triphenylmethyl hydroxylamine with PCl_5 [94]. The compound was converted into triphenyl imine. The reaction is similar to the Wagner–Meerwein rearrangement. An aryl group shifts from the carbon atom to the adjacent positively charged nitrogen atom. The migration center, the nitrogen atom, is divalent and sp^2 hybridized.

Gassman and Fox demonstrated that alkyl groups could also be involved in the Stieglitz rearrangement. They first synthesized 2-chloro-2-azabicycio[2.2.2]octane, which was reacted with silver nitrate in methanol to give 2-methoxy-1-azabicyclo[3.2.1]-octane in 60% yield [95, 96].

When the Stieglitz rearrangement is applied to N-chloraminocyclopropanol derivatives, they are converted into β-lactams in the presence of silver tetrafluoroborate. The reaction mechanism is similar to that of the pinacol–pinacolone rearrangement [97].

7.6 Rearrangement to Electron-deficient Oxygen

When oxygen compounds containing two bonds, such as alcohols and ethers, are treated with strong acids, tricoordinate oxonium ions are formed. Oxonium ions are positively (+) charged oxygen compounds having eight electrons in their valence shell, following the octet rule. Such compounds cannot undergo rearrangement, as they do not have an electron deficiency as discussed for ammonium salts. An oxygen atom in an oxonium ion cannot bind a group migrating with its bonding electrons. In order for a nucleophilic group to migrate onto the oxygen atom, the oxygen atom must have an electron deficiency in its valance shell, as well as a positive (+) charge.

To induce migration, a unicoordinate oxacation must first be generated. The most suitable starting compounds for the formation of an oxacation are those containing functional groups such as ROOH and ROX (X = a good leaving group). One of the best examples of this rearrangement is the industrial procedure for synthesizing phenol and acetone. The air oxidation of cumene (isopropyl benzene) leads to the formation of the hydroperoxide. The reaction of the hydroperoxide with an acid generates an oxacation. Removal of water produces an oxacation, followed by the phenyl group's migration to create a tertiary carbocation. The reaction of the cation with water followed by hydrolysis produces phenol and acetone.

7.6.1 Baeyer–Villiger Rearrangement (or Oxidation)

The Baeyer–Villiger oxidation is a reaction of ketones and aldehydes with peroxyacids to convert them into esters. The reaction was discovered in 1899 by German chemist Johann Friedrich Wilhelm Adolf von Baeyer (1835–1917) and his student Victor Villiger (1868–1934) and the reaction was named after them. Adolf von Bayer was awarded the Nobel Prize in Chemistry in 1905. One of the reactions they performed was the conversion of menthone to the corresponding lactone with peroxymonosulfuric acid [98].

The Baeyer–Villiger oxidation is especially useful for the synthesis of lactones starting from cyclic ketones [99, 100]. For this conversion, generally, alkyl peracids, aromatic peracids, or hydrogen peroxide are used as the oxidant. In peracids, there are two oxygen atoms in the peroxide part attached to the carbonyl group. The nucleophilic power of the oxygen atom that is directly attached to the carbonyl group is weak because nonbonding electrons on oxygen are conjugated with the carbonyl group. Because the other oxygen atom cannot conjugate with the carbonyl group, the nonbonding electrons are localized on that oxygen atom, making it more nucleophilic. Therefore, the second oxygen atom shows more nucleophilic properties and easily attacks electrophilic centers such as carbonyl carbon. Free acids are always present in the reaction medium. They increase the polarization of the carbonyl group to be reacted and facilitate the attack of peracid oxygen.

The reaction mechanism involves two separate steps: addition of the peracid to the carbonyl group and rearrangement. The reaction mechanism was always a subject of debate, as whether it is stepwise or concerted. Recent calculations support the concerted process [101]. According to this mechanism, the reaction starts with the peroxide oxygen atom attack on the carbonyl group, forming a tetrahedral Criegee intermediate (see the ozonolysis reaction), followed by proton transfer. In the next step, as the C=O double bond reforms, a 1,2-shift from the carbon atom to the oxygen atom occurs. Cleavage of the C—C bond, migration, and cleavage of the O—O bond take place in a concerted manner, forming an ester. Then deprotonation gives the neutral ester.

When a ketone is unsymmetrical, competitive migration of the substituents will occur. The migration aptitudes of the substituents in the Baeyer–Villiger oxidation are as follows:

tert-alkyl > cyclohexyl ≈ *sec*-alkyl > benzyl ≈ phenyl > *pri*-alkyl > methyl

For example, when an asymmetric cyclopentane derivative used in the synthesis of a natural product, sarracenin, is reacted with *m*-chloroperbenzoic acid, a lactone is formed in 78% yield [102]. Here, two different alkyl groups can shift. As can be seen from the structure, the secondary alkyl group shifts regiospecifically. Furthermore, the migrating group retains its configuration during the course of rearrangement.

The oxidation of ketones can also be carried out with hydrogen peroxide. However, the carbonyl group must be reactive. In the presence of a catalyst, carbonyl groups with low reactivity are converted into lactones with hydrogen peroxide with high efficiency. For example, when a bicyclobutanone derivative is reacted with hydrogen peroxide alone, its conversion into lactone does not exceed 5%. However, when a catalytic amount of *N*-alkylated derivative of flavin is added to the reaction mixture, the reaction's yield increases to 85% [103]. In this reaction, the catalyst, *N*-alkylated flavin, first reacts with hydrogen peroxide to form a hydroperoxide, which converts the ketone into the lactone. The *t*-alcohol formed as a result of oxidation is converted back into the catalyst.

How do aldehydes behave in the Baeyer–Villiger oxidation reaction? By a similar mechanism, the peracid adds the aldehyde carbonyl group. Two groups can migrate to the oxygen atom. One of them is a hydride and the other is an alkyl group. Because the hydride is a better migratory group, it first migrates to the oxygen atom, forming the corresponding carboxylic acids. Thus, aldehydes are generally converted into related acids.

Problems

7.1 Write the mechanisms for forming the products formed in the rearrangement reactions given in the equations below.

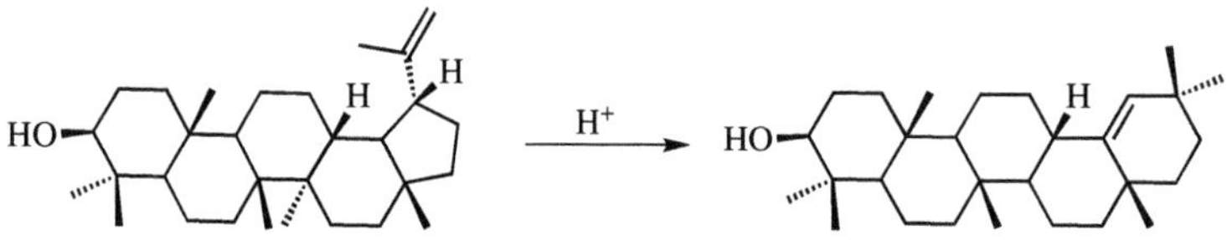

7.2 Please write the product's structure and suggest a mechanism for the formation of the product.

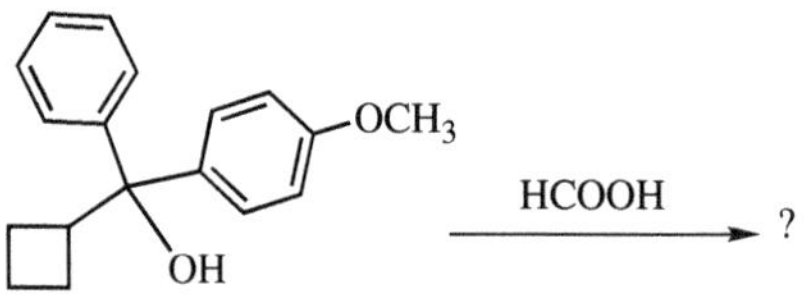

7.3 When two different systems (diol and aldehyde) labeled with C-13 are treated with acid, the same product is produced. It was observed that the C-13 label was distributed over two carbon atoms at a ratio of 1 : 1. Write the product and give the mechanism for distribution of the label.

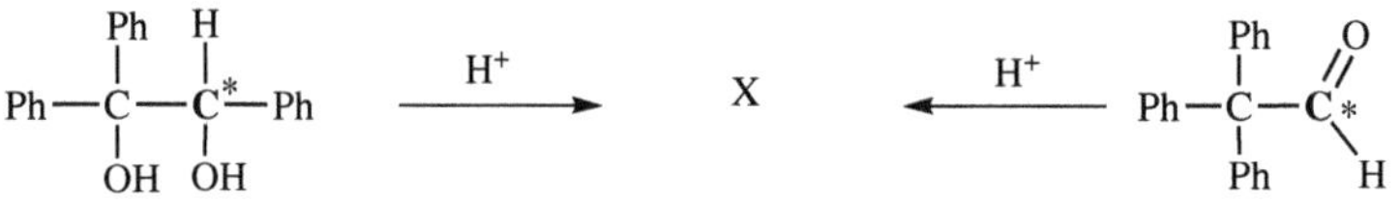

7.4 Write down the main product and by-product that may occur in the reaction below and discuss the reason for forming the main product.

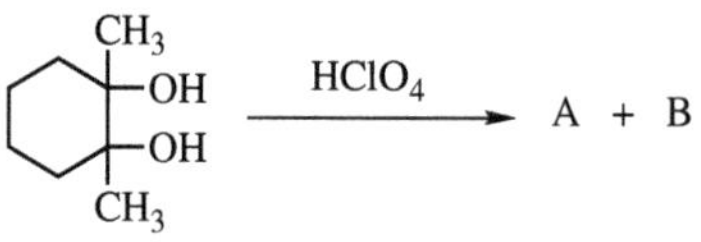

7.5 Write the products that are theoretically expected to occur in the reaction below.

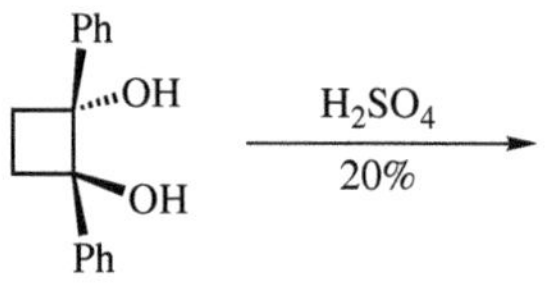

7.6 Write the products and their mechanisms of the products formed by the following reactions.

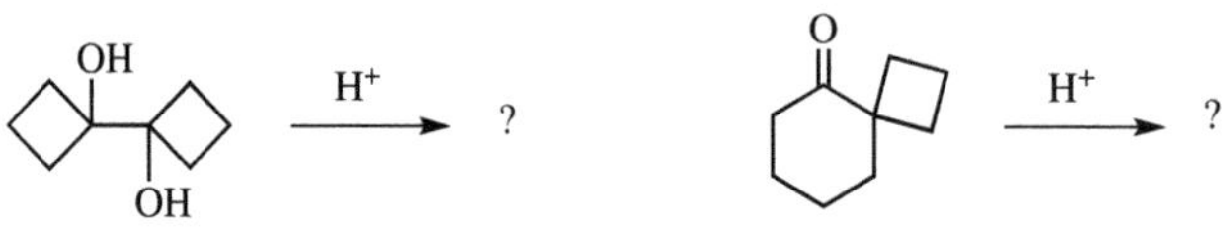

7.7 *erythro* and *threo*-Aminoalcohols rearrange by the reaction with HNO_2. Write the structures of the rearrangement products and discuss their mechanisms for formation.

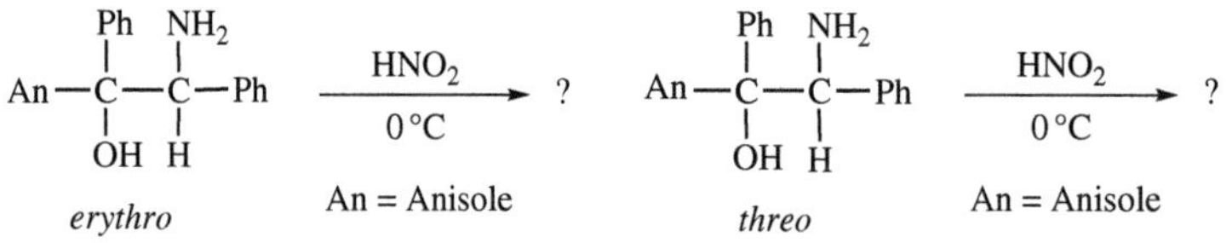

7.8 Please explain why the isomeric tosylates B isomer undergoes solvolysis in formic acid 1000 times faster than the isomer A.

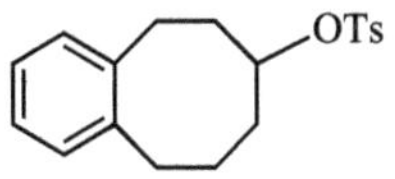

7.9 Tertiary amine rearranges in an aqueous medium. Explain the mechanism of this reaction.

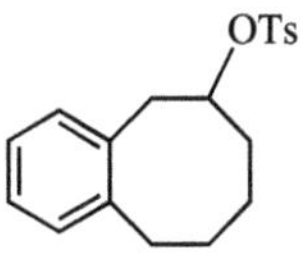

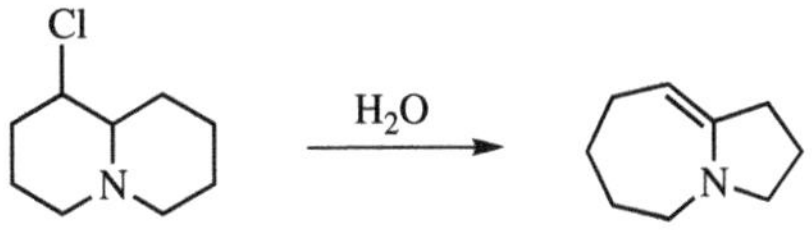

7.10 Write down the product resulting from the reaction below and explain its formation.

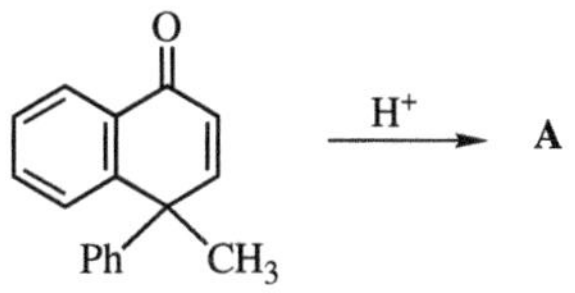

7.11 *exo*-Epoxide undergoes a ring-opening reaction. Please write down the product and give its configuration.

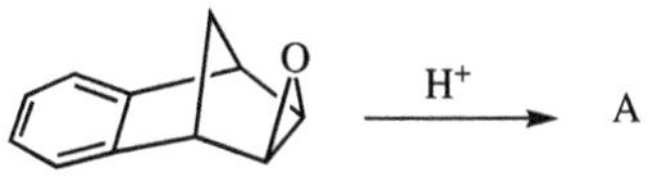

7.12 Explain the mechanism of formation of the products formed as a result of the following reaction.

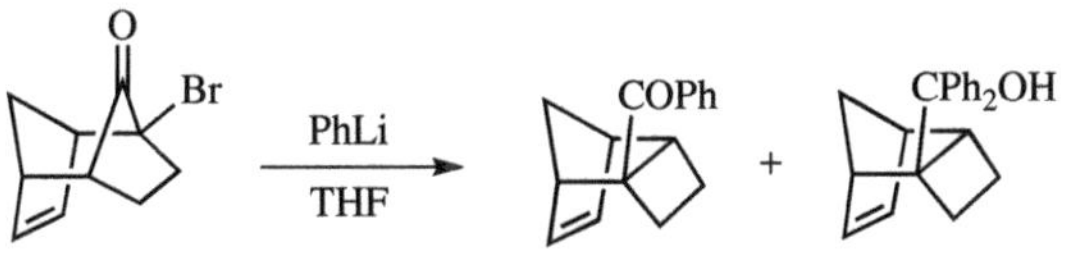

7.13 When the oxime given its structure below is treated with acid, two different products are formed. Write the structure and formation mechanism of the products.

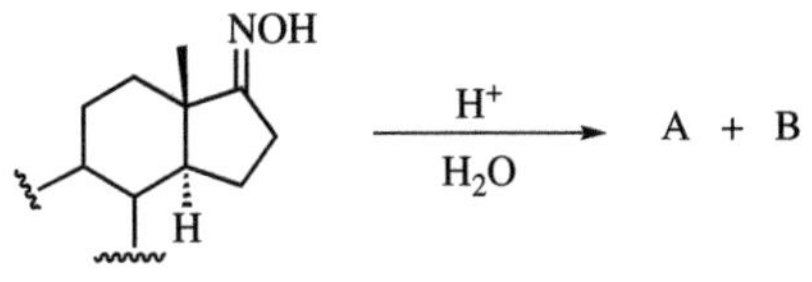

References

1. Traynham, J.G. (1986). *J. Chem. Educ.* 63: 930.
2. Olah, G. and Surya Prakash, G.K. (eds.) (2004). *Carbocation Chemistry*. Wiley.
3. Yannoni, C.S., Kendrick, R.D., Myhre, P.C. et al. (1989). *J. Am. Chem. Soc.* 111: 6440.
4. Hollenstein, S. and Laube, T. (1993). *J. Am. Chem. Soc.* 115: 7240.
5. Kim, C.K., Lee, K.A., Bae, S.Y. et al. (2004). *Bull. Korean Chem. Soc.* 25: 311.
6. Carrol, F.A. (1998). *Perspectives on Structure and Mechanism in Organic Chemistry*, 286. Brooks/Cole Publishing Company.

7 White, J.C., Cave, R.J., and Davidson, E.R. (1988). *J. Am. Chem. Soc.* 110: 6308.
8 Alabugin, I.V., Gilmore, K.M., and Peterson, P.W. (2011). *WIRES Comput. Mol. Sci.* 1: 109.
9 Olah, G.A., Baker, E.B., Kuhn, S.J., and Tolgyesi, W.S. (1962). *J. Am. Chem. Soc.* 84: 2733.
10 Olah, G.A. (1962). *Rev. Roum. Chim.* 7: 1129.
11 Olah, G.A. (1995). *Angew. Chem. Int. Ed.* Nobel Leceture 34: 1393.
12 Olah, G.A. and Schlosberg, R.H. (1968). *J. Am. Chem. Soc.* 90: 2726.
13 Olah, G.A., Baker, E.B., Evans, J.C. et al. (1964). *J. Am. Chem. Soc.* 86: 1360.
14 Olah, G.A. and Donovan, D.J. (1977). *J. Am. Chem. Soc.* 99: 5026.
15 Balcı, M. (2005). *Basic ^{1}H- and ^{13}C-NMR Spectroscopy*, 325. Elsevier.
16 Ausloos, P., Rebbert, R.E., Sleek, L.W., and Tiernan, T.O. (1972). *J. Am. Chem. Soc.* 94: 8939.
17 Andrei, H.-S., Solcà, N., and Dopfer, O. (2008). *Angew. Chem. Int. Ed.* 47: 395.
18 Saunders, M. and Hagen, E. (1968). *J. Am. Chem. Soc.* 90: 6881.
19 Olah, G.A. and White, A.M. (1969). *J. Am. Chem. Soc.* 91: 5801.
20 Olah, G.A. and White, A.M. (1969). *J. Am. Chem. Soc.* 91: 3954.
21 Olah, G.A., Mateescu, G.D., and Riemenschneider, J.L. (1972). *J. Am. Chem. Soc.* 94: 2529.
22 Myhre, P.C., Kruger, J.D., Hammond, B.L. et al. (1984). *J. Am. Chem. Soc.* 106: 6079.
23 Wagner, G. (1899). *J. Russ. Phys. Chem. Soc.* 31: 690.
24 Meerwein, H. (1914). *Justus Liebigs Ann. Chem.* 405: 129.
25 Collins, Q. (1960). *Rev. Chem. Soc* 14: 357.
26 Bachmann, W.E. and Ferguson, J.W. (1934). *J. Am. Chem. Soc.* 56: 2081.
27 Zaczek, N.M., Ruff, J.C., Jackewitz, A.H., and Roswell, D.F. (1971). *Chem. Educ.* 48: 257.
28 Tiffeneau, M., Weill, P., and Tchouba, B. (1937). *Compt. Rend.* 205: 54.
29 Kohlbacher, S.M., Ionasz, V.-S., Ielo, L., and Pace, V. (2019). *Monatsh Chem.* 150: 2011.
30 Smith, P.A.S. and Baer, D.R. (1960). *Org. React.* 11: 157.
31 Fattori, D., Hennry, S., and Vogel, P. (1993). *Tetrahedron* 49: 1649.
32 Kirmse, W. and Gruber, W. (1973). *Chem. Ber.* 106: 1365.
33 Goodyear, G. and Waring, A.J. (1990). *J. Chem. Soc. Perkin Trans. II*: 103.
34 Frimer, A.A., Marks, V., Sprecher, M., and Gilinsky-Sharon, P. (1994). *J. Org. Chem.* 59: 1831.
35 Arnold, R.T., Buckley, J.S. Jr., and Dodson, R.M. (1950). *J. Am. Chem. Soc.* 72: 3153.
36 Winstein, S., Grunwald, E., Buckles, R.E., and Hanson, C. (1948). *J. Am. Chem. Soc.* 70: 816.
37 Winstein, S. and Buckles, R.E. (1942). *J. Am. Chem. Soc.* 64: 2780.
38 Gash, K.B. and Yuen, G.U. (1966). *J. Org. Chem.* 31: 4234.
39 Winstein, S. and Lucas, H.J. (1939). *J. Am. Chem. Soc.* 61: 1576.
40 Tsuno, Y., Funatsu, K., Maeda, Y. et al. (1982). *Tetrahedron Lett.* 23: 2879.
41 Winstein, S. and Trifan, D.S. (1949). *J. Am. Chem. Soc.* 71: 2953.
42 Winstein, S. and Trifan, D. (1952). *J. Am. Chem. Soc.* 74: 1159.
43 Brown, H. and Spec, C. (1962). *Publ. Chem. Soc.* 16: 140–174.
44 Fischer, W., Grob, C.A., Hanreich, R. et al. (1980). *Helv. Chim. Acta* 63: 2298.
45 Grob, C.A. and Herzfeld, D. (1982). *Helv. Chim. Acta* 65: 2443.
46 Bielmann, R., Fuso, F., and Grob, C.A. (1988). *Helv. Chim. Acta* 71: 312.
47 Geier, H., Kautz, C.B., and Kirmse, W. (1993). *Chem. Ber.* 126: 739.
48 Olah, G.A., Prakash, G.K.S., and Saunders, M. (1983). *Acc. Chem. Res.* 16: 440.
49 Olah, G.A. (2001). *J. Org. Chem.* 66: 5943.
50 Olah, G.A., Prakash, G.K.S., Arvanaghi, M., and Anet, F.A.L. (1982). *J. Am. Chem. Soc.* 104: 7105.
51 Yannoni, C.S., Macho, V., and Myhre, P.C. (1982). *J. Am. Chem. Soc.* 104: 7380.
52 Scholz, F., Himmel, D., Heinemann, F.W. et al. (2013). *Nature* 341: 62.
53 Winstein, S., Lewin, A.H., and Pande, K.C. (1963). *J. Am. Chem. Soc.* 85: 2324.
54 Brown, H.C. and Bell, H.M. (1963). *J. Am. Chem. Soc.* 85: 2324.
55 Winstein, S. and Stafford, E.T. (1957). *J. Am. Chem. Soc.* 79: 505.
56 Winstein, S. and Shatavsky, M. (1956). *J. Am. Chem. Soc.* 78: 592.
57 Wilt, J.W. and Chenier, P.J. (1970). *J. Org. Chem.* 35: 1562.
58 Dastan, A., Demir, U., and Balcı, M. (1994). *J. Org. Chem.* 59: 6534.
59 Nametkin, S.S. (1923). *Justus Liebigs Ann. Chem.* 432: 207.

60 Starling, S.M., Vonwiller, S.C., and Reek, J.N.H. (1998). *J. Org. Chem.* 63: 2262.
61 Cope, A.C., Keough, A.H., Peterson, P.E. et al. (1957). *J. Am. Chem. Soc.* 79: 3900.
62 Prelog, V. and Kung, W. (1956). *Helv. Chim. Acta* 39: 1394.
63 Favorskii, A.E. (1894). *J. Russ. Phys. Chem. Soc.* 26: 559.
64 Bordwell, F.G., Scamhorn, J.G., and Springer, W.R. (1969). *J. Am. Chem. Soc.* 91: 2087.
65 Bordwell, F.G. and Strong, J.G. (1973). *J. Org. Chem.* 38: 579.
66 Tsuchida, N., Yamazaki, S., and Yamabe, S. (2008). *Org. Biomol. Chem.* 6: 3109.
67 Loftfield, R.B. (1951). *J. Am. Chem. Soc.* 73: 4707.
68 Stevens, C.L. and Farkas, E. (1952). *J. Am. Chem. Soc.* 74: 5352.
69 Barborak, J.C., Watts, L., and Petit, R. (1966). *J. Am. Chem. Soc.* 88: 1328.
70 Lou, X. (2015). *Mini Rev. Org. Chem.* 12: 449.
71 Taylor, R.J.K. and Casy, G. (2003). *Org. React.* 62: 357.
72 Corey, E.J. and Block, E. (1969). *J. Org. Chem.* 34: 1233.
73 Paquette, L.A. and Philips, J. (1971). *J. Am. Chem. Soc.* 93: 4516.
74 Sutherland, A.G. and Taylor, R.J.K. (1989). *Tetrahedron Lett.* 30: 3267.
75 Taylor, R.J.K. (1999). *Chem. Commun.*: 217.
76 Paquette, L.A., Wingard, R.E. Jr., Philips, J.C. et al. (1971). *J. Am. Chem. Soc.* 93: 4508.
77 Paquette, L.A., Philips, J.L.A., and Wingard, R.E. Jr. (1971). *J. Am. Chem. Soc.* 93: 4516.
78 Bowden, K. and Williams, K.D. (1994). *J. Chem. Soc. Perkin Trans.* 2: 77.
79 Selman, S. and Eastham, J.F.Q. (1960). *Rev. Chem. Soc.* 14: 221.
80 Bowden, K. and Fabian, W.M.F. (2001). *J. Phys. Org. Chem.* 14: 794.
81 Comisar, C.M. and Savage, P.E. (2005). *Green Chem.* 7: 800.
82 Donaruma, L.G. and Heldt, W.Z. (1960). *Org. React.* 11: 1.
83 Gawley, R.E. (1988). *Org. React.* 35: 14.
84 Yamabe, S., Tsuchida, N., and Yamazaki, S. (2005). *J. Org. Chem.* 70: 10638.
85 Kenyon, J. and Young, D.P. (1941). *J. Chem. Soc.*: 263.
86 Suginorne, H., Furukawa, K., and Orito, K. (1987). *J. Chem. Soc. Chem. Commun.*: 1004.
87 Drahl, M.A., Manpadi, M., and Williams, L. (2013). *J. Angew. Chem. Int. Ed.* 52: 11222.
88 Kirihara, M., Niimi, K., and Momose, T. (1997). *Chem. Commun.* 6: 599.
89 Neber, P.W. and Friedolsheim, A. (1926). *Justus Liebigs Ann. Chem.* 449: 109.
90 O'Brien, C. (1964). *Chem. Rev.* 64: 81.
91 House, H.O. and Berkowitz, W.F. (1963). *J. Org. Chem.* 28: 2271.
92 Ooi, T., Takahashi, M., Doda, K., and Maruoka, K. (2002). *J. Am. Chem. Soc.* 124: 7640.
93 Chung, J.Y.L., Ho, G.-J., Chartrain, M. et al. (1999). *J. Tetrahedron Lett.* 40: 6739.
94 Stieglitz, J. and Leech, P.N. (1913). *Ber. Dtsch. Chem. Ges.* 46: 2147.
95 Gassman, P.G. and Fox, B.L. (1967). *J. Am. Chem. Soc.* 89: 338.
96 Hoffman, R.V., Kumar, A., and Buntain, G.A. (1985). *J. Am. Chem. Soc.* 107: 4731.
97 Campomanes, P., Menendez, M.I., and Sordo, T.L. (2003). *J. Org. Chem.* 68: 6685.
98 Baeyer, A. and Villiger, V. (1899). *Chem. Ber.* 32: 3625.
99 Renz, M. and Meunier, B. (1999). *Eur. J. Org. Chem.*: 737.
100 Ten Brink, G.-J., Arends, I.W.C.E., and Sheldon, R.A. (2004). *Chem. Rev.* 104: 4105.
101 Alvarez-Idaboy, J.R., Reyesa, L., and Mora-Diezb, N. (2007). *Org. Biomol. Chem.* 5: 3682.
102 Chang, M.-Y., Chang, C.-P., Yin, W.-K., and Chang, N.-C. (1997). *J. Org. Chem.* 62: 641.
103 Mazzini, C., Lebreton, J., and Furstoss, R.J. (1996). *Org. Chem.* 61: 8–9.

8
Reactive Intermediates Carbanions, Carbenes, and Nitrenes

8.1 Carbanions: Electrophilic Rearrangements

In nucleophilic rearrangements, we have seen that carbocations tend to rearrange to more stable carbocations through 1,2-shifts of various groups to the carbocation center. During this rearrangement, the migrating group shifts with the bonding electrons. Therefore, we called this type of rearrangement *nucleophilic rearrangement*.

In sharp contrast to carbocations, carbanions undergo 1,2-shifts very rarely. In some cases, a shift to the carbanion center can be observed. As carbanions lack an electron-deficient center, the migrating groups shift without bonding electrons as an electrophile. This type of rearrangement is called *electrophilic rearrangement*.

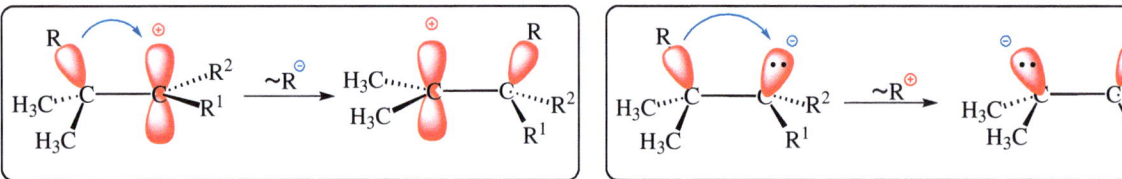

Nucleophilic rearrangement   Electrophilic rearrangement

What is the reason for a strong preference for cationic rearrangement? This question can be addressed by analyzing the transition states formed during the rearrangements. During the rearrangement, a cyclic system is generated and so the number of electrons involved is critical. In carbocations, just two electrons are involved; therefore, this transition state is aromatic. In contrast, in the rearrangement to the carbanion center, four electrons are involved, making the transition state antiaromatic.

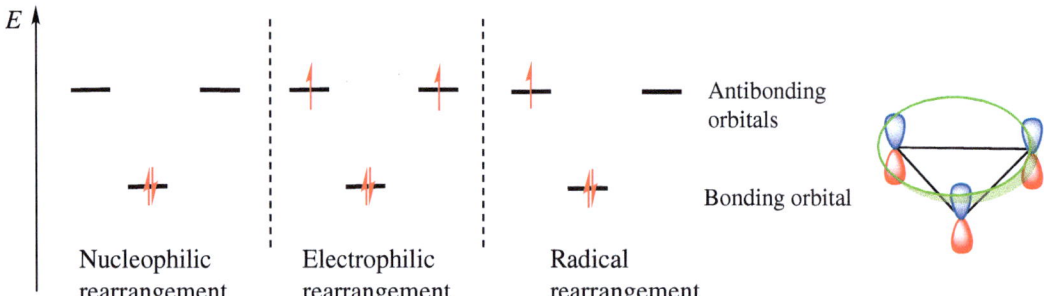

Figure 8.1 The population of orbitals during nucleophilic, electrophilic, and radical rearrangements.

The energy levels of orbitals in a delocalized three-membered ring formed during rearrangement are given in Figure 8.1. As shown, two electrons involved during nucleophilic rearrangement populate the bonding orbital. In the case of electrophilic rearrangement, two of four electrons populate the antibonding orbitals, making the transition state unstable and increasing the activation barrier. In radical rearrangement, one of three electrons populates the antibonding orbital.

Reaction Mechanisms in Organic Chemistry, First Edition. Metin Balcı.
© 2022 WILEY-VCH GmbH. Published 2022 by WILEY-VCH GmbH.

Electrophilic rearrangements can be mechanistically compared to intramolecular S_N2 reactions. In the S_N2 reaction mechanism, the nucleophile attacks the carbon atom from the backside, while the migrating group can leave the atom to which it is attached, leaving its electrons on that atom. During such a reaction, a steric strain is created in the transition complex. This is one of the reasons electrophilic rearrangements are prevented.

Generally, in electrophilic rearrangements, alkyl groups bonded to a heteroatom such as nitrogen, sulfur, or oxygen migrate to the carbanion center. The driving force of the reaction is to transfer the negative charge to the more electronegative element.

8.1.1 Stevens Rearrangement

Thomas Stevens, a Scottish chemist (1900–2000), discovered in 1928 a novel 1,2-shift involving ammonium ylides. Converting quaternary ammonium salt and sulfonium salt into the corresponding amines or sulfides in the presence of a strong base is called the *Stevens rearrangement*. For this reaction to occur, one of the groups attached to the nitrogen or sulfur atom must contain a CH or CH_2 group for proton abstraction. Additionally, the presence of an electron-withdrawing group attached to the CH or CH_2 group is required.

Stevens and coworkers reacted an ammonium salt formed by the reaction of 1-phenyl-1-(*N*, *N*-dimethyl)ethanone with benzyl bromide with NaOH in an aqueous medium, and a rearranged tertiary amine was formed in good yield [1].

After the formation of the quaternary ammonium salt, the next step of the reaction is the abstraction of the acidic proton by a strong base to form an ylide. The carbanion formed can be stabilized inductively by the nitrogen atom. However, since nitrogen contains four bonds, the carbanion cannot be stabilized by resonance by the nitrogen atom. For this reason, an electron-withdrawing group that can stabilize the carbanion by resonance must be attached to the CH_2 group. In that case, the carbonyl group can increase the acidity of the methylene protons and stabilize the carbanion by resonance [2].

By contrast, the third step, the 1,2-migration, has long been a subject of debate. Various mechanisms have been proposed for the Stevens rearrangement. First, it was proposed that the R group migrates as a carbanion, forming an ion pair. In addition to the ion-pair mechanism, a concerted mechanism and a radical pair mechanism were also proposed. A concerted mechanism through a bridge structure would require an inversion of configuration because of the orbital symmetry principle. Experiments showed the retention of the configuration of the migrating group. Therefore, a concerted mechanism was ruled out.

A third mechanism is the radical-pair mechanism. Most evidence suggests that the shift occurs through a radical-pair mechanism. According to this mechanism, the carbon–nitrogen bond undergoes cleavage homolytically to form a radical pair. The solvent lattice protects this pair of radicals. The radicals combine rapidly and so racemization cannot occur. Cross experiments show intramolecular migration. There is no radical transfer from one group to the other. This finding proves that the radicals formed combine, without leaving the solvent cage, to form the C—C bond. Recent calculations also support the radical-pair mechanism [3].

Quaternary ammonium salts are required for the Stevens rearrangement. They can be easily synthesized by the reaction of tertiary amines with alkyl bromides containing electron-withdrawing groups.

When the diallyl diethyl ammonium salt is treated with potassium *t*-butoxide in acetonitrile, the Stevens rearrangement results in diethylamino-1,5-diene [4].

The Stevens rearrangement has also been successfully applied to sulfonium salts. Cyclophanes (aromatic compounds bridged between two nonadjacent carbon atoms by a carbon chain) can be easily synthesized by applying the Stevens rearrangement. Treatment of 1,3-bis(bromomethyl)benzene dibromide with Na_2S gives 2,11-dithia[3.3]metacyclophane. Methylation with dimethoxymethylium tetrafluoroborate gives a sulfonium salt. The Stevens rearrangement is performed highly efficiently with NaH in tetrahydrofuran (THF). Metacyclophane is obtained by the elimination of the methyl sulfide formed with *t*-BuOK [5, 6]. The Stevens rearrangement followed by the Hofmann elimination is a method of choice to generate a double bond in highly strained molecules.

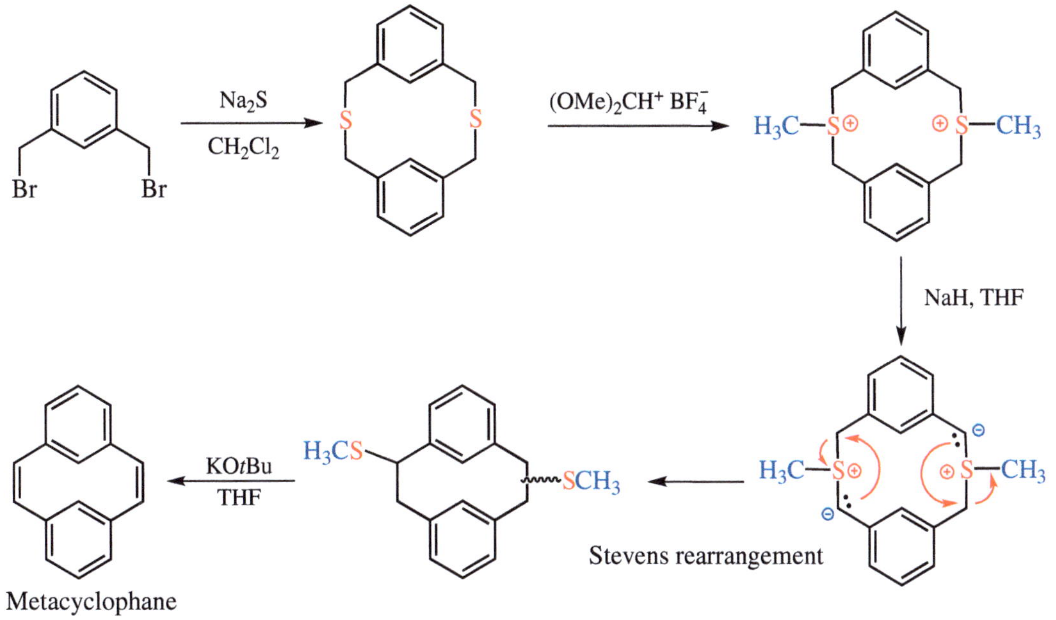

During the Stevens rearrangement, some by-products may occur if the sulfonium salt has two or more different acidic protons. To solve this problem, Vedejs [7, 8] used onium salts substituted with silyl groups as starting compounds to generate carbanions regioselectively as shown below. The removal of silyl groups with fluoride can easily form the carbanion intermediate required for the Stevens rearrangement. The major product is derived from the initially formed methylide, while the minor product is formed from the ester-stabilized ylide.

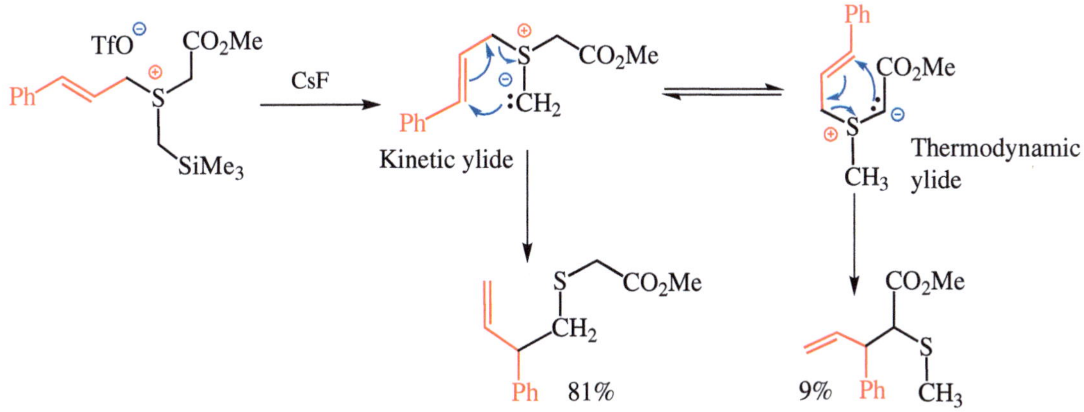

8.1.2 Sommelet–Hauser Rearrangement

The Sommelet–Hauser rearrangement involves intramolecular conversion of benzyl quaternary ammonium salts to *ortho*-substituted benzyl dialkyl amines with strong bases. Mechanistically, it is similar to the Stevens rearrangement and it was first discovered by Sommelet in 1937 [9–11].

Let us examine the mechanism of the reaction given below. In the first step, a strong base, such as an amide, abstracts the acidic benzylic methylene proton to produce an ylide. At this stage, no rearrangement takes place. Although the benzylic carbanion formed is thermodynamically stable, it is in equilibrium with a second ylide formed by deprotonation of one of the methyl protons. Although the concentration of this kinetic ylide is not very high, it undergoes a [2,3] sigmatropic shift and attacks the aromatic ring in its *ortho*-position, forming an exocyclic double bond. Subsequent aromatization by a 1,3-hydrogen shift produces the final product.

The Sommelet–Hauser rearrangement can be applied to ring enlargement reactions of cyclic amines. For example, the reaction of the 1,1-dimethyl-2-phenylpiperidinium ion with NaNH$_2$ in liquid ammonia forms a new nine-membered nitrogen-containing ring system [12, 13].

Because the benzylic proton is more acidic, it can be easily abstracted. However, the reaction does not proceed through the benzyl carbanion formed. To further strengthen this accepted mechanism, Sato and coworkers [14] regiospecifically generated the carbanion on the methyl group bonded directly to the nitrogen atom using the silylation method. Fluoride ion-induced desilylation formed the carbanion, which underwent rearrangement. The carbanion can attack the benzene ring in two ways (attacks indicated by the red and blue arrows). Indeed, the products formed as a result of both attacks can be isolated. One of the products underwent aromatization.

8.1.3 Wittig Rearrangement

In 1942, Georg Wittig (Nobel Prize winner, 1979) observed 1,2-alkyl migration from an oxygen center to a carbanion center by the reaction of benzyl ethers with phenyl lithium [15, 16]. This reaction should not be confused with the conventional Wittig reaction.

The reaction mechanism is analogous to that of the Stevens rearrangement. The driving force of the reaction is to transfer the negative charge to the more electronegative element. Three different mechanisms were proposed for the Wittig rearrangement, as discussed for the Stevens rearrangement: a concerted mechanism, a two-step ion-pair mechanism, and a two-step radical-pair mechanism [17]. It is well accepted that among these, the reaction proceeds through a radical dissociation–recombination mechanism. After carbanion formation, the bond between the oxygen and the alkyl group is cleaved homolytically and a ketyl radical is formed. At the same time, one of the electrons on the carbon is transferred to the oxygen atom. The rearranged product is formed by combining the ketyl radical and alkyl radical in the same solvent cage. In the case of allyl-substituted compounds, the 2,3-rearrangement competes with the 1,2-rearrangement. At lower temperatures, the 2,3-rearrangement is dominant.

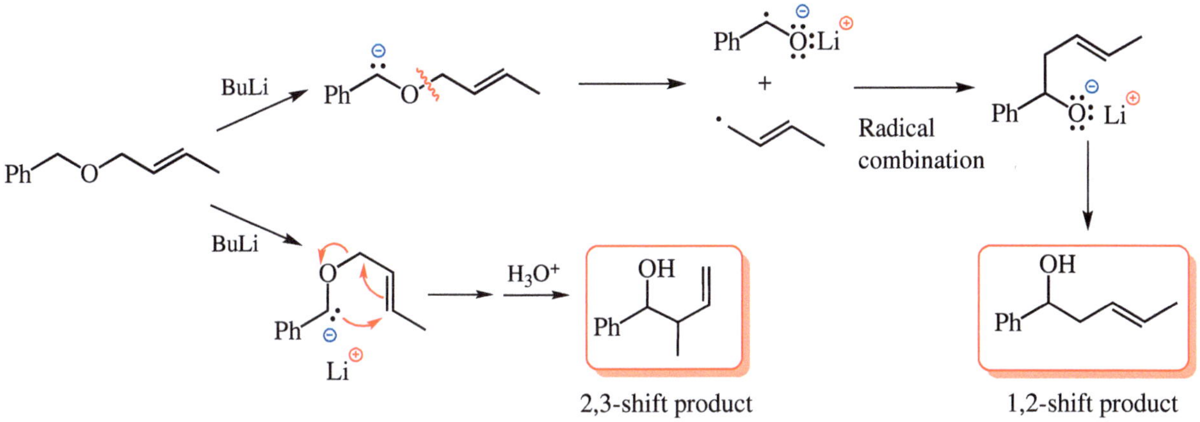

The 2,3-Wittig rearrangement has been successfully applied to a cyclohexenylmethyl propargyl derivative that is a key intermediate in the synthesis of a natural product, phomactins [18].

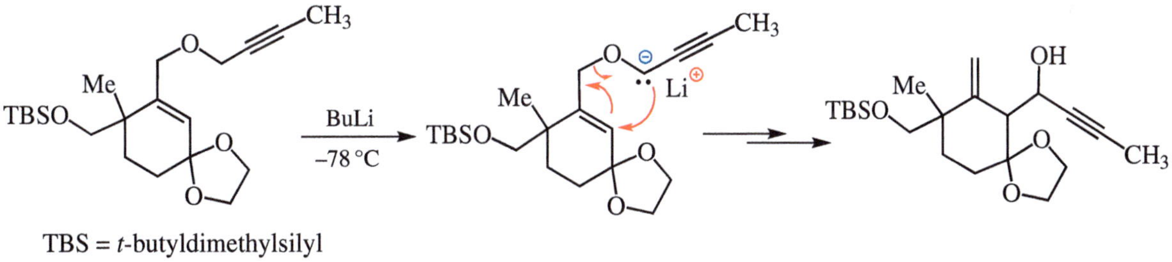

TBS = *t*-butyldimethylsilyl

8.2 Carbenes

A *carbene* is a neutral compound containing a divalent carbon atom with six electrons in the valence shell. The four electrons bond two substituents and two electrons are unshared. Because of the electron deficiency in the valence shell, carbenes are short-lived and highly reactive intermediates.

The atoms attached to the central atom can be carbons, heteroatoms, or halogens. The simplest carbene is the CH_2 compound, and it is called methylene carbene or methylene for short. The word "carbene" was first used in 1951 by the famous chemists Woodward, Doering, and Winstein.

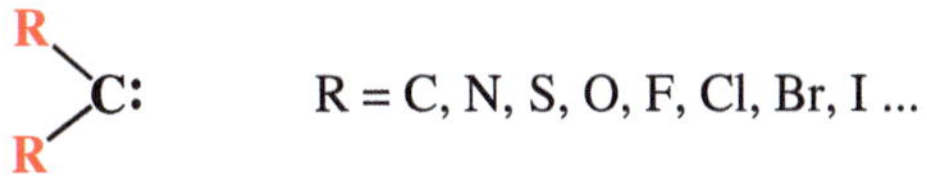

R = C, N, S, O, F, Cl, Br, I ...

Carbene

8.2.1 Naming Carbenes

The names of carbenes generally include the substituents. Substituents are given first and the word carbene is added at the end. If the carbene carbon atom is a part of a ring, the corresponding carbene is denoted by the suffix *ylidene*, as shown below.

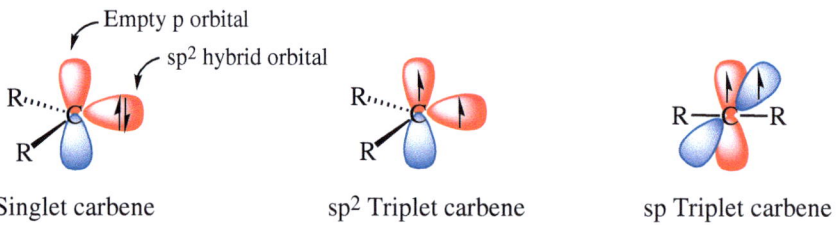

Methylene carbene   Ethylene carbene   Dibromocarbene   Ethylidene   Cyclopentylidene   Tropylidene

8.2.2 Structure and Reactivity of Carbenes

Carbene carbon atoms are generally sp^2-hybridized. Carbenes are called singlet or triplet depending on their electronic structures. In *singlet carbenes*, the nonbonding electrons occupy one of the sp^2 hybrid orbitals with the opposite spins as the energy level of the sp^2 orbitals is lower than that of the p orbital. The geometry of singlet carbenes is planar. According to Hund's law, if electrons are placed in different orbitals with parallel spins, the carbene is called a *triplet carbene*. The triplet carbene structure may be either linear (sp-hybridization) or bent (sp^2-hybridization) (Figure 8.2). Most carbenes have a nonlinear triplet structure. As triplet carbenes are paramagnetic, they can be observed by electron spin resonance spectroscopy (EPR). The experiments and the theoretical calculations show that the H—C—H angle for the triplet state of CH_2 is 136°. The H—C—H angle for the singlet state is 102° [19–21].

Figure 8.2 Singlet and triplet carbene structures.

Spin multiplicity is determined according to the formula $M = 2s + 1$. Because the spins are in the opposite direction in a singlet carbene, the spin multiplicity is $M = (2 \bullet 0) + 1 = 1$, as the total spin $s = 0$. This is why these carbenes are called singlet carbenes. In the case of parallel spins, the total spin will be $\frac{1}{2} + \frac{1}{2} = 1$; $M = (2 \bullet 1) + 1 = 3$ and the corresponding carbene is called a triplet carbene.

8.2.3 Inductive Effect

If the substituents attached to the carbene carbon atom are electron-withdrawing groups, the carbene prefers the singlet structure. The electron-withdrawing groups increase the energy difference between the σ and p orbital by stabilizing the σ orbital attached to the carbene carbon atom. The electrons prefer to occupy the σ orbital, leaving the p orbital empty. In contrast, if the substituents attached to the carbene carbon atom are electron-donating groups (Li—C—Li) via a σ bond, the carbene prefers the triplet structure [22–25].

8.2.4 Mesomeric Effect

The mesomeric effect plays a more significant role than the inductive effect in the structure of carbenes. Substituents interacting with the carbene center can be classified into two types, namely [26]:

1. *π Electron-donating groups*: Compounds with nonbonding electron pairs, such as –OR, –SR, –NR₂, –F, –Cl, –Br, or –I, can donate the electron pairs and fill one of the empty orbitals. Then, the carbene electrons are forced to occupy a single orbital. In that case, the carbenes prefer the singlet configuration in the ground state and become more stable. For example, dibromocarbene is a singlet carbene. Electrons on the bromine atom conjugate with the carbene carbon and increase the carbene stability.

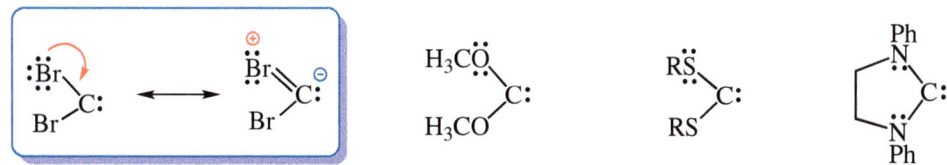

The electronic configuration of tropylidene is that of a singlet. Tropylidene is isoelectronic with pyridine. The empty p orbital on the carbene atom enables the electrons' delocalization on the seven-membered ring (as in the tropylium ion), and so the carbene electrons necessarily occupy the sp^2 hybrid orbital on the ring plane. Therefore, tropylidene is both a singlet and one of the rare carbenes with nucleophilic properties. Because carbenes are generally electron-deficient compounds, they react primarily with double bonds having higher electron density. However, tropylidene prefers to react with electron-poor double bonds (Figure 8.3).

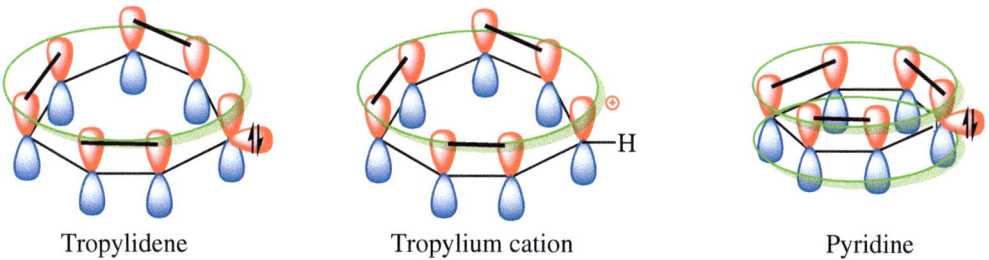

Figure 8.3 Electronic configuration of tropylidene, tropylium cation, and pyridine.

2. *π Electron-withdrawing groups*: Carbenes containing π electron-withdrawing groups, such as –COR, –CN, –SiR$_3$, or –PR$_3^+$, are generally linear singlet carbenes. In such compounds, the p_y orbital containing carbene electrons interact with the empty orbitals of the substituents. This interaction does not affect the p_x orbital on the carbene carbon atom. Therefore, the (p_x, p_y) degeneracy is broken, making these carbenes have a singlet ground state even though they are linear.

In the past three decades, some stable carbenes have been synthesized and isolated. The first stable carbene, 1,3-diadamantyl-imidazol-2-ylidene, was reported in 1991 by Arduengo and coworkers [27]. The nitrogen atoms attached to carbene carbon stabilize carbene electronically and bulky adamantane groups sterically. The carbene is a colorless and crystalline compound with a melting point of 240 °C. The X-ray crystal structure confirmed the proposed structure. Various stable even distillable carbenes stabilized by phosphine and trimethylsilane groups are also synthesized [26–28].

8.2.5 Carbene Generation

There are several ways of generating carbenes. They are generally formed from their precursors by elimination or fragmentation reactions. The carbene-forming reactions can be classified into four groups depending on the reaction type.

1. Thermal or metal-catalyzed decomposition of diazo compounds

2. α-Elimination reactions

3. Fragmentation

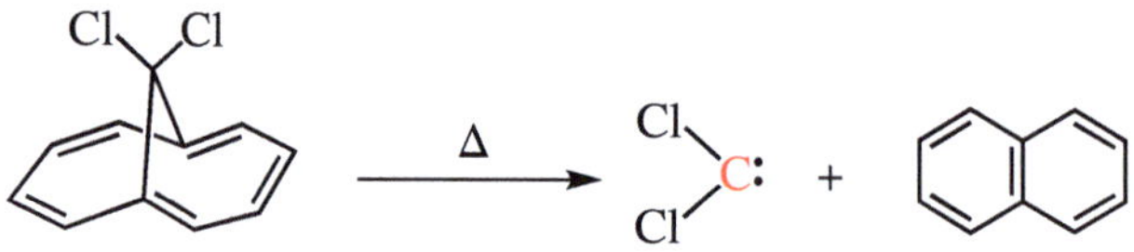

4. Miscellaneous methods

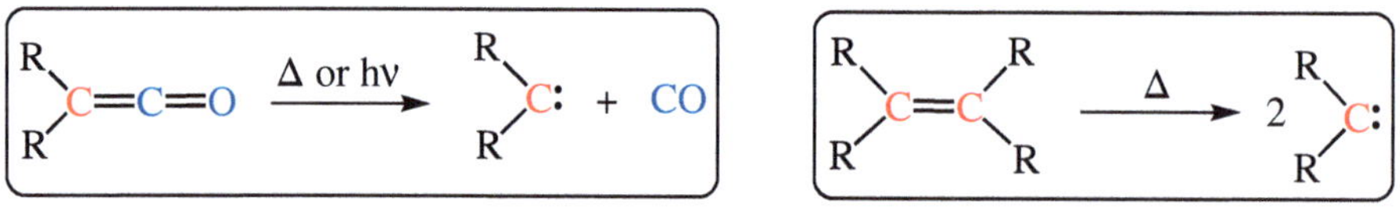

Although there are many different carbene synthesis methods, the two most common are thermal decomposition of diazo compounds and α-elimination. Other methods are not synthetically suitable but are mechanistically important.

8.2.5.1 Carbene Precursors: Synthesis of Diazo Compounds

Thermal or photochemical decomposition of diazo compounds in aprotic solvents to carbenes is a general method for the generation of carbenes. Various metal complexes are also used extensively for carbene formation. Because diazo compounds are easily degradable and generally unstable, they are often not available for purchase and they need to be synthesized each time before the reaction. Various methods are reported in the literature for the synthesis of diazo compounds.

Ketones and aldehydes react easily with hydrazine to form related hydrazones. Diazo compounds are formed because of oxidation of hydrazones with metal salts such as Ag_2O, HgO, MnO_2, and $Pb(OAc)_4$. This method works in cases in which one of the substituents is phenyl.

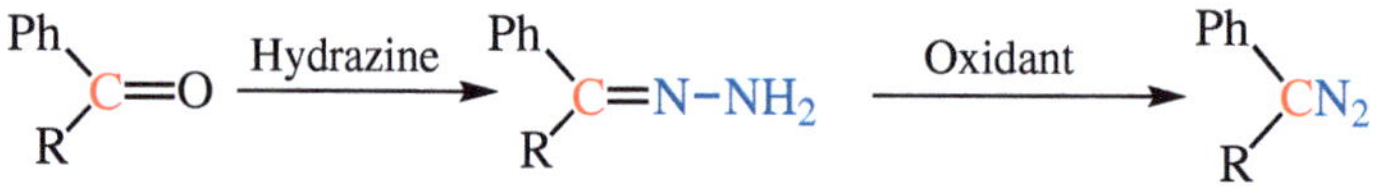

8.2.5.2 Bamford–Stevens Reaction

In 1952, Bamford (William Randall Bamford, British chemist) and Stevens (Thomas Stevens, Scottish chemist) observed that tosylhydrazones form alkenes via treatment of tosylhydrazones with strong bases [29, 30]. The base abstracts the more acidic NH hydrogen atom, forming a salt. This salt can be isolated and kept for a long time. If desired, the salt can be converted to a carbene by heating or photolysis at any time. Tosylhydrazone anions undergo thermal decomposition to give diazoalkanes via α-elimination.

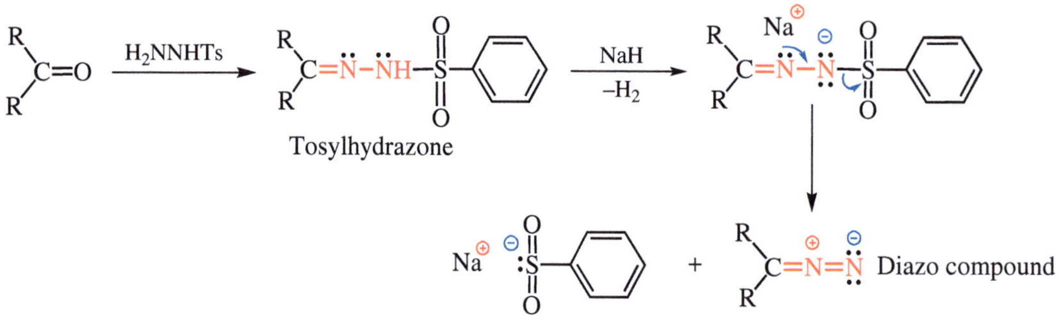

The fate of diazoalkane intermediates mainly depends on the nature of the solvent. In aprotic solvents, the Bamford–Stevens reaction gives carbenes, while in protic solvents, protonation of the diazoalkanes occurs, leading to carbocations after the removal of nitrogen gas. Diazonium salts, especially the aliphatic ones, are unstable. The carbocations formed can undergo various reactions (elimination, substitution, or rearrangement).

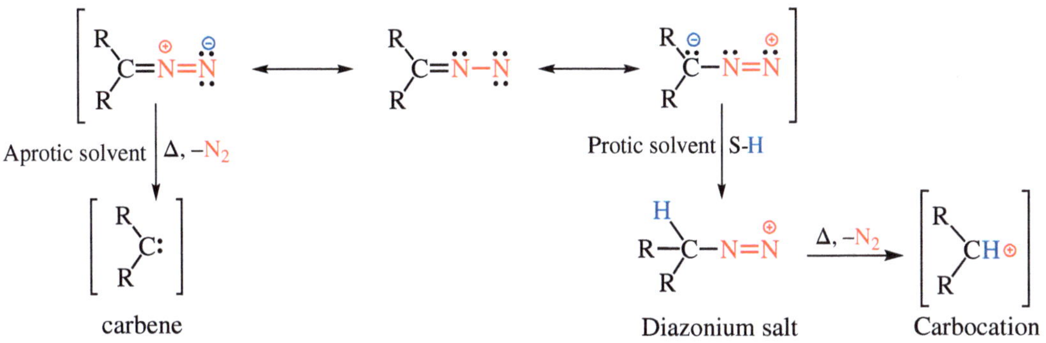

The resulting carbene generally forms an alkene because of a 1,2-hydrogen shift (see the insertion section). For this reason, the Bamford–Stevens reaction is successfully applied for the conversion of carbonyl compounds into the corresponding alkenes.

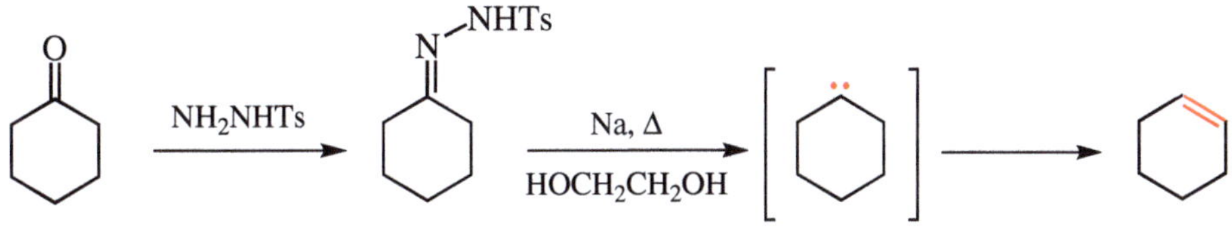

The *Shapiro reaction*, a variation of the Bamford–Stevens reaction, is a convenient and straightforward method for converting tosylhydrazones to alkenes [31–33]. The significant difference between these two reactions is that the Shapiro reaction forms a kinetic product, less substituted alkenes, while the Bamford–Stevens reaction yields a thermodynamic product. If the carbonyl group has a hydrogen atom in the α-position, two moles of MeLi are used as the base, alkenes are formed, and no other by-products are encountered. While one of the bases used abstracts the acidic proton on the nitrogen atom, the other removes the α-proton. The general mechanism is as shown below. In the last step, a vinyl carbanion is formed, which is then protonated or reacted with an electrophile.

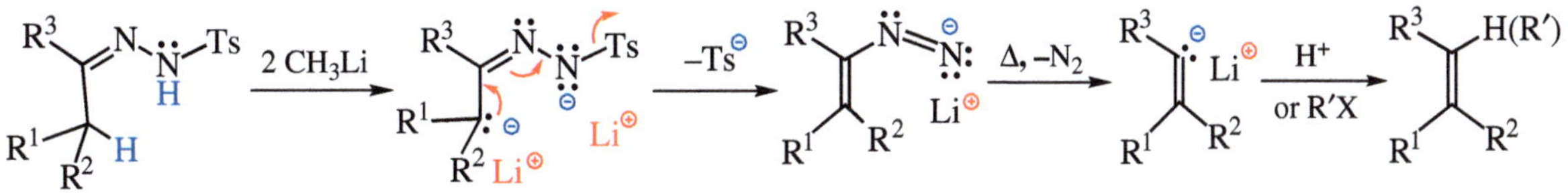

8.2.5.3 N-Nitrosoalkyl Urea Compounds

The base-catalyzed decomposition of *N*-nitrosoalkyl ureas, amides, or sulfonamides furnishes diazoalkanes. These compounds can be used for a variety of purposes. The *N*-nitrosoalkyl compounds can be synthesized and stored in advance. For their synthesis, potassium isocyanate is reacted with alkylammonium hydrochloride salts. In an acidic environment, potassium isocyanate transforms into isocyanic acid. The free alkyl amine adds to the carbonyl group to form *N*-nitrosoalkyl urea derivatives. A diazoalkyl compound is prepared by hydrolysis of ethereal solution of the urea derivative with aqueous NaOH.

434 | *8 Reactive Intermediates Carbanions, Carbenes, and Nitrenes*

Diazomethane is the compound synthesized and used most in laboratory conditions [34]. It is a toxic, explosive, and light yellow gas prepared as needed in laboratories. It is used as a solution in diethyl ether. As can be seen from diazomethane's resonance structures, the carbon atom behaves both electrophilically and nucleophilically.

Diazomethane reacts with acid chlorides to form α-diazoketones [35]. Diazocarbonyl compounds are much more stable than diazomethane because of the conjugation with the carbonyl group.

Diazomethane is a potent methylating agent in the laboratory for carboxylic acids. When it reacts with acids at room temperature, methyl esters are formed in one step. This esterification method works in a neutral medium; it provides a significant advantage over esterification reactions in acidic and basic mediums in yield and selectivity. The yield is often quantitative and the ester formed is very easy to isolate [35]. Esterification with diazomethane has been significantly reduced by the introduction of the safer and equivalent reagent trimethylsilyldiazomethane under very mild conditions [36].

8.2.5.4 Diazirines

An increasingly important method in carbene synthesis is the conversion of diazirines into reactive carbenes upon irradiation by ultraviolet light [37].

One of the points making carbene chemistry and carbenes synthetically important is the easy synthesis of carbene precursors. Synthetic methods developed for diazirines in recent years have expanded the application area of these intermediates.

There are various methods for the synthesis of azirines. In one of these methods, ketones are used as the starting materials. Ketones are first treated with NH_3 in methanol, followed by reaction with hydroxylamine-*O*-sulfonic acid to give diaziridines. Diaziridines are readily oxidized to diazirines by iodine [38].

8.2.5.5 α-Elimination Method

Modern carbene chemistry first started with the work by J. Hine. The formation of carbon monoxide during hydrolysis of chloroform in basic medium has been explained by dichlorocarbene formation as an intermediate [39].

In 1954, Doering and Hoffman [40] reacted chloroform under anhydrous conditions and noted that the dichlorocarbene generated from it adds to the alkenes, forming geminal dichlorocyclopropane derivatives. With this reaction, the first milestone of carbene chemistry was laid. A strong base (t-BuOK, n-BuLi, etc.) first abstracts the hydrogen atom from chloroform to form a trichloromethyl carbanion, which can be stabilized by electron-withdrawing chlorines. In the rate-determining step, one of the chlorine atoms leaves the carbanion with the bonding electrons to generate the neutral dichlorocarbene. This type of reaction is called an *α-elimination reaction* [41] because both groups (hydrogen and chlorine) eliminated were bonded to the same carbon atom.

Solvent = benzene, toluene, n-hexane, THF

Subsequent studies showed that α-elimination could be successfully applied to other halogen compounds. Dibromocarbene and diiodocarbene can also be synthesized in the same way. In the case of mixed haloform compounds, the leaving group determines which carbene is to be formed. The halogens' leaving ability is in the following order: I ~ Br > Cl ≫ F. For example, bromochlorocarbene (CBrCl), chlorofluorocarbene (CClF), and difluorocarbene (CF_2) are formed from $CHBr_2Cl$, CHBrClF, and $CHClF_2$, respectively.

Carbene reactions have been carried out in aprotic solvents with strong bases for years. Before starting the reaction, the solvents used must be well dried because H_2O in the reaction medium can easily react with the carbene and base and hinder the reaction. However, in 1969, the Polish chemist Mieczyslaw Makosza (1934–) showed that carbene generation and addition to a double bond could be performed in a two-phase system using concentrated aqueous NaOH as a base in the presence of a quaternary ammonium salt acting as a phase transfer catalyst [42]. Chloroform is used as both a reactant and a solvent. The alkene is dissolved in chloroform. A 50% aqueous NaOH solution is prepared and a quaternary ammonium salt, $PhCH_2NEt_3^+ Cl^-$, is added to this solution. The prepared organic and inorganic phases are combined and mixed at room temperature. Because the reaction occurs between two phases, efficient mixing plays a significant role in this reaction [43].

According to the mechanism, the base abstracts the proton from chloroform in the interfacial region. The sodium salt of the trichloromethyl carbanion undergoes ion exchange with the quaternary ammonium salt to form a new ion pair, which enters the organic phase because of the catalyst's organic groups. The ion pair dissociates immediately to dichlorocarbene, which then adds to the alkene present in the solvent.

This method has tremendous advantages over carbene synthesis in an anhydrous medium. Cheap NaOH is used instead of an expensive base. The solvent does not need to be of absolute purity. Various phase transfer catalysts can also be used. Crown ethers, for example, are also useful as phase transfer catalysts.

Dihalocarbenes are generally formed in a basic medium. If the acceptor compounds have base-sensitive groups, the molecule can undergo undesirable reactions with the base. In that case, the carbene must be generated in a neutral medium. The thermal decarboxylation of trihalomethyl carboxylate salts in aprotic solvents generates the trihalomethyl carbanion, which decomposes to the carbene. The carbene adds to the alkene [44].

Thermolysis of phenyl(trihalomethyl)mercury compounds is ideal for generating carbenes in a neutral medium. The mercury reagents are stable at room temperature. They decompose to the corresponding carbenes on heating. The carbene formed adds to an alkene. This method, developed by Seyferth, is generally applied to base-sensitive or low-reactivity olefins [45].

Under normal conditions, the most common carbene generation method is α-elimination. The yields obtained by applying the three different methods discussed above to the same compound are given in the following equation.

Method	Yield %
$CHCl_3$, 50% NaOH, Phase transfer catalyst	81
$Cl_3CCOONa$	47
$CHCl_3$, t-BuOK	27

Carbenes obtained by the elimination method are generally dihalocarbenes. Monohalocarbenes or methylene carbene are difficult to generate by α-elimination. For the synthesis of compounds derived from monohalocarbenes or methylene carbene, a dihalocarbene is first formed and added to the alkene. Then, one or both of the halogen atoms are reduced. Thus, the desired compounds are obtained indirectly.

Removal of one or both halogen atoms from geminal dihalocyclopropanes can be easily accomplished with tri-*n*-butyltin hydride, LiAlH$_4$, or *n*-BuLi or with alkali metals in protic solvents. Tri-*n*-butyltin hydride is the most frequently employed hydrogen atom donor. The reaction is performed in the presence of a radical initiator (AIBN = azabisisobutyronitrile) and only one of the halogens is reduced. If the reaction is carried out with sodium in liquid ammonia, both halogen atoms are reduced. If dihalocyclopropane is treated with organolithium reagents such as *n*-BuLi, the halogen will be exchanged for lithium in the first step. The lithium compound formed can be reacted with electrophiles to make the molecule more functional. Very recently, an efficient and highly stereoselective method for the reduction of a geminal dibromocyclopropane to the corresponding monobromocyclopropane was developed using dimethyl phosphite and potassium carbonate [46]. If geminal dihalogen compounds are dehalogenated with metals or alkylmetals in the presence of an alkene, cyclopropane compounds are formed.

There is evidence that no free carbenes are involved during the cyclopropanation. The addition is accomplished with a metal-complexed reagent with carbene-like reactivity. Such compounds are called *carbenoids*. These compounds are very stable at lower temperatures and can be observed by spectroscopic methods.

$$R_2CX_2 + RLi \text{ (or Li)} \longrightarrow R_2C\begin{smallmatrix}Li\\X\end{smallmatrix} \longleftarrow R_2CHX + R'Li$$

Carbenoid

8.2.5.6 Simmons–Smith Reaction

It is not possible to synthesize methylene carbene by α-elimination. When diiodomethane is reacted with zinc–copper couple, (iodomethyl)zinc iodide, a carbenoid, is formed. This carbenoid can transfer the methylene group (CH$_2$) to a double bond, furnishing a cyclopropane. For example, cyclohexene is easily converted to norcarane according to this method. This process for preparing nonhalogenated cyclopropane derivatives is called the *Simmons–Smith reaction* [47–49].

If the two faces of the double bond are not equal, the Simmons–Smith reaction is significantly affected by the steric effect. Cyclopropanation usually takes place on the less hindered face. However, when a hydroxyl group is present in proximity to the double bond, the hydroxyl group directs cyclopropanation *cis* to the hydroxyl group. For example, when the cyclopent-3-enol reacts under Simmons–Smith conditions, only the syn-product is formed. The addition of free methylene carbene cannot explain the exclusive formation of the syn-product. The zinc compound interacts with the hydroxyl group in the transition complex and transfers the methylene group from the same face to the double bond [50, 51].

The Simmons–Smith reaction is stereospecific, and the transfer of the methylene group from the iodomethylzinc iodide to the double bond is concerted and so the new bonds are formed simultaneously. For example, *trans*-2-butene gives only *trans*-1,2-dimethylcyclopropane, whereas *cis*-2-butene results in the formation of *cis*-1,2-dimethylcyclopropane.

The double bonds conjugated with a carbonyl group do not react with the carbenes. The mesomeric effect of the carbonyl group attracts electrons from the double bond, making the double bond electrophilic. Sulfur ylides are particularly effective in the cyclopropanation of electron-deficient alkenes.

Dimethylsulfoxonium methylide can be generated in situ by deprotonation of trimethylsulfoxonium iodide with strong bases. Because of its nucleophilic nature, the sulfoxonium anion easily adds to the β-carbon atom of the α,β-unsaturated carbonyl group in a Michael-type manner. By reforming of the carbonyl group, the double-bond electrons act as a nucleophile and attack the methylene carbon atom, removing the dimethylsulfoxide group and forming a cyclopropane compound. This reaction is called the *Corey–Chaykovsky reaction*. Considering this entire reaction, it should be noted that although it appears to be a methylene carbene addition reaction to a double bond, no free methylene carbene is formed in any step of the reaction.

8.2.6 Carbene Reactions

Carbenes are intermediates. They possess only a sextet of electrons and are highly reactive and electrophilic. Therefore, they react in various ways instantly in the medium in which they are formed. Carbene reactions are classified in four main groups.

1. *Addition Reactions*: Because carbenes are generally electrophilic species, they easily add to olefins to form cyclopropanes.

$$\text{cyclopentene} + :CR_2 \longrightarrow \text{cyclopropanated product} \quad \text{Cycloaddition}$$

2. *Carbene Dimerization*: When carbenes cannot find any reactants in the environment in which they are formed, they can undergo a dimerization reaction to form alkenes and fill the electron deficiency in their outer shell.

$$R_2C: + :CR_2 \longrightarrow R_2C=CR_2 \quad \text{Carbene dimerization}$$

3. *Carbene Insertion*: Carbene insertion is another common reaction of carbenes. A carbene can form stable structures by interposing itself into an existing bond (C—H or C—C).

$$\text{cycloalkane} \xrightarrow{\text{C-H insertion}} \text{bicyclic product} \quad \text{Carbene insertion}$$

4. *Carbene Rearrangement*: Another way to stabilize carbenes is by migration of the neighboring group to the carbene center. As a result of the rearrangement, double bonds are formed, and such reactions' synthetic potential is relatively high.

$$R-\underset{\underset{}{\overset{O}{\|}}}{C}-\ddot{C}-R \longrightarrow O=C=C\underset{R}{\overset{R}{\diagup}} \quad \text{Carbene rearrangement}$$

8.2.6.1 Carbene Cycloaddition Reactions

The addition of carbenes to alkenes to form cyclopropane is the most studied reaction of carbenes. Because carbenes are electron deficient, they behave as electrophiles and react with nucleophilic double bonds. The mechanism for a singlet carbene addition is a concerted process in which all bonds are broken and formed simultaneously. On the other hand, triplet carbenes cannot add to the double bonds in one step because this addition is prohibited from the point of view of orbital symmetry. Triplet carbenes form cyclopropanes by adding to double bonds as a result of a multistep reaction.

Although the singlet carbene addition reaction is a one-step process, the direct carbene approach to the double bond is forbidden because there is always an antibonding interaction. During the addition reaction, the highest occupied molecular orbital (HOMO) of one of the reactants interacts with the other's lowest unoccupied molecular orbital (LUMO) in the formation of the transition complex. If the empty p orbital of a singlet carbene, the LUMO, directly approaches the HOMO of the double bond, an antibonding interaction occurs between those orbitals, as shown in Figure 8.4. This interaction prevents the carbene from directly approaching the double bond. A similar situation occurs when the carbene HOMO approaches the LUMO of the double bond. In both cases, the carbene is forbidden from approaching the double bond directly.

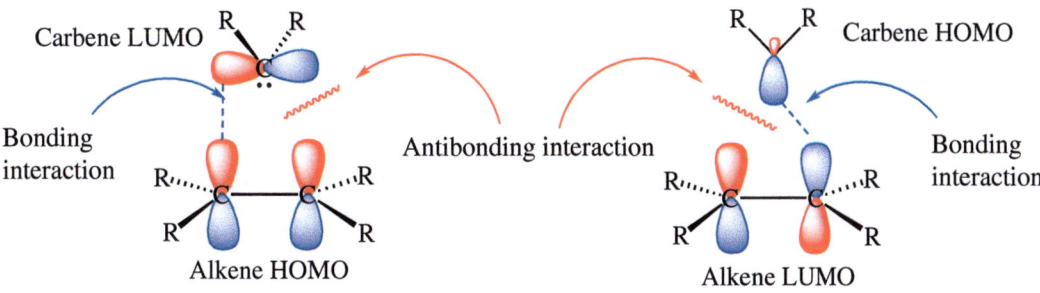

Figure 8.4 Possible HOMO–LUMO interactions in the direct approach of carbene to the double bond.

Therefore, carbenes approach the molecule in a sideways-on manner to eliminate these antibonding interactions that occur in the direct approach. This approach can accommodate HOMO–LUMO interactions in either combination (Figure 8.5).

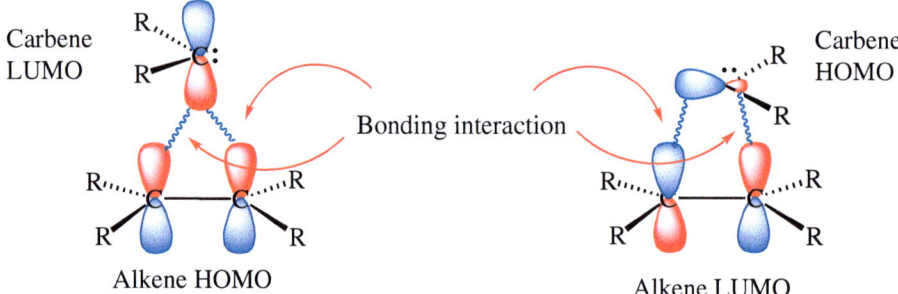

Figure 8.5 HOMO–LUMO interactions in the sideway approach of carbene to the double bond.

However, this approach would produce an addition product with an unusual geometry. Therefore, the carbene twists 90° during the course of the addition process as shown below.

According to the Skell rule, while singlet carbenes add stereospecifically to double bonds in a single step, no stereospecificity is observed in the addition of triplet carbenes. The addition occurs in several steps. In singlet carbene addition, the spins of the π electrons in cis-butene are paired. One of the bonds is formed by using one electron from the π electrons and one electron from the carbene. The spins of these two electrons are paired. The second bond will be formed at the same time by the remaining electrons. As the remaining electrons' spins are antiparallel, bond formation will take place without violating the Pauli exclusion principle, forming cis-configurated cyclopropane.

Let us now discuss triplet carbene addition. The spins of the triplet carbenes are unpaired. The spins of double bond and carbene electrons are not suitable for forming both bonds simultaneously. Once the first bond is formed according to the Pauli exclusion principle, the remaining electrons have the same spin. Before the formation of the second bond, one of the electron spins must flip to generate a singlet diradical via *intersystem crossing* (ISC, a transition between the two electronic states with different states spin multiplicity), which then closes to the cyclopropane. However, before undergoing ISC, we have to wait for a while. This indicates that the triplet diradical has an appreciable lifetime. During this period, rotation about the various carbon–carbon bonds is now possible. In such a way, the addition is not stereospecific. As a result, both cis- and trans-dimethylcyclopropane derivatives are formed.

Halogenated compounds (solvents) accelerate the ISC process. Fluorenylidene, an aryl carbene, can be generated by photolysis of 9-diazofluorene. The triplet ground state of fluorenylidene is only 1.1 kcal/mol (4.6 kJ/mol) lower in energy than the singlet state [52, 53]. Reactions involving fluorenylidene proceed through either the triplet or singlet states, and the products formed depend on the relative concentration of the carbene and experimental conditions. Fluorenylidene formed by the photochemical method has a singlet ground state. It reacts with *cis*-2-butene to form *cis*-cyclopropane. If the carbene concentration is kept too low, the rate of the addition reaction decreases. Hence, the carbene formed finds more time to spin-flip, the singlet carbene is partially transformed into a triplet carbene, and then a mixture of *cis*- and *trans*-adducts is formed. When the reaction is performed in a solvent such as dibromomethane or hexafluorobenzene that also accelerates the ISC process, the singlet carbene turns into a triplet carbene and again a product mixture is formed.

Carbene Addition to Alkenes Because carbenes are generally electrophilic species, they prefer double bonds with higher electron density when adding to compounds having different double bonds. In particular, singlet carbenes (dihalocarbenes) show more regiospecific properties as they are more stable. For example, isotetraline has two different double bonds. Although the steric barrier is more prominent in adding in the middle double bond, dihalocarbenes add to this double bond selectively. [10]Annulene was first synthesized by Vogel in 1964 by reduction of halogens, followed by bromination and dehydrobromination of the compound (see Section 6.3.2) [54].

442 | *8 Reactive Intermediates Carbanions, Carbenes, and Nitrenes*

On the other hand, carbethoxycarbene does not add to the central double bond of isotetraline. Because carbethoxycarbene is a reactive carbene and has a short lifetime, the carbene cannot be selective and prefers to add to the terminal double bonds. Monoadducts undergo aromatization under the reaction conditions [55].

We emphasized that carbenes generally add to double bonds with higher electron density. Under favorable conditions, carbenes also add to double bonds to which electron-withdrawing groups are attached. The examples below show that carbenes can readily add to α,β-unsaturated double bonds under Makosza conditions. The trichloromethyl anion generated by proton abstraction of chloroform with a base acts as a nucleophile and adds to the β-carbon atom of α,β-unsaturated double bonds. During the reformation of the C=O double bond, the C=C double bond electrons attack the trichloromethyl group and substitute the chloride to form the cyclopropanation product [56].

TEBA = triethylbenzylammonium salt

Carbene Addition to Alkynes The addition of carbenes to acetylene derivatives is as crucial as their addition to double bonds because cyclopropenes, building blocks for a wide array of interesting compounds, are generated by adding dihalocarbenes to alkynes. Hydrolysis of dihalocyclopropenes furnishes cyclopropenones [57, 58].

Carbene Addition to Allenes Carbenes can readily add to allenes to form alkylidenecyclopropanes. One of the most interesting reactions to allenes is the reaction between dimethylallene and carbenes. In general, singlet carbenes add to the double bond substituted with the methyl groups, while triplet carbenes preferably add to the less substituted double bond [59].

Carbenes can also add to some stable cyclic allenes to form alkylidenecyclopropanes. Treatment of 1,2-cyclononadiene with dibromocarbene generated by refluxing of phenyltribromomethyl mercury in benzene gives a cyclopropane derivative in almost quantitative yield [60].

Carbene Addition to Aromatic Compounds Because the compounds formed by the addition of carbenes to aromatic systems undergo ring expansion, these addition reactions are synthetically important. The first carbene addition reaction to aromatics was carried out by Curtius and Buchner in 1885 by the reaction of diazoacetate with benzene [61]. The first step is the generation of ethoxycarbonyl carbene, which adds to the benzene to form a seven-membered ring. The product obtained was estimated to have a norcaradiene structure. Studies carried out in later years showed that norcaradiene was formed first during the addition, and then, this system was transformed into cycloheptatriene by an electrocyclic process. Therefore, an acid having a norcaradiene structure is called a *Buchner acid* in reference to this study.

The addition of carbenes to aromatic compounds is often a problematic reaction because with the addition of carbene, the aromaticity of the benzene ring is destroyed. Some bonds in aromatic compounds have a more olefinic character. For example, the double bond at the 9,10-position in phenanthrene undergoes dichlorocarbene addition to give a cyclopropane derivative, which rearranges into the tropylium cation at 140 °C [62].

Carbenes quickly add to electron-rich aromatic compounds containing electron-donating substituents such as alkoxy groups. The reaction of 2-methoxynaphthalene with dichlorocarbene generated from ethyl trichloroacetate and sodium methoxide gives cyclopropane as the intermediate, which readily undergoes a ring-opening reaction, followed by hydrolysis, resulting in the formation of a benzotropone derivative [63]. Carbenes also add to the electron-rich heteroaromatic system. The reaction of pyrrole with dichlorocarbene furnishes a ring enlarged product, a pyridine derivative [64].

Naphthalene reacts with carbenes more easily than benzene does. When the resonance energies of both systems are compared, the different reactivities are easily understood. The resonance energy of benzene is 36.0 kcal/mol, while naphthalene's resonance energy is around 60.0 kcal/mol. Because the aromaticity disappears completely when carbenes are added to benzene, the energy loss will be 36.0 kcal/mol in the case of benzene. When a carbene is added to naphthalene, one ring's aromaticity will be lost, while the aromaticity will remain on the other ring. The energy loss will be around 60−36 = 24 kcal/mol. Because this value is less than the loss in benzene, carbenes more readily add to naphthalene. Theoretically, carbenes can add to four different double bonds in naphthalene: C1–C2, C2–C3, C8–C9, and C9–C10. The experiments show that carbenes add to the double bond C1–C2 carbons. Both theoretical and experimental studies show that the region with the highest electron density is between the C1–C2 bond. In naphthalene, there is a particular bond localization, which can also be seen from the relevant *ortho* coupling constants. When the resonance structures of naphthalene are examined, it is understood that the structure in which aromaticity is preserved in both rings is the dominant one (for more information, see Section 6.9.1).

The decomposition of ethyl diazoacetate in naphthalene leads to the formation of a benzonorcaradiene derivative. The addition takes place exclusively at the C1—C2 bond. While most of the norcaradiene system rearranges to cycloheptatrienes, the benzonorcaradiene formed is stable and does not rearrange. Isomerization into the cycloheptatriene is unfavorable, as this involves dearomatization of the benzene ring. Isomerization takes place at 260 °C, followed by a hydrogen shift to give a benzocycloheptatriene derivative [65].

Carbenes do not add to the central double bond (C9–C10) of naphthalene because both benzene rings' aromaticity will be lost. In rare cases, carbenes add to the C9—C10 bond. Dihalocarbenes add preferentially to the central double bond of the naphthalene unit in corannulene, which destroys the cyclic conjugation in two adjacent benzene rings. The product formed is a kinetic product [66].

Carbenes generated from diazo compounds by photolysis or thermolysis react with aromatic compounds giving low yields. Diazoketone synthesized in quantitative yield from dihydrocinnamic acid by standard methods undergoes intramolecular addition when exposed to copper(I) chloride to give a bicyclic ketone in 40–50% yields [67]. However, the transformation proceeds nearly quantitatively by reaction with a catalytic amount of rhodium(II) acetate at room temperature [68].

8.2.6.2 Carbene Insertion Reactions

Another characteristic reaction of carbenes is the insertion reactions in which a carbene interposes itself into an existing bond. For example, if a methylene carbene does not have a double bond to react with in the reaction medium, it attacks random C—H bonds in the liquid phase. Methylene behaves selectively in the gas phase and preferably inserts at tertiary C—H bonds. When carbene insertion reactions are selective, they are synthetically important. Otherwise, carbene insertion reactions are worth investigating in terms of mechanism. The path of the carbene insertion reaction differs according to whether the carbene is a singlet or triplet. Many singlet carbenes' C-H insertion processes are concerted, involving a three-center cyclic transition state, and they occur in one step with complete retention of the configuration [69].

The situation is different with triplet carbenes. For a triplet carbene, the C–H insertion process proceeds in two steps. Because of their diradical nature, triplet carbenes first abstract hydrogen from the carbon–hydrogen bonds, forming two new carbon radicals. The recombination of these radicals results in the formation of a new carbon–carbon bond. Meanwhile, the resulting radicals can undergo isomerization depending on the rate of recombination.

This is called the abstraction–recombination mechanism. According to this mechanism, in addition to the product given above, different dimerization products are expected. Generally, insertion products are formed because the radicals formed are in a solvent lattice. For different combinations, radicals have to move outside of the cage. However, the lifetime of the radical is too short to allow it to break the solvent cage. The triplet carbenes prefer insertion at tertiary C—H bonds because the tertiary radicals formed are the most stable ones.

Besides the intermolecular insertion reactions, intramolecular insertion reactions are frequently encountered. When an intramolecular insertion is possible, no intermolecular insertions are observed. For example, when cyclononanone tosylhydrazone is treated with NaOMe, the carbene formed inserts at various positions [70, 71].

Insertion reactions are also encountered frequently and with high efficiency in bicyclic systems. The carbene formed from the norbornane tosylhydrazone by treatment with a base undergoes 1,2- and 1,3-insertion reactions in yields of 30% and 70%, respectively. Insertion at the bridgehead C—H bond was hindered because of bridgehead double bond formation (see Bredt's rule) [72].

In saturated systems, carbenes can also insert into various bonds such as aromatic C—H, C—O, and C—N. Thermal or photochemical decomposition of ethyl diazoacetate in 2-phenyloxetane produces a mixture of *cis* and *trans* C—O insertion products, 2-carbethoxy-3-phenyltetrahydrofuran, in 72–80% yield [73]. Rh(I)-catalyzed carbene insertion into the C—C bond of benzocyclobutenols has been realized using diazoesters as a carbene precursor [74].

8.2.6.3 Carbene Rearrangements

The chemistry of singlet carbenes is similar to that of carbocations because both have an empty p orbital. Thus, a widespread reaction of both carbenes and carbocations yields an olefin. In the case of carbocations, a proton is eliminated, whereas in carbenes, a hydride ion migrates in what is termed a hydrogen insertion reaction. Another well-known reaction of carbocations is the rearrangement reaction. Carbenes also undergo rearrangement in which the carbon skeleton of the carbene is changed. In this part, we will discuss some carbene rearrangements.

Wolff Rearrangement In 1902, Ludwig Wolff, a German chemist (1857–1919), reported that heating an α-diazoketone with water generated carboxylic acid. At the time, Wolff did not predict the mechanism of this transformation. The first step is the removal of nitrogen gas from the diazo compound to give an electron-deficient α-ketocarbene. Like a carbocation, the carbene would be susceptible to a 1,2-shift. As the –R group attached to the carbonyl group migrates to the carbene center with its electrons, the carbene electrons form a double bond between the carbonyl carbon and the carbene carbon and a ketene is formed. The subsequent reaction of the ketene with water gives carboxylic acids. All these reactions are called the *Wolff rearrangement* [75, 76].

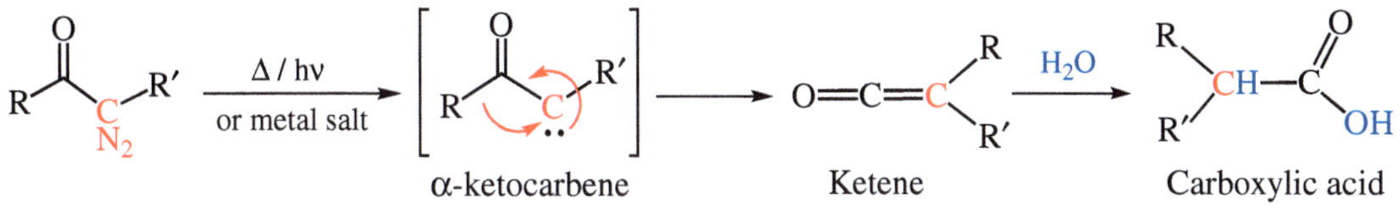

For the Wolff rearrangement, a general method for the synthesis of α-diazoketones is required. At the time this rearrangement was discovered, no general method was known. Therefore, this reaction's synthetic potential was not well understood until efficient methods for the successful synthesis of α-diazoketone were developed. Then, interest in this field grew enormously.

In particular, according to the method developed by Arndt and Eistert, α-diazoketones can be easily synthesized by the reaction of acyl chlorides with diazomethane (see the Arndt–Eistert reaction). Another critical method for synthesizing α-diazoketones is direct transfer of the diazo group to the α-carbon atom. An activated methylene group with two electron-withdrawing substituents reacts with tosyl azide in the presence of a weak base and gives the corresponding α-diazocarbonyl compounds. The base first abstracts one of the methylene protons, forming an enolate, which attacks the tosyl azide to form an α-diazo-1,3-dicarbonyl compound as shown below.

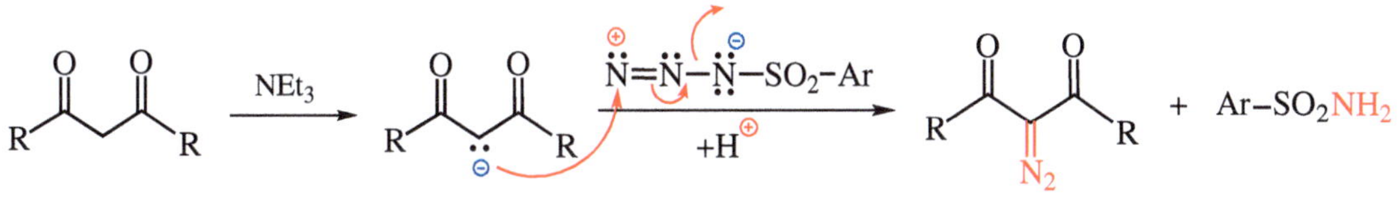

However, in cyclic systems, such as cyclohexanone, a stronger base must be used to achieve good results. The Wolff rearrangement is highly efficient for ring contraction of cyclic α-diazoketones. This strategy has proven useful for the synthesis of strained ring systems.

The Wolff rearrangement can proceed by either a concerted mechanism or a stepwise mechanism. The concerted mechanism seems to require a *cis* conformation of the α-diazo carbonyl compound (referred to the carbonyl group and the diazo group) in which the migrating group lies *trans* to the departing nitrogen moiety. The rearrangement of a cyclic α-diazoketone having a locked *cis* conformation generally proceeds via the concerted mechanism. Factors such as ring strain and carbene configuration also influence the mechanism and the fate of the reaction. Photolysis of the following cyclic diazoketone gives the ring-contracted product in 96% yield and the minor product formed by the 1,2-methyl shift. This reaction's driving force is the antiperiplanar orientation of the migrating group and the leaving group, nitrogen [77].

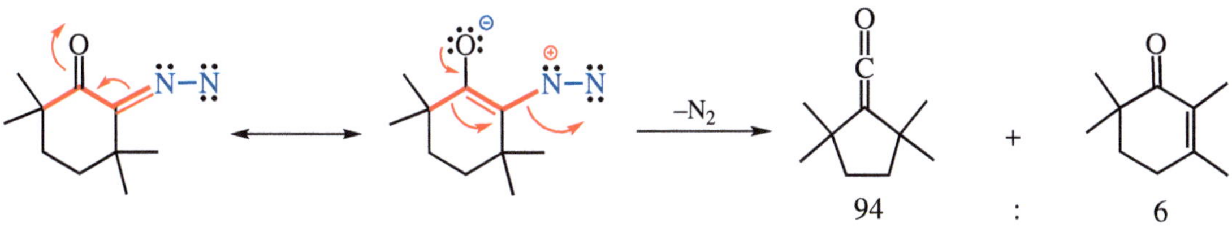

In contrast to this reaction, photolysis or heat of an acyclic diazoketone substituted with tertiary butyl groups gave only traces of the Wolff rearrangement product. Instead, two ketones were formed, arising from a 1,2-methyl shift and a intramolecular 1,2-insertion. The lack of the Wolff rearrangement product was attributed to the *trans* conformation of the carbonyl group and the leaving group, nitrogen. This steric repulsion between the bulky tertiary butyl groups hinders the formation of the *cis* conformation of the carbonyl group and the leaving group required for the Wolff rearrangement.

A fundamental part of the rearrangement mechanism is whether the configuration of the migrating group is preserved or not. This is another question that needs to be answered. Certain optically active compounds in which the migrating carbon atom is asymmetric have been subjected to the rearrangement. Experiments showed that the Wolff rearrangement of all tested α-diazoketones proceeded with complete retention of the migrating group's configuration. This indicates that the migrating group cannot exist in a free state, even for a short time. The retention of the configuration strictly supports the intramolecular rearrangement [75–78].

The Wolff rearrangement can also proceed in a stepwise process via a singlet state of α-ketocarbene. The α-ketocarbene can either undergo a 1,2-alkyl shift to form a ketene or form an antiaromatic oxirene, which can then rearrange to an isomeric α-ketocarbene. If the substituents bonded to the carbonyl group are equal, the same α-ketocarbene will be formed. In the case of different R groups, an isomeric α-ketocarbene will be formed. This process is called *carbene–carbene rearrangement*.

In the early experiments, no evidence could be obtained for the formation of the oxirene intermediate. Conclusive proof of oxirene formation was obtained with compounds labeled at the carbonyl carbon atom with ^{13}C-isotope. In the case of a concerted Wolff reaction, avoiding the oxirene intermediate, the label will be recovered exclusively in the carbonyl carbon. However, when the oxirene intermediate is involved, which can reopen in two different ways, the label will be scrambled between the carbonyl carbon and α-carbon of the ketene [79, 80]. Likewise, the distribution of the labeled carbon atom reveals that oxirene is formed as an intermediate product. This observation also proves that a carbene–carbene rearrangement takes place before the Wolff rearrangement.

The oxirene intermediate is not always observed during Wolff rearrangements. Especially in experiments conducted with conformationally constrained carbocycles, it was determined that oxirene did not occur. The antiaromatic properties of oxirene and the further increase in the molecule's strain in cyclic systems revealed that the labeled carbon atoms are not distributed in the molecule.

Ring Contraction The Wolff rearrangement has been widely applied to cyclic ketones, affording ring contraction products. For example, the Wolff rearrangement of 2-azo-1,3-cyclohexadiones can form the ring-contracted products β-dicarbonyl compounds, which are essential starting materials in the total synthesis of various natural products [81]. Irradiation of a diazo compound generates α-ketocarbene by releasing nitrogen gas. Rearrangement of the carbene results in a ketene. Ketenes may undergo further reactions with various nucleophiles, giving a wide variety of molecules with interesting functional groups.

There are two principal and most general synthetic routes to α-diazocarbonyl compounds: acylation of diazoalkanes and diazo transfer reactions to carbonyl compounds with sulfonyl azides. For the acylation of diazoalkanes, see the Arndt–Eistert reaction. When 2,4,6-triisopropylbenzene sulfonyl azide (TIBSA) is used, α-aryl ketones can easily be converted into diazo compounds [82].

The ring contraction reaction is one of the most common reactions in organic chemistry. Cyclopropanes can easily be synthesized by adding a carbene to double bonds. However, general methods applicable to four-membered and five-membered rings are somewhat limited. Therefore, ring contraction reactions become even more critical. For example, benzocyclopentanone is easily converted into the corresponding cyclobutene carboxylic acid [83]. Photolysis of the diazoketone derived from bishomocubene affords the ring-contracted homocubyl-9-carboxylic acid [84].

The strain of the resulting compounds determines the boundaries of ring contraction reactions. If there is an excessive amount of strain in the molecule, there will be no Wolff rearrangement. Then, the diazoketone may react in another way instead of ring construction. Photolysis of a tricyclodiazoketone does not lead to a ring contraction product; an interesting fragmentation product is formed instead [85].

Ring Expansion If there is an electron deficiency (carbene or carbocation) on a carbon atom in the ring, these systems are prone to ring contraction. On the other hand, if the carbon atom containing the electron deficiency is directly attached to the ring, these systems are prone to ring expansion. Ring expansion occurs when diazomethane is treated with cyclic ketones under neutral conditions. This reaction involves the addition of diazomethane to the carbonyl group. The carbon atom of diazomethane acts as a nucleophile, attacking the carbonyl group, forming the carbon–carbon bond, and opening the carbonyl group. The intermediate product formed is not stable. As the carbonyl group is reformed, one of the C—C bonds attacks the diazomethane carbon. Elimination of nitrogen triggers the migration of one of the neighboring carbon groups.

Cyclobutanones are useful synthetic intermediates undergoing ring expansion reactions forming five-membered rings. If the reacting ketone is asymmetric, differences in regioselectivity are expected in ring expansion reactions with diazomethane. The migration of the more substituted α-carbon is favored. As shown in the reaction below, two different cyclopentanone derivatives are formed, and the carbon atom carrying more alkyl groups preferably migrates [86].

A ring expansion reaction can also be carried out with ethyl diazoacetate. The reaction of cyclooctanone provides a keto ester after treatment with ethyl diazoacetate [87].

Arndt–Eistert Reaction: One-Carbon Homologation of Carboxylic Acids The Arndt–Eistert reaction allows the conversion of carboxylic acids to their homologs by the reaction of activated carboxylic acids with diazomethane and subsequent Wolff rearrangement of the intermediate diazoketones in the presence of water [88]. The reaction was discovered by the German chemist Fritz Arndt (1885–1969) and his PhD student Bernd Eistert (1902–1978). Carboxylic acids are first activated by conversion into acid chlorides or anhydrides.

α-Diazoketones are also obtained by the reaction of anhydrides formed by treatment of acids with ethyl chloroformate in the presence of a base with diazomethane. The anhydrides possess two different carbonyl groups. Diazomethane attacks the more reactive carbonyl group to which the alkyl group is attached. When the diazoketone is heated, photolyzed, or exposed to moist silver oxide, carboxylic acids are formed. If alcohol is used instead of water, an ester is formed. Similarly, by using ammonia, amides are formed.

When the Arndt–Eistert reaction is applied to acids containing optically active carbon atoms in the α-position, the optical activity is not lost during migration. Preservation of the configuration shows that the migrating group is not entirely separated from the molecule and the reaction is intramolecular.

Cyclopropylidene–Allene Rearrangement Cyclopropylidenes can undergo reactions such as addition, dimerization, and rearrangement to allenes. In particular, geminal dihalocyclopropanes are converted to allenes by the reaction with Mg or alkyl lithium.

Because dibromocyclopropanes are readily synthesized by adding dihalocarbenes to olefins, this reaction gains more synthetic significance. Treatment of the dibromocarbene adduct derived from cyclooctatetraene with Mg gives the allene cyclonona-1,2-diene. Treatment of the allene with PhHgCBr$_3$ in refluxing benzene yields the dibromo adduct 10,10-dibromobicyclo[7.1.0]dec-1-ene, which yields the cumulene (a hydrocarbon with three or more consecutive double bonds) cyclodeca-1,2,3-triene in high yield upon treatment with MeLi in Et$_2$O at −80 °C [60].

The reaction of 7,7-dibromobicyclo[5.1.0]octane with MeLi in ether was more interesting because of the formation of a highly strained allene, 1,2-cyclooctadiene. Four products were isolated as a result of this reaction. The expected allene is a more strained and unstable compound than cyclonona-1,2-diene. Therefore, most of the compound is dimerized by [2 + 2] cycloaddition reaction. Some intramolecular and intermolecular insertion products were also formed as side products [89].

Enormous effort has been devoted to the synthesis of cyclohepta-1,2-diene and cyclohexa-1,2-diene. They were generated as intermediates, while experimental evidence for 1,2-cyclopentadiene has remained elusive. For the generation of a five-membered ring allene, Balcı and coworkers first added fluorobromocarbene to a cyclobutene derivative. Treatment of an adduct with MeLi in ether at −25 °C in the presence of furan as the trapping reagent afforded an allene trapping product as the major product [90].

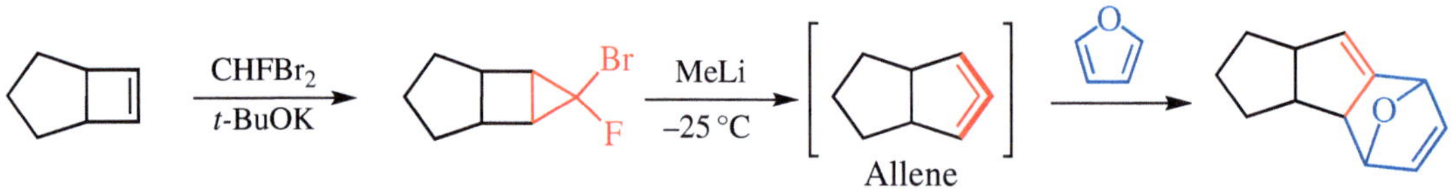

Cyclopropylcarbene–Cyclobutene Rearrangement Cyclopropylcarbene generated by the reaction of cyclopropanecarboxaldehyde tosylhydrazone with NaOMe in aprotic solvents gave mainly cyclobutene by ring expansion. Acetylene, ethylene, and butadiene were formed by fragmentation as the side products [91].

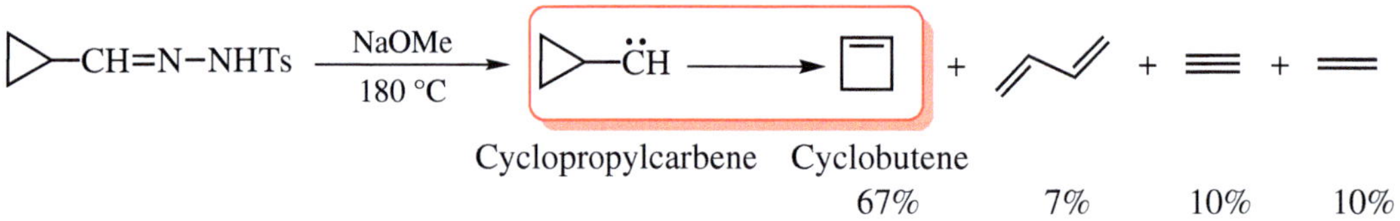

Cyclopropylcarbene reacts differently depending on the formation conditions. It was established that cyclobutene formation was increased when the decomposition was carried out with an equivalent amount of base in a protic solvent, usually ethylene glycol. Furthermore, the amounts of fragmentation products were suppressed if the cyclopropane was attached to a ring.

The cyclopropylcarbene–cyclobutene rearrangement has been successfully applied to the synthesis of essential compounds. For example, one of the most critical steps in octalene synthesis was the synthesis of the tetracyclic intermediate resulting from the application of the cyclopropylcarbene–cyclobutene rearrangement reaction of a dialdehyde with *p*-toluenesulfonyl hydrazide in boiling ethanol, which smoothly afforded a ditosylhydrazone. Heating of a bistosylhydrazone with sodium methoxide at 140 °C gave the desired biscyclobutene derivative, which was converted to octalene (for details of the synthesis, see Section 6.4.) [92].

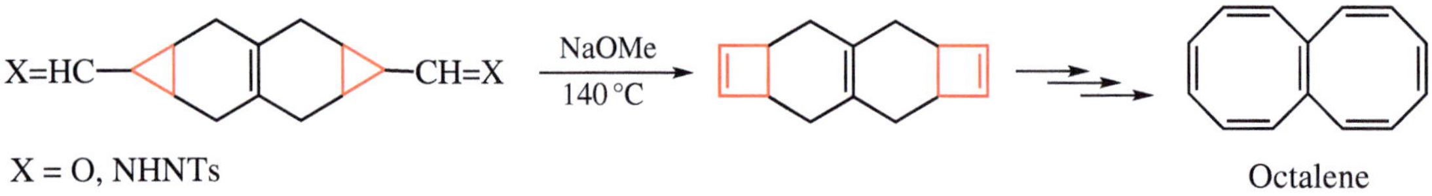

Starting from spiroketones, the synthesis of bicyclic olefins containing double bonds between the bridging carbons is also possible by application of the cyclopropylcarbene–cyclobutene rearrangement [93].

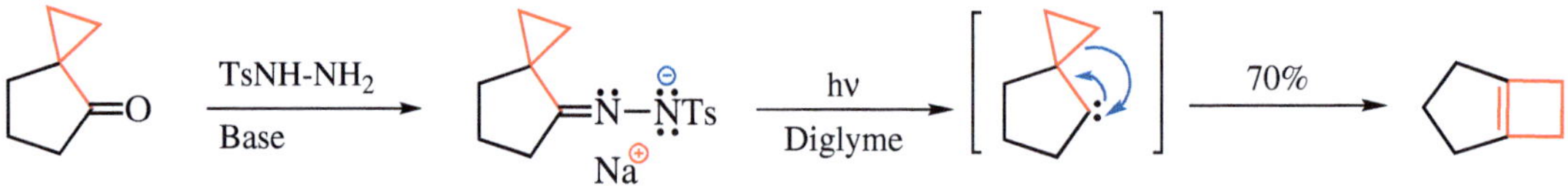

Vinylidene–Alkyne Rearrangement Vinylidenes easily undergo dimerization and addition reactions to double bonds. If vinylidenes are directly bonded to a ring, they can undergo rearrangement to form alkynes. This is a method of choice to synthesize elusive compounds such as strained cyclic alkynes. Generally, relevant ketones are first converted into exocyclic dibromomethylene derivatives or the corresponding diazo compounds as vinylidene precursors, as shown below [94–97].

Dibromomethylenecyclobutane is treated with MeLi or *n*-BuLi. The resulting organolithium compound converts into vinylidene by removing LiBr at room temperature or below, and the unstable vinylidene rearranges to a cycloalkyne. As cycloalkynes with smaller rings (ring carbon atom number less than 8) are reactive intermediates, they will be unstable under the reaction conditions. They are trapped with various dienophiles such as cyclohexene, furan, or dihydrofuran.

For the generation of vinylidenes, monohalomethylene compounds are also used instead of exocyclic dihalomethylene compounds. A base, such as potassium *t*-butoxide, is used to produce vinylidene intermediates. Balcı and coworkers have applied this method to the synthesis of a six-membered ring alkyne known as the most strained. The alkyne was trapped with 1,3-diphenylbenzoisofuran [98].

8.3 Azides and Nitrenes

The first organic azide, phenyl azide, as an energy-rich compound, was discovered by Griess in 1864, and later Tiemann proposed in 1891 the formation of nitrenes during decomposition of azides. After these findings, interest in this field has continued to grow. Nitrenes are nitrogen compounds that are analogous to carbenes. In nitrenes, only one substituent is attached to the nitrogen atom, and there are two pairs of nonbonding electrons on the nitrogen atom. Because the nitrogen atom has six electrons in its outer shell, nitrenes also fall into the class of electron-deficient compounds, like carbenes. Because of these electronic properties, nitrenes act as electrophiles, like carbenes; they are unstable and react very quickly. A nitrene has the following general formula.

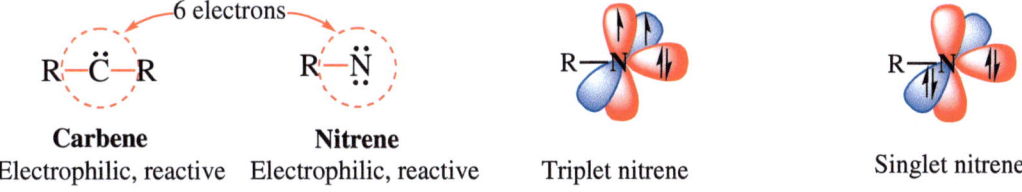

Nitrenes can have two different electronic configurations, like carbenes. The simplest nitrenes have a linear structure. If the nitrogen atom's hybridization is sp, one of the hybrid orbitals is used to bond the R group attached to the nitrogen, while the other sp hybrid orbital has a pair of electrons. Because there are two p orbitals, the two remaining electrons occupy two degenerate orbitals according to Hund's rule; electrons fill these orbitals one by one. For this reason, nitrenes in the ground state prefer the triplet configuration. In the singlet state, an electron pair populates one p orbital and the other one is empty.

While the energy difference between singlet and triplet carbenes in hydrocarbons is about 8–10 kcal/mol (33–42 kJ/mol), this difference in nitrenes is about 16–18 kcal/mol (63–75 kJ/mol) [99]. Moreover, calculations show that nitrenes are more stable than carbenes, with an energy difference of 25–26 kcal/mol (105–109 kJ/mol) [100]. Depending on the substituent's electronic structure attached to the nitrogen atom, the electronic configuration of nitrenes changes. Electron-donating groups force nitrenes to be singlets by filling one of the empty orbitals via electron transfer. For example, when a nitrogen atom is attached to the nitrene atom, nitrenes become so stable that they can be isolated.

8.3.1 Nitrene Synthesis

There are several ways of generating nitrenes. Nitrenes are generally formed from their precursors by elimination. The most common nitrene synthesis methods are given below [101].

1. Thermal or photochemical decomposition of alkyl azides and acyl azides.

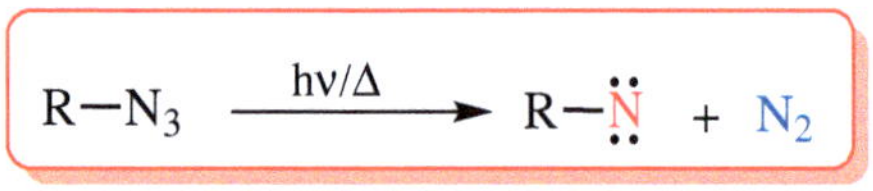

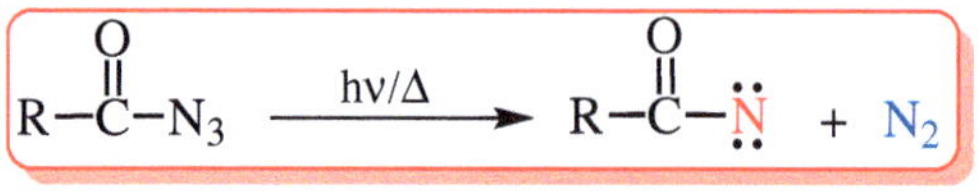

$$R-N_3 \xrightarrow{h\nu/\Delta} R-\ddot{N} + N_2 \qquad R-\overset{O}{\underset{\|}{C}}-N_3 \xrightarrow{h\nu/\Delta} R-\overset{O}{\underset{\|}{C}}-\ddot{N} + N_2$$

2. α-Elimination reaction.

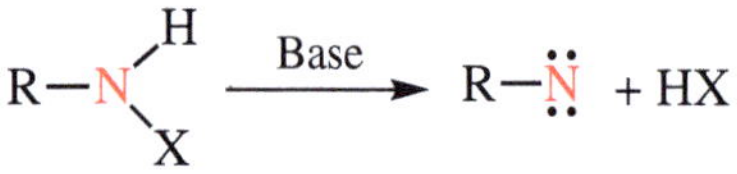

$$R-N\overset{H}{\underset{X}{\diagup}} \xrightarrow{Base} R-\ddot{N} + HX \qquad X = \text{Good leaving group}$$

3. Oxidation of amines and reduction of nitro and nitroso compounds.

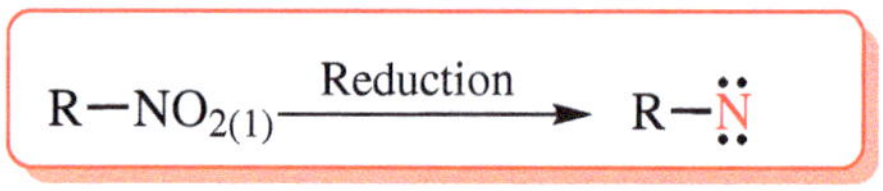

$$R-NH_2 \xrightarrow{Oxidation} R-\ddot{N} \qquad R-NO_{2(1)} \xrightarrow{Reduction} R-\ddot{N}$$

Because nitrenes are compounds with electron deficiency, like carbenes, they react similarly. Addition to the double bonds, rearrangement, and insertion are essential nitrene reactions [102, 103]. In this section, nitrene reactions will not be examined in separate groups.

Azides are highly polar compounds because of their different resonance structures shown below. These polar structures allow azides to undergo 1,3-dipolar cycloaddition reactions to double bonds. According to Pauling, the c and d structures are resonance structures that facilitate nitrogen gas's extrusion. Suppose the azide group is directly attached to an aromatic ring. In that case, the azide group will be in conjugation with the aromatic system, which makes the aromatic azides more stable than the alkyl azides.

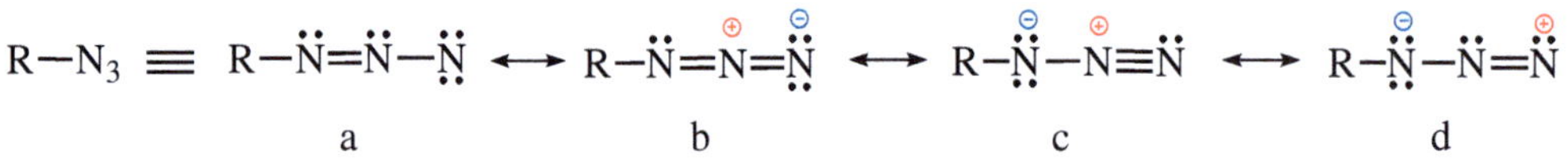

$$R-N_3 \equiv \underset{a}{R-\ddot{N}=\ddot{N}-\ddot{N}} \leftrightarrow \underset{b}{R-\ddot{N}=\overset{\oplus}{N}=\ddot{N}^{\ominus}} \leftrightarrow \underset{c}{R-\ddot{N}^{\ominus}-\overset{\oplus}{N}\equiv\ddot{N}} \leftrightarrow \underset{d}{R-\ddot{N}^{\ominus}-\ddot{N}=\overset{\oplus}{N}}$$

Most azides are explosive substances that decompose with nitrogen release by pressure or heat, forming nitrenes. The organic azides, particularly methyl azide, often decompose explosively. Therefore, great care must be taken when working with any azide. NaN$_3$ is used in airbags. The most common method for synthesizing alkyl azides is the substitution reaction with the azide anion (N$_3^-$).

8.3.1.1 Synthesis of Acyl Azides

Acyl azides are readily converted to α-ketonitrenes by thermal or photolytic decomposition with nitrogen gas release. For this reaction, acyl azides must first be synthesized. The first step in their synthesis is activation of the acid functional group. One of the conventional acyl azide synthesis methods is converting acids first to acyl chlorides, followed by the reaction with NaN_3 or trimethylsilyl azide [104]. Some functional groups (acid-sensitive groups) in the molecule may prevent the application of this method. Then, the acid functional group is first converted to an anhydride with ethyl chloroformate in the presence of a base, followed by the reaction with NaN_3. Anhydrides possess two different carbonyl groups. The azide anion attacks the more reactive carbonyl group to which the alkyl group is attached.

In addition to these methods, $SOCl_2$/DMF combination [105] is also a commonly used method. Diphenyl phosphoryl azide is another reagent used to synthesize acyl azides. The acid carbonyl group can also be activated by trifluorophenyl boric acid [106]. The reaction of the intermediate formed with NaN_3 also furnishes acyl azides.

8.3.2 Nitrene Rearrangements

The neutral monovalent nitrogen species with six electrons in the outer shell are highly reactive and generally electrophilic. Nucleophilic migration from a carbon atom to a nitrogen atom can take place. The reactions are used to prepare amines from carboxylic acids. We will discuss various rearrangement reactions in which acyl azides are involved.

8.3.2.1 Curtius Rearrangement

The Curtius rearrangement is the thermal decomposition of acyl azides to furnish isocyanates. This reaction was first discovered by the German chemist Julius Wilhelm Theodor Curtius (1857–1928). The isocyanates formed are stable and they can be isolated when the reaction is carried out in aprotic solvents. The isocyanates can be readily transformed into a variety of derivatives. Reaction of isocyanates with water results in the unstable carbamic acid, which undergoes spontaneous decarboxylation to give primary amines. The conversion of carboxylic acids to amines is named the *Curtius rearrangement* [107].

When acyl azides are heated in an aprotic solvent, they first release N_2 gas and form unstable acyl nitrenes in which the R group connected to the carbonyl group migrates to the electron-deficient nitrogen atom to form isocyanates in a similar way to that discussed for the Wolff rearrangement. Efforts to trap the acyl nitrene intermediate under the thermal condition failed. Therefore, a concerted mechanism without involving the acyl nitrene intermediate was suggested. Migration of the R group and removal of nitrogen take place at the same time. The migration proceeds with complete retention of stereochemistry.

When the decomposition of acyl azides is carried out in an alcohol, the alcohol molecules add to the carbonyl group of isocyanate to form urethanes (carbamic acid esters), which are stable compounds compared with carbamic acids. When isocyanates react with amines, urea derivatives are formed. This method has been successfully applied to the synthesis of asymmetric urea derivatives. Isocyanates also react with different nucleophiles and turn into compounds containing various heteroatoms.

A triacid, a cyclohexane derivative, has served as a useful framework for the design of enzyme models. Recently, Menger and coworkers [108] succeeded in conversion of this triacid to the corresponding tyramine by Curtius rearrangement.

There have been a wide range of applications of the Curtius rearrangement in the synthesis of natural products and their derivatives. This reaction has also been extensively utilized in the synthesis and application of a variety of interesting compounds [109].

Several methods are described for the synthesis of pyrimidoindolone skeletons. Balcı and coworkers [110] recently developed a straightforward synthetic methodology for indole-fused pyrimidone derivatives using two successive Curtius rearrangements. The construction of the target compound was based on the intramolecular cyclization of diacyl azides. One of the acyl azide functionalities was regiospecifically converted to the corresponding isocyanate by heating at 40 °C in benzene. The compound was trapped by amines to afford urea derivatives. A further Curtius rearrangement applied to the remaining acyl azide at higher temperatures followed by intramolecular cyclization gave the target compound. Because the acyl azide group that is attached to the double bond is more stable because of conjugation, it turns into an isocyanate at higher temperatures; however, the alkyl acyl azide group turns into an isocyanate at lower temperatures.

8.3.2.2 Schmidt Rearrangement

The Schmidt rearrangement is similar to the Curtius rearrangement both in terms of the starting compound used and mechanistically. The only difference between the two reactions is the generation of acyl azide [111]. Amines are formed starting from a carboxylic acid and hydrazoic acid (HN_3). When this method is applied, there is no way to isolate isocyanates.

The organic acid is first treated with mineral acids (H_2SO_4), and then HN_3 is added to the reaction medium. The reaction begins with the formation of an acylium ion, which reacts with hydrazoic acid to form a protonated azido ketone. The R group migrates to the nitrogen atom and displaces the nitrogen molecule to form a protonated isocyanate. The addition of water followed by decarboxylation of the carbamic acid produces an amine.

8.3.2.3 Lossen Rearrangement

The Lossen rearrangement involves the conversion of hydroxamic acid into an isocyanate [112, 113]. The hydroxamic acids used in the reaction are generated by treatment of the respective esters with hydroxylamine hydrochloride in the presence of a base. For H_2O elimination from hydroxamic acids, the hydroxyl group attached to nitrogen must first be activated through O-acylation or O-sulfonation. For this, the molecule is first reacted with tosyl chloride to obtain the corresponding tosylate. Treatment of the tosylate with a base generates a nitrene as the intermediate, which undergoes rearrangement to furnish an isocyanate as discussed for the Curtius rearrangement. The isocyanate can be subsequently converted with nucleophiles into various functional groups.

Because of some of the complications that occurred during the Lossen reaction, this procedure did not attract much attention among the scientific community because of the competing formation of self-condensation by-products. The rate-determining step is the activation of the hydroxamic acid. As the isocyanate formation is much faster than the activation step, the isocyanate formed can react with free hydroxamic acid to form the corresponding O-carbamoyl hydroxymate, which under heating affords a symmetric urea derivative.

In order to overcome these problems, various activated hydroxycarbamate synthesis methods have been developed in recent years [114–117]. Meier and coworkers used dimethyl carbonate as a green activation reagent in the presence of tertiary amines and methyl carbamates were obtained in good yields.

8.3.2.4 Hofmann Rearrangement

The Hofmann rearrangement, also known as the Hofmann degradation, is another well-known reaction to convert primary amides into primary amines [118, 119]. An amide is treated with bromine or chlorine in the presence of a strong base (NaOH or KOH). In the first step, the base abstracts one of the NH protons, forming an anion, which reacts with bromine to form N-bromoamide. The abstraction of the remaining proton forms a bromoamide anion. At this stage, the R group migrates to the nitrogen atom and substitutes the bromide ion simultaneously, forming an isocyanate. The reaction of isocyanate with water gives a primary amine. Several variations of the Hofmann reaction are available in the literature. For example, N-bromosuccinimide is also used instead of the bromine molecule to attach the bromine atom to nitrogen.

Hu and coworkers [120] developed a mild and efficient method for the synthesis of 2-oxazolone derivatives via the Hofmann rearrangement using bis(trifluoroacetoxy)iodobenzene. This method is superior to earlier ones in terms of yield. Zhdankin and coworkers [121] also converted alkyl carboxamides to the corresponding amines using hypervalent iodine species generated in situ from PhI and the commercial oxidant oxone ($2KHSO_5 \cdot KHSO_4 \cdot K_2SO_4$) in aqueous acetonitrile.

8.3.3 Nitrene Cycloaddition Reactions

Because nitrenes exhibit electrophilic properties, they easily add to double bonds with high electron density, forming aziridines. The configuration of the resulting product depends on the electronic structure of the nitrene. As in carbene chemistry, singlet nitrenes add simultaneously to double bonds and the configuration of the double bonds is preserved. In contrast, triplet nitrenes' addition proceeds in the same manner as triplet carbene addition and the addition is nonstereospecific.

Like acyl azides, alkyl azides are also easily transformed into alkyl nitrenes by heat (40–200 °C) or photolysis. While most acyl nitrenes prefer rearrangement to form isocyanate, alkyl nitrenes undergo insertion or cycloaddition to double bonds [122]. Although elimination is one of the most commonly used methods for carbene synthesis, it is rarely applied to alkyl nitrene synthesis because alkyl N-halogen compounds are not generally stable compounds. There must be a good leaving group on the nitrogen atom to apply α-elimination to nitrene synthesis. As seen in the example below, alkyl nitrene is formed by α-elimination of the sulfonyl group attached to the nitrogen atom.

As with carbenes, the reaction conditions (dilute solutions provide the formation of triplet carbenes) or how nitrenes are formed affect the structure of the products. In the formation of nitrenes by photochemical method, triplet nitrenes are formed directly; therefore, stereospecificity is not observed in addition reactions.

Kwart and Kahn [123] reported in a seminal paper that the copper-bronze-catalyzed aziridination reaction of cyclohexene with benzenesulfonyl azide gives the addition product in 15% yield. In contrast, insertion products are formed as the main product. Subsequently, Mansuy and coworkers [124] succeeded in the synthesis of N-substituted aziridines by transfer of nitrene (R—N) from PhI=NR in the presence of catalytic amounts of Fe(III) and Mn(III)tetrafenylporphyrin complexes.

Evans and coworkers applied the Cu-catalyzed aziridination of both electron-rich and electron-deficient olefins, employing PhI=NTs as the nitrene precursor, and they synthesized N-tosylaziridines in yields between 55% and 95% [125, 126].

Enantioselective addition is observed in aziridination reactions when an optically active ligand that can complex with copper is added to the reaction medium. Jabonsen and coworkers [127] reported the first enantioselective alkene aziridination using 1,2-diaminocyclohexane derivatives as excellent ligands for the Cu(I)-catalyzed asymmetric aziridination of olefins by PhI=NTs. The yields are high and the enantioselectivity is over 97%.

It is possible to synthesize nitrenes by removing the hydrogen from primary amines by oxidation. This method cannot be generally applied to nitrene synthesis. However, it has been successfully applied to the synthesis of aminonitrenes. Substituted hydrazine derivatives are oxidized with reagents such as HgO, MnO_2, $Pb(OAc)_4$, and $PhI(OAc)_2$ to form aminonitrenes [128].

Aminonitrenes are classified as stable nitrenes because of the free electrons on the adjacent nitrogen atom. The conjugation stabilizes the electron-poor nitrogen atom with the nonbonding electrons on the neighboring nitrogen atom. These electrons fill one of the empty p orbitals and force the nitrene to be in singlet state. Therefore, these nitrenes acquire nucleophilic properties, readily adding to electron-poor double bonds and regular double bonds.

The reaction of N-aminophthalimide with Pb(OAc)$_4$ in the presence of cyclohexene gives the corresponding aziridines in 40% yield. Because of lead tetraacetate toxicity, diacetoxyiodobenzene has been used and aziridination yield was increased as much as 71% [129]. The electrochemical nitrene transfer in acetonitrile in the presence of ammonia and trifluoroacetic acid increases the aziridination efficiency up to 92% [130].

Oxidant : Pb(OAc)$_4$/CH$_2$Cl$_2$, 40%
: PhI(OAc)$_2$/CH$_2$Cl$_2$, 71%

When unacylated azides are used as nitrene precursors, a cycloaddition reaction between the dipolar structure of azide and the double bond can generate the corresponding 1,2,3-triazolines. In many cases, these intermediates can be isolated. The intermediate can eliminate N$_2$ to produce aziridine derivatives.

8.3.4 Nitrene Insertion Reactions

Another significant reaction of nitrenes is insertion reactions like carbenes undergo [131]. A nitrene can insert into a carbon–hydrogen bond, forming an amine. While singlet nitrene insertions processes are concerted, involving a three-center cyclic transition state, triplet nitrenes form products because of multistage reactions. A singlet nitrene reacts with retention of configuration.

Pyrolysis of two different optically active azides, an *o*-azidoalkylbenzene and an alkyl azidocarbonate, led in each case to cyclization at the asymmetric carbon atom to give an indoline and oxazolone derivatives [132]. The products were shown to be still optically active. It was concluded that the insertion occurs by one process involving a singlet-state nitrene. For such insertion reactions to occur, the insertion center (C—H bond) must be close to the nitrogen atom and there must also be no other C—H bonds in the α-position.

For a triplet nitrene, the C–H insertion process proceeds in two steps. Because of its diradical nature, the triplet nitrene first abstract hydrogen from the C—H bonds, forming two new radicals. The combination of radicals results in the product of insertion.

Selectivity is observed in nitrene inversion and nitrenes prefer tertiary carbon atoms over secondary and primary ones in inversion. The example below clearly supports this. Insertion is observed more in tertiary carbon atoms than in secondary and primary ones.

Nitrene insertion can be applied to the synthesis of essential heterocycles. For example, vicinal amino alcohols are a common structural unit in various natural products and pharmaceutical agents. The metal-catalyzed insertion process of nitrene makes it possible to synthesize compounds having amino alcohol substructures. It is possible to synthesize heterocyclic compounds in high yields if nitrene insertion is controlled. For example, insertion of nitrene generated from the carbamate forms oxazolidone derivatives in high yields. This reaction does not work with $PhI(OAc)_2$. When 5% $Rh_2(OAc)_4$ is added to the medium as a catalyst, the inversion product is formed in 86% yield [133].

Problems

8.1 Write the product and show its mechanism for formation.

8 Reactive Intermediates Carbanions, Carbenes, and Nitrenes

8.2 When the ammonium salts given below are heated, two products are formed. Propose a mechanism of formation for each of the following products.

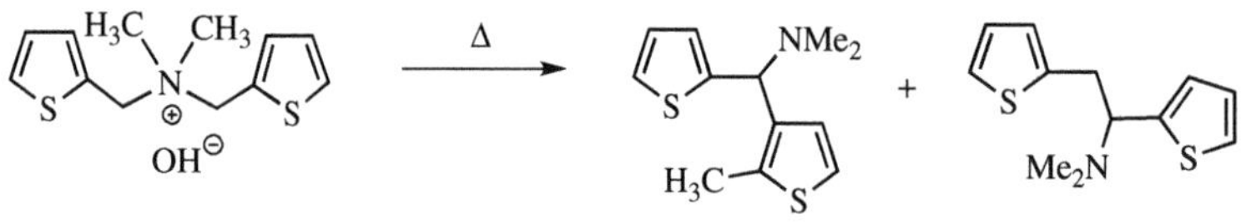

8.3 Propose a mechanism for the following conversion.

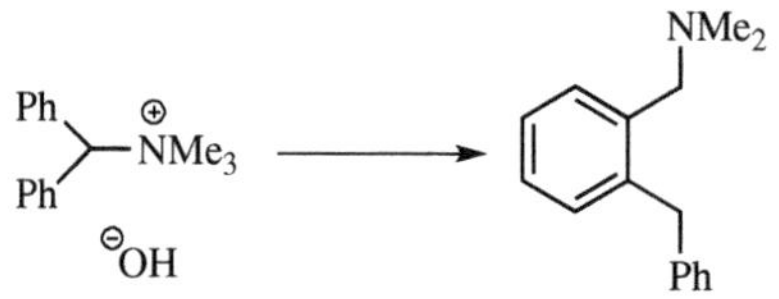

8.4 Write the structure of the product and explain its mechanism of formation.

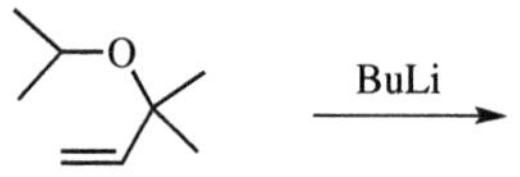

8.5 Give the products of the following reactions and discuss the mechanism of formation by giving the intermediate structures.

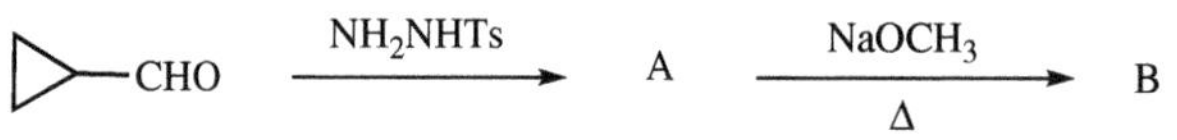

8.6 Describe the synthetic routes for the preparation of the following compound starting from the ketone.

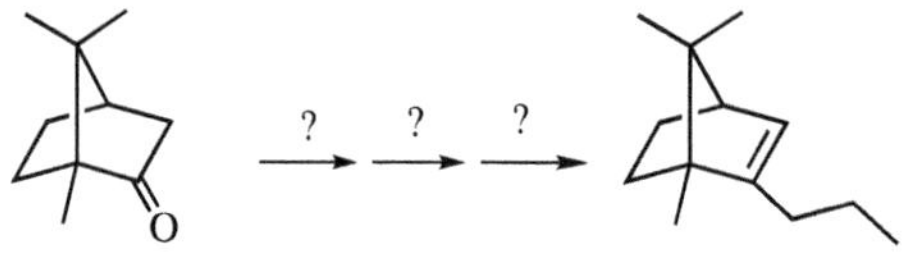

8.7 Show how the following acetyl compound could be prepared using the Shapiro reaction starting from cyclohexanone derivative.

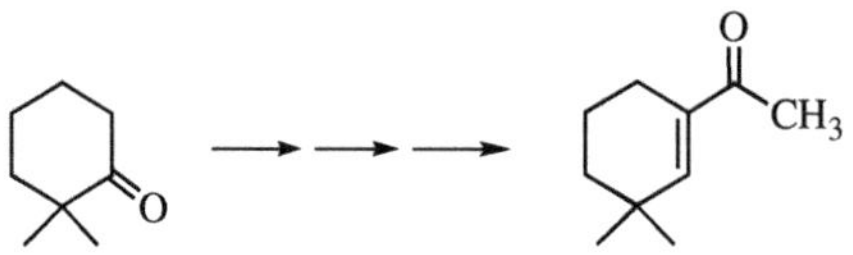

8.8 Give the product of the following reaction.

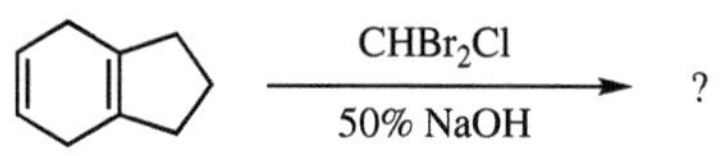

8.9 In the following reaction, chloroform and base are taken in excessive amounts. Write down the products that can occur.

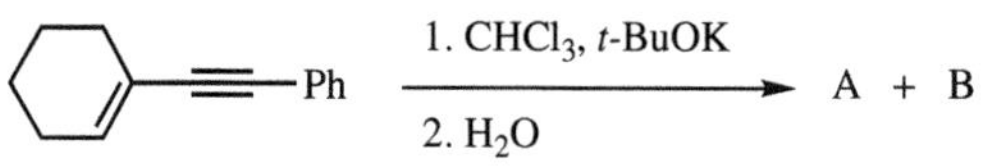

8.10 Identify the compounds A-E and discuss the mechanisms of formation of the products.

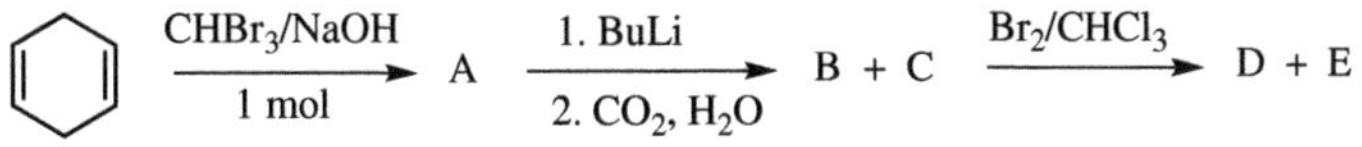

$$\text{cyclohexadiene} \xrightarrow[\text{1 mol}]{\text{CHBr}_3/\text{NaOH}} A \xrightarrow[\text{2. CO}_2, \text{H}_2\text{O}]{\text{1. BuLi}} B + C \xrightarrow{\text{Br}_2/\text{CHCl}_3} D + E$$

8.11 In the following reactions, fill the blanks and discuss the mechanisms of formation.

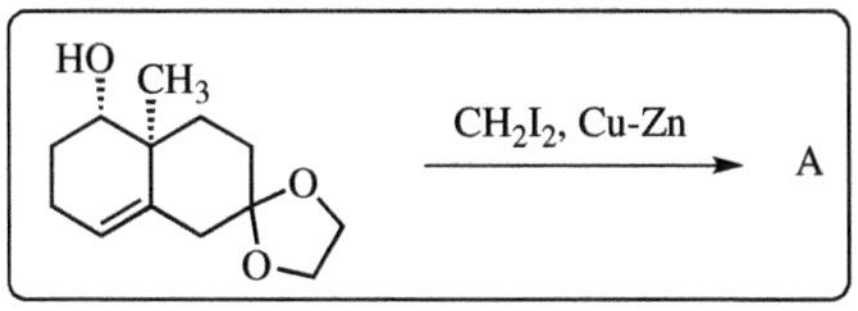

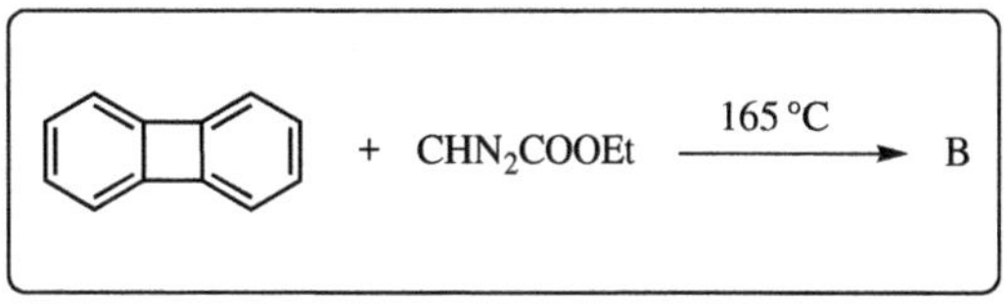

$$\xrightarrow{\text{CH}_2\text{I}_2, \text{Cu-Zn}} A$$

$$+ \text{CHN}_2\text{COOEt} \xrightarrow{165\,°\text{C}} B$$

8.12

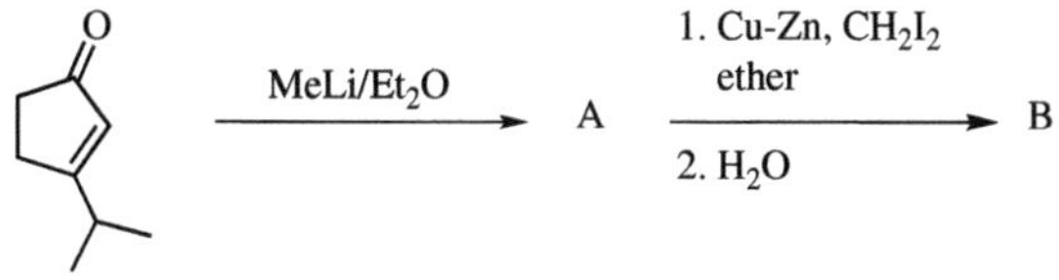

$$\xrightarrow{\text{MeLi/Et}_2\text{O}} A \xrightarrow[\text{2. H}_2\text{O}]{\text{1. Cu-Zn, CH}_2\text{I}_2 \text{ ether}} B$$

8.13 Indicate how the following compound can be synthesized from the given starting material.

8.14 Write down the products that may be formed by heating tosyl hydrazones in a basic medium.

$$\text{Ar-CH=NNHTs} \xrightarrow{\text{Base/heat}} ? \qquad \xrightarrow{\text{Base/heat}} ?$$

8.15 As a result of the photolysis of the following diazo compound in methanol, three products are formed. Explain the mechanisms of their formation.

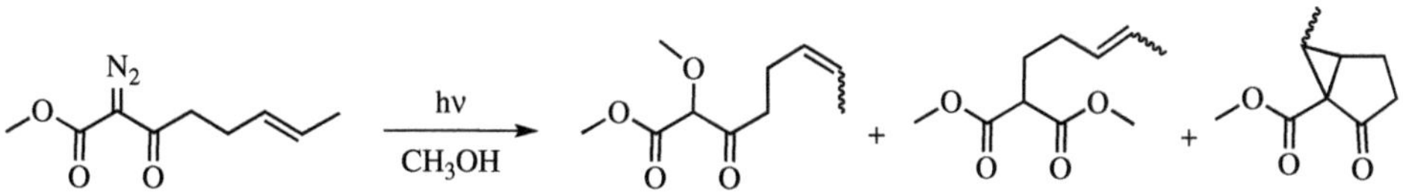

$$\xrightarrow[\text{CH}_3\text{OH}]{h\nu}$$

8.16 Isomeric α-diazo ketones transform into the same product when heated. Discuss the mechanism of formation of the product.

$$\xrightarrow{140\,°\text{C}} \xleftarrow{140\,°\text{C}}$$

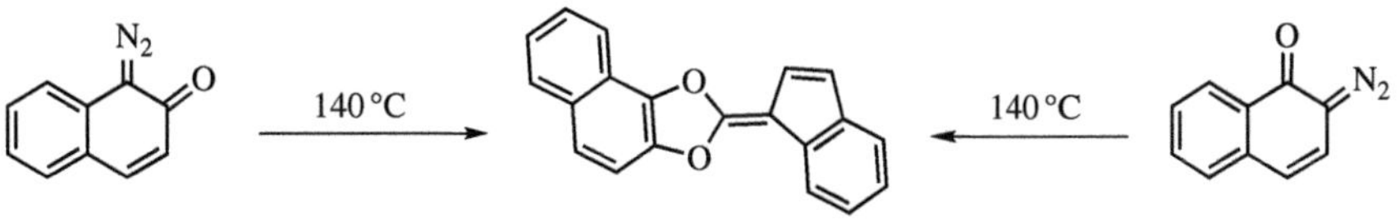

8.17 2-Diazo-1-3-diketone transforms into a tricyclic compound when heated. Discuss the formation mechanism of the product.

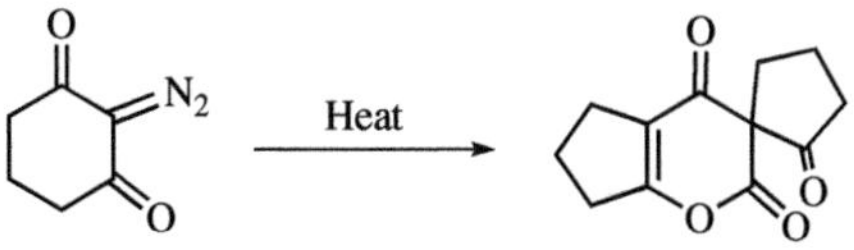

$$\xrightarrow{\text{Heat}}$$

8.18 Show how the following compound can be synthesized starting from the 1,3-diketone.

8.19 When the α-diazoketone is heated, it transforms into a dimer product with high efficiency. Explain its formation and give a reasonable answer to why the expected Wolff arrangement did not occur.

8.20 When the diazo-1,3-dicarbonyl compound is heated, it converts to products A and B. Write the structure of products A and B. How do the yields of products A and B change qualitatively when the substituents H, OCH$_3$, and NO$_2$ attached to the benzene ring are changed?

$$X = H, OCH_3, NO_2$$

8.21 Explain the mechanism of the following transformation below.

8.22 The following epoxide undergoes a ring-opening reaction upon irradiation in methanol. Suggest a rational mechanism for this transformation.

8.23 When the bicyclic diazo compound is heated to 420 °C, it undergoes fragmentation and turns into furan and compound A. Give the structure of the product and discuss the mechanism of formation.

8.24 Fill in the blanks in the table below and discuss how products are formed.

8.25 Explain the mechanism of the transformation below.

8.26 Explain the mechanism of the following transformation.

References

1. Stevens, T.S., Creighton, E.M., Gordon, A.B., and MacNicol, M. (1928). *J. Chem. Soc.*: 3193.
2. Vanecko, J.A., Wan, H., and West, F.G. (2006). *Tetrahedron* 62: 1043.
3. Ghigo, G., Cagnina, S., Maranzana, A., and Tonachini, G. (2010). *J. Org. Chem.* 75: 3608.
4. Allin, S.M., Button, M.A.C., and Shuttleworth, S.J. (1997). *Synlett*: 725.
5. Bodwell, G.J., Houghton, T.J., Koury, H.E., and Yarlagadda, B. (1995). *Synlett*: 751.
6. Mitchell, R.H. and Boekelheide, V. (1974). *J. Am. Chem. Soc.* 96: 1547.
7. Vedejs, E. and West, F.G. (1986). *Chem. Rev.* 86: 941.
8. Vedejs, E. and Martinez, G.R. (1979). *J. Am. Chem. Soc.* 101: 6452.
9. Sommelet, M. (1937). *Compt. Rend.* 205: 56.
10. Tayama, E., Watanabe, K., and Sho, S. (2017). *Org. Biomol. Chem.* 15: 6668.
11. Roy, T., Gaykar, R.N., Bhattacharjee, S., and Biju, A.T. (2019). *Chem. Commun.* 55: 3004.
12. Lednicer, D. and Hauser, C.R. (1957). *J. Am. Chem. Soc.* 79: 4449.
13. Kitano, T., Shirai, N., Motoi, M., and Sato, Y. (1992). *Chem. Soc. Perkin Trans. I*: 2851.
14. Narita, K., Shirai, N., and Sato, Y. (1997). *J. Org. Chem.* 62: 2544.
15. Wittig, G. and Löhmann, L. (1942). *Justus Liebigs Ann. Chem.* 550: 260.
16. Wittig, G. (1958). *Experientia* 14: 389.
17. Tomooka, K., Yamamoto, H., and Nakai, T. (1997). *Liebigs Ann./Recl.*: 1275.
18. Marsden, A. and Thomas, E.J. (2002). *Arkivoc* ix: 78.
19. Bunker, P.R. and Jensen, P. (1983). *J. Chem. Phys.* 79: 1224.
20. Bender, C.F. and Schaefer, H.F. III, (1970). *J. Am. Chem. Soc.* 92: 4984.
21. Tomioka, H. (1997). *Acc. Chem. Res.* 30: 315.
22. Schuster, G.B. (1986). *Adv. Phys. Org. Chem.* 22: 311.
23. Irikura, K.I., Goddard, W.A. III,, and Beauchamp, J.L. (1992). *J. Am. Chem. Soc.* 114: 48.
24. Schoeller, W.W. (1980). *J. Chem. Soc. Chem. Commun.*: 124.
25. Pauling, L. (1980). *J. Chem. Soc. Chem. Commun.*: 688.
26. Bourissou, D., Guerret, O., Gabbai, F.P., and Bertrand, G. (2000). *Chem. Rev.* 100: 39.
27. Arduengo, A.J., Harlow, R.L., and Kline, M. (1991). *J. Am. Chem. Soc.* 113: 361.
28. Präsang, C., Donnadieu, B., and Bertrand, G. (2005). *J. Am. Chem. Soc.* 127: 10182.
29. Bamford, W.R. and Stevens, T.S. (1952). *J. Chem. Soc.*: 4735.
30. Fulton, J.R., Aggarwal, V.K., and de Vicente, J. (2005). *Eur. J. Org. Chem.*: 1479.
31. Shapiro, R.H. and Heath, M.J. (1967). *J. Am. Chem. Soc* 89: 5734.
32. Adlington, R.M. and Barrett, A.G.M. (1983). *Acc. Chem. Res.* 16: 55.
33. Funes-Ardoiz, I., Losantosa, R., and Sampedro, D. (2015). *RCS Adv.* 5: 37292.
34. de Boer, T.J. and Backer, H.J. (1963). *Org. Syn. Coll. Vol.* 4: 250.
35. Doyle, M.P., McKervey, M.A., and Te, Y. (1998). *Modern Catalytic Methods for Organic Synthesis with Diazo Compounds*, 1–60. New York: Wiley-Interscience.

36 Presser, A. and Hüfner, A. (2004). *Monatsh. Chem.* 135: 1015.
37 Moss, R.A. (2006). *Acc. Chem. Res.* 39: 267.
38 Krois, D. and Brinker, U.H. (2001). *Synthesis*: 379.
39 Hine, J. (1950). *J. Am. Chem. Soc.* 72: 2438.
40 Doering, W.E. and Hoffmann, A.K. (1954). *J. Am. Chem. Soc.* 76: 6162.
41 Kirmse, W. (1965). *Angew. Chem. Int. Ed.* 4: 1.
42 Makosza, M. and Wawrzyniewicz, M. (1969). *Tetrahedron Lett.* 10: 4659.
43 Fedorynski, M. (2003). *Chem. Rev.* 103: 1099.
44 Wagner, W.M. (1959). *Proc. Chem. Soc.*: 229.
45 Seyferth, D. (1972). *Acc. Chem. Res.* 5: 65.
46 Chen, T., Wang, X.-B., and Han, L.-B. (2015). *Phosphorus Sulfur Silicon Relat. Elem.* 190: 1820.
47 Simmons, H.E. and Smith, R.D. (1958). *J. Am. Chem. Soc.* 80: 5323.
48 Simmons, H.E. and Smith, R.D. (1959). *J. Am. Chem. Soc.* 81: 4256.
49 Denis, J.M., Girard, C., and Conia, J.M. (1972). *Synthesis*: 549.
50 Furukawa, J., Kawabata, N., and Nishimura, J. (1968). *Tetrahedron* 24: 53.
51 Shitama, H. and Katsuki, T. (2008). *Angew. Chem. Int. Ed.* 47: 2450.
52 Grasse, P.B., Brauer, B.E., Zupancic, J.J. et al. (1983). *J. Am. Chem. Soc.* 105: 6833.
53 Moss, R.A. and Joyce, M.A. (1978). *J. Am. Chem. Soc.* 100: 4475.
54 Vogel, E. and Böll, W.A. (1964). *Angew. Chem. Int. Ed.* 3: 642.
55 Menzek, A., Saracoglu, N., Krawiec, M. et al. (1995). *J. Org. Chem.* 60: 829.
56 Tanabe, Y., Seko, S., Nishii, Y. et al. (1996). *J. Chem. Soc., Perkin Trans. 1*: 2157.
57 Dehmlow, E.V. (1968). *Chem. Ber.* 101: 427.
58 García, J.L., Salvatella, L., Pires, E. et al. (2014). *Comprehensive Organic Synthesis II*, vol. 4 (eds. P.K. Gary and A. Molande), 1081. Elsevier.
59 Ball, W.J., Landor, S.R., and Punja, N. (1967). *J. Chem. Soc. (C)*: 194.
60 Moore, W.R. and Ozretich, T.M. (1967). *Tetrahedron Lett.* 33: 3205.
61 Buchner, E. and Curtius, T. (1885). *Ber. Dtsch. Chem. Ges.* 18: 2377.
62 Nguyen, J.M. and Thamattoor, D.M. (2007). *Synthesis*: 2093.
63 Parham, W.E., Bolon, D.A., and Schweizer, E.E. (1961). *J. Am. Chem. Soc.* 83: 603.
64 Jones, R.L. and Rees, C.W. (1969). *J. Chem. Soc. (C)*: 2249.
65 Huisgen, R. and Juppe, G. (1961). *Chem. Ber.* 64: 2332.
66 Preda, D.V. and Scott, L.T. (2000). *Tetrahedron Lett.* 41: 9633.
67 Scott, L.T. and Sumpter, C.A. (1990). *Org. Syntheses* 69: 180.
68 Kennedy, M., McKervey, M.A., Maguire, A.R. et al. (1990). *J. Chem. Soc. Perkin Trans.* 1: 1047.
69 Bach, R.D., Su, M.-D., Aldabbagh, E. et al. (1993). *J. Am. Chem. Soc.* 7: 10237.
70 Casanova, J. and Waegell, B. (1977). *Bul. Soc. Chim. France* 5–6: 560.
71 Casanova, J. and Waegell, B. (1975). *Bul. Soc. Chim. France* 3–4: 598.
72 Nickon, A. and Werstiuk, N.H. (1972). *J. Am. Chem. Soc.* 94: 7081.
73 Nozaki, H., Takaya, H., and Noyori, R. (1966). *Tetrahedron* 22: 3393.
74 Xia, Y., Liu, Z., Liu, Z. et al. (2014). *J. Am. Chem. Soc.* 136: 3013.
75 Kirmse, W. (2002). *Eur. J. Org. Chem.*: 2193.
76 Meier, H. and Zeller, K.-P. (1975). *Angew. Chem. Int. Ed.* 14: 32.
77 Kaplan, F. and Mitchell, M.L. (1979). *Tetrahedron Lett.*: 759.
78 Rodina, L.L. and Korobitsyna, I.K. (1967). *Russ. Chem. Rev.* 36: 260.
79 Zeller, K.-P. (1977). *Angew. Chem. Int. Ed.* 16: 781.
80 Zeller, K.-P., Blocher, A., and Haiss, P. (2004). *Mini Rev. Org. Chem.* 1: 291.
81 Silva, L.F. Jr., (2002). *Tetrahedron* 58: 9137.
82 Taber, D.F. and Tian, W. (2007). *J. Org. Chem.* 72: 3207.
83 Horner, L., Kirmse, W., and Muth, K. (1958). *Chem. Ber.* 91: 430.
84 Dauben, W.G., Schallhorn, C.H., and Whalen, D.L. (1971). *J. Am. Chem. Soc.* 93: 1446.
85 Wiberg, K.B. and Snoonian, J.R. (1998). *J. Org. Chem.* 63: 1390.
86 Reeder, L.M. and Hegedus, L.S. (1999). *J. Org. Chem.* 64: 3306.
87 Mock, W.L. and Hartman, M.E. (1977). *J. Org. Chem.* 42: 459.

88 Arndt, F. and Eisterdt, F. (1935). *Ber. Dtsch. Chem Ges. B.* 68: 200.
89 Marquis, E.T. and Gardner, P.D. (1966). *Tetrahedron Lett.*: 2793.
90 Algı, F., Özen, R., and Balcı, M. (2002). *Tetrahedron Lett.* 43: 3129.
91 Friedman, L. and Shechter, E. (1960). *J. Am. Chem. Soc.* 82: 1002.
92 Vogel, E., Runzheimer, H.V., Hogrefe, F. et al. (1977). *Angew. Chem.* 89: 909.
93 Kirmse, W. and Pook, K.H. (1966). *Angew. Chem. Int. Ed.* 5: 594.
94 Gilbert, J.C. and Baze, E. (1984). *J. Am. Chem. Soc.* 106: 1885.
95 Gilbert, J.C., Hou, D.R., and Grimme, J.W. (1999). *J. Org. Chem.* 64: 1529.
96 Fitjer, L. and Modaressi, S. (1983). *Tetrahedron Lett.* 24: 5495.
97 Gilbert, J.C. and Hou, D.-R. (2003). *J. Org. Chem.* 68: 10067.
98 Taskesenligil, Y., Kashyap, R.P., Watson, W., and Balcı, M. (1993). *J. Org. Chem.* 58: 3216.
99 Winkler, M. (2008). *J. Phys. Chem. A* 112: 8649.
100 Kemnitz, C.R., Karney, W.L., and Borden, W.T. (1998). *J. Am. Chem. Soc.* 120: 3499.
101 Scriven, E.F.V. and Turnbull, K. (1988). *Chem. Rev.* 88: 297.
102 Labbe, G. (1969). *Chem. Rev.* 69: 345.
103 Braese, S., Gil, C., Knepper, K., and Zimmermann, V. (2005). *Angew. Chem. Int. Ed.* 44: 5188.
104 Prakash, G.K.S., Arvanaghi, M., and Olah, G.H. (1983). *J. Org. Chem.* 48: 3358.
105 Arrieta, A., Aizpurua, J.M., and Palomo, C. (1984). *Tetrahedron Lett.* 25: 3365.
106 Tale, R.H. and Patil, K.M. (2002). *Tetrahedron Lett.* 43: 9715.
107 Balcı, M. (2018). *Synthesis* 50: 1373.
108 Menger, F.M., Bian, J., and Azov, V.A. (2002). *Angew. Chem. Int. Ed.* 41: 2581.
109 Ghosh, A.K., Brindisi, M., and Sarkar, A. (2018). *ChemMedChem.* 13: 2351.
110 Kapti, T., Dengiz, C., and Balcı, M. (2017). *Synthesis* 49: 1898.
111 Lang, S. and Murphy, J.A. (2006). *Chem. Soc. Rev.* 35: 146.
112 Thomas, M., Alsarraf, J., Araji, N. et al. (2019). *Org. Biomol. Chem.* 17: 5420.
113 Bauer, L. and Exner, O. (1974). *Angew. Chem. Int. Ed.* 13: 376.
114 Kreye, O., Wald, S., and Meier, M.A.R. (2013). *Adv. Synth. Catal.* 355: 81.
115 Anilkumar, R., Chandrasekhar, S., and Sridhar, M. (2000). *Tetrahedron Lett.* 41: 5291.
116 Stafford, J.A., Gonzales, S.S., Barrett, D.G. et al. (1998). *J. Org. Chem.* 63: 10040.
117 Jasikova, L., Hanikyrova, E., Skriba, A. et al. (2012). *J. Org. Chem.* 77: 2829.
118 Debnath, P. (2019). *Curr. Org. Chem.* 23: 2402.
119 Shioiri, T. (1991). *Comp. Org. Syn.* 6: 800.
120 Yu, C., Jiang, Y., Liu, B., and Hu, B. (2001). *Tetrahedron Lett.* 42: 1449.
121 Zagulyaeva, A.A., Banek, C.T., Yusubov, M.S., and Zhdankin, V.V. (2010). *Org. Lett.* 12: 4644.
122 Singh, G.S., D'hooghe, M., and De Kimpe, N. (2007). *Chem. Rev.* 107: 2080.
123 Kwart, H. and Kahn, A.A. (1967). *J. Am. Chem. Soc.* 89: 1951.
124 Mansuy, D., Mahy, J.P., Dureault, A. et al. (1984). *J. Chem. Soc. Chem. Commun.*: 1161.
125 Evans, D.A., Faul, M.M., and Bilodeau, M.T. (1991). *J. Org. Chem.* 56: 6744.
126 Evans, D.A., Faul, M.M., and Bilodeau, M.T. (1994). *J. Am. Chem. Soc.* 116: 2742.
127 Li, Z., Conser, K.R., and Jacobsen, E.N. (1993). *J. Am. Chem. Soc.* 115: 5326.
128 Anderson, D.J., Gilchrist, T.L., Horwell, D.C., and Rees, C.W. (1970). *J. Chem. Soc. (C)*: 576.
129 Li, J., Liang, J.-L., Chan, P.W.H., and Che, C.-M. (2004). *Tetrahedron Lett.* 45: 2685.
130 Siu, T., Picard, C.J., and Yudin, A.K. (2005). *J. Org. Chem.* 70: 932.
131 Müller, P. and Fruit, C. (2003). *Chem. Rev.* 103: 2905.
132 Smolinsky, G. and Feuer, B.I. (1964). *J. Am. Chem. Soc.* 86: 3085.
133 Du Bois, J. and Espino, C.G. (2001). *Angew. Chem. Int. Ed.* 40: 598.

9

Reactive Intermediates: Radicals and Singlet Oxygen

Compounds in which three groups are attached to the central carbon atom (trivalent) have unpaired electrons called radicals. They are also called *free radicals*. Typical examples are methyl, *t*-butyl, vinyl, allyl, phenyl, and benzyl radicals, the structures of which are shown below.

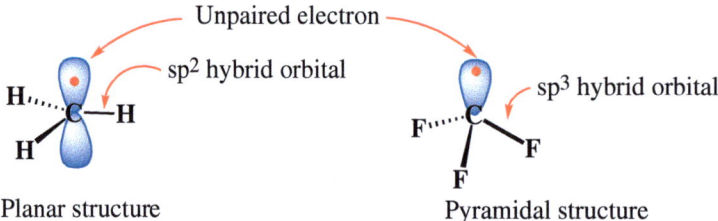

Because of the electron deficiency in the outer shell, radicals are classified as reactive intermediates like carbocations and carbenes. Radicals are highly reactive and involved in many chemical reactions to increase the number of electrons in the outer shell from seven to eight.

9.1 Structure of Radicals and Their Stability

In general, it is possible to examine radicals by using electron spin resonance (ESR) spectroscopy. The working principle of ESR spectroscopy is similar to that of nuclear magnetic resonance (NMR) spectroscopy. The difference between the two is that while NMR works according to the spin created by the nucleus, ESR is associated with the electron spin. Experiments show that radicals have either planar or pyramidal structures. ESR spectroscopy can be used to distinguish between planar and pyramidal structures. It is well established that the methyl radical has a planar structure [1]. In methyl radicals, the carbon atom is sp^2-hybridized and the unpaired electron occupies a 2p orbital. Fessenden and Schuler determined that the trifluoromethyl radical has a pyramidal structure and the F–C–F angle is about 111.1°, indicating sp^3 hybridization [2, 3].

Based on the results of experiments, it is possible to comment on the structures of radicals. For example, the chlorination of a pure enantiomer of 2-bromobutane produces a racemic mixture of 2-bromo-2-chlorobutane. The abstraction of a hydrogen atom with a chlorine radical generates a planar, sp^2 hybridized bromobutane radical. Because of the planar structure, the chlorine radical attacks from two enantiotopic faces with the same probability. As a consequence, two enantiomers are formed in equal amounts. This observation can be explained in two ways. Firstly, a radical formed as an intermediate product has a plane structure. Secondly, a radical with a pyramidal structure is formed, but the energy barrier required for inversion is very low. Some spectroscopic measurements and theoretical calculations indicate that the first possibility is responsible for racemization.

Reaction Mechanisms in Organic Chemistry, First Edition. Metin Balcı.
© 2022 WILEY-VCH GmbH. Published 2022 by WILEY-VCH GmbH.

Radicals generated at the bridgehead atom in bicyclic systems give information about the three-dimensional structure. Because of the high rigidity of these systems, their radicals cannot have a planar structure. Remember, the formation of a carbocation at the bridgeheads leads to a more significant steric strain (see substitution at bridgeheads) than in the reactant. Therefore, bicyclic systems at the bridgeheads are generally resistant to S_N1 reactions. However, it is possible to perform radical reactions at the bridgeheads because the intermediate radical formed has a pyramidal structure; the system's strain does not increase excessively compared to the starting compound. It is possible to generate radicals at the bridgeheads in norbornyl and adamantyl systems [4].

Radical intermediates are stabilized by electron-donating alkyl groups, like carbocations. The more substituted radicals are more stable than the less substituted ones. The stability of radicals can be determined by the dissociation energies of the C—H bonds that must be homolytically cleaved to obtain the radicals. The decrease in the C—H bond dissociation energies from methyl, ethyl, *i*-propyl, and *t*-butyl is in agreement with the related radicals' stability.

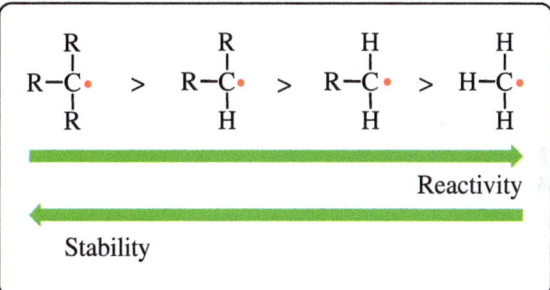

The differences in the relative radical stability are significantly lower than those in carbocations. The alkyl groups cannot stabilize radicals as well as they stabilize carbocations. In Section 7.1, we established that carbocations are stabilized by hyperconjugation. The interaction of a p orbital containing an electron with a filled C–H σ orbital results in the formation of two new molecular orbitals. Two electrons settle in the newly formed molecular orbital, while the third electron goes into an antibonding molecular orbital, SOMO (singly occupied molecular orbital). The three-electron system will stabilize the radical because of more electrons in the bonding orbitals (Figure 9.1). However, in the case of carbocations, there is no electron in the antibonding orbital. That is why, the alkyl groups cannot stabilize radicals as well as they can carbocations.

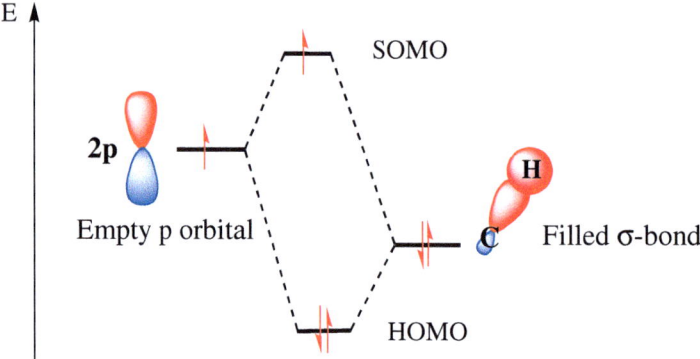

Figure 9.1 Presentation of hyperconjugation by orbital interaction.

Thus, the formation of a tertiary radical is favored over that of secondary or primary ones. Stability does not mean that a radical can be isolated under normal conditions. The more stable the tertiary radical, the longer the lifetime of that radical. The stability of radicals can vary depending on the molecular structure. There is generally a connection between stability and reactivity. The more unstable a radical is, the more reactive it is. However, this statement cannot be generalized. It is not always correct to say that secondary radicals are more reactive than tertiary radicals. This reveals that other factors affect reactivity.

For example, in the allyl system, different protons can be abstracted by a radical initiator. The strongest C—H bond is the vinyl C=C—H bond, and this bond is practically never broken. The allylic hydrogen can be easily abstracted because the C—H bond is relatively weak. Furthermore, abstraction of this hydrogen atom generates a radical that is stabilized by delocalization.

$$H_2C=CH-CH(H)-R \xrightarrow[-InH]{In^\bullet} H_2C=CH-\overset{\bullet}{C}H-R \longleftrightarrow H_2\overset{\bullet}{C}-HC=CH-R$$

In$^\bullet$ = radical initator

The stability of radicals can also be increased by aromatic groups attached to the radical carbon atom. In benzylic systems, the abstracted hydrogen will be on the α-carbon of the alkyl side chain because such a bond is relatively weak compared to the aromatic C—H bond. The resulting radical is stabilized by delocalization. The hydrogens attached to the aromatic ring cannot be abstracted. The phenyl radical formed cannot delocalize on the ring.

Because of electron delocalization, allyl and benzyl radicals are both more stable than the other alkyl radicals. The relative stabilities of various radicals are given below.

$$Ph-\overset{\bullet}{C}H_2 > H_2C=CH-\overset{\bullet}{C}H_2 > R-\overset{R}{\underset{R}{\overset{\bullet}{C}}} > R-\overset{R}{\underset{H}{\overset{\bullet}{C}}} > R-\overset{H}{\underset{H}{\overset{\bullet}{C}}} > H_2C=\overset{\bullet}{C}H > H-\overset{H}{\underset{H}{\overset{\bullet}{C}}}$$

← Increased stability

9.1.1 Generation of Radicals

Radicals are generally formed by the homolytic cleavage of covalent bonds, a process that requires significant amounts of energy. The energy required is the bond dissociation energy and each molecule has a different value. For example, the energy required for the H—H bond's cleavage is 104 kcal/mol (435 kJ/mol), while 58 kcal/mol (243 kJ/mol) is required for cleavage of the Cl—Cl bond. The electronegativities of the elements forming an σ bond must be equal or very close to each other to break the bonds homolytically by heat or photolysis. For example, organic peroxides and diazo compounds are easily broken homolytically by heat or irradiation. The oxygen–oxygen bond (RO—OR) is one of the weakest bonds in organic chemistry and its bond energy is around 34.9 kcal/mol (146 kJ/mol). This value decreases to 30.0 kcal/mol in acyl peroxides because of resonance.

Diacyl peroxides are sources for alkyl radicals. When diacyl peroxides are heated, acyloxy radicals are formed first. Decarboxylation of acyloxy radicals generates phenyl or alkyl radicals. These radicals react in various ways as radical initiators. Another general source of free radicals is the decomposition of diazo compounds. One of the most widely used radical initiators is azobisisobutyronitrile (AIBN). The stabilizing effect of the cyano substituent is responsible for the easy decomposition of AIBN.

Homolytic bond cleavage can be achieved in some compounds. If any molecule has heat-sensitive functional groups, photolysis is preferred to thermolysis. For example, when the bromine molecule is irradiated with UV light, it breaks down homolytically into bromine radicals, which can undergo various reactions.

$$Br-Br \xrightarrow{h\nu} 2\ Br\bullet$$

In addition to the homolytic cleavage of σ bonds, radicals can also be formed by excitation of carbonyl groups. For example, when benzophenone is irradiated, one of the nonbonding electrons goes into the antibonding π* orbital. The molecule's excited state shows a diradical property, which can abstract hydrogen from the environment, forming a highly stable ketyl radical. Ketyl radicals are easily dimerized into pinacol.

Radicals are also formed by the chemical or electrochemical oxidation or reduction of stable molecules. If an electron is removed from the molecule, a radical cation is formed, and if a single electron transfer is made to the molecule, a radical anion is formed. For example, when sodium and naphthalene react in an aprotic solvent, sodium transfers an electron to the naphthalene ring to form sodium naphthalenide, which is used successfully as a reductant in synthetic organic chemistry.

9.1.2 Radical Reactions

Radical reactions are unique in comparison with other types that we have seen up to this point. The significant difference is that the many radical reactions are chain reactions and they are applied to polymerization in industry. In this section, we will focus on the general reactions of radicals and five different reaction types will be discussed.

1. Atom-Abstraction Reaction
2. Radical Combination
3. Radical Disproportionation
4. Radical Addition
5. Electron Transfer Reaction.

9.1.2.1 Atom-Abstraction Reaction

One of the simplest atom-abstraction reactions is the removal of a hydrogen atom from methane by a radical. In these reactions, the lower the bond dissociation energy of the abstracted bond (in other words, the weaker the bond), the more easily the related bond is broken.

$$Cl\bullet + H-CH_3 \longrightarrow \bullet CH_3 + Cl-H$$

Furthermore, the stability of the new radical (methyl radical) and the formation energy of the new bond (H—Cl) formed after atom abstractions are factors affecting the reaction rate. For example, let us examine hydrogen abstraction reactions with fluorine, chlorine, bromine, and iodine radicals from the ethane molecule from a thermodynamic perspective. We have to compare the $\Delta H°$ values for forming ethyl radical by halogen radicals. The $\Delta H°$ values can be calculated using the bond dissociation energies in Table 9.1. $\Delta H°$ values refer to the energy of the bond broken minus the energy of the bond formed.

$$Hal-Hal \xrightarrow{h\nu} Hal\bullet \longleftarrow Hal-Hal + In\bullet \quad In\bullet = \text{radical initiator}$$

$$F\bullet + H-CH_2-CH_3 \longrightarrow F-H + \bullet CH_2-CH_3 \quad \Delta H° = -38 \text{ kcal/mol}$$
98 kcal/mol, 136 kcal/mol — exothermic

$$Cl\bullet + H-CH_2-CH_3 \longrightarrow Cl-H + \bullet CH_2-CH_3 \quad \Delta H° = -5 \text{ kcal/mol}$$
98 kcal/mol, 103 kcal/mol — exothermic

$$Br\bullet + H-CH_2-CH_3 \longrightarrow Br-H + \bullet CH_2-CH_3 \quad \Delta H° = 10 \text{ kcal/mol}$$
98 kcal/mol, 88 kcal/mol — endothermic

$$I\bullet + H-CH_2-CH_3 \longrightarrow I-H + \bullet CH_2-CH_3 \quad \Delta H° = 27 \text{ kcal/mol}$$
98 kcal/mol, 71 kcal/mol — endothermic

Table 9.1 Bond dissociation energies in kcal/mol and kJ/mol.

Bond	kcal/mol	kJ/mol	Bond	kcal/mol	kJ/mol	Bond	kcal/mol	kJ/mol
H—H	104	435	C_2H_5—Br	68	285	CH_2=CH—CH_3	97	406
H—F	136	571	C_2H_5—I	53	222	CH≡C—H	125	523
H—Cl	103	432	C_2H_5—OH	91	381	HO—H	119	498
H—Br	88	368	$(CH_3)_2$CH—H	95	397	HO—OH	51	213
H—I	71	298	$(CH_3)_2$CH—Cl	80	335	CH_3O—H	102	426
Cl—Cl	58	242	$(CH_3)_2$CH—Br	68	285	NC—H	130	543
Br—Br	46	192	$(CH_3)_3$C—H	91	381	H_2N—H	103	431
I—I	33	138	$(CH_3)_3$C—Cl	79	331	CH_3CO—H	86	360
CH_3—H	104	435	$(CH_3)_3$C—Br	65	272	Ph—H	112	469
CH_3—Cl	84	350	CH_2=CH—H	108	451	Ph—Cl	97	406
CH_3—Br	70	293	CH_2=CH—Cl	88	368	Ph—Br	82	343
CH_3—I	56	243	CH_2=CH—CH_2—H	87	364	Ph—CH_2—H	85	356
CH_3—OH	91	381	CH_2=CH—CH_2—Cl	69	289	Ph—CH_2—Cl	70	293
C_2H_5—H	98	410	CH_3—CH_3	88	368	Ph—CH_3	102	427
C_2H_5—Cl	81	339	C_2H_5—CH_3	85	356			

The first step in the radical substitution reactions of hydrocarbons is the removal of a hydrogen atom by the radical attack on the C—H bond. The H—halogen bond occurs as a result of breaking. When we look at the formation enthalpies of the bonds formed and broken, we see that the reaction of fluorine with alkanes is too violent to be useful or to be controlled. The $\Delta H°$ values for chlorine and bromine show why alkenes undergo bromination and chlorination reactions. On the other hand, the $\Delta H°$ value for iodination explains why the iodine radical is the least reactive of the halogen radicals and cannot abstract hydrogen from alkanes.

The chlorination of higher molecular alkanes gives many monosubstituted halogenated products. The chlorine atom is so reactive that it is not selective in the substitution reaction. For example, the chlorination of isopentane gives four different products. Isopentane contains primary, secondary, and tertiary carbon atoms. Hydrogens do not all exhibit equal reactivity. The C—H bonds' reactivity decreases in the following order: 3 > 2 > 1. The expected product, 2-chloro-2-methyl-butane, is formed in only 22% yield. However, the same compound's bromination is much more selective and even more surprising, and 2-bromo-2-methylbutane is formed in 92% yield as the major product. These results strongly suggest that tertiary hydrogens are inherently more reactive than secondary and primary hydrogens.

Aromatic compounds bearing alkyl groups react easily with halogens under radical conditions, and substitution products are formed at the side chains. The benzylic C—H bonds are weaker than the most sp³-hybridized C—H bonds (85 kcal/mol, 356 kJ/mol). This is because the radical formed is resonance stabilized. For example, when toluene is mixed with bromine and reacted under thermal or photochemical conditions, the methyl group is easily brominated. However, there is no substitution of the benzene ring under the radical bromination conditions.

First, heat or UV light breaks the bromine homolytically to generate two bromine radicals and start the chain reaction (initiation step). A bromine radical abstracts the benzylic proton to form a benzyl radical and HBr. The benzyl radical stabilized by resonance abstracts a bromine atom from another molecule of Br₂ to give benzyl bromide and another bromine radical, which continues the radical chain mechanism (propagation step). The radical chain mechanism of benzylic bromination can be terminated by various reactions between the possible pairs of radicals (termination step).

After examining the halogenation of aromatic compounds' benzylic positions, we turn to the halogenation of allylic systems. We cannot apply the same procedure to allyl systems. When the halogens are mixed with compounds having C=C double bonds, the expected reaction will be halogens' addition to the double bond, forming vicinal dihalides. *How can we solve the problem and favor the allylic substitution over halogenation?* When the halogen concentration is kept low enough, alkenes can be halogenated at the allylic position rather than undergo addition at the double bond. Then, we may consider the question, "*How can we generate a low concentration of bromine?*"

N-bromosuccinimide (NBS) is an essential reagent for bromination at the allylic positions [5–7]. Mechanistic investigations revealed that Br₂ is the activating reagent for bromination. NBS is always contaminated with trace HBr as well as bromine. When NBS is dissolved in a solvent with an alkene, a trace amount of bromine will be formed. Heat or light (or radical initiators) will dissociate the weak halogen bond to generate bromine radicals to start the chain reaction. In the propagation steps, a bromine radical formed abstracts a hydrogen from the allylic position of the alkene to form the resonance-stabilized allyl radical and HBr. The allyl radical abstracts a bromine atom from another molecule of Br₂ to form the desired product and generate a bromine radical, which continues the chain. Termination occurs between the reactions of possible pairs of radicals. These reactions, allylic bromination of alkenes using NBS and a radical initiator, are called *Wohl–Ziegler* reactions.

As the bromine concentration remains at a very low level, no addition to the double bond is observed. Furthermore, the stability of the allyl radical provided by delocalization of the radical is the reason that the substitution at the allylic position is favored over addition to the double bond. Supposing the bromine radical adds to the double bond, forming a carbon radical, it cannot take a hydrogen from HBr as its concentration is low, and the radical reverts to the alkene and the bromine radical. Because of the low concentration of HBr (it reacts immediately with NBS), polar addition of HBr to the double bonds is not observed either.

The carbon radicals generated by cleavage of carbon–halogen bonds can be useful in the synthesis of monohalocyclopropane derivatives. As discussed in the carbene section (Section 8.2.1), monohalocarbenes are difficult to synthesize by α-elimination. For the synthesis of compounds derived from a monohalocarbene, a dihalocarbene is first added to an alkene. Then, one of the halogen atoms is reduced. For this process, tri-n-butyltin hydride is used in the presence of a radical initiator. The radical initiator, the azoisobutyronitrile radical, starts the reaction by abstracting a proton from tri-n-butyltin hydride, generating tri-n-butyltin radical. The tin radical removes one of the bromine atoms from the dibromocyclopropane. The cyclopropyl radical formed abstracts a hydrogen from tri-n-butyltin hydride, generating bromocyclopropane and tri-n-butyltin radical, continuing the chain [8].

9.1.2.2 Radical Combination and Disproportionation

One of the most important steps of radical chain reactions is the termination reaction. The most common termination processes are radical combination and disproportionation. The collision of two radicals can result in two different reactions. Both radicals combined form a new bond; this reaction is called *radical combination*. A second alternative is that one radical is reduced by removing a hydrogen from the other, while two electrons in the other molecule combine to form a double bond. In this case, one molecule is reduced while the other is oxidized. This reaction is called *radical disproportionation*.

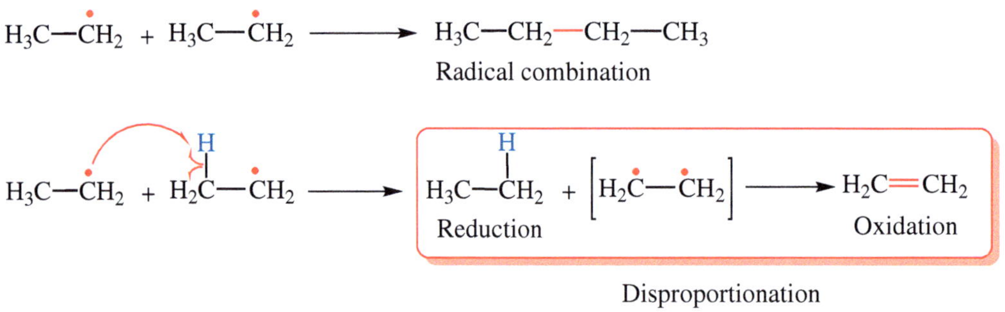

$$H_3C-\overset{\bullet}{C}H_2 + H_3C-\overset{\bullet}{C}H_2 \longrightarrow H_3C-CH_2-CH_2-CH_3$$
Radical combination

$$H_3C-\overset{\bullet}{C}H_2 + H_2\overset{\bullet}{C}-CH_2 \longrightarrow H_3C-CH_2 + [H_2\overset{\bullet}{C}-\overset{\bullet}{C}H_2] \longrightarrow H_2C=CH_2$$
Reduction   Oxidation

Disproportionation

9.1.2.3 Kolbe Electrolysis (Kolbe Reaction)

Radicals can also be obtained by the electrochemical oxidative decarboxylation of carboxylic acid salts. Carboxylate ions transfer an electron to the anode in which carboxylate radicals are formed. As a result of the decarboxylation of the carboxylate radicals formed, alkyl radicals are formed. Because this reaction takes place around the anode where the concentration of radicals formed is high, they easily dimerize. This process is called the *Kolbe electrolysis* or the *Kolbe reaction* [9–11]. It is best applied to the synthesis of symmetrical dimers; however, unsymmetrical dimers can also be synthesized.

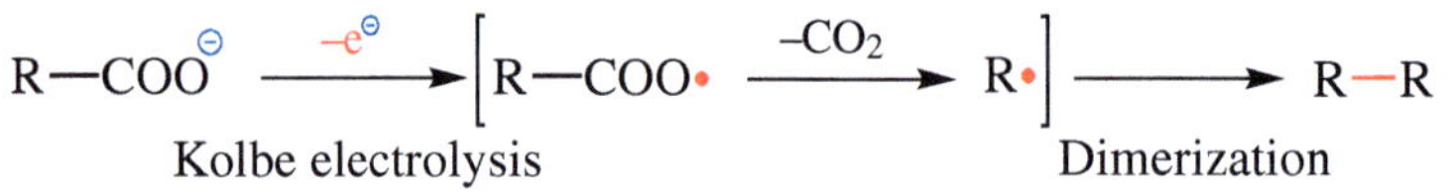

$$R-COO^{\ominus} \xrightarrow{-e^{\ominus}} [R-COO\bullet] \xrightarrow{-CO_2} R\bullet \longrightarrow R-R$$
Kolbe electrolysis   Dimerization

The radical that occurs as a result of the oxidative decarboxylation of carboxylic acid can undergo an intermolecular reaction as well as an intramolecular reaction, as shown below. The radical formed during the Kolbe electrolysis intramolecularly adds to the double bond, forming a new radical. This radical undergoes a reaction with the methyl radical, formed by the oxidative decarboxylation of acetic acid, by an intermolecular reaction to form a perhydroazulene skeleton [12].

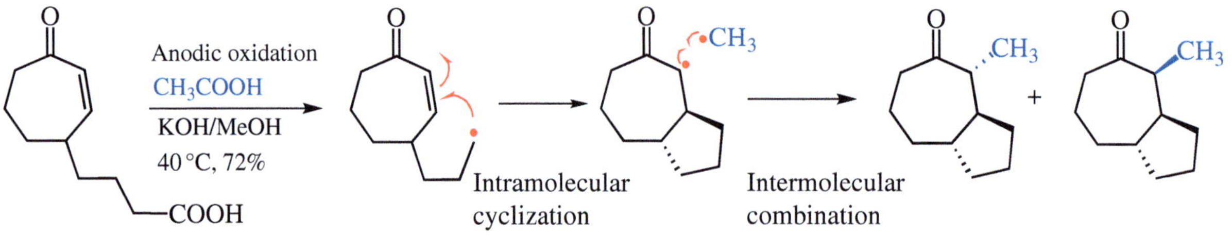

The electrochemical decarboxylation of acids, when applied to vicinal acids, causes the resulting radicals to be in the vicinal position and so the combination of radicals easily forms double bonds.

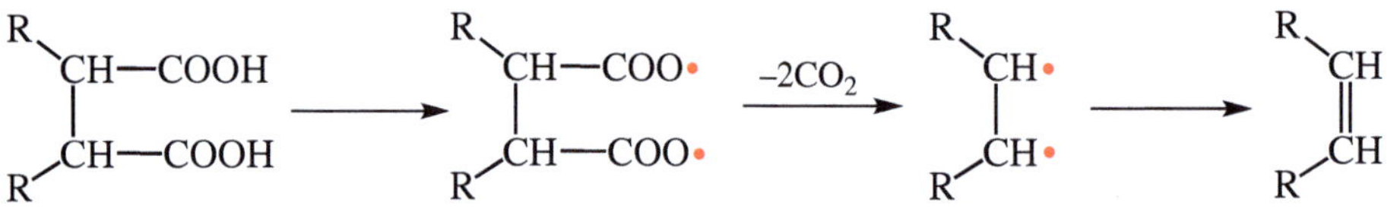

The vicinal diradical derived from dicarboxylic acids can also be generated starting from the corresponding peresters. For example, the treatment of the acyl chlorides of vicinal dicarboxylic acids with *t*-butyl hydroperoxide furnishes the corresponding di-*t*-butyl peresters, which are then decomposed either photochemically or thermally to provide the corresponding alkene. Because the peresters contain weak oxygen–oxygen bonds, they readily undergo homolytic oxygen–oxygen bond cleavage when heated and form biscarboxylate radicals. These radicals transform into the corresponding alkyl radicals by removal of CO_2. The combination of radicals gives the alkene.

Masamune succeeded in the synthesis of basketene by application of this methodology [13]. Cyclooctatetraene is in equilibrium with the valence isomer bicyclooctatriene. Although the equilibrium is mostly shifted to the cyclooctatetraene side, maleic anhydride adds to the diene unit of bicyclooctatriene to form a tricyclic compound. Photolysis, followed by hydrolysis of anhydride, gives a diacid. Treatment of the acyl chlorides of dicarboxylic acids with *t*-butyl hydroperoxide furnishes di-*t*-butyl peresters, providing basketene upon heating or photolysis. Furthermore, treatment of the dicarboxylic acids with lead tetraacetate (Pb(OAc)$_4$) [14] or anodic oxidation [15] also affords basketene.

9.1.2.4 Hunsdiecker Reaction

The *Hunsdiecker reaction* involves the decomposition of the silver(I) salts of carboxylic acids in the presence of halogens, such as bromine or iodine, in an inert solvent to form an unstable intermediate, which readily decarboxylates thermally to give aryl or alkyl halides.

$$R-COOAg \xrightarrow[CCl_4]{Br_2} R-Br + AgBr + CO_2$$

Hunsdiecker reaction

Alexander Porfiryevich Borodin (1833–1887), a Russian chemist and well-known composer, was the first to demonstrate the formation of methyl bromide from silver acetate in the presence of bromine [16]. Later, the German chemists Heinz Hunsdiecker (1904–1981) and his wife Cläre Hunsdiecker (1903–1995) improved Borodin's reaction and developed a general method to produce organic halides starting from carboxylic acids [17], and the reaction was named after them.

In terms of the reaction mechanism, the Hunsdiecker reaction is similar to the Kolbe electrolysis. The critical difference between the two reactions is that the Kolbe reaction is initiated electrochemically while the Hunsdiecker reaction is initiated chemically [18, 19]. Carboxylic acids must first be converted into their silver salts. Silver carboxylate is heated in CCl$_4$ along with bromine to start the reaction. The silver ions break the Br—Br bond heterolytically and the stable silver bromide precipitates. The remaining positively (+) charged bromine ion binds to the carboxylic acid's oxygen atom, forming acyl hypobromite. Homolysis of the weak oxygen–bromine bond results in the formation of a carboxylate radical and a bromine radical. Decarboxylation of the carboxylate radical forms an alkyl radical, which combines with the bromine radical to form alkyl bromide.

When the silver salt of picolinic acid reacts with bromine in hot nitrobenzene, the expected bromopyridine is formed along with 2,2′-dipyridine [20]. The formation of this dimer indicates that the reaction proceeds through radical intermediates because 2,2′-dipyridine can only occur as a result of dimerization of the pyridyl radicals.

Various modifications of the classical Hunsdiecker reaction have been developed. However, it remains challenging to apply the Hunsdiecker reaction to carboxylic acid attached to a vinyl group or an alkyne. Vinyl bromides are obtained by microwave irradiation of α,β-unsaturated carboxylic acids in the presence of NBS and a catalytic amount of lithium acetate for 1–2 minutes [21].

Recently, a practical method for decarboxylative bromination of sterically hindered carboxylic acids was developed using commercially available (diacetoxyiodo)benzene, (PhI(OAc)$_2$), and potassium bromide [22, 23].

9.1.2.5 Radical Addition to Alkenes

Synthetic polymers are essential in daily life. They can be synthesized by polymerization of monomers. Chain growth polymerization proceeds via different mechanisms: cationic, anionic, and radical polymerization mechanisms.

In radical polymerization, a radical generated by a radical initiator adds to a double bond of a monomer, forming a new radical. This radical adds to the double bond of another monomer to propagate the chain. This process is repeated over and over. The polymerization stops if the radical is terminated. For termination, two chains can combine, the radical can undergo disproportionation somehow, or the radical can abstract a hydrogen from the environment.

Anti-Markovnikov addition of HBr to olefins in the presence of alkyl peroxides proceeds through radical intermediates. This topic was discussed in detail in the section "Addition Reactions." Therefore, we will not discuss this type of addition here.

In recent years, intramolecular radical addition reactions have attracted enormous interest as a synthetically useful method for constructing ring systems. If the radical formed and the double bond to which the radical will add are in close proximity, it is possible to control the ring closure reactions with very high efficiency.

The hex-5-enyl radical is one of the most studied radicals in recent years in terms of cyclization reactions. Once a radical is generated, it can react with a double bond in an intramolecular fashion to yield cyclized radical intermediates. If the double bond is asymmetric, there are two paths to follow in the intramolecular cyclization. In that case, the formation of either a five-membered or a six-membered ring is possible. When the radical attacks the terminal carbon atom, C1, forming a secondary radical in a six-membered ring, this process is called *endo cyclization*; however, when the radical attacks the carbon atom C2, forming a five-membered ring with a primary radical, this cyclization is called *exo cyclization*. Instead of a thermodynamically stable six-membered ring, a five-membered ring is generally produced [24–26].

At first glance, it is thought that *exo* cyclization may not be possible for two reasons. The occurrence of a five-membered ring and the formation of a less stable primary radical may be sufficient reasons to prevent this addition. *Endo* cyclization can be preferred because of the formation of both a strain-free six-membered ring and a secondary radical. However, experiments show that this is not the case; *exo* cyclization is preferred over *endo* cyclization, even though it results in less stable primary radical formation.

In the transition complex, the radical center, the SOMO, interacts with the double bond's antibonding π* orbital. An ideal interaction occurs when the molecule forms a transition complex in the chair conformation, in which case the radical prefers an exocyclic addition. For the formation of a six-membered ring, a boat-shaped transition complex must be formed, which increases the system's energy. These results are in agreement with the theoretical calculations, which suggest that *exo* cyclization proceeds through a chair-like transition state, while *endo* cyclization involves a boat-like transition state.

Steric factors play an important role in radical cyclization reactions. Two different quinolone derivatives give *endo* or *exo* cyclization products depending on the alkyl chain's substituents. In the following example, in the presence of dimethyl groups, the *exo* cyclization product is formed in a yield of 76%, while in the absence of methyl groups, the *endo* cyclization product occurs in 99% yield [27]. When this reaction is carried out in the presence of an optically active catalyst, chiral compounds with an optical yield of 94–99% from the prochiral starting compound can also be synthesized.

Radicals can also intramolecularly add to the carbonyl groups, forming the corresponding cycloalkanols. However, if an aldehyde group is present in the molecule beside the vinyl group, cycloalkanol and methyl cycloalkane formation will be competitive. In the system given below, the radical adds to the vinyl group and the carbonyl group (C=O), forming 5-exocyclic products in a ratio to 3 : 1 [28].

The example below involves cyclization of the enamine group with an aromatic ring. The radical formed by treatment of enamide with Bu₃SnH in the presence of 1,1-azobis(cyclohexanecarbonitril) (ABCN) in toluene generates an aryl radical, which undergoes cyclization to give an *exo* cyclization product along with a small amount of *endo* cyclization product. On the other hand, the second enamide undergoes *endo* aryl radical cyclization to give a benzazepine derivative exclusively [29, 30]. These results indicate that the position of the carbonyl group directs the course of the cyclization.

9.1.2.6 Manganese(III)-Mediated Oxidative Radical Additions to Alkenes

It has been known for a long time that manganese(III) acetate (Mn(OAc)₃ in acetic acid at reflux temperature converts olefins to γ-lactones [31–38].

The manganese radical formed in this reaction first abstracts a proton from acetic acid and forms the carboxymethyl radical, which adds to the double bond, forming a new radical. The fate of this radical has been a topic of debate for a long time. This radical can be reductively or oxidatively terminated. Heiba and Dessau [37] proposed oxidation of the radical, followed by cyclization to the tetrahydrofuran (THF) derivative (route B). On the other hand, Fristad and coworkers [39, 40] proposed an alternative route A, in which the radical formed undergoes a cyclization reaction first, followed by oxidation.

To address this question, namely, at which stage the oxidation occurs, possible intermediates were incorporated in a bicyclic system, benzobarrelene. It is well established that a radical type **A** does not undergo rapid rearrangement. However, a carbocation incorporated in a benzobarrelene system, structure **B**, has a great tendency to form rearranged products. Therefore, the formed products' structures would give information about the stage at which oxidation takes place.

9.1 Structure of Radicals and Their Stability | 481

No tendency for rearrangement   Great tendency for rearrangement

Balcı and coworkers [41, 42] reacted benzobarrelene with a 1,3-dicarbonyl compound because removing a hydrogen atom is much easier than using monocarbonyl compounds. Because rearranged products are generally formed, it was concluded that any radical formed after the addition is oxidized before cyclization occurs. Likely, a nonclassical carbocation is first produced, which induces Wagner–Meerwein rearrangement in the bicyclic system, leading to the formation of the rearranged products.

Corey and Kang [43] applied the intramolecular cyclization reaction to synthesize a tricyclic lactone in a single step. The diacyl radical formed in the first step adds to the double bond in cyclohexene. It generates a new radical, which undergoes further cyclization with carboxylic acid to form the tricyclic lactone.

9.1.2.7 Birch Reduction

The reduction of benzene (and its aromatic relatives) to cyclohexa-1,4-diene and its derivatives using sodium (lithium) in liquid ammonia is called the *Birch reduction* [44, 45]. This reaction was first discovered by the Australian chemist Arthur John Birch (1915–1995) in 1944 [46].

Cyclohexa-1,4-diene

When sodium or lithium is dissolved in liquid ammonia at −78 °C, a dark blue solution of cations and solvated electrons is formed. The reaction mixture also contains an alcohol required as a proton donor that protonates the anionic intermediate. In the first step, an electron is transferred into the antibonding π orbital of the benzene ring, yielding a radical anion. The alcohol present in the reaction medium protonates the anion, generating a cyclohexadienyl radical. This radical is stabilized by resonance. By transferring a second electron, a cyclohexadienyl anion is formed, which can be protonated by the alcohol to give cyclohexa-1,4-diene. Surprisingly, the thermodynamically more stable compound, cyclohexa-1,3-diene, is not formed and by considering the resonance structures we can understand why. The electron–electron repulsion in the radical anion will be minimal when separated as much as possible in a 1,4-relation. The product is formed from that intermediate.

When substituents are attached to the aromatic ring, they control the position of the 1,4-double bonds. The determining factor for the substituents is the electron-donating or electron-withdrawing properties of the substituents [47].

For example, if an electron-withdrawing group, such as carboxylic acid, is attached to the aromatic ring, the result would theoretically be two different products. The acid group can be attached to a saturated carbon atom or a double bond carbon atom in the reduced product. When we analyze these two products from a thermodynamic perspective, the product bearing the carboxyl group at the double bond should be more stable because of the conjugation. However, this product is not formed. The carboxylic acid in the resulting product is attached to a saturated carbon atom. The thermodynamic stability of possible products does not always determine the product that will result from the reaction. The stability of the radical anion, formed from electron transfer to the aromatic ring, is essential. It is possible to write various resonance structures for radical anions. Among them, it is necessary to consider the structure of the intermediate product, which is the most stable. If the carbanion is found on the carbon atom to which the carboxylic acid is attached, it is stabilized because it will conjugate with the carbonyl group. This means that the intermediate will remove hydrogen from the alcohol molecule at this location. Thus, it is understood that the C3-substituted product will preferably occur.

With electron-donating groups, such as alkoxy or alkyl groups, the protonation occurs on the carbon atom adjacent to the carbon bearing the alkoxy or alkyl group; the substituent will be attached to one of the double bonds. The resonance structures of radical anions again will determine which product will preferably be formed. The resonance structures of the radical anion formed by the reduction of anisole are given below. Suppose the carbanion electrons are on the carbon atom to which the OCH$_3$ group is attached. In that case, there will be repulsion between the nonbonding electrons on the oxygen atom and the carbanion electrons. Therefore, the resonance structure in which the carbanion electrons are far from the oxygen atom will be more stable. The hydrogen abstraction will take place from that resonance structure. The reduction product becomes a vinyl-substituted product.

The Birch reaction conditions cannot reduce isolated double bonds. However, if a double bond is conjugated with an aromatic ring, the alkene double bond can be reduced.

There are several modifications to the Birch reduction in the literature. One of them is the reduction of aromatic compounds at room temperature with Li metal in THF or dioxane under ammonia atmosphere [48].

In 1954, Benkeser reported an alternate and simplified version of the Birch reduction. It was observed that instead of liquid ammonia, amines with lower molecular weights could be used to perform the Birch reduction. Lithium or calcium metal was used as a safer alternative to sodium. This modification of the Birch reduction is called *the Benkeser reduction reaction* [49, 50]. Monoamines with lower molecular weight, such as methylamine or ethylamine, are excellent solvents for reduction with lithium. Calcium metal has also been applied extensively besides lithium as a reducing reagent.

For example, when naphthalene is reduced in liquid ammonia, both rings are reduced to form tetrahydronaphthalene (isotetraline). On the other hand, application of the Benkeser conditions reduces naphthalene mainly to 9,10-octalin. Benzene was reduced in ethylamine to a mixture of cyclohexene and cyclohexane [49].

The Birch reduction also plays an essential role in the total synthesis of natural products. The radical carbanion that occurs as an intermediate product in the Birch reduction can be captured with protons. Interesting cyclohexadiene derivatives can be synthesized if this carbanion is captured with other electrophiles instead of a proton. For example, in the biaryl compound given below, one of the aromatic rings is reduced under the Birch condition. The electrophile $I(CH_2)_3N_3$ regiospecifically traps the resulting carbanion. The product formed is used in the synthesis of the alkaloid vindoline [51].

9.1.2.8 Di-π-methane Rearrangement

The di-π-methane rearrangement is a photochemical reaction of molecules that contain two vinyl groups separated by an sp³ hybridized carbon atom to form vinyl-substituted cyclopropanes. This reaction was first discovered by an American chemist Howard E. Zimmerman (1926–2012) in 1976 [52].

Later studies showed that this reaction is general; even when one of the vinyl groups is incorporated into a benzene ring, the molecule is rearranged to form benzene-substituted cyclopropanes [53, 54].

484 | *9 Reactive Intermediates: Radicals and Singlet Oxygen*

The mechanism of formation of vinyl cyclopropanes is depicted below. Upon irradiation of the molecule, one of the double bonds of the diene system is first excited. The excited electrons show a diradical property. One of the electrons adds to the opposite double bond, forming a 1,4-cyclopropane diradical. Now, one of the C—C bonds in the cyclopropane ring can be opened homolytically. If the newly formed bond (C2—C4) is opened, the molecule reverts to its former state. However, the situation is different if one of the other bonds (C2—C3 or C3—C4) is opened. One of these bonds will preferably be opened depending entirely on the electronic structures of the substituents attached to the cyclopropane ring. In the case of a symmetrical 1,4-diradical, it is not so important which bond will be opened. As the bond opens homolytically, electrons on neighboring carbon atoms instantly form a double bond, generating a new 1,3-diradical. The combination of electrons forms vinyl cyclopropane and the reaction ends.

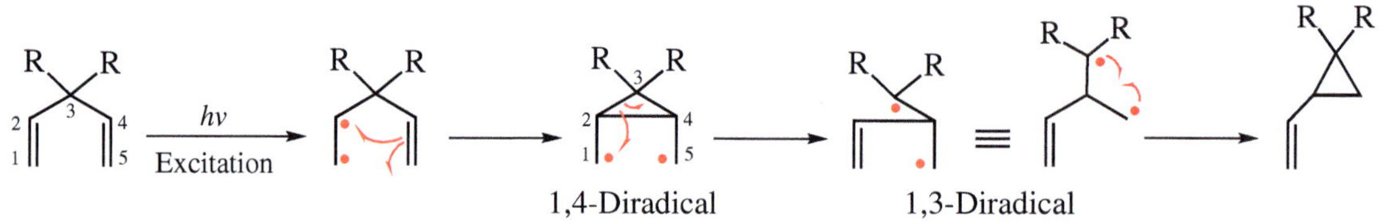

Bicyclic systems with rigid geometry, such as barrelene, are ideal for di-π-methane rearrangement. The barrelene system rearranges efficiently to semibullvalene upon irradiation. One of the most interesting properties of barrelene is that it has two sp³-hybridized carbon atoms and three C=C double bonds attached to sp³-hybridized carbon atoms. Theoretically, there can be six different types of vinyl–vinyl bridging required for di-π-methane rearrangement.

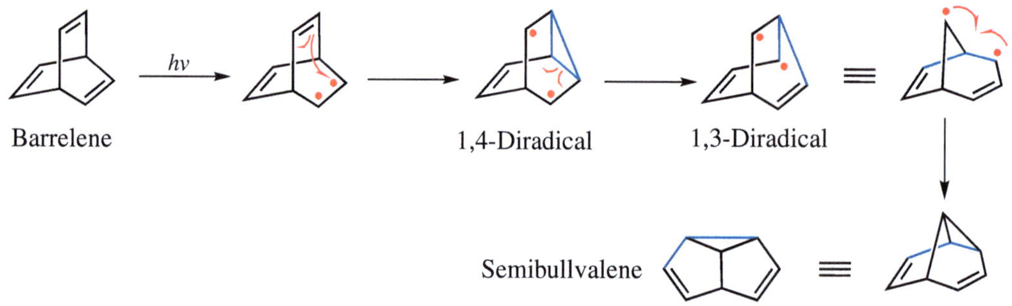

It is not possible to distinguish between these different bridging modes. When the symmetry in the barrelene system is distorted, different bridging will result in different products. Two types of bridging are possible in benzobarrelene: vinyl–vinyl bridging and benzo–vinyl bridging. Benzosemibullvalene is formed as a result of both different types. Zimmermann and coworkers replaced the hydrogen atoms at the bridgehead with deuterium. Photolysis of the labeled benzobarrelene gave deuterated benzosemibullvalene with deuterium labels in the benzylic positions, strongly suggesting that vinyl–vinyl bridging is the main pathway in the photorearrangement of benzobarrelenes [55].

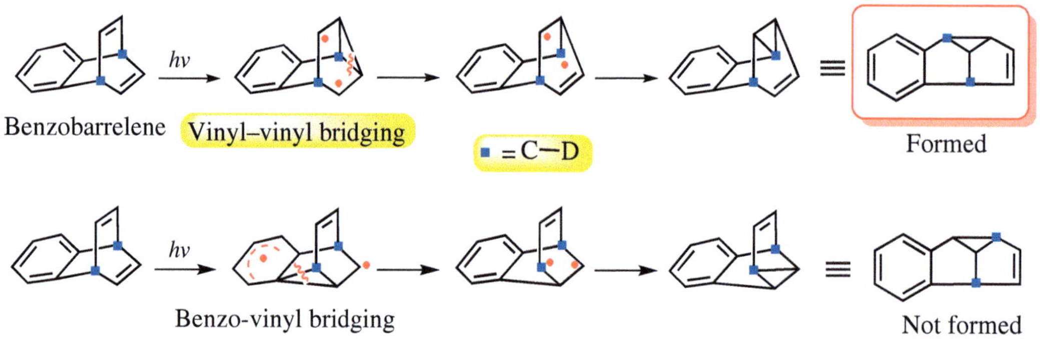

The introduction of a substituent on the vinyl group in benzobarrelene destroys the symmetry, increasing the possible initial bonding from two to six. Nevertheless, the di-π-methane rearrangement generally shows excellent regioselectivity. For example, the photolysis of cyanobenzobarrelene furnishes the only benzosemibullvalene in which the cyano group is bonded to the cyclopropane ring [56, 57]. Mechanistic studies revealed that this reaction proceeds via vinyl–vinyl bridging between the less substituted carbon atoms because of the good stabilization of the diradical intermediate by the cyano group.

However, the destabilization of the intermediate by the substituent (above on the left side) has not been considered. In a recent study, Balcı and coworkers [58, 59] showed that the destabilization of the cyclopropane ring plays an essential role in determining the reaction mode as much as the stabilization effect of the diradical intermediates.

Photolysis of dicyanobenzobarrelene gave two different di-π-methane rearrangement products [60]. The major product (82% yield) is formed by benzo–vinyl bridging. Zimmerman and coworkers proved by labeling experiments that vinyl–vinyl bridging is the major process for transforming benzobarrelene to benzosemibullvalene. However, contrary to this finding, the major product is formed via benzo–vinyl bridging. This can be explained by the destabilization effect of the adjacent bond in an initially formed cyclopropane ring. It is likely that the destabilization effect of the cyclopropane ring by the cyano group overwhelms the radical stabilization effect of the cyano group.

How can a cyano group destabilize the formation of a cyclopropane ring? When an electron-withdrawing group such as a nitrile is attached to the cyclopropane ring, electron transfer occurs from the cyclopropane highest occupied molecular orbital (HOMO) (Walsh orbital) to the nitrile lowest unoccupied molecular orbital (LUMO). Because the bond between C2 and C3 in the cyclopropane ring is antibonding, electron transfer to the nitrile group strengthens this bond, while the bonds between C1 and C2 and C1 and C3 weaken after electron transfer (Figure 9.2) [61–63]. This shows that the cyano group will weaken the vicinal bonds of the cyclopropane ring formed by vinyl–vinyl bridging.

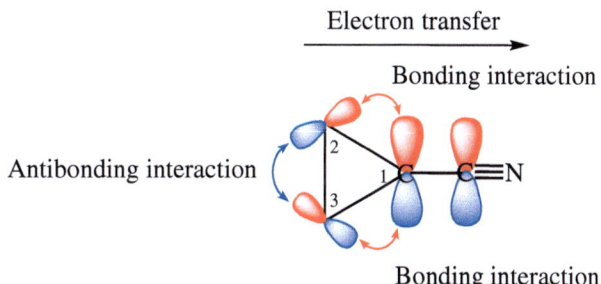

Figure 9.2 Interaction between the HOMO of cyclopropane (Walsh orbital) and the LUMO of nitrile.

In di-π-methane systems, one of the vinyl groups attached to the sp^3 carbon atom can also be replaced by a carbonyl group. For example, when a benzocyclohexenone derivative is irradiated, it transforms into a cyclopropane compound with high efficiency [64]. A similar mechanism can explain the formation of this compound. The rearrangement reactions of systems in which one of the vinyl groups is the carbonyl group are called *oxa-di-π-methane rearrangements*.

9.1.2.9 Diradicals Derived from Diazo Compounds

Diradicals [65] can be synthesized by various methods. Photochemical or thermal decomposition reactions of cyclic diazo compounds yield hydrocarbons via diradical intermediates. The thermal and photochemical eliminations of nitrogen are considered to be general reactions of cyclic diazo compounds. In this section, we will briefly focus on the radicals that are synthetically important and formed as a result of irradiation or heating of diazo compounds. For example, pyrazole and similar compounds can easily generate 1,3-, 1,4-, and 1,5-diradicals depending on the structure of the diazo compound [66, 67].

1,3-, 1,4-, and higher diradicals can easily be synthesized starting from 4,5-dihydro-3H-pyrazole or higher homologs, which can be obtained by the cycloaddition reaction of 4-phenyl-1,2,4-triazoline-3,5-dione (PTAD) with dienes or bicyclic alkenes. Adducts are hydrolyzed with KOH in isopropyl alcohol, followed by oxidation to give cyclic diazo compounds. The diazo compounds can be very stable as well as extremely unstable depending on the structure of the molecule. Diradicals formed can undergo various reactions, such as fragmentation, coupling, and cyclization.

The diradicals obtained by this method have various synthetic applications and play an important role in elucidating the mechanistic details of the di-π-methane rearrangement. For example, the photolysis of benzonorbornadiene affords a benzotricyclic hydrocarbon [68, 69]. First, a 1,4-diradical is formed by initial benzo–vinyl bridging, which rearranges to a 1,3-diradical. The collapse of this radical furnishes the benzotricyclic hydrocarbon.

Erden and Balcı [70, 71] succeeded in generating the 1,3-diradical and transformed it into the same benzotricyclic hydrocarbon obtained by the photolysis of benzonorbornadine.

PTAD undergoes a cycloaddition reaction with benzonorbornadiene to give a rearranged cycloadduct. Reflux of the adduct with KOH in isopropyl alcohol followed by treatment with $CuCl_2$ in methanol affords a diazoalkane. Direct photolysis of the diazoalkene gives mainly the benzotricyclic hydrocarbon and some benzonorbornadiene. The result of this experiment supported the formation of a 1,3-diradical in the di-π-methane rearrangement of benzonorbornadiene.

The addition of PTAD to a diene system, followed by di-π-methane rearrangement and generation of a diazo compound, was successfully applied to the synthesis of benzopinene derivatives by Balcı and coworkers [72]. The diene system was reacted with PTAD to give a cycloaddition product. Photolysis of the resulting benzonorbornadiene derivative furnished the desired rearranged product. Hydrolysis of the rearranged product, followed by oxidation and denitrogenation, gave a benzopinene derivative.

9.2 Singlet Oxygen

Although singlet oxygen was first recognized in 1924, its chemistry started to develop after the 1960s. Singlet oxygen is a high-energy oxygen molecule and is classified as a reactive intermediate. It is vital in terms of both chemical reactions and biochemical transformations. To better understand the reactivity and reactions of singlet oxygen, the electronic structures of the oxygen molecule should be carefully studied.

9.2.1 The Electronic Configuration of the Oxygen Molecule

The properties and behavior of oxygen reflect its unique electronic structure. Unlike many molecules, the electronic configuration of the oxygen molecule in the ground state is a triplet. To better understand this configuration, let us examine oxygen's molecular orbitals in the ground state. The atomic and molecular orbitals of oxygen are shown in

Figure 9.3. In this diagram, only the s and p orbitals of the second shell are represented. The 2s atomic orbital is lower in energy than the 2p atomic orbitals. The combination of 2s atomic orbitals generates two molecular orbitals: the bonding σ_{2s} and antibonding σ_{2s}^*. The oxygen molecule (O_2) has 16 electrons. The four electrons of the total of 12 in the second shell populate these two σ molecular orbitals. The two p orbitals can overlap in a parallel fashion, generating the bonding orbitals π_{2px} and π_{2py} and the antibonding orbitals π_{2px}^* and π_{2py}^*. The remaining pair of p orbitals generates the bonding orbital σ_{2pz} and the antibonding orbital σ_{2pz}^*. As the interaction between the $2p_z$ orbitals (head-on interaction) is stronger than that between the other p orbitals, the energy of the formed molecular orbital σ_{2pz} is lower than that formed by a combination of the p_x and p_y orbitals.

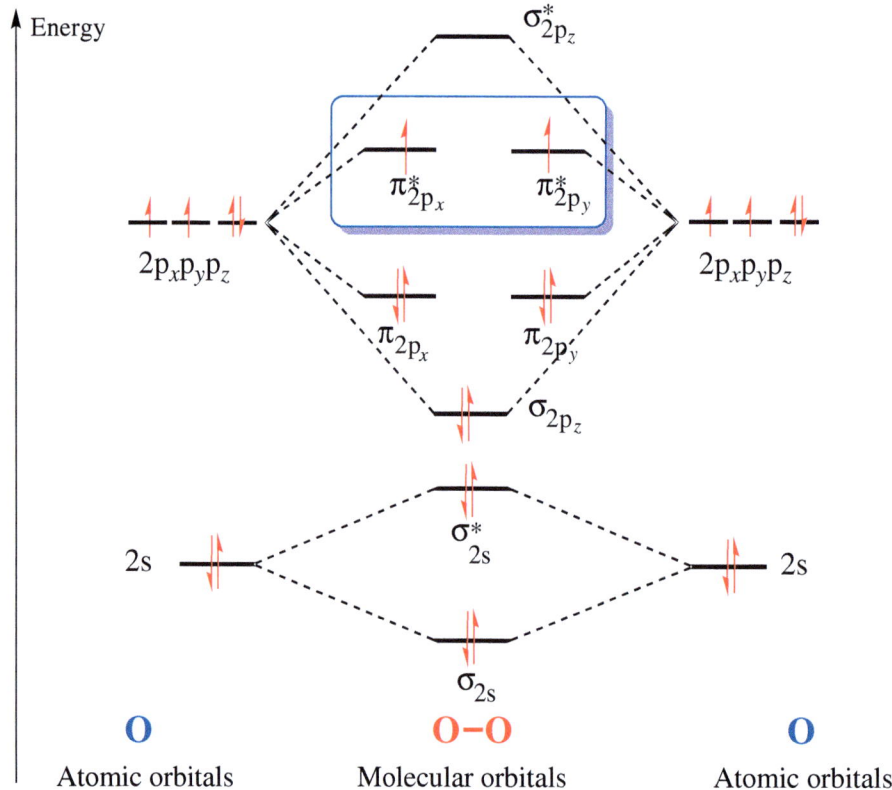

Figure 9.3 Schematic presentation of atomic and molecular orbitals of oxygen.

A total of eight electrons in the p orbitals of the two atomic oxygens must be placed in the six molecular orbitals that are formed. If starting from the orbital with the lowest energy, two electrons populate the σ_{2pz} orbital and four electrons the π_{2px} and π_{2py} orbitals. There are two degenerate antibonding orbitals. The last two electrons populate the degenerate π_{2px}^* and π_{2py}^* antibonding orbitals with parallel spins and qualify ground-state oxygen as a diradical. Because the spins are in the parallel direction in the ground state, the spin multiplicity will be ½ + ½ = 1; the ground state configuration of oxygen molecule is a triplet (as the multiplicity, $M = (2\times 1)+1 = 3$) and it is called *triplet oxygen* ($^3O_2$). The oxygen molecule is highly reactive, and it exhibits radical-like behavior in chemical reactions.

Let us take a closer look at the orbitals shown in the blue frame in Figure 9.3 to examine other electronic configurations of oxygen more closely. The lowest electronic state of oxygen is a triplet ground state ($^3O_2$) with two unpaired electrons with parallel spin distributed in the π* orbitals. The oxygen molecule has two excited electronic configurations (Figure 9.4). If two electrons are paired in a single orbital, leaving the other degenerate orbital vacant, the corresponding oxygen molecule is called *singlet oxygen* ($^1O_2$), denoted by $^1\Delta$ (the preceding superscript "1" indicates a singlet state). The total spin will be $I = ½ + -½ = 0$; $M = (2\times 0)+1 = 1$. Therefore, it is called singlet oxygen and it is 22 kcal/mol (94 kJ/mol) higher in energy than the triplet ground state of oxygen. In the liquid phase, energy transfer occurs faster because of the collision of molecules and the lifetime of singlet oxygen decreases to 10^{-3} seconds. This time is sufficient for chemical reactions to take place. Oxygen with a higher energy level and having electrons in different orbitals with antiparallel spins is $^1\Sigma$ singlet oxygen that is 37.5 kcal/mol (157 kJ/mol) higher energy than the ground state. Because the lifetime of $^1\Sigma$ singlet oxygen is very short, 10^{-9} seconds, this time is not sufficient for chemical reactions.

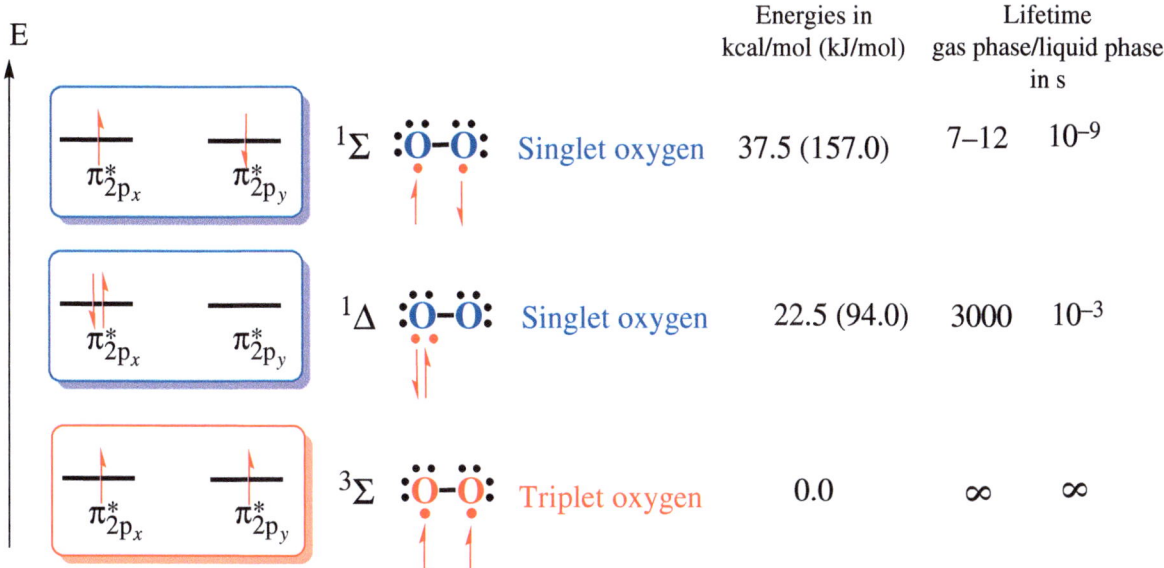

Figure 9.4 Electronic configuration of singlet and triplet oxygens and some physical properties.

9.2.2 Singlet Oxygen Generation

Singlet oxygen can be produced by chemical methods and photosensitization. If singlet oxygen is to be used as a reactant in a chemical reaction, it must be generated in the reaction medium because of the very short lifetime of singlet oxygen.

9.2.2.1 Generation of Photosensitized Singlet Oxygen

Direct irradiation of the oxygen molecule cannot generate singlet oxygen. Singlet oxygen can be produced in the presence of photosensitizers [73–76]. Irradiation of the oxygen molecule in the presence of organic dyes as sensitizers results in the generation of singlet oxygen. These sensitizers are generally compounds with triplet energies between 30 and 70 kcal/mol. The structures of some sensitizers are given below.

Tetraphenylporphine   Rose bengal   Eosin Y   Methylene blue

For generation of singlet oxygen, it requires only oxygen, light of an appropriate wavelength, and a photosensitizer capable of absorbing the light. While oxygen gas is passed through the reaction medium containing the sensitizer and reactant, the reaction apparatus is irradiated from the inside or outside. Upon absorption of light by the sensitizer, one of the electrons in the uppermost orbital passes to the next upper orbital, maintaining its spin direction, forming an excited singlet state of the sensitizer ($^1S^*$). The excited sensitizer can return to the ground state by two different mechanisms. The first one is the emission of light. This process is an allowed transition because there is no spin change. Emission of an electromagnetic wave is called *fluorescence*. As the lifetime of the electron is around 10^{-8} to 10^{-9} seconds, it returns almost immediately to the ground state.

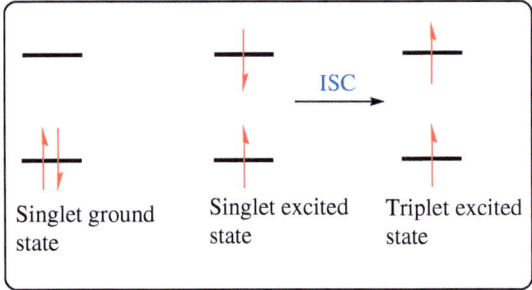

By the second mechanism, an excited singlet state of a molecule will form an excited triplet state. Actually, this process is a forbidden process because of the spin transition, and it is called *intersystem crossing*. Such a transition is possible if heavy atoms are present in the molecule, environment, or solvent molecules. A molecule excited at the triplet state can revert to the ground state by emitting light. This phenomenon is called *phosphorescence*. Sensitizers in the triplet state are chemically highly reactive because of their unpaired electrons and long relaxation times. The excited sensitizer in the triplet state can react with ground-state triplet oxygen ($^3O_2$), which forms the excited singlet oxygen by an electron transfer mechanism (Figure 9.5). The energy transfer occurs during collision of the excited state of the sensitizer with triplet oxygen.

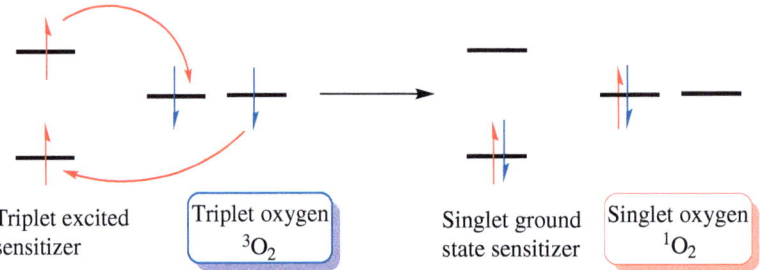

Figure 9.5 Singlet oxygen generation by electron transfer mechanism.

9.2.2.2 Chemical Sources for Singlet Oxygen

There are several methods for singlet oxygen generation by chemical reactions. Because the photosensitized generation of singlet oxygen is generally useful in singlet oxygen production, we will not argue about chemical methods. However, we will not continue without mentioning a few of those methods.

Hydrogen Peroxide–Hypochlorite System Chemically singlet oxygen can be generated in large concentrations (>10%) using a reaction of hydrogen peroxide with hypochlorite ions. In 1927, Mallet [77] described weak *chemiluminescence* that occurred in the reaction of hydrogen peroxide with hypochlorite ion. Later, Khan and Kasha [78] and then Arnold et al. [79] determined that the emission bands arose from the electronically excited oxygen molecule. Foote and Wexler [80] determined that the products obtained by reacting the oxygen formed as a result of this reaction with dienes were the same products obtained with singlet oxygen by photosensitization. Labeling experiments showed that singlet oxygen arose from the hydrogen peroxide, as shown below [81].

$$H_2O_2^* + ClO^\ominus \longrightarrow HOO^{*\ominus} + HOCl \qquad \boxed{O = {}^{18}O}$$
$$HOO^{*\ominus} + HOCl \longrightarrow HOOCl^* + HO^\ominus$$
$$HOOCl^* + HO^\ominus \longrightarrow ClOO^{*\ominus} + H_2O$$
$$ClOO^{*\ominus} \longrightarrow Cl^\ominus + {}^1O_2^* \qquad \text{Singlet oxygen}$$

Decomposition of the Triphenyl Phosphite–Ozone Complex The triphenyl phosphite–ozone complex can be synthesized by passing ozone into a methylene chloride solution of triphenyl phosphite at −78 °C [82, 83]. This complex is not stable and decomposes at temperatures over −20 °C and releases singlet oxygen according to the mechanism given below. The advantage of this method is that it is possible to obtain a higher concentration of singlet oxygen.

$$(PhO)_3P \xrightarrow[-70\,°C]{O_3} (PhO)_3P\begin{array}{c}O-O\\|\ \ \ |\\O-O\end{array} \xrightarrow{>-20\,°C} (RO)_3P\begin{array}{c}O^\oplus\\|\\O-O^\ominus\end{array} \longrightarrow (RO)_3P{=}O + {}^1O_2^*\ \text{Singlet oxygen}$$

Decomposition of Aromatic Endoperoxides It has been known for a long time that some aromatic compounds, such as rubrene and anthracene derivatives, undergo cycloaddition reactions with singlet oxygen to furnish the corresponding bicyclic aromatic endoperoxides. It is also well established that these peroxides undergo dissociation on heating to release singlet oxygen and the parent aromatic hydrocarbon. In 1926, Moreau and coworkers reported that heating of rubrene endoperoxide generates singlet oxygen and hydrocarbon [84].

Rubrene endoperoxide $\xrightarrow{\Delta}$ Rubrene + ${}^1O_2^*$ Singlet oxygen

Later, Wassermann and coworkers [85] showed that a similar decomposition of endoperoxides derived from anthracene furnishes singlet oxygen. This method was used as a singlet oxygen source in many chemical reactions. Anthracene endoperoxide is stable and when heated under reflux in benzene, it releases singlet oxygen. Naphthalene endoperoxide [86] behaves the same way and also releases singlet oxygen at lower temperatures. However, naphthalene endoperoxide cannot be obtained by adding singlet oxygen to naphthalene.

Aromatic endoperoxides can also be considered as singlet oxygen-storing compounds. The advantage of the method is that singlet oxygen can be generated in high concentration without using a light source. As will be discussed in the next chapter, the oxygen–oxygen bond is one of the weakest bonds in organic chemistry. While the oxygen–oxygen bond is expected to break homolytically at first upon heating of these endoperoxides, the carbon–oxygen bond, which is preferably more stable, is broken. This behavior of endoperoxides can only be observed in systems in which stable aromatic compounds are formed as a result of decomposition.

Calcium Peroxide Diperoxohydrate Recent studies show that calcium peroxide diperoxyhydrate ($CaO_2 \cdot 2H_2O_2$) is an environmentally friendly generator of singlet oxygen. The compound to be reacted with singlet oxygen is dissolved in a solvent such as methanol or THF, and $CaO_2 \cdot 2H_2O_2$ is added. When the reaction mixture is heated to 50 °C, the resulting singlet oxygen forms [2 + 2], [4 + 2] cycloaddition and/or ene-reaction products, depending on the system to be reacted [87].

[Bis(trifluoroacetoxy)iodo]benzene and Hydrogen Peroxide Recently, Kılıç and coworkers demonstrated that treating hypervalent iodine with hydrogen peroxide produces singlet oxygen ($^1O_2$) [88, 89]. Mechanistic studies revealed hydroperoxyl radicals produced from hydrogen peroxide, as the intermediate dimerizes to tetraoxidane. The tetraoxidane decomposes to singlet oxygen and hydrogen peroxide. Chemical reactions established the generation of singlet oxygen.

9.2.3 Reactions of Singlet Oxygen

The chemical reactions of singlet oxygen ($^1\Delta$) are very different from those of triplet oxygen ($^3\Sigma$). In triplet oxygen, two electrons are located in different orbitals with parallel spin, whereas in singlet oxygen, electrons are located in a single orbital with antiparallel spin. The HOMO of singlet oxygen can be compared with the HOMO of a double bond. Therefore, singlet oxygen often acts as a double bond in chemical reactions. Because of its electrophilic nature, singlet oxygen readily undergoes ene and cycloaddition reactions with C=C double bonds.

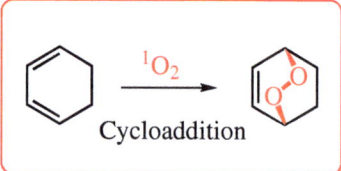

Because the lifetime of singlet oxygen is very short in the liquid phase (10^{-3} seconds), it must be formed in the reaction medium. The solvent to be chosen must be capable of dissolving both the compound to be reacted and the sensitizer to be used. The lifetime of singlet oxygen varies according to the solvent chosen [90]. The most extended lifetimes are observed in perhalogenated solvents. On the other hand, the lifetime of singlet oxygen decreases on increasing the number of hydrogen atoms in the solvent molecule, having OH groups, and heavy atoms. Solvent deuteration increases the lifetime of singlet oxygen (Table 9.2).

Table 9.2 Lifetime of singlet oxygen in various solvents.

Solvent	t (µs)	Solvent	t (µs)
H_2O	4.5	C_6H_{12}	17
D_2O	58	$CHCl_3$	60 ± 15
CH_3OH	23	$CDCl_3$	300 ± 100
CD_3OD	224	C_6H_6	31
C_2H_5OH	10	C_6D_6	700
C_2D_5OD	230	CCl_4	700 ± 200
CH_3COCH_3	26	CS_2	200 ± 60

9.2.3.1 Ene Reactions

The ene reaction is a chemical reaction between an alkene having allylic hydrogen and a compound having multiple bonds in which the double bond shifts and a new C—H and C—C σ bonds are formed [91–93]. This reaction is also known as the *Alder-ene reaction* and was discovered by Alder in 1943 [94].

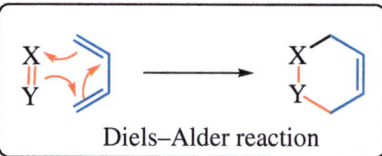

Singlet oxygen also undergoes the ene reaction. The ene reaction with singlet oxygen was initially discovered by Schenck in 1953 and named the *Schenck reaction* [95]. For example, the reaction of singlet oxygen with 1-methyl-cyclopent-1-ene gives two ene products, as major products, when the proton abstraction takes place from the allylic protons located in the five-membered ring [96, 97]. The exocyclic olefin was only formed in 4% yield. Steric and electronic factors directly affect regioselectivity.

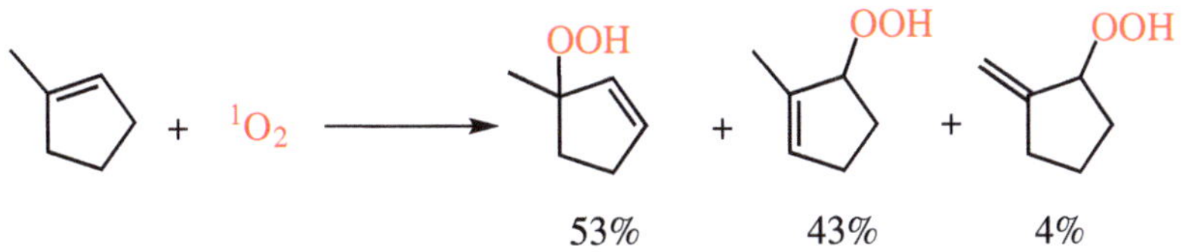

A wide variety of reaction mechanisms are proposed for the ene reaction. The mechanism may differ depending on the nature of the substrate. Four different mechanisms are generally considered: concerted, perepoxide, zwitterion, and diradical mechanisms. The concerted mechanism and the perepoxide mechanism are the most common. The geometry of the allylic hydrogen abstracted by the singlet oxygen molecule is very important. If the hydrogen to be abstracted is perpendicular to the double bond plane, it is very likely to react.

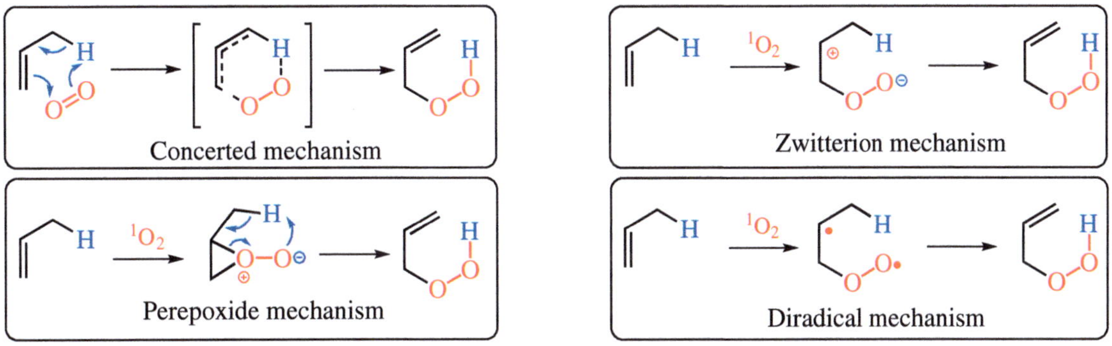

Photooxygenation of asymmetric olefins shows that the hydrogen abstraction mainly occurs from the groups in the *cis* position. The numbers on the formulas shown below give the yields of the products formed as a result of the hydrogen removed from that location. The preferred cleavage of hydrogens from the *cis* position is called a *cis effect* [93].

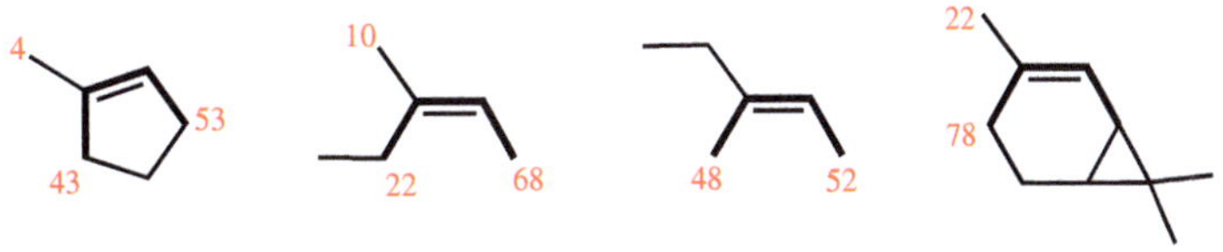

9.2.3.2 [2 + 2] Cycloaddition Reactions

Singlet oxygen undergoes a [2 + 2] cycloaddition reaction with alkenes not containing allylic hydrogen to give 1,2-dioxetane [98]. If there are electron-donating groups such as –OR, –SR, and –NR$_2$ attached to the double bond, dioxetane formation is preferred because of the increased electron density. However, the first 1,2-dioxetane was synthesized by treatment of β-halohydroperoxide either with a silver salt or with a base. β-Halohydroperoxide can be obtained by the reaction of alkenes with 1,3-dibromo-5,5-dimethylhydantoin in the presence of hydrogen peroxide [99].

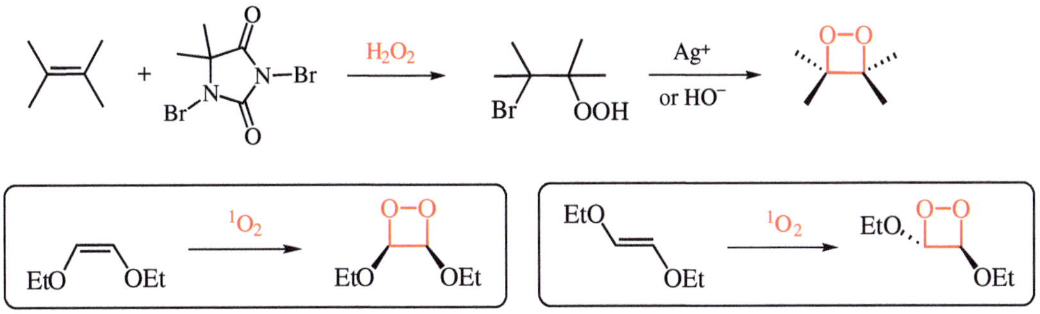

The stereoselectivity observed in addition reactions and the fact that radical inhibitors do not inhibit the reactions reveal that no radicals are involved during the addition reaction. A concerted cycloaddition mechanism is consistent with the retention of the substituents' configuration by the addition of singlet oxygen to *cis*- and *trans*-1,2-diethoxyethane [100].

The thermal suprafacial–suprafacial addition is known to be prohibited for the [2 + 2] cycloaddition reaction. According to the Woodward–Hoffmann rules, the oxygen molecule must approach the double bond in a suprafacial–antrafacial fashion, as shown in Figure 9.6. This topic is discussed in detail in the Woodward–Hoffmann rules section.

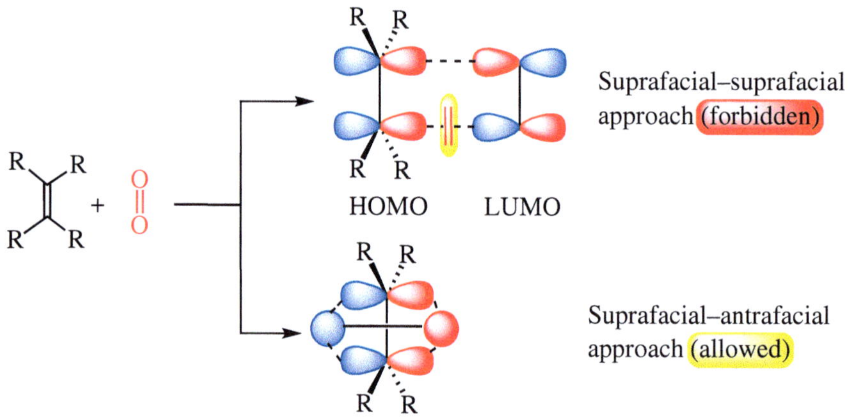

Figure 9.6 Singlet oxygen's approach to an alkene double bond.

In addition to a concerted dioxetane formation mechanism, the perepoxide mechanism is also emphasized. For example, adamantylideneadamantane reacts with singlet oxygen to afford the unusually stable 1,2-dioxetane (melting point 174–176 °C) [101]. The mechanism for the formation of the dioxetane was a subject of considerable controversy. For the mechanism of formation, it was predicted that a perepoxide is formed first and it rearranges to dioxetane. When the photooxygenation reaction was carried out in the presence of phenyl methyl sulfoxide in a ratio of 1 : 1, it was observed that the epoxide and methyl phenyl sulfoxide were formed with very high yields without dioxetane. This finding shows that the rearrangement of the initially formed perepoxide is slowed by adamantyl groups' steric hindrance to allow time for oxygen transfer to methyl phenyl sulfoxide.

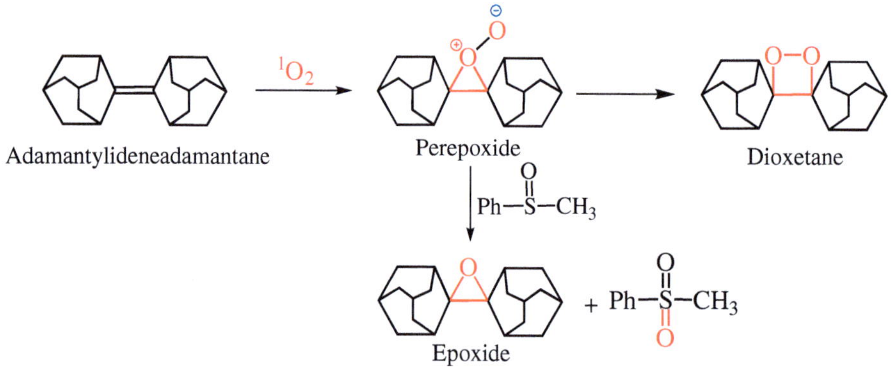

Chemiluminescence is the production of light from a chemical reaction. The most spectacular types of chemical reactions of dioxetanes are those that produce light upon decomposition in carbonyl compounds. If the process takes place in a living organism, it is called *bioluminescence*. The method by which fireflies produce light is perhaps the best-known example of bioluminescence. Dioxetanes are generally converted into ketones or esters quantitatively when heated to 60–70 °C. It was observed that 3,3,4,4-tetramethyl-1,2-dioxetane, as pale yellow crystals, when heated to 50 °C decomposed smoothly with the emission of blue light [101–103]. We may raise the question *Why do dioxetanes emit light*? To emit light, there must be sufficient energy to generate an electronically excited molecule. The energy of the lowest excited states of simple carbonyl compounds lies in the range 75–85 kcal/mol (314–356 kJ/mol) for singlet and 60–78 kcal/mol (251–326 kJ/mol) for triplet excitation. Therefore, at least 75–85 kcal/mol energy must be available in the transition state during the decomposition of the dioxetane so that one of the carbonyl groups can be excited [104]. The reaction enthalpy for the decomposition of dioxetane is around 63 kcal/mol (264 kJ/mol) and the activation energy 27 kcal/mol (113 kJ/mol). That means the total energy (63 + 27 = 90 kcal/mol (377 kJ/mol)) released in the thermolysis of dioxetanes is enough to excite one of the carbonyl groups. The excited singlet and triplet acetone emit their excess energy as electromagnetic waves. This phenomenon is called chemical irradiation, chemiluminescence (Figure 9.7).

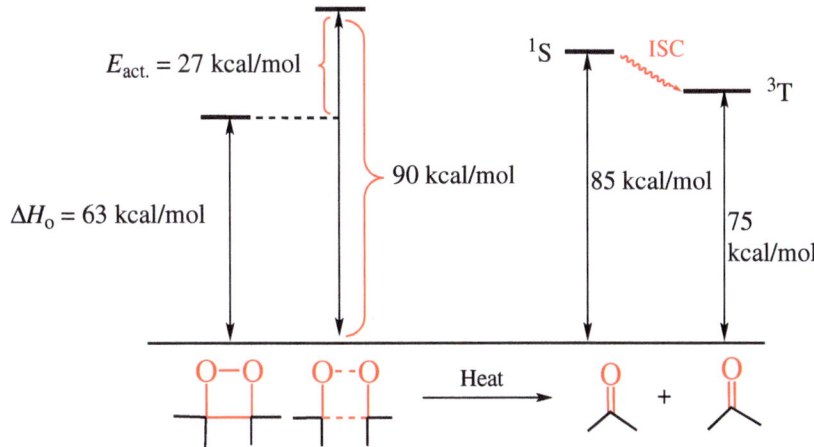

Figure 9.7 Energy diagram for the generation of electronically excited singlet and triplet acetone in the thermolysis of tetramethyldioxetane.

9.2.3.3 [4 + 2] Cycloaddition Reactions

The 1,4 cycloaddition of singlet oxygen to a diene results in the formation of a cyclic endoperoxide. If the reacting diene is a cyclic diene, such as 1,3-cyclohexadiene, then *bicyclic endoperoxides* are formed. As the singlet oxygen approaches the diene unit from one side of the double bond, the configuration of the oxygen atoms is *cis*.

Bicyclic unsaturated endoperoxides are molecules with very high synthetic potential [105–108]. With the transformation of the peroxide functional group, compounds of various functionalities can be synthesized. One of the common reactions of unsaturated bicyclic endoperoxides is the cleavage of the weak oxygen–oxygen bond upon thermolysis or photolysis, followed by the addition of the oxygen radicals to the adjacent double bond to furnish bisepoxides with *syn* configuration [109].

Another feature of endoperoxides is that the peroxide linkage undergoes a ring-opening reaction in the presence of a base to give unsaturated hydroxy ketones. Although there is no certainty about the mechanism of formation, an accepted mechanism is suggested for some systems [110, 111]. The first step of the reaction includes abstraction of the bridgehead hydrogen to give an intermediate carbanion, which cleaves the oxygen–oxygen bond, forming a carbonyl group. Proton transfer to the alkoxide completes the reaction. During this reaction, the configuration of the hydroxyl groups is preserved.

The peroxide linkage in endoperoxides is highly susceptible to reductive cleavage by a variety of reductants. Therefore, it is not surprising that in the catalytic hydrogenation of unsaturated endoperoxides, the peroxide bond and the double bond are reduced. These functional groups can also be reduced in a controlled manner. Reagents such as LiAlH$_4$ or thiourea ((H$_2$N)$_2$C=S) selectively reduce the O–O bond, forming 1,4-unsaturated diols in the *cis* configuration [112]. As a result of the reaction, the configuration of the carbon–oxygen bonds is preserved. Diimide selectively reduces the carbon–carbon double bond, resulting in saturated bicyclic endoperoxides [113]. Interestingly, the weak oxygen–oxygen bond is preserved during this reaction. In the reaction of endoperoxides with triphenylphosphine, one of the oxygen atoms is removed and epoxy olefins are formed [114].

9.2.3.4 Bicyclic Endoperoxides in Synthesis

Even though singlet oxygen is a short-lived metastable excited state of molecular oxygen, it is a useful reagent for the synthesis of various compounds with complicated structures. By using the controlled reaction of singlet oxygen, it can be applied to the synthesis of various natural products [106] and especially to the synthesis of polyhydroxy compounds [115].

4,5-Dimethylenecyclohex-1-ene was subjected to a photooxygenation reaction to introduce oxygen functionalities. The endoperoxide obtained underwent with a second equivalent singlet oxygen an ene reaction to form a mixture of hydroperoxides with 1,3-diene structures. Further addition of singlet oxygen to the diene units resulted in the formation of tricyclic hydroperoxides having three oxygen molecules in their structures. Cleavage of the oxygen–oxygen bonds furnished a carbasugar [116].

Tropone and tropolone represent the key structural elements in a wide range of natural products, many of which display essential biological activities. Recently, bicyclic endoperoxides derived by the cycloaddition of singlet oxygen to cycloheptatriene derivatives have been applied to synthesize some tropone and tropolone derivatives [117].

References

1 Karplus, M. (1959). *J. Chem. Phys.* 30: 15.
2 Fessenden, R.W. and Schuler, R.H. (1963). *J. Chem. Phsy.* 39: 2147.
3 Bernardi, F., Cherry, W., Shaik, S., and Epiotis, N.D. (1978). *J. Am. Chem. Soc.* 100: 1352.
4 Walton, J.C. (1992). *Chem. Soc. Rev.*: 105.
5 Bloomfield, G.F. (1944). *J. Chem. Soc.*: 114.
6 Gosselain, P.A., Adam, J., and Goldfinger, P. (1956). *Bull. Soc. Chim. Belg.* 65: 533.
7 Day, J.C., Lindstrom, M.J., and Skell, P.S. (1974). *J. Am. Chem. Soc.* 96: 5616.
8 Menapace, L.W. and Kuivila, H.G. (1964). *J. Am. Chem. Soc.* 86: 3047.
9 Moeller, K.D. (2000). *Tetrahedron* 56: 9527.
10 Kurihara, H., Fuchigami, T., and Tajima, T. (2008). *J. Org. Chem.* 73: 6888.
11 Vijh, A.K. and Conway, B.E. (1967). *Chem. Rev.* 67: 623.
12 Matzeit, A., Schaefer, H.J., and Amatore, C. (1995). *Synthesis* 11: 1432.
13 Cain, E.N., Vukov, R., and Masamune, S. (1969). *Chem. Commun.*: 98.
14 Masamune, S., Cuts, H., and Hogben, M.G. (1966). *Tetrahedron Lett.* 10: 1017.
15 Radlick, P., Klem, R., Spurlock, S. et al. (1968). *Tetrahedron Lett.* 12: 5117.
16 Borodin, A. (1861). *Justus Liebigs Ann. Chem.* 119: 121.
17 Hunsdiecker, H. and Hunsdiecker, C. (1942). *Chem. Ber.* 75: 291.
18 Johnson, R.G. and Ingham, R.K. (1956). *Chem. Rev.* 56: 219.
19 Kong, Z., He, L., Shi, Y. et al. (2020). *Heliyon* 6: e03446.
20 Kuffner, F. and Russo, C. (1954). *Monatsh. Chem.* 85: 1097.
21 Das, J.P. and Roy, S. (2002). *J. Org. Chem.* 67: 7861.
22 Watanabe, A., Koyamada, K., Miyamoto, K. et al. (2020). *Org. Process Res. Dev.* 24: 1328.
23 Moriarty, R.M. (2005). *J. Org. Chem.* 70: 2893.
24 Beckwith, A.L.J. (1981). *Tetrahedron* 37: 3073.
25 Beckwith, A.L.J. and Zimmerman, J. (1991). *J. Org. Chem.* 56: 5791.
26 Gilmore, K. and Alabugin, I.V. (2011). *Chem. Rev.* 111: 6513.
27 Dressel, M. and Bach, T. (2006). *Org. Lett.* 14: 3145.
28 Walton, R.A. and Fraser-Reid, B. (1991). *J. Am. Chem. Soc.* 113: 5791.
29 Taniguchi, T., Ishita, A., Uchiyama, M. et al. (2005). *J. Org. Chem.* 70: 1922.
30 Kamimura, A., Taguchi, Y., Omata, Y., and Hagihara, M. (2003). *J. Org. Chem.* 68: 4996.
31 Snider, B.B. (1996). *Chem. Rev.* 96: 339.
32 Iqbal, J., Bhatia, B., and Nayyar, N.K. (1994). *Chem. Rev.* 94: 519.
33 Melikyan, G.G. (1993). *Synthesis*: 833.
34 Mondal, M. and Bora, U. (2013). *RSC Adv.* 3: 18716.
35 Pan, X.-Q., Zou, J.-P., and Zhang, W. (2009). *Mol. Divers.* 13: 421.
36 Demir, A.S. and Emrullahoglu, M. (2007). *Curr. Org. Synth.* 4: 321.
37 Heiba, E.I., Dessau, R.M., and Koehl, W.J. Jr., (1968). *J. Am. Chem. Soc.* 90: 5905.
38 Bush, J.B. Jr., and Finkbeiner, H. (1968). *J. Am. Chem. Soc.* 90: 5903.
39 Fristad, W.E., Peterson, J.R., Ernst, A.B., and Urbi, G.B. (1986). *Tetrahedron* 42: 3442.
40 Fristad, W.E. and Peterson, J.R. (1985). *J. Org. Chem.* 50: 10.
41 Ali, M.F., Çalışkan, R., Şahin, E., and Balcı, M. (2009). *Tetrahedron* 65: 1430.
42 Çalışkan, R., Ali, M.F., Şahin, E. et al. (2007). *J. Org. Chem.* 72: 3353.
43 Corey, E.J. and Kang, M.-C. (1984). *J. Am. Chem. Soc.* 106: 5384.
44 Birch, A. (1996). *J. Pure Appl. Chem.* 68: 553.
45 Rabideau, P.W. and Marcinow, Z. (1992). *Org. React.* 42: 1–334.
46 Birch, A.J. (1944). *J. Chem. Soc.*: 430.
47 Hook, J.M. and Mander, L.N. (1986). *Nat. Prod. Rep.* 3: 35.
48 Altundaş, A., Menzek, A., Demirci-Gültekin, D., and Karakaya, M. (2005). *Turk. J. Chem.* 29: 513.
49 Benkeser, R.A., Robinson, R.E., Sauve, D.M., and Thomas, O.H. (1955). *J. Am. Chem. Soc.* 77: 3230.
50 Dhatrak, N.R. (2019). *Resonance* 24: 735.
51 Casimiro-Garcia, A. and Schultz, A.G. (2006). *Tetrahedron Lett.* 47: 2739.

52 Zimmerman, H.E., Binkley, R.W., Givens, R.S., and Sherwin, M.A. (1967). *J. Am. Chem. Soc.* 89: 3932.
53 Hixon, S.S., Mariano, P.S., and Zimmerman, H.E. (1973). *Chem. Rev.* 73: 531.
54 Zimmerman, H.E. and Armesto, D. (1996). *Chem. Rev.* 96: 3065.
55 Zimmerman, H.E., Givens, R.S., and Pagni, R.M. (1968). *J. Am. Chem. Soc.* 90: 6096.
56 Bender, C.O. and King-Brown, E.H.J. (1976). *J. Chem. Soc. Chem. Commun.*: 878–879.
57 Bender, C.O. and O'Shea, S.F. (1979). *Can. J. Chem.* 57: 2804.
58 Altundaş, R., Daştan, A., Ünaldı, N.S. et al. (2002). *Eur. J. Org. Chem.*: 526.
59 Altundaş, R. and Balcı, M. (1997). *Aust. J. Chem.* 50: 787.
60 Ünaldi, N.S. and Balcı, M. (2001). *Tetrahedron Lett.* 42: 8365.
61 Hoffmann, R. (1970). *Tetrahedron Lett.* 11: 2907.
62 Günther, H. (1970). *Tetrahedron Lett.* 11: 5173.
63 Balcı, M. (1992). *Turk. J. Chem.* 16: 42.
64 Cheung, E., Netherton, M.R., Scheffer, J.R., and Trotter, J. (1999). *Tetrahedron Lett.* 40: 8737.
65 Abe, M. (2013). *Chem. Rev.* 113: 7011.
66 Berson, J.A. (1978). *Acc. Chem. Res.* 11: 446.
67 Meier, H. and Zeller, K.-P. (1977). *Angew. Chem.* 89: 876.
68 Edman, J.R. (1966). *J. Am. Chem. Soc.* 88: 3454.
69 Paquette, L.A., Cottrell, D.M., and Snow, R.A. (1977). *J. Am. Chem. Soc.* 99: 3723.
70 Erden, I. and Balcı, M. (1980). *Tetrahedron Lett.* 21: 1825.
71 Adam, W., De Lucchi, O., and Erden, I. (1980). *J. Am. Chem. Soc.* 102: 4806.
72 Altundaş, A., Akbulut, N., and Balcı, M. (1998). *Helv. Chim. Acta* 81: 828.
73 Wasserman, H.H. and Murray, R.W. (1979). *Singlet Oxygen*, A Series of Monographs, vol. 40. New York: Academic Press.
74 Kearns, D.R. (1971). *Chem. Rev.* 71: 395.
75 Schmidt, R. (2006). *Photochem. Photobiol.* 82: 1161.
76 Nosaka, Y. and Nosaka, A.Y. (2017). *Chem. Rev.* 17: 11302.
77 Mallet, L. and Hebd, C.R. (1927). *Seances Acad. Sci. Ser. C* 185: 352.
78 Kahn, A.M. and Kasha, M.L. (1963). *J. Chem. Phys.* 39: 2105.
79 Arnold, S.J., Ogryzole, E.A., and Witzke, H. (1964). *J. Chem. Phys.* 40: 1769.
80 Foote, C.S. and Wexler, S. (1964). *J. Am. Chem. Soc.* 86: 3880.
81 Maetzke, A. and Jensen, S.J.K. (2006). *Chem. Phys. Lett.* 425: 40.
82 Murray, R.W. and Kaplan, M.L. (1969). *J. Am. Chem. Soc.* 91: 5358.
83 Stephenson, L.M. and McClure, D.E. (1973). *J. Am. Chem. Soc.* 95: 3074.
84 Moreau, C., Dufraisse, C., and Dean, P.M. (1926). *Compt. Rend.* 182: 1440.
85 Wasserman, H.H. and Schefter, J.R. (1967). *J. Am. Chem. Soc.* 89: 3073.
86 Schaefer-Ridder, M., Brocker, U., and Vogel, E. (1976). *Angew. Chem. Int. Ed.* 27: 262.
87 Pierlot, C., Nardello, V., Schrive, J. et al. (2002). *J. Org. Chem.* 67: 2418.
88 Catir, M., Kilic, H., Nardello-Rataj, V. et al. (2009). *J. Org. Chem.* 74: 4560.
89 Kalay, E., Kilic, H., Catir, M. et al. (2014). *Pure Appl. Chem.* 86: 945.
90 Hurst, J.R. and Schuster, G.B. (1983). *J. Am. Chem. Soc.* 105: 5756.
91 Clennan, E.L. and Pace, A. (2005). *Tetrahedron* 61: 6665.
92 Clennan, E.L. (2000). *Tetrahedron* 56: 9151.
93 Stratakis, M. and Orfanopoulos, M. (2000). *Tetrahedron* 56: 1595.
94 Alder, K., Pascher, F., and Schmitz, A. (1943). *Ber. Dtsch. Chem. Ges.* 76: 27.
95 Schenck, G.O., Eggert, H., and Denk, W. (1953). *Justus Liebigs Ann. Chem.* 584: 177.
96 Schulte-Elte, K.H. and Rautestrauch, V. (1980). *J. Am. Chem. Soc.* 102: 1738.
97 Yardımcı, S.D., Kaya, N., and Balcı, M. (2006). *Tetrahedron* 62: 10633.
98 Bartlett, P.D. and Landis, M.E. (1979). *Singlet Oxygen*, A Series of Monographs, vol. 40 (eds. H. Wasserman and R.W. Murray), 173. New York: Academic Press.
99 Kopecky, K.K. and Mumford, C. (1969). *Can. J. Chem.* 47: 709.
100 Schaap, A.P. and Bartlett, P.D. (1970). *J. Am. Chem. Soc.* 92: 6055.
101 Schaap, A.P., Recher, S.G., Faler, G.R., and Villasenor, S.R. (1983). *J. Am. Chem. Soc.* 105: 1691.
102 Turro, N.J. and Lechtken, P. (1972). *J. Am. Chem. Soc.* 94: 2886.

103 Turro, N.J. and Lechtken, P. (1973). *Pure Appl. Chem.* 33 (2–3): 363.
104 Adam, W., Reinhardt, D., and Saha-Möller, C.R. (1996). *Analyst* 121: 1527.
105 Balcı, M. (1981). *Chem. Rev.* 81: 91.
106 Ghogare, A.A. and Greer, A. (2016). *Chem. Rev.* 116: 9994.
107 Balcı, M. (1997). *Pure Appl. Chem.* 69: 97.
108 Gültekin, M.S., Celik, M., and Balcı, M. (2004). *Curr. Org. Chem.* 8: 1159.
109 Adam, W. and Balcı, M. (1980). *Tetrahedron* 36: 833.
110 Kornblum, N. and de La Mare, H.E. (1951). *J. Am. Chem. Soc.* 73: 880.
111 Mete, E., Altundaş, R., Seçen, H., and Balcı, M. (2003). *Turk. J. Chem.* 27: 145.
112 Kaneko, C., Sugimoto, A., and Tanaka, S. (1974). *Synthesis*: 876.
113 Coughlin, D.J. and Salomon, R.G. (1977). *J. Am. Chem. Soc.* 99: 655.
114 Bartlett, P.D., Landis, M.E., and Shapiro, M.J. (1977). *J. Org. Chem.* 42: 1661.
115 Balcı, M., Sütbeyaz, Y., and Seçen, H. (1990). *Tetrahedron* 46: 3715.
116 Baran, A., Aydın, G., Savran, T., and Balcı, M. (2013). *Org. Lett.* 15: 4350.
117 Daştan, A. and Balcı, M. (2006). *Tetrahedron* 62: 4003.

10

Pericyclic Reactions

In organic chemistry, reactions can be classified according to their mechanisms into three groups.

1. Ionic or polar reactions
2. Radical reactions
3. Pericyclic reactions

Ions and radicals occur as intermediates in polar and radical reactions. Solvent polarity significantly affects the reaction. Pericyclic reactions have characteristic features compared to polar and radical reactions.

- The reactions proceed in a concerted manner and bond cleavage and bond forming occur simultaneously in a single step.
- Generally, a cyclic transition complex is formed.
- No intermediates are formed in pericyclic reactions.
- Heat or light is required to start the reactions.
- Reactions are stereoselective.
- Solvent polarity generally does not affect the reaction rate or product distribution.

We will consider four categories of pericyclic reactions:

Electrocyclic reactions: An electrocyclic reaction is a unimolecular process in which a new σ bond across the end of a conjugated π system is formed. During this process, the double bonds are displaced and a cyclic compound is formed containing one more σ bond and one fewer π bond than the reactant. The conversion of 1,3,5-hexatriene to 1,3-cyclohexadiene is one of the most typical examples of these reactions.

The reverse or retroelectrocyclic reaction can also occur. In the reverse reaction, a σ bond of a cyclic compound breaks to form a conjugated product with one more π bond, such as the conversion of 1,3-cyclohexadiene to 1,3,5-hexatriene. An additional example is the opening of a cyclobutene ring by heat or photochemical reaction. As the σ bond in the ring opens, a new double bond is formed.

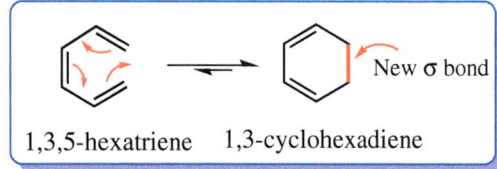
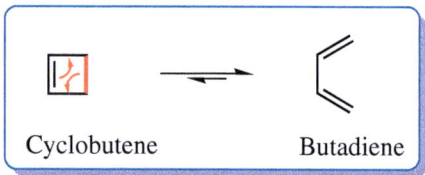

Cycloaddition reactions: The reaction between a conjugated diene and an alkene or alkyne results in the formation of a cyclic compound. Two new σ bonds are formed at each end of both compounds. The number of double bonds in molecules that react is variable. The reaction between butadiene and ethylene is one of the best examples of cycloaddition reactions.

Butadiene ethylene → Cyclohexene (New σ bonds)

Sigmatropic reactions: A sigmatropic reaction is an intramolecular reaction in which one or more σ bonds are broken in the reactant and an equal number of new σ bonds are formed in the product. The π bonds are rearranged, but the number of the double bonds is not changed. For example, the shift of an R group in the alkylmethylene group attached to the butadiene system from the C1 carbon atom to the C5 carbon atom is a sigmatropic reaction. As can be seen from the examples below, the position of the double bonds also changes. The rearrangement of 3-methylhexa-1,5-diene to 1-methylhexa-1,5-diene is also a sigmatropic reaction.

Cheletropic reactions: Small molecules, such as carbon monoxide, sulfur dioxide, and singlet carbenes, having a filled and vacant orbital on the same atom can bond two other atoms. In cycloaddition reactions across the terminal atoms, two new σ bonds will form a single atom. In the reverse reaction, two σ bonds will break from the same atom. Two examples of cheletropic addition and elimination reactions are given below.

Birth of the Woodward – Hoffman rules: Havinga and Schlatmann previously noted that precalciferol underwent electrocyclic ring closure differently depending on the activation mode (photochemical or thermal). According to model studies, heating of the compound resulted in the formation of an expected *trans* ring-closure product. Contrary to what was expected, they determined that irradiation causes a ring-closure reaction with a *cis* configuration of a methyl group and hydrogen atom. These results were puzzling [1].

R = (R)-6-methylhept-2-yl

Precalciferol
previtamin D$_3$

A classic example of a pericyclic reaction is the ring opening of dimethyl *cis*-cyclobut-3-ene-1,2-dicarboxylate, which is an electrocyclic reaction. The stereochemistry of the product formed in a reaction carried out by Vogel [2] was the opposite of what was expected. The *cis* diester isomerizes almost quantitatively to *cis,trans*-muconic acid dimethyl ester already at 120 °C within a few minutes.

Satisfactory rationalization of the striking stereospecificity of these reactions was lacking until the Woodward–Hoffmann rules were formulated. These reactions and some others were fundamental findings for the birth of these rules [3–6].

In electrocyclic ring closures, p orbitals at the ends of a π system interfere and form a new σ bond. In electrocyclic ring-opening processes, the opening of the σ bond must generate the π bond of the new system. As the hybridization of the terminal carbon atoms transforms from sp^2 to sp^3 (in the opening, from sp^3 to sp^2) during the ring-closure process, the orbitals' phases at the ends must also be compatible. The symmetry of the orbitals in the starting compound should also be preserved in the resulting compound. For that reason, the *Woodward–Hoffmann rules* are also called *conservation of orbital symmetry*.

In 1965, American chemist and experimentalist Robert Burns Woodward (1917–1979) and Polish – American theoretical chemist Roald Hoffmann (1937–) published a paper about the conservation of orbital symmetry theory to explain the relationships between the structure and configuration of products depending on the reaction conditions (thermal or photochemical) [7]. Woodward was awarded the Nobel Prize in Chemistry in 1965, while Hoffmann received it in 1981. Kenichi Fukui, a Japanese chemist, was the corecipient of the 1981 Nobel Prize in Chemistry with Hoffmann. Fukui interpreted these concerted reactions in terms of their symmetry properties.

In 2004, American chemist Elias James Corey (1928–), Nobel Prize winner in 1990, published an article in which he wrote: *On May 4, 1964, I suggested to my colleague R. B. Woodward a simple explanation involving the symmetry of the perturbed (HOMO) molecular orbitals for the stereoselective cyclobutene → 1,3-butadiene and 1,3,5-hexatriene → cyclohexadiene conversions that provided the basis for the further development of these ideas into what became known as the Woodward–Hoffmann rules* [8, 9]. (reprinted with permission of American Chemical Society from the Ref. 8.

Furthermore, he states that his name was not included in the first article even though his contribution was significant. Thereupon, a defense article was published by Hoffmann wondering why Corey waited 40 years to make this public [10].

Before moving on to the Woodward–Hoffman rules, we will explain the orbitals, orbital symmetries, and how symmetry is preserved during the ring opening and closing of the bonds.

10.1 Frontier Molecular Orbitals

Chemical reactions take place as a result of electron exchange between molecules. In general, one of the reactants transfers electrons from the highest energy-filled orbital (HOMO) to the other component's empty orbital (LUMO). As new bonds are formed, the phases of the orbitals in the terminal atoms must be compatible. No bond is formed by the interference of orbitals in different phases.

To understand the stereochemistry during ring-closure or ring-opening reactions, we must focus on the p orbitals on the terminal carbons of the HOMO and determine whether like phases of the orbitals are on the same side or opposite sides of the molecule. Remember, the orbitals' arbitrary coloring represents different phases of the orbitals. In the example given below, the like phases of orbitals must interact to generate a σ bond. Therefore, the respective orbitals must rotate around an axis before the σ bond is formed. These rotations can vary according to the phases of the orbitals. If the p orbitals are in the same phase, either the orbital lobes in red or the orbital lobes in blue must intervene for the formation of a σ bond. For this, one p orbital must rotate clockwise and the other one counterclockwise. Two modes of rotation are possible. Both orbitals can rotate in or out. This type of rotational motion, one of the orbitals clockwise and the other counterclockwise, is called *disrotatory rotation* (Figure 10.1).

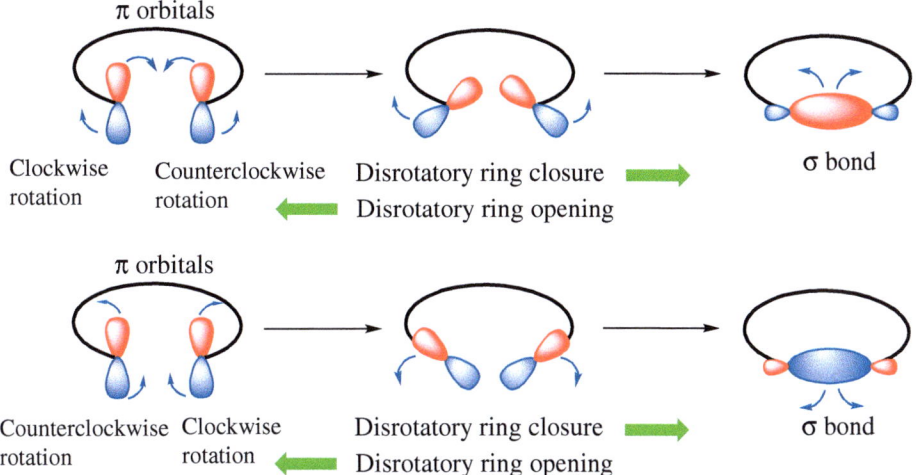

Figure 10.1 Presentation of disrotatory ring-closure and ring-opening reactions using the terminal p orbitals.

When like phases of the p orbitals are on the opposite sides of the molecule, the two p orbitals must rotate either clockwise or counterclockwise in the same direction to form the σ bond. This type of rotational motion is called *conrotatory rotation* (Figure 10.2).

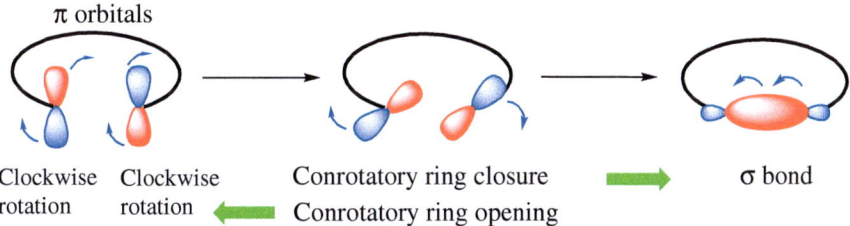

Figure 10.2 Presentation of conrotatory ring-closure and ring-opening reactions using the terminal p orbitals.

These rotations will determine the configurations of the substituents attached to the terminal carbon atoms in the newly formed compound. Therefore, to determine the product's configuration in an electrocyclic reaction, the phases of the orbitals at the ends must be known. Let us try to explain how to determine the phases of these orbitals in the HOMO and LUMO. First, we have to examine the molecular orbitals and their formation.

The molecular orbitals are formed by a combination of atomic orbitals. The number of molecular orbitals is equal to the number of atomic orbitals combined. Let us look at the sideways combination of two atomic p orbitals. The combination of two p orbitals generates two new molecular orbitals, one a low-energy π bonding orbital and the other a high-energy π* antibonding orbital. In the bonding orbital, electrons are concentrated between two nuclei and the lobes of the p orbitals interact constructively with each other (Figure 10.3). Thus, the energy of the system decreases and this makes the molecule stable. In the antibonding π* orbital, one lobe (the red one) of one p orbital interacts destructively with the lobe (the blue one) of the other p orbital, leading to a node between the two nuclei. At a node, there is zero probability of finding an electron. Therefore, the repulsion between the carbon nuclei occurs because of the phase difference and the energy of the system increases.

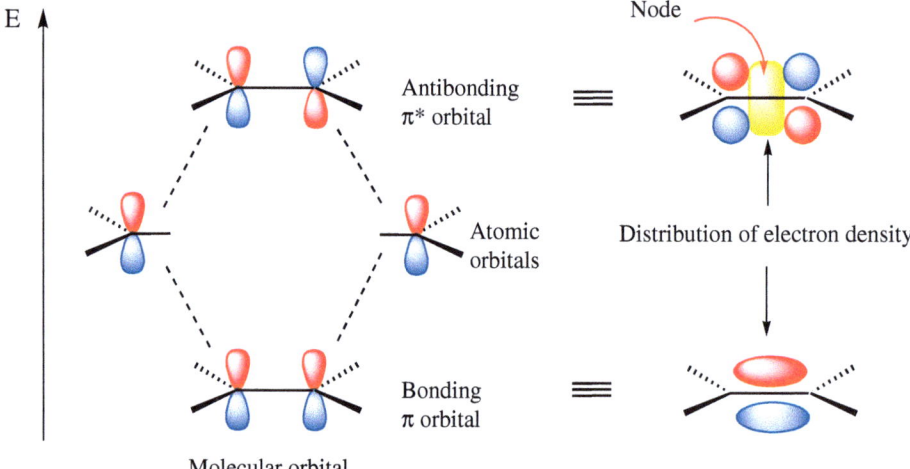

Figure 10.3 Molecular orbitals of ethylene and distribution of electron density.

It is quite simple to determine the orbitals of butadiene, hexatriene, octatetraene, etc. In butadiene, there are four carbon atoms involved in the π system. Combining a p orbital from each atom will generate four molecular orbitals: two bonding orbitals, π_1 and π_2, and two antibonding orbitals, π_3^* and π_4^*. When determining these orbitals, it is necessary to consider the number of nodes. The lowest energy molecular orbital π_1 will have p orbitals with phases in complete alignment with each other and there will be no node. The second-lowest-energy molecular orbital, π_2, will have one node. The trick is in knowing where to place it. In systems with one node, we always place the node in the middle. This requires the orbitals on the left to be in the same phase; the orbitals on the right are also in the same phase (Figure 10.4). These two orbitals, π_1 and π_2, are filled orbitals. Because butadiene has four π electrons, they settle in these orbitals.

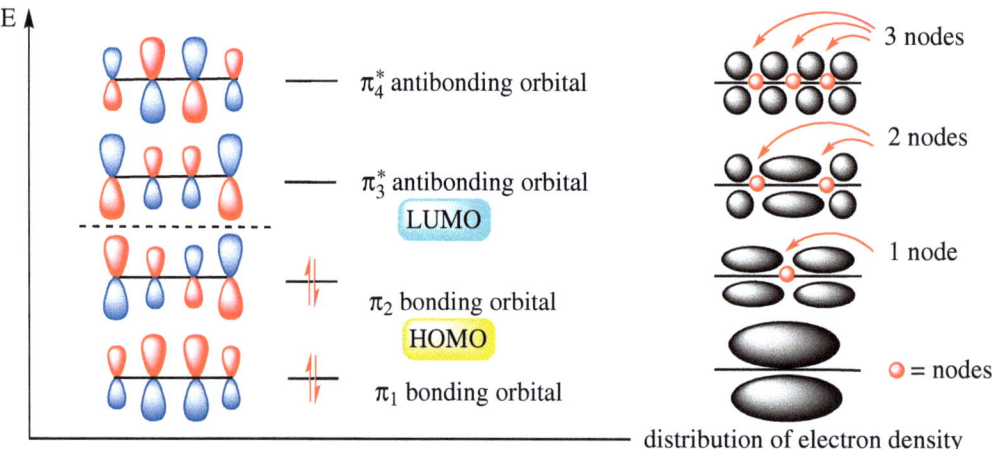

Figure 10.4 Molecular orbitals of butadiene and representation of the electron density distribution.

The empty antibonding orbital π_3^* with the lowest energy level is the LUMO and it has two nodes. *Where to place the nodes?* The general principle is that they are placed symmetrically to the center, as shown in Figure 10.4. The orbital with the highest energy level, π_4^*, has three nodes.

When the molecular orbitals in Figure 10.4 are examined carefully, it can be seen that the orbital lobes are of different sizes. The size of the lobes indicates the probability of finding an electron there. It is also possible to express lobe sizes in numbers (lobe coefficient). These are one of the most critical factors determining regioselectivity in addition reactions. Detailed information about orbital coefficients is given in the Diels–Alder addition section.

The molecular orbitals belonging to cyclohexa-1,3,5-triene are shown in Figure 10.5. In hexatriene, there are six adjacent carbon atoms involved in the π system. The combination of these six p orbitals results in six molecular orbitals. Three of these orbitals are bonding orbitals (π_1, π_2, and π_3), while the other three are antibonding orbitals (π_4^*, π_5^*, and π_6^*) Figure 10.5.

Orbital coefficients are not taken into account in the representation of these orbitals. Of course, the lobe sizes of these orbitals are different, as in butadiene. As can be seen, the orbitals' energies increase in parallel with the increasing number of nodes. Considering the number of nodes, it is possible to draw the orbitals of conjugated systems such as octatetraene and compounds with more π electrons.

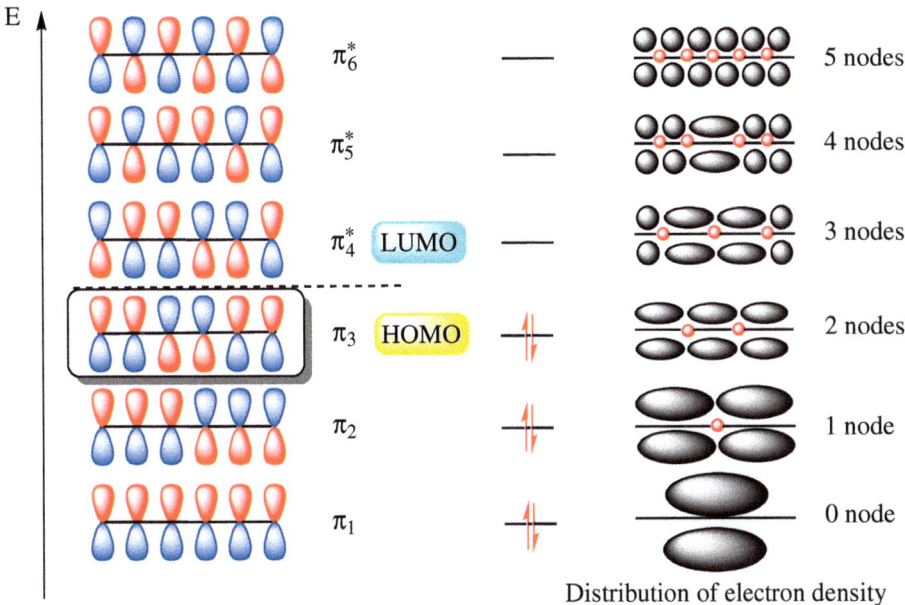

Figure 10.5 Molecular orbitals of hexa-1,3,5-triene and representation of the electron density distribution.

10.2 Electrocyclic Reactions

10.2.1 Thermal Electrocyclic Reactions

The concerted cyclization of a conjugated π electron system to a cycloalkene by forming a σ bond and the reverse reaction, the ring opening of an alkene to a conjugated polyene, are called *electrocyclic reactions*. Two examples, the ring closure of an octatriene system and the ring opening of a cyclobutene derivative, are shown below. To decide whether such a reaction can occur under thermal conditions, we need to look at the orbitals' phases on the terminal atoms. Because chemical reactions proceed through the electrons located in the HOMO, we must consider the phases of the HOMO.

Electrocyclic ring closure   Electrocyclic ring opening

Now let us start to examine the ring-closure reaction of butadiene to cyclobutene. First, we apply a disrotatory rotation of the terminal p orbitals to convert butadiene to cyclobutene. To make a new σ bond, the rotating p orbitals must be in phase. As shown in Figure 10.6, the application of a disrotatory rotation will generate the σ bond out of phase. This kind of overlapping of orbitals will be an antibonding interaction. This process is symmetry forbidden. To achieve in-phase overlapping, we have to apply a conrotatory motion, so that the orbitals making the σ bond will be in phase.

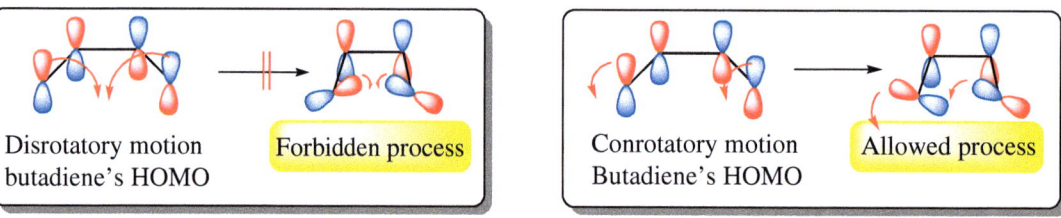

Figure 10.6 Disrotatory and conrotatory ring closure of butadiene.

After examining the disrotatory and conrotatory closure of butadiene, let us explore how these closures affect the substituents' configurations bonded to the terminal carbon atoms. We have seen that the p orbitals on the terminal carbon atoms must rotate in a conrotatory manner to generate a σ bond. For example, *trans,trans*-hexa-2,4-diene and *cis,cis*-hexa-2,4-diene transform into *trans*-3,4-dimethylcyclobut-1-ene as a result of a thermal electrocyclic reaction.

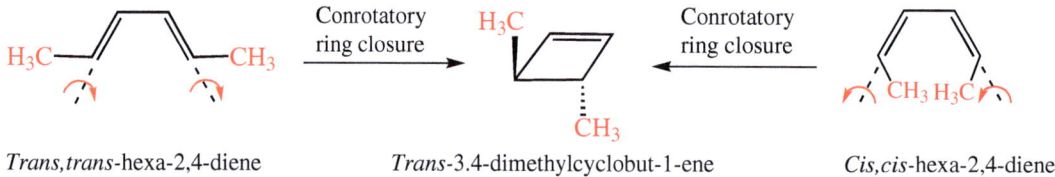

Trans,trans-hexa-2,4-diene   *Trans*-3.4-dimethylcyclobut-1-ene   *Cis,cis*-hexa-2,4-diene

10.2.2 Photochemical Electrocyclic Reactions

If the reaction is carried out under photochemical conditions, electrocyclic reactions follow principles similar to those discussed in thermal reactions. However, there is a significant difference. In a photochemical reaction, the molecule absorbs light, and one of the electrons is promoted from the HOMO to the LUMO to form a higher energy configuration (Figure 10.7). Now, butadiene is in excited state; one electron is in the π_2 orbital and one electron in the π_3 orbital. The HOMO filled with two electrons in the ground state is no longer the HOMO. The LUMO in the ground state is now the HOMO of the excited state. Therefore, we now have to consider π_3 as the HOMO (Figure 10.7).

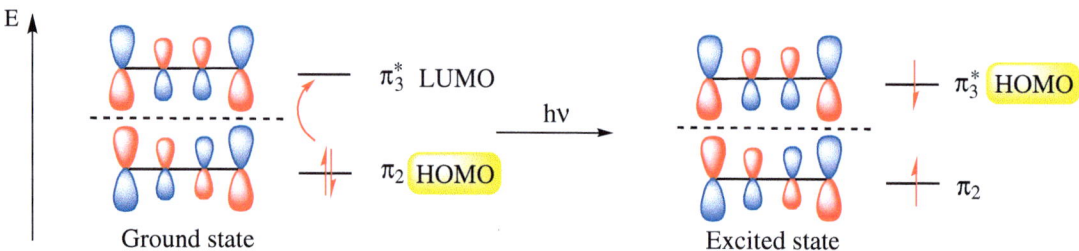

Figure 10.7 HOMO and LUMO of butadiene in the ground state and the excited state.

Now let us examine the ring-closure process in butadiene. The excited state HOMO of butadiene has like phases of the outermost p orbitals on the same side of the molecule. For the generation of a σ bond, disrotatory ring closure is a symmetry-allowed process. As a result, the method of ring closure of a photochemical electrocyclic reaction is the opposite of that of a thermal electrocyclic reaction. On the other hand, a conrotatory motion of the outermost p orbitals would lead to antibonding orbital formation. Therefore, this process is forbidden (Figure 10.8).

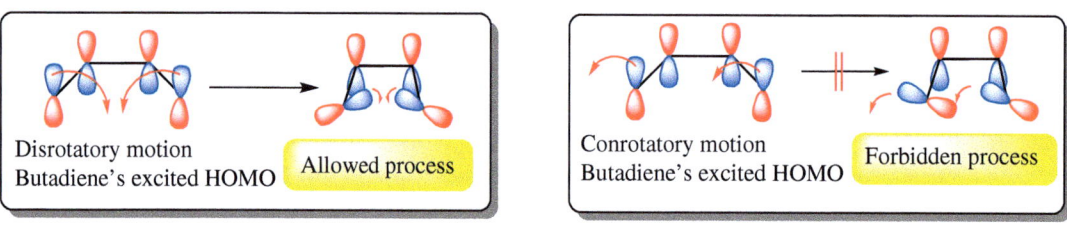

Figure 10.8 Disrotatory and conrotatory ring closure of butadiene in photochemical reactions.

When the *trans,trans*-hexa-2,4-diene is irradiated, the p orbitals located at the ends of the HOMO are in the same phase. Therefore, cyclization occurs in a disrotatory manner to generate a σ bond. Of course, these motions can be inward and outward, as shown below. In both cases, *cis*-3,4-dimethylcyclobut-1-ene is formed.

As the excited state HOMO of *trans,cis*-hexa-2,4-diene is symmetric, the ring closure is disrotatory. Disrotatory ring closure leads to the *trans* product. On the other hand, the ground state HOMO is antisymmetric. The ring closure must be conrotatory. Conrotatory ring closure leads to the formation of the *cis* product.

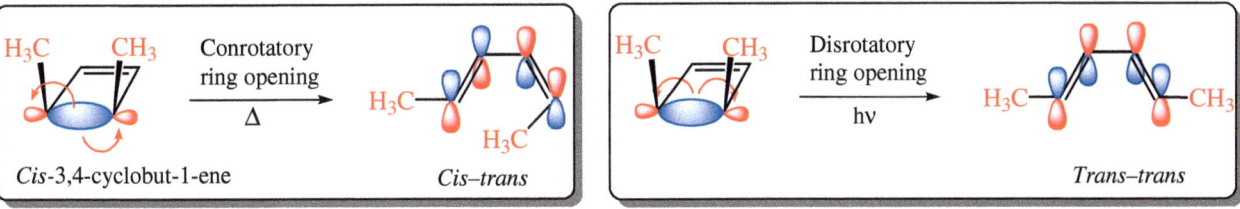

Because the electrocyclic reactions are reversible, the Woodward–Hoffman rules can also be applied to electrocyclic ring-opening reactions. The σ bond in the ring will open in such a way that the p orbitals formed will have the same symmetry as the HOMO of the product. The thermolysis of cyclobutenes leads to ring-opening products. What is particularly significant about these reactions is that they are stereospecific. *cis*-3,4-Dimethylcyclobut-1-ene is converted to *cis,trans*-hexa-2,4-diene, while photolysis yields the *trans,trans*-isomer.

After examining the thermal and photochemical ring closure of butadiene and the ring opening of cyclobutene systems, let us examine the hexa-1,3,5-triene system. The hexa-1,3,5-triene likewise forms cyclohexa-1,3-diene by heat or irradiation. The ground state HOMO π_3 (see Figure 10.5) with three conjugated π bonds is symmetric. The outermost p orbitals have like phases on the same side of the triene system. This means that we have to apply disrotatory rotation under thermal conditions. Thus, thermolysis of *trans,cis,trans*-octa-2,4,6-triene gives rise to the *cis* ring-closure product. This means that ring closure under thermal conditions is disrotatory. Remember, in the case of butadiene, we applied the conrotatory rotation under thermal conditions.

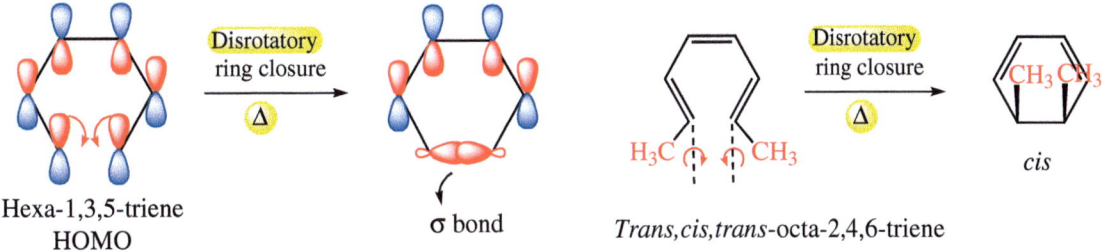

If the same reaction is carried out under photochemical conditions, the π_4^* orbital becomes the HOMO because an electron from π_3 will settle into the π_4^* orbital. If the phases of the outermost p orbital are different, this time the orbital lobes must perform a conrotatory rotation to generate a σ bond. Irradiation of *trans,cis,trans*-octa-2,4,6-triene results in the formation of the *trans* isomer.

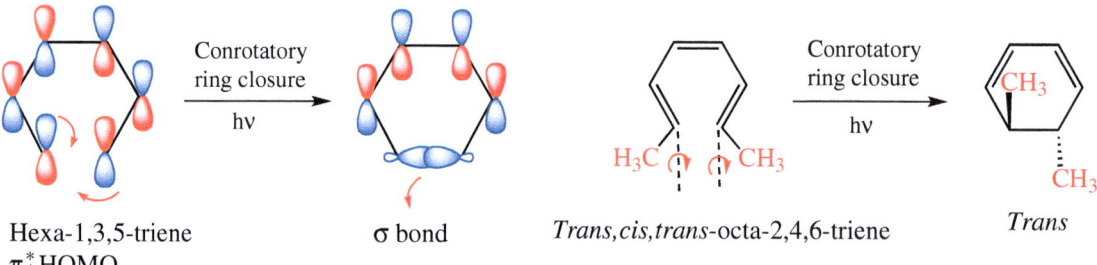

As a result of these two systems (butadiene and hexatriene) examined, we have seen that the stereochemistry of ring-opening and ring-closure reactions depends on the mode of ring closure and ring opening (disrotatory or conrotatory). This depends on the total number of π electrons undergoing reorganization. The number of electrons can be either $4n + 2$ or $4n$ (where n is an integer). Once the number of electron is known, the following generalizations can be made.

- Systems containing $4n$ ($n = 1, 2, 3, 4 \ldots$) π electrons undergo ring-closure reactions in a conrotatory manner (either clockwise or counterclockwise) in thermal electrocyclic reactions, while in photochemical reactions, they undergo disrotatory ring-closure reactions.
- System containing $4n+2$ ($n = 0, 1, 2, 3 \ldots$) π electrons undergo ring-closure reactions in a disrotatory manner in thermal reactions, while in photochemical reactions, they undergo conrotatory ring-closure reactions. These selection rules are also known as the *Woodward–Hoffmann rules* and are listed in Table 10.1.

Table 10.1 Woodward – Hoffmann rules for electrocyclic reactions.

Number of conjugated π bonds	Thermal reaction (Δ)	Photochemical reaction (hν)
($4n$) electrons, $n = 1, 2, 3, 4,\ldots$	Conrotatory	Disrotatory
($4n + 2$) electrons, $n = 0, 1, 2, 3, 4,\ldots$	Disrotatory	Conrotatory

10.2.3 Application of the Woodward–Hoffman Rules to Electrocyclic Reactions

10.2.3.1 Neutral Compounds

Cycloheptatriene is in equilibrium with its valence isomer, norcaradiene [11, 12]. Cycloheptatriene with six π electrons ($4n + 2$ system) undergoes a thermally allowed disrotatory ring-closure reaction to form the norcaradiene, which undergoes a thermally allowed disrotatory ring-opening reaction. Because the ring closure is disrotatory, the hydrogen atoms attached to the cyclopropane are in the *cis* configuration.

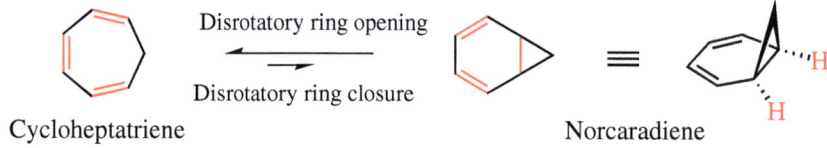

Similarly, cyclooctatriene [13] and cyclooctatetraene [14] are also in equilibrium with their valence isomers, bicyclooctadiene and bicyclooctatriene. Cyclooctatriene isomerizes thermally to bicyclo[4.2.0]cyclooocte-2,4-diene in the expected disrotatory manner. For cyclooctatetraene and bicyclo[4.2.0]octa-2,4,7-triene, the equilibrium strongly favors cyclooctatetraene. Although there are eight π electrons, only six electrons are involved in the disrotatory ring-closure and ring-opening reactions. The disrotatory ring-closure reaction of these compounds causes the hydrogen atoms to be *cis* in the bicyclic compounds.

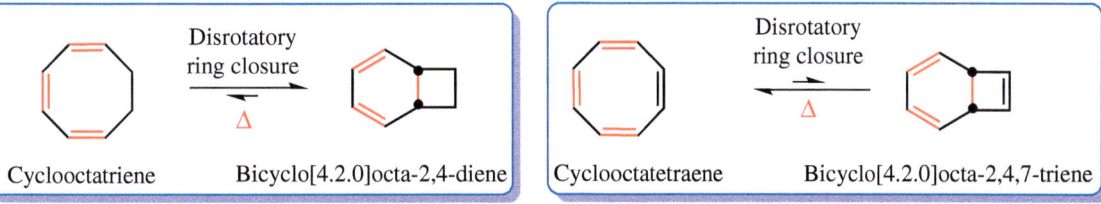

UV irradiation of a solution of α-pyrone in ether affords a cyclic isomer in almost quantitative yield [15]. Cyclization occurs in a disrotatory manner because the excited state HOMO of a conjugated diene is involved during the cyclization reaction. The hydrogen atoms attached to the cyclobutene ring are in the *cis* configuration. Because the ring closure in the thermal reaction will be conrotatory, a highly strained compound will be formed. Therefore, α-pyrone does not undergo ring closure under thermal conditions.

cis,cis,cis-1,3,5-Cyclononatriene has three conjugated double bonds and a $(4n + 2)$ system and undergoes disrotatory ring closure under thermal conditions to give *cis*-bicyclo[4.2.0]nona-2,4-diene. Irradiation of this diene ($4n + 2$ system, we also count the σ bond involved in the opening reaction) at low temperature ($-20\,°C$) causes conrotatory ring opening to furnish *cis,cis,trans*-triene, which undergoes disrotatory cyclization at room temperature (thermal conditions) to give *trans*-bicyclo[4.2.0]nona-2,4-diene [16].

Cyclononatetraene is transformed into bicyclo[4.1.0] nona-2,4,6-triene upon irradiation (photochemical conditions). As can be seen, the product is formed by disrotatory ring closure. The HOMO of the conjugated tetraene system in its ground state has lobes of like sign on the opposite side of the molecule. Therefore, disrotatory ring closure of the eight π system in the ground state cannot take place. However, the lobes of the HOMO in the excited state have like signs on the same side of the molecule and so ring closure takes place. In a photochemical reaction, the tetraene system uses all eight π electrons available for cyclization, while in the thermal ring-closure reaction, only six π electrons are involved in the disrotatory cyclization process.

Irradiation of the tricyclic hydrocarbon causes a disrotatory ring-opening reaction, forming the tetraene. When the resulting bicyclic compound is heated, it undergoes ring opening by a disrotatory path, forming monocyclic hexaene. This process is ring opening of a 10 π system. Remember that we also count the σ bond electrons involved in the opening process. Only six π electrons of the hexaene system participate in the disrotatory ring-closure process forming the final product, bicyclopentaene [17].

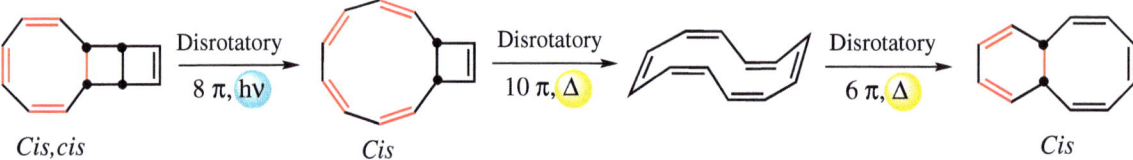

10.2.3.2 Ionic Compounds

Application of the Woodward–Hoffmann rules is not limited to neutral compounds. They can also be applied to ionic compounds. The conversion of a cyclopropyl cation to an allyl cation represents the simplest possible case of electrocyclic transformation since it involves only two π electrons. First, let us examine the orbitals of allyl systems formed by opening the cyclopropyl cation and anion.

An allyl cation and an allyl anion have similar molecular orbitals. The lowest orbital π_1 is the HOMO of the allyl system and the orbital lobes are in the same phase (Figure 10.9). In the ground state, two π electrons occupy the bonding molecular orbital. The lowest energy orbital π_2, the nonbonding orbital that contains no electrons, is called the LUMO. This molecular orbital has a single node. The lobes of the outermost orbitals in the LUMO are in different phases. The allyl anion has four π electrons and they populate π_1 and π_2 orbitals. The LUMO of the allyl cation is the HOMO of the allyl anion.

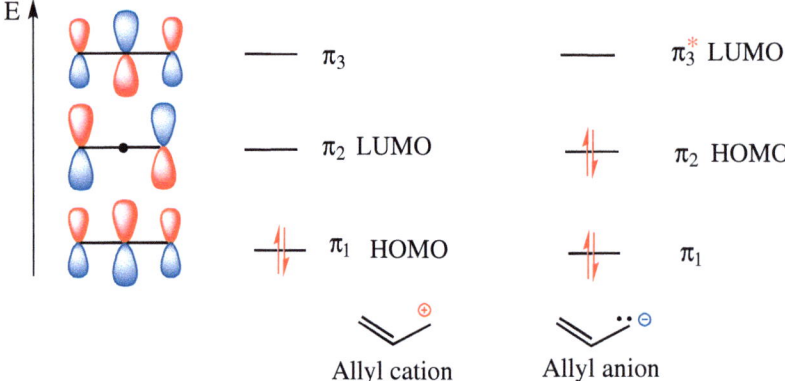

Figure 10.9 HOMO and LUMO of an allyl cation and anion in the ground state and the excited state.

Likewise, orbitals belonging to a pentadienyl cation and a pentadienyl anion are shown in Figure 10.10. In this system, the LUMO of the pentadienyl cation is the HOMO of the pentadienyl anion.

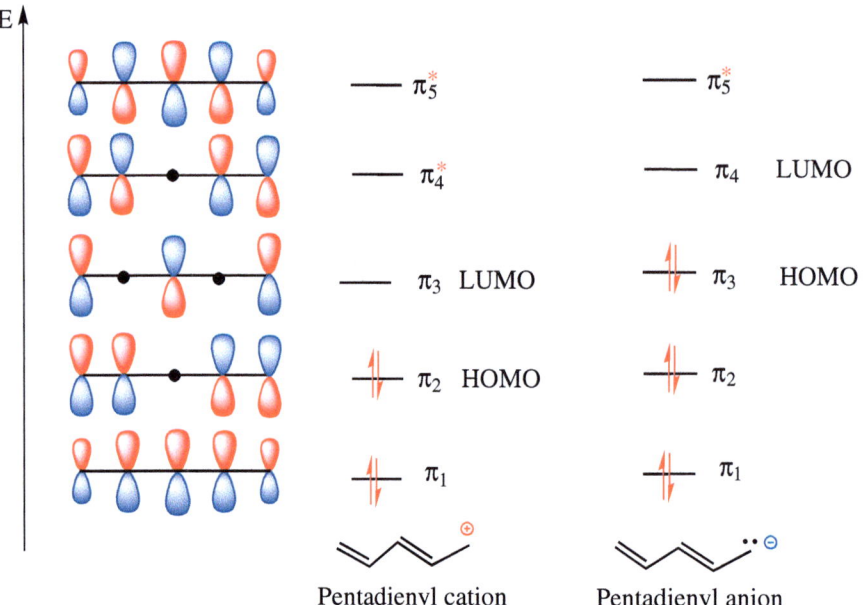

Figure 10.10 HOMO and LUMO of a pentadienyl cation and anion in the ground state and the excited state.

The allyl cation is formed by ring opening of the cyclopropyl cation. We will examine only the opening of the cyclopropyl cation as the probability of the allyl cation forming a cyclopropyl cation is small.

Monohalocyclopropanes are readily accessible compounds and they represent an excellent class of compounds for effecting ring expansion reactions. The ring opening (retroelectrocyclic reaction) proceeds in a disrotatory manner, as predicted by Woodward–Hoffmann rules (Table 10.1). The strain in the three-membered ring is relieved and resonance stabilization of the allyl system is accomplished.

Let us analyze the ring-opening process of (1*r*,2*R*,3*S*)-1-bromo-2,3-dimethylcyclopropane. It was shown that the cyclopropyl cation is not an intermediate and the ring opening proceeds simultaneously with displacement of the leaving group. The thermally allowed disrotatory ring-opening reaction proceeds according to a concerted S_N2 mechanism. During the opening process, the electron density from the breaking σ bond increases at the back away from the bromine atom and substitutes the bromine atom. During the ring-opening process, the methyl groups arranged *trans* to the bromine atom rotate outward, forming the *trans,trans* allyl system (Figure 10.11).

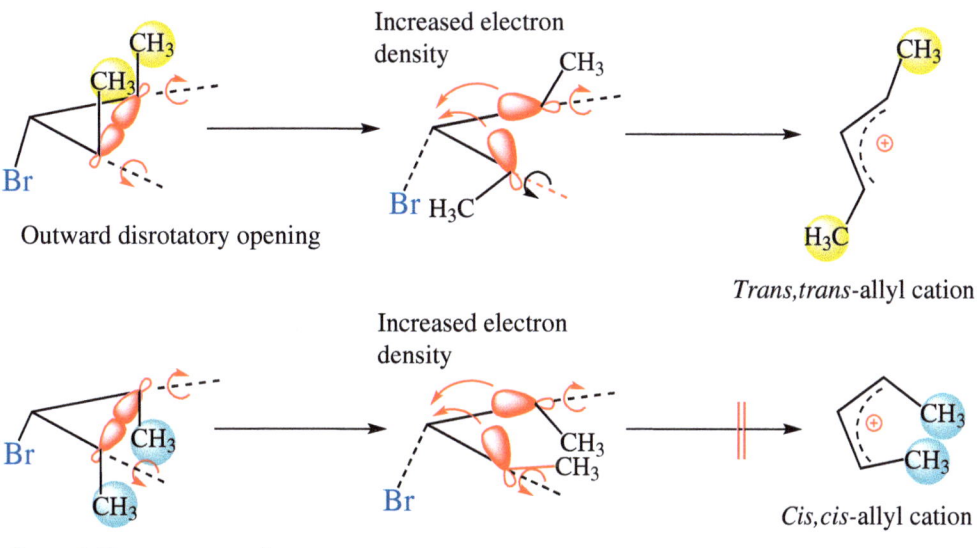

Figure 10.11 Disrotatory ring opening of isomeric *cis*- and *trans*-1-bromo-2,3-dimethylcyclopropanes.

The ring-opening process of the isomeric compound, (1*s*,2*R*,3*S*)-1-bromo-2,3-dimethylcyclopropane, in which the bromine atom is in the *cis* position to the methyl groups is different. Because of the two π system, the compound can undergo disrotatory ring opening under thermal conditions. Again, the electron density must increase at the back of the bromine atom for an S_N2-like substitution reaction. As shown in Figure 10.11, by applying disrotatory motion, the methyl groups that are arranged *cis* to the leaving group rotate inward and approach each other during the rotation. These groups will be in the *cis,cis* position in the final product. In such an opening, a steric barrier arises because of the groups repelling each other.

The difference that these two openings will reveal is much more pronounced, especially in compounds with a bicyclic structure. For example, the reactions of the isomeric 7-chlorobicyclo[4.1.0]heptanes (norcarane) are quite different. While *endo* chloronorcarane is easily solvolyzed in acetic acid at 125 °C, the *exo* isomer does not react, even at 210 °C [18, 19].

In these isomeric bicyclic compounds, the *endo* isomer can undergo disrotatory ring opening, while the *exo* isomer is geometrically constrained against ring opening. There is no steric obstacle in the opening of the *endo* isomer. As shown below, the strain in the molecule is further reduced as a result of the disrotatory ring opening. The situation is very different in the *exo* isomer. During disrotatory rotation, the electron density must increase behind the chlorine atom. In this case, the allyl cation that would be produced would be quite highly strained as the product will be a *trans*-cycloheptenyl cation. Therefore, this ring opening does not occur. Two examples show us that the ring opening of these systems is completely controlled by the Woodward–Hoffman rules.

Balcı and coworkers have applied these ring-opening reactions to the synthesis of an elusive compound, a five-membered ring allene. The addition of fluorobromocarbene (CFBr), generated from $CHFBr_2$ and NaOH under phase-transfer conditions, to bicyclo[3.2.0]hept-6-ene affords the ring-opened product and fluorobromocyclopropane [20, 21]. The expected isomer **B** was not among the products. The initially formed isomer **B** likely undergoes disrotatory ring opening under the reaction conditions. For the disrotatory ring-opening reaction, bromide must be in the *endo* position. The isomer **A** can also undergo disrotatory ring opening. However, in this case, the departing group is the fluorine atom. As the fluorine is a bad leaving group, the compound does not undergo ring opening.

Treatment of fluorobromocyclopropane with MeLi in ether at −25 °C in the presence of furan as the trapping reagent furnishes the trapping product, which is the first example of a five-membered cyclic allene [20, 21].

10.3 Correlation Diagrams

The Woodward–Hoffman rules were first published under the title *orbital symmetry conservation*. According to this theory, during the transformation of the molecular orbitals of reactants into those of products, the orbitals' symmetry must remain unchanged as the orbitals of the reactants are smoothly converted into the orbitals of the products. When the symmetry is maintained, the reaction will occur. If the symmetry is not maintained, the reaction is forbidden. In the reactions we have seen so far, we analyzed the phases of the HOMOs to decide whether a reaction can proceed or not, which was the correct course of action. In this section, we will examine the conservation of orbital symmetry in more detail via a few examples. There is no need to know about correlation diagrams to apply the Woodward–Hoffman rules. Those who need more detailed information can read this section. Otherwise, they can skip this section and move on to the next.

10.3.1 Symmetry Elements

Because the symmetries of the orbitals are taken as the basis in applying the Woodward–Hoffman rules, let us first examine which symmetry elements are used and how. There are two main symmetry elements applied to determine the orbital symmetries (Figure 10.12).

A plan of symmetry, designated by the symbol Σ, is the mirror plane element perpendicular to the molecule plane and divides the molecule into two parts. *A rotational axis* is designated C_2, which shows the degree of rotation that restores the object after 180° rotation.

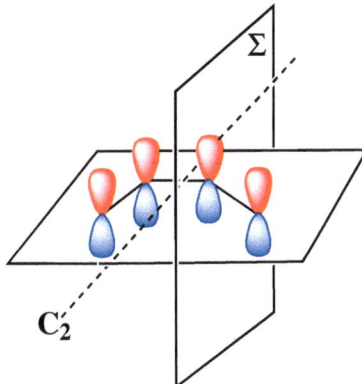

Figure 10.12 Representation of symmetry elements Σ and C_2.

Let us apply symmetry elements to ethylene orbitals. The symmetry element Σ passes through the middle of the HOMO, and, according to the plane of symmetry, ethylene's HOMO is symmetric. Now let us apply the C_2 axis element to the same orbital. The C_2 axis passes through the middle of the C—C bond. The orbitals are rotated by 180° about this axis, and the resulting orbitals are compared with the initial state. As can be seen from the figure, the symmetry is lost. The phases of the orbitals are changed. In summary, we can say that the HOMO of ethylene is symmetric according to the symmetry element Σ and antisymmetric according to the C_2 axis (Figure 10.13).

When the same symmetry elements are applied to the LUMO of ethylene, it can be seen that the orbital is antisymmetric with respect to the plane of symmetry and symmetric with respect to the C_2 axis. No orbital can be symmetric or antisymmetric in terms of both symmetry elements.

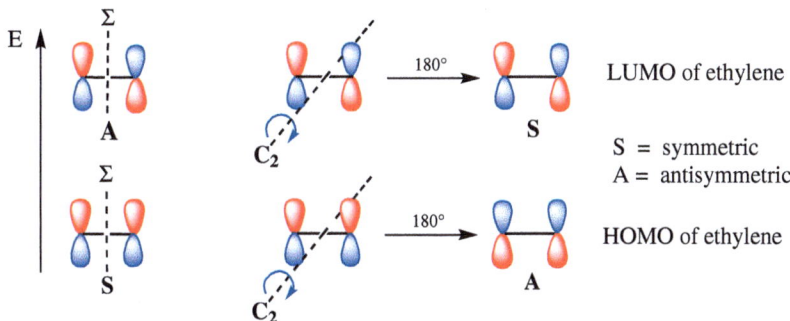

Figure 10.13 Application of symmetry elements Σ and C_2 to ethylene's HOMO and LUMO.

The symmetry of the butadiene orbitals according to the plane of symmetry and the C_2 axis is given in Figure 10.14.

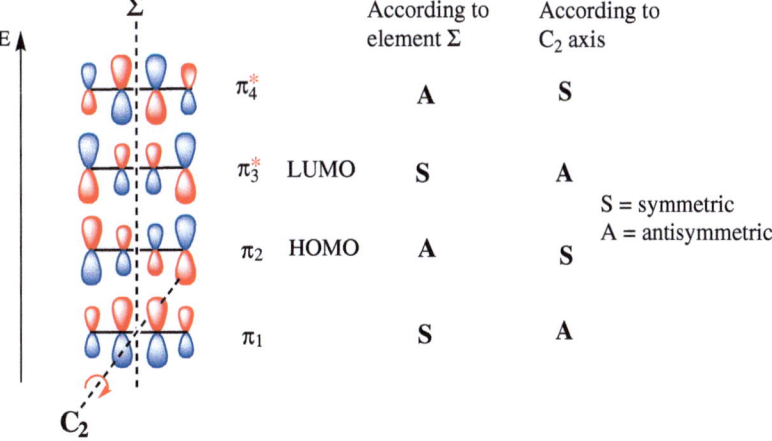

Figure 10.14 Application of symmetry elements Σ and C_2 to butadiene's HOMO and LUMO.

After examining the symmetry elements, it is relatively easy to determine whether a reaction is allowed or forbidden by drawing correlation diagrams of it. The orbitals in the starting compounds form the orbitals of the product formed due to the electrocyclic reaction. Therefore, the orbitals' symmetries (including the transition complex) must be maintained during ring opening and ring closing. The point to be taken into consideration here is that the symmetry of the reactants' orbitals should correlate with the symmetry of the interacting bonding molecular orbitals of the product. Similarly, the antibonding orbitals should also correlate. Otherwise, the reaction is thermally not possible.

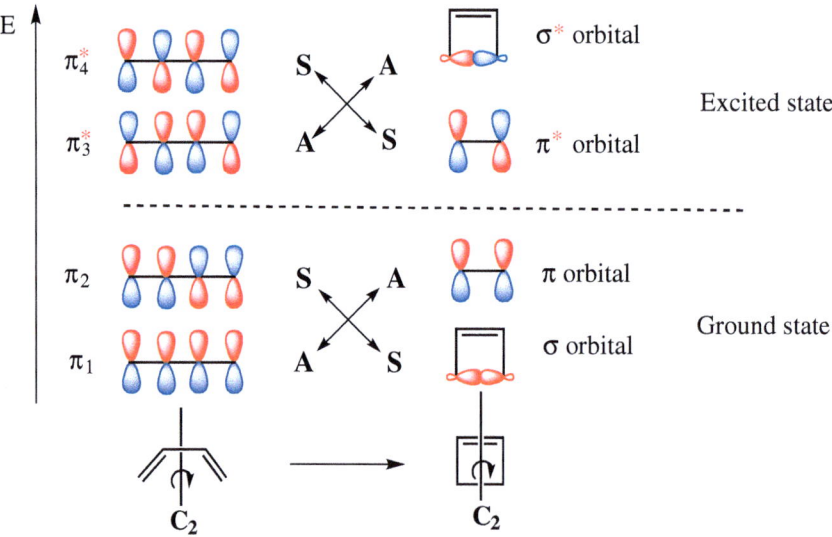

Figure 10.15 Correlation diagram for conrotatory conversion of butadiene to cyclobutene.

The electrocyclic ring closure of 1,3-butadiene can proceed through either a conrotatory or a disrotatory reaction mechanism. We have to identify a symmetry element maintained throughout the reaction. In the conrotatory ring closure, we apply the C_2 axis as the symmetry element and in the disrotatory ring closure the plane of symmetry Σ. After ring closure, a new σ bond and a new π bond are formed. We also have to consider the newly formed orbitals of these bonds. To correlate orbitals of the starting material and product, we have to determine whether the molecular orbitals are symmetric or antisymmetric with respect to these symmetry elements. Then, we list the orbitals in the usual order of increasing energy. Filled and empty orbitals are separated by a dashed line (Figure 10.15). In this diagram, the orbital π_1 correlates with the bonding π orbital of the cyclobutene double bond. The π_2 orbital of butadiene correlates with the newly formed σ orbital of cyclobutene. Because the orbitals in the ground state correlate with each other, this process is thermally allowed. The four electrons occupying π_1 and π_2 in the ground state of butadiene go into the π and σ bonding orbitals of cyclobutene without involving an excited state. Moreover, all orbitals in the excited state also correlate with each other. In short, butadiene can undergo a ring-closure reaction under thermal conditions by conrotatory motion.

When we examine the disrotatory ring closure of butadiene, we must use the plane of symmetry element Σ. Next, we determine whether the molecular orbitals are symmetric or antisymmetric with respect to this symmetry element Σ. The molecular orbitals of butadiene and cyclobutene are shown in Figure 10.16. The lowest energy molecular orbital π_1 is symmetric and correlates with the symmetric σ orbital of cyclobutene. However, the remaining orbitals, π_2 and π orbitals in the ground state, do not correlate. This process is thermally not feasible; it is a symmetry-forbidden process. Correlation with antibonding orbitals is required to ensure orbital conservation. This means that the antibonding orbitals must be involved during this reaction. Irradiation can induce the promotion of an electron to the next level. Now, the π_2 orbital of butadiene can correlate with the π^* orbital of cyclobutene. As a result of these correlations, we can conclude that the disrotatory ring closure of butadiene can occur under photochemical conditions. Note that the frontier molecular orbital approach, analyzing only the HOMO of the diene, provides precisely the same prediction.

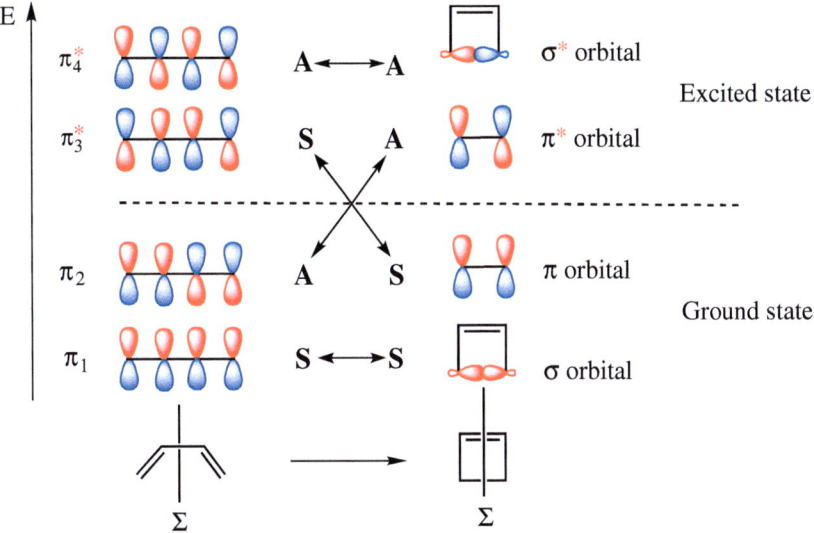

Figure 10.16 Correlation diagram for disrotatory conversion of butadiene to cyclobutene.

A symmetry correlation diagram for the disrotatory ring closure of hexatriene can also be constructed. We will use the plane symmetry element Σ maintained during the reaction. The correlation diagram showing the disrotatory closure of hexatriene is given in Figure 10.17. When the reactant and product orbital symmetries are compared, it is seen that the correlation is possible between the orbitals in the ground state. This reveals that the disrotatory ring closure of the hexatriene molecule under thermal conditions is an allowed process according to the Woodward–Hoffmann rules.

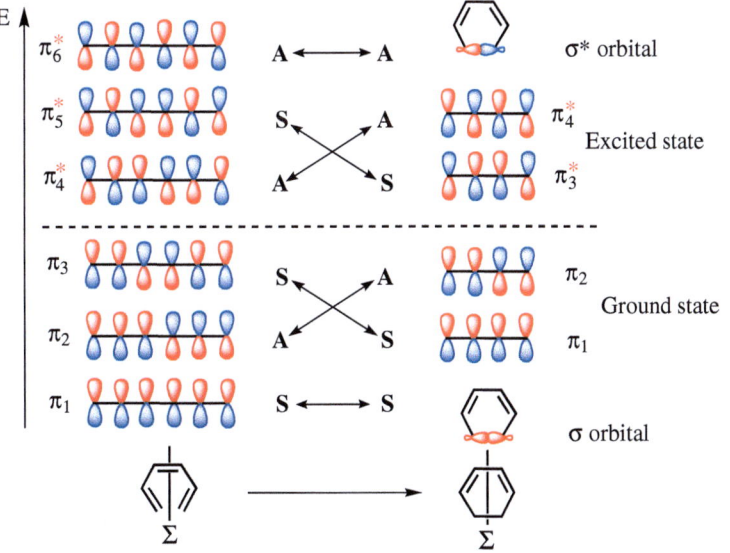

Figure 10.17 Correlation diagram for the disrotatory conversion of hexatriene to cyclohexadiene.

10.4 Cycloaddition Reactions

As discussed, cycloaddition reactions are reactions in which two compounds having π bonds add to one another, yielding new cyclic compounds with two new σ bonds and two fewer double bonds. Cycloadditions are among the reactions in organic chemistry that have been well known for many years. Reactions are classified according to the number of π bonds present in the reactants and directly involved in the reaction. The numbers of participating π electrons in both components are given in brackets. For example, the most common cycloaddition reaction is the [4 + 2] cycloaddition reaction, known as the Diels–Alder cycloaddition [22]. The German chemists Otto Paul Hermann Diels (1876–1954) and Kurt Alder

(1902–1958) were awarded the Nobel Prize in Chemistry in 1950 for their work in this field. Cycloaddition reactions, like electrocyclic reactions, may be thermally or photochemically allowed or forbidden. In this section, we will focus on the application of the Woodward–Hoffmann rules to these reactions.

One of the most classic examples of cycloaddition reactions is the conversion of butadiene to cyclohexene by reacting with ethylene. In this reaction, two new σ bonds are formed as a result of the reaction of the three π bonds, while the location of one π bond is displaced.

Chemical reactions are electron exchange reactions. Electron transfer occurs from the HOMO of one of the reactants to the LUMO of the other. In cycloaddition reactions, we need to consider the phases of the outermost p orbitals of the HOMO and LUMO to determine whether any addition reaction will occur or not. In a [4 + 2] cycloaddition reaction, mainly electron transfer occurs from the HOMO of the diene system to the LUMO of the dienophile. For a successful cycloaddition reaction, the terminal π lobes of the reactants must have the same phases for bonding to occur. As shown below, a HOMO–LUMO interaction with the same symmetry results in the stabilization of bonding orbitals.

While investigating this reaction, we use the HOMO of butadiene and the LUMO of ethylene. One may raise the following question: *What would be the result if we used the LUMO of butadiene and the HOMO of ethylene?* The answer to this question is: *The result would not change, as shown below.*

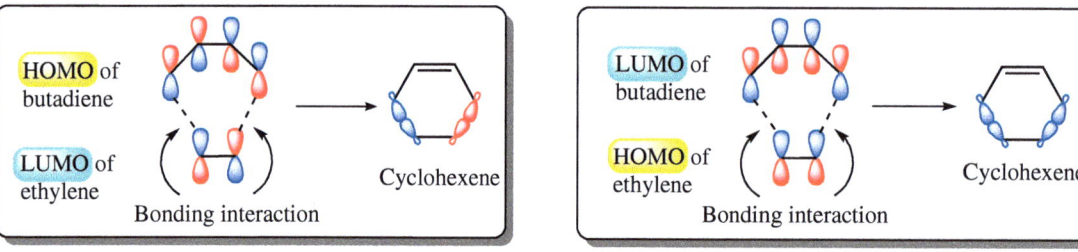

There is no obstacle in terms of orbital symmetry for the formation of the product in both interactions. *Which orbitals are used in these reactions?* The energy difference (ΔE) between the energy levels of the HOMO and LUMO determines which orbitals will be used. Suppose the energy difference ΔE_1 between the diene's HOMO and the dienophile LUMO is smaller than the ΔE_2 between the dienophile's HOMO and the diene's LUMO. In that case, the cycloaddition reactions will take place between the diene's HOMO and the dienophile's LUMO (Figure 10.18). There are also cases in which the diene's LUMO and the dienophile's HOMO are used (see Inverse Diels–Alder Reactions).

Finally, we can conclude that a [4 + 2] cycloaddition reaction between a diene (butadiene) and a dienophile (ethylene) is a thermally allowed process.

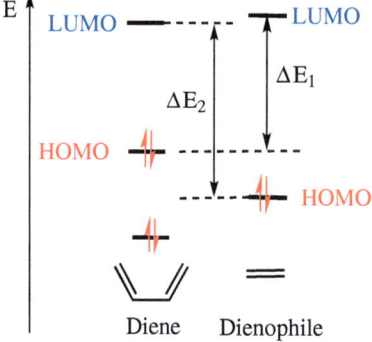

Figure 10.18 Energy levels of the frontier orbitals of a [4 + 2] cycloaddition reaction.

The cycloaddition reactions, proceeding according to the Woodward–Hoffmann rules, are concerted processes. The formation of σ bonds can occur in two ways. Considering that the diene and dienophile have plane structures, if the bonding interaction occurs between the lobes on the same sides of both reactants, this is termed *suprafacial bond formation*. However, if the bonding interaction occurs between the lobes on the same face of one reactant and the lobes on opposite faces of the other reactant, this is called *antarafacial bond formation* and is shown in Figure 10.19.

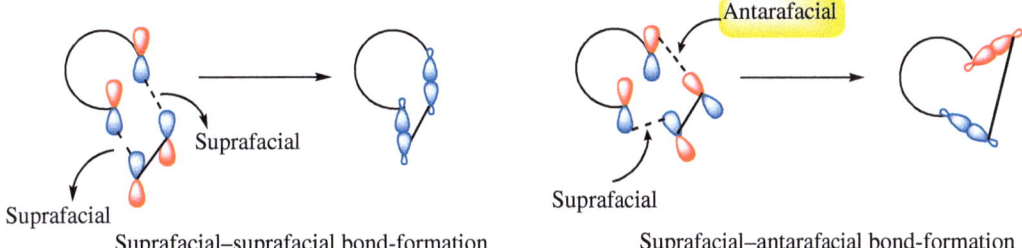

Figure 10.19 Suprafacial – suprafacial and suprafacial – antarafacial bond formation.

After examining [4 + 2] cycloaddition reactions, let us now investigate [2 + 2] cycloaddition reactions. Under thermal conditions, the HOMO of ethylene (or substituted ethylene) reacts with the LUMO of a second ethylene molecule. As seen in Figure 10.20, while the phases of two orbitals are suitable for bond formation, the other orbitals' phases are not suitable to form a σ bond.

It is apparent that a thermal [2 + 2] cycloaddition can take place by an antarafacial pathway. However, the geometric constraints increase excessively in the transition complex and so antarafacial addition cannot occur. Suprafacial–antarafacial [2 + 2] cycloaddition reactions are limited to large rings where the strain in the transition state is partially reduced. In general, a thermal [2 + 2] cycloaddition is not observed between alkenes.

In contrast to the thermal [2 + 2] cycloaddition, photochemical [2 + 2] cycloaddition reactions can occur and they are allowed processes. When the alkene is irradiated, an electron from the alkene HOMO is promoted to the alkene LUMO, which becomes the excited state HOMO. The phases of an alkene LUMO and excited state HOMO are appropriate for interacting and making σ bonds (Figure 10.20). This is one of the most frequently used methods in the synthesis of four-membered rings in organic chemistry.

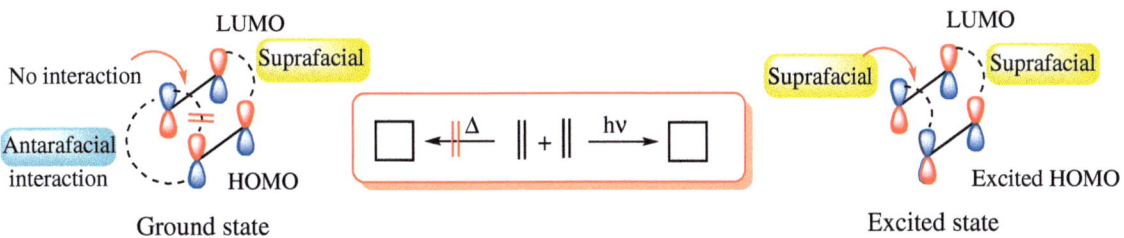

Figure 10.20 HOMO and LUMO interactions in [2 + 2] cycloadditions under thermal and photochemical conditions.

The cycloaddition reactions are classified according to the number of π electrons involved in each reacting compound, $4n$- and $(4n + 2)$ π electrons ($n = integer$). As the total number of π electrons in the reaction of butadiene with ethylene is six, this reaction falls into the $(4n + 2)$ π ($n = 1$) class. Table 10.2 summarizes the Woodward–Hoffmann rules applied to cycloaddition reactions.

Table 10.2 Woodward – Hoffmann rules for cycloaddition reactions.

Number of π bonds	Thermal reactions	Photochemical reactions
$(4n + 2)$-π system $n = 0, 1, 2, 3,\ldots$	Suprafacial	Antarafacial
$4n$-π system $n = 1, 2, 3, 4\ldots$	Antarafacial	Suprafacial

We discussed that a thermal antarafacial [2 + 2] cycloaddition does not occur because of the geometric constraints in the transition complex. However, it is remarkable that ketenes (heteroanalogs, allenes) undergo a rapid [2 + 2] cycloaddition rather readily at or below room temperature in a concerted manner [23, 24]. The orbital phases of the ketene double bond are not suitable for a concerted cycloaddition. However, the orbital of the low lying C=O π^* orbital is involved. Ketene has a second π bond orthogonal to the C=C double bond. An interaction takes place with this double bond, as shown in Figure 10.21. One end of the C=C double bond interacts with the p_z orbital at the terminal carbon atom of the ketene moiety. The other end of the double bond interacts with the C=O double bond's p_y orbital.

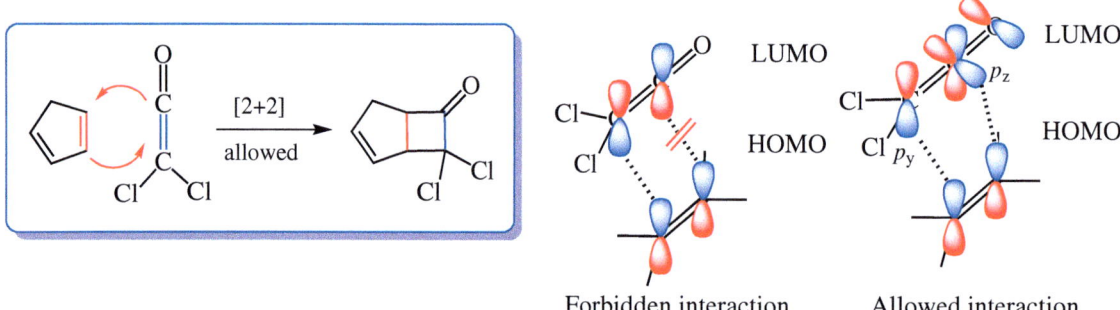

Figure 10.21 Orbital overlap in the cycloaddition of an olefin to a ketene.

When a diene system, for example, butadiene, reacts with itself or with another diene system under thermal conditions, a [4 + 4] cycloaddition reaction never occurs. This result can be explained by examining the HOMO and LUMO of butadiene. In contrast, a suprafacial [4 + 2] cycloaddition reaction occurs under thermal conditions. The like phases of the orbitals on the outermost carbon atoms can overlap (Figure 10.22).

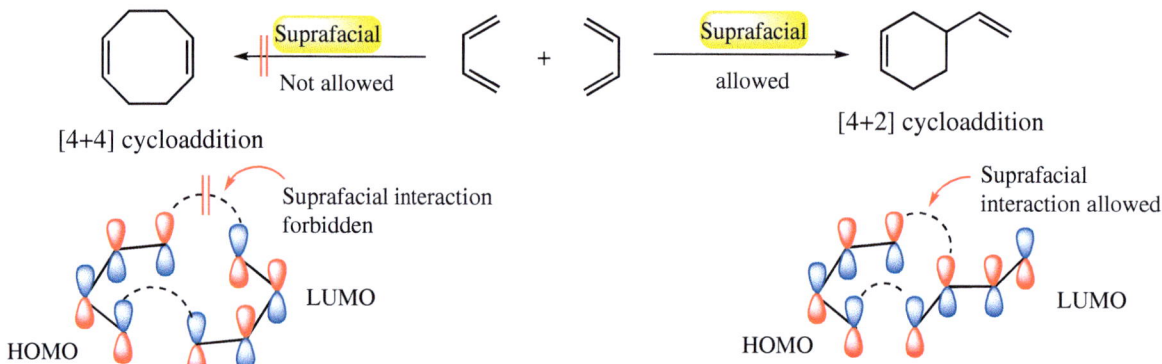

Figure 10.22 Analysis of [4 + 4] and [4 + 2] cycloaddition reactions.

Some dienes behave as both dienes and dienophiles and they undergo [4 + 2] cycloaddition reactions under thermal conditions. For example, cyclopentadiene is unstable at room temperature and its dimerization proceeds to form dicylopentadiene.

The [8 + 2] cycloaddition reactions are thermally allowed; however, they are rarely observed. Because the total number of electrons involved in the cycloaddition reactions is 10, these reactions are classified as a $(4n + 2)\,\pi$ system and suprafacial addition occurs. When heptafulvene is reacted with dimethyl acetylenedicarboxylate, dimethyl azulene-1,2-dicarboxylate is formed [25].

10.4 Cycloaddition Reactions | 519

One of the best examples of [6 + 4] cycloaddition is the reaction between tropone and cyclopentadiene. According to the Woodward–Hoffmann rules, the reaction is thermally allowed, and it takes place at room temperature. In boiling xylene, the adduct decomposes into its original components [26, 27]. One of the exciting aspects of this reaction is that the reactants can also form a [4 + 2] cycloaddition product. An [4 + 2] adduct has been observed as a by-product [28] and for the reason why it is formed, see the section "orbital coefficients."

The intramolecular [6 + 4] cycloaddition of tropone with a diene is a potentially powerful tool for direct access to 10-membered carbocycles, prominently displayed in a number of interesting and biologically significant natural product families [29].

Tropone undergoes dimerization when irradiated in an aqueous or organic solution to give the [6 + 6] cycloaddition product. The addition is suprafacial and proceeds under photochemical conditions [30].

The reaction of heptafulvalene with tetracyanoethylene (TCNE) under thermal conditions gives an adduct, whose structure has been firmly established by X-ray analysis. According to the data obtained by X-ray analysis, the hydrogens attached to sp³-hybridized carbon atoms are in the *trans* position [31]. This reaction is a [14 + 2] cycloaddition reaction and a total of 16 π electrons are involved. A suprafacial addition is not allowed. One of the new σ bonds is formed with the upper face of the molecule and the other σ bond with the lower face. Thus, it shows that the heptafulvalene underwent an antarafacial addition. The most important part of this reaction is that the orbital symmetry controls the mode of the reaction.

10.4.1 Stereoselectivity in Cycloaddition Reactions: Secondary Orbital Interactions

Endo selectivity is observed in most of the cycloaddition reactions occurring between cyclic systems. For example, the Diels–Alder reaction of cyclopentadiene with maleic anhydride forms two products: *endo* and *exo* addition products. The ratio of *endo* to *exo* products in this reaction is about 4 : 1. The preference for *endo* over *exo* is curious as the *exo* product is thermodynamically more stable than the *endo* product. Furthermore, the *endo* product is more sterically hindered.

Two orientations of the reactants are possible to form *exo* and *endo* adducts. Both *endo* and *exo* additions are symmetrically allowed processes. Experiments show that although the *exo* adduct is thermodynamically more stable, the *endo* adduct is formed faster than the *exo* adduct. In some cases, the initially formed *endo* adduct can isomerize to the less sterically hindered and more stable *exo* adduct by a retro-Diels–Alder reaction followed by recombination.

The phases of the orbital lobes at the outermost carbon atoms are suitable for forming σ bonds. To explain the preferential formation of the *endo* adduct, we need to examine the orbitals' interactions, which are not directly involved in forming new bonds.

In Figure 10.23, the transition complexes required for the formation of *endo* and *exo* adducts are presented. The interactions of the outermost orbitals leading to the formation of σ bonds are shown by bold green lines. The C2–C3 orbitals of the diene HOMO are close to the carbonyl group (C=O) orbitals of maleic anhydride and so they can interact. There is no similar interaction in the case of the *exo* transition state. The interactions shown with dotted lines are not bond-forming. Although these interactions do not lead directly to the formation of new bonds, they lower the transition state's energy. It is a stabilizing interaction, nonetheless. This orbital interaction is called a *secondary orbital interaction* [32, 33]. Although the product formed by the *exo* transition complex is more stable, preferably the thermodynamically less stable product is formed as the major product. This stabilizing interaction compensates for the steric hindrance.

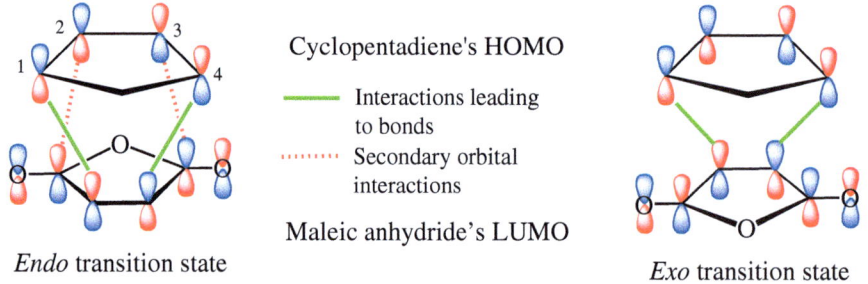

Figure 10.23 Transition states formed by the reaction of cyclopentadiene with maleic anhydride.

The reaction of tropone with cyclopentadiene gives exclusively the *exo* adduct rather than the *endo* adduct. The secondary orbital interaction does not favor the formation of the *endo* product.

Endo adduct   Intermolecular [6+4] cycloaddition   *Exo* adduct

To understand the preferential formation of the *exo* adduct, let us analyze the secondary orbital interactions. Both transition complexes are shown in Figure 10.24. Of course, this addition is an allowed process according to the Woodward–Hoffmann rules. The like phases of the orbitals at the outermost carbon atom interact to form σ bonds. However, during the *endo* approach, the phases of some orbitals, which do not participate in the reaction, are not the same. Therefore, the destructive interaction between these orbitals increases the energy of the *endo* transition complex. Because there is no such repulsive interaction between the secondary orbitals in the *exo* transition complex, it is preferred and the *exo* adduct is formed.

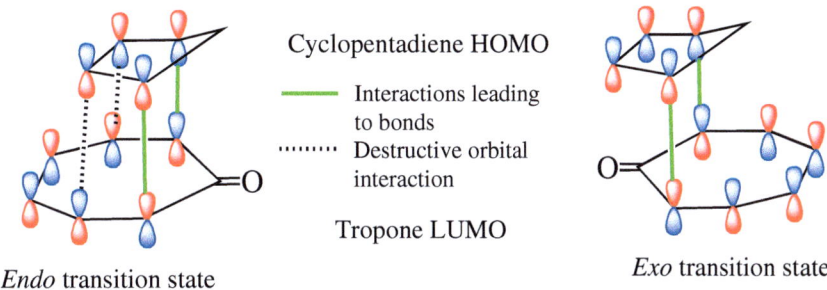

Figure 10.24 Transition states formed by the reaction of cyclopentadiene with tropone.

10.4.2 Factors Affecting the Rates of Cycloadditions

As discussed in the introduction part of this section, cycloaddition reactions occur as a result of electron transfer from the HOMO of one of the reactants to the LUMO of the other. The difference between HOMO and LUMO energy levels determines which component's HOMO and which component's LUMO will be used. In most cycloaddition reactions, the HOMO of the diene reacts with the LUMO of the dienophile. The substituents attached to the diene and dienophiles strongly affect the HOMO's and LUMO's energy levels depending on their electron-donating or electron-withdrawing properties.

For example, the energy difference between the HOMO of butadiene and the LUMO of ethylene is relatively high. It is a well-known fact that the smaller the energy gap between the HOMO and LUMO, the faster the reaction. Therefore, butadiene does not react efficiently with ethylene. On the other hand, it reacts readily with maleic anhydride. The electron-withdrawing carbonyl groups in maleic anhydride reduce the LUMO energy level of the molecule; thus, the energy difference between the LUMO of the dienophile and the HOMO of the diene is decreased, which leads to an increase in the rate of the reaction. On the other hand, substituents with electron-donating abilities attached to the diene or dienophile increase the energy levels of their HOMO's and LUMO's (Figure 10.25) [34, 35].

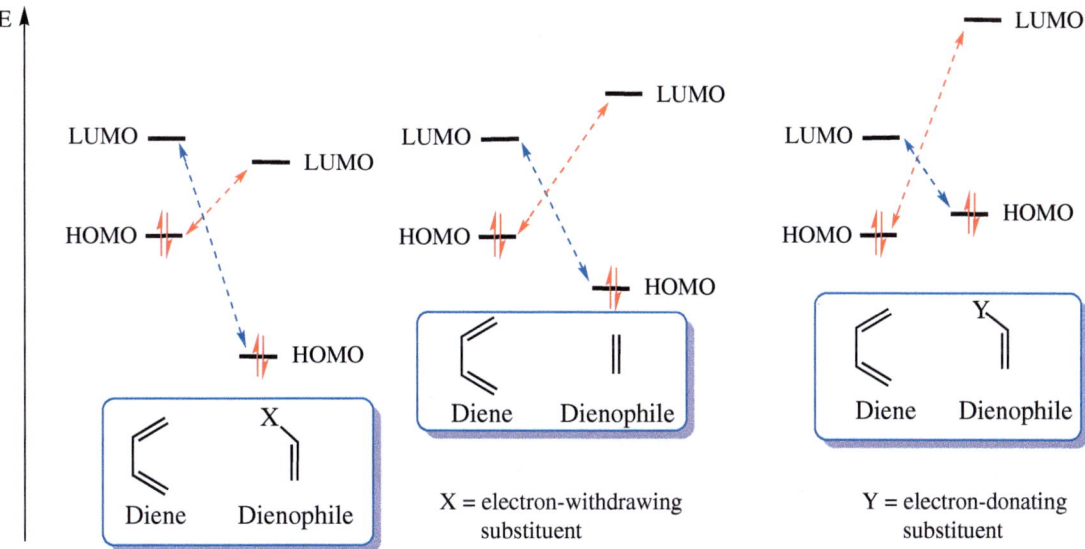

Figure 10.25 Energy levels of the HOMO and LUMO for Diels – Alder reactions.

Suppose that the electron-donor substituent is attached to the dienophile, the HOMO and LUMO energy levels of the dienophile increase. Thus, the energy difference between the HOMO of the diene and the LUMO of the dienophile increases even more than in butadiene and ethylene (Figure 10.25). This means that the reaction rate will decrease. Nevertheless, a diene substituted with an electron-donating group and a dienophile substituted with an electron-withdrawing group react at a lower temperature giving high yield. For example, *Danishefsky's diene*, *trans*-1-methoxy-3-trimethylsilyloxy-buta-1,3-diene, named after Samuel J. Danishefsky, is an electron-rich diene that reacts rapidly with electrophilic alkenes [36].

Danishefsky's diene
electron-rich diene   Electron-poor dienophile

An alternative version of this reaction is the *inverse demand Diels–Alder reaction*, in which an electron-rich alkene reacts with an electron-poor diene [37]. In the inverse demand Diels–Alder reaction, the HOMO of the dienophile reacts with the LUMO of the diene.

Let us examine the addition reaction between tetrazine and acetylene. Tetrazine is an electron-poor compound because of four nitrogen atoms in the ring and the electron-withdrawing ester groups. The ethoxy group is a mesomeric electron-donating group; it increases the electron density on the triple bond, making the dienophile electron rich. The reaction between these two compounds is the inverse demand Diels–Alder reaction [38]. The LUMO of the tetrazine reacts with the HOMO of the dienophile.

Electron-poor   Electron-rich   Inverse Diels–Alder   Retro Diels–Alder
diene   dienophile   cycloaddition   reaction

Lewis acids used as catalysts in Diels–Alder cycloaddition reactions increase the rate of the cycloaddition [39]. Lewis acid catalysis not only increases the reaction rate; it also makes it more selective. The increase in reaction rate also varies depending on the groups in the reactants. Lewis acid catalysts such as $AlCl_3$, BF_3, $B(OAc)_3$, $ZnCl_2$, $SnCl_4$, and $TiCl_4$ coordinate at the Lewis base side of the dienophile. For example, at the carbonyl oxygen of an α,β-unsaturated carbonyl group, it makes the carbonyl group even more electron-withdrawing and, therefore, more reactive.

The complexation of the Lewis acid with a carbonyl group increases the polarization of the carbonyl group, which reduces the electron density at the double bond. This effect lowers the carbonyl substrate's LUMO energy and contributes to reducing the energy gap between the HOMO of the diene and the LUMO of the dienophile. This increases the rate of the cycloaddition reaction. How Lewis acids affect selectivity (regioselectivity) will be discussed in the next section.

10.4.3 Coefficients of the Frontier Orbitals: Application to Regioselectivity in Diels–Alder Reactions

First of all, let us shortly discuss: *What is the orbital coefficient "c"?* The linear combination of atomic orbitals is an approximate method for representing molecular orbitals. The following equation describes a molecular orbital. The molecular orbital ψ is described as the sum of n atomic orbitals, each multiplied by a coefficient. The coefficient c describes the amount each atomic orbital contributes to the molecular orbital. The higher the value of c^2, the more significant the contribution.

$$\psi = c_1 \phi_1 + c_2 \phi_2 + \cdots$$

ψ is the molecular orbital, ϕ is the atomic orbital, and c is the orbital coefficient.

If there is an electron in a molecular orbital, that electron can be in various parts of the orbital. The square of these coefficients indicates the probability that the electron will be in that region. Because there will be only one electron in the same spin in a molecular orbital, the sum of the squares of the atomic orbital coefficients within a single molecular orbital must be equal to 1. Let us examine the HOMO and LUMO orbitals of the ethylene molecule. The orbital coefficients give us an idea about the probability of finding an electron at a particular place. As can be seen, the sum of the squares is equal to 1.

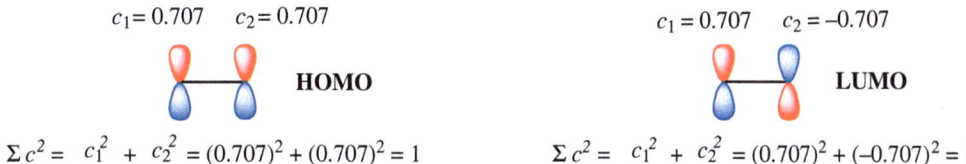

While drawing molecular orbitals, it is noteworthy that some of the atomic orbitals are large and some are small. For example, the HOMO and LUMO of butadiene are given below. The different orbital coefficients show that the electron density is not equally distributed within a molecular orbital. The atomic orbital coefficients found at the ends of butadiene are greater than those found in the interior. This means that the electron density in this orbital is higher at the ends. Of course, the sum of the atomic orbital coefficients' squares must also be equal to 1 here. The orbital coefficients of some conjugated systems are given in Table 10.3.

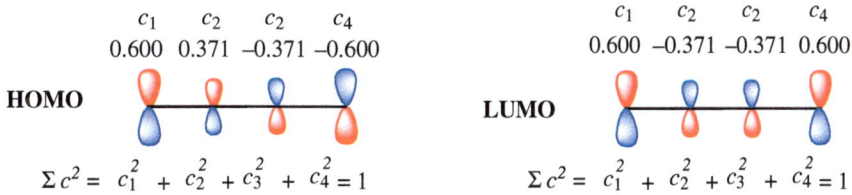

Table 10.3 The coefficients of the atomic orbitals in the molecular orbitals of some conjugated systems.

Molecule	Frontier orbitals	Coefficients of atomic orbitals							
Ethylene	LUMO	0.707	−0.707						
	HOMO	0.707	0.707						
Allyl cation	LUMO	0.707	0.0	−0.707					
	HOMO	0.500	0.707	0.500					
Allyl anion	LUMO	0.500	−0.707	0.500					
	HOMO	0.707	0.0	−0.707					
Butadiene	LUMO	0.600	−0.371	−0.371	0.600				
	HOMO	0.600	0.371	−0.371	−0.600				
Pentadienyl cation	LUMO	0.576	0.0	−0.576	0.0	0.576			
	HOMO	0.500	−0.500	0.0	0.500	−0.500			
Pentadienyl anion	LUMO	0.500	0.500	0.0	−0.500	−0.500			
	HOMO	0.576	0.0	−0.576	0.0	0.576			
Hexatriene	LUMO	0.521	−0.232	−0.418	0.418	0.232	−0.521		
	HOMO	0.521	0.232	−0.418	−0.418	0.232	0.521		
Octatetraene	LUMO	0.464	−0.161	−0.408	0.303	0.303	−0.408	−0.161	0.464
	HOMO	0.464	0.161	−0.408	−0.303	0.303	0.408	−0.161	−0.464

After this preliminary information about orbital coefficients, let us examine a specific reaction given below. 1-Methoxybutadiene forms a cycloaddition product with acrylaldehyde. As can be seen, the Diels–Alder reaction is regioselective. When an unsymmetric dienophile reacts with an unsymmetric diene, two regioisomeric products can be obtained, depending on the substituents' orientation. While the aldehyde group bonded to the dienophile decreases the energy level of the LUMO of the dienophile, the electron-donating methoxy group attached to butadiene increases the energy level of the HOMO of the diene. As the energy difference between the HOMO and LUMO is getting smaller, the product is easily formed [40]. In this addition reaction, acrylaldehyde can approach the diene in two different ways to form two different regioisomeric products. One of these products is a 1,3-substituted product and the other is a 1,2-substituted product.

If the steric factors are taken into account, at first glance, it can be thought that the 1,3-substituted product can be formed more easily. Contrary to this expectation, experiments show that only the 1,2-adduct is formed. Therefore, steric factors definitively have no effect on product formation. Other factors should be at the forefront for this observed site selectivity. The orbital coefficients determine the regioselectivity in this reaction.

The atomic orbital coefficients change from orbital to orbital. The sizes of the coefficients of the frontier orbitals entering the reaction are significant in terms of the reaction's direction. The sizes of the orbital coefficients control the regioselectivity in Diels–Alder reactions. Suppose the magnitudes of the orbital coefficients in the diene or dienophile within a molecular orbital are different. In that case, the orbitals with the larger coefficients of the diene prefer to interfere with the orbital of the dienophile with a larger coefficient, while the smaller ones of the diene interfere with the smaller ones of the dienophile. This is the most energy-efficient interaction (Figure 10.26). The circles show the lobes of the p orbital above the plane of the paper.

Figure 10.26 Interaction of orbitals with different orbital coefficients.

Finally, we come to the following conclusion: The orbital coefficients determine the regioselectivity. The orbital coefficients of conjugated systems are given in Table 10.3. The substituents attached to the diene and dienophile change the orbital coefficients. These coefficients are determined as a result of theoretical calculations. However, qualitatively, it is possible to determine the orbital coefficients (large or small) in some systems without doing any calculations. This gives us information about how a reaction will proceed.

Let us look at how to determine these coefficients for acrylaldehyde qualitatively. Acrylaldehyde has a butadiene structure if we consider the carbonyl oxygen atom as a carbon atom. On the other hand, because of the carbonyl group's polarization, acrylaldehyde also has an allyl cation structure. Therefore, we can consider the electronic structure of acrylaldehyde as a structure formed by mixing butadiene and allyl cation.

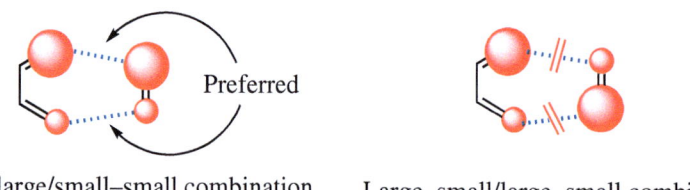

If we add the orbital coefficients of the butadiene and allyl cation using the values given in Table 10.3, the following result will be obtained. If we sum these coefficients, as shown in Figure 10.27, we approximate the orbital coefficients of acryl aldehyde in the HOMO and LUMO. The blue color shows the opposite sign of the orbitals.

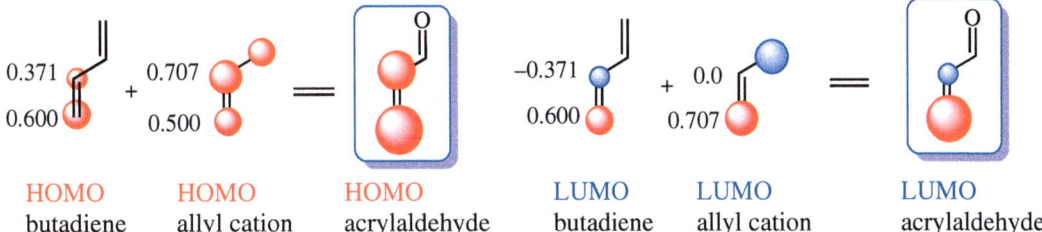

Figure 10.27 Qualitatively estimated orbital coefficients of the HOMO and LUMO of acrylaldehyde.

Similarly, we can qualitatively determine the orbital coefficients of the HOMO of 1-methoxybutadiene. Here, as the methoxy group is electron donating, we consider the molecule to be composed of butadiene and pentadienyl anions to determine the orbital coefficients (Figure 10.28).

We already know that in the reaction of 1-methoxybutadiene with acrylaldehyde, the HOMO of butadiene will interact with the dienophile's LUMO. The magnitude of the orbital coefficient at the α-carbon atom in acrylaldehyde is smaller than that at the β-carbon atom (Figure 10.28). On the other hand, the magnitude of the orbital coefficient at the C4 carbon atom in 1-methoxybutadiene is larger than that at the C2 carbon atom. Thus, the large–large/small–small combinations of the diene and dienophile will furnish a 1,2-disubstituted product.

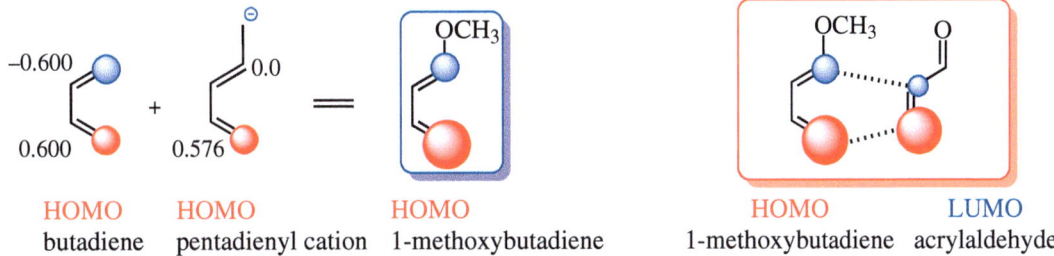

Figure 10.28 Qualitatively estimated orbital coefficients of the HOMO of 1-methoxybutadiene of acrylaldehyde and the interaction between the frontier orbitals of the diene and dienophile.

We saw that the orbital coefficients affect regioselectivity. The orbital coefficients also affect *periselectivity* [41, 42]. First, let us briefly explain what periselectivity is. The differentiation between two symmetry-allowed processes, for example, the [6 + 4] vs. [4 + 2] cycloaddition of dimethylfulvene to tropone, is called periselectivity. Tropone can undergo a [4 + 2] cycloaddition reaction with dimethylfulvene acting as a diene. According to the Woodward–Hoffmann rules, this addition is an allowed process as the number of electrons entering the reaction is 4 + 2 = 6. As an alternative, the diene part of the fulvene can add to tropone's hexatriene system, forming an [6 + 4] adduct. However, this product is not a thermodynamically preferred one.

The frontier orbitals indicate that the longer conjugated system of tropone is more reactive than the shorter one. The largest coefficients of the LUMO of tropone are at C2 and C7. On the other hand, the larger coefficients of the HOMO of dimethylfulvene are at C1 and C2 (Figure 10.29) [43]. Therefore, the interaction occurs between these orbitals, forming the [6 + 4] cycloadduct. As one perimeter preferably occurs, the reaction is called periselective.

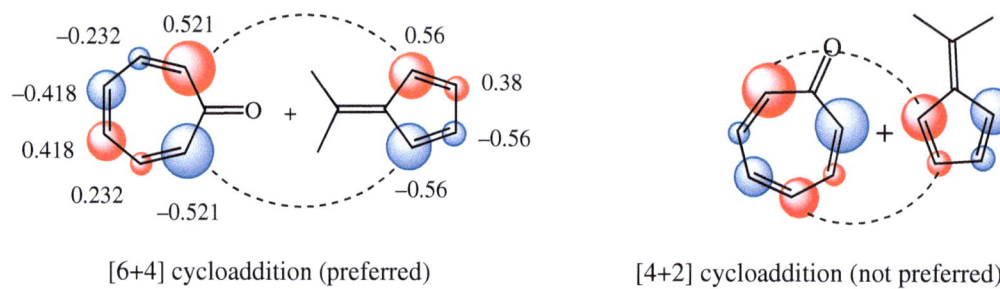

[6+4] cycloaddition (preferred) [4+2] cycloaddition (not preferred)

Figure 10.29 Orbital coefficients of the LUMO and HOMO of dimethylfulvene and tropone.

10.5 Sigmatropic Reactions

The intramolecular migration of an allylic σ bond (hydrogen or alkyl group) across a conjugated system to the terminus of the system through a cyclic transition state is called a *sigmatropic rearrangement*. Sigmatropic reactions occur under thermal and photochemical conditions, and they do not need any catalyst. During this rearrangement, the total numbers of σ bonds and π bonds remain unchanged.

A general example of a sigmatropic shift is given below. Here, the hydrogen atom, together with the sigma electrons (as a hydride), shifts from the C1 carbon atom to the C5 atom through the conjugated system. This shift is called a sigmatropic [1,5] shift or [1,5] sigmatropic rearrangement. To cause a sigmatropic shift, the π system must be conjugated.

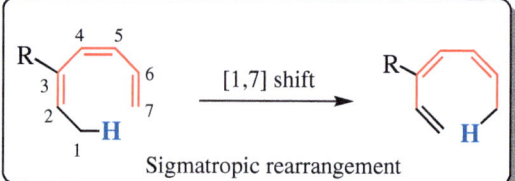

According to the number of π electrons in the conjugated system involved in the reaction, sigmatropic shifts can also be [1,3], [1,5], or [1,7] shifts. The process is concerted, involving simultaneous cleavage of a σ bond and the formation of a new σ bond while the π bonds rearrange. Thus, a cyclic transition state structure is formed by combining both ends of the π system through a migrating group. Of course, the orbitals must have like phases for the formation of such a transition state.

Let us assume that the π system has a planar structure; then, there can be two types of shifts. If the cleavage of the σ bond and its bonding to another carbon atom occurs on the same side of the plane, this shift is called a *suprafacial shift*. If the bonding occurs on the other side of the plane, this shift is called an *antarafacial shift* (Figure 10.30).

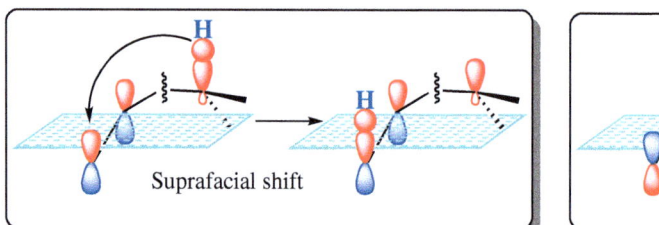 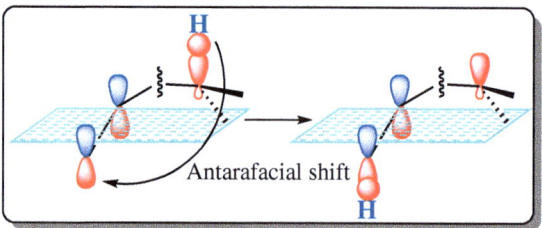

Figure 10.30 Presentation of suprafacial and antarafacial shifts.

It is possible to determine which shift will occur using the Woodward–Hoffman rules. In the sigmatropic rearrangements, besides the π bond electrons, the migrating group's σ bond electrons also participate in the reaction. Therefore, we need to consider the [1,3] shift as an allyl anion and the [1,5] shift as a pentadienyl anion. To understand the site selectivity and stereochemistry observed in these reactions, it is necessary first to examine the HOMO–LUMO interaction of the system. The HOMO of the system (allyl anion) and the LUMO of the migrating group or vice versa have to be considered to find the symmetry-allowing process.

For a [1,3] shift, we take the HOMO of the allyl system. As shown below, the orbital phase to which the hydrogen atom is bonded is not in the same phase as the orbital at the C3 carbon atom (Figure 10.31). According to the orbital symmetry, such a suprafacial shift is a thermally symmetry forbidden process. However, the Woodward–Hoffmann rules dictate that an antarafacial shift is an allowed process, but it is geometrically impossible (Figure 10.31). A suprafacial photochemical [1,3] shift is an allowed process according to the Woodward–Hoffmann rules. However, as the photochemical processes proceed through the diradical mechanism, the Woodward–Hoffmann rules cannot be applied.

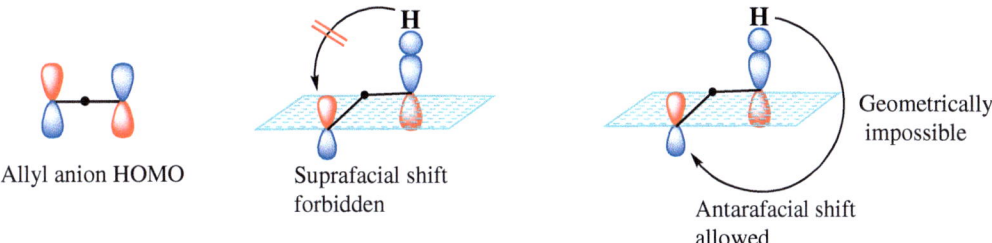

Figure 10.31 Illustration of a thermal 1,3-hydrogen shift.

Similar to [1,3] hydrogen shifts, suprafacial thermal [1,3] alkyl shifts are also forbidden. They must also proceed antarafacially. Again, the geometry of the transition state hinders an antarafacial shift. However, the alkyl groups can use the back lobe of the sp³ hybrid orbital to cause a suprafacial shift (Figure 10.32). However, the configuration at the migrating carbon atom undergoes an inversion during this rearrangement.

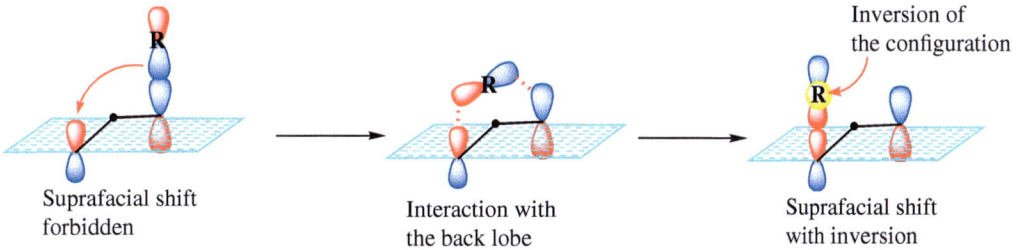

Figure 10.32 Presentation of a thermal suprafacial [1,3] shift with configuration inversion.

Pyrolysis of bicyclo[2.1.1]hex-2-ene at about 150–200 °C effects conversion to [3.1.0] hex-2-ene [44]. Normally, this reaction is a forbidden process. The product resulting from the [1,3] shift shows that the configuration of the carbon atom bearing the methyl group is inverted. This can only be explained by a suprafacial shift of the alkyl group using the back lobe of the migrating group, causing configuration isomerization.

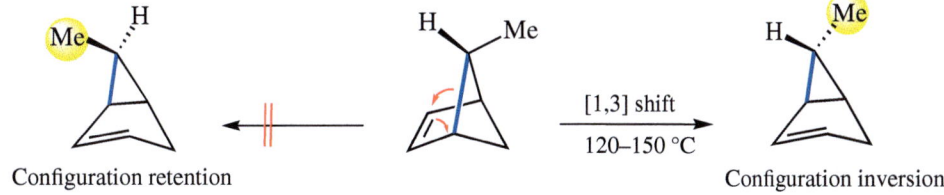

A remarkable stereospecific [1,3] alkyl shift was observed when *endo*-bicyclo[3.2.0]- hept-2-en-6-yl acetate was pyrolyzed at 300 °C to give *exo*-norbornyl acetate [45, 46]. Mechanistic studies reveal that this rearrangement can only proceed via a suprafacial [1,3] alkyl shift with configuration inversion at the migrating carbon atom where the carbon atom uses its back orbital lobe.

Among all of the sigmatropic migrations, a [1,5] shift is one of the most common reactions in cyclic systems because of both orbital symmetry and suitable geometry [47]. Under thermal conditions, the [1,5] shift proceeds suprafacially with retention of the configuration. Hydrogen can undergo a [1,5] shift in cyclic and open-chain systems at high temperatures (Figure 10.33). The transition structure for such processes is relatively easy to obtain and such reactions are common.

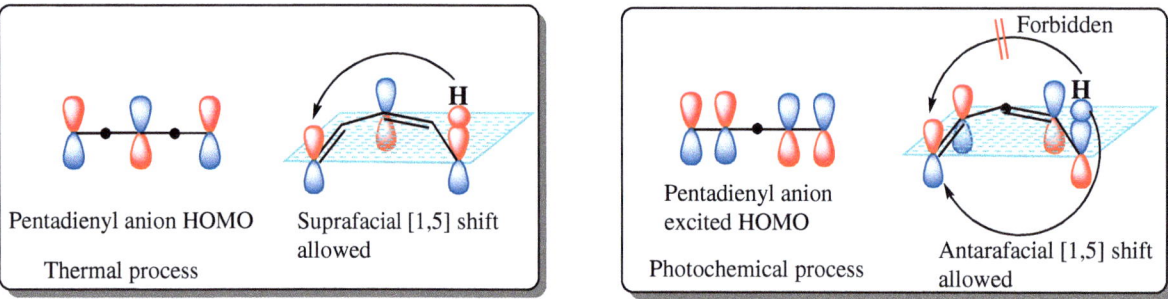

Figure 10.33 Illustration of thermal and photochemical [1,5] hydrogen shifts.

For example, substituted cyclopentadienes readily undergo a [1,5] hydrogen shift, resulting in isomeric compounds, because the geometry of the compound is suitable for interactions. Similar shifts occur in open diene systems; however, high temperatures such as 250 °C are required.

(S)-cis,trans-3-Methyl-7-deutero-octa-4,6-diene rearranges to a mixture of (R)-trans-cis-3-methyl-7-deutero-octa-3,5-diene and (S)-3-cis,cis-3-methyl-7-deutero-octa-3.5-diene. In both cases, the reaction proceeds by a suprafacial transition state, which is at least 8.0 kcal/mol lower in energy than the antarafacial one [48].

A [1,5] hydrogen shift is also encountered in substituted cycloheptatriene derivatives. Because of the cyclic nature of the structure, the formation of the required transition complex occurs at lower temperatures than with open systems. Temperatures of 100–150 °C are required for a [1,5] hydrogen shift. The monosubstituted cycloheptatriene derivatives undergo successive suprafacial [1,5] hydrogen shifts reasonably slower than those of cyclopentadienes [49].

Thermolysis of the spirocyclopentadiene derivative at 230–280 °C forms a [1,5] alkyl shift product, reducing the molecule's strain. The stereospecific rearrangement involves a [1,5] sigmatropic shift of the alkyl group with complete retention of the configuration [50].

Cycloheptatriene is in equilibrium with its valence isomer, norcaradiene. When 3,7,7-trimethylcycloheptatriene is heated at 300 °C in the gas phase, a mixture containing the recovered starting material and 2,7,7- and 1,7,7-trimethylcycloheptatriene was obtained [51, 52]. In the form of norcaradiene, the cyclopropyl group undergoes a suprafacial [1,5] alkyl shift following the Woodward–Hoffmann rules and isomeric cycloheptatriene derivatives are formed.

Thermally, a [1,7] shift is a suprafacial symmetry forbidden process. Unlike a [1,3] shift, the [1,7] sigmatropic shift proceeds in an antarafacial manner because the geometrical restrictions on the antarafacial transition state are not as great as those in the [1,3] shift case because of the flexibility of the π system (Figure 10.34).

Figure 10.34 Illustration of thermal suprafacial and antarafacial [1,7] hydrogen shifts.

The thermal equilibrium between precalciferol (previtamin D_3) and calciferol (vitamin D_2) is illustrated by an antarafacial [1,7] hydrogen shift.

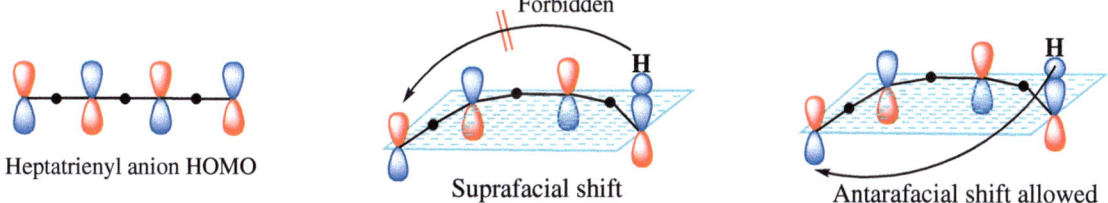

Based on this information, we can give selection rules from the above analysis according to the number of electrons involved in the transition structure (Table 10.4). Shortly, we count the π electrons and include the σ electrons from the migrating group.

Table 10.4 Woodward – Hoffmann rules for sigmatropic shifts.

Electrons in transition state	Reaction condition	
$4n$ π electrons	Thermal	Antarafacial
[1,3], [1,7] shifts	Photochemical	Suprafacial
$(4n+2)$ π electrons	Thermal	Suprafacial
[1,5] shifts	Photochemical	Antarafacial

10.6 Cope and Claisen Rearrangements

A [3,3] sigmatropic rearrangement of 1,5-dienes under thermal conditions leading to a regioisomeric 1,5-diene is called the *Cope rearrangement* [53–56]. The Claisen rearrangement is a [3,3] sigmatropic rearrangement of allyl vinyl ethers leading to homoallyl carbonyl compounds [53–58]. The Claisen rearrangement is also called the oxo-Cope rearrangement.

[3,3] Cope rearrangement

[3,3] Claisen rearrangement

Arthur C. Cope, an American chemist (1909–1966), discovered that ethyl (*E*)-2-allyl-2-cyanopent-3-enoate rearranges readily on heating at temperatures between 150 and 260 °C, giving ethyl (*Z*)-2-cyano-4-methylhepta-2,6-dienoate [59].

To test this reaction's generality, 1,5-hexadiene was heated to 300 °C and it rearranged back to 1,5-hexadiene. Because the initial and final products are the same here, it may not be understood if there is such a rearrangement at the first stage. If a substituent is attached to any of the carbon atoms, it can easily be determined that a rearrangement occurs as the resulting product will be different. Attachment of a substituent lowers the energy of the transition complex and rearrangement occurs at lower temperatures (150–200 °C).

Because a [1,3] shift occurs at both terminal carbon atoms, this reaction falls into the group of sigmatropic reactions and the overall reaction is classified as a [3,3] sigmatropic shift. The Cope rearrangement is a reversible reaction and the equilibrium is mainly shifted to the side of the thermodynamically stable isomer. The reaction proceeds through a six-membered cyclic transition state. A chair-like transition state is energetically preferred that minimizes steric interaction between the substituents in the transition state. The reaction is not affected by solvents in general, and it is a stereoselective reaction. The Cope rearrangement is an intramolecular reaction as a mixture of two different hexa-1,5-dienes does not form any cross product.

Doering and Roth [60] showed that heating of *meso*-3,4-dimethyl-hexa-1,5-diene at 225 °C gives exclusively *cis,trans*-octa-2,4-diene. The formation of this compound is only consistent with the formation of a chair-like conformation of the transition state. In the case of a boat-like conformation in the transition state, either *cis,cis* or *trans,trans*-octa,2,4-diene would be formed.

Chair-like conformation

In 1960, Vogel discovered that 1,2-divinylcyclopropane [61] and divinylcyclobutane [62, 63] rearrange to cyclohepta-1,4-diene and cycloocta-1,4-diene. Mechanistic studies show that the rearrangement proceeds through a concerted pathway. A boat-like transition state has been proposed as the chair-like transition state is sterically impossible because of the steric constraint. Both reactions proceed smoothly. The relief of the strain in the small rings is the driving force of these rearrangements. *cis*-Divinylcyclopropane undergoes rapid Cope rearrangement and so it can only be isolated at low temperatures.

The Cope rearrangement can proceed with even greater ease when the vinyl groups can be incorporated in a ring by connecting them with a methylene linkage. The compound formed, homotropilidene, undergoes a degenerate Cope rearrangement [64]. The structure is not changed by valence isomerization and it is identical to the starting material.

Homotropilidene *Cisoid* conformation

To increase the Cope rearrangement's reaction rate, it is necessary to ensure that the molecule is in the *cisoid* conformation. If the cyclopropane ring is connected in some way to the methylene group in homotropilidene, the conformation will be more rigid and the rate of the Cope rearrangement will increase. For this purpose, many exciting compounds have been synthesized. The most important ones among these are dihydrobullvalene, bullvalene, barbaralane, barbaralone, and semibullvalene [65].

Dihydrobullvalene   Bullvalene   Barbaralane   Barbaralone   Semibullvalene

All of these systems have a 1,5-hexadiene structure, and they are rapidly transformed into the corresponding valence isomers. They have two identical valence isomers, and they are degenerate systems. However, bullvalene has a threefold axis of symmetry, and it does not have only two valence isomers; it has 1 209 600. The ^{1}H NMR spectrum recorded at 120 °C shows a singlet, indicating the equivalency of all carbon atoms. Semibulvalene represents a Cope system with the lowest activation barrier (6.4 kcal/mol) for isomerization [65].

10.6.1 Oxy-Cope Rearrangement

When a hydroxyl group is attached to one of the sp^3-hybridized carbon atoms in a Cope system, the system will also undergo a [3,3] sigmatropic shift. Because the hydroxyl group will be attached to the double bond in the final product, the enol formed will be subsequently converted to the corresponding carbonyl group. This reaction is called *oxy-Cope rearrangement* [66, 67].

[3,3] oxy-Cope rearrangement

If the oxy-Cope systems are treated with a base that will remove the proton on the alcohol, then the oxy-Cope rearrangement rate will be 10^{17} times faster and the reaction can be conducted at room temperature. These reactions are generally referred to as *anionic oxy-Cope reactions* and they have interesting synthetic applications [68]. Taking advantage of the oxy-Cope rearrangement, it is possible to synthesize medium-sized ring systems, as shown below.

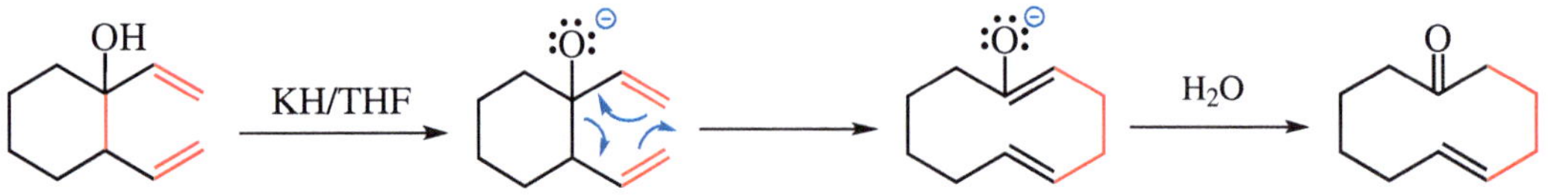

Interesting tricyclic compounds can be synthesized in one step by applying the anionic oxy-Cope rearrangement to a norbornene system [69].

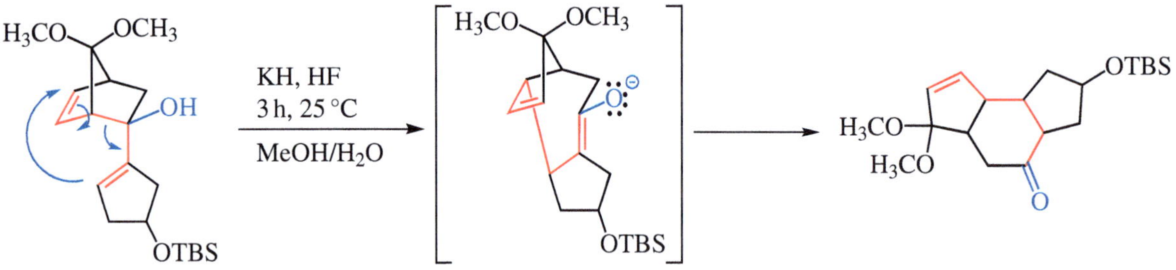

10.6.2 Claisen Rearrangement

The thermal [3,3] sigmatropic rearrangement of a vinyl allyl ether to an unsaturated carbonyl compound is called the *Claisen rearrangement* [53–58]. The reaction was first discovered by the German chemist Rainer Ludwig Claisen (1851–1930) in 1912 by heating a substituted vinyl allyl ether [70, 71].

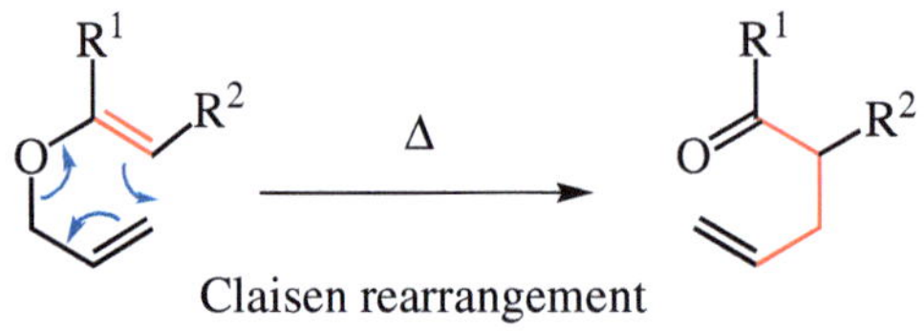

The Claisen reaction can occur even if one of the double bonds is incorporated in a benzene ring. One of the first examples is the [3,3] sigmatropic rearrangement of an allyl phenyl ether to an intermediate, which quickly tautomerizes to an *ortho*-substituted phenol. When substituents block the *ortho* positions in an allyl phenyl ether, the *para*-substituted product is formed by tandem Claisen and Cope rearrangement.

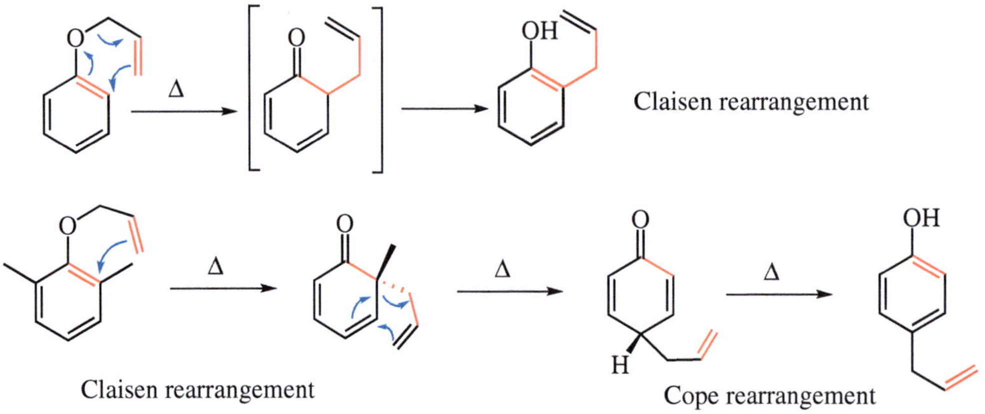

Several variations of the Claisen rearrangement have been reported. One of them is the *Ireland–Claisen rearrangement*. This variation uses the allyl ester of a carboxylic acid instead of allyl vinyl ethers. The reaction starts with deprotonation of an allylic ester, followed by trimethylsilyl chloride, to give the corresponding silyl-protected enolate, which undergoes [3,3] sigmatropic rearrangement at temperatures below 100 °C. Unsaturated acids are formed as a result of hydrolysis [72].

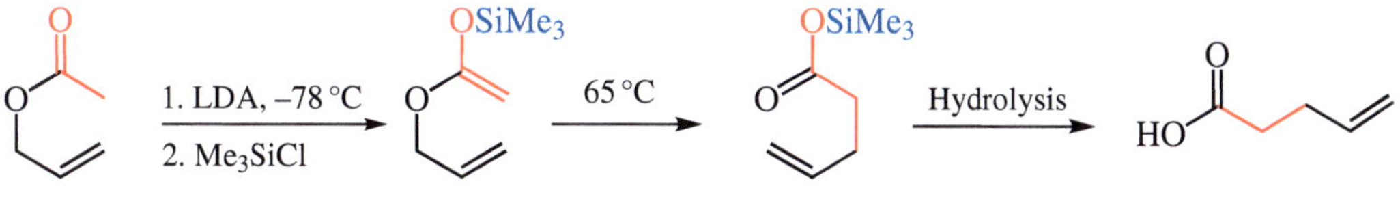

Ireland–Claisen rearrangement

Problems

10.1 Determine the configurations of hydrogen atoms in products formed under thermal conditions.

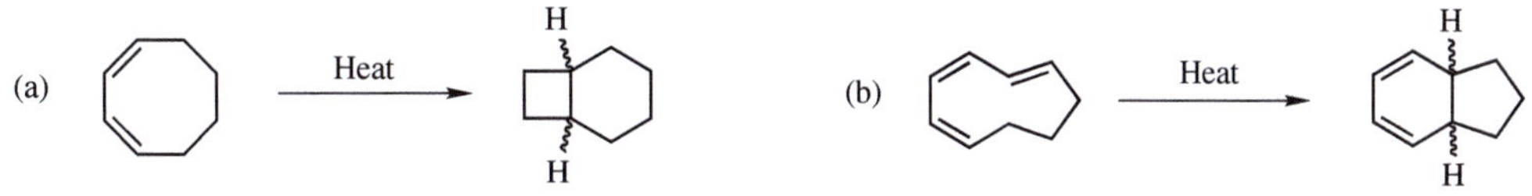

10.2 Give the products of each of the following reactions and give the exact configurations.

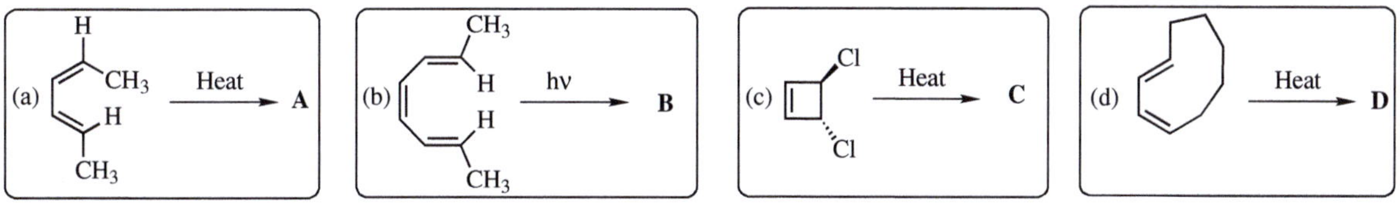

10.3 Give the reaction conditions under which ring closure will occur.

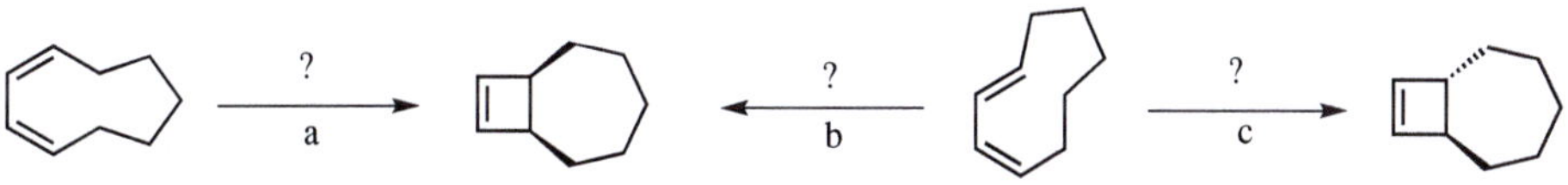

10.4 When bicylododecatetraene is heated, *trans*-dihydronaphthalene is formed. Give the structure of the intermediate and discuss its mechanism for formation.

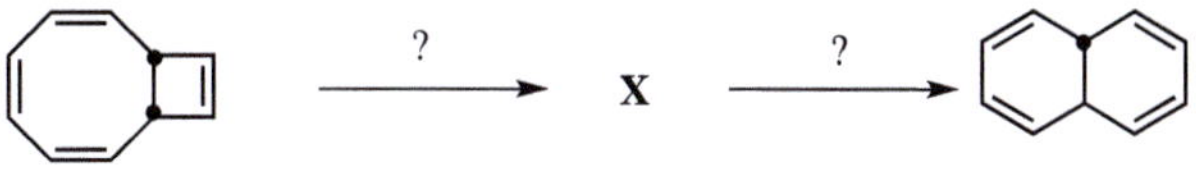

10.5 How can you perform the following transformations using only heat or light?

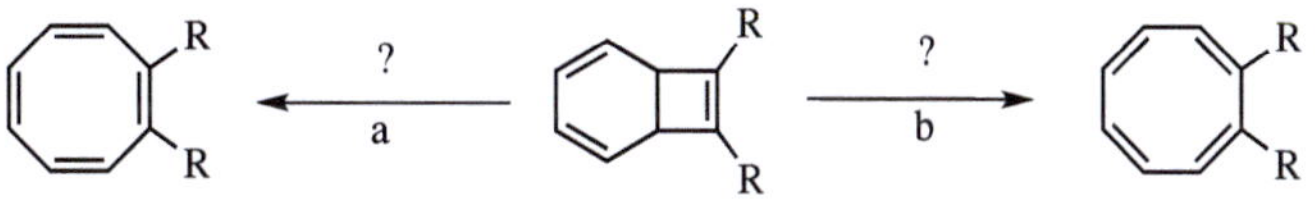

10.6 Which of the following compounds undergoes easier solvolysis in acetic acid? Give the structures of the products formed.

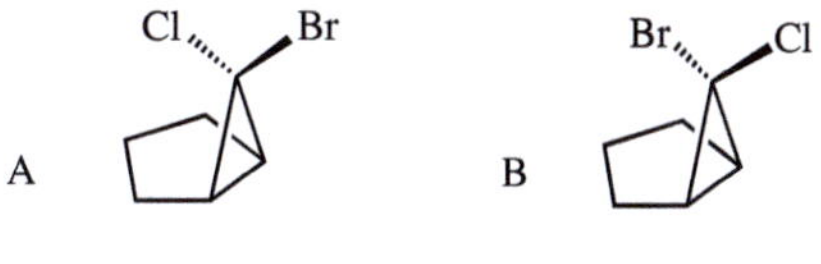

534 | 10 Pericyclic Reactions

10.7 Give the products formed when the following compound undergoes an electrocyclic reaction under thermal and photochemical conditions.

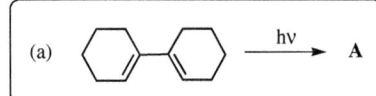

10.8 Give the product of each of the following reactions.

(a) [cyclohexenyl-cyclohexene] $\xrightarrow{h\nu}$ A

(b) [chloro-oxepine] $\xrightarrow{h\nu}$ B

10.9 Give the structures of the expected products and their correct configurations.

(a) butadiene + acrolein (CHO)

(b) furan + dimethyl maleate (COOCH₃, COOCH₃)

(c) H₃CO-vinyl ether + NC-acrylonitrile (CN)

(d) (CH₃, CH₃)-diene + dimethyl fumarate (COOCH₃, COOCH₃)

10.10 Give the products formed by the following reactions.

(a) cyclopentadiene + cyclopentadiene

(b) 2,3-dicyano-cyclohexadienone + butadiene

(c) anthracene + maleic anhydride

(d) cycloheptatriene + maleic anhydride

10.11 Propose a mechanism for the formation of the following compound.

tropone + cycloheptatriene → A → [cage product]

10.12 Write the products and discuss the mechanism for the formation of the products.

(a) isopropylidenecyclopentane $\xrightarrow{h\nu}$ A + B

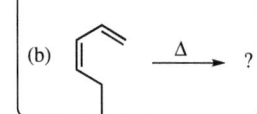

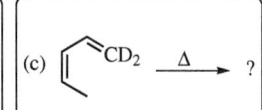

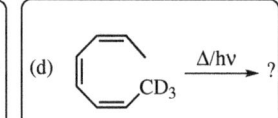

10.13 Give the product of each of the following reactions.

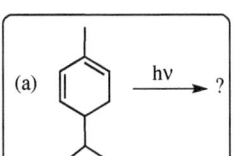

(b) 2-methylphenyl cyclopentadienylmethyl ether $\xrightarrow{\Delta}$?

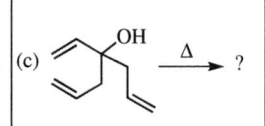

(d) H₃C-CN, CO₂Et diene $\xrightarrow{\Delta}$?

10.14 Give the product of each of the following sigmatropic rearrangements.

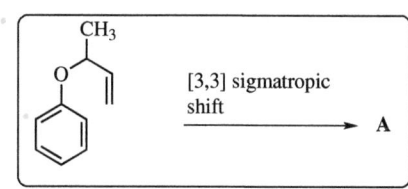
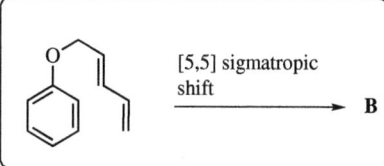

References

1. Havinga, E. and Schlattmann, J.L.M.A. (1961). *Tetrahedron* 16: 146.
2. Vogel, E. (1958). *Justus Liebigs Ann. Chem.* 615: 14.
3. Woodward, R.B. and Hoffmann, R. (1969). *Angew. Chem. Int. Ed.* 8: 781.
4. Woodward, R.B. and Hoffmann, R. (1970). *Die Erhaltung der Orbital Symmetrie*. Verlag Chemie.
5. Gilchrist, T.L. and Storr, R.C. (1972). *Organic Reactions and Orbital Symmetry*. Cambridge University Press.
6. Fleming, I. (1979). *Grenzorbitale und Reaktionen Organischer Verbindungen*. Verlag Chemie.
7. Woodward, R.B. and Hoffmann, R. (1965). *J. Am. Chem. Soc.* 87: 395.
8. Corey, E.J. (2004). *Impossible Dreams, J. Org. Chem.* 69: 2917.
9. Corey, E. (2004). *J. Chem. Eng. News* 82: 42.
10. Hoffmann, R. (2004). *Angew. Chem. Int. Ed.* 43: 6586.
11. Balcı, M. (1992). *Turk. J. Chem.* 16: 42.
12. McNamara, O.A. and Maguire, A.R. (2011). *Tetrahedron* 67: 9.
13. Cope, A.C., Haven, A.C., Ramp, F.S., and Trumbull, E.R. (1952). *J. Am. Chem. Soc.* 74: 4867.
14. Reppe, W., Schlichting, O., Klager, K., and Toepel, T. (1948). *Justus Liebigs Ann. Chem.* 560: 1.
15. Corey, E.J. and Streith, J. (1964). *J. Am. Chem. Soc.* 86: 950.
16. Boche, G., Bernheim, M., Lawaldt, D., and Ruisinger, B. (1979). *Tetrahedron Lett.* 20: 4285.
17. Paquette, L.A. and Stowell, J.C. (1971). *J. Am. Chem. Soc.* 93: 5735.
18. Sliwinski, W.F., Su, T.M., and Schleyer, P.v.R. (1972). *J. Am. Chem. Soc.* 94: 133.
19. Cristol, S.J., Segueira, R.M., and De Puy, C.H. (1965). *J. Am. Chem. Soc.* 87: 4007.
20. Özen, R. and Balcı, M. (2002). *Tetrahedron* 58: 3079.
21. Kilbas, B., Azizoglu, A., and Balcı, M. (2009). *J. Org. Chem.* 74: 7075.
22. Diels, O. and Alder, K. (1929). *Ber. Dtsch. Chem. Ges.* 62: 2087.
23. Huisgen, R. and Otto, P. (1968). *Tetrahedron Lett.* 9: 4491.
24. Şengül, M.E., Şimşek, N., and Balcı, M. (2000). *Eur. J. Org. Chem.*: 1359.
25. Schenk, W.K., Kyburz, R., and Neuenschwander, M. (1975). *Helv. Chim. Acta* 58: 1099.
26. Cookson, R.C., Drake, B.V., Hudec, J., and Morrison, A. (1966). *Chem. Commun.*: 15.
27. Ito, S., Fujise, Y., Okuda, T., and Inoue, Y. (1966). *Bull. Chem. Soc. Japan* 39: 1351.
28. Isakovic, L., Ashenhurst, J.A., and Gleason, J.L. (2001). *Org. Lett.* 3: 4189.
29. Rigby, J.H., Rege, S.D., Sandanayaka, V.P., and Kirova, M. (1996). *J. Org. Chem.* 61: 842.
30. Reingold, I.D., Kwong, K.S., and Menard, M.M. (1989). *J. Org. Chem.* 54: 708.
31. Ref. 1. Page 816. A private communication from W. von E. Doering to R. B. Woodward and R. Hoffmann
32. Ginsburg, D. (1983). *Tetrahedron* 39: 2095.
33. Garcia, J.I., Mayoral, J.A., and Salvatella, L. (2000). *Acc. Chem. Res.* 33: 658.
34. Fleming, I. (1976). *Frontier Orbitals and Organic Reactions*. Wiley.
35. Gilchrist, T.L. and Storr, R.C. (1972). *Organic Reactions and Orbital Symmetry*. Cambridge at the University Press.
36. Danishefsky, S. and Kitahara, T. (1974). *J. Am. Chem. Soc.* 96: 7807.
37. Png, Z.M., Zeng, H., Ye, Q., and Xu, J. (2017). *Chem. Asian J.* 12: 2142.
38. Sauer, J., Mielert, A., Lang, D., and Peter, D. (1965). *Chem. Ber.* 98: 1435.
39. Vermeeren, P., Hamlin, T.A., Fernández, I., and Bickelhaupt, M. (2020). *Angew. Chem. Int. Ed.* 59: 6201.
40. Chen, I.-H., Young, J.-N., and Yu, S. (2004). *Tetrahedron* 60: 11903.
41. Paddon-Row, M.N. and Warrener, R.N. (1974). *Tetrahedron Lett.* 15: 3797.
42. Paddon-Row, M.N. (1974). *Austr. J. Chem.* 27: 299.

43 Mandal, D.K. (2018). *Pericyclic Chemistry: Orbital Mechanisms and Stereochemistry*. Elsevier.
44 Roth, W.R. and Friedrich, A. (1969). *Tetrahedron Lett.* 10: 2607.
45 Berson, J.A. and Nelson, G.L. (1967). *J. Am. Chem.* 89: 5503.
46 Berson, J.A. (1968). *Acc. Chem. Res.* 1: 152.
47 Hess, B.A. Jr., and Baldwin, J.E. (2002). *J. Org. Chem.* 67: 6025.
48 Roth, W.R., König, J., and Stein, K. (1970). *Chem. Ber.* 103: 426.
49 Egger, K.W. (1967). *J. Am. Chem. Soc.* 89: 3688.
50 Boersma, M.A.M., De Haan, J.W., Kloosterziel, H., and Van de Ven, L.J.M. (1970). *J. Chem. Soc. Chem. Commun.*: 1168.
51 Berson, J.A. and Willcott, M.R. (1965). *J. Am. Chem. Soc.* 87: 2751.
52 Berson, J.A. and Willcott, M.R. (1966). *J. Am. Chem. Soc.* 88: 2494.
53 Rhoads, S.J. and Raulins, N.R. (1975). *Org. React.* 22: 1.
54 Nowicki, J. (2000). *Molecules* 5: 1033.
55 Lutz, R.P. (1984). *Chem. Rev.* 84: 205.
56 Baird, R.D. (2001). *Curr. Org. Chem.* 5: 395.
57 Ziegler, F.E. (1988). *Chem. Rev.* 88: 1423.
58 Wipf, P. (1991). *Compr. Org. Synth.* 5: 827.
59 Cope, A.C. and Hardy, E.M. (1940). *J. Am. Chem. Soc.* 62: 441.
60 Doering, W.v.E. and Roth, W.R. (1963). *Angew. Chem. Int. Ed.* 2: 115.
61 Vogel, E. (1960). *Angew. Chem.* 72: 4.
62 Vogel, E., Ott, K.-H., and Gajek, K. (1961). *Justus Liebigs Ann. Chem.* 644: 172.
63 Vogel, E. (1956). *Angew. Chem.* 68: 413.
64 Schröder, G., Oth, J.F.M., and Merenyi, R. (1965). *Angew. Chem. Int. Ed.* 4: 752.
65 Maier, G. (1972). *Valenzisomerisierungen*. Verlag Chemie.
66 Berson, J.A. and Jones, M. (1964). *J. Am. Chem. Soc.* 86: 5019.
67 Paquette, L.A. (1990). *Angew. Chem. Int. Ed.* 29: 609.
68 Paquette, L.A. (1997). *Tetrahedron* 53: 139.
69 Paquette, L.A., Gao, Z., Ni, Z., and Smith, G.F. (1998). *J. Am. Chem. Soc.* 120: 2543.
70 Claisen, L. (1912). *Chem. Ber.* 45: 3157.
71 Castro, A.M.M. (2004). *Chem. Rev.* 104: 2939.
72 Chai, Y., Hong, S.-P., Lindsay, H.A. et al. (2002). *Tetrahedron* 58: 2905.

11

Carbon–Carbon Coupling Reactions

11.1 History

The coupling reactions of organometallic reagents with organic electrophiles are widely used in the industry for the synthesis of agrochemicals, pharmaceuticals, and polymers. Among these tools, Pd-catalyzed coupling reactions are of great interest because of the favorable working conditions, excellent functional group tolerance, and chemoselectivity as well as their wide applicability. These cross-coupling processes have a rich and intriguing history commencing in the nineteenth century.

The history of acetylenic coupling began in 1869 with the observation by Carl Glaser that copper(I) phenylacetylide when exposed to air underwent smooth oxidative dimerization to diphenyldiacetylene [1]. Later, the Glaser coupling was extended to various organic compounds possessing a terminal ethynyl group.

In 1905, Straus observed that heating copper(I) phenylacetylide in acetic acid under inert gas (CO_2) gave enynes instead of the expected diphenyldiacetylene. This Straus coupling has even found industrial application in the production of vinylacetylene and divinyl-acetylene [2].

$$HC\equiv CH \xrightarrow[60\,°C]{CuCl/NH_4OH} H_2C=CH-C\equiv CH \;+\; H_2C=CH-C\equiv C-CH=CH_2$$

An important modification was reported in 1956 by Eglinton and Galbraith, who performed oxidative acetylenic couplings with a copper(II) salt in methanolic pyridine to construct macrocyclic compounds [3] including the pioneering annulene syntheses by Sondheimer [4].

The advantages of this new sp—sp bond forming reaction were successfully used for the construction of various acetylenic compounds. An impressive example of the use of the Glaser coupling was the synthesis of indigo by Baeyer in 1882 [5].

Following the development of C(sp)—C(sp) homocoupling, the methodology was extended to C(sp^2)—C(sp^2) bond formation. In 1901, Ullmann reacted 2-chloro- and 2-bromonitrobenzene with copper salts and obtained symmetric biaryls at elevated temperatures (200 °C) [6]. When dimethylformamide (DMF) is used as a solvent, the reaction can be carried out at lower temperatures. The reaction has most often been performed on aryl iodides. Aryl bromides and chlorides do not usually react unless there is an electron-withdrawing group like nitro or ester at the *ortho* and/or *para* positions to the halogen atom. The active species is a copper(I) compound that undergoes oxidative addition with the second equivalent of halide, followed by reductive elimination and the formation of the aryl—aryl bond.

The mechanism for the Ullmann reaction is not fully understood. The proposed mechanism begins with oxidative addition of the aryl halide to the copper, followed by a single electron transfer to form an organocuprate reagent. The organocuprate undergoes another oxidative addition on an aryl halide and after reductive elimination results in the final biaryl product.

The Wurtz coupling [7] is another of the oldest organic coupling reactions, wherein sodium metal is reacted with two alkyl halides to generate a simple dimer leading to C(sp^3)—C(sp^3) bond formation. The intramolecular Wurtz reaction has also found application in the preparation of strained ring compounds [8].

The reaction starts with electron transfer from sodium to halogen to produce sodium halide and an alkyl radical. The alkyl radical formed accepts an electron from another sodium atom to generate an alkyl anion. The nucleophilic carbon of the alkyl anion then substitutes the halide in an S_N2 reaction, forming a new carbon–carbon covalent bond.

$$R-X \xrightarrow[-NaX]{Na} \dot{R} \xrightarrow{Na} Na^{\oplus} R^{\ominus} \quad R-X \longrightarrow R-R + NaX$$

Grignard reagents are commonly prepared by reaction of an organohalogen with magnesium in a nitrogen atmosphere, and they are represented by the general formula RMgX [9]. For example, alkyl iodides generally react very rapidly. They are called Grignard reagents after their discoverer, French chemist Victor Grignard, who received the 1912 Nobel Prize for Chemistry for this work. Bennett and Turner in 1914 reacted phenylmagnesium bromide with a stoichiometric amount of chromium(III) chloride and they obtained C—C coupling product [10]. A few years later, it was shown that this homocoupling reaction can also be carried out with $CuCl_2$ [11].

$$Ph-MgBr \xrightarrow[Ether]{CrCl_3} Ph-Ph$$

In 1955, Cadiot and Chodkiewicz reported the first selective copper-catalyzed cross-coupling of alkynes with bromoalkynes (C(sp)—C(sp) coupling) under mild conditions [12]. A few years later, Castro and Stephens reported the C(sp)—C(sp^2) coupling involving aryl or vinyl halides with copper acetylides [13]. The Castro–Stephens method demanded elevated temperatures.

Cadiot–Chodkiewicz coupling:
$$Ph-C\equiv C-Br + H-C\equiv C-C(CH_3)_2OH \xrightarrow[EtNH_2/H_2O, rt]{CuCl\ (1-2\%\ mol),\ NH_2OH \cdot HCl} Ph-C\equiv C-C\equiv C-C(CH_3)_2OH \quad 92\%$$

Castro–Stephens coupling:
$$Ph-C\equiv C-Cu + I-C_6H_4-OMe \xrightarrow[120\,°C,\ 10\,h]{Pyridine} Ph-C\equiv C-C_6H_4-OMe \quad 99\%$$

11.2 Mizoroki–Heck Coupling Reaction

The cross-coupling reaction between an aryl/alkenyl halide and a terminal olefin in the presence of a Pd(0) catalyst to produce a substituted olefin is called the *Mizoroki–Heck reaction*. The most important catalytic reactions for the construction of carbon–carbon bonds involve the chemistry of Pd(0). The most stable palladium complexes are those in which the sum of the d electrons from the metal as well as the electrons donated by ligands are in total of 18. Pd(0) has a d^{10} electron configuration. Therefore, palladium can coordinate with four ligands where each of the ligands donates a pair of electrons to the metal, making an 18 electron configuration. Such complexes are saturated (all s, p, and d orbitals are filled), stable, and relatively unreactive. Dissociation of one or two ligands produces a 16 or 14 electron complexes that are highly reactive.

In 1968, Heck reported that the reaction of alkenes with an equal amount of phenylpalladium acetate prepared from phenylmercuric acetate and palladium acetate results in the arylation of alkenes. The reaction of propylene with phenylpalladium acetate gave an olefin mixture (total yield 66%), consisting of 60% *trans*-1-phenyl-1-propene, 9% *cis*-1-phenyl-1-propene, 15% allylbenzene, and 16% α-methylstyrene [14, 15].

11 Carbon–Carbon Coupling Reactions

$$Ph-Hg-OAc + Pd(OAc)_2 \longrightarrow Ph-PdOAc + Hg(OAc)_2$$
Phenylpalladium acetate

[Reaction scheme showing Ph–PdOAc reacting with H₂C=CH(CH₃) to form two intermediates Ph–H₂C–CH(CH₃)–PdOAc and AcOPd–H₂C–CH(CH₃)–Ph, each losing HPdOAc to give four products:]

- (H)(Ph)C=C(CH₃)(H) — 60%
- (Ph)(H)C=C(CH₃)(H) — 9%
- H₂C=CH–H₂C–Ph — 15%
- H₂C=C(CH₃)(Ph) — 16%

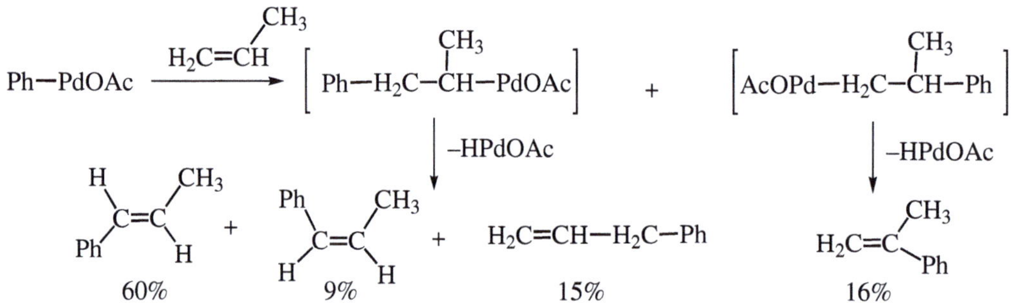

This pioneering work by Heck paved the way for a new reaction later called the *Mizoroki–Heck reaction*. In 1971, Mizoroki and coworkers reported the PdCl$_2$-catalyzed arylation of alkenes by iodobenzene in the presence of potassium acetate as a base [16]. In 1972, Heck and Nolley used Pd(OAc)$_2$ as a catalyst and *n*-Bu$_3$N as a base. The reactions were performed in a solvent (*N*-methylpyrrolidone [NMP]) at 100 °C or without any solvent [17]. The Nobel Prize for Chemistry in 2010 was awarded jointly to Richard F. Heck, an American chemist (1931–2015), Ei-ichi Negishi, a Japanese chemist (1935–), and Akira Suzuki, a Japanese chemist (1930–) "for palladium-catalyzed cross-couplings in organic synthesis." The untimely death of Mizoroki led to the development of this reaction by Heck.

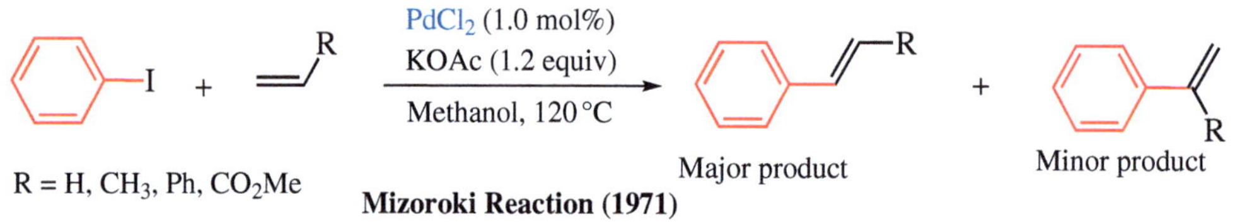

Mizoroki Reaction (1971)

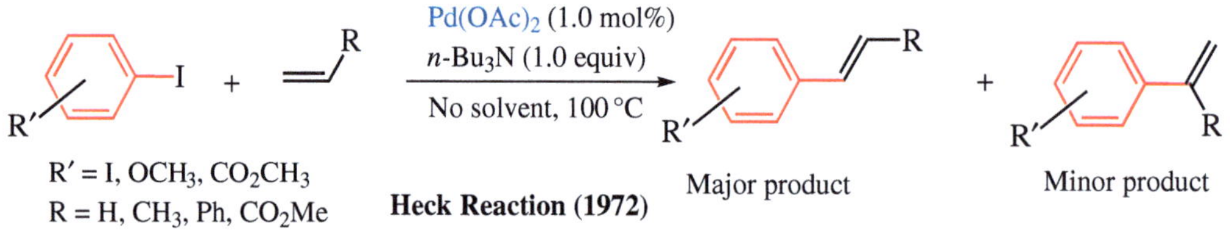

Ph–I + CH₂=CH–R → (PdCl$_2$ (1.0 mol%), KOAc (1.2 equiv), Methanol, 120 °C) → Ph–CH=CH–R (Major product) + Ph–C(=CH₂)–R (Minor product)

R = H, CH$_3$, Ph, CO$_2$Me

Heck Reaction (1972)

R′–C$_6$H$_4$–I + CH₂=CH–R → (Pd(OAc)$_2$ (1.0 mol%), *n*-Bu$_3$N (1.0 equiv), No solvent, 100 °C) → R′–C$_6$H$_4$–CH=CH–R (Major product) + R′–C$_6$H$_4$–C(=CH$_2$)–R (Minor product)

R′ = I, OCH$_3$, CO$_2$CH$_3$
R = H, CH$_3$, Ph, CO$_2$Me

More importantly, they proposed for the first time a full mechanism for the catalytic insertion reaction. At that time, it was well established that tetrakis(triphenylphosphine)-palladium(0) (Pd(PPh$_3$)$_4$) reacts readily with a variety of organic halides to form oxidative addition products as shown below [18].

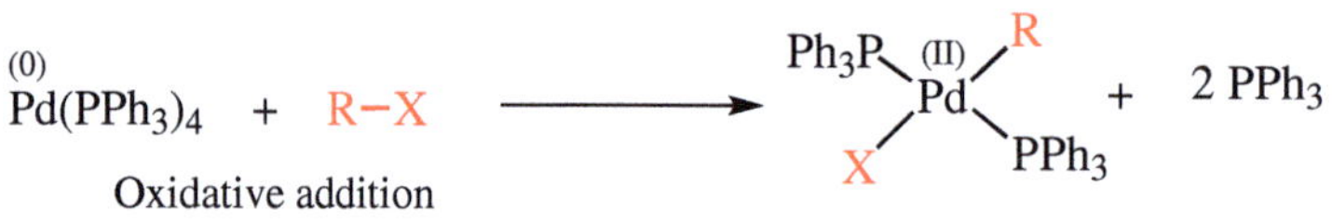

$$\text{Pd}^{(0)}(\text{PPh}_3)_4 + R-X \longrightarrow \text{Ph}_3P-\text{Pd}^{(II)}(R)(X)(\text{PPh}_3) + 2\,\text{PPh}_3$$
Oxidative addition

Heck and Nolley proposed the following reaction mechanism. First, the Pd(II) complex must be reduced to Pd(0) in order to enter the catalytic cycle. Pd(OAc)$_2$ undergoes reduction in situ by the olefin added to the reaction mixture to generate palladium metal. The first step of the coupling reaction is the oxidative addition of the aryl halide RX (or pseudohalide) to the catalytically active L$_n$Pd^0 species to generate RPdX, which initiates the catalytic cycle.

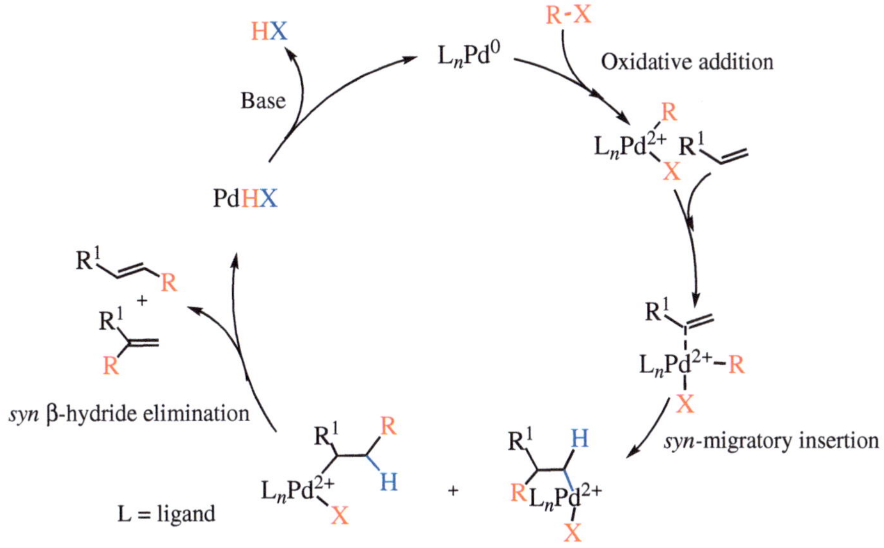

Scheme 11.1 The catalytic cycle for the Heck reaction.

When this reaction is carried out in the presence of an alkene, the reaction progresses by coordination of an alkene to the Pd^{II} species, followed by its syn migratory insertion. The regioselectivity of this insertion depends on the nature of the alkene and the catalyst, as well as the reaction conditions employed. The adduct formed, an organopalladium species, then decomposes by *syn* β-hydride elimination of hydridopalladium halide, forming the substituted olefin. For this elimination, an internal C—C bond rotation in the σ-alkyl–palladium(II) halide is necessary in order to bring the β-hydrogen into a syn position relative to the palladium atom. Subsequently, base-assisted elimination of H–X from the HPdX regenerates the L_nPd^0 catalyst (Scheme 11.1).

Actually, the first step of the Heck reaction is not the oxidative addition of the substrate [19–22]. The palladium salt, taken as a catalyst, must be first reduced to Pd(0) in order to enter the catalytic cycle. There are different mechanisms involving alkene-, amine-, and phosphine-mediated reduction. Heck proposed that alkenes may play a role in reducing agents according to the mechanism depicted below [15]. The catalytic precursor Pd(OAc)$_2$, associated with phosphine ligands such as PPh$_3$, is normally used to catalyze the reaction. Jutand and coworkers discovered that a Pd(0) complex can be generated by reacting Pd(OAc)$_2$ and PPh$_3$ [23]. In this process, PPh$_3$ is oxidized to phosphine oxide, thereby attesting to the reduction of Pd(II) to Pd(0) by the phosphine. In the case of having too many monophosphine ligands, the catalyst can be inhibited because a coordinatively saturated metal complex will be formed via ligand association. Furthermore, it is commonly accepted that another role of phosphine ligands is to prevent the formation of palladium black by acting as a stabilizing ligand. The reduction of Pd(II) catalysts can also be achieved without the assistance of phosphines. Amines used as bases in Mizoroki–Heck reactions have also been proposed as reducing agents. Indeed, β-hydride elimination may take place in the amine coordinated Pd(II) complex (Scheme 11.1).

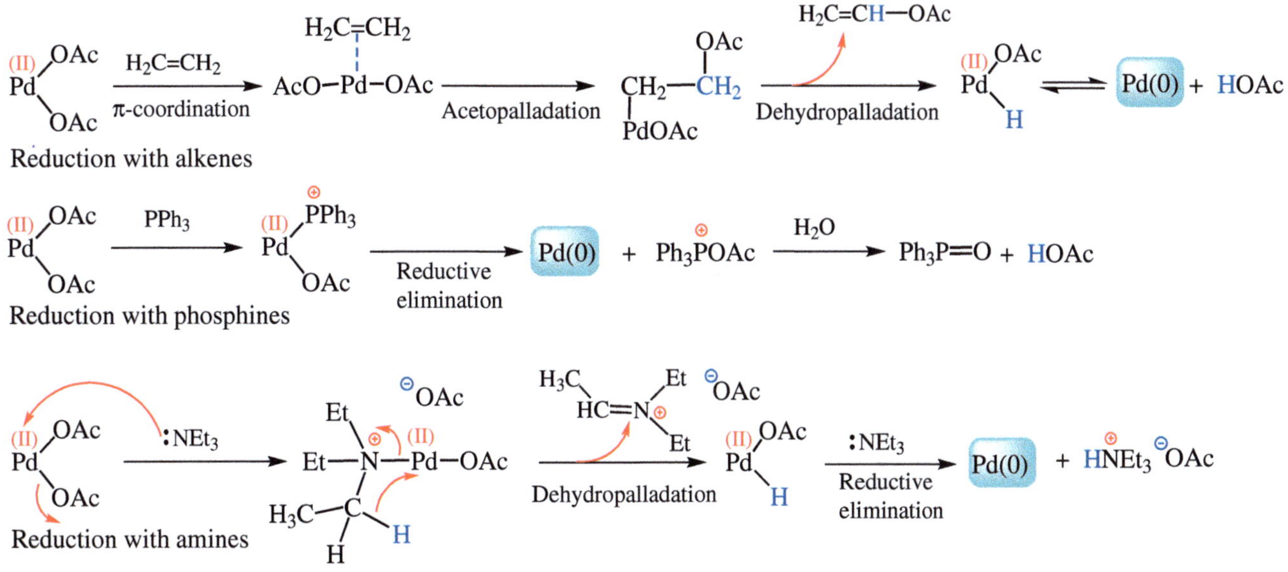

Oxidative addition is the most difficult step of the entire catalytic cycle. The Pd(0) catalyst can be more activated by the presence of electron-donating groups on the phosphine ligands so that the R—X bond can be easily broken along with the formation of Pd—R and Pd—X bonds. The rate of oxidative addition also depends on the chemical property of halides. Under the same experimental conditions, the reactivity order of aryl halides in Mizoroki–Heck reactions is usually ArI > OTf > ArBr ≫ ArCl [24]. Aryl chlorides are easily available substrates, but they have been rarely used for palladium-catalyzed cross-coupling reactions because their oxidative addition to palladium(0) is too slow to develop the catalytic cycle. The oxidative addition step is the rate-determining step for the less reactive halides. For the most reactive ones, for example, aryl iodides, the complexation/insertion of the alkene is the rate-determining step.

To prove syn-elimination, several cyclic alkenes such as cyclopentene, cyclohexene, and cycloheptene were subjected to phenylation. Indeed, little or no conjugated products were formed. For example, phenylation with cycloheptene gave an allylic compound. The expected conjugated product was not formed [25]. As mentioned above, the palladium and the hydride attached must be syn-coplanar for the initiation of elimination. Because of the restricted rotation about the C—C bond, no *syn* β-hydride is available. Therefore, the elimination takes place with the *syn* β′-hydride, leading to the formation of a nonconjugated product.

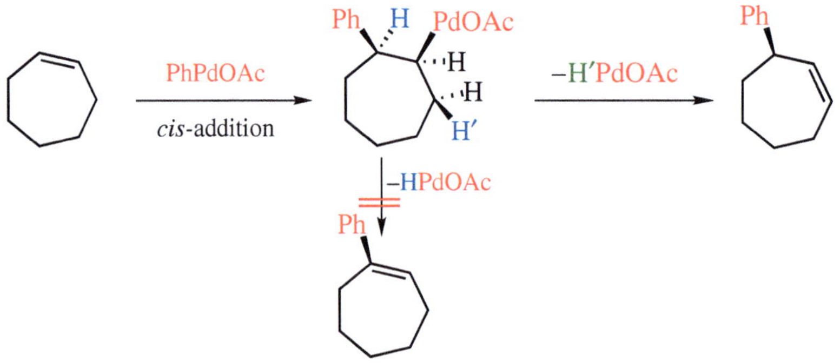

11.2.1 Regioselectivity

The regioselectivity of the Mizoroki–Heck reaction is determined by two major factors, namely, electronic and steric effects. The nature of the alkene, the catalyst, and the reaction conditions employed also affect the regioselectivity. After oxidative addition of the R–X group, the olefin must first associate onto the palladium complex, which requires the dissociation of one of the existing ligands. During the insertion process, the organic group R can add either to the α-carbon atom of the monosubstituted alkene, forming an α-product, or to the β-carbon, providing *trans*- or *cis*-β-products as shown below. Generally, electron-deficient alkenes react smoothly, forming mainly *trans*-isomers of the β-substitution products. On the other hand, alkenes with an electron-donating substituent, such as a heteroatom, generally form a mixture of the α-product and the β-product in the presence of monodentate ligands.

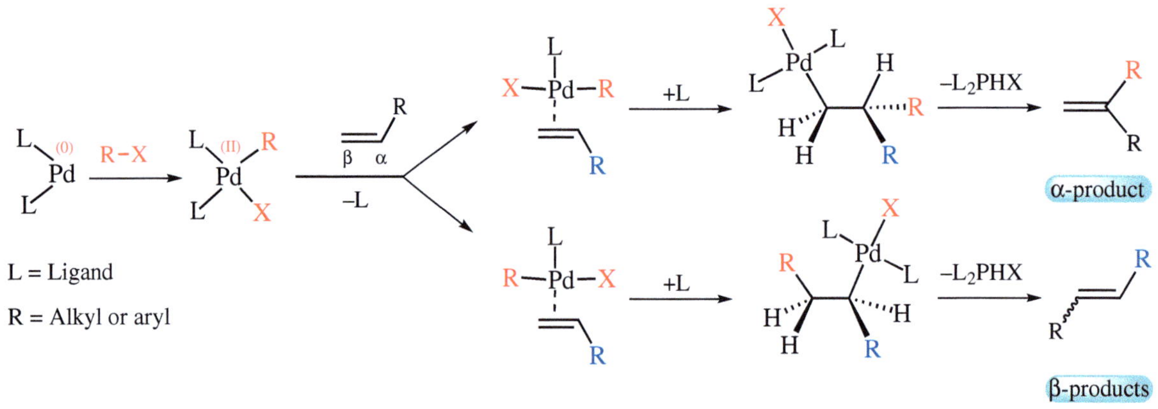

We mentioned that one of the existing ligands must leave the palladium complex after oxidative addition. There are two different leaving groups on palladium: the leaving group X and the ligand L. Mainly, phosphine ligands such as PR₃ are used. Palladium makes a strong Pd—X bond where the Pd—PR₃ bond is weak. Therefore, an oncoming group, the olefin, attacks the palladium complex and substitutes one of the ligands and forms a neutral π-complex (Path A). In a neutral π-complex, palladium coordinates one ligand and one anionic counter ion, usually a halide (Path A) [26–28]. If the X group in R–X is a very good leaving group, such as –OTf, –OTs, or –OAc, the oncoming group, the olefin, will substitute the leaving group X because these groups are weakly associated with palladium. Finally, a charged palladium complex will be formed (Path B) [29, 30].

Generally, monodentate phosphine ligands lead to the occurrence of both neutral and cationic mechanisms of course depending on the nature of the leaving group. On the other hand, bidentate phosphorus or nitrogen ligands along with triflates generate a cationic complex [31].

Now, the regioselectivity for electron-rich alkenes is determined by the nature of the π-complex. In a cationic π-complex, the organopalladium moiety is generally stabilized by an uncharged bidentate phosphine or nitrogen ligand. The cationic π-complex will lead to the α-product, while the neutral π-complex will give rise to a mixture of the α- and β-substituted products. For a neutral palladium complex, the regioselectivity is controlled by steric effects, which means nucleophiles attack on the less hindered site of the alkene, leading to a linear alkene. For cationic palladium complexes, the regioselectivity is controlled by electronic effects, which implies that nucleophilic attack occurs on the carbon atom with the least electron density of the alkene [32, 33].

For intermolecular Mizoroki–Heck reactions with neutral Pd complexes (coordination of the olefin via dissociation of one neutral ligand) and unactivated or electron-poor alkenes, the regioselectivity for R′ insertion is controlled by steric effects, leading to the formation of products substituted at the less sterically hindered position. However, in the case of neutral Pd complexes with electron-rich alkenes, the regioselectivity is controlled by electronic effects, resulting in substitution of α to the electron-donating group.

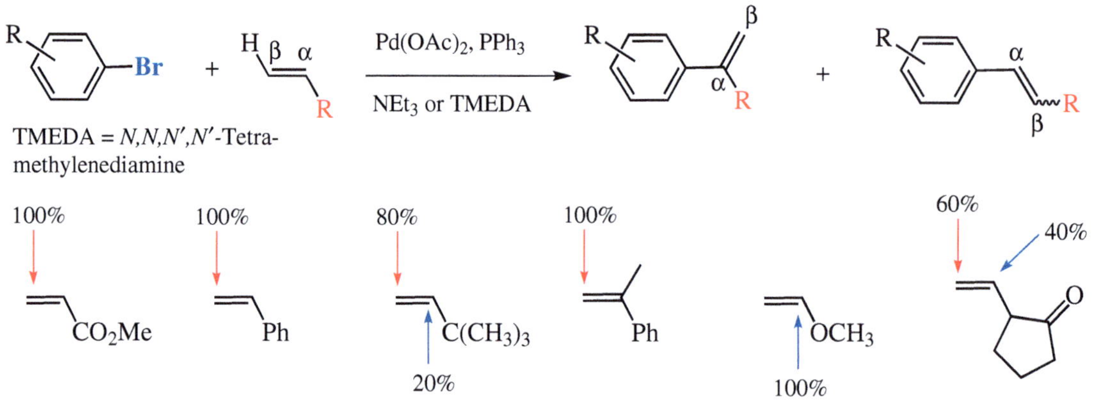

Regioselectivity of migratory insertion with neutral Pd complexes

However, when the reaction proceeds via dissociation of the counterion (the reaction of aryltriflate with various alkenes), electronic factors predominate. In fact, the coordination of the π-system in a cationic complex increases the polarization of the C=C bond; then, the aryl moiety migrates onto the carbon with lower charge density. Some examples are given below.

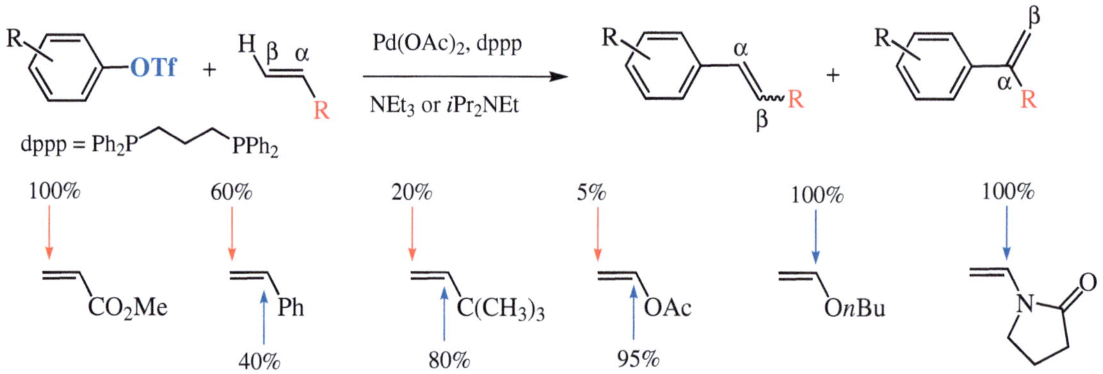

Regioselectivity of migratory insertion with cationic Pd complexes

As discussed, the Heck reaction is one of the most significant approaches for the construction of C—C linkages. Unlike other C—C bond-forming reactions that involve a polar addition, the Heck reaction tolerates almost any sensitive functionality such as unprotected amino, hydroxyl, aldehyde, ketone, carboxyl, ester, cyano, and nitro groups. Recently, the reaction was successfully applied to the synthesis of 2-methylbenzofuran by an intramolecular coupling of 1-iodo-2-(prop-1-en-2-yloxy)benzene. Different ligands were used for this cyclization reaction. However, the best result was obtained using [1,1′-bis(diphenylphosphino)ferrocene]dichloropalladium(II) (Pd(dppf)Cl$_2$) as the catalyst [34].

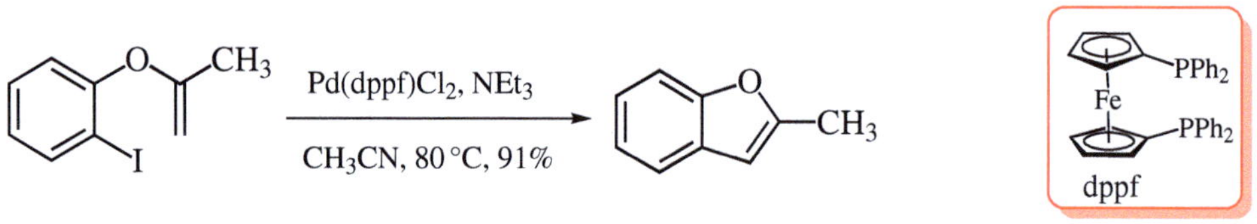

The Mizoroki–Heck reaction has also been successfully applied to sugar chemistry. The results gave a new insight into palladium-mediated cyclization in sugar chemistry. Cyclization of unsaturated carbohydrates stereospecifically forms the corresponding tricyclic heterosystems as illustrated below. The mechanism of this cyclization starts with the formation of a vinyl–palladium complex **A**. The migratory insertion of **A** generates a new σ-alkylpalladium complex **B**, which inserts into the double bond present in the side chain via a 5-*exo*-trig process to generate compound **C**. The β-hydride elimination results in the formation of a tricyclic compound [35].

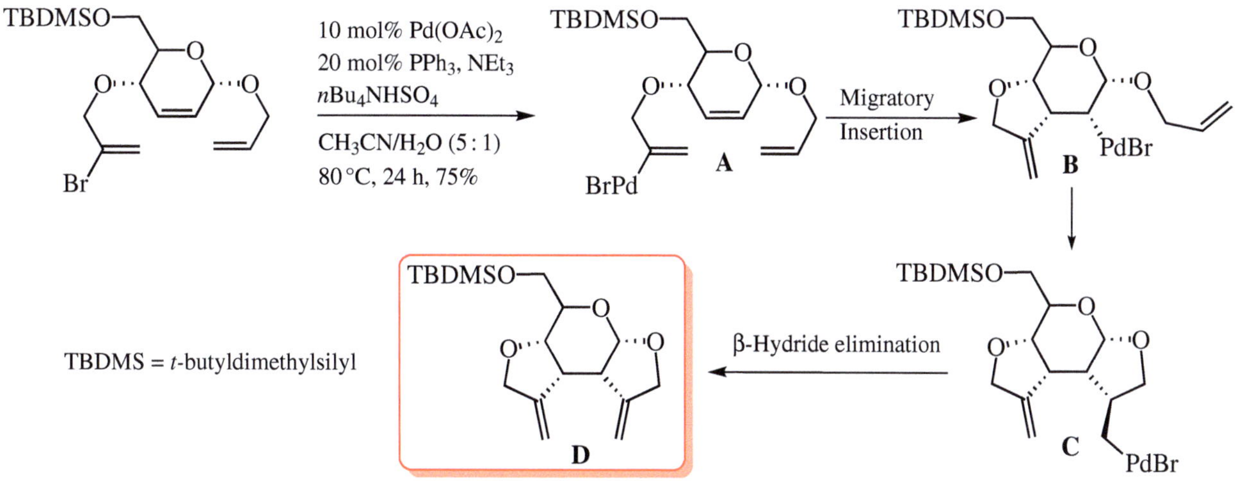

11.2 Mizoroki–Heck Coupling Reaction

TBDMS = *t*-butyldimethylsilyl

Abscisic acid is a natural product, and it is a well-known phytohormone showing interesting biological activities. The palladium-catalyzed Mizoroki–Heck cross-coupling reaction has been successfully applied to the synthesis of this natural product [36]. The protection of one of the carbonyl groups was performed with (S,S)-hydrobenzoin in the presence of a milder catalyst, pyridinium *p*-toluenesulfonate (PPTS). The Grignard reaction with an excess of vinylmagnesium bromide gave the desired addition product in quantitative yield. The Mizoroki–Heck coupling was performed with palladium acetate and silver carbonate under air without any ligand or solvent. Hydrolysis of the ester group followed by acidic treatment enabled the formation of abscisic acid.

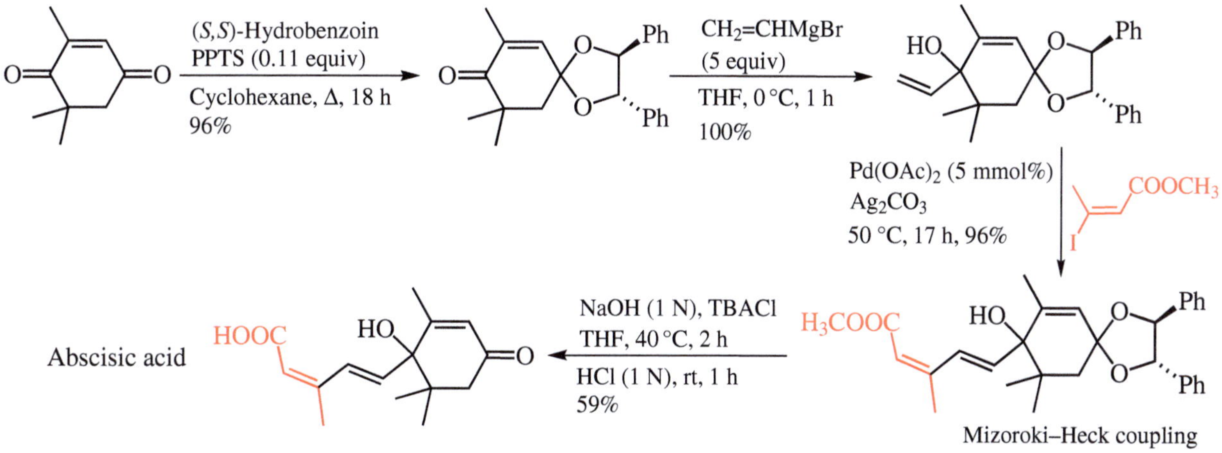

The asymmetric Heck reaction is a powerful method for the synthesis of both tertiary and quaternary chiral carbon centers, with an enantiomeric excess up to 99%. Of course, the chiral ligands are necessary to generate an asymmetric center in the molecule. Various new chiral ligands were discovered and they are applied to the synthesis of optically active compounds [37].

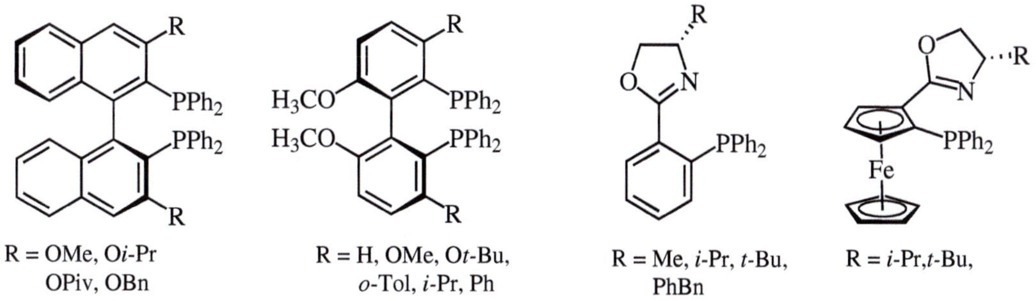

R = OMe, O*i*-Pr, OPiv, OBn

R = H, OMe, O*t*-Bu, *o*-Tol, *i*-Pr, Ph

R = Me, *i*-Pr, *t*-Bu, PhBn

R = *i*-Pr, *t*-Bu,

Optically active catalysts

These ligands can be mainly classified into two groups: BINAP (2,2′-bis(diphenylphosphino)-1,1′-binaphthyl) derivatives and oxazoline-containing ligands. The first example of the asymmetric intermolecular Heck reaction was reported by Hayashi and coworkers in 1991, where they used (R)-BINAP as a catalyst [38, 39]. t-Butyldihydrooxazole derivative was used as the catalyst for the asymmetric intermolecular Heck reaction of cyclohexenyl triflate with 2,3-dihydrofuran. Excellent conversion as well as regioselectivity was observed. The enantiomeric excess was found up to 99% [40].

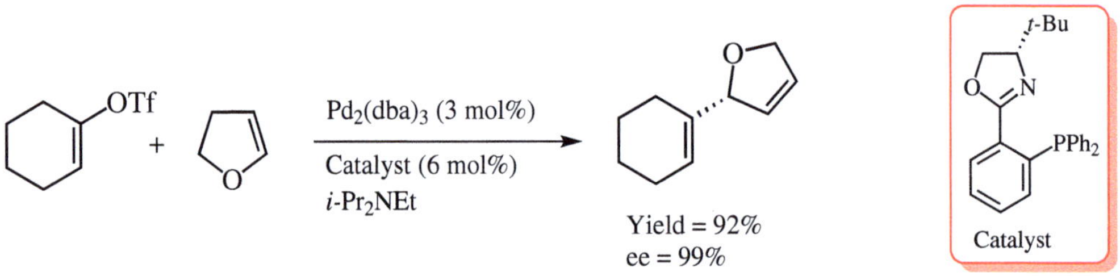

Overman and coworkers applied the asymmetric cascade Heck reaction to the synthesis of spiro-cyclic systems. This is a powerful method for catalytic asymmetric construction of quaternary carbon stereocenters. For example, cyclization of an enol triflate of a 1,3-diketone using Pd(OAc)$_2$ and (−)-DIOP proceeded via such a tandem Heck process forming the tricyclic hydrocarbon in 90% yield and 45% ee via an intermediate, which is shown below [41].

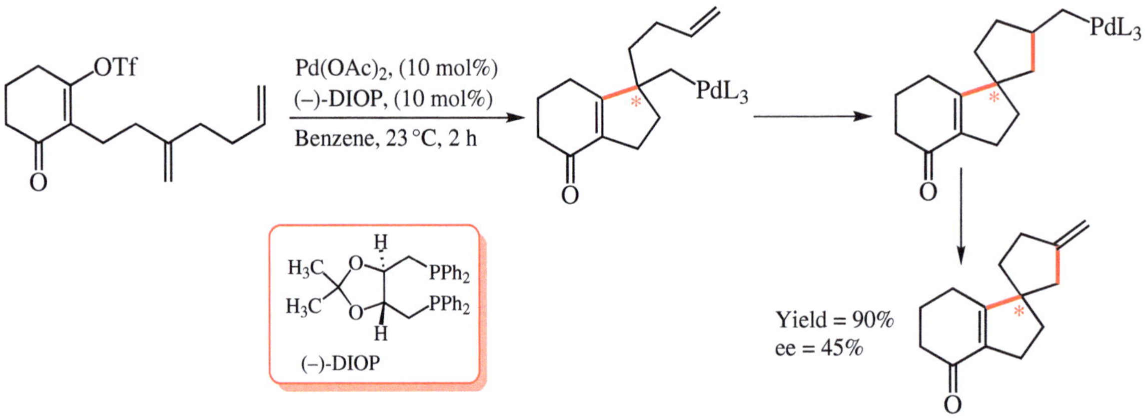

It has been shown that the Heck reaction of aryl or vinyl halides with electron-rich olefins such as vinyl ethers or enamides provides an indirect method for the synthesis of alkyl aryl (vinyl) ketones, where the aryl or vinyl group inserts at the α-carbon atom next to the heteroatom. Hydrolysis of the C—C coupling product provides the corresponding ketones.

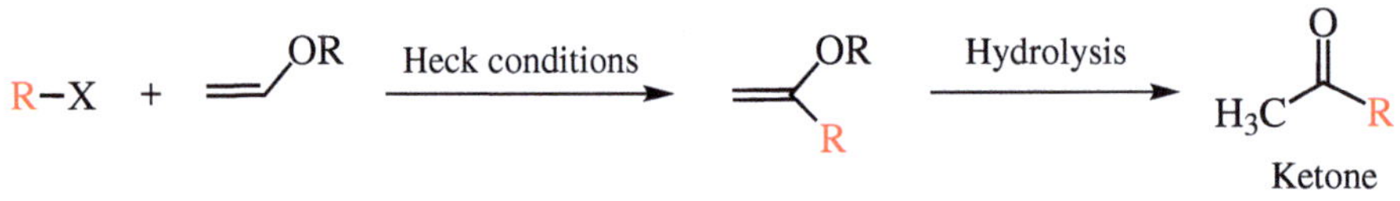

Under the basic conditions, an aldehyde can generate an enolate having a structure similar to that used in the Heck coupling reaction. Starting from this point, Xiao and coworkers developed a methodology for the acylation of aromatic compounds using various aldehydes [42]. The reaction of 1-bromo-4-methoxybenzene with pentanal in the presence of a base, using Pd(dba)$_2$-phosphine as a catalyst precursor and dppp, gives the acylation product in good yield.

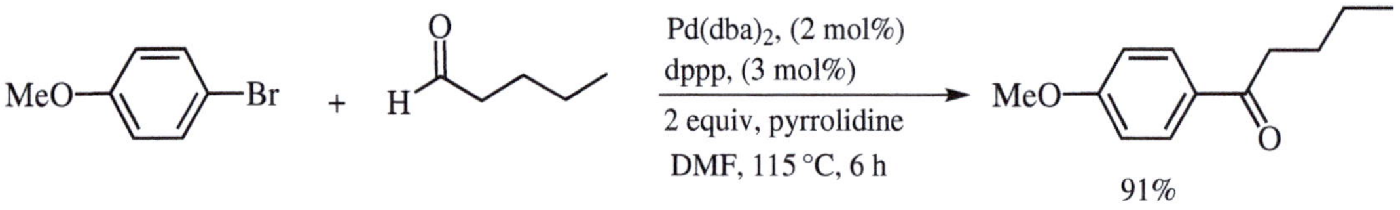

11.3 Stille Coupling Reaction

Over the past several decades, palladium-catalyzed coupling reactions have proved to be indispensable in organic synthesis. In particular, the Stille cross-coupling reaction is widely used in organic synthesis, and it involves the coupling of an organotin compound with a variety of organic electrophiles via a palladium-catalyzed coupling reaction.

$$R-X + R^1SnR_3 \xrightarrow[\text{Stille coupling}]{\text{Pd(0)}} R-R^1 + XSnR_3$$

The groundwork for the Stille reaction was carried out in 1976 and 1977 by two different groups. They explored numerous palladium-catalyzed couplings involving organotin reagents. Eaborn and coworkers explored the reaction of organic halides with R_3M-MR_3 (M = metal) in the presence of tetrakis(triarylphosphine)palladium(0). The coupling products were formed in 7–53% yields [43]. In 1977, Kosugi and coworkers expanded this reaction to acyl chlorides, and they reported catalytic alkylation, arylation, and vinylation of acyl chlorides with organotin compounds in the presence of the same catalyst [44].

X—C₆H₄—Br + Bu₃Sn—SnBu₃ →[Pd(PPh₃)₄, 7–53%] X—C₆H₄—C₆H₄—X + BrSnBu₃

X = H, OMe, NO₂   Eaborn and coworkers

R^1COCl + SnR₄ →[Pd(PPh₃)₄, 120–140 °C, 53–87%] R^1COR + ClSnR₃

R^1 = Me, Ph
R = Me, Bu, Ph, CH=CH₂   Kosugi and coworkers

In 1978, Stille and Milstein developed a much milder and more broadly applicable procedure. They reported that organotin compounds readily undergo palladium-catalyzed coupling with acid chlorides, providing a general and simple method for the preparation of ketones [45]. They have synthesized various ketones and found that the yields are high and in many cases quantitative. This coupling reaction can tolerate a wide range of functional groups on each coupling partner. Furthermore, it is regioselective. Stille made an amazing contribution to the mechanism of the different steps involved in the catalytic cycle [46–48]. John Kenneth Stille (1930–1989), an American chemist, would have shared the Nobel Prize for "Palladium-Catalyzed Cross-Coupling Reactions" given in 2010 to Richard Heck, Ei-ichi Negishi, and Akira Suzuki. However, his untimely death at the age of 59 in a plane crash in 1989 deprived him of that honor.

X—C₆H₄—COCl + R₄Sn →[Pd(PPh₃)₄, HMPA, 60 °C] X—C₆H₄—COR + R₃SnCl

X = 4-Cl, 4-Br, 4-OCH₃
4-NO₂, 4-CHO, 4-CN   R = Me, n-Bu, Ph, CH₂Ph
2-NO₂, 2-OCOCH₃

R—COCl + Me₄Sn →[Pd(PPh₃)₄, HMPA, 60 °C] R—COMe + Me₃SnCl

R = CH₂CH₂Ph, CH=CH₂
CH=CHPh, furanyl

The Stille reaction uses a palladium catalyst. An 18- or 16-electron Pd(0) complex can be used as the source catalyst, such as Pd(PPh$_3$)$_4$ or Pd(dba)$_2$. Then, through ligand dissociation, a 18-electron Pd(0) can be formed into a 14-electron Pd(0), an active catalyst.

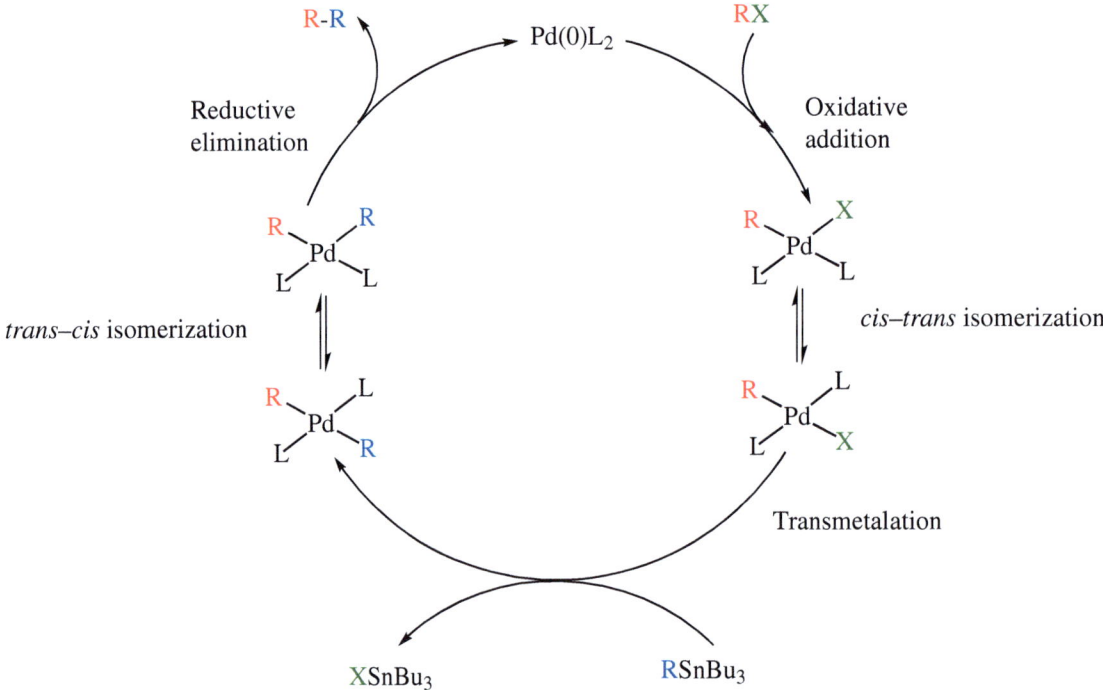

Scheme 11.2 The catalytic cycle for the Stille reaction.

In the first step, as discussed in the Heck coupling reaction, an oxidative addition of Pd(0) across a C—X bond to form an organopalladium species takes place. Transmetalation between the organopalladium(II) species and organotin generates a diorganopalladium(II) complex, which is followed by reductive elimination to form the coupling product and generate the active palladium(0) catalyst for the next catalytic cycle (Scheme 11.2).

While Pd(0) typically undergoes facile oxidative addition across iodine-, bromine-, and triflate-carbon bonds to generate the a *cis*-complex, chlorine–carbon bonds tend to be more difficult to insert. This fact can be explained on the basis of the dissociation energies between the halogens and carbon atoms. For sp^2-hybridized organohalides, a concerted three-center oxidative addition to Pd(0) complex is proposed. This kind of addition should form first a *cis*-complex as shown below. In the case of complexes with monodentate phosphine ligands, the initially formed *cis*-complex can isomerize to form the more stable *trans*-complex. The driving force for *cis–trans* isomerization is mainly the presence of bulky phosphine ligands, which tend to adopt the *trans*-configuration. All the systems with bidentate ligands investigated are less reactive. Complexes with Pd(L–L)$_2$ were found to not undergo oxidative addition with phenyl iodide [49].

The next step, the transmetalation, proceeds via different mechanisms depending on the substrates and reaction conditions. The most common accepted transmetalation involves an associative mechanism. First, when the organostannane initially adds to the *trans* metal complex, the X group can coordinate to the tin, in addition to the palladium, producing a cyclic transition state. Breakdown of this adduct results in the loss of Bu$_3$Sn-X and a trivalent palladium complex with R and R′ present in a *cis* configuration. The second mechanism is the open and ionic mechanism observed in the case of badly coordinating anionic ligands (e.g. triflate). In this case, the X group does not coordinate to the tin, producing an open transition state, which creates a highly electrophilic Pd center, leading to fast transmetalation, where the R group attacks the palladium atom. The transmetalation step is generally assumed to be the rate-determining step [50, 51]. The accelerative effect of additives is thought to be due to promotion of the transmetalation process. The order of the rate of transmetalation from tin is alkynyl > alkenyl > aryl > allyl, benzyl > α-alkoxyalkyl > alkyl. Therefore, the butyl and methyl groups from the organotin reagents rarely participate in the transmetalation reaction.

The reductive elimination is thought to occur from the *cis*-complex [52]. Therefore, two *trans* R groups must first isomerize into a *cis*-configuration. Then, it can undergo concerted reductive elimination. This step involves a cyclic three-center transition state [53]. There is another proposed mechanism, which involves the association or dissociation of ligands. An extra ligand can be bonded to palladium, forming a trigonal bipyramidal structure and this forces the two R groups into an equatorial position, which is a suitable configuration for the formation of C—C bonds. If the ligands are bulky enough, like phosphines with large cone angles, it is possible to push the two R groups closer to each other into an appropriate coordination angle, and this can accelerate the rate of the reductive elimination process. The activation energy of coupling depends on the hybridization of the carbon atoms to be coupled, following the trend sp^3-C—C-sp^3 > sp^3-C—C-sp^2 > sp^2-C—C-sp^2 [54].

It has been well documented that the inclusion of a Cu(I) salt and fluoride ion in the Stille coupling reactions accelerates the reaction rate as well as increasing the yield [55, 56]. It is thought that the function performed by the CuI salt in the catalytic cycle is solvent dependent [57]. In solvents such as tetrahydrofuran (THF) and dioxane, the CuI acts as a scavenger of the free neutral ligand that otherwise causes retardation of the transmetalation. On the other hand, in polar solvents such as NMP and DMF, Cu(I) ions are directly involved in the transmetalation reaction. It is assumed that a more reactive organocopper intermediate is formed from the organotin compound, which then takes part in the catalytic cycle (Scheme 11.3).

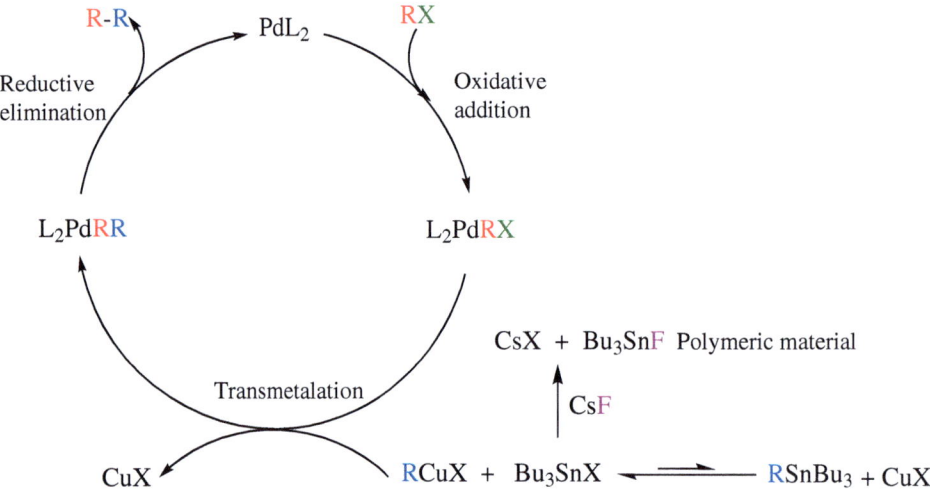

Scheme 11.3 Mechanism of the CsF/CuI-promoted Stille coupling reaction.

Stille and Scott observed previously that the fluoride ion increases the rate of the palladium-catalyzed coupling reaction between vinyl triflates and organostannanes [58]. CsF reacts with Bu_3SnX to form the insoluble polymeric material Bu_3SnF, which can be removed from the reaction mixture by filtration [59]. If this transmetalation reaction is reversible, then the removal of the by-product Bu_3SnX from the reaction will shift the equilibrium toward the formation of the more reactive organocopper intermediate.

Baldwin and coworkers investigated the Stille coupling reaction of 4-iodotoluene and aryl stannane, 4-(tri-n-butylstannyl) nitrobenzene, under different conditions [60]. When the reaction was carried out in the absence of CsF and CuI, most of the starting material was recovered. The addition of CsF to the reaction medium had little effect on the yield. A significant improvement in the yield was observed by adding CuI. However, when both CuI and CsF were used together, the coupling reaction proceeded to completion with 98% yield.

Reagents	Yield (%)
$Pd(PPh_3)_4$	2
$Pd(PPh_3)_4$, CsF	8
$Pd(PPh_3)_4$, CuI	46
$Pd(PPh_3)_4$, CuI and CsF	98
CuI, CsF	0

11.3.1 Synthesis of Organotin Compounds

In the palladium-catalyzed coupling reactions of organic electrophiles with organotin compounds, essentially only one of the substituents on tin is involved in the coupling reaction. This is not a problem if a relatively simple organic group, like a methyl or tetra-n-butyl group, is transferred. Then, tetramethyltin or tetra-n-butyltin can be used.

$$R-X + R-SnR_3 \xrightarrow{PdL_n} R-R + X-SnR_3$$

However, if the group to be transferred is more difficult to synthesize, then the utilization of only one of four identical groups would be a disadvantage. Fortunately, the rate of transfer of groups varies. Methyl and *n*-butyl groups are often used as the other three R groups that bind to stannane, as these alkyl groups are not as reactive in the transmetalation process. Therefore, these groups are called *non-transferable* groups. The order of reactivity of the R groups bonded to tin is as follows:

Alkynyl > Alkenyl > Aryl > Allyl = Benzyl > α-Alkoxyl > Alkyl

A number of methods for the synthesis of unsymmetrical organotins are known. Reaction of a triorganotin halide with an organometallic compound is widely used. However, its scope of application is limited because the organometallic compounds undergo reactions with various functional groups. The reaction of stannyl anions with electrophilic substrates allows the introduction of functional groups [61, 62]. The addition of R_3SnH to C—C double or triple bonds is also an efficient way to synthesize substituted organotin compounds. The addition starts by initiators such as azobisisobutyronitrile (AIBN) or by ultraviolet (UV) irradiation. Palladium complexes are also used for the synthesis of organotin compounds.

E = COOR, $CONH_2$, CN

dba = dibenzylideneacetone

NMP = *N*-Methyl-2-pyrrolidone

One of the reasons for the popularity of the Stille reaction in modern organic synthesis is the fact that trialkyl organotin species are readily available and tolerate many functional groups. Moreover, these reagents are not particularly oxygen or moisture sensitive. Because of their value in coupling reactions, a few examples are presented below.

Macrocyclic lactones occur widely in nature and possess a variety of biological activities. Therefore, the construction of large-ring compounds, particularly in relatively high yields, is challenging. The Stille coupling has been successfully applied to the synthesis of various macrocyclic lactones and various natural products. Piers et al. [63] synthesized fused bicyclic systems by an intramolecular cross-coupling reaction. Stille and coworker [64] used the same method to obtain macrolidic lactones of different sizes.

Piers (1985)

Stille (1987)

n = 5–8

Heliolactone, a natural product, has a synthetically challenging structure, with a doubly vinylogous stereogenic center at C6 and a configurationally defined cross-conjugated diene system. The coupling partner **A** was synthesized from the known chiral iodide using a palladium-catalyzed stannylation reaction. The Stille cross-coupling of **A** and **B** delivered heliolactone [65, 66].

11.4 Suzuki–Miyaura Coupling Reaction

The Suzuki coupling reaction [67–69] (also called Suzuki–Miyaura coupling) is a palladium-catalyzed reaction in which the coupling partners are alkenyl, aryl, or alkynyl organoboranes (boronic acid or boronic ester) and organohalides under basic conditions. This reaction is used to produce carbon–carbon bonds. It was first published in 1979 by Suzuki [67] and he shared the 2010 Nobel Prize in Chemistry with Richard F. Heck and Ei-ichi Negishi for the discovery of palladium-catalyzed cross-couplings in organic synthesis.

Suzuki and coworkers reported the reaction of alkenylboranes readily obtained by hydroboration of alkynes with alkenyl halides in the presence of a catalytic amount of Pd(PPh$_3$)$_4$ and base to provide conjugated dienes. The reaction requires a base for the activation of the boron compound and proceeds under mild conditions [70].

In 1981, Suzuki and Miyaura and coworkers made a breakthrough in the synthesis of biaryl compounds using aryl boronic acids and aryl bromide under homogeneous palladium-catalyzed conditions in the presence of bases. Under mild conditions, biaryl compounds were formed with high selectivities and in excellent yields [71].

When Suzuki and Stille coupling reactions are compared, it is seen that the difference is the metal used in transmetalation. Stille coupling uses tin as metal while Suzuki coupling uses boron as boron has a similar electronegativity to tin. Both couplings have a similar reaction scope and proceed via a similar mechanistic cycle.

The general catalytic cycle for Suzuki cross-coupling involves three fundamental steps: (i) oxidative addition of an organohalide to Pd(0)L$_2$ to form an organopalladium compound, (ii) transmetalation of a nucleophilic carbon from boron to RPdX, and (iii) reductive elimination of the cross-coupling product with the generation of the original palladium(0) catalyst, which is used for the next cycle (Scheme 11.4).

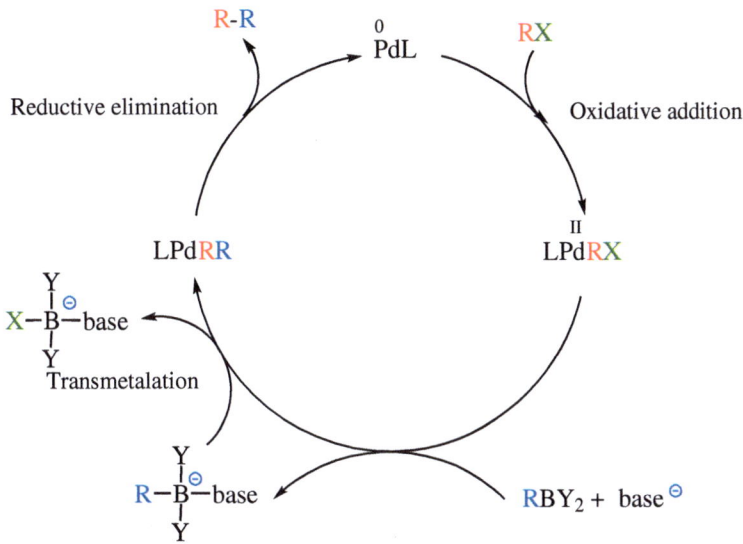

Scheme 11.4 The catalytic cycle for the Suzuki–Miyaura reaction.

Oxidative addition is the rate-determining step in the catalytic cycle and the order of activity decreases from iodine to chlorine, I > OTf > Br > Cl. Chlorine electrophiles are the least likely to react. The reactivity of chloride electrophiles can be increased by using bulky, electron-donating (electron-rich) phosphine ligands on the Pd catalyst and stronger bases to support the dissociation of chloride [67]. Mechanistic studies have revealed that the initially formed *cis*-complex isomerizes to the more stable *trans*-complex as discussed in the case of the Stille coupling reaction. The transmetalation process between RPdX and organoborane BRY_2 does not occur readily because of the weak nucleophilic character of the organic group R to be transferred in the organoboranes. On the other hand, tetraorganoborate complexes, BR_4^-, do undergo a cross-coupling reaction while the organoboranes BR_3 themselves do not normally couple without the added base. Mechanistic studies show that the added base forms a complex with the boron compound and increases the nucleophilicity of the R group attached to the boron [72]. Gropen and Haaland demonstrated that the methyl group in tetramethylborate [$(B(CH_3)_4^-$] is 5.5 times more electronegative than the methyl group in trimethylborane, $B(CH_3)_3$ [73].

Suzuki and coworkers examined the optimum conditions for the coupling reaction between *E*-1-hexenylborane and *trans*-styryl bromide. The results are given below, which shows that high yields of the diene were obtained when the relatively strong base NaOEt was used. Similar results were also obtained when using phenoxide and hydroxide. When weak bases such as sodium acetate and triethylamine were used, they were not effective in increasing the yields [74]. Furthermore, they showed that the coupling reactions are highly regio- and stereospecific and take place while retaining the original configurations of both of the coupling reagents. The isomeric purity of the products generally exceeds 97%. A base is required to carry out successful coupling.

Catalyst	Base	Solvent	Reaction time (h)	Yield
Pd(PPh$_3$)$_4$	None	THF	6	0
Pd(PPh$_3$)$_4$	None	Benzene	6	0
Pd(PPh$_3$)$_4$	NaOEt	THF	2	73
Pd(PPh$_3$)$_4$	NaOEt	Benzene	2	86

They proposed two different mechanisms for the transmetalation step. According to the first mechanism, the base initially reacts with the borane to form a more nucleophilic tetrahedral boronate, from which the transfer of an R group on palladium can be achieved more easily. According to the second mechanism, the base displaces the halide from the palladium complex and forms a more reactive alkoxo–palladium species. Such a palladium complex then reacts with an alkenylborane and transfers the R group on palladium to provide the diorganopalladium complex as shown in Scheme 11.5 [69, 74]. These two pathways are named the *boronate pathway* and the *oxo–palladium pathway*, respectively. Recent rapid injection nuclear magnetic resonance (NMR) spectroscopic studies at low temperatures established the identity of three different species containing palladium–oxygen–boron linkages [75]. The transmetalation step of the Suzuki coupling reaction proceeds with retention of stereochemistry.

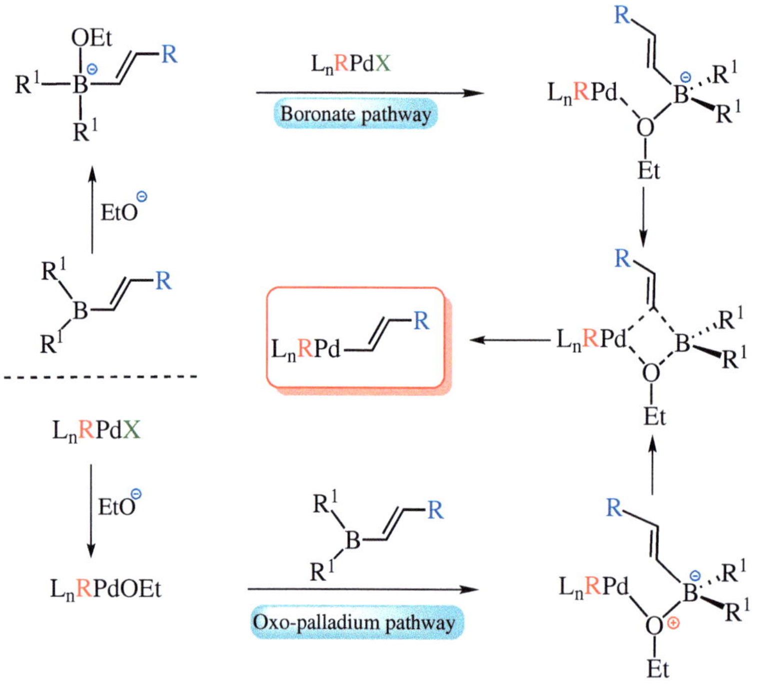

Scheme 11.5 Two mechanisms suggested for the transmetalation reaction.

The final step is the reductive elimination, which occurs from the *cis*-complex: therefore, the *trans*-complex formed after the transmetalation reaction isomerizes to *cis*-complex to undergo reductive elimination [75]. Ridgway and coworkers have shown, using deuterium labeled compounds, that the reductive elimination proceeds with retention of stereochemistry [76].

The advantages of Suzuki coupling over other similar reactions include the availability of common boronic acids and vinyl or alkylborono compounds and mild reaction conditions. Boronic acids are easily synthesized by two common methods: (i) reaction of Grignard reagents with borates, followed by hydrolysis, and (ii) reaction of arylsilanes with boron tribromide, followed by acid hydrolysis.

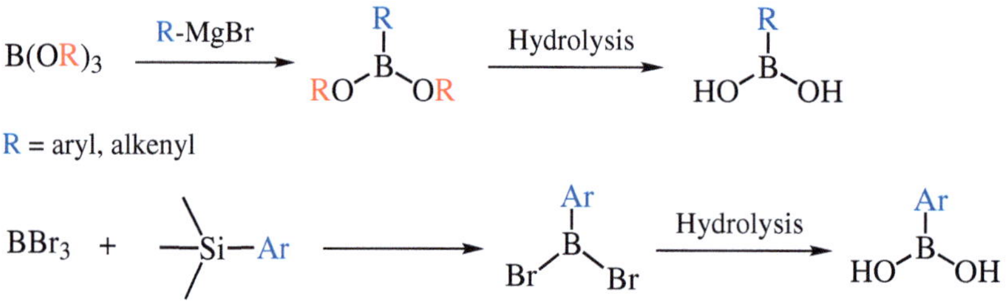

Organoborane reagents can also be easily synthesized by the hydroboration reaction (see Section 4.5). The addition of dialkylboranes such as 9-borabicyclo-[3.3.1]nonane (9-BBN), disiamylborane, or dicyclohexylborane to an alkene or alkyne gives the corresponding alkyl- or alkenylboranes. The reaction is essentially quantitative, proceeds through *cis* anti-Markovnikov addition from the less hindered side of the double bond, and can tolerate various functional groups.

11.4 Suzuki–Miyaura Coupling Reaction

9-BBN= 9-Borabicyclo[3.3.1]nonane   BH(Sia)$_2$ =Disiamylborane   Dicyclohexylborane

The 9-alkyl- and alkenyl-9-BBN derivatives thus obtained are particularly useful for the transfer of primary alkyl and vinyl groups by the palladium-catalyzed cross-coupling reaction.

Within the context of "green chemistry," some modifications have been made to the Suzuki–Miyaura cross-coupling reaction. Badone and coworkers reported the "ligandless" palladium acetate-catalyzed Suzuki–Miyaura coupling reaction of arylboronic acids with substituted aryl derivatives in water without using organic cosolvent in the presence of tetra-n-butylammonium bromide [77]. They found that n-Bu$_4$N$^+$Br$^-$ enhances the rate of the coupling reaction and a wide variety of functional groups, including the base-sensitive ones (i.e. esters), can be tolerated.

As mentioned above, that chloride electrophiles are reluctant to participate in oxidative addition in the Suzuki–Miyaura coupling reaction because of their higher bond dissociation energies (the approximate bond dissociation energies for Ph—X are C—Cl = 96 kcal/mol, C—Br = 81 kcal/mol, and C—I = 65 kcal/mol).

Nickel complexes are known to activate arylchlorides [78]. Percec and coworkers [79] recently developed a Suzuki coupling reaction for various aryl sulfonates including mesylates with arylboronic acid using a nickel catalyst. The Ni(0) catalyst is generated in situ from NiCl$_2$(dppf) and Zn. This novel reaction with Zn, as a catalyst, is highly regiospecific and tolerates various functional groups.

In 1996, Miyaura and coworkers applied for the first time the nickel-catalyzed cross-coupling reaction to the reluctant aryl chlorides, fluorides, and unreactive bromides with boronic acids [80–82]. The highly active nickel(0) catalysts were obtained in situ by treating the nickel chloride complexes with 4 equiv of n-BuLi or DIBAL-H at room temperature. The reaction was carried out under heating in the presence of a base, K$_3$PO$_4$.

The disadvantage of this reaction is that *n*-BuLi is used to form the nickel(0) catalyst because *n*-BuLi reacts with a wide variety of functional groups. Indolese [83] adopted the reaction conditions of Percec to the reaction of various aryl chlorides and boronic acids without using the external reductant Zn. The corresponding coupling products were formed in good yields. Moreover, even 1 mol% of catalyst loading was also workable in some cases. Later, Miyaura and Inada reported a more efficient and cheaper catalyst. They used 2 equiv of PPh_3 as external supporting ligands to achieve the coupling of arene chlorides and boronic acids [84].

R^1, R^2 = Alkyl, F, OR, SR, NH_2, CN, CHO, CF_3

In the first step of the Ni-catalyzed Suzuki–Miyaura coupling reaction, oxidative addition of Ni(0) across an Ar—X bond takes place to form an organonickel species. Transmetalation between the organonickel(II) species and boronic acid generates a diorganonickel(II) complex, which is followed by reductive elimination to form the coupled product and generate the active nickel(0) catalyst for the next catalytic cycle (Scheme 11.6) [85].

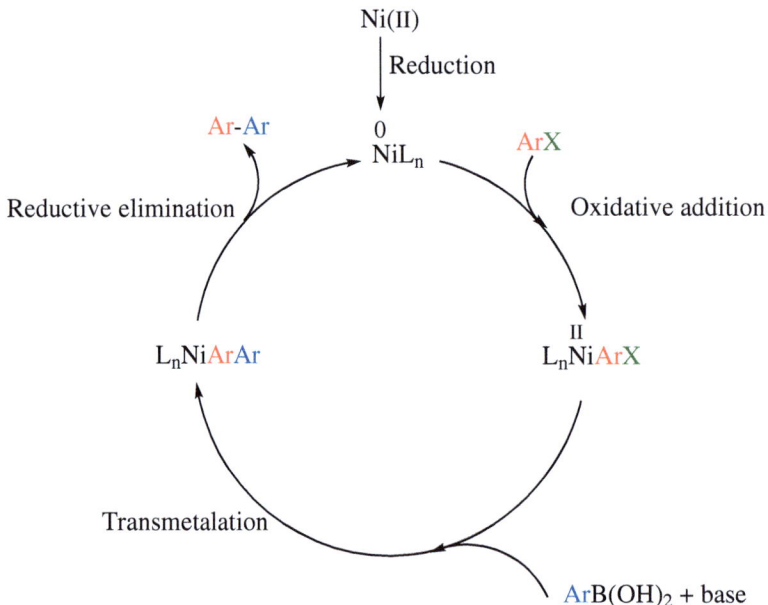

Scheme 11.6 Ni-catalyzed Suzuki–Miyaura reaction. Source: Based on Zhang et al. [85].

The natural product ribisin A, initially isolated from the fruiting bodies of *Phellinus ribis*, was used as a folk medicine in East Asia for treating several diseases and enhancing body immunity. The first total synthesis of ribisin A was accomplished starting from D-glucopyranoside, which was converted to α-iodoenone in several steps. The key step in the synthesis was the Suzuki–Miyaura coupling reaction between 2-hydroxyphenylboronic acid and α-iodoenone. The reaction was carried out in the presence of $Pd_2(PCy_3)_2$ as a catalyst and Na_2CO_3. After several steps, the desired natural product, ribisin A, was obtained [86, 87].

Within the framework of the Suzuki–Miyaura coupling reaction, significant progress has been made in the field of Suzuki acyl-coupling in recent years [88]. The Suzuki acyl cross-coupling involves the coupling of an organoboron reagent with an acyl electrophile (acyl halide, anhydride, ester, and amide). Suzuki acyl cross-coupling also constitutes an alternative to Friedel–Crafts acylation [89, 90]. Suzuki acyl cross-coupling often proceeds in the absence of an external base, which is an advantage. The biaryl derivatives were synthesized in high yields. Over the past few years, significant contributions have been made to Suzuki–Miyaura coupling reactions. In recent years, it has been demonstrated that even amides, acids, and esters can also be involved in the Suzuki–Miyaura coupling reaction [91].

More recently, the high cost, toxicity, and instability of phosphine complexes have prompted researchers to search for phosphine-free catalysts. Therefore, in the past two decades, N-heterocyclic carbenes (NHCs) have attracted great interest [92–94]. Initially, NHC complexes were perceived as some kind of imitation of phosphine compounds, but later, it was found that they are strong σ-donors and the metal–carbene bond in them is stronger and shorter than that in metal–PR_3 bonds. Therefore, NHC complexes are more stable than phosphine complexes. This stability is also seen against moisture and oxygen. A selection of commonly used NHC ligands is given below.

Recently, Ozdemir and coworkers [95] synthesized new *NHCs* that are resistant to moisture and they successfully applied these Pd-PEPPSI (Pyridine Enhanced Precatalyst, Preparation, Stabilization and Inhibition) catalysts to Suzuki cross-coupling reactions. The reaction was carried out with arylchlorides, which generally tend not to undergo coupling reactions, in aqueous medium. The corresponding coupling products were formed in high yields.

11.5 Negishi Coupling Reaction

In 1976, Negishi initiated a series of studies to search for more chemoselective organometallic species in the palladium-catalyzed couplings with organohalides. They reported the formation of C—C coupling products by the reaction of organoaluminum reagents with aryl bromides in the presence of a nickel catalyst, Ni(PPh$_3$)$_4$. Namely, the reaction of an alkenylaluminum intermediate, synthesized by hydroalumination of acetylene, and 1-bromonaphthalene provided a simple route to the synthesis of *trans* aryl-coupled alkene [96]. The same reaction was carried out with the palladium catalyst Pd(PPh$_3$)$_4$; however, it was found that the reaction is much slower than the corresponding nickel-catalyzed reaction and does not appear to offer any advantage over the latter.

Shortly after this work, Negishi and coworkers reported the use of zinc reagents in cross-coupling reactions. They synthesized unsymmetrical biaryls using organozinc compounds instead of aluminum compounds in the presence of nickel or palladium catalysts. Unsymmetrical biaryls and diarylmethanes were obtained chemo- and regioselectively in high yields [97, 98].

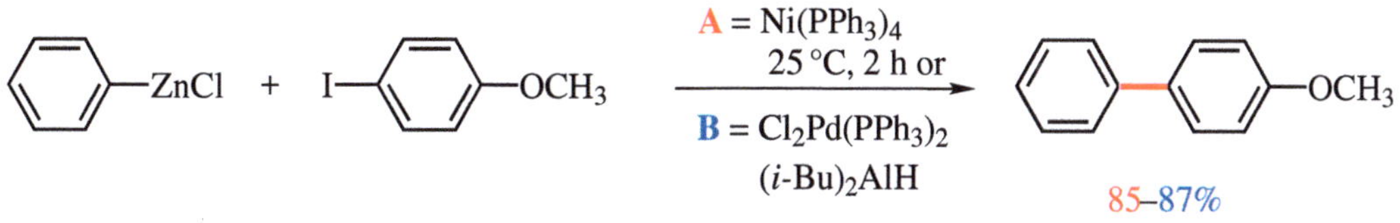

11.5 Negishi Coupling Reaction

Fawarque and Jutand showed that organometallic compounds such as RMgX can undergo cross-coupling reactions with aryl bromides in the presence of palladium(0) complexes to give the coupling products R-Ar [99]. This reaction was extended to the Reformatsky reagent, BrZnCH$_2$COOEt. They reacted the Reformatsky reagent with various aromatic halides in the presence of Pd(PPh$_3$)$_4$ or Ni(PPh$_3$)$_4$ to give aryl acetic acid esters in good to moderate yields [100].

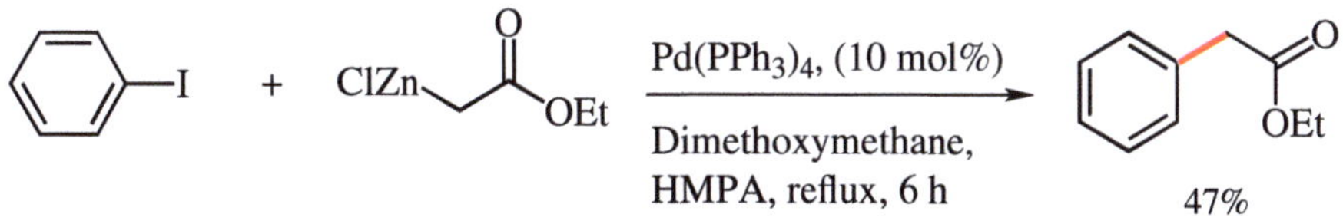

Negishi cross-coupling reaction is the organic reaction of an organohalide with an organozinc compound to give the coupled product using a palladium or nickel catalyst.

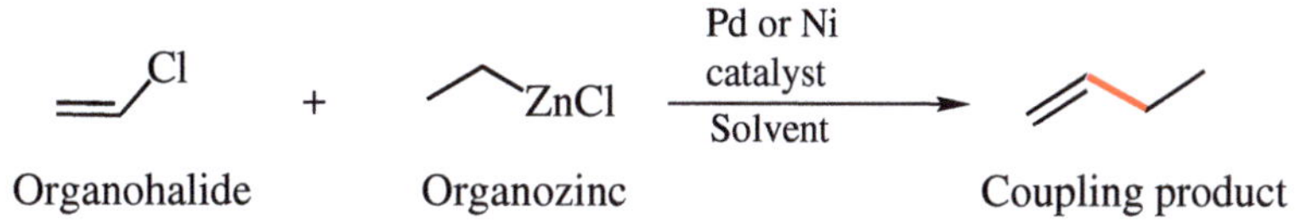

Organohalide   Organozinc   Coupling product

Negishi coupling [101–104] was applied for versatile nickel- or palladium-catalyzed coupling of organometals with various aryl, vinyl, benzyl, or allyl halides. The scope and potential use of the Negishi reaction were also extended to alkyl–zinc compounds beyond regular C(sp^2)—C(sp^2) couplings. The formation of C-sp^2—C-sp^3 bonds by cross-couplings has been generally performed by the reaction of an alkyl group with an aryl or alkenyl halide. The alternative coupling of an alkyl halide with an aryl or alkenylmetal reagent has not been used because of the potential β-hydrogen elimination in the intermediate alkylmetal complexes. Giovannini and Knochel [105] developed a Ni-catalyzed cross-coupling reaction of various polyfunctional alkyl iodides with arylzinc bromides in the presence of Ni-(acac)$_2$ (nickel(II) acetylacetonate) and 4-(trifluoromethyl)styrene. It is assumed that 4-(trifluoromethyl)styrene facilitates reductive elimination from the intermediate. As shown below, electron-rich and electron-poor arylzinc species are suitable cross-coupling partners.

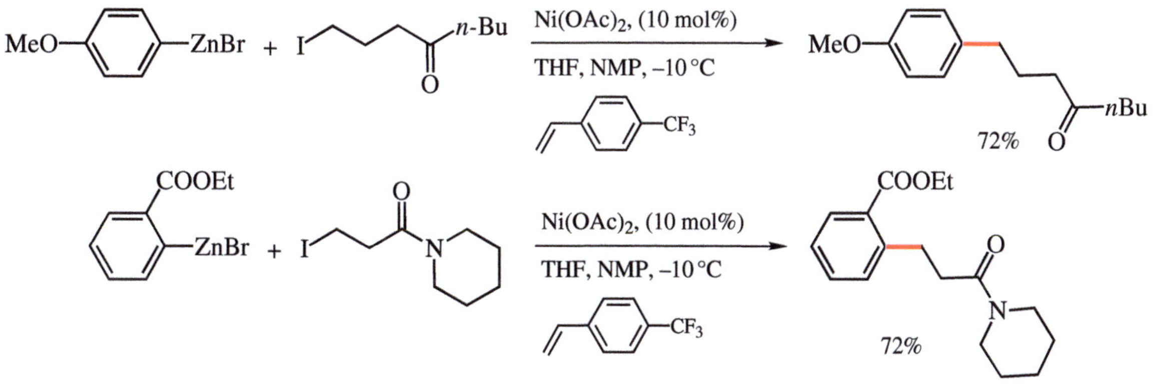

Knochel and coworkers [106, 107] pioneered the development of Ni-catalyzed alkyl–alkyl Negishi cross-coupling reactions between primary iodoalkanes and primary diorganozinc compounds in the presence of *m*-trifluoromethylstyrene or acetophenone. These promoters allow efficient and fast coupling at lower temperatures. Iodoalkanes contain functional groups such as esters, amides, ketones, and vinyl couple smoothly. The key to the success of this reaction is suggested to be the formation of intermediate Ni(II)–olefin complexes, which accelerate the transmetalation and reductive elimination reactions.

The Negishi cross-coupling reaction involves three key steps. The palladium-catalyzed reaction begins with oxidative addition of the organohalide or pseudohalide (e.g. triflate) to the Pd(0) to form a Pd(II) complex. Transmetalation of the palladium complex with the organozinc reactant forms the diorganopalladium complex where the R group of the organozinc reactant replaces the halide anion on the palladium complex and makes a zinc(II) halide salt. Subsequent reductive elimination then affords the final coupling product and regenerates the catalyst and the catalytic cycle can begin again (Scheme 11.7).

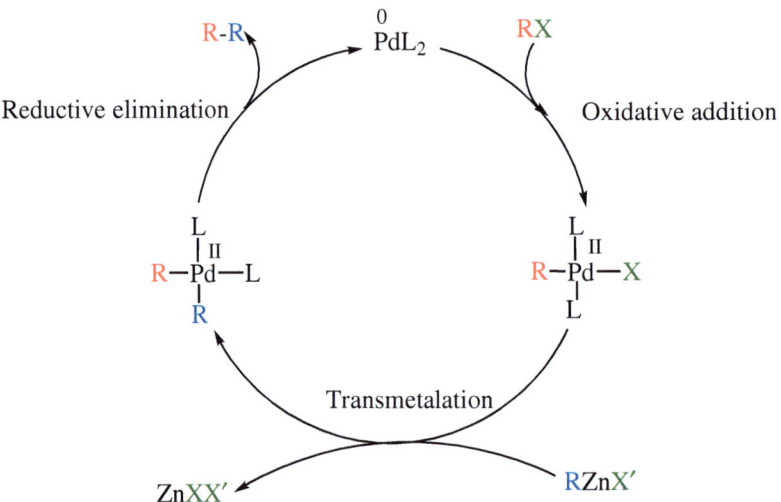

Scheme 11.7 Palladium-catalyzed cycle for the Negishi reaction.

In reactions where nickel is used as a catalyst, the reaction mechanism differs from that of palladium. For example, for Pd-catalyzed reactions, it is well established that only Pd(0) and Pd(II) intermediates are involved in the cycle. However, in the case of nickel-catalyzed reactions, the oxidation states of the intermediates are different. Kochi [108] proposed a radical chain process for nickel-catalyzed aryl–aryl couplings of aryl halides in the presence of phosphine ligands. Recently, Schley and Fu [109] conducted a systematic mechanistic investigation of a Ni-catalyzed cross-coupling reaction. According to this mechanism, first a Ni(II) complex is formed by the oxidative addition of aryl halide to Ni(0). The ArNiL$_2$ formed reacts with an additional aryl halide by a single electron transfer to form a Ni(III) species, which decomposes to the catalytically active species Ni(I)X. The radical chain propagation starts with the transmetalation process between the Ni(I) complex and ZnRX to give a Ni(I)R species. The oxidative addition followed by the reductive elimination process gives the coupling product RR and regenerates the Ni(I) active complex for a further cycle (Scheme 11.8).

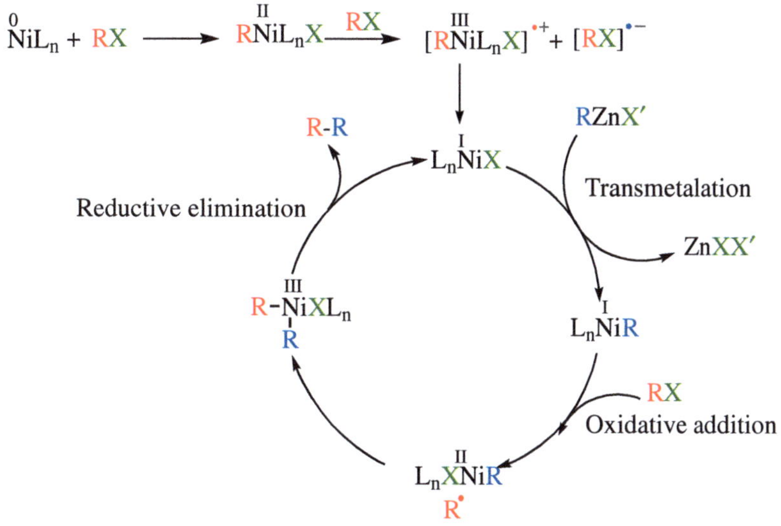

Scheme 11.8 Nickel-catalyzed cycle for the Negishi reaction.

The major disadvantage of the Negishi coupling, at least by comparison with the Stille and Suzuki reactions, is that the organozinc compounds are relatively moisture and air sensitive. These drawbacks lead to the use of excess Zn organometals, which can be considered a limitation to the applications of the Negishi coupling reaction. On the other hand, the low toxicity of zinc salts, as well as the growing number of zinc reagents, has increased their use in cross-coupling reactions.

Organozinc reagents can be prepared directly by the insertion of zinc powder to various aromatic and heterocyclic electrophiles. Commercial zinc powder can be used in the presence of lithium chloride. LiCl removes the impurities on the surface of zinc, such as oxide, increases the solubility, and activates the metal. For example, zinc regioselectively inserts into the C—Br bond in 2,4-dibromothiazole in the presence of LiCl to produce (4-bromothiazol-2-yl)zinc(II) chloride, which easily undergoes the Negishi cross-coupling reaction with 2-iodobenzaldehyde [110].

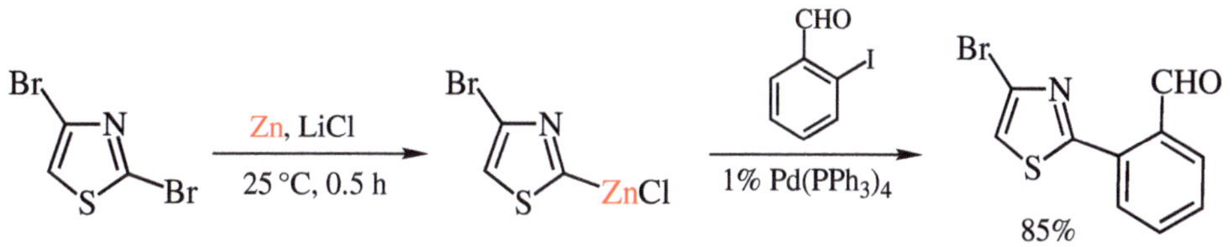

Although, in recent years, the catalysts used in the Negishi cross-coupling reactions have been further developed, the coupling of simple aryl halides with aryl zinc reagents is often less successful. The use of palladocycle precatalysts enables the performance of Negishi cross-couplings with a broad range of substrates. Buchwald and coworkers [111] recently developed a new class of easily prepared air- and moisture-stable palladacycle precatalysts, which can rapidly generate the Pd(0) allowing Negishi cross-couplings to proceed at ambient temperature with low catalyst use. The 2-zincated benzofuryl-derived reagent was coupled with a chloropyrazole, providing a complex heterocyclic compound in 90% yield [112].

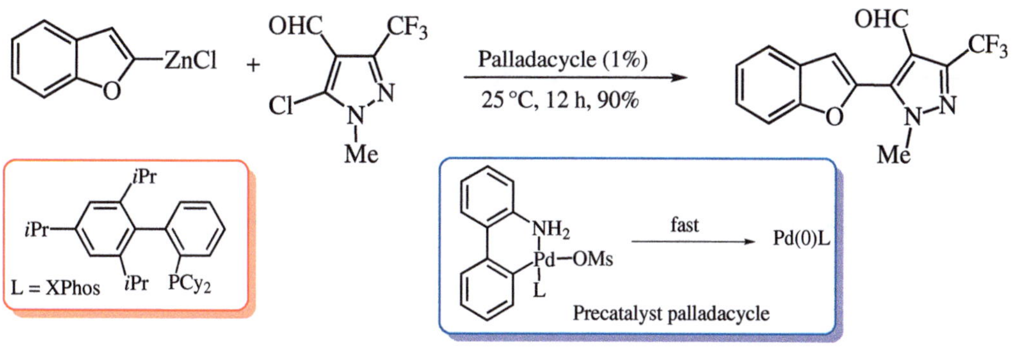

11.6 Sonogashira Coupling Reaction

The palladium-catalyzed cross-coupling between terminal acetylenes and sp^2-hybridized carbon atoms (aryl, heteroaryl, and vinyl halides) is a modern and extremely powerful synthetic method for the synthesis of important organic intermediates. Conjugated acetylenic compounds have found application in the synthesis of natural products, pharmaceuticals, and nanomaterials. In 1975, three different groups independently reported the alkynylation reaction of aryl halides using aromatic acetylenes.

Cassar [113] converted acetylene or monosubstituted acetylene derivatives into aryl- or vinyl-substituted acetylenes by the reaction of aryl or vinyl halides with aryl acetylenes in the presence of a palladium or nickel complex and a base under mild conditions. Some palladium(II) compounds were also used, but they were probably reduced to Pd(0) complexes under the reaction conditions.

Dieck and Heck [114] reacted monosubstituted acetylenes with aryl, heterocyclic, or vinylic bromides or iodides at 100 °C and they also obtained disubstituted acetylenes in the presence of an amine and $Pd(OAc)_2(PPh_3)_2$. Various acetylenes and organic halides were reacted under the same reaction conditions. It was noted that the alkylacetylenes were less reactive than phenylacetylene. With triethylamine as a base, only reactive halides such as iodobenzene underwent a coupling reaction. The alkylacetylenes, however, were involved in the coupling reaction when more basic secondary amines were used. Both procedures required harsh reaction conditions, such as high temperatures.

Kenkichi Sonogashira, a Japanese chemist (1931–) and coworkers [115] reacted acetylene derivatives with various iodoarenes, bromoalkenes, or bromopyridines in the presence of a catalytic amount of $PdCl_2(PPh_3)_2$ and CuI as a cocatalyst in diethylamine under very mild conditions and they obtained disubstituted alkynes. Addition of a catalytic amount of copper(I) iodide greatly accelerates the reaction, thus enabling the alkynylation reaction at room temperature.

This coupling reaction developed by Sonogashira and coworkers was related to the already known coupling between phenyl or vinyl halides and copper acetylides, known as the Castro–Stephens reaction [13]. This reaction was carried out in refluxing pyridine under a nitrogen atmosphere. Under these conditions, aryl iodides bearing an *ortho*-nucleophilic substituent are converted exclusively to the corresponding heterocycles in high yields.

The use of a Cu⁺ cocatalyst in the Sonogashira reaction [1, 116–119] increases the reactivity of the reagents and allows the reaction to proceed at room temperature.

Langer and coworkers [120] synthesized 4-bromo-2,3,5-trichloro-6-iodopyridine containing three different halides as leaving groups, and they tested their chemoselectivity in the Sonogashira reaction. They studied the selective monoalkynylation reaction using standard conditions ($PdCl_2(PPh_3)_2$ (5 mol%) and CuI (5 mol%) in NEt_3 with 1.1 equiv of phenylacetylene. The starting material was converted to the desired product with high chemoselectivity and in a yield of 90%, where only iodide was substituted. The best yields were obtained with arylalkynes having electron-donating groups. By using 2 equiv of phenylacetylene, 2,3,5-trichloro-4,6-dialkynyl-pyridines were formed in 67% yield. The selectivity is exclusively based on the different leaving groups. The positions C3 and C5 in pyridine are considerably less reactive and require the use of more reactive catalysts. For this reason, XPhos was used as the catalyst and pentaalkynylpyridines were obtained in good yields.

The exact mechanism of the homogeneous copper-cocatalyzed Sonogashira reaction is unknown. The copper-cocatalyzed Sonogashira reaction is thought to proceed via two independent catalytic cycles. Before oxidative addition, the Pd(II) complex must be reduced to the active catalyst Pd(0). The Pd(0) catalyst is formed by the reduction of palladium(II) complexes under the employed reaction conditions. It is well known that *n*-electron donors such as amines, phosphanes, and ethers used as solvents and ligands reduce palladium(II) species typically via σ-complexation–dehydropalladation–reductive elimination (see Section 11.2) [117].

The Pd(0) complex undergoes oxidative addition with RX (R = aryl, heteroaryl, and vinyl; X = I, Br, Cl, and OTf) to form a four-coordinated palladium complex. The palladium cycle intersects with the copper cycle. First, a π-alkyne–copper complex is formed to increase the acidity of the alkyne for deprotonation. The base abstracts the acetylenic proton to form the copper acetylide. Then, transmetalation takes place by the transfer of copper acetylide to palladium, forming a four-coordinated palladium complex. This step is the rate-determining step. *Trans/cis* isomerization is then followed by reductive elimination to afford the final product and regenerates the Pd(0) catalyst for the next cycle (Scheme 11.9).

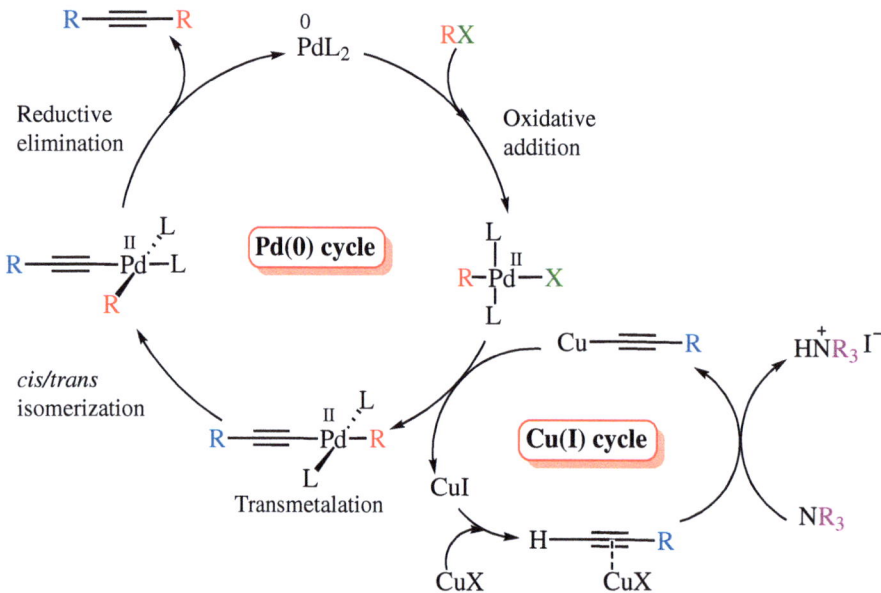

Scheme 11.9 Palladium-copper-catalyzed cycle for the Sonogashira reaction.

The most widely employed cocatalysts in the Sonogashira coupling reaction are copper salts, which form homocoupling products (the so-called Glaser coupling) from the terminal alkynes when the reaction medium is exposed to air. Therefore, chemists searched for reaction conditions to carry out the Sonogashira coupling reaction in the absence of copper. These reactions can be carried out under aerobic conditions because the copper-mediated oxidative homocoupling of acetylene is prevented. On the other hand, the use of environmentally unfriendly copper is another drawback.

A mild protocol for the copper-free Sonogashira coupling of terminal acetylenes with aryl iodides in water under aerobic conditions has been developed. The use of 1 mol% $PdCl_2$ in the presence of pyrrolidine gives the coupling products at room temperature or 50 °C in good yields [121].

All these copper-free methodologies are usually called copper-free Sonogashira couplings. Actually, copper-free coupling reactions were discovered for the first time by Heck and Cassar. It is unfair to call these reactions "*copper-free Sonogashira coupling*." These reactions use excess amine, often acting as a solvent.

Very recently, Košmrlj and coworkers [122] hypothesized that the Cu-free Sonogashira reaction proceeds through a tandem Pd/Pd double cycle as shown in Scheme 11.10. This pathway is practically identical to the Pd/Cu-catalyzed mechanism shown in Scheme 11.9. The only difference between the two mechanisms is that the role of copper is taken by a Pd complex. The first step begins with the oxidative addition of the aryl halide to the Pd(0) catalyst, forming complex **A** and activating aryl halide substrate for the reaction. Acetylene is activated in the second, Pd(II) cycle. It was recently proven that phenylacetylene forms Pd monoacetylide complex **C** as well as Pd bisacetylide complex **D** under mild reaction conditions. The complexes **A** and **D** are involved in the transmetalation process. Reductive elimination of the complex **B** generates the product as well as the catalyst [116].

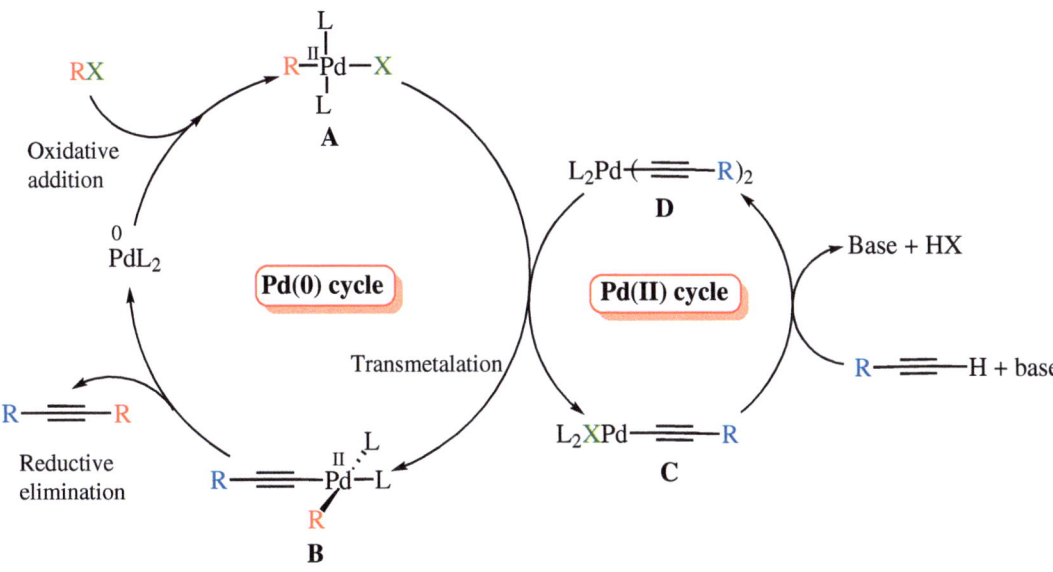

Scheme 11.10 Copper-free palladium-catalyzed cycle for the Sonogashira reaction.

The versatility of the Sonogashira coupling reaction makes it widely used in the synthesis of various compounds. Recently, Basceken and Balcı [123] applied the Sonogashira coupling reaction as a key step in the synthesis of pyrazolo–pyrrolo–diazepine skeletons. The key features of this method include the (i) synthesis of pyrrole-derived α,β-alkynyl ketones, (ii) introduction of various substituents into the alkyne functionality, and (iii) cyclization reaction. The synthesis of pyrazole units by the reaction of α,β-alkynyl compounds with hydrazine followed by gold-catalyzed cyclization of pyrazoles with alkyne units results in the formation of the target compound.

Dissymmetrically functionalized anthracene derivatives can be used as active semiconducting layers for organic field-effect transistors (OFETs). Schweizer and coworkers [124] synthesized highly π-conjugated dissymmetric anthracenes starting from 9-bromo-10-iodoanthracene by two successive Sonogashira coupling reactions.

11.7 Kumada Coupling Reaction

The Kumada cross-coupling reaction is useful for generating carbon-carbon bonds by the reaction of an organohalide with a Grignard reagent to give the coupled product using a nickel or palladium catalyst. Unlike other coupling reactions, alkyl groups also undergo coupling reactions successfully. The first coupling reaction of a Grignard reagent with an organic halide using cobalt as a catalyst was reported in 1941 by Kharasch and Fields [125]. They reacted an aryl Grignard reagent with an organic halide (aryl or alkyl bromide) in the presence of 3–10 mol% with cobalt or nickel halides, and the coupling products were formed in excellent yields.

Corriu and Masse [126] reacted aromatic Grignard reagents with olefinic halides. Thus, stilbene was synthesized by treatment of trans-Cl—CH=CH—Cl with phenylmagnesium bromide in the presence of a catalytic amount of nickel halide. They found that the most effective catalyst was nickel(II)acetylacetonate when used in 0.1–0.5 mol%. Independently from Corriu and Masse, in the same year, Kumada and coworkers [127] also reported that Grignard reagents undergo cross-coupling reactions in a similar manner with alkyl and aryl halides in the presence of Ni–phosphine complexes.

The formation of C-(sp^3)—C-(sp^3) bonds by cross-coupling reactions has undergone an important development in recent years. The limitations associated with these kinds of couplings are mainly the more difficult oxidative addition of an alkyl electrophile and the fast β-elimination of hydrogen. This has been overcome in a variety of ways by using bulky and highly electron-donating ligands such as heterocyclic carbenes (NHCs) and trialkyl phosphines. Recently, Cardenas reported the Ni-catalyzed alkyl–alkyl Kumada reaction, which tolerates several functional groups [127, 128].

R = COOMe

The proposed nickel-catalytic cycle for the Kumada coupling reaction involves both nickel(0) and nickel(II) oxidation states. Initially, the electron-rich Ni(0) catalyst inserts into the R—X bond of the organic halide. This oxidative addition forms a Ni(II) complex. Subsequent transmetalation with the Grignard reagent, followed by *trans-cis* isomerization and finally reductive elimination, forms a carbon–carbon bond and releases the cross-coupled product while regenerating the Ni(0) catalyst for the next cycle (Scheme 11.11) [129, 130].

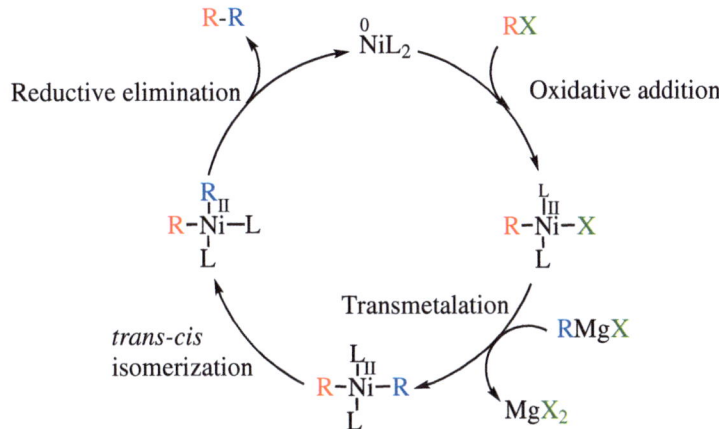

Scheme 11.11 Nickel-catalyzed cycle for the Kumada reaction. Source: Adapted from Guisán-Ceinos et al. [129]; Iffland et al. [130].

Although the first cross-coupling reactions of Grignard reagents described by Kumada and Corriu utilized nickel catalysts, the most efficient and commonly employed catalysts for such C—C coupling reactions nowadays are based on palladium catalysts. The palladium-catalyzed Kumada coupling was first reported by Murahashi and coworkers in 1975 [131, 132]. They reacted bromobenzene with vinyl and alkyl Grignard reagents in the presence of a palladium complex, Pd(PPh$_3$)$_4$, and the coupling products were formed in high yields.

Alkyl halides are often inactive substrates for metal-catalyzed cross-coupling reactions because of their reluctance to undergo oxidative addition [132]. However, palladium complexes were used as effective catalysts for the cross-coupling reaction of Grignard reagents with unactivated alkyl chlorides [132].

11.8 Hiyama Coupling Reaction

The Hiyama coupling is a palladium- or nickel-catalyzed cross-coupling reaction of organosilanes with organic halides or triflates to generate carbon–carbon bonds with chemo- and regioselectivity. The Hiyama coupling was first reported by Hatanaka and Hiyama in 1988 [133]. They reacted various aromatic iodides or bromides with tris(diethylamino)-sulfonium difluorotrimethylsilicate (TASF) in the presence of a palladium catalyst, allylpalladium chloride dimer, to form the corresponding methylated aromatic compounds in moderate to high yields.

568 | 11 Carbon–Carbon Coupling Reactions

Organosilicon reagents are in general highly stable because of the less polarized carbon–silicon bonds. Therefore, the organosilicon reagents are less reactive toward electrophiles than the other organometallic nucleophiles. As with the other coupling reactions, the Hiyama coupling [134–137] also proceeds through the usual oxidative addition, transmetalation, *trans–cis* isomerization, and reductive elimination steps. Generally, transmetalation, transferring an organic group from silicon to a transition metal such as palladium, is the key step. Because the organosilicon compounds are not reactive, they must first be activated for transmetalation. Activation of silane can be done with a base or fluoride ions (TASF and tetrabutylammonium fluoride [TBAF]) leading to pentavalent silicon compounds, which are more prone to transmetalation. Without the added fluorine, the organosilicon compound is simply too stable.

Denmark and coworkers [138] applied the Hiyama coupling reaction to the construction of medium-sized rings by activating the silicon compound using TBAF (2.0 equiv). The organosilane is activated by forming a pentavalent silicon center that is labile enough to allow for the breaking of a C—Si bond for smooth transmetalation of organosilicon reagents to complete the catalytic cycle of the cross-coupling reaction [139].

The proposed mechanism for the palladium-catalytic cycle for the Hiyama coupling reaction is given in Scheme 11.12. First, Pd(0) inserts into the R—X bond of the organic halide. Pentacoordinated silicate generated by the nucleophilic attack of a fluoride ion on silicon is nucleophilic enough to participate in transmetalation with the palladium complex to form the *trans*-complex. *Trans–cis* isomerization and finally reductive elimination forms a carbon–carbon bond and release the cross-coupled product while regenerating the Pd(0) catalyst for the next cycle. This reaction is comparable to the Suzuki coupling, which also requires an activating agent such as a fluoride ion or a base.

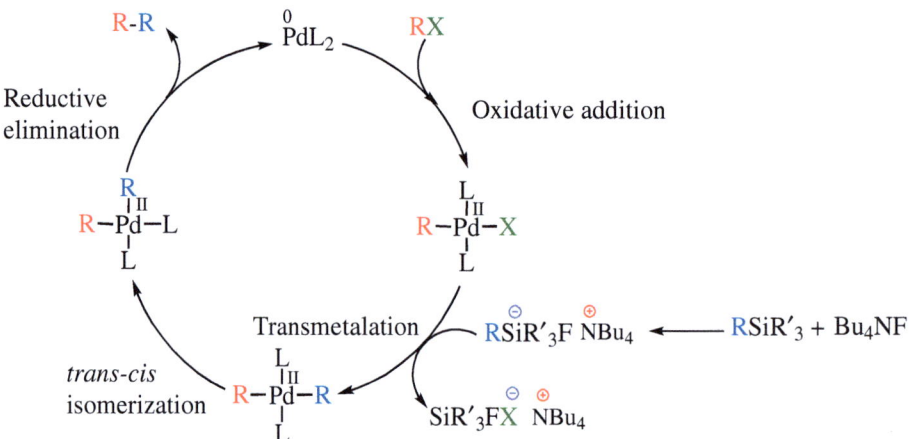

Scheme 11.12 Palladium-catalyzed cycle for the Hiyama reaction.

Tamao and Ito [140] demonstrated that alkoxy groups can serve as an alternative to the fluoro or chloro substituents on silicon. Dialkoxymethylsilanes undergo a palladium-catalyzed cross-coupling reaction with alkenyl and aryl halides in the presence of tetra-*n*-butylammonium fluoride.

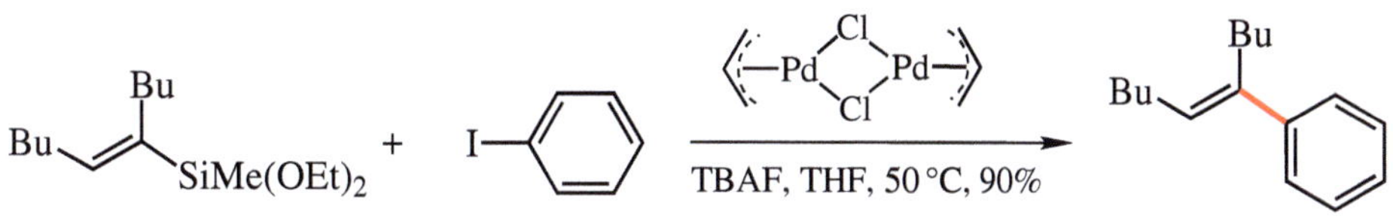

Later, Hiyama and coworkers reported that NaOH is a better promoter than fluoride ions [141]. The reactivity of the silicon reagents is significantly increased by using NaOH so that the palladium-catalyzed cross-coupling reaction of even aryl and alkenylchlorosilanes with organic halides proceeds under much milder reaction conditions compared to the fluoride ions. In contrast to the fluoride-promoted coupling reaction of Ar–SiRCl$_2$ with aryl bromides, which requires rather drastic conditions (120–150 °C, 20 hours, DMF), the NaOH-promoted reaction occurs smoothly at 60–80 °C in THF or in benzene.

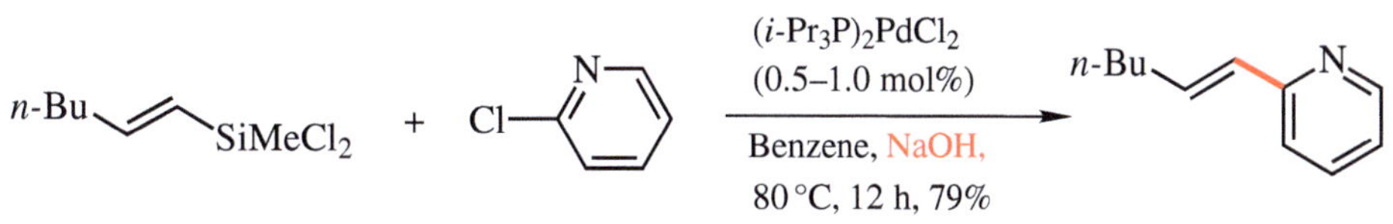

11.8.1 Hiyama–Denmark Coupling

For the transmetalation in the Hiyama coupling, fluoride activation and the formation of a pentavalent silicon is required. Using fluoride ions as a promoter precludes the employment of this reaction for the synthesis of complex products where the coupling substrates contain silyl protective groups. Moreover, TBAF is expensive to use on a large scale. Denmark and coworkers used silanols bearing the group to be transferred instead of fluoride ions. For the Hiyama–Denmark coupling [142], most of these steps are similar to those of the Hiyama coupling. The in situ-generated silanolate **B** first adds to the oxidative addition product forming an organopalladium complex **C**. Now, this complex is further activated by adding a second silanolate before transmetalation to give **D**. The transmetalation proceeds directly from an organopalladium(II) silanolate complex **D** to form the coupling product **E**, which undergoes reductive elimination to form the final coupling product **F** (Scheme 11.13).

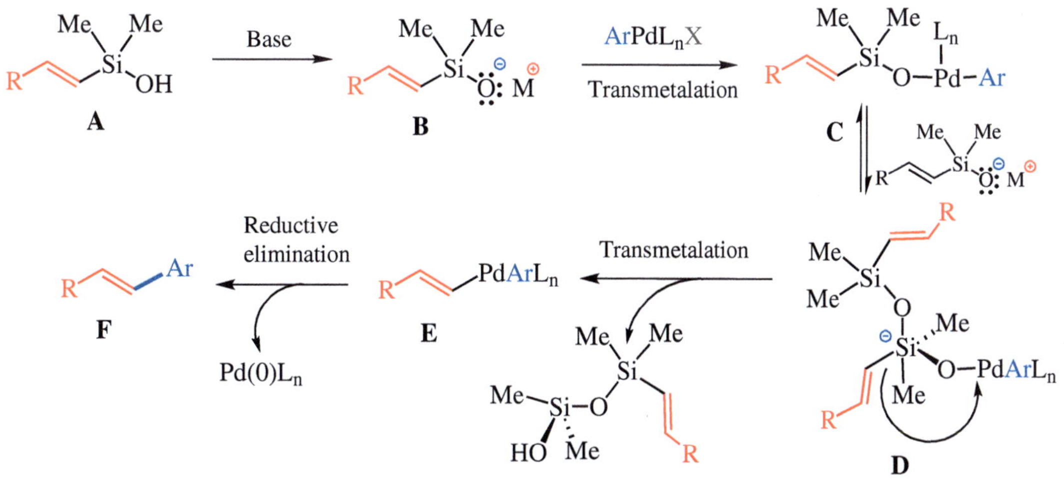

Scheme 11.13 Palladium-catalyzed cycle for the Hiyama–Denmark coupling reaction.

One of the most widely used methods for the construction of conjugated acetylenes is the Sonogashira reaction. Denmark and Tymonko [143] used the palladium-catalyzed cross-coupling of aryl iodides with aliphatic alkynylsilanols to synthesize conjugated alkyne derivatives. Potassium trimethylsilanolate (TMSOK) as the coupling promoter and copper(I) iodide as a cocatalyst were used. The cross-coupling proceeds at room temperature and produces good yields. The addition of copper(I) iodide was essential in order to achieve clean reactions with reasonable rates.

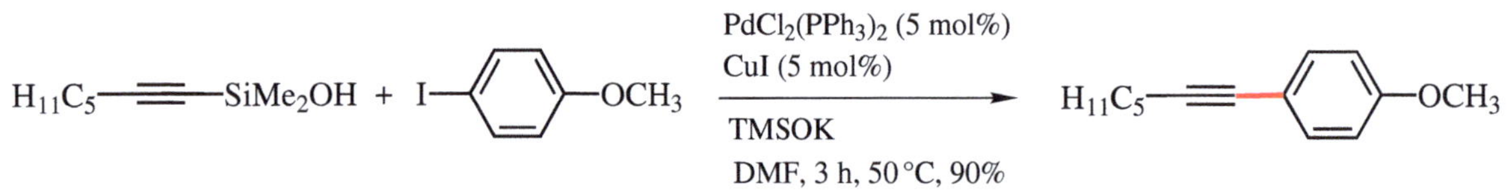

11.9 Buchwald–Hartwig Coupling Reaction

The Buchwald–Hartwig coupling reaction is a cross-coupling reaction of an aryl halide with an amine using palladium as a catalyst and a strong base to generate a carbon–nitrogen bond [144–147]. In 1993, Migita and coworkers published an example of a palladium-catalyzed C—N cross-coupling reaction. They reacted aryl bromides with N,N-diethylamino-tributyltin in the presence of a palladium catalyst to give N,N-diethylaminobenzene derivatives [148].

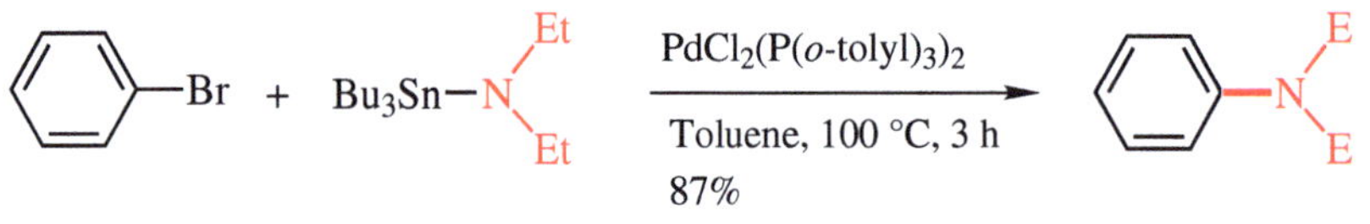

After a decade, Buchwald and coworkers improved the original work by Migita. They generated aminostannanes in situ via a simple transamination reaction. Aminostannane derivatives with a volatile amine were reacted with a higher boiling amine (in toluene), concomitantly with the removal of the volatile amine [149]. Then, the in situ generated aminostannane underwent Pd-catalyzed reactions with aryl bromides substituted with electron-donating or electron-withdrawing substituents. In the same year, Hartwig and coworkers [150] identified and characterized several intermediates formed in the palladium-catalyzed C—N bond formation.

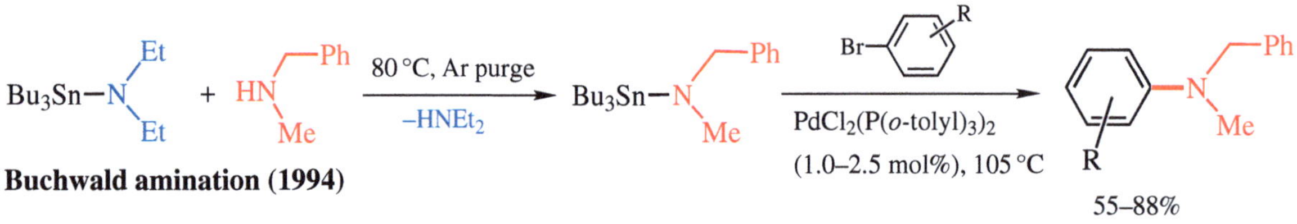

Buchwald amination (1994)

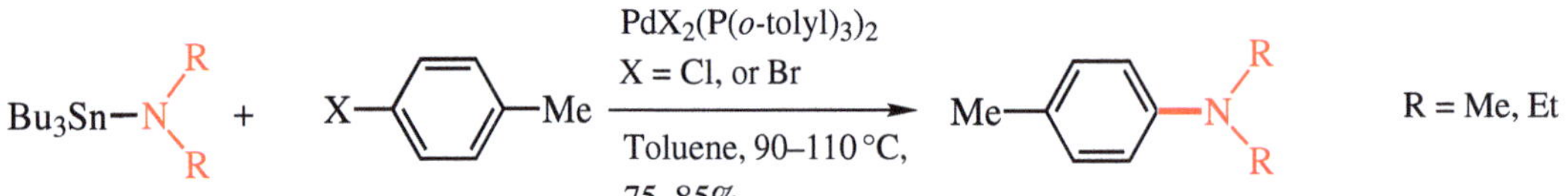

Hartwig amination (1994)

One year later, in 1995, Buchwald [151] as well as Hartwig [152] and coworkers independently reported that coupling can be performed with free secondary amines in the presence of a bulky base (t-BuONa, Buchwald; LiN(SiMe$_3$)$_2$, Hartwig). These processes provide a convenient method for performing these C—N coupling reactions without the necessity for forming tin amides and disposing of tin halides.

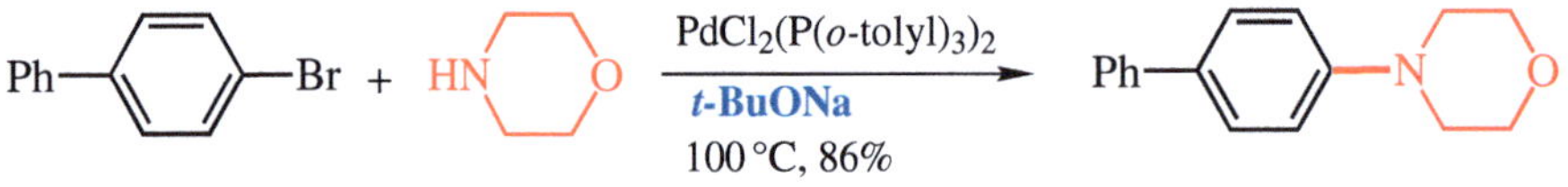

Buchwald amination without Sn (1995)

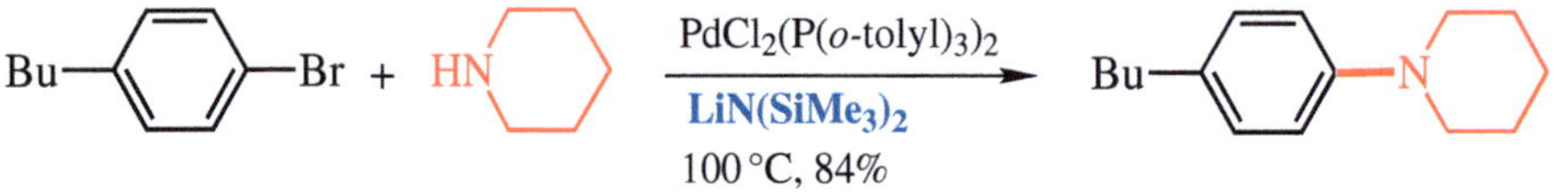

Hartwig amination without Sn (1995)

The reaction starts by insertion of Pd(0) into the R—X bond of the organic halide, which is in equilibrium with the dimer, whose structure has been determined by X-ray analysis [150] (Scheme 11.14). The stability of this dimer decreases in the following order: X = I > Br > Cl. The stability of the dimer formed with an iodine ion is responsible for the slow reaction of aryl iodides. The next step is the coordination of the amine to the palladium, followed by deprotonation by a base, forming an amide. Reductive elimination then produces the final aryl amine product and regenerates the catalyst.

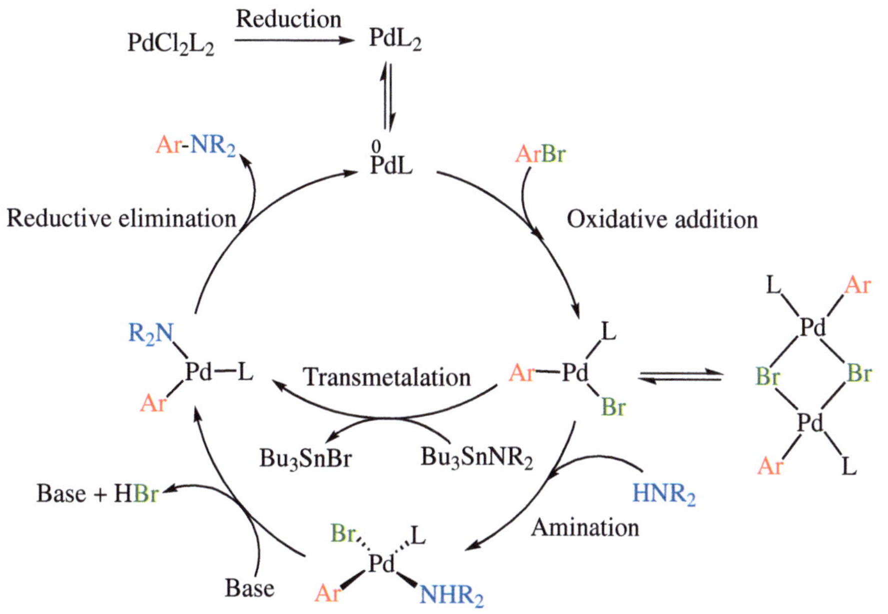

Scheme 11.14 Palladium-catalyzed cycle for the tin-free and with tin Buchwald–Hartwig amination reaction.

It is critical to choose the correct coordinating ligands to the palladium in the Buchwald–Hartwig amination. The catalyst $Pd_2(dba)_3$ in the presence of $P(o\text{-tolyl})_3$ as a ligand successfully cyclizes enantiomerically enriched amine and amide substrates [153]. The intramolecular coupling reaction of aryl bromides with amines having stereocenters at the α-position to the nitrogen atom does not give a racemic mixture using $Pd_2(dba)_3/P(o\text{-tolyl})_3$. For example, cyclization of (R)-N-phenethyl-1-phenylethan-1-amine produces (R)-1-(1-phenylethyl)indoline in 96% enantiomeric purity [154].

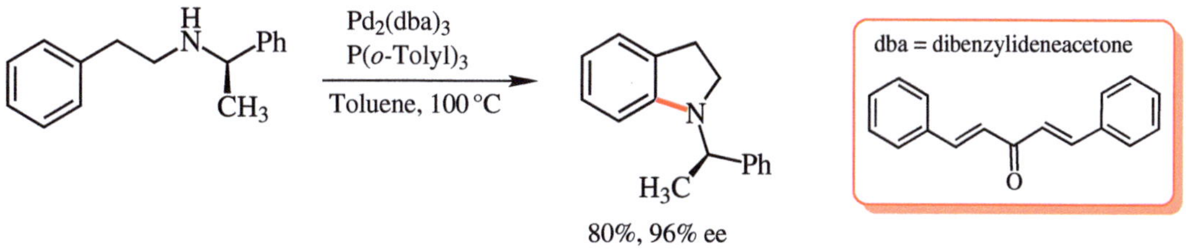

11.10 Tsuji–Trost Coupling Reaction

The Tsuji–Trost reaction is a palladium-catalyzed nucleophilic substitution reaction involving a substrate that contains a leaving group in an allylic position. It is a powerful method for C—C and C—heteroatom bond formation that has gained considerable recognition in organic synthesis.

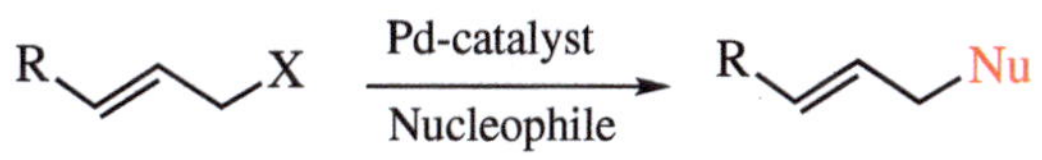

11 Carbon–Carbon Coupling Reactions

In 1965, Tsuji reported that π-allylpalladium chloride dimer reacts smoothly with the sodium salt of ethyl malonate or acetoacetate to give a mixture of mono- and dialkylated products [155]. It was established that the carbanion attacks the carbon atom of the palladium complex, giving allyl derivatives in high yields.

However, the lack of catalytic use of palladium and regioselectivity problems have hindered the widespread applicability of this transformation. In 1973, Barry M. Trost, an American chemist (1941–), and Fullerton [156] reported the first allylic alkylation of an alkene using Hüttel's Pd(II)-catalyzed stoichiometric C—H activation [157] and Tsuji's allylic alkylation process [155] with the introduction of phosphine ligands.

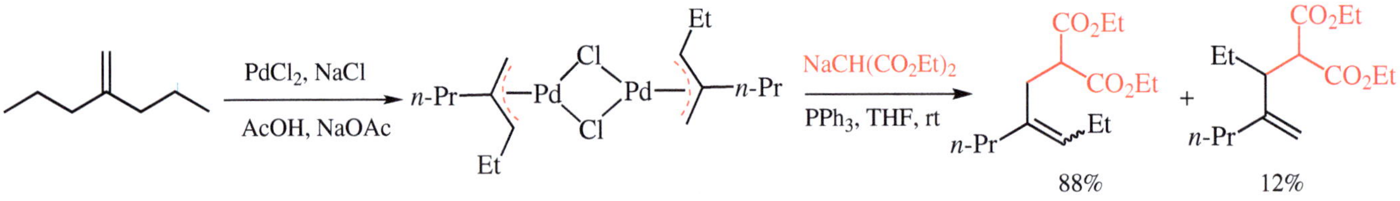

Later, Trost and coworkers developed the asymmetric version of allylic alkylation [158, 159]. Further, the reaction was improved using allylic acetates and performing the reaction catalytically under mild conditions at room temperature. A variety of nucleophiles can be used, such as alkali metal enolates or heteroatom nucleophiles, but the most commonly used are soft stabilized carbon nucleophiles, such as malonate. For example, the reaction of a cyclic allylic acetate catalyzed by palladium proceeds with complete retention of configuration at the carbon undergoing displacement [160].

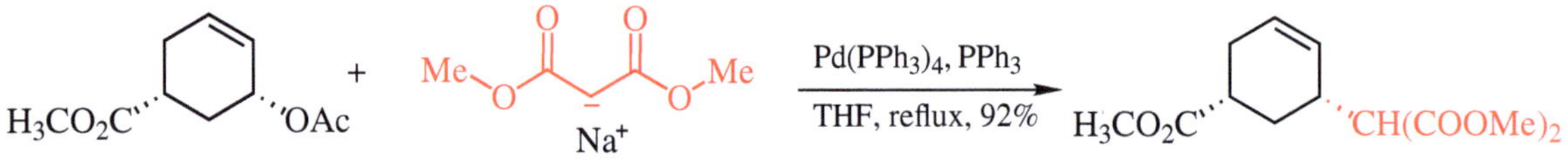

The mechanism for the Tsuji–Trost allylic alkylation reaction [161–164] begins with coordination of the palladium(0) complex to the allylic substrate, forming a Pd complex (Scheme 11.15). The next step is oxidative addition in which the

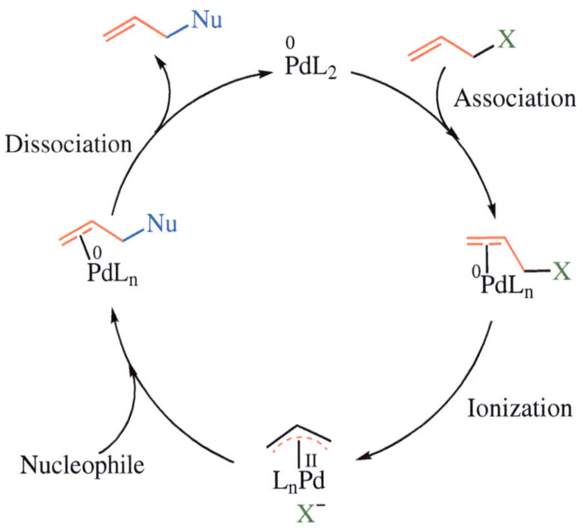

Scheme 11.15 Palladium-catalyzed cycle for the Tsuji–Trost allylic substitution reaction.

leaving group is removed by inversion of configuration, generating a π-allyl–Pd(II) complex. The oxidative addition in these types of complexes is also called ionization. Now, the nucleophile attacks the allyl group, regenerating the π-allyl–Pd(0) complex. In the final step, the product is released after dissociation from the Pd(0), which can start again in the catalytic cycle.

Trost and coworkers described a highly stereoselective method for the synthesis of oxazolidinone starting from cyclic biscarbamate via a palladium-catalyzed reaction, as shown below [165].

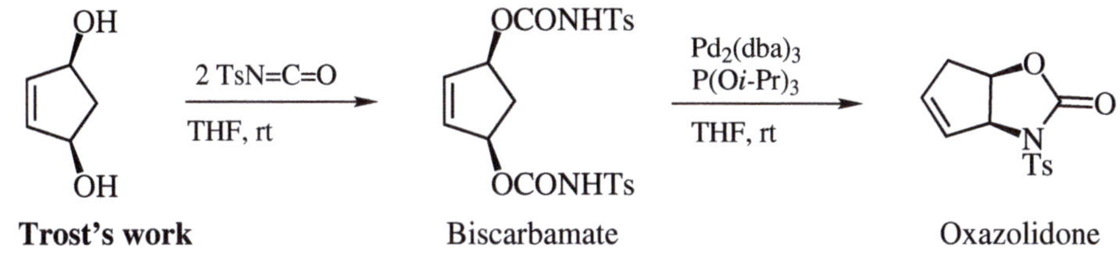

Trost and coworkers [165] and Fiaud and coworkers [166] proposed that only metal–olefin complexation anti to the leaving group will lead to the product.

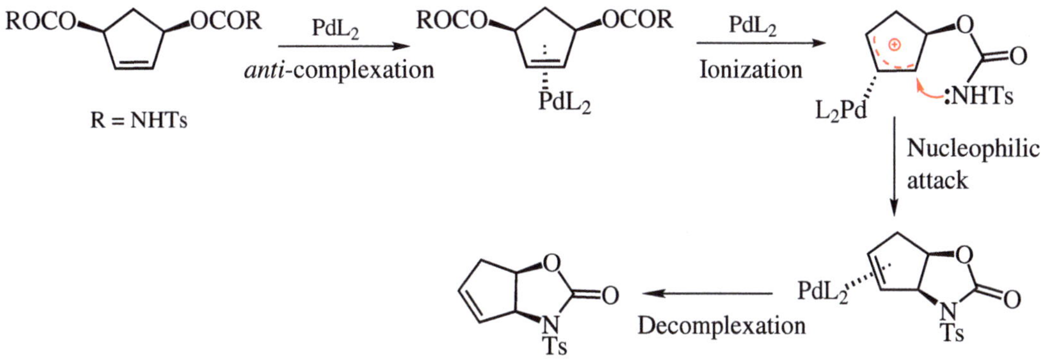

However, Balcı and coworkers [167] synthesized the norcarane diol derivative with a nitrile group located over the six-membered ring. This nitrile blocks the syn-face of the double bond and may hinder the approach of the palladium complex from the side of the cyclopropane ring. The corresponding oxazolidone was formed in 58% yield upon treatment of the biscarbamate with a palladium complex. Probably, removal of the leaving group and complexation take place at the same time.

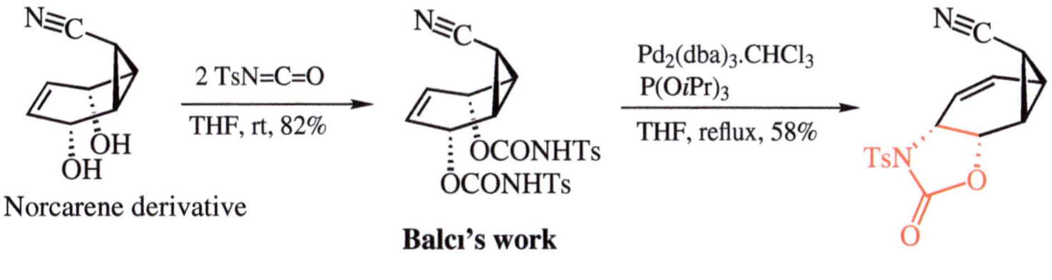

11.11 Palladium-Catalyzed Carbonylation Reactions

In parallel with carbon–carbon coupling reactions, palladium-catalyzed carbonylation reactions have also gained importance [168–172]. For example, carbonylation of olefins is extremely important for the synthesis of more valuable products such as aldehydes, alcohols, and carboxylic acids. Palladium-catalyzed carbonylation was first reported by Heck and coworkers. They described that the reaction of vinyl bromides, vinyl iodides, and benzyl chlorides with carbon monoxide in the presence of a catalytic amount of palladium complex and a tertiary amine forms the corresponding esters [173]. The reaction tolerates a variety of functional groups and shows appreciable stereospecificity with cis and trans vinylic halides, producing esters with retained configuration. When this reaction is performed in the presence of primary or secondary amines instead of alcohols, the corresponding amides are formed, and this is called carboamidation [174]. On the other hand, a combination of carbon monoxide and hydrogen converts vinyl and aromatic halides into aldehydes, and this reaction can be used to produce aldehydes from aryl, heterocyclic, and vinylic halides catalytically [175].

Aryl halides are widely used as substrates for coupling processes. The corresponding alcohols can also be used as substrates. However, they must first be converted into a more reactive derivative such as a sulfonate or triflate, which undergoes carbonylation with activity comparable with aryl iodides. In 2006, Cai and coworkers [176] reported the synthesis of aryl carboxylic esters by the reaction of aryl p-fluorobenzenesulfonates or -tosylates with Pd(OAc)$_2$ and carbon monoxide in the presence of a Josiphos ligand.

Cacchi and coworkers published a CO-free protocol in the presence of an acetic anhydride and lithium formate, which are a source for carbon monoxide, for the hydroxycarbonylation of aryl and vinyl halides or triflates [177]. The transformations tolerate a wide range of functional groups, including ether, ester, ketone, and nitro groups. The labeling experiments clearly demonstrate that the formate anion is the source of the carbonyl group. The reaction involves the intermediacy of formic acetic anhydride (generated in situ by the reaction of the formate anion with acetic anhydride). Because of the instability of formic anhydride, it decarbonylates under the reaction conditions to release carbon monoxide.

Bocelli and coworkers synthesized a benzodiazepine-1,3-dione derivative by the reaction of l-butyl-l(o-iodobenzyl)-3-phenylurea and carbon monoxide via palladium-catalyzed intramolecular cyclization at atmospheric pressure in 91% yield [178].

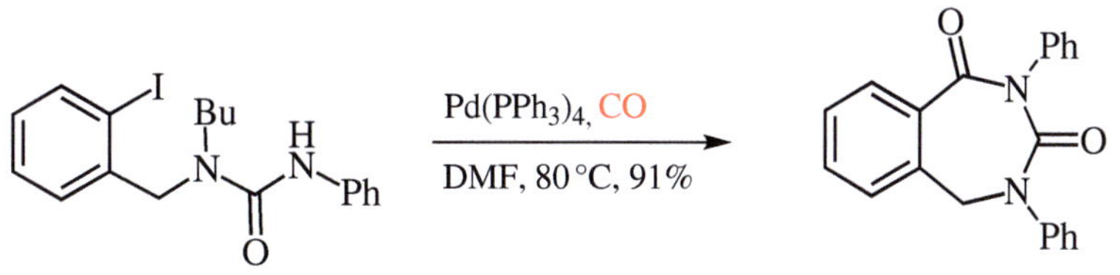

11.11 Palladium-Catalyzed Carbonylation Reactions

The mechanism for the carboalkoxylation and amidation reactions is similar to that of the C—C cross-coupling reactions as discussed above. The first step of the catalytic cycle is the oxidative addition of the aryl or vinyl compound to a Pd(0) species. Next, the palladium complex is obtained after the coordination and insertion of CO. Afterward, the ligand exchange affords the acyl palladium complex. There are different possibilities for the attacking nucleophile, one of which is coordination with palladium, followed by reductive elimination or direct attack of the nucleophile on the acyl–carbon atom. The latter is thought to be the dominant pathway in amino carbonylation reactions (Scheme 11.16).

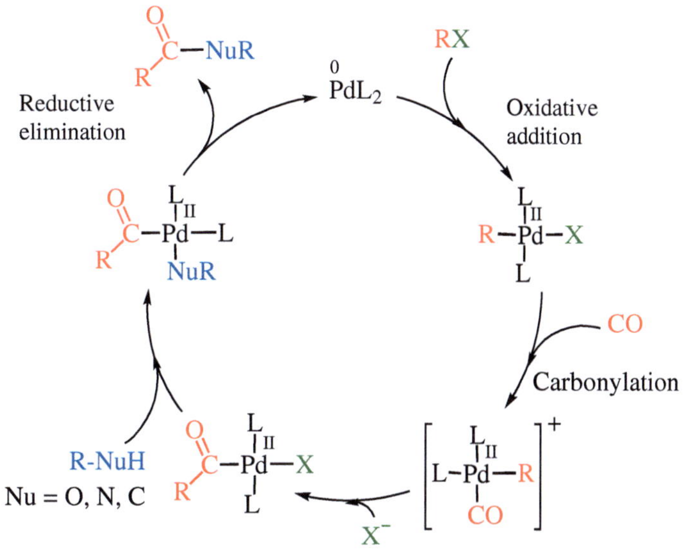

Scheme 11.16 Palladium-catalyzed cycle for the carbonylation and amidation reactions.

In reductive carbonylation reactions, aldehydes are produced as the terminal products. The palladium-catalyzed reductive carbonylation reaction was originally discovered by Schoenberg and Heck [175]. In the presence of a relatively large amount of palladium catalyst and CO at 80–150 °C, aryl and vinyl bromides or iodides were converted into the corresponding aldehydes. One decade later, Stille and coworkers used tributyltin hydride (Bu_3SnH) as a reducing agent under very mild conditions [179]. Because of the toxicity and waste generation of tin hydrides, it is no longer used. Beller and coworkers have developed an efficient palladium-catalyzed formylation procedure using $Pd(OAc)_2$/cataCXium® and TMEDA at 100 °C [180, 181].

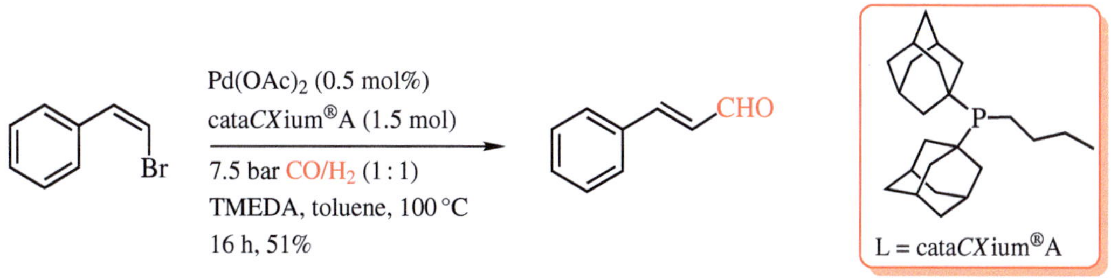

11.11.1 Carbonylative Coupling Reactions with Organometallic Reagents

The palladium-catalyzed cross-coupling of organoboranes with organohalides is known as the Suzuki reaction. The carbonylative Suzuki reaction is the formation of ketones by carrying out the classical Suzuki reaction in an atmosphere of CO and insertion of one carbonyl group into the two coupling partners. Occhiato and coworkers reported the carbonylative coupling reactions of various enol triflates with boronic acids in the presence of Pd(OAc)$_2$ [182].

A novel bulky thiourea ligand has been successfully applied to Suzuki-type carbonylative coupling reactions. As the metal–sulfur bond in the thiourea complexes is stronger than the metal–phosphorus bond in the phosphine complexes, thiourea ligands cannot easily dissociate from the metal center under catalytic conditions. This means that the thiourea palladium complexes are effective catalysts for palladium-catalyzed cross-coupling reactions [183].

The Stille cross-coupling reaction can also be carried out in an atmosphere of CO. Very recently, Dai and coworkers successfully applied the Stille carbonylation reaction to the total synthesis of a natural product, *trans*-resorcylide, which is a 12-membered macrocycle [184].

Jatrophone is a macrocyclic diterpene that exhibits significant inhibitory activity in vivo and in vitro against various carcinomas. The total synthesis of jatrophone was completed with the use of a palladium-catalyzed carbonylative reaction. A Stille-carbonylative cross-coupling reaction of a vinylic triflate with an organostannane in the final step formed an 11-membered macrocycle [185].

If the Sonogashira reaction is carried out in a carbon monoxide atmosphere, alkynones are formed, which are interesting structural motifs found in numerous biologically active molecules. The alkynone structure formed serves as a platform for the synthesis of various heterocyclic structures [186].

Wu and coworkers developed a convenient palladium-catalyzed Sonogashira carbonylative coupling. They used formic acid as the CO source. The desired alkynones can be isolated in moderate-to-good yields under mild conditions [187].

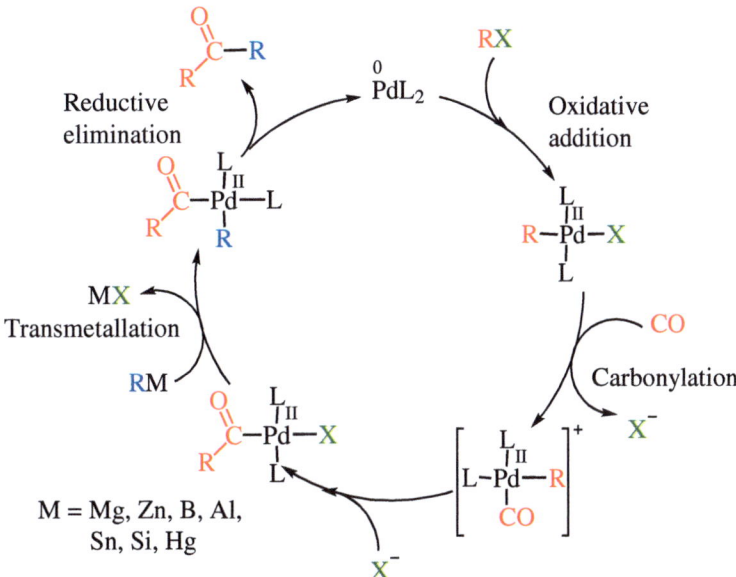

Palladium(0)-catalyzed reactions are initiated by the oxidative addition of an electrophile, typically an aryl/vinyl halide. The next step in the carbonylation reactions is the coordination of CO to palladium, followed by migration of the organyl group to the CO to form an acylpalladium complex. The transmetalation step involves the reaction of an organometallic or organometalloid nucleophile denoted M—R (M = Mg, Zn, B, Al, Sn, Si, and Hg) with the Pd complex. This process delivers the organonucleophile to the Pd center. Finally, the reductive elimination step forms the product and releases the catalyst for the next cycle (Scheme 11.17).

Scheme 11.17 Palladium-catalyzed cycle for the carbonylative coupling reactions with organometallic reagents.

11.11.2 Mo(CO)$_6$-Mediated Carbonylation

Incorporation of the carbonyl group into a molecule is an important process in the industry for converting bulk chemicals into more valuable products. Despite industrial applications, synthetic organic chemists are not willing to use carbon monoxide, which is a highly toxic and flammable gas, and high pressure reactors are required. Therefore, they are seeking alternatives. The solution to this problem is the use of solid chemicals capable of releasing carbon monoxide. Examples of these are Cr(CO)$_6$, W(CO)$_6$, Co$_2$(CO)$_8$, and especially Mo(CO)$_6$. Various alternative compounds have been used to date instead of carbon monoxide gas, for example, alkyl- and arylformates, aldehydes, formic acid, formamides, and metal carbonyls. However, transition metal carbonyls are preferred. The metal should easily release carbon monoxide. In addition, the metal carbonyl used must act catalytically [188].

Larhed and coworkers reported a palladium-catalyzed CO-free carbonylative Sonogashira/cyclization reaction of 4-quinolones starting from 2-iodoaniline derivatives and alkynes. Mo(CO)$_6$ was used as a convenient solid source of CO [189].

The Stille coupling of phenylstannane with 2-bromoindene under these carbonylative conditions in the presence of Mo(CO)$_6$ as a CO source gives an unsaturated ketone. The reaction was carried out with 1 equiv of Mo(CO)$_6$ and 10 mol% of DBU in DMF at 100 °C for 16 hours [190].

References

1. Siemsen, P., Livingston, R.C., and Diederich, F. (2000). *Angew. Chem. Int. Ed.* 39: 2632.
2. Straus, F. (1905). *Justus Liebigs Ann.Chem.* 342: 190.
3. Eglinton, G. and Galbraith, A.R. (1956). *Chem. Ind. (London)*: 737.
4. Sondheimer, F. (1963). *Pure Appl. Chem.* 7: 363.
5. Baeyer, A. (1882). *Ber. Dtsch. Chem. Ges.* 15: 50.
6. Ullmann, F. and Bielecki, J. (1901). *Ber. Dtsch. Chem. Ges.* 34: 2174.
7. Hassan, J., Sevignon, M., Gozzi, C. et al. (2002). *Chem. Rev.* 102: 1359.
8. Wurtz, A. (1855). *Ann. Chim. Phys.* 44: 275.
9. Grignard, V. (1900). *C. R. Hebd. Seances Acad. Sci.* 130: 1322.
10. Bennett, G.M. and Turner, E.E. (1914). *J. Chem. Soc. Trans.* 105: 1057.
11. Krizewsky, J. and Turner, E.E. (1919). *J. Chem. Soc.* 115: 559.
12. Chodkiewicz, W. and Cadiot, P. (1955). *C. R. Hebd. Seances Acad. Sci.* 241: 1055.
13. Stephens, R.D. and Castro, C.E. (1963). *J. Org. Chem.* 28: 3313.
14. Heck, R.F. (1968). *J. Am. Chem. Soc.* 90: 5518.
15. Heck, R.F. (1969). *J. Am. Chem. Soc.* 91: 6707.
16. Mizoroki, T., Mori, K., and Ozaki, A. (1971). *Bull. Chem. Soc. Jpn.* 44: 581.
17. Heck, R.F. and Nolley, J.P. Jr., (1972). *J. Org. Chem.* 37: 2320.
18. Fitton, P., Johnson, P., and McKeon, J.E. (1968). *J. Chem. Soc., Chem. Commun.*: 6.
19. Beletskaya, I.P. and Cheprakov, A.V. (2000). *Chem. Rev.* 100: 3009.
20. Felpin, F.-X., Nassar-Hardy, L., Le Callonnec, F., and Fouquet, E. (2011). *Tetrahedron* 67: 2815.
21. Jutand, A. (2009). Mechanisms of the Mizoroki–Heck reaction. In: *The Mizoroki-Heck Reaction* (ed. M. Oestreich), 1–50. Wiley.
22. De Meijere, I.A. and Meyer, F.E. (1994). *Angew. Chem. Int. Ed.* 33: 2379.
23. Amatore, C., Jutand, A., and M'Barki, M.A. (1992). *Organometallics* 11: 3009.
24. Tambar, U.K. (2003). The Heck reaction: mechanistic insight into a synthetically useful reaction. https://www.yumpu.com/en/document/view/37786536/the-heck-reaction-mechanistic-insight-into-a-the-stoltz-group (accessed October 2019)
25. Heck, R.F. (1971). *J. Am. Chem. Soc.* 93: 6896.
26. Heck, R.F. (1982). *Org. React.* 27: 345.
27. Dieck, H.A. and Heck, R.F. (1974). *J. Am. Chem. Soc.* 96: 1133.
28. Larherd, M., Anderson, C.-M., and Hallberg, A. (1994). *Tetrahedron* 50: 285.
29. Cabri, W., Candiani, I., Bedeschi, A., and Santi, R. (1991). *Tetrahedron Lett.* 32: 1753.
30. Cabri, W., Candiani, I., Bedeschi, A. et al. (1992). *J. Org. Chem.* 57: 1481.

31 Cabri, W. and Candiani, I. (1995). *Acc. Chem. Res.* 28: 2.
32 Jagtap, S. (2017). *Catalysts* 7: 267.
33 Von Schenck, H., Akermark, B., and Svensson, M. (2003). *J. Am. Chem. Soc.* 125: 3503.
34 Zhou, L., Shi, Y., Zhu, X., and Zhang, P. (2019). *Tetrahedron Lett.* 60: 2005.
35 Nguefack, J.F., Bolitt, V., and Sinou, D. (1997). *J. Org. Chem.* 62: 6827.
36 Dumonteil, G., Hiebel, M.-A., and Berteina-Raboin, S. (2018). *Catalysts* 8: 115.
37 Mc Cartneya, D. and Guiry, P.J. (2011). *Chem. Soc. Rev.* 40: 5122.
38 Ozawa, F., Kubo, A., and Hayashi, T. (1991). *J. Am. Chem. Soc.* 113: 1417.
39 Ozawa, F. and Hayashi, T. (1992). *J. Organomet. Chem.* 428: 267.
40 Loiseleur, O., Meier, P., and Pfaltz, A. (1996). *Angew. Chem. Int. Ed.* 35: 200.
41 Carpenter, N.E., Kucera, D.J., and Overman, L.E. (1989). *J. Org. Chem.* 54: 5846.
42 Ruan, J., Saidi, O., Iggo, J.A., and Xiao, J. (2008). *J. Am. Chem. Soc.* 130: 10510.
43 Azarian, D., Dua, S.S., Eaborn, C., and Walton, D.R.M. (1976). *J. Organomet. Chem.* 117: C55–C57.
44 Kosugi, M., Shimizu, Y., and Migita, T. (1977). *Chem. Lett.* 6: 1423.
45 Milstein, D. and Stille, J.K. (1978). *J. Am. Chem. Soc.* 100: 3636.
46 Stille, J.K. (1986). *Angew. Chem. Int. Ed.* 25: 508.
47 Espinet, P. and Echavarren, A.M. (2004). *Angew. Chem. Int. Ed.* 43: 4704.
48 Cordovilla, C., Bartolome, C., Martínez-Ilarduya, J.M., and Espinet, P. (2015). *ACS Catal.* 5: 3040.
49 Amatore, C., Broeker, G., Jutand, A., and Khalil, F. (1997). *J. Am. Chem. Soc.* 119: 5176.
50 Farina, V. and Krishnan, B. (1991). *J. Am. Chem. Soc.* 113: 9585.
51 Nova, A., Ujaque, G., Maseras, F. et al. (2006). *J. Am. Chem. Soc.* 128: 14571.
52 Brown, J.M. and Cooley, N.A. (1988). *Chem. Rev.* 88: 1031.
53 Gillie, A. and Stille, J.K. (1980). *J. Am. Chem. Soc.* 102: 4933.
54 Ananikov, V.P., Musaev, D.G., and Morokuma, K. (2005). *Organometallics* 24: 715.
55 Liebeskind, L.S. and Fengi, R.W. (1990). *J. Org. Chem.* 55: 5359.
56 Casado, A.L. and Espinet, P. (2003). *Organometallics* 22: 1305.
57 Farina, V., Kapadia, S., Krishnan, B. et al. (1994). *J. Org. Chem.* 59: 5905.
58 Scott, W.J. and Stille, J.K. (1986). *J. Am. Chem. Soc.* 108: 3033.
59 Lee, V. (2019). *Org. Biomol. Chem.* 17: 9095.
60 Mee, S.P.H., Lee, V., and Baldwin, J.E. (2004). *Angew. Chem. Int. Ed.* 43: 1132.
61 San Flipo, J. Jr., and Silbermann, J. (1981). *J. Am. Chem. Soc.* 103: 5588.
62 Trost, B.M. and Tanigawa, Y. (1979). *J. Am. Chem. Soc.* 101: 4743.
63 Piers, E., Friesen, R.W., and Keay, B.A. (1985). *J. Chem. Soc., Chem. Commun.*: 809.
64 Stille, J.K. and Tanaka, M. (1987). *J. Am. Chem. Soc.* 109: 3785.
65 Woo, S. and McErlean, C.S.P. (2019). *Org. Lett.* 21: 4215.
66 Farina, V. and Hauck, S.I. (1991). *J. Org. Chem.* 56: 4317.
67 Suzuki, A. *Proc. Jpn. Acad. Ser. B.* 2004, 80: 359.
68 (a) Miyaura, N. and Suzuki, A. (1995). *Chem. Rev.* 95: 2457. (b) Beletskaya, A., Alonso, F., and Tyurin, V. (2019). *Coord. Chem. Rev.* 385: 137.
69 Lennox, A.J.J. and Lloyd-Jones, G.C. (2014). *Chem. Soc. Rev.* 43: 412.
70 Miyaura, N., Yamada, K., and Suzuki, A. (1979). *Tetrahedron Lett.* 20: 3437.
71 Miyaura, N., Yanagi, T., and Suzuki, A. (1981). *Synth. Commun.* 11: 513.
72 Matos, K. and Soderquist, J.A. (1998). *J. Org. Chem.* 63: 461.
73 Gropen, O. and Haaland, A. (1973). *Acta Chim. Scan.* 27: 521.
74 Miyaura, N., Yamada, K., Suginome, H., and Suzuki, A. (1985). *J. Am. Chem. Soc.* 107: 972.
75 Thomas, A.A. and Denmark, S.E. (2016). *Science* 352: 329.
76 Ridgway, B.H. and Woerpel, K.A. (1998). *J. Org. Chem.* 63: 458.
77 Badone, D., Baroni, M., Cardamone, R. et al. (1997). *J. Org. Chem.* 62: 7170.
78 Grushin, V.V. and Alper, H. (1994). *Chem. Rev.* 94: 1047.
79 Percec, V., Bae, J.-Y., and Hill, D.H. (1995). *J. Org. Chem.* 60: 1060.
80 Saito, S., Sakai, M., and Miyaura, N. (1996). *Tetrahedron Lett.* 37: 2993.
81 Saito, S., Oh-tani, S., and Miyaura, N. (1997). *J. Org. Chem.* 62: 8024.
82 For a review on Ni see: Han, F.-S. (2013). *Chem. Soc. Rev.* 42: 5270.

83 Indolese, A.F. (1997). *Tetrahedron Lett.* 38: 3513.
84 Inada, K. and Miyaura, N. (2000). *Tetrahedron* 56: 8657.
85 For detailed mechanistic studies see:Zhang, K., Conda-Sheridan, M., Cooke, S.R., and Louie, J. (2011). *Organometallics* 30: 2546.
86 For a review on Applications of Suzuki–Miyaura coupling reaction to synthesis of natural products see:Koshvandi, A.T.K., Heravi, M.M., and Momeni, T. (2018). *Appl. Organomet. Chem.* 32: e-4210.
87 Zhang, C., Liu, J., and Du, Y. (2014). *Tetrahedron Lett.* 55: 959.
88 Buchspies, J. and Szostak, M. (2019). *Catalysts* 9: 53.
89 Haddach, M. and McCarthy, J.R. (1999). *Tetrahedron Lett.* 40: 3109.
90 Rafiee, F. and Hajipour, A. (2015). *Appl. Organomet. Chem.* 29: 181.
91 Osumi, Y., Liu, C., and Szostak, M. (2017). *Org. Biomol. Chem.* 15: 8867.
92 Gürbüz, N., Karaca, E.Ö., Özdemir, İ., and Çetinkaya, B. (2015). *Turk. J. Chem.* 39: 1115.
93 Zhang, D. and Wang, Q. (2015). *Coord. Chem. Rev.* 286: 1.
94 Yus, M. and Pastor, I.M. (2013). *Chem. Lett.* 42: 94.
95 Yaşar, S., Çağlar, Ş., Arslan, M., and Özdemir, İ. (2015). *J. Organomet. Chem.* 776: 107.
96 Negishi, E. and Baba, S. (1976). *J. Chem. Soc., Chem. Commun.* 15: 596b.
97 Negishi, E., King, A.O., and Okukado, N. (1977). *J. Org. Chem.* 42: 1821.
98 Okukado, N., Negishi, E., and King, A.O. (1977). *J. Chem. Soc., Chem. Commun.* 19: 683.
99 Fauvarque, J.F. and Jutand, A. (1976). *Bull. Soc. Chim. Fr.*: 765.
100 Fauvarque, J.F. and Jutand, A. (1977). *J. Organomet. Chem.* 132: C17–C19.
101 Haas, D., Hammann, J.M., Greiner, R., and Knochel, P. (2016). *Catalysts* 6: 1540.
102 Heravi, M.M., Hashemi, E., and Nazari, N. (2014). *Mol. Divers.* 18: 441.
103 Phapale, V.B. and Cárdenas, D.J. (2009). *Chem. Soc. Rev.* 38: 1598.
104 Negishi, E. (1982). *Acc. Chem. Res.* 15: 340.
105 Giovannini, R. and Knochel, P. (1998). *J. Am. Chem. Soc.* 120: 11186.
106 Giovannini, R., Stüdemnann, N., Devasagayaraj, A. et al. (1999). *J. Org. Chem.* 64: 3544.
107 For "Mechanisms of Nickel-Catalyzed CrossCoupling Reactions" see:Diccianni, J.B. and Diao, T. (2019). *Trends Chem.* 1: 830–844. https://doi.org/10.1016/j.trechm.2019.08.004.
108 Kochi, J. (1980). *Pure Appl. Chem.* 52: 571.
109 Schley, N.D. and Fu, G.C. (2014). *J. Am. Chem. Soc.* 136: 16588.
110 Boudet, N., Sase, S., Sinha, P. et al. (2007). *J. Am. Chem. Soc.* 129: 12358.
111 Yang, Y., Oldenhuis, N.J., and Buchwald, S.L. (2013). *Angew. Chem. Int. Ed.* 52: 615.
112 Tollefson, E.J., Dawson, D.D., Osborne, C.A., and Jarvo, E.R. (2014). *J. Am. Chem. Soc.* 136: 14951.
113 Cassar, L. (1975). *J. Organomet. Chem.* 93: 253.
114 Dieck, H.A. and Heck, F.R. (1975). *J. Organomet. Chem.* 93: 259.
115 Sonogashira, K., Tohda, Y., and Hagihara, N. (1975). *Tetrahedron Lett.* 15: 4467.
116 Chinchilla, R. and Nájera, C. (2011). *Chem. Soc. Rev.* 40: 5084.
117 Chinchilla, R. and Nájera, C. (2007). *Chem. Rev.* 107: 874.
118 Doucet, H. and Hierso, J.-C. (2007). *Angew. Chem. Int. Ed.* 46: 834.
119 Sonogashira, K. (2002). *J. Organomet. Chem.* 653: 46.
120 Rivera, R.P., Ehlers, P., Ohlendorf, L. et al. (2019). *Tetrahedron* 75: 130559.
121 Liang, B., Dai, M., Chen, J., and Yang, Z. (2005). *J. Org. Chem.* 70: 391.
122 Gazvoda, M., Virant, M., Pinter, B., and Košmrlj, J. (2018). *Nat. Commun.* 9: 1.
123 Basceken, S. and Balcı, M. (2015). *J. Org. Chem.* 80: 3806.
124 Schweizer, S., Erbland, G., Bisseret, F. et al. (2015). *Turk. J. Chem.* 39: 1180.
125 Kharasch, M.S. and Fields, E.K. (1941). *J. Am. Chem. Soc.* 63: 2316.
126 Corriu, R.J.P. and Masse, J.P. (1972). *J. Chem. Soc., Chem. Commun.*: 144a.
127 Tamao, K., Sumitani, K., and Kumada, M. (1972). *J. Am. Chem. Soc.* 94: 4374.
128 Cardenas, D.J. (2003). *Angew. Chem. Int. Ed.* 42: 384.
129 Guisán-Ceinos, M., Soler-Yanes, R., Collado-Sanz, D. et al. (2013). *Chem. Eur. J.* 19: 8405.
130 Iffland, L., Petuker, A., Gastel, M.V., and Apfel, U.-P. (2017). *Inorganics* 5: 78.
131 Yamamura, M., Moritam, I., and Murahashi, S.-I. (1975). *J. Organomet. Chem.* 91: C39.
132 Dahadha, A.A. and Aldhoun, M.M. (2018). *Arkivoc* 6: 234.

133 Hatanaka, Y. and Hiyama, T. (1988). *Tetrahedron Lett.* 29: 97.
134 Nakao, Y. and Hiyama, T. (2011). *Chem. Soc. Rev.* 40: 4893.
135 Foubelo, F., Najera, C., and Yus, M. (2016). *Chem. Rec.* 16: 2521.
136 Monfared, A., Mohammadi, R., Ahmadi, S. et al. (2019). *RSC Adv.* 9: 3185.
137 Denmark, S.E. and Regens, C.S. (2008). *Acc. Chem. Res.* 41: 1486.
138 Denmark, S.E. and Yang, S.M. (2002). *J. Am. Chem. Soc.* 124: 2102.
139 Hiyama, T. (2002). *J. Organomet. Chem.* 653: 58.
140 Tamao, K., Kobayashi, K., and Ito, Y. (1989). *Tetrahedron Lett.* 30: 6051.
141 Hagiwara, E., Gouda, K., Hatanaka, Y., and Hiyama, T. (1997). *Tetrahedron Lett.* 38: 439.
142 Denmark, S.E. and Sweis, R.F. (2001). *J. Am. Chem. Soc.* 123: 6439.
143 Denmark, S.E. and Tymonko, S.A. (2003). *J. Org. Chem.* 68: 9151.
144 Dorel, R., Grugel, C.P., and Haydl, A.M. (2019). *Angew. Chem. Int. Ed.* 58: 17118.
145 Heravi, M.M., Kheilkordi, Z., Zadsirjan, V. et al. (2018). *J. Organomet. Chem.* 861: 17.
146 Ruiz-Castillo, P. and Buchwald, S.L. (2016). *Chem. Rev.* 116: 12564.
147 Louillat, M.-L. and Patureau, F.W. (2014). *Chem. Soc. Rev.* 43: 901.
148 Kosugi, M., Kameyama, M., and Migita, T. (1983). *Chem. Lett.* 12: 927.
149 Guram, A.S. and Buchwald, S.L. (1994). *J. Am. Chem. Soc.* 116: 7901.
150 Paul, F., Patt, J., and Hartwig, J.F. (1994). *J. Am. Chem. Soc.* 116: 5969.
151 Guram, A.S., Rennels, R.A., and Buchwald, S.L. (1995). *Angew. Chem. Int. Ed.* 34: 1348.
152 Louie, J. and Hartwig, J.F. (1995). *Tetrahedron Lett.* 36: 3609.
153 Wolfe, J.P., Rennels, R.A., and Buchwald, S.L. (1996). *Tetrahedron* 52: 7525.
154 Wagaw, S., Rennel, R.A., and Buchwald, S.L. (1997). *J. Am. Chem. Soc.* 119: 8451.
155 Tsuji, J., Takahashi, H., and Morikawa, M. (1965). *Tetrahedron Lett.* 49: 4387.
156 Trost, B.M. and Fullerton, T.J. (1973). *J. Am. Chem. Soc.* 95: 292.
157 Hüttel, R. and Bechter, M. (1959). *Angew. Chem.* 71: 456.
158 Trost, B.M. and Verhoeven, T.R. (1976). *J. Org. Chem.* 41: 3215.
159 Trost, B.M. and Dietsche, T.J. (1973). *J. Am. Chem. Soc.* 95: 8200.
160 Trost, B.M. and Strege, P.E. (1977). *J. Am. Chem. Soc.* 99: 1649.
161 Fernandes, R.A. and Nallasivam, J.L. (2019). *Org. Biomol. Chem.* 17: 8647.
162 Le Bras, J. and Muzart, J. (2016). *Eur. J. Org. Chem.*: 2565.
163 Trost, B.M. and Crawley, M.L. (2003). *Chem. Rev.* 103: 2921.
164 Trost, B.M. and Van Vranken, D.L. (1996). *Chem. Rev.* 96: 395.
165 Trost, B.M., Van Vranken, D.L., and Bingel, C. (1992). *J. Am. Chem. Soc.* 114: 9327.
166 Fiaud, J.C. and Aribi-Zouioueche, L. (1986). *J. Chem. Soc., Chem. Commun.*: 390.
167 Kilbas, B. and Balcı, M. (2011). *Beilstein J. Org. Chem.* 7: 246.
168 Mancuso, R., Ca, N.D., Veltri, L. et al. (2019). *Catalysts* 610: 1–32.
169 Beller, M. and Wu, X.F. (2013). *Transition Metal Catalyzed Carbonylation Reactions*. Berlin, Heidelberg: Springer Verlag.
170 Wu, X.F., Neumann, H., and Beller, M. (2011). *Chem. Soc. Rev.* 40: 4986.
171 Barnard, C.F. (2008). *J. Organomet.* 27: 5402.
172 Brennführer, A., Neumann, H., and Beller, M. (2009). *Angew. Chem. Int. Ed.* 48: 4114.
173 Schoenberg, A., Bartoletti, I., and Heck, R.F. (1974). *J. Org. Chem.* 39: 3318.
174 Schoenberg, A. and Heck, R.F. (1974). *J. Org. Chem.* 39: 3327.
175 Schoenberg, A. and Heck, R.F. (1974). *J. Am. Chem. Soc.* 96: 7761.
176 Cai, C., Rivera, N.R., Balsells, J. et al. (2006). *Org. Lett.* 8: 5161.
177 Cacchi, S., Fabrizi, G., and Goggiamani, A. (2003). *Org. Lett.* 5: 4269.
178 Bocelli, G., Catellani, M., Cugini, F., and Ferraccioli, R. (1999). *Tetrahedron Lett.* 40: 2623.
179 Baillargeon, V.P. and Stille, J.K. (1986). *J. Am. Chem. Soc.* 108: 452.
180 Klaus, S., Neumann, H., Zapf, A. et al. (2006). *Angew. Chem. Int. Ed.* 45: 154.
181 Sergeev, A.G., Zapf, A., Spannenberg, A., and Beller, M. (2008). *Organometallics* 27: 297.
182 Bartali, L., Guarna, A., Larini, P., and Occhiato, E.G. (2007). *Eur. J. Org. Chem.*: 2152.
183 Mingli, D., Liang, B., Wang, C. et al. (2004). *Adv. Synth. Catal.* 346: 1669.
184 Luo, Y., Yin, X., and Dai, M. (2019). *J. Antibiot.* 72: 482.

185 Gyorkos, A.C., Stille, J.K., and Hegedus, L.S. (1990). *J. Am. Chem. Soc.* 112: 8465.
186 Neumann, K.T., Laursen, S.R., Lindhardt, A.T. et al. (2014). *Org. Lett.* 16: 2216.
187 Qi, X., Jiang, L.B., Li, C.L. et al. (2015). *Chem. Asian J.* 10: 1870.
188 Akerbladh, L., Odell, L.R., and Larhed, M. (2019). *Synlett* 30: 141.
189 Akerbladh, L., Nordeman, P., Wejdemar, M. et al. (2015). *J. Org. Chem.* 80: 1464.
190 Sävmarker, J. (2012). Palladium-Catalyzed Carbonylation and Arylation Reactions. PhD Thesis. Uppsala University.

Solutions

Chapter 1 (Structure and Bondings)

1

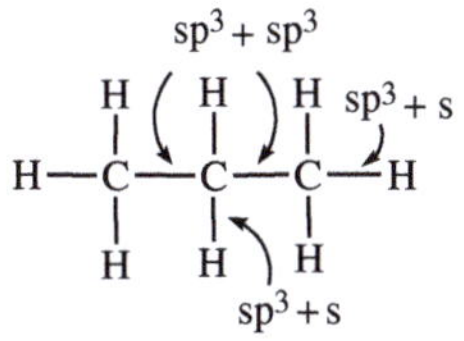

2

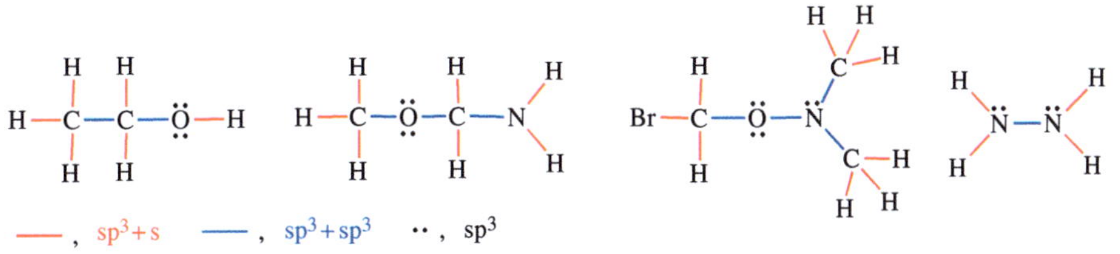

3

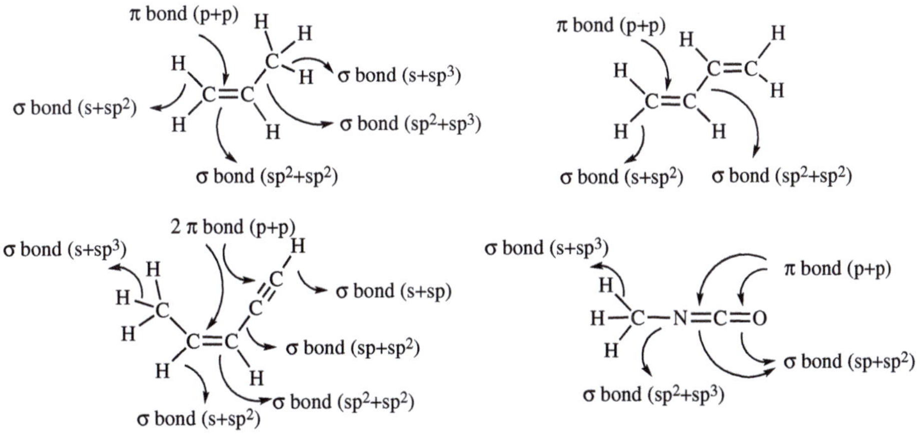

4

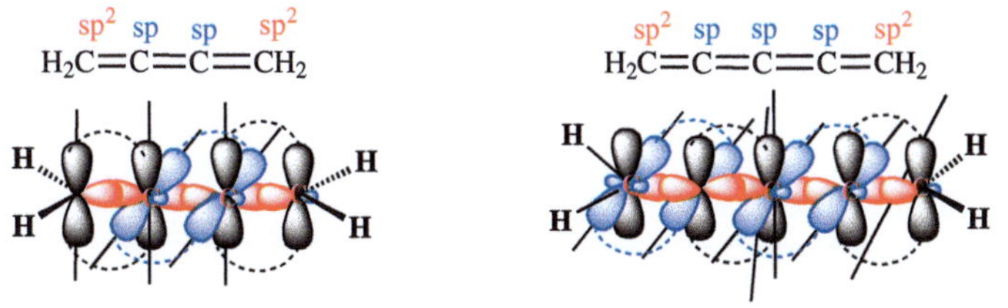

While the terminal hydrogen atoms of the butatriene molecule are on the same plane, the terminal hydrogen planes in the pentatetraene are perpendicular to each other.

Reaction Mechanisms in Organic Chemistry, First Edition. Metin Balcı.

5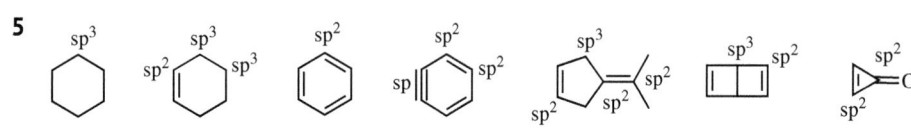

6

7 Because the hybridization of the spiro carbon atom of the first molecule is sp³, the planes formed by these rings are perpendicular to each other. On the other hand, because the hybridization of the carbon atoms connecting the two rings of the second molecule is sp², all the atoms in both rings are on a plane.

8 The π bond connecting two double bonds in butadiene is formed by the overlap of sp² hybrid orbitals. The molecule can rotate freely around this σ bond. For this reason, there are two different conformations *cisoid* and *transoid* (not to be confused with the configuration).

transoid conformation   *cisoid* conformation

9 Because four carbon atoms are sp-hybridized in the first molecule, these four carbon atoms and hydrogen and methyl carbon form a linear structure. In the second molecule, the triple-bond carbon atoms and the double-bond carbons directly attached to them are in a line. The vinyl groups can freely rotate to have two conformations: *cisoid* and *transoid*.

10 The hybridization of olefinic carbons in cyclohexadiene is sp². The bond angles are 120°. Therefore, there is no strain. The second compound has an allene structure. Hybridization of the C2 carbon atom is sp. Accordingly, C1–C2–C3 should form a linear structure. However, there is a serious bending here. This molecule is unstable under normal conditions and is an extremely reactive molecule. There is a similar situation with alkyne. Alkyne carbons and atoms directly attached to them must be in a line C1–C2–C3–C4. Because the structure does not allow this, the molecule is extremely unstable.2

11 [structures shown]

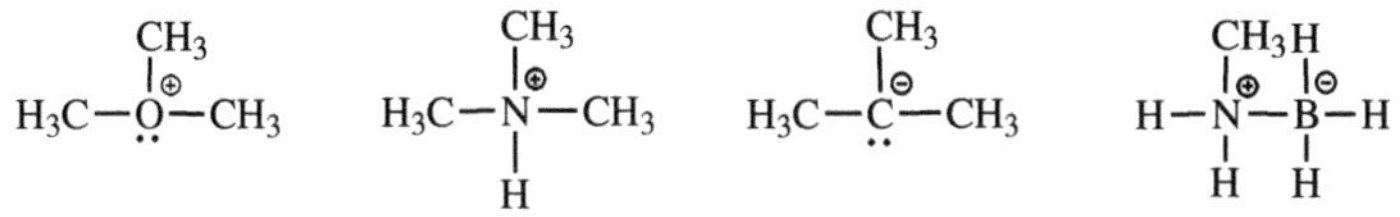

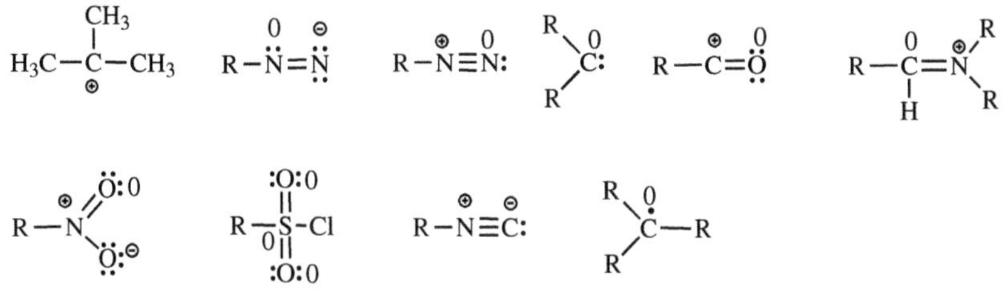

12 [structures shown]

13 [resonance structures shown]

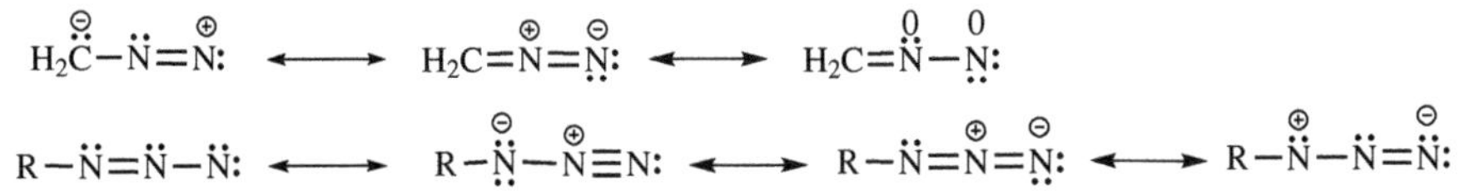

14 In both reactions, double-bond carbon atoms are oxidized. Bromine is reduced.

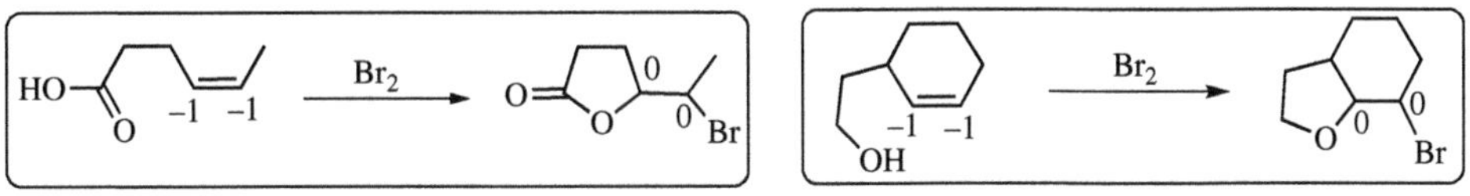

[reaction schemes shown]

15 Here, four elements, two oxygen atoms and two carbon atoms, are involved in this reaction. The oxygen atoms are reduced while one of the carbon atoms does not change oxidation number, the other is oxidized. This rearrangement reaction is in itself a reduction and oxidation reaction.

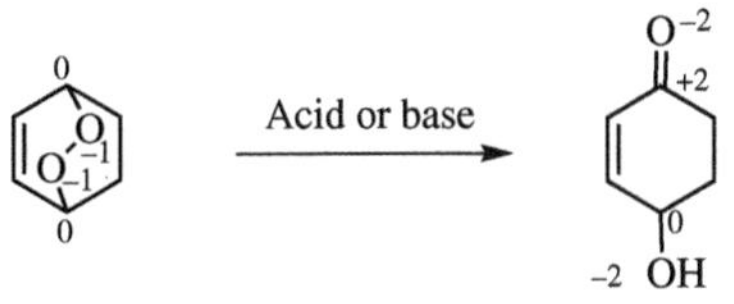

[reaction scheme shown]

16 The first reaction involves the removal of nitrogen gas from the molecule, an oxidation and reduction reaction in itself. In the second reaction, a ketene is formed by the elimination of two chlorine atoms from the molecule. The carbon atoms are reduced. Zn is used as a reductant for the reaction to take place.

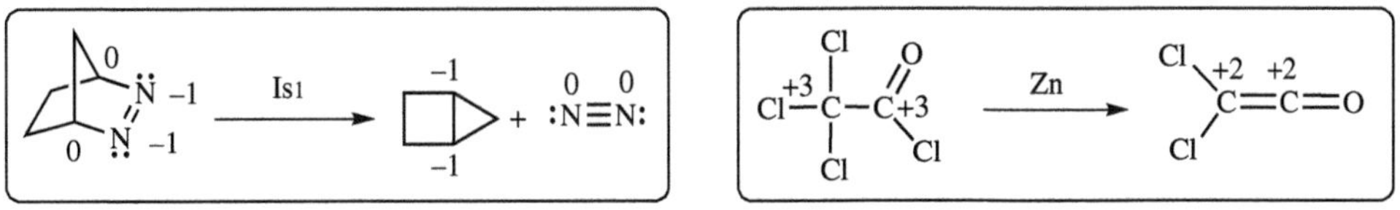

[reaction schemes shown]

17 1. The molecule (a) is stronger because of the inductive effect of oxygen in the chain.
2. The molecule (b). The first one is an alcohol. The hydroxyl group cannot conjugate with the carbonyl group.
3. The molecule (b). In the second molecule, the ether oxygen is closer to the hydroxyl group.

18 Halogens increase acidity by the inductive effect. The fluorine-containing molecule is the strongest acid.
b > c > d > a

19 d > b > a > c

20 Methylsulfonic acid is more acidic. This is because the anion formed after the dissociation of the proton can distribute the negative charge on more oxygen atoms.

21 Because chlorine is more electron attracting, it makes the acetate anion more stable. HBr is more acidic because of the larger volume of bromine, which can better stabilize the negative charge.

22 d > b > c > a

23 NaOH can react with b, c, and d.

24 t-BuOK is more basic than water. Therefore, it would abstract a proton from water.

25 Acetic acid dissolves in sodium bicarbonate solution but phenol does not.

Chapter 2 (Nucleophilic Substitution)

1 a. Because the configuration of the starting compound is not given, it should be considered to be a racemic mixture. The product will be a racemic mixture of 1-methoxy-2-methylbutan-2-ol.
   b. In the second product, it will be the same, however with the inverted configuration R.

2 (a) The first compound reacts faster. Nucleophile HO⁻ is a stronger nucleophile than H_2O. (b) The first compound reacts faster as there is less branching in the carbon atom attacked by the nucleophile. (c) The second compound reacts faster because sulfur is a stronger nucleophile.

3 Because the reaction mechanism is S_N1, the system that forms the most stable carbocation reacts fast.
   c > a > b > d

4 Because the reaction proceeds according to the S_N1 mechanism, the electronic structure of the nucleophile has no effect on the reaction rate. The rate-determining step is the separation of chloride from the molecule.

5 Because the reaction proceeds according to the S_N1 mechanism, carbocation will be formed in the first stage. The most stable among the carbocations will be the cation a. The charge distributed over the allyl system and the mesomeric effect of oxygen will stabilize the carbocation. In the case of the structure b, the oxygen atom will inductively reduce the stability of the carbocation. a > c > b.

6 The compound (a) is more reactive than the other as it will form an allyl cation.

7 The rate of the first reaction increases. Because of this, the reaction works according to the S_N2 mechanism. The concentration of the nucleophile does not affect the rate of the second reaction as it will proceed according to the S_N1 mechanism.

8 The benzilic bromine atoms are more reactive and easier to substitute.

9 The systems that form the most stable carbocations react faster. a. i-Propyl cation; b. –OMe group will stabilize the carbocation by mesomeric effect. c. benzyl cation; d. The norbornyl system reacts very hard because the carbocation formed is a bridgehead cation.

10 In both systems, a primary carbocation occurs first and then they rearrange into a more stable carbocation.

11 The bromide anion attacks the double bond according to the S$_N$2' reaction, and the double-bond electrons shift and substitute the bromide (allyl rearrangement).

12 PhS−C(=O)−CH$_2$−OH ⟵[NaSPh] β-propiolactone ⟶[NaOEt] HO−CH$_2$−C(=O)−OEt

13 (CH$_3$)$_2$C=CH−CH$_2$−NC ⟵[AgCN, DMF] (CH$_3$)$_2$C=CH−CH$_2$−Br ⟶[NaSPh, THF] (CH$_3$)$_2$C=CH−CH$_2$−SPh

14 2-methyl-1-(trifluoromethylsulfonyloxy)cyclohexene ⟵[LDA, PhNH(SO$_2$CF$_3$)] 2-methylcyclohexanone ⟶[LDA, EtI] 2-ethyl-6-methylcyclohexanone

15 (a) H$_3$C−CH$_2$−CH(CH$_3$)−OH + Br−CH$_2$−CH$_2$−CH$_3$ (b) H$_3$C−CH$_2$−CH(CH$_3$)−CH$_2$−OH + Br−CH$_2$−CH$_3$

(c) cyclohexyl-OH + CH$_3$Br   (d) PhCH$_2$−Br + HO−cyclohexyl

Chapter 3 (Elimination)

1 The carbocation formed after the removal of the halogen eliminates hydrogen. After its rearrangement in two different ways, three more products are formed by elimination again.

(H$_3$C)(H$_3$CH$_2$C)C=C(CH$_3$)$_2$ > (H$_3$C)(H$_3$CHC(CH$_3$))C=C(CH$_3$)(H) > (H$_3$C)(H$_3$CHC(CH$_3$))C=C(H)(CH$_3$) > (H$_3$CH$_2$C)(H$_3$CHC(CH$_3$))C=C(H)(H)

2 Compound a does not have a β-proton to remove. A more stable olefin will be formed from compound b. Bromine is a better leaving group than chlorine.

b. CH$_3$−C(CH$_3$)$_2$−CH$_2$−CH$_2$Cl   b. H$_3$C−CH$_2$−CHBr−CH$_3$ b. cyclohexyl−Br

3 Zaitsev products are given below.

H$_3$C−C(CH$_3$)=CH−CH$_3$ H$_3$C−C(CH$_3$)=CH−CH$_3$ H$_3$CH$_2$C−C(CH$_3$)$_2$−CH=CH$_2$

4 *cis*- and *trans*-Stilbene are formed. In the formation of both products, the desired antiperiplanar conformation is provided. The *trans* product is the main product, as the energy of the transition complex required for *trans* product formation is lower.

Ph−CH=CH−Ph

5 The reaction proceeds according to the E1 mechanism. The initially formed cation undergoes a rearrangement, followed by elimination.

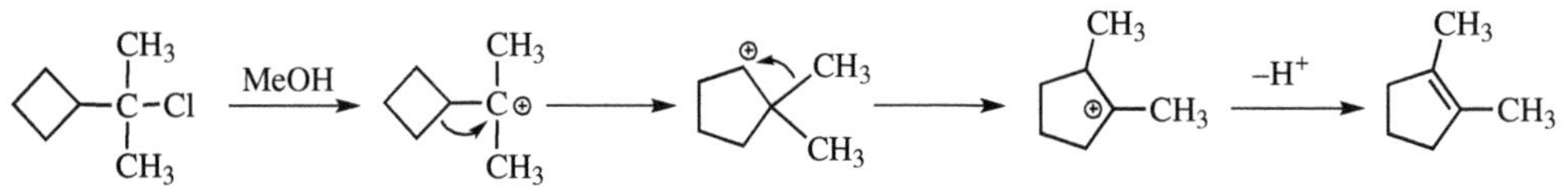

6 *trans*-Olefin if formed from the erythro compound and *cis*-olefin from threo compound.

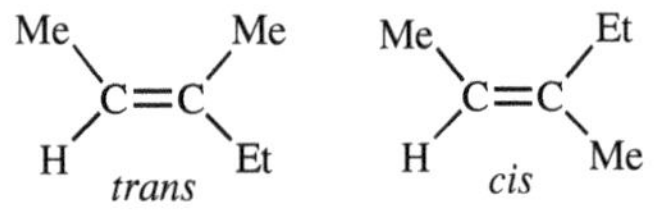

7 Compounds A and B are produced from the secondary carbocation. The secondary carbocation rearranges into the tertiary carbocation from which A and C are formed.

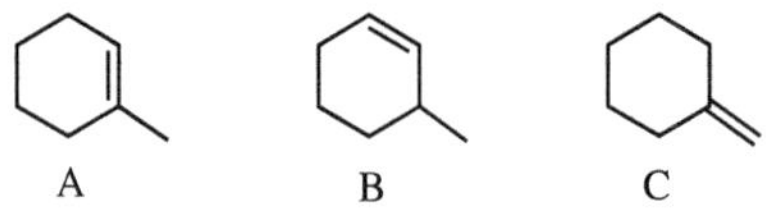

8 *trans* Alkene is formed from the erythro tosylate and *cis* alkene from the threo tosylate.

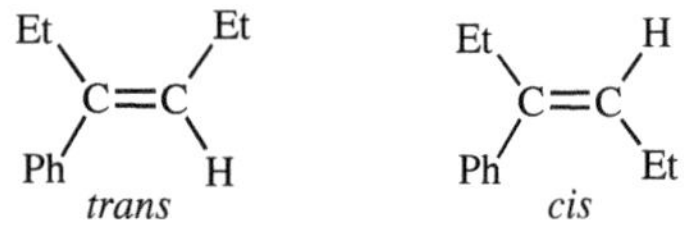

9 The less-stable olefin is formed. There is no hydrogen atom in the *trans* position for the Satyzeff product to form.

10 Zaitsev product cannot be formed. DBr is eliminated.

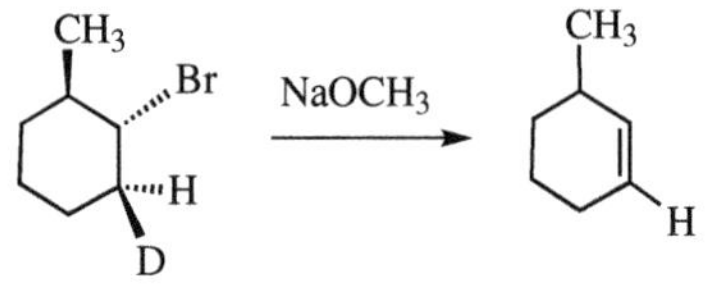

11 As there is no proton in *trans* position to the bromine atom, no elimination product will be formed. However, a substitution can take place.

12

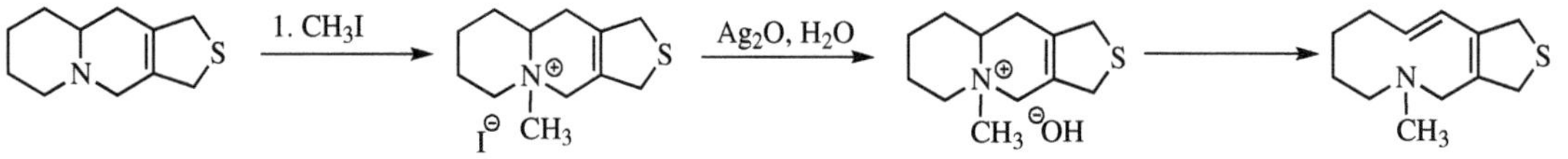

13

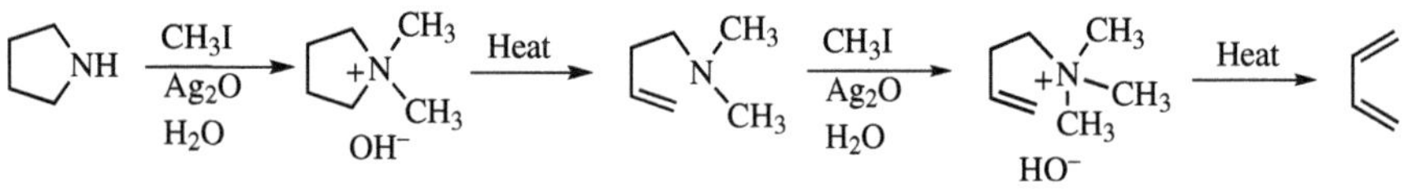

Chapter 4 (Addition)

7 [reaction scheme: H₃C-C(CH₃)₂-CH=CH₂ → H₃C-C(CH₃)₂-⁺CH-CH₃ → H₃C-C(CH₃)-⁺CH-CH₃ (methyl shift) → H₃C-C(CH₃)(OH)-CH(CH₃)-CH₃]

8 [reaction scheme showing methylcyclobutyl vinyl compound → carbocation → ring-expansion to cyclopentyl cation → tertiary alcohol]

9 [reaction scheme: furanose with allyl and BnO groups, Hg(OAc)₂, THF, 25 °C, NaCl → cyclized tetrahydrofuran with CH₂HgCl]

10 Water will add according to the anti-Markovnikov rule. Because BH₃ will approach the molecule from the *exo* face, the hydroxyl group will be in the *exo* position. Both groups will be *trans* to each other.

[norbornene-CH₃ + BH₃ / H₂O₂, NaOH → norbornane with H, OH, CH₃ exo]

11 Compound A can be synthesized by the addition of H₂O in an acidic medium or by oxymercuration. Compound B can be synthesized by hydroboration.

12 [cyclopentene → (RCO₃H) → epoxide → (H⁺/H₂O) → trans-1,2-cyclopentanediol]

13 [methylenecyclopentane → (HBr) → 1-bromo-1-methylcyclopentane → (Base) → 1-methylcyclopentene → (O₃, oxidative work-up) → keto-acid]

14 [cyclization sequence of geranyl-type chain: epoxide formation, H⁺-mediated cyclization through protonated epoxide, cation intermediate, –H⁺ to give cyclohexanol with isopropylidene group]

15 [1,3-cyclohexadiene → (RCO₃H) → monoepoxide → (H⁺/H₂O) → trans-diol with remaining alkene → (OsO₄, NMO) → tetraol]

16 [octahydronaphthalene alkene → (H⁺/H₂O) → tertiary alcohol at ring junction → (H⁺) → alkene isomer → (O₃) → cyclodecane-1,6-dione]

Chapter 5 (Carbonyl Compounds)

1

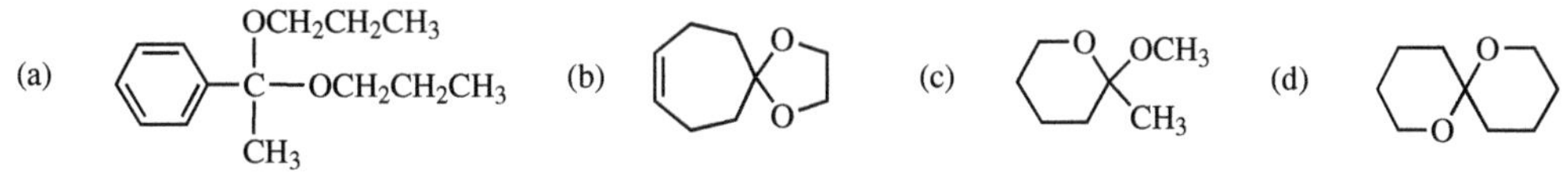

(a) Ph-C(OCH₂CH₃)(OCH₂CH₃)(CH₃) (b) cycloheptene-fused 1,3-dioxolane spiro (c) 2-methoxy-2-methyltetrahydropyran (d) cyclohexanone bis-ethylene ketal spiro

2

$$R-\overset{O}{\underset{\|}{C}}-CH_2-CH_2-COOCH_3 \xrightarrow[H^+]{\substack{CH_2-OH \\ CH_2-OH}} R-\underset{\text{(dioxolane)}}{C}-CH_2-CH_2-COOCH_3$$

$$\xrightarrow{LiAlH_4, H_2O}$$

$$R-\underset{\text{(dioxolane)}}{C}-CH_2-CH_2-CH_2OH \xrightarrow{H^+} R-\overset{O}{\underset{\|}{C}}-CH_2-CH_2-CH_2OH$$

3

$$CH_3CH_2CHO \xrightarrow[H^+]{HS(CH_2)_3SH} \text{(dithiane-H,Et)} \xrightarrow{BuLi} \text{(dithianyl anion-Et)} \xrightarrow{\text{styrene oxide (PhCH-CH_2-O)}} Ph-CH(OH)-CH_2-C(dithiane)(CH_2CH_3) \xrightarrow{Hg^{++}}$$

4

(bromo-diol cyclohexene) $\xrightarrow{m\text{-CPBA}}$ (bromo-diol epoxide) $\xrightarrow{H^+}$ (tetraol dibromide) $\xrightarrow[p\text{-TsOH}]{\text{Dimethoxypropane}}$ (bis-acetonide dibromide)

$\xrightarrow[\text{DMSO}]{Zn}$ (bis-acetonide alkene) $\xrightarrow{H^+}$ Conduritol-E

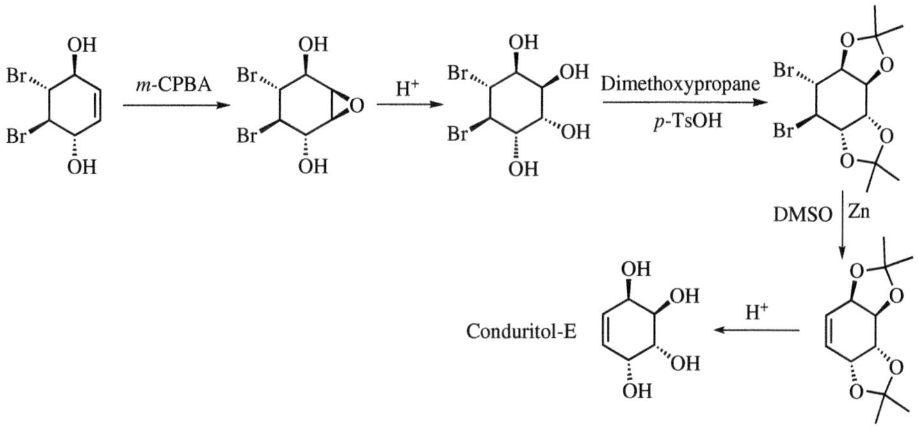

5

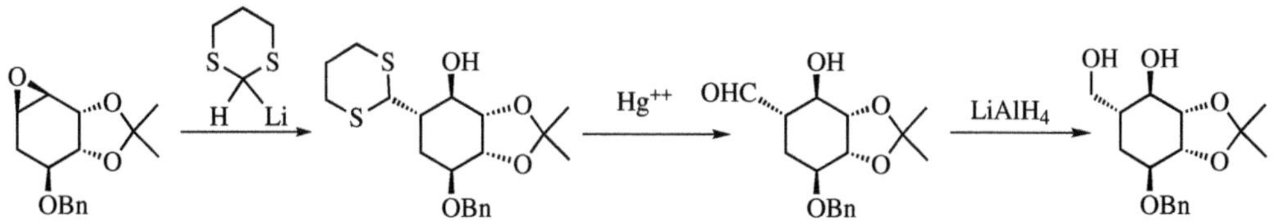

(epoxy-acetonide-OBn) $\xrightarrow{\text{dithianyl-Li}}$ (dithiane-OH-acetonide-OBn) $\xrightarrow{Hg^{++}}$ (CHO-OH-acetonide-OBn) $\xrightarrow{LiAlH_4}$ (CH₂OH-OH-acetonide-OBn)

6

(5-oxohexanal) $\xrightarrow[H^+]{\text{Ethylene glycol}}$ (keto-dioxolane aldehyde) $\xrightarrow{\text{Wolff–Kishner reduction}}$ (dioxolane hydrocarbon) $\xrightarrow{H^+}$ (hexanal)

7

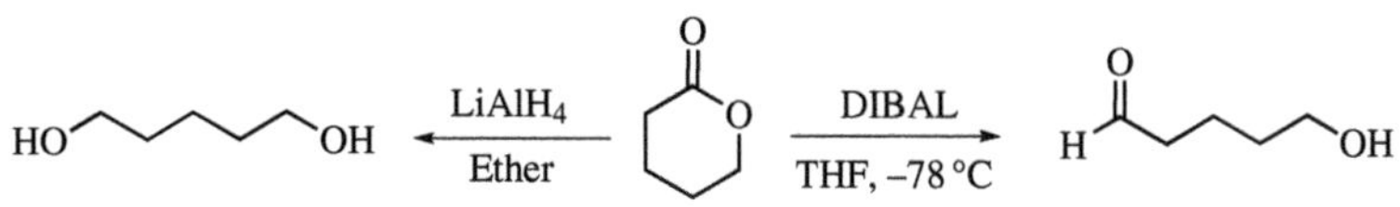

$HO-(CH_2)_5-OH \xleftarrow[\text{Ether}]{LiAlH_4} \text{δ-valerolactone} \xrightarrow[\text{THF}, -78\,°C]{\text{DIBAL}} OHC-(CH_2)_4-OH$

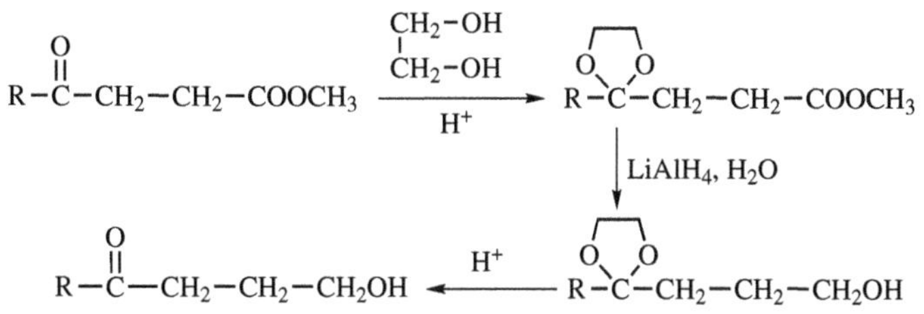

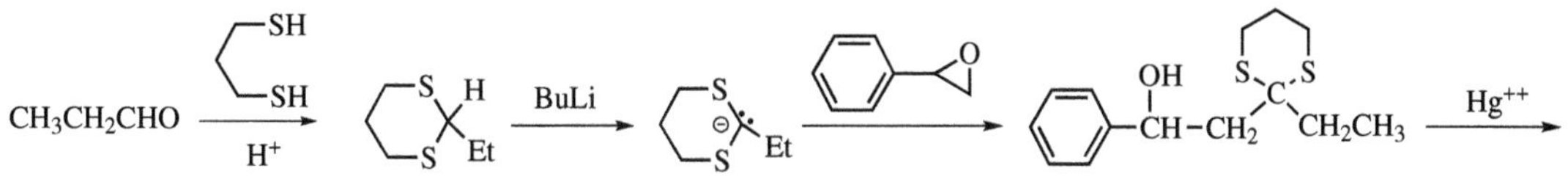

8

(cyclohexenone-CH₂CH₂COOH) $\xrightarrow[\text{MeOH}, 0\,°C]{NaBH_4/CeCl_3}$ (cyclohexenol-CH₂CH₂COOH) $\xrightarrow{HCl}$ [allylic cation resonance structures] $\longrightarrow$ (spirolactone)

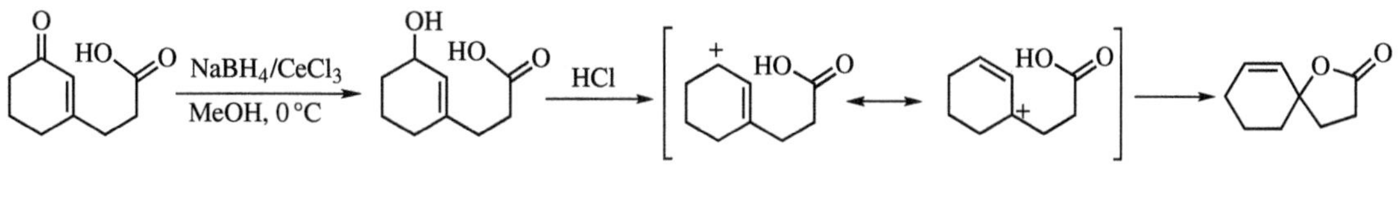

9

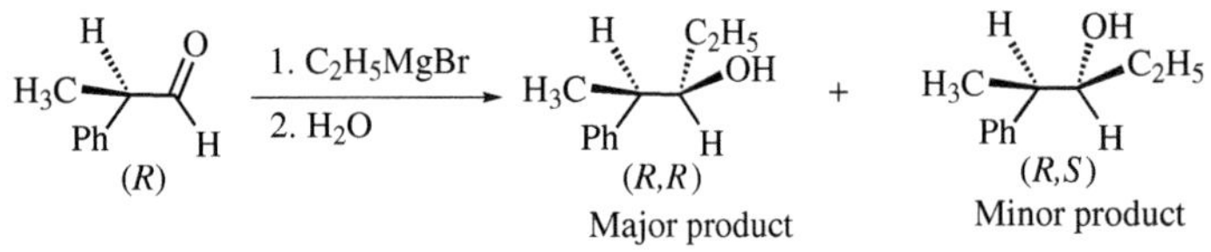

[Structure: (R)-PhCH(CH3)CHO reacts with 1. C2H5MgBr, 2. H2O to give (R,R) major product and (R,S) minor product of PhCH(CH3)CH(OH)C2H5]

10

A = benzophenone (Ph-CO-Ph)

B = acetophenone (Ph-CO-CH3)

C = CH3-CO-OCH3

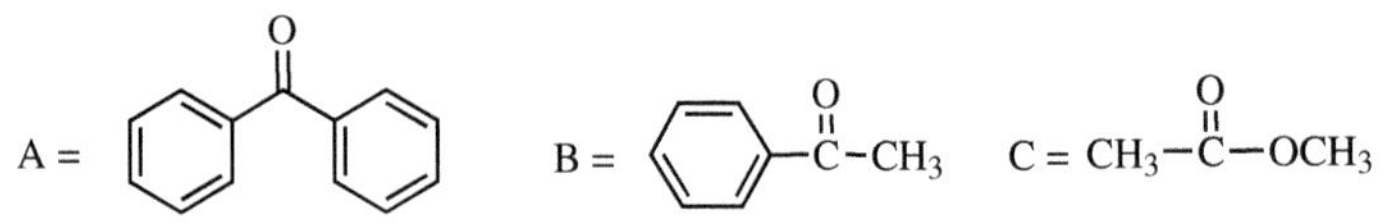

11 Magnesium first forms the expected Grignard compound with the alkyl bromide. However, because the compound contains an acidic hydrogen, the alkyne Grignard compound is formed by the exchange.

HC≡C—CH$_2$CH$_2$Br + Mg ⟶ [HC≡C—CH$_2$CH$_2$MgBr] ⟶ BrMg—C≡C—CH$_2$CH$_3$

12

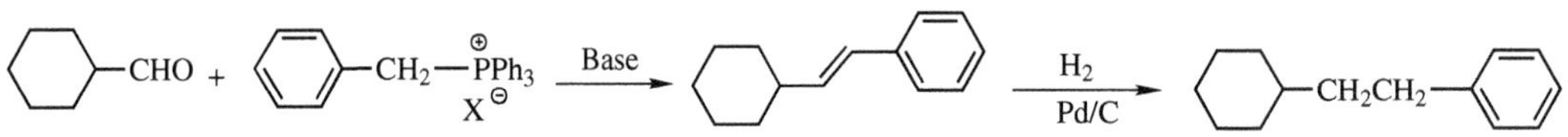

Cyclohexyl-CHO + Ph-CH$_2$-PPh$_3$⁺ X⁻ —Base→ cyclohexyl-CH=CH-Ph —H$_2$/Pd/C→ cyclohexyl-CH$_2$CH$_2$-Ph

13 1,4-Addition product is formed. The ylide formed behaves like a sulfoxonium ylide as the carbanion formed after proton abstraction is stabilized by the adjacent ester group.

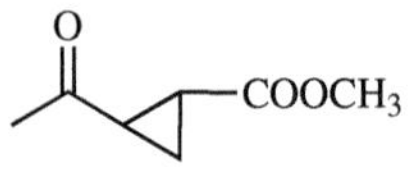

[Structure: methyl ketone with cyclopropane bearing COOCH3 group]

14 The more substituted double bonds are the most stable ones.

[Structures showing Major enol (more substituted) ⇌ less substituted enol for two cases]

15

[Four structures: 2-formylcyclohexanone, 2-hydroxycyclohexanone, cyclohex-3-enone, cyclohex-2-ene-1,4-dione]

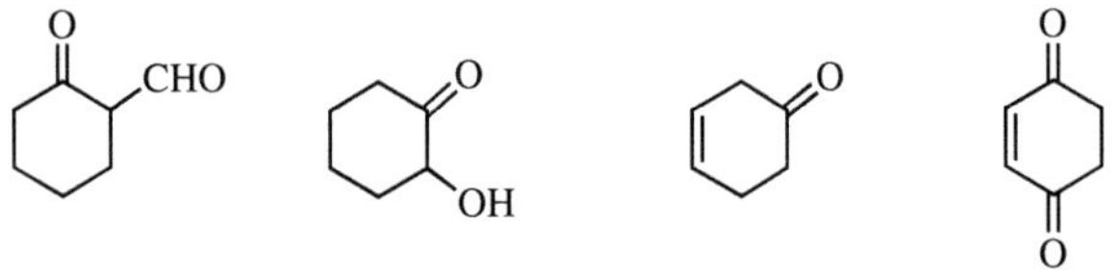

16 Two products will be formed. However, these products are diastereoisomers.

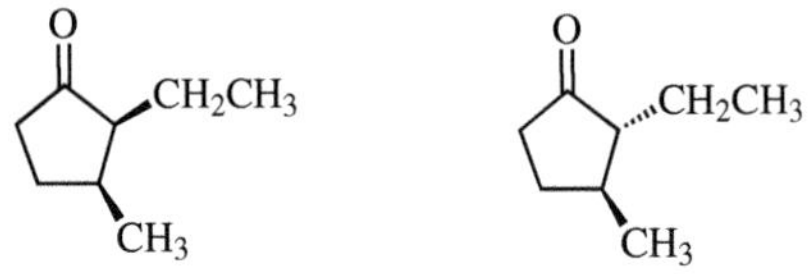

[Two cyclopentanone structures with CH$_2$CH$_3$ and CH$_3$ substituents as diastereomers]

17

CD$_3$—C(=O)—CD$_3$ CH$_3$—CD$_2$—C(=O)—CD$_2$—CH$_3$ [cyclic ketone with CD$_2$, CD positions and CH$_3$ group]

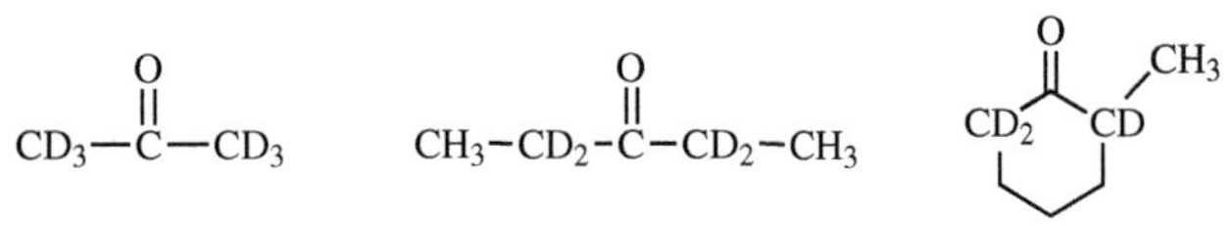

18 Both molecules are treated with D_2O either in an acidic (D^+) or a basic (OD^-) medium. While in the first molecule, all methylene protons will be exchanged by deuterium, in the second molecule, all methylene protons except benzylic protons will be exchanged.

19 A = [Cl–C₆H₄–COOH + $CHCl_3$]   B = Cl–C₆H₄–C(O)–CH_2Br

20 A = neopentyl-like diacid with two –COOH groups + $CHCl_3$ B = tetrachloro diketone structure

21
- α-tetralone + LDA/THF → enolate + Cl–CH₂–C(Cl)=CH₂ → α-alkylated tetralone
- Cl–CH₂CH₂–C(O)–CH₂CH₂–Cl + NaOH, CH_3OH/H_2O → dicyclopropyl ketone

22 $H_3C–C(O)–CH_2–C(O)OC_2H_5$ + Br–CH₂CH₂CH₂–C(CH₃)=CH₂ $\xrightarrow{NaOEt}$ alkylated β-ketoester $\xrightarrow[\text{Heat}]{\text{Hydrolysis}}$ methyl ketone with alkenyl chain

23
- A: cyclopentanone enolate (two resonance structures) + benzyl electrophile → B: 2-(benzyl ketone)-cyclopentanone
- A: $H_2C^-–C(O)–S(O)–CH_3$ enolate → B: alkylated product $CH_2–S(O)–CH_3$

24 Theoretically, three cyclization products can be formed during this reaction. The formation of the first product is not suitable for entropy. When the other two products are compared, the probability of product number 2 is higher because of the addition of an enolate to the aldehyde.

1: cycloheptenone
2: 1-acetyl cyclopentene (CH_3)
3: cyclopentene carbaldehyde (CH_3, H)

25 A = octahydronaphthalenone with angular CH_3 B = bicyclic ketone with CH_3 and OH substituents

26

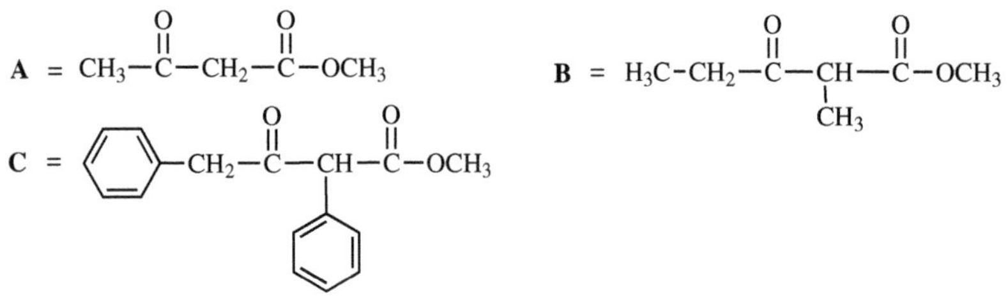

27

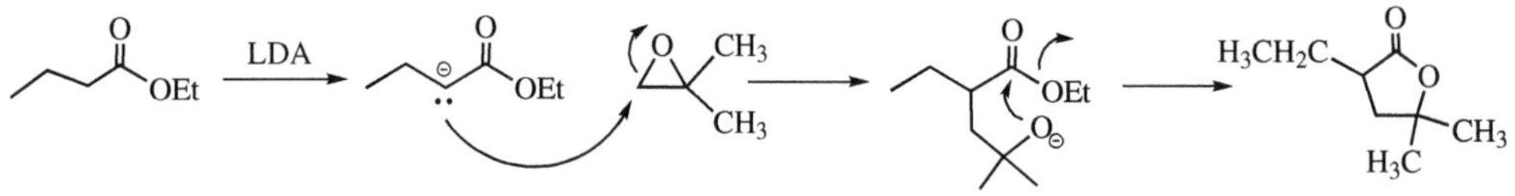

28

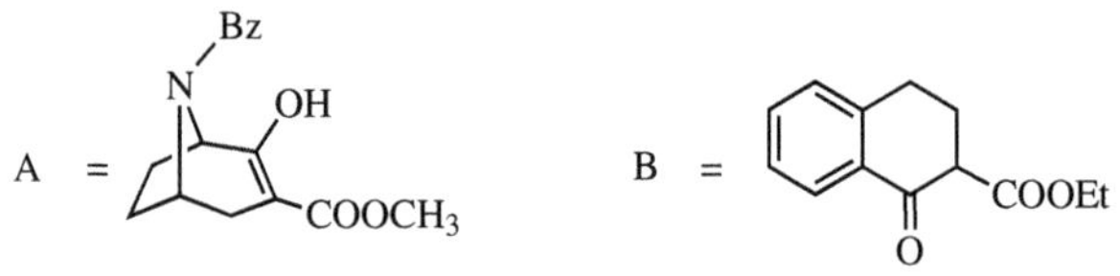

Chapter 6 (Aromatic Compounds)

1

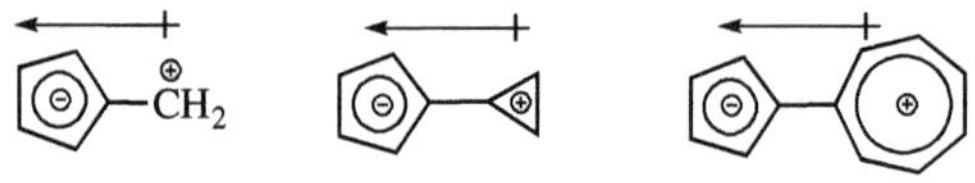

2

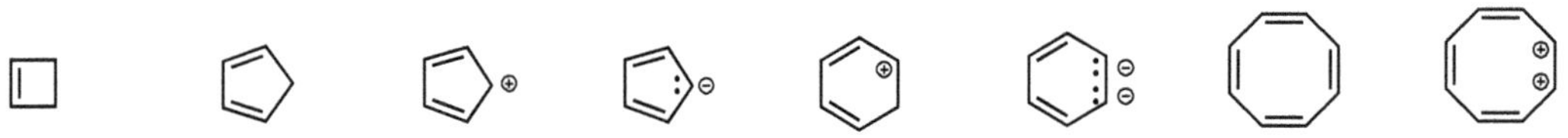

No conjugation No conjugation Antiaromatic Aromatic No conjugation Antiaromatic Nonaromatic Aromatic

3 Cyclopentadienone shows antiaromatic character and cyclooheptatrienone shows aromatic character.

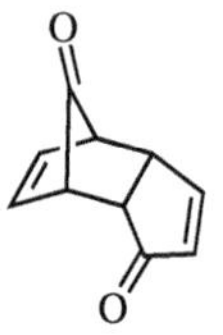

4 Compound a undergoes faster solvolysis. Solvolysis of compound b generates an antiaromatic intermediate.

5 Pentalene (8π) is antiaromatic. Pentalene dianion (10π) is an aromatic compound.

6 Protonation of the carbonyl oxygen atom generates an aromatic cationic product.

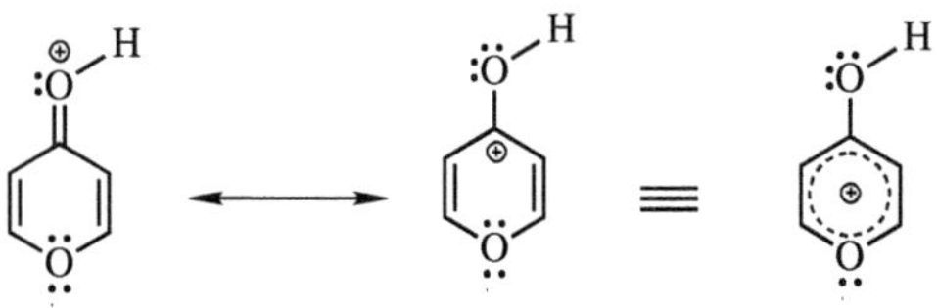

7 HBr elimination can form two isomeric benzynes, which can undergo addition reactions.

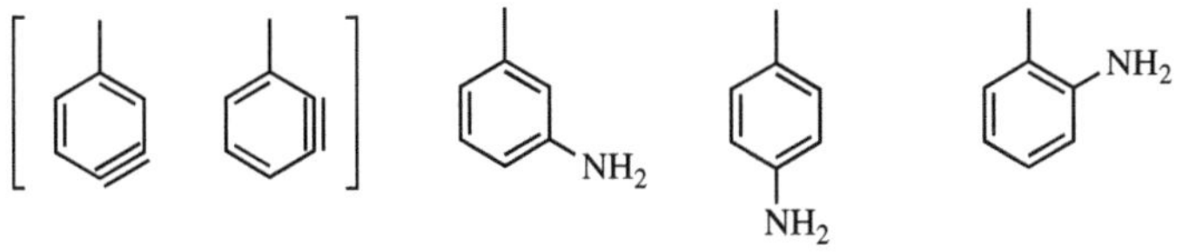

8

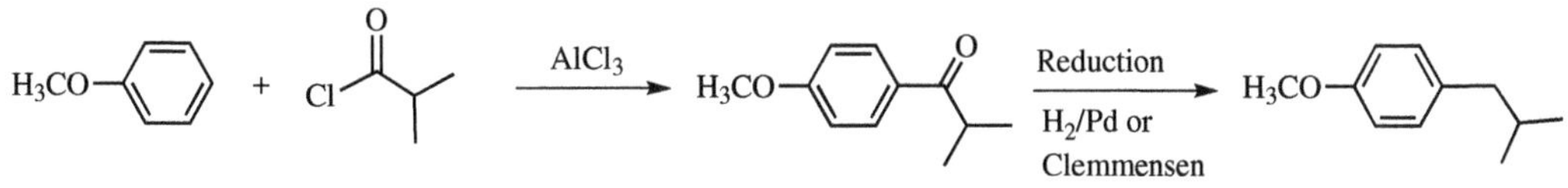

9 Electron-withdrawing group / Electron-donating group

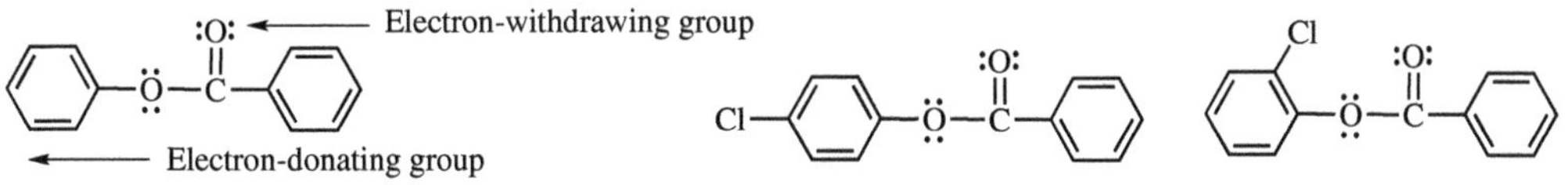

10 The more electronegative halogen, fluorine, pulls the electron density out of the ring and activates it toward the attack so that the reaction of 1-fluoro-4-nitromethane is faster.

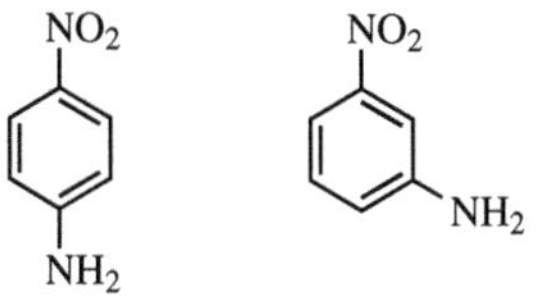

11

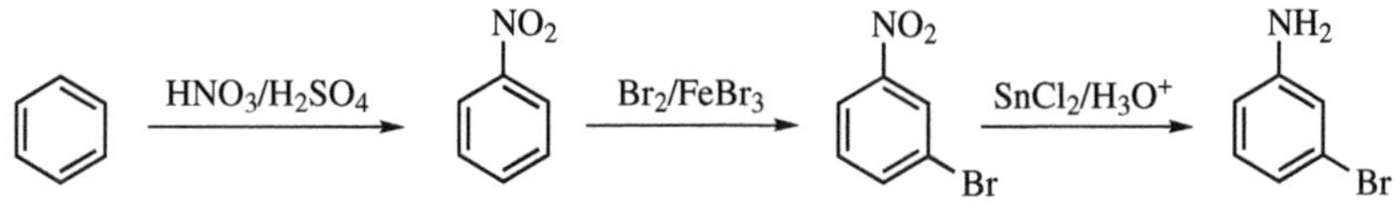

12

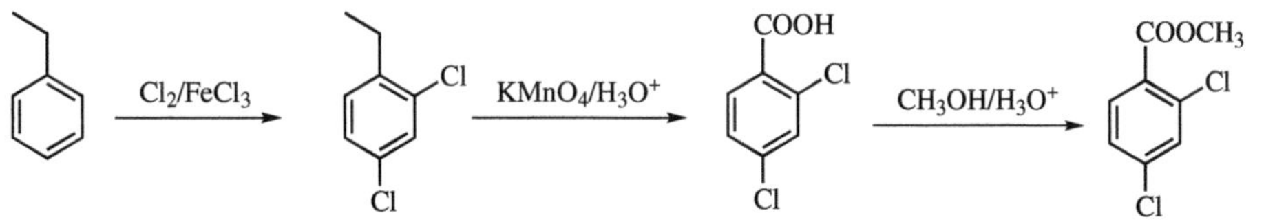

13

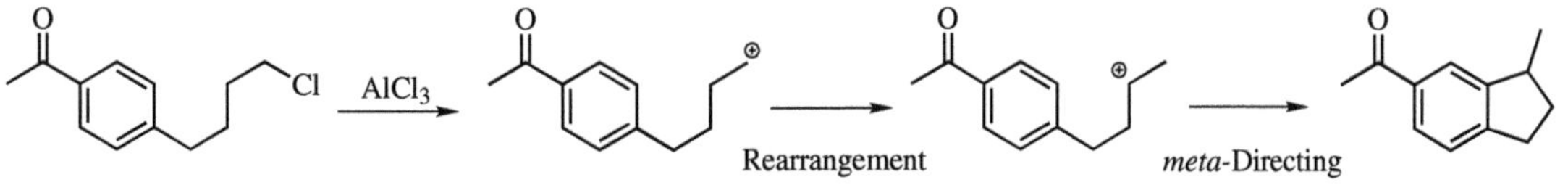

14

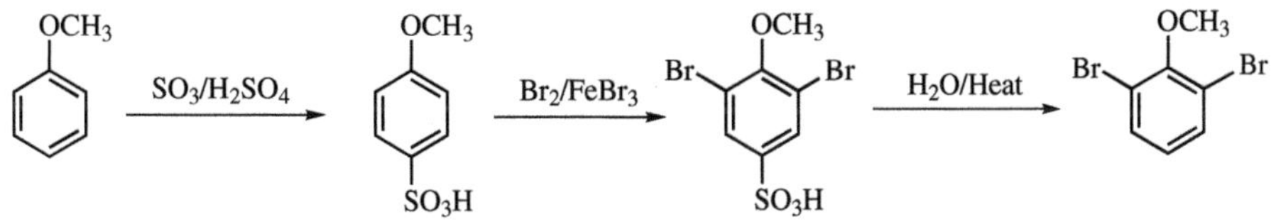

15

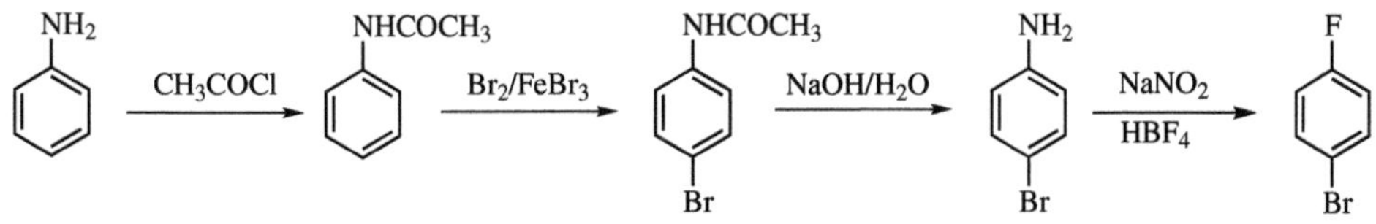

16

COO⊖
N2⊕
Heat

17

CH3Cl
AlCl3
CH3
KMnO4
H3O+
COOH
Br2/FeBr3
COOH
Br

18

HNO3
H2SO4
NO2
SnCl2
H3O+
NH2
NaNO2/H3O+
N2⊕
H2O/H3O+
OH
N2⊕
+
OH
N=N
OH

19

H3CO
O
HOOC
H3CO
HOOC
H3CO
ClOC
H3CO
O
O2N
H3CO
O

20

n-PrCl
AlCl3
HNO3
H2SO4
NO2
KMnO4
COOH
NO2
Br2
FeBr3
COOH
H3C
Br
NO2
CH3Cl
AlCl3
COOH
Br
NO2
KMnO4
Br
NO2
SnCl2/H3O+
Br
NH2
NaNO2
HBF4
Br
F
iso-Amyl nitrit
Br
N2⊕
Benzene
N=N
Br

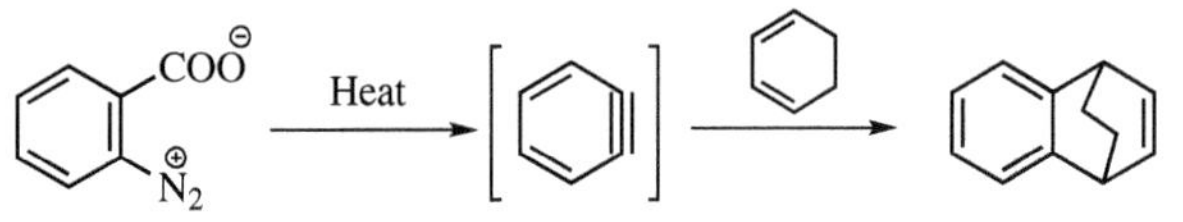

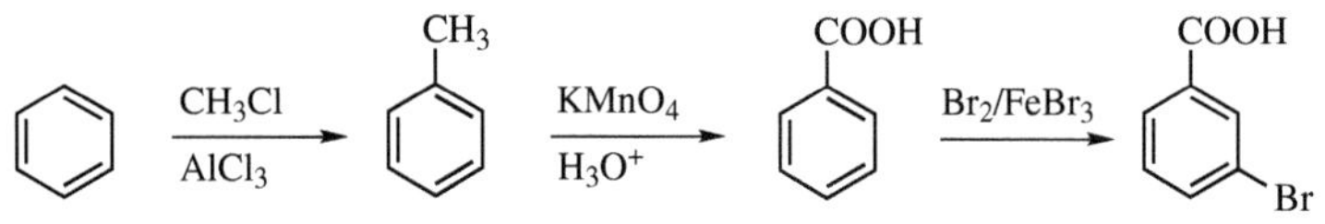

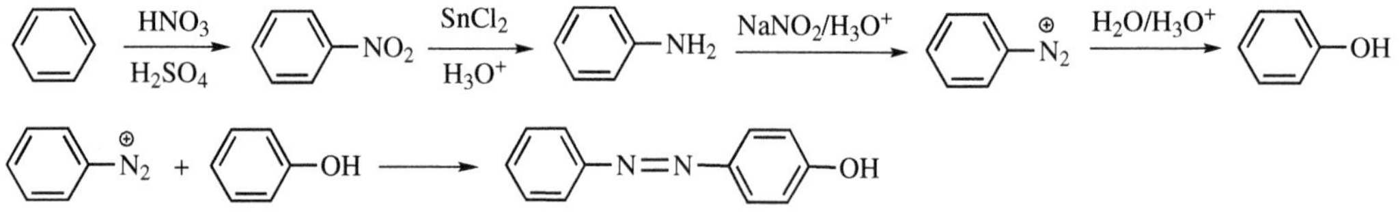

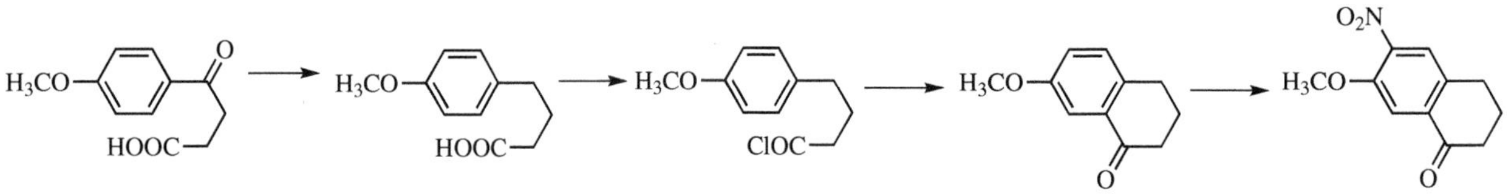

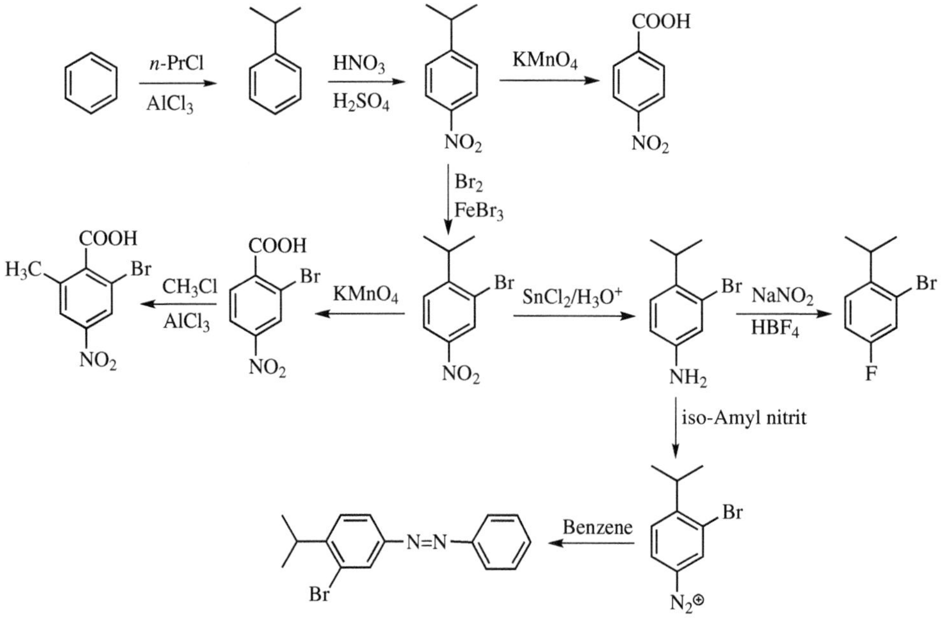

Chapter 7 (Carbocations)

1

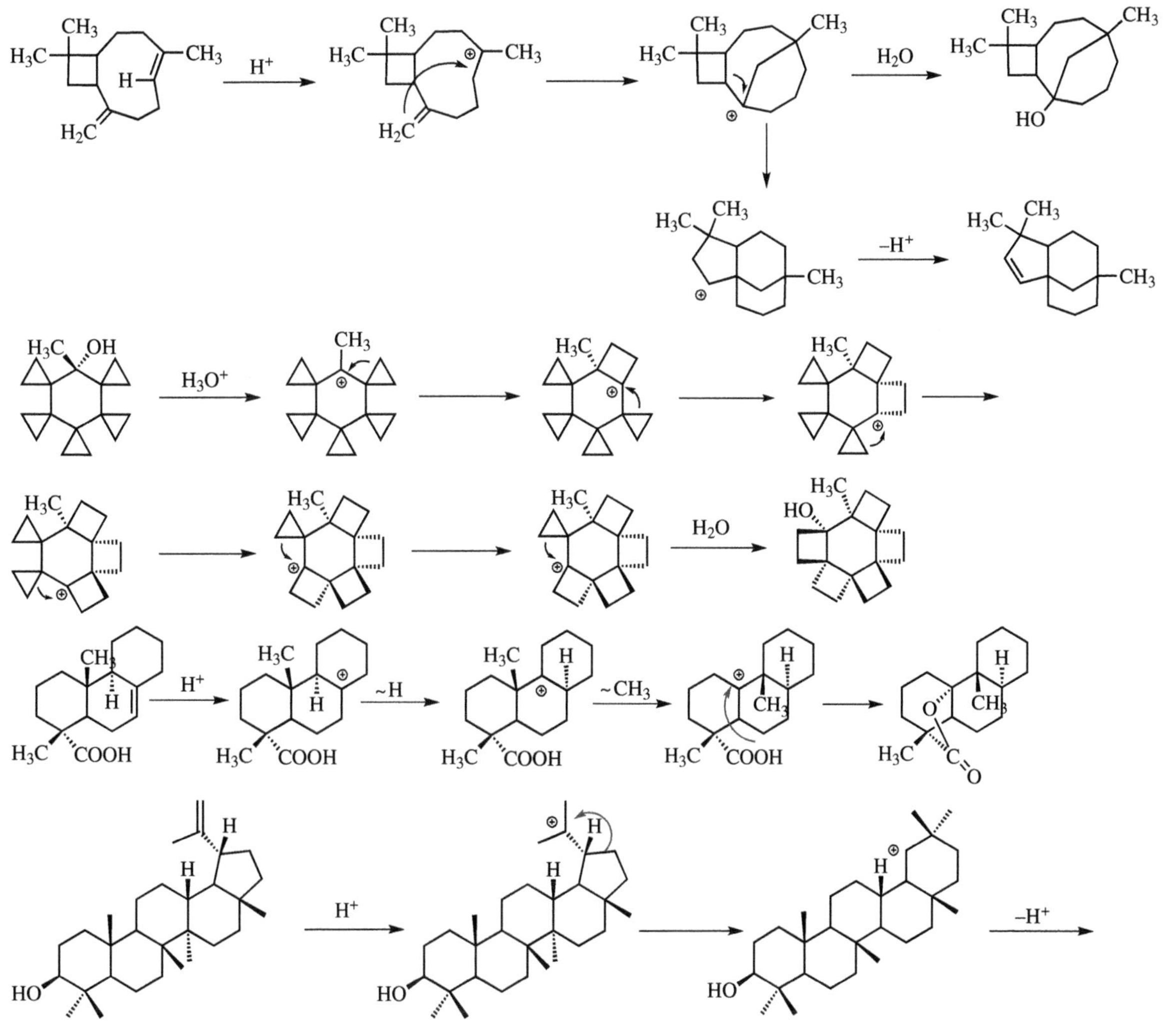

2

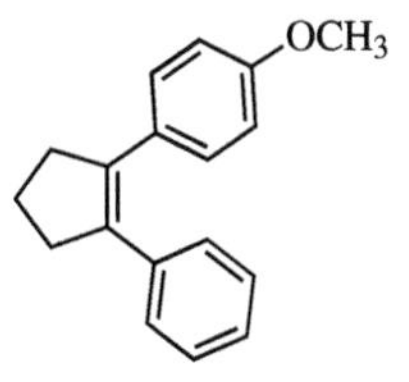

3

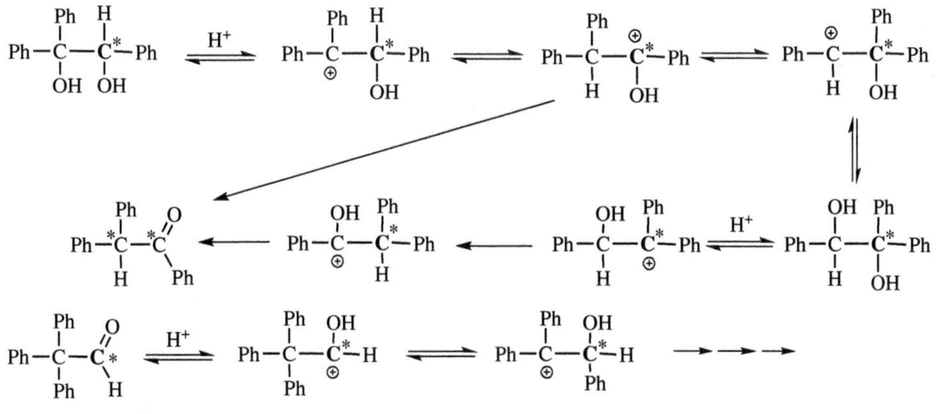

598 | Solutions

4 Two groups can migrate after carbocation occurs. Because the methylene group migrates better than the methyl group, the cyclopentane derivative is the main product.

Major product 90%   Minor product 6%

5 Theoretically, two products are expected in this reaction. The product with phenyl migration does not occur because of steric repulsion.

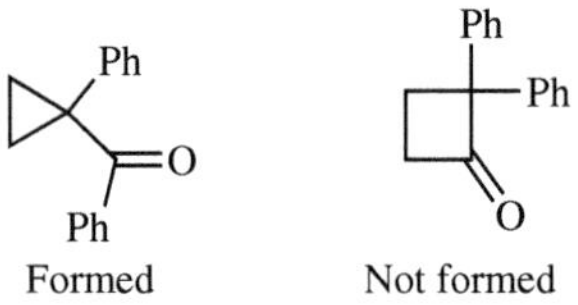

Formed   Not formed

6

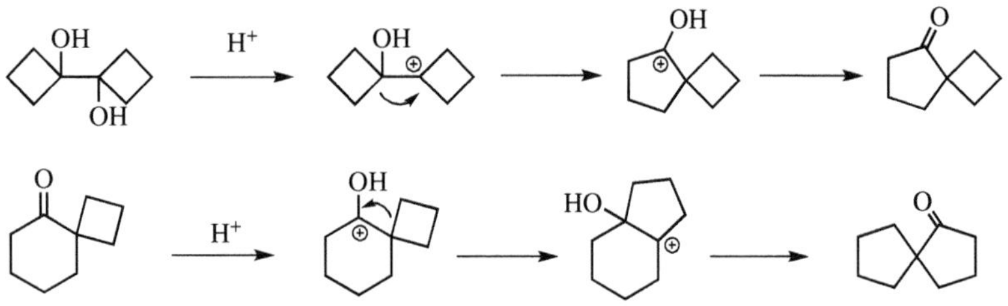

7

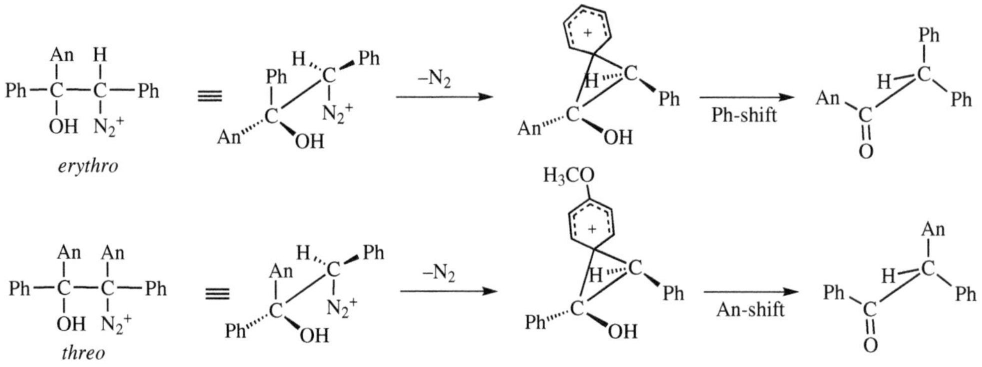

8 In isomer B, the aromatic ring supports the removal of the tosyl group by the neighboring group participation, but this is not possible in the other isomer.

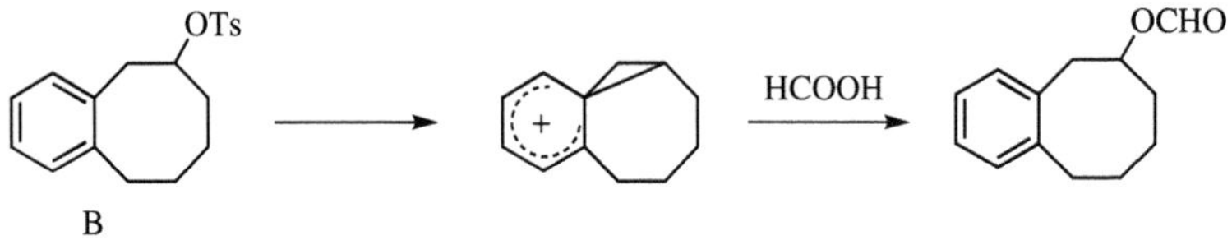

9

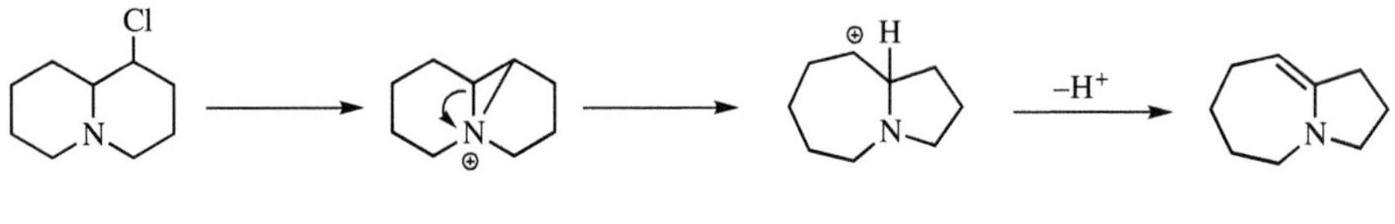

Chapter 8 (Carbanions, Carbenes, and Nitrenes)

600 | Solutions

13

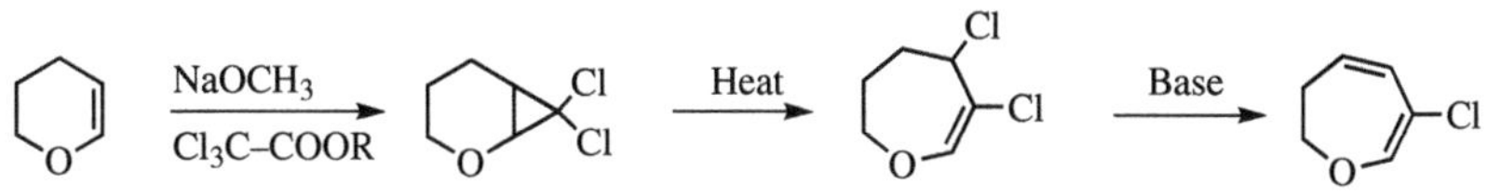

14

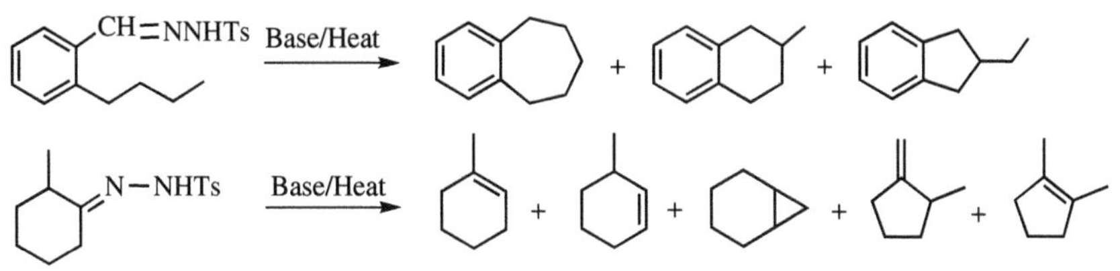

15 The first compound is formed by the insertion of the formed carbene into a solvent molecule, methanol. The second compound is the product of the normal Wolff arrangement, followed by methanol addition to the ketene. The third product is formed by the intramolecular addition of carbene to the double bond.

16 Both carbenes formed in the first stage are in equilibrium through the intermediate, oxirene. The dimerization product is formed by 1,3-dipolar addition of carbene to the Wolff rearrangement product.

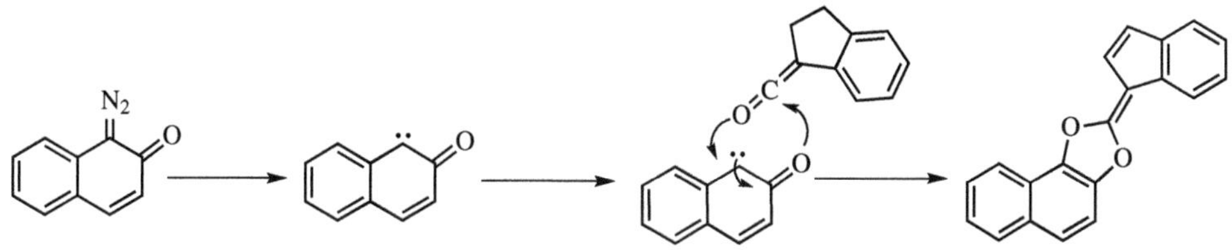

17

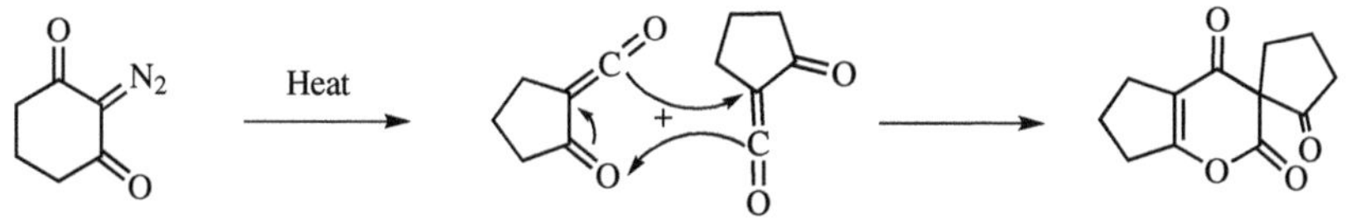

18

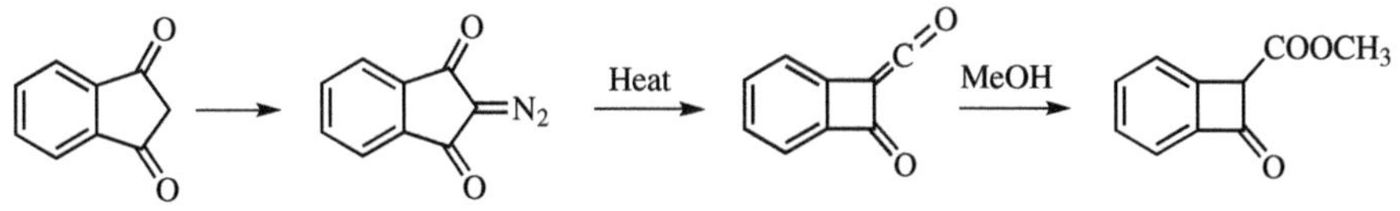

19 Carbene dimerizes in this reaction. The reason for the lack of Wolff rearrangement is that the product, which will be formed as a result of ring contraction, is highly strained because of the rigid structure.

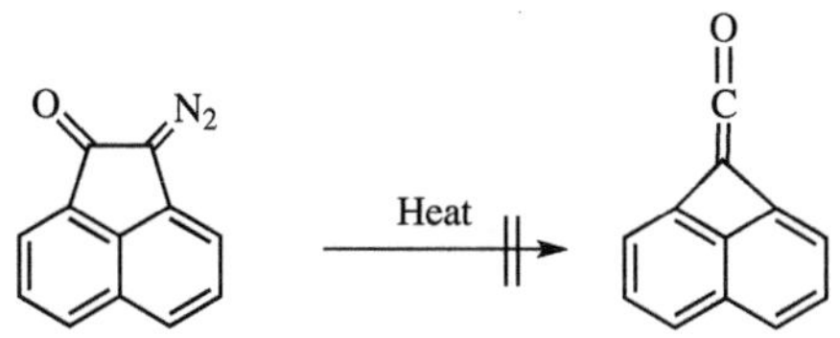

20 After the carbene is formed, two different groups (phenyl or methyl) can migrate. The migration of these groups forms A and B products. When R = OCH$_3$, more product A will be formed because the OCH$_3$ group forces the phenyl group to migrate as it is an electron-donating group. When the nitro group is attached, the phenyl group's migration is reduced or even prevented, then the ratio of product B increases.

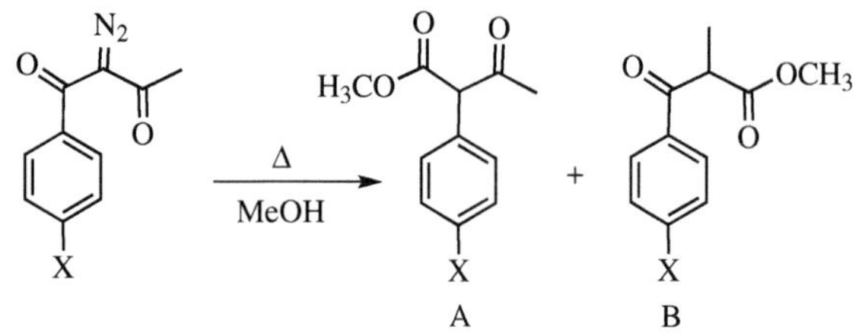

21

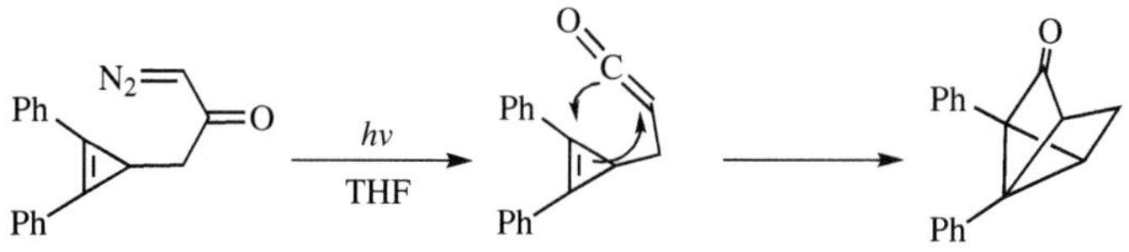

22

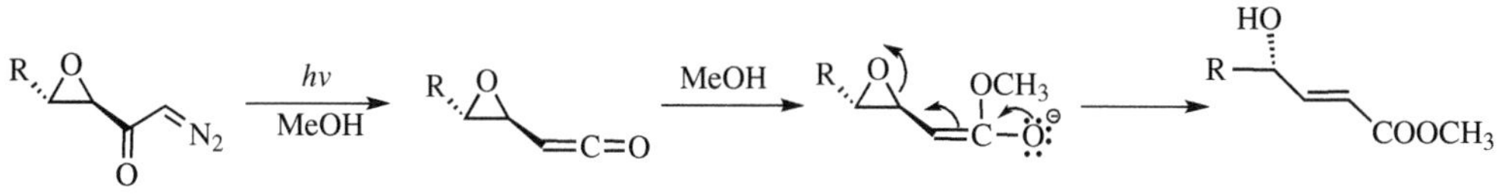

23 Following the Wolff arrangement, the molecule undergoes a retro-Diels–Alder reaction.

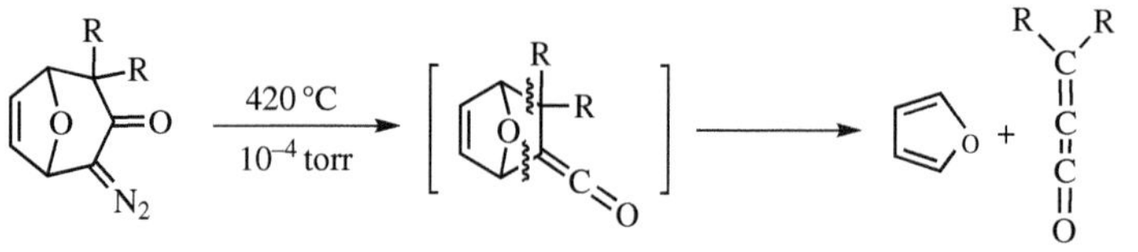

24

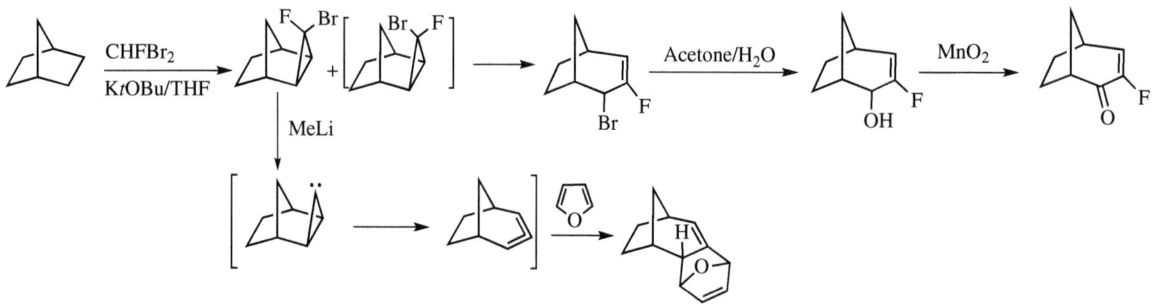

25

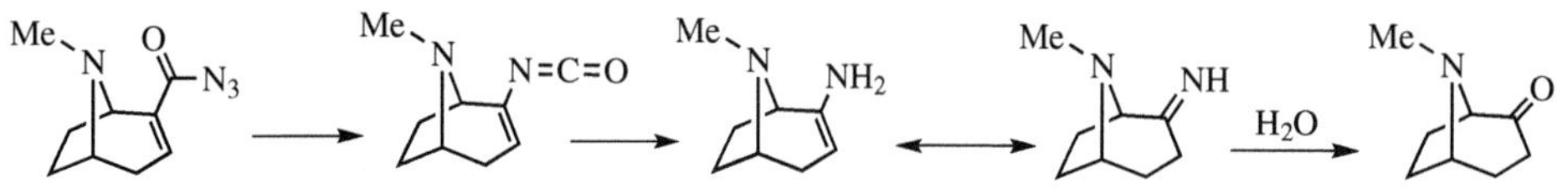

26

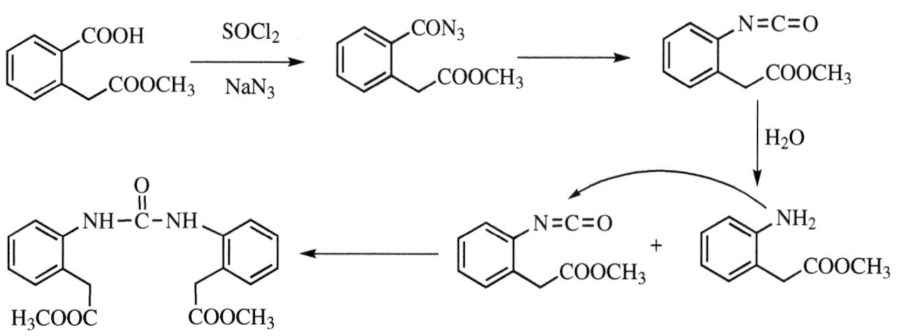

Chapter 10 (Pericyclic Reactions)

1 (a) *trans*, conrotatoric ring closure (b) *trans*, disrotatoric ring closure

3 a. *hv*, disrotatoric, b. heat, conrotatoric, and c. *hv*, disrotatoric

4 conrotatoric ring opening (4π or 8π system); disrotatoric ring closure (6π system)

The structure of X

5 a. heat, disrotatoric (6π system); b. *hv*, disrotatoric, (4π system)

6 B undergoes solvolysis more easily than A because the bromine is a better leaving group than the chlorine.

11 In the first step, a [6 + 4] cycloaddition takes place to form the product A. The final product is formed by an intramolecular [4 + 2] cycloaddition.

A =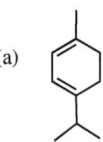

12

(a) 1.3-suprafacial H-shift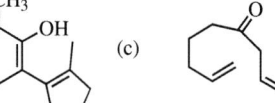

(b) 1.5-suprafacial H-shift

(c) 1.5-suprafacial H-shift

(d) 1.5-suprafacial or 1.3 (1.7) suprafacial H-shift

13

(a) 1.5-suprafacial H-shift

(b) 3.3-sigmatropic rearrangement

(c) 3.3-sigmatropic rearrangement

(d) 3.3-sigmatropic rearrangement

14

A =

B =

Index

a

α-aminoketones 415
abnormal bimolecular substitution 72
α-bromocarboxylic acid 241
abscisic acid 545
α,β-unsaturated carbonyl compounds 208, 246
α,β-unsaturated carbonyl groups 184, 226
α-carbon atom
   acid-catalyzed enolization 232
   acidity of α-hydrogens 228–229
   α-alkylation reactions of carbonyl compounds 241–251
   base-catalyzed enolization 233–234
   C-alkylation/O-alkylation 249–251
   Claisen ester condensation 257–259
   crossed aldol condensation 254–256
   crossed Claisen ester condensation 259–260
   Dieckmann condensation 260–261
   electron-deficient nitrogen 412–416
   electron-deficient oxygen 416–419
   enol and enolate reactions 235–241
   keto–enol tautomerism 229–232
   kinetic and thermodynamic enolates 234–235
   Knoevenagel condensation 261–264
   Perkin condensation 264–265
   Robinson annulation 256–257
   Stobbe condensation 265
acetoacetic ester synthesis 248
acetylacetone 30, 230, 245, 246
acid–base reactions 28, 30, 31, 34, 189
acid-catalyzed enolization 232
acids and bases
   acidity strength of organic compounds 35–38
   Arrhenius acid–base theory 29
   Brønsted–Lowry acid–base theory 29–30
   hard acids and bases 32–33
   Lewis acid–base theory 30–31
   nitrogen-containing compounds 38–40
   Pearson hard and soft acid–base theory 32
   pKa values of acids 34–38
   soft acids and bases 33–34
acrylaldehyde 524

acyl azides 455
adamantane carboxylic acid 207
addition–elimination mechanism 271
addition–elimination reactions 196
addition reactions
   alkenes reduction 165–171
   anti-Markovnikov addition, hydrogen halides 141–143
   conjugated dienes 171–175
   halogenation 133–138
   halohydrins formation of 145–146
   hydration 143–144
   hydration of alkenes 146–148
   hydroboration of alkenes 148–152
   oxidation of alkenes 152–165
   water and alcohols, alkoxylation 144–145
α-deprotonation 238
α-elimination method 435
α-epimerization 235
α-halogenation 237
α-haloketones 410
alcohol dehydration 91–92
aldehydes 193, 196, 230
Alder-ene reaction 492
aldol condensation reaction 113, 252
aliphatic carboxylic acid 206
alkene oxidation 152
alkenes 441
alkenes reduction 165–171
alkoxyborohydrides 209
alkoxylation 144–145
alkylation reactions 242
alkyl bromides 97
alkyl chlorides 97
alkyl Grignard reagents 212
alkyl iodides 97
alkynyl Grignard reagents 217
allenes 442
allylic alcohols 156
allylic epoxides 158
allylic rearrangement 72
allylic substitution reactions 73

Reaction Mechanisms in Organic Chemistry, First Edition. Metin Balcı.
© 2022 WILEY-VCH GmbH. Published 2022 by WILEY-VCH GmbH.

allylic systems   71–73
1,2-aminoalcohols   396
1-aminobenzotriazole   364
ammonia   38
anchimeric assistance   75, 398
anhydrides   338
aniline   39
annulenes
    [10]annulene   305–307
    [12]annulenes   307
    [14]annulenes   308
    azulene   311–312
    cyclobutadiene   303–305
    defined   303
    dioxo-anti[14]annulene   308–310
    fulvalenes   313–314
    fulvenes   313
    heptalene   312–313
    1,7-methano[12]annulene   307–308
    octalene   312
    pentalene   311
    trans-cis-cis-trans-cis   305
    X-ray structural analysis   308
antarafacial bond formation   517
antarafacial shift   526
anthracene endoperoxide   491
antiaromatic   285
antibonding orbitals   514
anti-elimination   112
anti-Markovnikov addition   478
anti-periplanar conformation   103, 118
anthracene   368
α-pyrone   509
arenediazonium ion   356–357
aromatic azo-compounds   357
aromatic carboxylic acids   206
aromatic compounds   281, 284
    acyclic unsaturated compounds   285
    antiaromatic   287
    benzene, structure and discovery of   281–284
    cyclooctatetraene   285
    defined   284
    heteroaromatic compounds
        five-membered ring   315–323
        seven-membered ring   330–333
        six-membered ring   323–330
        three-membered ring   314–315
    homoaromaticity   291–292
    Li-NMR   290–291
    magnetic susceptibility   289–290
    Möbius aromaticity   293–294
    molecular orbitals of   287–288
    NICS method   291
    nonaromatic   287
    nonaromatic compounds   286–287
    nucleophilic aromatic substitution reactions   357
    polycyclic aromatic compounds   365
    structural evidence for   289
    thermodynamic and aromatic resonance stabilization energy   288–289
    triple bond electrons   285
aromatic endoperoxides   491
aromatic ions   294
    antibonding molecular orbitals   300
    cyclobutene   296
    cyclobutenyl dianion   296
    1,3,5-cycloheptatriene   298
    cycloheptatrienyl cation and cycloheptatetraenyl anion   299
    cyclooctatetraene (COT)   300
    cyclooctatetraene dianion   302
    cyclopentadiene   298
    cyclopentadienone   297
    cyclopropene   295
    1,2-diphenyl-3,4-bis(trimethylsilyl)-cyclobutenyl dianion   296
    1,3-diphenyl-2,4-bis(trimethylsilyl)-cyclobutenyl dianion   296
    hydride exchange reactions   299
    5-iodocyclopentadiene   297
    tetraphenylyclopentadienone   298
aromatic stability   283
aromatic sulfonation   44
Arrhenius acid–base theory   29
aryl halides   574
α-silylcarbanions   227
α-substitution product   44
asymmetric environment   58
asymmetric ester hydrolysis   273
asymmetric hydroboration   151
asymmetric induction   155, 215
atom-abstraction reaction   472–475
axial hydrogens   106
azides   449, 453
1,1-azobis(cyclohexanecarbonitril) (ABCN)   480
azobisisobutyronitrile (AIBN)   471
azoisobutyronitrile radical   475
azulene   311–312

b

Baeyer–Villiger oxidation   27, 417, 419
Baeyer–Villiger rearrangement   417
$B_{AL}2$ mechanism   271–272
Bamford–Stevens Reaction   432
base-catalyzed halogenation   238
β-carbon atom   116

Beckmann fragmentation   414
β-elimination   87
benzene
   activating groups   344–349
   Clemmensen reduction   341
   deactivating groups   349–352
   electrophilic addition reactions   282
   electrophilic aromatic substitution   333–334, 352
   Friedel–Crafts acylation   338–339
   Friedel–Crafts alkylation   339–341
   halogenation of   335–336
   heat of hydrogenation   283
   monosubstituted benzene derivatives   341–344
   nitration   336–337
   resonance energy of   283
   side chain substituents
      of alkylbenzenes   354–356
         arenediazonium ion   356–357
   structure   281–284
   sulfonation   337–338
benzenoid aromatic compounds   368–373
benzodiazepine-1,3-dione derivative   574
β-esters   249
β-halohydroperoxide   493
β-hydroxyalkenes   121
β-hydroxyketones   243
bicyclic endoperoxides   496
bioluminescence   494
Birch reaction conditions   483
Birch reduction   481–483
β-ketoacid   246
β-ketoaldehyde   249
boat conformation   106
bond lengths   12
borane   150
borirene   314
boron trifluoride   31
β-protons   91, 117, 228
Bredt's rule   123–126
bromine   239
1-bromobicyclo[2.2.2]octane   68
2-bromobutane   468
3-bromobutan-2-ol   399
2-bromo-2-chlorobutane   468
bromocyclohexane   109
1-bromocyclopentane-1,3-dicarboxylic acid   124
1-bromo-1,2-diphenylpropane   103
1-bromo-4-methoxybenzene   546
2-bromo-2-methylbutane   89, 96, 97
3-bromopyridine   325
3-bromopyridine N-oxide   328
Brønsted–Lowry acid–base theory   29–30

Brown, H. C.   403
Brown's theory   403
β-substitution product   44
Buchwald–Hartwig coupling reaction   570
butadiene   171, 506, 514
1,3-butadiene   514

C

calciferol   529
carbanions
   population of orbitals   424
   S_N2 reaction mechanism   425
   Sommelet–Hauser rearrangement   427–428
   Stevens rearrangement   425–427
   Wittig rearrangement   429
carbenes   429, 444
   Arndt–Eistert reaction   450–451
   carbene insertion   439, 445
   cycloaddition reactions   439
   cyclopropylcarbene–cyclobutene rearrangement   452
   cyclopropylidene–allene rearrangement   451–452
   generation   431–438
   inductive effect   430
   insertion reactions   445
   mesomeric effect   430–431
   naming carbenes   430
   π electron-withdrawing groups   431
   rearrangement   446
   ring contraction   449–450
   ring expansion   450
   Simmons–Smith reaction   437–438
   structure and reactivity of   430
   vinylidene–alkyne rearrangement   452–453
   Wolff rearrangement   446–448
carbenoids   437
carbocations   31, 90, 381, 384
   base-induced nucleophilic rearrangements   408
   benzil–benzilic acid rearrangement   412
   detection of   388–389
   electrophilic addition to π bonds   388
   hydride ion affinity (HIA)   385
   hydride shift   406–407
   hyperconjugation   386
   ionization mechanism   387–388
   Nametkin rearrangement   406
   neighboring group participation, in molecular rearrangement   398–401
   nonclassical carbocations   401–405
   reactions of   389–412
   sp^2 hybrid orbitals   384
   structure and stability of   384
   t-butyl carbocation   384

carbon
   sp-hybridization   10–12
   sp² hybridization   7–10
   sp³ hybridization   4–7
carbon-carbon coupling reaction   537
   Grignard reagents   539
   Mizoroki-Heck coupling reaction   539
   Ullmann reaction   538
   Wurtz coupling   538
carbon nanotubes   371
carbonyl compounds   14, 181
carbonyl groups
   Clemmensen reduction   201
   diisobutyl aluminum hydride (DIBAL)   203
   hemiacetal formation   190–196
   hydration   187–189
   Meerwein-Ponndorf-Verley reduction   209–210
   metal hydride reduction   201–210
   Oppenauer oxidation   209–210
   with organometallic compounds reaction   210–218
   reactivity of   181–186
   sodium borohydride (NaBH₄) reduction   203–205
   Wolff–Kishner reduction   200
   with ylides reaction   218–227
carboxylic acids   205, 206
cations   14
chair conformation   105
cheletropic reactions   501
chemical reactions   50
   bond polarization   2
   covalent bond   2–13
   effective collisions   41
   inductive effect   2
   Lewis acid catalysts   1
   mesomeric effect   2
   reaction coordinates   41
   reaction kinetics   41
chemiluminescence   494
Chichibabin reaction   326, 327
chlorine   239
3-chlorocyclopropene   295
3-chloro-3-methylbut-1-ene   73
3-chloro-2-methyl-2-phenylbutane   90
Chugaev elimination   121–123
cine-substitution   364
cis-1,2-dimethylcyclohexane   108
Claisen ester condensation   257–259
Claisen rearrangement   533
Claisen–Schmidt condensation   255
classical carbocations   384
classical resonance theory   386
Clemmensen reduction   201, 341

condensation reactions   196
configuration retention   77
conjugated dienes   171–175
conjugated double bonds   287
conrotatory rotation   503
Cope rearrangement   126, 530
coronene   370
correlation diagrams   512
Criegee mechanism   164
crossed aldol condensation   254–256
crotyl chloride (1-chlorobut-2-ene)   72
cumulenes   12
cyclic 1,3-dithianes   195
cyclic ethers   80
cycloaddition reactions   494, 500
   cyclopentadiene   518
   Diels–Alder reactions   521
   frontier orbital coefficient   522–526
   inverse demand Diels–Alder reaction   522–526
   secondary orbital interactions   520–521
   suprafacial bond formation   517
   tetrazine and acetylene   522
   tropone   519
[2+2] cycloaddition reactions   493–495
[4+2] cycloaddition reactions   495
cyclobutadiene   285, 303–305
cyclobutanones   184, 450
cyclobutene   296
cyclodeca-1,3,5,7,9-pentaene   286
cycloheptatriene   298, 529
1,3,5-cycloheptatriene   298
cyclohexa-3,5-diene-1,2-diol   193
cyclohexane-1,3-dione   231
cyclohexane system   105
   conformation and configuration   105–111
   syn-elimination   112
cyclohexanone   231, 243
   enolate anion   251
   oxime   199
cyclohexa-1,3,5-triene   504
2-cyclohexyl-2-propanol   92
cyclononatetraene   509
1,5-cyclooctadiene   150
cyclooctatetraene (COT)   119, 300, 477, 508
cyclooctatriene   508
cyclopentadiene   297, 298
cyclopentadienone   297
cyclopentanone   231
cyclopentyl carbocation   391
cyclopropanone   188, 408
cyclopropanone mechanism   408
cyclopropene   295

d

deactivating substituents 342
degenerate systems 107
dehydrohalogenation 90
delocalization energy 283
deprotonation 241, 258
diacyl peroxides 471
diamagnetic ring current 290
diastereomers 100
1,3-diaxial interaction 107, 108
diazirines 434
diazo compounds 432
diazomethane 434
1,2-dibromobenzene 282
2,5-dibromobenzoquinone 410
2,3-dibromobutane 101
2,2-dibromocyclohexane-1,3-dione 239
1,2-dibromo-1,2-dichloroethane 101
2,3-dibromo-2,3-dimethylbutane 100
1,2-dibromo-1,2-diphenylbutane 104
1,2-dibromo isomers 282
dibromomethylenecyclobutane 453
1,3-dicarbonyl compounds 230, 245, 246, 481
1,3-dicarbonyl substrates 250
2,3-dichloro-5,6-dicyano-1,4-benzoquinone (DDQ) 306
2,2-dichloro-1,1,1-trifluoroethane 114
Dieckmann condensation 260–261, 267
Diels–Alder adduct 410
Diels–Alder reactions 27, 50, 297, 521–526
dienone–phenol rearrangement 398
diethyl tartrate 155
dihalocarbenes 436, 437, 475
dihydroaromates 383
2,3-dihydroxybutane 101
dihydroxylation via pifa 161
Diisobutyl Aluminum Hydride (DIBAL) 203
1,3-diketones 243
dimethoxyethane 250
1,6-dimethoxynaphthalene 266
2,2-dimethoxypropane 193
dimethylamino group 40
1,3-dimethylbenzene 353
2,2-dimethyl-1-bromopropane 392
3,3-dimethylbutan-2-ol 146
3,3-dimethyl-1-butene 146
1,2-dimethylcyclohexane 108
1,2-dimethyl-cyclohex-1-ene 150
dimethyldioxirane (DMDO) 156
4,5-dimethylenecyclohex-1-ene 496
dimethylfulvene 525
dimethylsulfonium methylide 438
2,4-dinitrophenylhydrazine 198
dioxetanes 494

dioxirane 156
1,2-diphenyl-3,4-bis(trimethylsilyl)-cyclobutenyl dianion 296
1,3-diphenyl-2,4-bis(trimethylsilyl)cyclobutenyl dianion 296
1,1-diphenylethane-1,2-diol 395
di-π-methane rearrangement 483–485
di-π-methane systems 485
1,3-dipolar cycloreversions 164
diradicals 486–487
disrotatory ring closure
    butadiene 514
    hexatriene 515
disrotatory rotation 502
1,3-disubstituted benzene derivatives 353
1,3-dithianes 195
double bonds 183

e

eclipsed conformation 99
effective collisions 41
electrocyclic reactions 500, 505
    thermal electrocyclic reactions 505–506
    Woodward–Hoffman rules 508–512
electron deficiency 468
electron-deficient carbon 383
electron-deficient nitrogen 412
    Beckmann rearrangement 413–414
    Neber rearrangement 415
    Stieglitz rearrangement 416
electron-deficient oxygen 416
electron density 504
electron donation 385
electron spectroscopy for chemical analysis (ESCA) 391
electron spin resonance (ESR) spectroscopy 468
electron-withdrawing effect 238
electrophilic addition reactions 282
electrophilic compounds 14
electrophilic rearrangement 424
electrophilic substitution 328, 330
electrophilic substitution reactions 324
elimination-addition mechanism 363–365
elimination reactions 87
    biomolecular elimination reactions 94
        anti-periplanar/syn-periplanar conformation 94
        cyclohexane system 105
        erythro and threo-configurations 100
        i-propyl chloride and sodium ethoxide 95
        kinetic isotope effect 98
        and product distribution 95
        second-order reactions and transition state 94
        stereochemistry 98–105
    Bredt's rule 123–126
    α-elimination 87
    γ elimination 87

elimination reactions (contd.)
    Grob fragmentation   126–128
    halogen elimination   114–116
    Hofmann elimination   116–119
    pyrolytic elimination   119–123
    unimolecular conjugate base elimination   112–114
    unimolecular elimination reactions
        alcohol dehydration   91–92
        E1 reaction mechanism   88–91
        factors   92–94
enamines   196, 198
enantiomers   59
endocyclic   92
endo cyclization   478, 479
ene reaction   492–493
enolate anions   34
enzymatic dihydroxylation   162
epoxidation   152, 155
epoxide ring-opening reactions   157
epoxides   80, 82
1,2-epoxy-4,5-dibromocyclohexane   330
equatorial hydrogens   106
erythro-1-bromo-2,3-diphenylbutane   104
erythro-1-bromo-1,2-diphenylpropane   103
ester hydrolysis   268, 270, 272
ester pyrolysis   120–121
1,2-ethanedithiol   194
ether cleavage   81
ethers   81
ethyl carbocation   390
exo cyclization   479

f

Favorskii rearrangement   408, 410, 412
Finkelstein reaction   67
first-order reactions   45
flagpole hydrogens   106
fluorescence   489
formal charge   24
free radicals   468
Friedel–Crafts acylation   338–339
Friedel–Crafts alkylation   339–341
Friedel–Crafts formylation   339–341
Friedel–Crafts reaction   406
frontier molecular orbitals   502–505
fullerenes   372
fulvalenes   313–314
fulvenes   313
furan   319, 320

g

gauche interaction   108
geminal diols   189

generation of carbocations   387
Glaser coupling   537
graphene   371
graphite   371, 372
Grignard reagents   210, 213, 216, 539
Grob fragmentation   126–128

h

half-chair conformation   106
halo-2,4-dinitrobenzene derivatives   359
haloform reaction   239
halogenation   335–336
    bromination reactions   134
    bromine–bromine bond   134
    electron-donating groups   135
    stereospecificity of   136
halogens   351
halopyridines   327
hard reagents   209
1H-azepine   331
heliolactone   552
Hell–Volhard–Zelinsky reaction   240
hemiacetal formation
    acid-catalyzed acetal formation mechanism   190
    cyclic acetals and synthetic application   191–192
    defined   190
    in glucose   191
    4-hydroxybutanal and 5-hydroxypentanal   191
    protecting group   192
Henbest rule   154
heptalene   312–313
heterogeneous catalysts   166
heterolytic bond cleavage   50
hexaphenyltriapentafulvalene   314
hexatriene   504, 515
hexa-1,3,5-triene   505, 507
highest energy-filled orbital (HOMO)   502
Hiyama-coupling reaction   567
    Hiyama–Denmark coupling   569–570
    organosilicon reagents   568
    palladium-catalyzed cycle for   568
Hofmann degradation   116, 119
Hofmann elimination   116–119
Hofmann reaction   116
Hofmann rearrangement   458
homogeneous catalytic reduction   168
homolytic bond cleavage   50, 471
Horner–Wadsworth–Emmons Reaction   223
Hückel's molecular orbital theory   302, 303
Hückel's rule   284, 285, 294, 296, 301, 308
Hunsdiecker reaction   477
hydration   143, 187–189
hydration of alkenes   146–148

hydrazone derivatives 199
hydride ion affinity (HIA) 385
hydride shift 406–407
hydrogenation 165
hydrogen atom 216
hydrogen–deuterium exchange 236–237
1-hydroxycyclopropane-1-carboxylic acids 412
hyperconjugation 386

i

inductive effect 15
  bond polarization 16
  carbon–carbon double bonds 18
  carbon–carbon triple bonds 18
  pKa values 17
  symmetrical σ bonds 15
intermediate product 382
intersystem crossing 490
iodine 336
iodobenzene derivatives 361
Ireland–Claisen rearrangement 533
isodesmic reaction 288
isodesmotic to isodesmic 288
isopropyl benzene 406
isopropyl carbocation 390
3-isopropyl-6-methylcyclohex-1-ene 90

j

Jahn–Teller theorem 304
Julia–Kocienski olefination 226
Julia olefination 226

k

keto–enol tautomerism 229–232
ketones 193, 196, 230, 397
ketoximes 199
kinetic controlled reactions 43
kinetic isotope effect 98
Knoevenagel condensation 261–264
Köbrich's rules 126
Kolbe electrolysis 476–477
Kumada cross-coupling reaction 566

l

l-butyl-l(*o*-iodobenzyl)-3-phenylurea 574
leaving group 50, 54
Le Chatelier's principle 67
Lewis acid–base theory 30–31
Lewis acid catalysts 1
Lewis acids 14, 193
  lithium diisopropylamide (LDA) 233
Luche reduction 209

m

macrocyclic lactones 551
magnesium alkyl hydroperoxides 217
malonic ester 248
malonic ester synthesis 247
Markovnikov's rule 138
m-chloroperbenzoic acid 27, 153, 418
Meerwein, Hans 392
Meerwein-Ponndorf-Verley Reduction 209–210
Meisenheimer complex 358
menthyl chloride 111
meso compounds 101, 102
mesomeric effect 18
  α,β-unsaturated carbonyl compounds 20
  aromatic compounds 23
  carbonate anion 19
  1,2-dibromobenzene 19
  and inductive effect 20
  oxygen atom 22
  phenol 21
1,7-methano[12]annulene 307–308
1-methoxybutadiene 524, 525
methyl bromide 251
2-methylbutadiene 118
methylcyclohexane 107
2-methylcyclohexanone 242, 243
1-methylcyclohexene 110
2-methyl-1,1-diphenyl-propane-1,2-diol 394
5-methylenebicyclo[2.2.1]hept-2-ene 157
3-methylpyrrolidine 118
2-methylpyrrolidine 117
methyl trimethylacetate 272
1,2-migration 426
Mizoroki–Heck coupling reaction 539
  amines 541
  catalytic cycle 541
  catalytic insertion reaction 540
  palladium acetate and silver carbonate 545
  phosphine ligands 541
  regioselectivity of 542–546
m-nitroaniline 351
Möbius aromaticity 293–294
moderately activating substituents 343
moderately deactivating substituents 344
molecular orbital (MO) 53, 296, 503
molecular rearrangement 78
Molozonide 163
monoalkylborane compounds 151
monobromination 237, 240
monobromoacetone 237
monobromophenol 347
monoepoxide 157
monohalocarbenes 437, 475

monohalocyclopropanes 511
monosubstituted cyclohexanes 108
monosubstituted malonic ester 248
Mozingo reduction 194

n

Nametkin rearrangement 406
naphthalene 365–367, 444
naphthalene endoperoxide 491
N-bromosuccinimide (NBS) 474
N-chloraminocyclopropanol derivatives 416
Neber reaction 415
Neber rearrangement 415
Negishi-coupling reaction 558
   iodoalkanes 560
   Nickel-catalyzed cycle 561
   organozinc reagents 561
   Palladium-catalyzed cycle 560
   polyfunctional alkyl iodides 559
   unsymmetrical biaryls and diarylmethanes 558
neighboring groups 75–78
neomenthyl chloride 111
N-heterocyclic carbenes (NHCs) 557
nitration 336–337
nitrenes
   acyl azides 455
   Curtius rearrangement 455–456
   cycloaddition reactions 458
   Hofmann rearrangement 458
   insertion reactions 460
   Lossen rearrangement 457–458
   synthesis 454
   rearrangements 455
   Schmidt rearrangement 457
nitriles 216
nitrogenous compounds 11
3-nitropyridine 325
N,N-dimethylneomentylamines 123
N-nitropyridinium nitrate 325
N-Nitrosoalkyl Urea Compounds 433
nonaromatic 287
nonclassical carbocations 384
nonnucleophilic lithium tetramethylpiperidide (LTMP) 330
norbornane system 231
norbornyl cation 405
2-norbornyl cation 404
nucleophiles 389
nucleophilic aromatic substitution 326
   addition-elimination mechanism 358–360
   arenediazonium salts 360–363
   elimination-addition mechanism 363–365
nucleophilic compounds 15

nucleophilic rearrangement 392, 424
nucleophilic substitution reaction 50, 328, 330
   in allylic systems 71–73
   ambident nucleophiles 78–80, 249
   and basicity 51–53
   epoxides 82
   internal nucleophilic substitution reaction 74–75
   neighboring group participation 75–78
   S_N1 reaction
      optical activity 59–61
      stereochemistry in 58–59
      steric factors 61–62
   S_N2 reaction
      factors 66–71
      stereochemistry of 64–66
   unsaturated systems 83
   Williamson ether synthesis 80–81
nucleus-independent chemical shift (NICS) 291
nylon-6 414

o

O-alkylation 250
octalene 312
oppenauer oxidation 209
optically inactive compounds 60
orbital symmetry conservation 512
organic field-effect transistors (OFETs) 565
organoborane reagents 554
organomagnesium halides 210
organotin compounds 550–552
osmium tetroxide (OsO_4) 160
overlapping orbitals 2
oxaphosphetane 220
oxepine 330
oxidation number 25
   Baeyer–Villiger oxidation 27
   bonding electrons 25
   carbon atom 26
   carbon–carbon bonds 27
   cyclization reactions 28
oxidative workup 164
oximes 199, 413
oxirene 314
oxymercuration–demercuration to 3,3-dimethyl-but-1-ene 146
oxymercuration reaction 147, 148
ozonolysis 162

p

palladium-catalyzed carbonylation reactions
   aryl halides 574
   $Mo(CO)_6$-mediated carbonylation 577–578
   organometallic reagents 576–577

paramagnetic ring current   290
partial racemization   58
p-bromobenzenesulfonate   400
p-chlorotoluene   352
2,4-pentadione   230
pentalene   311
pericyclic reactions   500
   characteristic features   500
   cheletropic reactions   501
   cycloaddition reactions   500
   electrocyclic reactions   500, 505
   frontier molecular orbitals   502–505
   sigmatropic reaction   501
   Woodward–Hoffmann rules   501–502
Perkin condensation   264–265
Peterson olefination   227
phenanthrene   368
3-phenyl-2-butanone   237
phenylhydrazine   198
4-phenyl-1,2,4-triazoline-3,5-dione (PTAD)   486
phosphonium ylides   219
phosphorescence   490
photochemical electrocyclic reactions   506
pinacolone   393
pinacol rearrangement   393
polarized bonds   31
polarizer   60
polar reactions   50
polycyclic aromatic compounds
   benzenoid aromatic compounds   368–373
     coronene   370
     fullerene   372
     graphite   370–371
     naphthalene   365–367
potassium *t*-butoxide   97
precalciferol   529
primary ozonide (molozonide)   163
protecting group   192
protection of diols   193
proton elimination   389
pyrazine   329, 330
pyrene   369
pyridazine   329, 330
pyridine   323
Pyridine Enhanced Precatalyst, Preparation, Stabilization and Inhibition catalyst   558
pyridine *N*-oxides   327
pyridine-sulfur trioxide complex   319
pyrimidine   329, 330
pyrolitic elimination   119
pyrolytic elimination
   Chugaev elimination   121–123
   ester pyrolysis   120–121

pyrrole   316–318
pyrrolidine   243
3-pyrrolidinethiophene   332

r

radical-pair mechanism   426
radical polymerization   478
radical reactions   50
radicals   12
   addition to alkenes   478
   antibonding orbital   469
   atom-abstraction reaction   472–475
   benzylic systems   470
   Birch reduction   481–483
   carbocations   469
   C–H bond dissociation energies   469
   di-π-methane rearrangement   483–485
   diradicals   486–487
   ESR spectroscopy   468
   generation   471
   Kolbe electrolysis   476–477
   manganese radical   480–481
   planar and pyramidal structures   468
   radical addition to alkenes   478–480
   radical combination   475–476
   radical disproportionation   476
   singlet oxygen   487–496
   structure and stability   468
Ramberg–Bäcklund reaction   411
reaction kinetics   41
reaction rate   44, 76
reductive deamination   360
reductive dediazonization   360
reductive workup   163
ring expansion   450
ring inversion   105
ring opening process   511
Robinson annulation   256–257

s

Schiff base   196
Schiff base formation   196
Schlenk equilibrium   211
secondary amines   197
semibenzilic rearrangement   408, 409
semicarbazide   198
semipinacol rearrangement   396
sesquifulvalene   314
Shapiro reaction   433
Sharpless asymmetric epoxidation   156
Sharpless asymmetric oxidation   155
Sharpless oxidation   155

sigmatropic reactions 501, 526
   antarafacial shift 526
   calciferol 529
   configuration isomerization 527
   cycloheptatriene 529
   precalciferol 529
   suprafacial shift 526
   thermal 1,3-hydrogen shift 527
   Woodward–Hoffmann rules 530
Simmons–Smith Reaction 437–438
singlet oxygen
   chemical sources 490–492
   [2+2] cycloaddition reactions 493–495
   [4+2] cycloaddition reactions 495, 496
   electronic configuration 487–489
   ene reaction 492–493
   photosensitized singlet oxygen 489, 490
single transition state 42
singly occupied molecular orbital (SOMO) 469
sitagliptin 155
sodium ethoxide 110
solvent effect 68
solvolysis 269
Sommelet–Hauser rearrangement 427–428
Sonogashira-coupling reaction 562
   acetylenes and organic halides 562
   4-bromo-2,3,5-trichloro-6-iodopyridine 563
   conjugated acetylenic compounds 562
   copper-free Sonogashira coupling 564, 565
   organic field-effect transistors (OFETs) 565
   Palladium-copper-catalyzed cycle 564
sp^2-hybridization 184
spirodienone systems 398
stable ylides 220
staggered conformation 99
stereochemistry 214
Stevens rearrangement 425–427
Stieglitz rearrangement 416
Stille cross-coupling reaction 547
   organotin compounds 550–552
   palladium catalyst 548
   sp^2-hybridized organohalides 548
Stobbe condensation 265
Stork enamine reaction 243
Straus coupling 537
strongly activating substituents 342
substituted anilines 39
substitution mechanism 271
sulfonation 337–338
sulfur ylides 224
suprafacial bond formation 517
suprafacial shift 526

Suzuki–Miyaura coupling reaction 552
   advantages of 554
   aryl boronic acids and aryl bromide 552
   biaryl derivatives 557
   boronate pathway and oxo–palladium pathway 554
   catalytic cycle 553
   E-1-hexenylborane and *trans*-styryl bromide 553
   N-heterocyclic carbenes (NHCs) 557
   nickel complexes 555
   organoborane reagents 554
   Pd-PEPPSI catalysts 558
   ribisin A 556
   transmetalation reaction 554
syn-elimination 112
syn-periplanar conformation 103, 120

t

tautomerism 229
t-butylhydroperoxide 155
t-butylmagnesium bromide 217
tertiary butylamine 245
tertiary carbocations 414
testosterone synthesis 267
tetrahedral structure 3
tetrahydrofuran 250
1,4,5,8-tetrahydronaphthalene 136
1,3,5,7-tetramethylcyclooctatetraene dianion 302
tetraphenylyclopentadienone 298
thermodynamic controlled reactions 43
thienothiophenes 322
thiepine 332
thiirene 315
thioacetals 193, 194
thioethers 93
thiophene 321
threo isomers 100, 120
Tiffeneau–Demjanov reaction 28, 395
trans alkenes 91
trans,cis,trans-octa-2,4,6-triene 507, 508
trans configuration 109
trans-1,2-dibromocyclohexane 109, 111
trans-1,2-dihalocyclohexanes 108
transesterification 273, 274
trans isomer 91
transition state 42
trans-olefin 120
trialkylamine 117
trialkyloxonium tetrafluoroborate 329
trichloromethyl carbanion 436
trifluoroacetic acid 37
trimethylamine 39
trimethylsilyl carboxylate 331
2,4,6-trinitroaniline 39

tri-*t*-butylcyclobutadiene   304
tropone   519
tropylium ion   299
Tsuji–Trost coupling reaction   571

u
umpolung   195
unimolecular conjugate base elimination   113
unsaturated systems   83

v
vicinal *cis*-dihydroxylation   159
vinylidene–alkyne rearrangement   452–453
vinylmagnesium bromide   545

w
Wagner–Meerwein rearrangements   391, 405, 481
weakly activating substituents   343
weakly deactivating substituents   344
Wilkinson's catalyst   169
Williamson ether synthesis   80–81
Wittig cyclization   126
Wittig–Horner reaction   222, 223
Wittig reactions   219, 220
Wittig rearrangement   429
Wittig–Schlosser Reaction   221, 222
Wohl–Ziegler reactions   474
Wolff–Kishner reduction
    aldehydes and ketones   200
    hydrazone derivatives   200
    reduction reaction at room temperature   200
Wolff rearrangement   446–448
Woodward–Hoffmann rules   501–502, 507–512
    cycloaddition reactions   515
    symmetry elements   512, 513
Wurtz coupling   538

x
X-ray crystallographic analysis   297, 299
X-ray diffraction   19, 149, 282
X-ray structural analysis   296, 301, 308

y
ylides reaction   218–227

z
Zaitsev product   89, 96, 116, 122, 123